T0338672

Principles of Planetary Climate

This book introduces the reader to all the basic physical building blocks of climate needed to understand the present and past climate of Earth, the climates of Solar System planets, and the climates of the newly discovered extrasolar planets. These building blocks include thermodynamics, infrared radiative transfer, scattering, surface heat transfer, and various processes governing the evolution of atmospheric composition. General phenomena such as Snowball Earth states, habitability zones, and the Runaway Greenhouse are used to illustrate the interplay of the basic building blocks of physics. The reader will also acquire a quantitative understanding of such key problems as the Faint Young Sun, the nature of Titan's cold liquid-methane hydrological cycle, and the warm, wet Early Mars climate, in addition to phenomena related to anthropogenic global warming on Earth, Earth's glacial–interglacial cycles, and their analogs on other planets. Exploration of simple analytical solutions is used throughout as a means to build the intuition needed to interpret the behavior of more complex phenomena requiring numerical simulation. Where numerical simulation is necessary, all necessary algorithms are developed in the text, and implemented in user-modifiable software modules supplied in the online supplement to the book. Nearly 400 problems are supplied to help consolidate the reader's understanding, and to lead the reader towards original research on planetary climate.

This textbook is invaluable for advanced undergraduate or beginning graduate students in atmospheric science, Earth and planetary science, astrobiology, and physics. It also provides a superb reference text for researchers in these subjects, and is very suitable for academic researchers trained in physics or chemistry who wish to rapidly gain enough background to participate in the excitement of the new research opportunities opening in planetary climate.

RAYMOND T. PIERREHUMBERT is the Louis Block Professor in the Geophysical Sciences at the University of Chicago, where he has taught and undertaken research on a wide variety of Earth and planetary climate problems for over 20 years. He shared in the Nobel Peace Prize as a lead author of the Intergovernmental Panel on Climate Change Third Assessment Report. He co-authored the US National Research Council report on Abrupt Climate Change, and is currently on the National Research Council panel on CO_2 stabilization targets, as well as being a member of their Board on Atmospheric Science and Climate. He has been a John Simon Guggenheim Fellow, is a Fellow of the American Geophysical Union, and was named Chevalier de l'Ordre des Palmes Academiques by the Republic of France. In addition to his research on planetary climate, he writes regularly for the popular climate science blog RealClimate.org.

PRINCIPLES OF PLANETARY CLIMATE

RAYMOND T. PIERREHUMBERT
University of Chicago

CAMBRIDGE
UNIVERSITY PRESS

CAMBRIDGE
UNIVERSITY PRESS

University Printing House, Cambridge CB2 8BS, United Kingdom

One Liberty Plaza, 20th Floor, New York, NY 10006, USA

477 Williamstown Road, Port Melbourne, VIC 3207, Australia

314-321, 3rd Floor, Plot 3, Splendor Forum, Jasola District Centre, New Delhi - 110025, India

79 Anson Road, #06-04/06, Singapore 079906

Cambridge University Press is part of the University of Cambridge.

It furthers the University's mission by disseminating knowledge in the pursuit of education, learning and research at the highest international levels of excellence.

www.cambridge.org
Information on this title: www.cambridge.org/9780521865562

First published 2010
8th printing 2019

A catalogue record for this publication is available from the British Library

Library of Congress Cataloging in Publication data
Pierrehumbert, Raymond T.
Principles of planetary climate / Raymond T. Pierrehumbert.
p.   cm.
Includes bibliographical references and index.
ISBN 978-0-521-86556-2
1. Planetary meteorology.   2. Climatology.   3. Paleoclimatology.   I. Title.
QB603.A85P54   2010
551.5–dc22
2010036193

ISBN 978-0-521-86556-2 Hardback

Additional resources for this publication at www.cambridge.org/pierrehumbert

For Arnold E. Ross

who taught us to think deeply of simple things

Contents

Preface

When it comes to understanding the whys and wherefores of climate, there is an infinite amount one needs to know, but life affords only a finite time in which to learn it; the time available before one's fellowship runs out and a PhD thesis must be produced affords still less. Inevitably, the student who wishes to get launched on significant interdisciplinary problems must begin with a somewhat hazy sketch of the relevant physics, and fill in the gaps as time goes on. It is a lifelong process. This book is an attempt to provide the student with a sturdy scaffolding upon which a deeper understanding may be built later.

The climate system is made up of building blocks which in themselves are based on elementary physical principles, but which have surprising and profound collective behavior when allowed to interact on the planetary scale. In this sense, the "climate game" is rather like the game of Go, where interesting structure emerges from the interaction of simple rules on a big playing field, rather than complexity in the rules themselves. This book is intended to provide a rapid entrée into this fascinating universe of problems for the student who is already somewhat literate in physics and mathematics, but who has not had any previous experience with climate problems. The subject matter of each individual chapter could easily fill a textbook many times over, but even the abbreviated treatment given here provides enough core material for the student to begin treating original questions in the physics of climate.

The Earth provides our best-observed example of a planetary climate, and so it is inevitable that any discussion of planetary climate will draw heavily on things that can be learned from study of the Earth's climate system. Nonetheless, the central organizing principle is the manner in which the interplay of the same basic set of physical building-blocks gives rise to the diverse climates of present, past, and future Earth, of the other planets in the Solar System, of the rapidly growing catalog of extrasolar planets, and of hypothetical planets yet to be discovered. A guiding principle is that new ideas come from profound analysis of simple models – thinking deeply of simple things. The goal is to teach the student how to build simple models of diverse planetary phenomena, and to provide the tools necessary to analyze their behavior.

This is very much a how-to book. The guiding principle is that the student should be able to reproduce every single result shown in the book, and should be able to use those skills as a basis for explorations that go wherever the student's curiosity may lead. Similarly, the student should have access to every dataset used to produce the figures in the book, and ideally to more comprehensive datasets that draw the student into further and even original analyses. To this end, I have set as a ground rule that I would not use reproductions of figures from other works, nor would I show any results which the student would not be able to reproduce. With the exception of a very few maps and images, every single figure and calculation in this book has been produced from scratch, using software written expressly for the purposes of this book and provided as an online software supplement. The computer implementations have pedagogy as their guiding principle, and readability of the implementation has been given priority over computational efficiency. A companion to this philosophy is what I call "freedom to tinker." The code is all in a form that can easily be modified for other purposes. The goal is to allow the student to first reproduce the results in the book, and then use the tools immediately as the basis for original research. In this, I have been much inspired by what the book *Numerical Recipes* did for numerical analysis. This book does not sell fish. Instead, it teaches students how to catch fish, and how to cook them. As gastronomical literature goes, the book before the reader is somewhat in the spirit of one of Elizabeth David's or Brillat-Savarin's extended pedagogical discourses on food (with recipes interspersed). My efforts to implement all the algorithms described in the present book and make them accessible to the reader also put me in mind of Julia Child's decade-long effort to winkle out and make explicit all the secret knowledge that makes the recipes in *Mastering the Art of French Cooking* actually work. Often, the most important things known to specialists and practitioners of such arts as radiative transfer or atmospheric escape are never committed to paper in a form that is recognizable to the uninitiated.

The software underlying this book was implemented in the open-source interpreted language Python, because it lends itself best to the design principles annunciated above. It has a versatile and powerful syntax but nonetheless is easy to learn. In my experience, students with no previous familiarity with the language can learn enough to make a substantial start on the computational problems in this book in only two weeks of self-study or computer labs. Python also teaches good programming style, and is a language the student will not outgrow, since it is easily extensible and provides a good basis for serious research computations. It will work on virtually any kind of computer, and because it is open-source the instructor does not have the bother and expense of dealing with licensing fees. I do hope that the student and instructor will fall for Python as madly as I have, but I emphasize that this book is not Python-specific. The text focuses on ideas that are independent of implementation. Specific reference to Python is confined to the online supplement and to the Workbook section of each chapter, where Python-specific advice is isolated in clearly demarcated *Python tips*. The instructor who wishes to make use of some other computer language in teaching the course will find few obstacles. The transparency and readability of Python is such that the Python implementations should provide a convenient aid to re-implementation in other languages. It is envisioned that MATLAB versions of most of the software will ultimately be made available.

In this book I have chosen to deal only with aspects of climate that can be treated without consideration of the fluid dynamics of the atmosphere or ocean. Many successful scientists have spent their entire careers productively in this sphere. The days are long gone when leading-edge problems could be found in planetary fluid dynamics alone, so even the student whose primary interests lie in atmosphere/ocean dynamics will need to know a considerable amount about the other bits of physics that make up the climate system. There are many excellent textbooks on what is rather parochially known as "geophysical fluid dynamics," from which the student can learn the fluid dynamics needed to address that aspect of planetary climate. That does not prevent me from entertaining a vision of adding one more at some point, as a sequel to the present volume. This sequel, subtitled *Things that Flow*, would treat the additional phenomena that emerge when fluid dynamics is introduced. It would continue the theme of taking a broad planetary view of phenomena, and of providing students with the computational tools needed to build models of their own. It would take a rather broad view of what counts as a "flow," including such things as glaciers and sea ice as well as the more traditional atmospheres and oceans. We shall see; for the moment, this is just a vision.

Remarks on notation and terminology

Since I have in mind the full variety of planets in our Solar System and in extrasolar systems, there is the question of what kind of terminology to use to emphasize the generality of the phenomena. Should we create new terminology that emphasizes that we are talking about an arbitrary system, at the risk of creating confusion by introducing new jargon? Or should we adopt terminology that emphasizes the analogy with familiar concepts from Earth and our own Solar System? For the most part, I have adopted the latter approach, which leads to a certain amount of Earth-centric terminology. For example, if I sometimes refer to "the sun" or "solar radiation," it is to be thought of as referring to whatever star the planet under discussion is orbiting, and not necessarily Earth's Sun or even a star like it. In the same spirit, the term *solar constant* will be used to refer to the rate at which a planet receives energy from its star (as defined precisely in Chapter 3), regardless of what that star may be and where the planet may be located. One may thus talk about the solar constant for Mars, for Earth, for Gliese 581d, or for that matter the difference between Earth's solar constant in June and July; from this remark, it is clear that the solar "constant" is a rather inconstant constant, but I will stick to the terminology since it has considerable familiarity within the field of climate physics. I will use the notation $L_\odot$ to refer to the value of this quantity for the planet under consideration, and use the same $\odot$ subscript to refer to all properties of a planet's star and the electromagnetic radiation which emanates from it. The new notation is necessary in order to distinguish the value pertinent to a given planet from the specific *number* $L_\odot$ which refers to the corresponding mean flux from the Sun itself, measured in the present era at a standard distance of 1 astronomical unit (roughly the Earth's mean orbit). The reader should also be cautioned that astronomers usually use the symbol L to refer to a star's *luminosity*, or net power output, rather than the flux as measured at some particular orbit. Since the luminosity of a star affects the climate of a planet only through the energy flux at the planet's own orbit, however, I have taken the liberty of co-opting the symbol for this flux. Astronomers are free to think of the quantity as a form of apparent luminosity, seen from the planet's orbit.

More proper terms would be *stellar radiation* and *stellar constant*, but those unfortunately call to mind starlight from the night sky and seem potentially confusing (though I will gradually break in the use of the terms to help the reader get used to the idea that there are a lot of stars out there, with a lot of planets with a lot of climates). The radiation from a planet's star will also sometimes be referred to as *shortwave radiation*, to emphasize that it is almost invariably of considerably shorter wavelength than the thermal radiation by which a planet cools to space.

In a similar vein, "air" will mean whatever gas the atmosphere is composed of on the planet in question – after all, if you grew up there, you'd just call it "air." When I need to refer to the specific substances that make up our own atmosphere, it will be called "Earth air." All this is a bit like the way one refers to Martian "geology" and "geophysics," so we don't need to refer to Areophysics on Mars and Venerophysics on Venus when we are really talking about the same kind of physics in all these cases. Eventually, we will all need to learn to get used to terms like "periastron" as a generalization of "perihelion," as the focus of the field shifts more to the generality of phenomena amongst planetary systems.

To improve the readability of inline equations, I will usually leave out parentheses in the denominator. For example, $a/2\pi$ is the same thing as $\frac{a}{2\pi}$, whereas I would write $(a/2)\pi$ or $\pi a/2$ if $\frac{a}{2}\pi$ was intended.

With few exceptions, SI units (based on kilograms, meters, and seconds) are used throughout this book. To avoid the baggage of miscellaneous factors of 1000 floating around, when counting molecules kilogram-moles are used, denoted with a capital, i.e. Mole. Thus 1 Mole of a substance is the number of molecules needed to make a number of kilograms equal to the molecular weight – one thousand times Avogadro's number. There are a few cases where common practice dictates deviations from SI units, as in the use of millibars (mb) or bars for pressure when pascals (Pa) involve unwieldy numbers, or the use of cm^{-1} for wavenumbers in infrared spectroscopy.

Throughout the book we will most commonly use pressure as a vertical coordinate, and this raises some possibilities for confusion regarding the meaning of the word "above" applied to vertical position in an atmosphere, since low pressures occur at high altitudes and high pressures occur at low altitudes. The word "above" should always be understood in this context to refer to altitude. Thus, the phrase "layers above 100 mb" is to be understood as a shortcut for the more precise phrase "Layers whose altitude lies above the altitude of the 100 mb pressure level."

How to use this book

The short exercises embedded in the text are meant to be done "on the spot," as an immediate check of comprehension. More involved and thought-provoking problems may be found in the accompanying Workbook section at the end of each chapter. The Workbook provides an integral part of the course. Using the techniques and tools developed in the Workbook sections, the student will be able to reproduce every single computational and data analysis result included in the text. The Workbook also offers considerable opportunities for independent inquiry launching off from the results shown in the text.

There are four basic kinds of problems in the Workbook. Some calculations are analytic, and require nothing more than pen and paper (or at most a decent pocket calculator). Others involve simple computations, data analysis, or plotting of a sort that can be done in a spreadsheet or even many commercially available graphing programs, without the need for any actual computer programming. Many of these problems involve analysis of datasets from observations or laboratory experiments, and all critical datasets are provided in the online supplement to this book in a tab-delimited text format which can be easily read into software of any type. Students who have competence in a programming language, either from prior courses or because the instructor has integrated programming instruction into the climate sequence (as is done at the University of Chicago), have the option of doing these problems in the programming language of their choice. They should be encouraged to do so, since these simpler problems make good warm-up exercises allowing students to consolidate and hone their programming skills. The third class of problems requires actual programming, but can easily be carried out from scratch by the student in the instructor's language of choice (perhaps with the assistance of some standard numerical analysis routines). While just about any language would do, I have found that interactive interpreted languages such as Python and MATLAB offer considerable advantages, since they provide instant feedback and encourage exploration and experimentation.

The fourth class of problem consists of major projects which would take a lot of time for the student to implement from scratch. Some of these become relatively straightforward, though, if the basic tools such as the moist adiabat routine of Chapter 2 or the "homebrew" exponential sum real gas radiation code of Chapter 4 are provided as tools the student can use in doing the problems. I have provided Python implementations of all such tools, but this is the class of problems that poses the greatest challenge for the instructor who has not adopted Python as the language of choice for instruction. It is highly recommended that the instructor take the time to re-code at least some of the critical tools in the language of choice if Python is not being used. I have provided algorithm descriptions in the text that are independent of the computer implementation, and examination of the Python code should also help. The object-oriented features and powerful list handling capabilities of Python mean that translations into languages that do not support these language features are apt to be more complicated and unwieldy, but still it is only the exponential sums radiation code that is likely to pose any real challenge to the instructor. It is an excellent training exercise to pursue the necessary code development as a team-project effort with a few enthusiastic graduate students, who can then serve as teaching assistants for the course.

There are just a handful of basic computational methods and computer skills needed to do the Workbook problems and to reproduce all the calculations in this book. None of the calculations requires any more computer power than is available on any decent laptop computer. The required numerical skills are outlined and exercised in the Chapter 1 Workbook section, which the student should master before proceeding to the rest of the book. I have not provided detailed discussions of basic algorithms like ordinary differential equation integration, interpolation, or numerical quadrature, since they are well described in the book *Numerical Recipes*, also available from Cambridge University Press. *Numerical Recipes* should be viewed as an essential companion to this book, though only a small part of the material in that opus is actually required for the problems that concern us here.

With very few exceptions, all the datasets needed to produce the figures in the text, or needed to do the data analysis problems in the Workbook sections, have been provided in the online supplement in plain text tabular form. These are organized into subdirectories according to the chapter to which the data is pertinent. These datasets can be plotted and analyzed using virtually any software. The only exceptions to the text format are a very few datasets used in making temperature and humidity maps in Chapters 7 and 9, which are too large to handle conveniently in text form. These datasets are in the machine-independent netCDF format, but they are not heavily used and it is not necessary to learn how to deal with this file format for the purposes of working through this book. The format is widely used for archiving numerical simulations and observations, so effort expended in this direction will be well repaid. It is very easy to read netCDF data using various packages written for use with Python, and not very hard to do it within MATLAB either.

For the foreseeable future, the online supplementary material can be downloaded from Cambridge University Press, at www.cambridge.org/pierrehumbert. The supplementary material, particularly the courseware, is meant to be somewhat viral, and so it should remain available over time (hopefully in improved and evolving forms) at various open-source repositories and individual websites. This will no doubt be a mutable territory, and so the best way to locate such resources is to run suitably chosen keywords through a search engine.

There are three groups of Python courseware that are provided as part of the online supplement. Even non-Pythonista should be aware of what is there, since it is recommended that the same basic organization be adopted regardless of the computer language used for instruction.

- First, there are the basic courseware modules ClimateUtilities.py, phys.py, and planets.py. They should be placed in a publicly available directory which the students' Python interpreter will look for when loading add-on software (called modules in Python). The student should be able to read these files so as to get a better understanding of what they are doing, but it is not expected that the student will need to modify them.

  ClimateUtilities.py provides basic graphics, input/output, data manipulation, and numerical analysis utilities. The modules phys.py and planets.py replace the data tables that typically appear on the endpapers and in the appendices of books such as this one. Being software rather than paper, they can include a more versatile level of data organization and can include functions as well as static data. However, since they are also human-readable text files, even the non-Pythonista can consult them in order to find needed data.

- Second, there are the *Chapter Scripts*, organized in subdirectories according to the chapter to which they pertain. These reproduce all the computations and figures appearing in the respective chapters, provide the means of further explorations, and also illustrate techniques needed to solve the Workbook problems. Some instructors will want to have the student refrain from examining the Chapter Scripts until they have had a go at implementing the ideas on their own, while others may want to make them immediately available as a study aid. Whenever they are made available, students are expected to have their own individual copies of these, as the basic intended use of these scripts is that the student will modify them and customize them, and re-run them as needed.

- Third, there are the *Solution Scripts*, which carry out solutions to selected Workbook problems. Access to these requires an instructor password, and they are intended to be doled out to students after they have turned in their own work.

The online supplement also includes Python tutorials, and various sample scripts illustrating numerical techniques and basic data analysis tasks. Software requirements and tips on installation are provided here as well. Solution write-ups for selected Workbook problems are also provided (instructor password required); these are for the most part language-independent, even where the calculations themselves were done in Python.

Although I have tried to rely primarily on calculations done with software that was written expressly for this book – and which the student can read, understand, and customize – at a few points I have found it necessary to make use of calculations carried out with a full-featured terrestrial radiation model. For this purpose I have used the ccm column radiation model from the National Center for Atmospheric Research. For the most part, I have designed the problems so that they can be done using various polynomial fits to calculations with this model; it is not strictly necessary for the student to have access to the model. Nonetheless, it is desirable that the student be able to reproduce the results on his or her own. A stand-alone FORTRAN version of the ccm radiation model can be downloaded from public sources and used to do the needed calculations, but to make life easier for the Pythonista, I have included as part of the courseware a user-friendly Python interface to the `ccm` model, which makes it easy to use the model in conjunction with other `Python` computations. More details are provided in the online supplement.

Every author likes to think that their book will occasionally be perused from time to time even in a century or two. I think of this myself, when reading through the crumbling though still-illuminating pages of Arrhenius' work, or the less famous though equally crumbling (and still-illuminating) papers of Frank Very. Whether the reader of the future (should I be so lucky) is absorbing this material through crumbling paper or neural implant, the text and equations will still in some sense be readable. One cannot dare to hope, however, that the associated software used in the computations for this book will still run on the hardware prevailing in the future, though one can hope that the underlying algorithms are eternal. Even the `Python` of 2050 is unlikely to look like the `Python` of 2010, if indeed the language still exists at all. For this reason, I have minimized the discussion of the detailed software implementation within the text, and left that to the supplementary online material. Surely, while Frank Very's ideas on radiative transfer are still of interest, any description of the charts and graphical techniques for doing the calculations at the time would have at most historical interest. The associated `Python` software is meant to be not a static thing, but a living entity, which will be adopted, ported, and mutated by the community of users as necessary. It could well be that quantum computers of the next century will allow direct line-by-line calculations to replace all the approximate radiative fanciness developed in this book, but even then people will need a good set of example programs in order to help build an understanding of the underlying physics. Computation is not understanding. The calculations embodied in the suite of software upon which this book is built are intended to provide the nucleation point for understanding.

Prerequisites and suggested syllabi

Before tackling the material in this book, the student should have attained a good mastery of classical mechanics, such as is typically provided by a first-year college physics course or an advanced secondary school course. This is not so much because classical mechanics itself is heavily exercised in this book, as because classical mechanics introduces the student to the necessary kind of problem-solving skills, as well as building foundational concepts such as energy conservation and its use in problem-solving. It would also be useful for the student to have some familiarity with the basics of electromagnetism, including the concept of electric and magnetic fields and the forces they exert on charged particles; it is certainly not necessary for the student to have dealt with Maxwell's Equations in their full glory, though. The treatment of thermodynamics in this book is designed to be self-contained, though a student with some prior exposure to the subject in a physics course will be able to approach the material on a deeper level. There is some lightweight use of chemical kinetics and equilibrium in Chapter 8, but a self-contained, if minimal, introduction to the subject has been provided.

As for mathematics, all that is really required is a thorough understanding of single-variable differential and integral calculus, including first order ordinary differential equations. There is some use of second order differential equations in Chapters 5 and 7, but with help most students will be able to grasp the generalization even if they haven't formally studied the subject. The discussion of the diffusion equation in Chapter 7 makes use of a one-dimensional partial differential equation, but this material does not require a very deep understanding of the subject and for the most part can be grasped intuitively from a physical basis. In fact, for students who haven't before dealt with partial differential equations, or who are rusty on the subject, this material and the associated problems serve as a good refresher. There is some optional material in the final chapter which exercises multivariable calculus more heavily, but the discussion is designed so that the details can be skipped if necessary.

It takes several courses to cover the material in this book, where by a "course" I mean 30 hours of lectures in a typical 10-week quarter, or 45 hours of lectures in a typical 15-week semester. Europeans using different quanta of instruction can calibrate the following accordingly.

For complete beginners, Chapters 2 and 3, with just a brief dip into the material of Chapter 1, plus all the necessary computer and algorithm preliminaries, fit in a one-quarter introductory course, though with little room for digressions. It is essential to introduce the students to some programming environment sophisticated enough to handle numerical differential equation integration. Even if most of the planetary history information in Chapter 1 is skipped, or left to self-study, the material in the Computational Toolkit section of the Chapter 1 Workbook should be covered, since it will be needed to do the rest of the problems. It usually works best to save lecture time by having the students learn programming and algorithms in lab or section meetings, supplemented by copious use of computational examples in the course of the lectures. When using Python, I find that students usually have enough basic skills to do the computational problems after about two weeks of such training; more advanced programming and numerical analysis skills can be introduced as needed. Besides covering the Computational Toolkit problems, it is a good idea to have the students work through the problems covering basic physics and chemistry, to get them up to speed on the fundamental concepts they will need to proceed; the "chemistry" covered in these problems is mostly a matter of understanding how to do problems involving molecular weight.

In a semester, or for students who already possess a good knowledge of either thermodynamics or basic computer skills, a bit more material can be fitted in. This could be a full coverage of the Earth and planetary history material of Chapter 1, a more thorough treatment of programming and numerical methods, or inclusion of the gray-gas portion of Chapter 4.

Chapters 4 and 5 can form the basis of a one-semester course for advanced undergraduates or beginning graduate students, but I have sometimes taken as much as a full quarter just to cover the material in Chapter 4 in depth, with plenty of time allowed for simulation projects using the material.

The material on surface energy balance in Chapter 6 is less fundamental to planetary climate than some of the other topics addressed, but it is something that everybody needs to know eventually. It could be paired naturally with Chapter 4 in place of scattering. My own preference is to truncate the material somewhat and teach it together with the material on the seasonal cycle and geographic variations in Chapter 7, which is truly fundamental and essential. One cannot begin to understand the issues surrounding Pleistocene ice ages without the material in this chapter.

Chapter 8, which deals with planetary formation and evolution of atmospheric composition – including feedbacks between climate and composition – can comfortably fit into a one-quarter course. It depends somewhat on material from previous chapters as a prerequisite, but the nature of the material is independent enough that with suitable presentation of background, it could be used in a stand-alone course. In a semester, the material could be supplemented with additional instruction on atmospheric chemistry, oceanic carbonate chemistry, or silicate weathering.

Chapter 9, which points the student toward an appreciation of the importance of fluid dynamical effects not considered as part of this book, can be worked in toward the end to fill out any of the above courses. It is a particularly good complement to the material in Chapter 7. Alternately, it can be left for the student to peruse at leisure, given that study of the other material has no doubt awakened considerable curiosity about what comes next.

The material is laid out in what seems to me to be the natural didactic order, but in view of the realities of teaching and the necessity of dividing the material up amongst multiple courses the instructor may wish to jump around a bit, so as to retain the students' interest. For example, Chapters 2 and 3 fit comfortably in a one-quarter course and provide the student with an introduction to thermodynamics and radiation, but adhering to that syllabus would leave the student without an appreciation of the important real gas aspects of CO_2 and water vapor. It is thus desirable to work in some of the more qualitative real gas material from Chapter 4, including the use of polynomial *OLR* fits in solving climate problems. These can be introduced as simply a drop-in replacement for σT^4, once the qualitative underlying physics is explained. Similar opportunities abound, and I have tried to organize things so as to help out the instructor who wishes to wander nonlinearly through the subject matter.

Acknowledgements

First and foremost, I would like to thank all the students at the University of Chicago who have participated in my various planetary climate courses over the years. Without them, I would have had little chance to understand what questions were the right ones to ask. The patience and compassion of the students in the early years when I was first fumbling through the business of turning myself from a dry fluid dynamicist into something more is especially appreciated. I also gratefully acknowledge the tireless efforts of David Crisp, who patiently responded to an endless flurry of emails from me when I was getting up to speed on real gas radiative transfer. I hope that in this book I have managed to pass on to the next generation some of the secrets known formerly only to the high priesthood.

Over the many years during which various drafts of this book have been in circulation, I have received a great deal of valuable feedback from readers, some of whom I have known for years and others of whom provided help out of the blue. These include Itay Halevy, Rodrigo Caballero, David Andrews, Colin Goldblatt, Jim Galasyn, Philip Machanik, Barton Paul Levenson, Ellen Thomas, Jack Barrett, and Samuel Odell Campbell. My apologies to any I may have inadvertently left out. I hope the long wait for the final product proves worthwhile. I would especially like to thank the readers who brought me through the home stretch and helped compile the index: Ari Solomon, Ian Williams, Mac Cathles, and Janet Pierrehumbert.

I am indebted to Ron Blakey, Professor Emeritus at Arizona State University, for kindly allowing me the use of his stunning paleogeography maps. And not least, I am indebted to my editor at Cambridge University Press, Matt Lloyd, and his capable co-workers. They surely have the patience of an Inuit waiting for days on end at an airhole in the ice, hoping at long last for the seal to put in an appearance.

Notation

This is not a completely exhaustive list of symbols used in the text, but most of the important and frequently used ones are included. Note that some symbols are used for multiple purposes, according to context. In addition, some symbols have incidental alternate uses besides the major usages listed below.

Units	
m	Length, meter
cm	centimeter (0.01 m)
mm	millimeter (0.001 m)
μm	micron (10^{-6} m)
nm	nanometer (10^{-9} m)
pm	picometer (10^{-12} m)
km	kilometer (1000 m)
A.U.	Astronomical unit (mean distance of Earth from Sun, $1.496 \cdot 10^{11}$ m)
Parsec	Parallax-second ($3.08 \cdot 10^{16}$ m)
Light year	$9.46 \cdot 10^{15}$ m
kg	Mass, kilogram
Gt	gigatonne (10^{12} kg)
Mole	kilogram-mole, i.e. $1000 \cdot N_{avo}$ molecules
ppmv	Parts per million by count of molecules
s	Time, second
d, da	Standard Earth day (86 400 s)
sol	Standard Earth day
Hz	Frequency, 1/s
N	Force, newton (kg m/s^2)
Pa	Pressure, pascal (N/m^2)
bar	10^5 Pa, approximately one Earth atmosphere
mb	millibar (0.001 bar or 100 Pa)

J	Energy, joule (Nm)
kJ	kilojoule, 1000 J
attoJ	attojoule, (10^{-18} J)
W	Power, watt (J/s)
mW	milliwatt (0.001 W)
K	Temperature, kelvin
°C	Temperature, degrees Celsius

Physical and mathematical constants

k	Boltzmann thermodynamic constant, $1.38 \cdot 10^{-23}$ J/K
N_{avo}	Avogadro's number, $6.022 \cdot 10^{23}$. Number of molecules in the number of grams of a substance equal to the substance's molecular weight. Sometimes called the "Loschmidt number"
μ	Mass of a proton, approx. $0.001/N_{avo}$ or $1.66 \cdot 10^{-27}$ kg
R^*	Universal gas constant, k/μ or 8314.5 J Mol^{-1} K^{-1}
c	Speed of light in vacuum, $3.0 \cdot 10^8$ m/s (Symbol also used for speed of sound in Chapter 8)
h	Planck's constant, $6.626 \cdot 10^{-34}$ J s
σ	Stefan-Boltzmann constant, $2\pi^5 k^4/(15c^2 h^3)$, or $5.67 \cdot 10^{-8}$ W/m^{2}K^4
G	Newton's gravitational constant, $6.674 \cdot 10^{-11}$ m^3/kg $\cdot$ s
$L_\odot$	Mean present Solar flux measured at 1 A.U. from the Sun, about 1365 W/m^2
K_{vk}	von Karman constant, 0.41
π	Ratio of circumference to diameter of circle, 3.14159...
e	Base of natural logarithms, 2.71828...

Variables and physical quantities

$B(\nu, T)$	Planck density in frequency space, expressed as function of frequency and temperature
C_D	Drag (bulk exchange) coefficient
C_H	Henry's law temperature constant
D	Diffusivity (thermal or mass)
H	Atmospheric density scale height (also used for weighting function in scattering calculation)
I	A general flux of radiation (sometimes per unit wavenumber or steradian, according to context)
$I(p, \hat{n}, \nu)$	Spectral irradiance at pressure p, in direction $\hat{n}$, at frequency ν
$I_+, \bar{I}_+$	Upward radiation flux; band-averaged upward flux
$I_-, \bar{I}_-$	Downward radiation flux; band-averaged down flux
J_ν	Photon flux at frequency ν
K_1	First carbonate equilibrium constant
K_2	Second carbonate equilibrium constant
K_H	Henry's law constant
K_{sp}	Solubility product constant
L	Latent heat

$L_\odot$	Stellar constant. Generalization of "solar constant" Incoming flux from a planet's star, measured at the orbit. Time-dependent or mean, according to context.
M	Mach number (ratio of fluid velocity to speed of sound)
M	Generic colliding molecule
M, M_A	Molecular weight, of substance A
$M_\odot$	Stellar mass
OLR	Outgoing Longwave Radiation
OLR_{all}	All-sky OLR
OLR_{clear}	Clear-sky OLR
P	Phase function (Chapter 5)
Q_{abs}	Absorption efficiency
Q_{sca}	Scattering efficiency
R	Gas constant for some particular gas (R^*/M)
R, R_A	Gas constant, for substance A
Ri	Richardson number
S	Generic incident shortwave (stellar) radiation, averaged over appropriate portion of planet's surface according to context
S	Line strength (intensity) (Chapter 4)
T	Temperature
T^*	Constant for radiative absorption temperature scaling
T_U	Temperature constant for Ebelmen–Urey reactions
$T_\odot$	Temperature of photosphere of star
T_s	Surface temperature
T_g	Ground temperature
T_{sa}	Surface air temperature
T_{skin}	Skin temperature
U	Wind speed at outer edge of surface layer
χ	Absorption or scattering cross-section, between molecules or between molecules and photons, according to context
χ_ν	Cross-section for collision with photon having frequency ν
Δ	Used to represent an increment of some quantity, as in ΔT for temperature difference
Γ	Gamma function
Ω	Angular rotation rate of planet
$d\Omega$	Differential of solid angle
Ω^+	Solid angle of upward hemisphere
Ω^-	Solid angle of downward hemisphere
Φ	Climate feedback factor (Chapter 3)
Φ	Escape flux, per unit surface area of planet (Chapter 8)
Φ^*	Limiting escape flux
Φ	Meridional heat transport (Chapter 9)
α	Albedo
α_g	Albedo of ground
χ	Collision cross-section
χ_{abs}	Absorption cross-section (per particle)
χ_{sca}	Scattering cross-section (per particle)
δ	Used with an isotope to indicate isotopic anomaly, e.g. $\delta^{18}O$
δ_{org}	Organic carbon isotopic anomaly
δ	Angle; subsolar latitude (Chapter 7)
δ	Angle; latitude of the Sun/star (subsolar/substellar latitude)

δ	Used to represent an increment of some quantity, usually presumed small, as in δr or δQ
δ	Depolarization factor (Chapter 5)
$\delta(x)$	Dirac delta "function" (Chapter 5)
ℓ	Characteristic length for impact erosion (Chapter 8)
ℓ	Mean free path (Chapter 8)
ℓ	Mass path for absorption
ℓ_s	Equivalent mass path for strong lines
η, η_A	Molar concentration, of substance A
γ	Angle; obliquity (Chapter 7)
γ	Line width (Chapter 4)
γ	Ratio of specific heats, c_p/c_v
γ, γ'	Numerical constants in two-stream scattering (Chapter 5)
$\gamma_B, \gamma_\pm$	Numerical constants in two-stream scattering (Chapter 5)
κ	Season angle (Chapter 7)
κ	Thermal conductivity
κ, κ_A	Absorption cross-section, of substance A
λ	Longitude in a geographically fixed coordinate system on a planet
λ	Wavelength
$\lambda_\odot$	Longitude of the subsolar/substellar point ("longitude of the sun") in a geographically fixed coordinate system (Distinct from astronomers' "ecliptic longitude")
λ_c	Jeans escape parameter
$\mathcal{L}_\circledast$	Stellar luminosity
$\mathcal{L}_\odot$	Luminosity of the Sun, (at present, if not qualified with a time argument)
$\mathfrak{H}$	Longwave radiative heating per unit optical depth
H_ν	Longwave radiative heating per unit mass
$\mathfrak{M}$	Moist static energy per unit mass
$\mathfrak{T}$	Transmission function
$\overline{\mathfrak{T}}$	Band-averaged transmission function
ν	Frequency
ν_{max}	Frequency of maximum emission in frequency space
ω	Angular frequency of insolation cycle (Chapter 7)
ω_0	Single-scattering albedo
ϕ	Latitude
ρ	Density (sometimes subscripted according to species)
τ	Optical depth or thickness, sometimes as a function of wavenumber
τ^*	Optical thickness in normal direction
τ_ν	Optical thickness at frequency ν
τ_∞	Optical thickness of the entire atmosphere
τ	Generic thermal response time (Chapter 7)
τ_1	Diffusive thermal relaxation time scale (Chapter 7)
τ_D	Mixed layer thermal relaxation time scale (Chapter 7)
θ	Potential temperature
θ	Also used for propagation angle in radiative transfer
Θ	Scattering angle
$\check{g}$	Asymmetry factor in scattering
$\hat{g}$	An alternate form of asymmetry factor
ξ	Molar mixing ratio
ζ	Zenith angle

ζ	Non-dimensional boundary layer coordinate (Monin-Obukhov theory, Chapter 6)
a	Generic radius of a spherical body
a	Absorptivity (Chapter 3)
a_{sw}	Shortwave absorptivity of atmosphere
b	Multiple uses as a coefficient, but often used as a coupling constant (derivative of flux with respect to temperature), as in b_{ir} in Chapter 6.
c_p	Specific heat at constant pressure
c_v	Specific heat at constant volume
e	Emissivity (Not to be confused with e as base of natural logarithms)
e_a	Effective atmospheric emissivity for back-radiation to ground (Chapter 6)
e^*	Surface cooling factor, $1 - e_a$ (Chapter 6)
e_g	Emissivity of ground (Chapter 6)
f_{org}	Organic carbon burial fraction
g	Acceleration of gravity (at surface, unless otherwise qualified)
g_s	Surface gravity
h	Hour angle (Chapter 7)
h	Relative humidity
h_t	Hour angle of the terminator
h_{sa}	Surface layer relative humidity
k	Generic equilibrium constant
n	Index of refraction (Chapter 5)
n	Particle number density (esp. in Chapter 8), often subscripted according to species or position, e.g. n_{ex} for exobase density.
n	Occasionally used for wavenumber ν/c. More commonly, ν is used and "wavenumber in 1/cm" is treated as an alternate unit of frequency
pH	$-\log_{10}[H^+]$ in aqueous solution.
p	Pressure
pCO_2	Partial pressure of CO_2, sometimes expressed as ppmv for Earth atmosphere
p_A	Partial pressure of substance A
p_s	Surface pressure
p_{rad}	Effective radiating pressure
p_{sat}	Saturation vapor pressure
p_{trop}	Tropopause pressure
q	Mass-specific concentration (also specific humidity)
q	Exponent in power law describing impactor distribution in Chapter 8
r, r_A	Mass mixing ratio, of substance A
r	Generic radius of a spherical object
$r_\odot$	Radius of star
r_c	Critical radius, transonic point (Chapter 8)
u	$h\nu/kT$
y	$\sin\phi$
z	Height of a pressure surface

The Big Questions

1.1 OVERVIEW

This chapter will survey a few of the major questions raised by observed features of present and past Earth and planetary climates. Some of these questions have been answered to one extent or another, but many remain largely unresolved. This will not be a comprehensive synopsis of Earth and planetary climate evolution; we will be content to point out a few striking facts about climate that demand a physical explanation. Then, in subsequent chapters, we'll develop the physics necessary to think about these problems. Although we hope not to be too Earth-centric in this book, in the present chapter we will perforce talk at greater length about Earth's climate than about those of other planets, because so much more is known about Earth's past climate than is known about the past climates of other planets. A careful study of Earth history suggests generalities that may apply to other planets, and also raises interesting questions about how things might have happened differently elsewhere, and it is with this goal in mind that we begin our journey.

1.2 CLOSE TO HOME

When the young Carl Linnaeus set off on his journey of botanical discovery to Lapland in 1732, he left on foot from his home in Uppsala. He didn't wait until he reached his destination to start making observations, but found interesting things to think about all along the way, even in the plant life at his doorstep. So it is with climate as well.

To the discerning and sufficiently curious observer, a glance out the window, a walk through the woods or town, a short sail on the ocean, all raise profound questions about the physics of climate. Even without a thermometer, we have a perception of "heat" or "temperature" by examining the physical and chemical transitions of the matter around us. In the summertime, ice cream will melt when left out in the sun, but steel cooking pots don't. Trees and grass do not spontaneously burst into flame every afternoon, and a glass of water left outdoors in the summer does not boil. Away from the tropical regions, it often gets cold enough for water to freeze in the wintertime, but hardly ever cold enough for alcohol to freeze. What is it that heats the Earth? Is it really the Sun, as seems intuitive from the perception of warmth on a sunny day? In that case, what keeps the Earth from just accumulating more and more energy from the Sun each day, heating up until it melts? For that matter, why don't temperatures plummet to frigid wintry values every night when the Sun goes down? Similarly, what limits how cold it gets during the winter?

With the aid of a thermometer, such questions can be expressed quantitatively. The first, and still most familiar, kinds of thermometers were based on one particular reproducible and measurable effect of temperature on matter – the expansion of matter as it heats up. Because living things are composed largely of liquid water, the states of water provide a natural reference on which to build a temperature scale. The *Celsius* temperature scale divides the range of temperature between the freezing point of pure water and the boiling point at sea level into 100 equal steps, with zero being at the freezing point and 100°C at the boiling point.[1]

Through observations of fire and forge, even the ancients were aware that conditions could be much hotter than the range of temperatures experienced in the normal course of climate. However, they could have had no real awareness of how much *colder* things could get. That had to await the theoretical insights provided by the development of thermodynamics in the nineteenth century, followed by the invention of refrigeration by Carl von Linde not long afterwards. By the close of the century, temperatures low enough to liquify air had been achieved. This was still not as low as temperatures could go. The theoretical and experimental developments of the nineteenth century consolidated earlier speculations that there is an *absolute zero* of temperature, at which random molecular motions cease and the volume of an ideal gas would collapse to zero; no temperature could go below this absolute zero. On the Celsius scale, absolute zero occurs at −273.15 °C. Most of thermodynamics and radiation physics can be expressed more cleanly if temperatures are given relative to absolute zero, which led to the formulation of the *Kelvin* temperature scale, which shifts the zero of the scale while keeping the size of the units the same as on the Celsius scale. On the Kelvin scale, absolute zero is at zero kelvin, the freezing point of water is at 273.15 K, and the sea-level boiling point of water is at 373.15 K. Viewed on the Kelvin scale, the temperature range of Earth's climate seems quite impressively narrow. It amounts to approximately a ±10% variation about a typical temperature of 285 K. A 20% variation in the Earth's temperature (as viewed on the Kelvin scale) would be quite catastrophic for life as we know it. This remark can be encapsulated in a saying: "Physics may work in degrees Kelvin, but Earth life works in degrees Celsius."

There is more to climate than temperature. Climate is also characterized by the amount and distribution of precipitation (rainfall and snowfall), as well as patterns of atmospheric winds and oceanic currents. However, temperature will do for starters. In this book we will discuss temperature at considerable length, and venture to a somewhat lesser degree into the factors governing the amount of precipitation. We will not say much about wind

---

[1] The scale is named for the Swedish astronomer Anders Celsius, who originally formulated a similar temperature scale in 1742. Celsius' scale was reversed relative to the modern one, putting 100 at the freezing point and zero at the boiling point. The Celsius scale is sometimes called *centigrade*, but Celsius is considered to be the preferred term. The official definition of the temperature scale is now based on standards that are more precise and unambiguous than the freezing and boiling points of water.

patterns, though some of their effects on the temperature distribution will be discussed in Chapter 9.

If you live outside the tropical zone, you will come to wonder why it is hotter in summer than in winter, and why the summer/winter temperature range has the value that it does (e.g. 30 °C in Chicago) and why the variation is generally lower over the oceans (e.g. 7 °C in the middle of the Pacific Ocean, at the same latitude as Chicago). If you communicate with friends living in the Arctic or Antarctic regions, and other friends living near the Equator, you will begin to wonder why, on average, it is warmer near the Equator than in the polar regions, and why the temperature difference has the value it does (e.g. 40 °C difference between the annual average around the Equator and the annual average at the north pole). The physics underlying the seasonal cycle and the pole to Equator temperature gradient is discussed in Chapters 7 and 9. If you climb a mountain (or even observe the snow-capped peaks of a mountain from the valley floor on a hot summer day), or if you go up in a hot-air balloon, or fly in an airliner which informs you of the outdoor temperature – you will notice that the air gets colder as one goes higher in altitude. Why should this be? This turns out to be a general feature of planetary atmospheres, and the basic physics underlying the phenomenon is discussed in Chapter 2.

The air that surrounds us is itself a matter of interest. We know that it is there because it has a temperature and exerts pressure, and because it is necessary that we breathe it in order to remain alive. But what is the air made of, and why does it have the composition it does? We can see water condense out of the air, but why don't other components condense in the course of natural weather and climate variations? How much air is there? And has it always been there with its present composition, or has it changed over time? If so, how much and how quickly?

We know that our planet journeys through the hard vacuum of outer space, clothed in a thin blanket of air – our atmosphere. It is natural to wonder how our atmosphere affects the Earth's climate. The airless Moon shares the orbit of the Earth, at the same distance from the Sun, so one can look to the Moon to get an idea of what the Earth's climate would be like if it had no atmosphere. We know the Moon is airless because a reasonably thick atmosphere would bend the light rays from the Sun and stars, just as objects appear displaced when viewed through the surface of a swimming pool. But how to measure its temperature?

Of course, one could go there with a thermometer (and this did eventually happen) but people became curious about lunar conditions long before it seemed likely that anybody would ever get there. Dante Alighieri himself, in the *Paradiso* written between 1308 and 1321, devoted fully one hundred cantos to a learned discussion between himself and Beatrice concerning the source of lunar light and the solidity of the lunar surface. By the mid nineteenth century, science had progressed to the point that the questions could be formulated more sharply, and the means for an answer had begun to emerge. With the discovery of infrared light by Sir William Herschel in 1800, astronomy opened a new window into the properties of planets and stars. Over the coming decades, it gradually became clear that all bodies emit radiation according to their temperature. This is known as *blackbody radiation* and will be discussed in detail in Chapter 3. Infrared light from the Moon was detected by Charles Piazzi Smyth in 1856, and the first attempt to use it to estimate temperature was by the Fourth Earl of Rosse in 1870. The instruments available at the time were not up to the task. In 1878, Langley invented the bolometer, which made good observations of lunar infrared radiation possible. However, while Langley made the first accurate observations of lunar infrared, theory was not quite up to the task of interpreting the observations. These issues were largely sorted out by 1913, though Langley gave up on his earlier estimates rather reluctantly. By 1913 it was pretty clear that the daytime temperature of the Moon at the point where the Sun is directly overhead is well in excess of 373 K (the sea-level temperature of boiling water on Earth). Night-time temperatures were harder to determine accurately, since the infrared emission from cold objects is weak; however, it was clear that temperatures at night dropped by well over 140 K relative to the daytime peak. Pettit and Nicholson observed the temperature of the Moon during the lunar eclipse of 1927, using the Mt. Wilson telescope. They found something even more remarkable: over the span of the few hours of the eclipse, the lunar temperature fell from 342 K at the point of observation to 175 K. Modern measurements show the daily average temperature at the lunar equator to be around 220 K, while the mean temperature at 85° N latitude is 130 K.

It appears that without an atmosphere or ocean, the Earth would be subject to extreme swings of temperature between day and night. The Moon's "day" is 28 Earth days, since it always shows the same face to the Earth; on that basis, one could imagine that the day/night extremes were due to a longer night offering more time to cool down, but the rapid cooling during an eclipse gives the lie to this idea. Given the rapid cooling of an airless body at night, it is likely that the Earth's summer/winter temperature difference would be far more extreme in the absence of an atmosphere. Further, a comparison of the pole to equator gradient in daily mean temperature with that on Earth suggests that the atmosphere significantly moderates this gradient, too. What is it about the atmosphere or ocean that damps down day/night or summer/winter swings in temperature? This subject will be taken up in Chapter 7, where we'll also learn why summer is warmer than winter and why the poles are on the average colder than the equator. How does an atmosphere or ocean moderate the temperature difference between pole and equator? We'll learn something of that in Chapter 9.

At its hottest the Moon gets much hotter than Earth, and at its coldest it gets much colder. But how does the Moon's mean temperature stack up against that of Earth? The 220 K mean equatorial temperature of the Moon is very much colder than the observed mean tropical temperature on Earth, which is on the order of 300 K. If the Earth's mean temperature were as low as that of the Moon, the oceans would be solidly frozen over. The cold mean temperature of the Moon does not come about because the Moon reflects more sunlight than the Earth; the Moon looks silvery but measurements show that it actually reflects less than Earth. Why is the Earth, on average, so much warmer than the Moon? Does this have something to do with our atmosphere, or is it the case that Earth is warmed by some internal heat source that the Moon lacks?

The search for the first stirrings of an answer to this problem takes us back to 1827, when Fourier published his seminal treatise on the temperature of the Earth. Fourier could not have known anything about the temperature of the Moon, but he did know a great deal about heat transfer – having in fact largely invented the subject. Using his new theory of heat conduction in solids, Fourier analyzed data on the rate at which average temperature increases as one descends deeper below the Earth's surface; he also analyzed the attenuation of day/night or summer/winter temperature fluctuations with depth. (Fourier's solution for the latter problem will be derived in Chapter 7.) Based on these analyses, Fourier concluded that the flow of heat outward from the interior of the Earth was utterly insignificant in comparison to the heat received from the Sun. We will see shortly that this situation applies to other rocky planets as well: dry rock is a good insulator, and doesn't let internal heat out very easily.

If the Earth is continually absorbing solar energy, it must also have some way of getting rid of it. Otherwise the energy would have accumulated over the past eons, leading to a molten, incandescent uninhabitable planet (see Problem 3.2) – which is manifestly not the case. Fourier seems to have known that there was little or no matter in the space through which planets plied their orbits, and so he posited that planets lose heat almost exclusively through emission of infrared radiation (called "dark heat" at the time).[2] He also knew that the rate of emission of "dark heat" increased with temperature, which provided a means for an equilibrium temperature to be achieved: a planet would simply heat up until it radiated infrared energy at the same rate as it received energy from the Sun. Finally, Fourier refers to experiments showing that something in the atmosphere emits infrared radiation downward toward the ground, and seems to have been aware also of the fact that something in the atmosphere absorbs infrared. Based on these somewhat sketchy observations, Fourier inferred that the Earth's atmosphere retards the emission of infrared to space, allowing it to be warmer than it would be if it were airless.

Fourier's treatise made it clear that the thermal emission of infrared light was not just useful for astronomical observations – it was in fact part and parcel of the operation of planetary climate. At Fourier's time the state of understanding of infrared radiation emission was not sufficiently developed to allow him to complete the calculation he set up. Nonetheless, he correctly formulated the problem of terrestrial temperature as one of achieving a balance between the rate at which solar radiation is absorbed and the rate at which infrared is emitted. With this great insight, the modern era of study of planetary temperature had begun. Fleshing out the "details," however, required major advances in several areas of fundamental physics. The basic principles of planetary energy balance, and of the manner in which an atmosphere increases planetary temperature, are introduced in Chapter 3 and elaborated on in the earlier parts of Chapter 4.

One of the many details that needed to be settled was the question of which components of the atmosphere affected the transmission of infrared radiation. In 1859 Tyndall found that the dominant components of the Earth's atmosphere – nitrogen (in the form of N_2) and oxygen (in the form of O_2) – are very nearly transparent to infrared radiation. He found instead that it was two relatively minor constituents – water vapor and CO_2 – which accounted for most of the infrared absorption and emission by Earth's air. Gases of this sort, which let solar energy through virtually unimpeded but strongly retard the outward loss of infrared radiation, are known as *greenhouse gases*. Their warming effect on the lower portions of a planet's atmosphere, and on its surface (if it has one), is called the "greenhouse effect." The term was not coined by Fourier, and in some ways is misleading, since real greenhouses do not work by blocking infrared emission. However, the glass or plastic enclosure of a real greenhouse does warm the interior by reducing heat loss to the environment while allowing solar heating, and in that sense – viewed as a broader metaphor for the implications of energy balance – the analogy is apt. Besides CO_2 and water vapor, we now know of a number of additional greenhouse gases, including CH_4 (methane), which may have played a very important role on the Early Earth, and plays some role even today. In fact, it turns out that in some very dense atmospheres such as that of Titan, even nitrogen can become a greenhouse gas. What determines whether a molecule is or is not a good greenhouse gas, and how do we characterize the effects of individual gases, and thus the

---

[2] Fourier also refers to the importance of heating from what he calls the "temperature of space." It is unclear whether he thought there was some substance in space that could conduct heat to the atmosphere, or whether he was referring to some invisible radiation which pervades space. His inferences regarding the importance of this factor were erroneous – the only real error in an otherwise remarkable paper.

influence of atmospheric composition on climate? These questions will be taken up in the latter half of Chapter 4.

In thinking about the effect of greenhouse gases on climate, it is important to distinguish between *long-lived greenhouse gases* which are removed slowly from the atmosphere on a time scale of thousands of years or more, and *short-lived greenhouse gases* which are removed on a time scale of weeks to years by condensation or rapid chemical reactions. The short-lived greenhouse gases act primarily as a feedback mechanism. Their concentration adjusts rapidly to other changes in the climate, serving to amplify or offset climate changes caused by other factors – including changes due to long-lived greenhouse gases. Long-lived greenhouse gases can also participate in feedbacks, but only on time scales longer than their typical atmospheric adjustment time. Whether a greenhouse gas is long-lived or short-lived depends on environmental conditions. On the Earth, CO_2 is a long-lived greenhouse gas but water vapor is a short-lived greenhouse gas; however, on Mars, which gets cold enough for CO_2 to condense, that gas can be considered short-lived.

Greenhouse gases are largely invisible, but the atmosphere also holds a readily visible component that exerts a profound influence over our planet's energy balance – the clouds. Clouds on Earth are composed of suspended droplets of condensed water, in the form of liquid or ice. Clouds, like water vapor, act as a short-lived greenhouse gas affecting the rate at which infrared can escape to space. The infrared opacity of clouds is used routinely in weather satellites, since this property makes cloud patterns visible from space even on the night side of the Earth. However, clouds affect the other side of the energy balance as well, because cloud particles quite effectively reflect sunlight back to space. The two competing effects of clouds are individually large, but partly offset each other, so that small errors in one or the other term lead to large errors in the net effect of clouds on climate. Moreover, the effect of clouds on the energy budget depends on all the intricacies of the physics that determine things like particle size and how much condensed water remains in suspension. For this reason, clouds pose a very severe challenge to the understanding of climate. This is the case not just for Earth, but for virtually any planet with an atmosphere. The physics underlying the effects of clouds on both sides of the radiation balance will be discussed in Chapters 4 and 5.

1.3 INTO DEEPEST TIME: FAINT YOUNG SUN AND HABITABILITY OF THE EARTH

The Solar System was not always as we see it today. It formed from a nebula of material collapsing under the influence of its own gravitation, and once the nebula began to collapse, things happened very quickly. The initial stage of formation of the Solar System was complete by about 4.6 billion years ago. By this time, the Sun had begun producing energy by thermonuclear fusion; the formation of the outer gas giant planets and their icy moons by condensation, and the formation of the inner planets by collision of smaller rocky planetesimals, were essentially complete. The last major event in the formation of the Earth was collision with a Mars-sized body 4.5 billion years ago, which formed the Moon and may have melted the Earth's primitive crust in the process. All these collisions left behind a great deal of heat that had to be gotten rid of before the crust could stabilize. To determine how long it takes to get rid of this heat, we must learn about the mechanisms by which planets lose energy, and about how the rate of energy loss depends on temperature and atmospheric composition; this will happen in Chapters 3 and 4. It turns out that a planet loses energy almost exclusively by radiation of infrared light to space. While the precise rate of loss depends on the nature of the atmosphere, all estimates show that the surface of the Earth quickly cools to 2000 K, at which point molten rock solidifies; in the absence of an atmosphere, this process takes a thousand years or less, while with a thick atmosphere it could take as long as two or three million years.

Once a solid crust forms, the flow of heat from the interior of the Earth to the surface is sharply curtailed, because heat diffuses very slowly through solid rock. In this situation, supply of heat from the interior becomes insignificant in comparison with the energy received from the Sun, and the Earth has settled into a state where the climate is determined by much the same processes that determine today's climate: a competition between the rate at which energy is received from the Sun and the rate at which energy is lost to space by radiation of infrared light. This is very likely to have been the case by 4.4 billion years ago, if not earlier. There are no actual rocks as old as this, but there are individual zircon crystals embedded in the Jack Hills formation of Western Australia which are 4.4 billion years old. Zircons of a similar age are also found within the 3.7-billion-year-old crustal rocks of the neighboring Narryer Gneiss Complex. These crystals provide indisputable evidence for the existence of at least some continental crust of a sort very like that we see today; they also provide convincing though less certain evidence of the existence of liquid water in contact with the early continental crust. The existence of liquid water does not in itself put much constraint on temperature, since water can be maintained in a liquid state even at temperatures in excess of 500 K, provided the pressure exerted by water vapor in the atmosphere is high enough. The thermodynamics needed to address this issue will be introduced in Chapter 2. Certain aspects of the chemical composition of the zircons, however, suggest that they interacted with near-surface water having a temperature of 100 °C or less. By 4.4 billion years ago, it would appear, the Earth was no longer a molten volcanic inferno.

The precise nature of the climate evolution between 4.5 and 3.8 billion years ago is obscure at present. Depending on the composition of the atmosphere, the surface temperature could have been as high as 200 °C or low enough to cause the ocean (if any) to freeze over completely, and the climate could well have swung wildly between the two extremes. In addition, the dates of lunar craters indicate that the Earth very likely underwent a period of heavy bombardment by interplanetary debris between 4.1 and 3.8 billion years ago; it is generally supposed that this *late heavy bombardment* affected the rest of the inner Solar System as well, though that is far from certain. The energy brought in by impacts during this period could easily have been sufficient to bring surface temperatures episodically to values well in excess of 100 °C, sterilizing any nascent ecosystems. Life, if any, may have waged and won a battle for survival in deep ocean refugia.

By 3.8 billion years ago, the veil begins to lift. This is the age of the oldest intact rocks, found today in the Isua Greenstone Belt of Greenland. The appearance of these rocks marks the end of the *Hadean eon*, and the dawn of the *Archean eon*. Remnants of 3.7-billion-year-old shales in the Isua formation show the unmistakable signs of deposition of sediments in open water. More intriguingly, these shales are rich in organic carbon, and this carbon preserves a chemical signature generally associated with microbial activity – life. The Barberton formation of South Africa and the Warrawoona formation of Australia, both about 3.5 billion years old, contain layered carbonate sedimentary structures known as *stromatolites*, which in later times are known to be laid down by microbial mats. This is not an unambiguous sign of life, since inorganic processes can also produce stromatolite-like features. Be that as it may, the early stromatolites certainly require ponds of open water evaporating into air. The Barberton and Warrawoona formations also contain microscopic features that are suggestive of bacterial fossils, though not unambiguously so.

The record of surface conditions during the subsequent billion years is hardly continuous, but preserved rocks dating to this period very commonly show a sedimentary character of a type most easily explained by deposition in an open, unfrozen ocean. The first truly unmistakable microbial fossils date to 2.6 billion years ago, where they are found in the Campbell formation of Cape Province, South Africa, and argue for open water conditions having a moderate temperature. At about this time, we bid farewell to the Archean eon, and enter the *Proterozoic eon*, which extends to the appearance of animal life 544 million years ago. Certain fine-grained silica based sedimentary rocks known as *cherts* preserve information about past temperatures, as well as a wealth of fossils. Very ancient cherts contain no unambiguous microbial fossils, but certain aspects of their chemical composition point to temperatures as high as 70 °C at 3.5 billion years ago, declining to 60 °C at 2 billion years ago, and declining further to 30 °C at 1 billion years ago. Well-preserved ancient cherts are rare, however, so this data by no means implies that temperatures were uniformly warm on the young Earth. It only indicates that the Earth attained high surface temperatures at least part of the time; there is ample room to hide lengthy cold periods within the gaps in the chert record, as we shall soon see.

The earliest geological indication of the presence of glaciers on Earth occurs in the upper part of the Pongola formation of South Africa, and dates to 2.9 billion years ago. The evidence consists of glacial sedimentary deposits called *diamictites*, material of a sort usually transported by floating ice, and even glacier-scratched rocks. This does not mean that there were no earlier glaciations, but in light of the chert record and widespread occurrence of marine sedimentary rock it seems fairly certain that the Earth did not spend the bulk of its earlier history locked in a deep-freeze. Still, the Pongola glaciation seems to mark the beginning of Earth's long flirtation with ice. The Makganyene glaciation beginning around 2.3 billion years ago, recorded in rocks of the Transvaal group of Southern Africa, was a big one, and may well have been global. We know this because a record of the Earth's magnetic field is preserved in the rocks, and this can be used to infer the latitude at which the rocks were located when the glacial deposits were laid down. This *paleomagnetic data* shows that there was ice within 12 degrees of the Equator, strongly suggestive of a global glaciation.

The first unambiguous bacterial microfossils (found in the Campbell group of South Africa) date to 2.6 billion years ago, shortly before the Makganyene glaciation. While earlier fossil and geochemical evidence is very strongly suggestive of life, the Campbell group fossils are the ocular proof that biology was well under way. These fossils mark a watershed in another important way, in that they are identifiable as *cyanobacteria* – the type of organisms that produce oxygen by photosynthesis. The issue of when cyanobacteria evolved is hotly debated, with some lines of indirect evidence putting their appearance early in the Archean and others dating their onset to the time of the Campbell group microfossils. Be that as it may, the appearance of these fossils speaks for a fairly benign environment, with open water and temperatures no more than about 40 °C. After the Makganyene glaciation, microbial fossils become quite abundant. The two-billion-year-old Gunflint Chert of Canada is one of many such marine sedimentary formations in which cyanobacterial microfossils are preserved.

So far, no glaciations have been reported in the period between two billion years ago and 800 million years ago, though there are abundant sedimentary rocks dating to this time. The record is far from continuous and the lack of glaciations in this period may be an artifact of preservation, but the evidence certainly indicates that icy climates were not dominant at this time, and were probably quite rare. The long hiatus in ice is terminated by the massive – and possibly global – Snowball Earth glaciations of the Neoproterozoic, about 700 million years ago. Thereafter, the climate alternated between fairly lengthy periods when the Earth was ice-free or nearly so, and periods when there was at least some ice in polar regions. The ice never again, however, reached the nearly global proportions it attained during the Neoproterozoic, suggesting that the Earth passed some new threshold of climate stability in the Neoproterozoic. What might that be? This is one of the central questions of climate science.

Our overall picture of Earth history is that liquid water and moderate temperatures appeared at least episodically very shortly after the Moon-forming collision, and that the next three billion years had widespread open water with temperatures probably not exceeding 70 °C and generally much less. These conditions were punctuated by occasional glaciations, only a very few of which may have been global in extent. It was an environment that could support the evolution and survival of life, including (by 2.6 billion years ago, if not before) photosynthetic life requiring moderate temperature conditions. Let's keep this picture of relative stability in mind as we go on to discuss long-term changes in the atmosphere and the Sun – the two principal ingredients determining the Earth's climate, or indeed that of any planet.

There are many processes at play that cause the composition of a planet's atmosphere to evolve over time. In the earliest times, bombardments can help supply atmosphere-forming volatiles such as water, nitrogen, and carbon dioxide. Equally, however, sufficiently energetic bombardments can cause loss of atmosphere through literally splashing it into orbit. On a volcanically active planet with a hot interior, such as the Earth or Venus, or the younger Mars, new atmosphere is continually being supplied by outgassing of volatile substances from the interior of the planet. The heat needed to keep the interior of the planet churning so it can recycle minerals formed at the surface and cook out volatile gases in the hot interior is supplied by leftover heat from formation of the planet and by radioactive decay. How long this process can continue before the planet freezes out and becomes tectonically inactive depends on the size of the planet and the stuff it is made of; the nature of the gases which come out of volcanoes and other types of vents depends on the chemistry of the planet. For example, the early segregation of iron in the Earth's core made it harder to bind up oxygen in minerals, and therefore resulted in fairly oxidized gases like carbon dioxide (CO_2) and sulfur dioxide (SO_2) being released in preference to gases like methane (CH_4) and hydrogen (H_2) – though some of the latter two do nonetheless escape. The interior Earth also outgasses water vapor (H_2O), which is cooked out of hydrated minerals; the volume of the oceans appears to have been in a steady state for a long time, though, indicating that the rate of release is balanced by the rate of formation and subduction of new hydrated minerals at the surface. Nitrogen (N_2) is fundamentally different from other current and past constituents of the Earth's atmosphere as it does not readily get incorporated into the minerals that form the bulk of Earth's crust and interior. Unlike, say, CO_2, nitrogen does not cycle through the Earth's interior. The bulk of the Earth's N_2 is in its atmosphere and stays there, where it has probably been for almost all of our planet's history. This is likely to be the case as well for any other rocky planet made of stuff similar to the Earth – iron, oxygen (mostly bound up in minerals), silicon, magnesium, and sulfur.

While atmosphere is being supplied by outgassing from the interior, other processes cause material to be lost from the atmosphere. Parts of a planet's atmosphere extend far out from the surface, where hot, fast-moving molecules can reach escape velocity and escape to space. Besides escape from random molecular motions, the solar-heated tenuous outer atmosphere can sustain fluid flows which cause atmospheric mass to fountain systematically into the void. In addition, the solar wind can literally blow away the outer portions of an atmosphere; the extent to which this happens is affected by the intensity of the planet's magnetic field, which shields the atmosphere from the solar wind. As outer parts of the atmosphere are eroded, new gases from lower altitudes well up to replace the lost material, sustaining the gradual loss of atmosphere. All three mass loss processes preferentially remove lighter molecules, either because lighter molecules move more swiftly for a given temperature, or because the outer atmosphere is enriched in gases having a lower molecular weight. For a given density, a smaller planet has lower surface gravity, and so binds its atmosphere less tightly; in consequence, escape of atmosphere to space proceeds more rapidly on a small planet. Impacts by large, swift bodies can impart sufficient energy to part of the atmosphere to blast it into space. This mechanism of atmosphere loss does not discriminate as to molecular weight, but as with the other mechanisms, it is easier for a small planet to lose atmosphere this way. Overall, the Moon or Mars is more prone to lose atmosphere than more massive bodies such as the Earth or Venus, to say nothing of Jupiter or Saturn. For Earth and Venus, escape to space is significant only for H_2 and He, and of these the latter is important only as an indicator of planetary history rather than as a physically or chemically active component of the atmosphere. Saturn's satellite Titan is an interesting case, as it maintains a mostly N_2 atmosphere more massive than that of Earth (per unit surface area) despite having a surface gravity lower than that of the Moon. The very cold temperature of Titan helps it retain its atmosphere, but it is nonetheless likely that the persistence of the atmosphere requires a substantial rate of resupply from the interior of the planet.

Some components of the atmosphere can also be lost through chemical reactions with rocks at the Earth's surface. A particularly important example of this is the class of reactions commonly known today as *Urey reactions*,[3] which remove CO_2 from the atmosphere. When CO_2 dissolves in water, it forms a weak acid (carbonic acid), which reacts with silicate minerals (e.g. $CaSiO_3$) to form carbonate minerals (e.g. $CaCO_3$, or "limestone"). The reactions that form carbonate take place only in the presence of liquid water, so if a planet becomes so hot that liquid rain never reaches the surface, or if it somehow loses its water altogether, then CO_2 outgassed from the interior of the planet will accumulate in the atmosphere until the interior source is depleted or the rate of supply is balanced by loss to space. On Earth, all of the CO_2 presently in the atmosphere could be removed by the Urey reactions within 5000 years, forming a layer of limestone a mere 5 millimeters thick; if all the carbon stored in ocean water were to outgas as CO_2 and react to form limestone, the process would take a half million years and form a layer a half meter thick.

Life itself, once it appears, has a profound effect on atmospheric composition. While little methane escapes directly from the Earth's interior, bacteria known as *methanogens* can synthesize it from H_2 and CO_2 or from organic material produced by other organisms. Methanogens may well have dominated the ecosystems of the Earth's first two billion years, potentially allowing a methane-dominated atmosphere to build up. The advent of life also had a profound effect on nitrogen cycling. The bonds holding together N_2 are so strong that

---

[3] The reactions are named after the University of Chicago geochemist Harold Urey, who discussed them in a 1952 book called *The Planets: Their Origin and Development*. Although modern science was made aware of the importance of these reactions through Urey's work, the reactions were first introduced by the French chemist and metallurgist J. J. Ebelmen more than a century earlier. Ebelmen also introduced the notion that the silicate/carbonate reactions play an important role in determining atmospheric CO_2 and hence Earth's climate. Similar ideas were independently rediscovered by the Swedish geochemist A. G. Högbom in 1894, and then finally by Urey. For more details of the history see Berner 1996: *Geochim. Cosmochim. Acta.* **60**, 1633–1637.

in the abiotic world only rare energetic events such as lightning strikes can form nitrogen compounds. In fact, though nitrogen is an essential ingredient of all living material, higher forms of life – including all plants and animals – are incapable of performing the chemical magic that makes N_2 available to organisms. This trick is accomplished by nitrogen-fixing bacteria, which can efficiently transform atmospheric N_2 into ammonium (NH_4^+), in turn transformed into nitrate (NO_3^-) which can be used by other organisms in the synthesis of living matter. Other bacteria, in oxygen-starved conditions, can make a living combining the oxygen in nitrate with carbon, returning N_2 to the atmosphere in the process. It was only in 1910, with the invention of the Haber process for turning atmospheric N_2 into ammonia (NH_3), that humans caught up with the bacteria. This innovation has become essential to the human population, as the demands of industrial-style agriculture have far outstripped the ability of the natural bacterial ecosystem to supply nitrate (which is not to deny that other forms of agriculture might be able to live within the means provided by our bacterial friends). Nitrogen-fixing bacteria are still way ahead of industry in terms of chemical sophistication, though, since the Haber process requires molecular hydrogen (made from fossil methane), an iron catalyst, temperatures exceeding $400\,^\circ$C, and an operating pressure over 200 times that of air at the Earth's surface; nitrogen-fixing bacteria can do the same trick, in contrast, in ambient temperature/pressure conditions and using materials found readily in their immediate environment.

In the absence of oxygen-producing photosynthetic life, only minuscule amounts of oxygen would be present in the atmosphere, since only a trickle could be produced through the breakdown of H_2O by exposure to sunlight. That there was very little oxygen in the early atmosphere (under 0.2%, compared with 20% today) is confirmed by the widespread presence of striking rock structures known as *banded iron formations* until 2.4 billion years ago. Banded iron formations can be laid down only when iron is very soluble in the ocean and can be transported long distances. This requires low oxygen, since in the presence of oxygen iron forms compounds that are not very soluble in water. Additional evidence stemming from the chemical composition of certain sulfur-containing minerals indicates that during at least some periods earlier than 2.6 billion years ago the atmospheric oxygen content might in fact have been orders of magnitude lower than 0.2%.

Note that the appearance of oxygen-producing photosynthesis is not synonymous with the rise of oxygen in the atmosphere. For oxygen to accumulate, a sufficient proportion of the organic matter must be buried before it is oxidized by other bacteria, taking the synthesized oxygen right back out of the atmosphere. Further, if the Earth had accumulated a great stock of available organic matter in the ocean during its anoxic phase, this backlog would have to be worked off before oxygen could begin to accumulate to any significant degree. Be that as it may, banded iron formations begin to falter somewhat after the time of the Campbell group cyanobacterial microfossils, and disappear entirely by around 2 billion years ago. By this time, oxygen may have made up at least 3% of the atmosphere. Once oxygen made its appearance in the atmosphere at significant concentration, it changed all the rules of atmospheric chemistry, since it is so powerfully reactive. In particular, it made it much harder for CH_4 and H_2 to accumulate in the atmosphere, since the former oxidizes readily to CO_2 and the later to H_2O. The rise of oxygen may also have fostered another great biological innovation – the *eukaryotic cell*, which has a complex internal organization with specialized structures, including a nucleus within which genetic material is segregated. We are made of eukaryotic cells, as are all animals and plants. Eukaryotic cells make their first unambiguous appearance in the fossil record about 1.5 billion years ago, in the Roper group shales of Australia, though biochemical molecules preserved in ancient sediments suggest strongly that eukaryotic life may have evolved much earlier. However the answer to that issue may shake out, it is certain that eukaryotic life – even of the single-celled variety – did not proliferate and diversify until much later in the Proterozoic.

There was a sporadic reappearance of banded iron formations at the time of the Neoproterozoic Snowball events (600–700 million years ago), but by 600 million years ago oxygen was approaching its present concentration, and banded iron formations disappeared for good (at least so far). This second pulse of oxygenation made the rise of multicellular animals possible, which happened in a great flurry of biological innovation known as the *Cambrian Explosion*, occupying a remarkably short span of years around 543 million years ago. It is even possible that the rise of animals helped to stabilize atmospheric oxygen levels, by providing a reliable means of transporting organic carbon to the sea floor where it can be buried and preserved.

All of this atmospheric evolution takes place against a backdrop of a gradually brightening Sun. The energy produced within a star leaves the star almost exclusively in the form of electromagnetic radiation – loosely speaking, light of all wavelengths. The net power output is called the *luminosity*, and is measured in watts (a measure of energy per unit time), just as if the star were a light bulb. Stars like the Sun get their energy by fusing hydrogen into helium, and as time goes on the proportion of helium in the Sun increases, thus increasing the mean molecular weight of the Sun. This in turn means that the core of the Sun needs to contract and heat up in order to maintain the pressure required to balance gravity. The increased density and temperature increases the rate of fusion more than the reduced availability of hydrogen reduces it, so the rate of production of energy – and hence the solar luminosity – *increases* with time. The resulting evolution of luminosity over time rests on fundamental aspects of solar physics that are not seriously in question, and does not depend greatly on the finer points of solar modeling. The results of the standard solar evolution model can be expressed as

$$\mathcal{L}_\odot(t) = \frac{\mathcal{L}_\odot(t_\odot)}{1 + \frac{2}{5}(1 - t/t_\odot)} \tag{1.1}$$

where $\mathcal{L}_\odot(t)$ is the luminosity of the Sun at time t and $t_\odot$ is the current age of the Sun, usually taken as 4.6 billion years. This formula was originally proposed to describe the younger Sun, but it continues to be reasonably accurate out to 4 billion years in the future as well.

It follows from Eq. (1.1) that 4 billion years ago the Earth received solar energy at only 75% of the rate it does today. All other things being equal – atmospheric composition in particular – that means the Earth would have been colder than it is today. How much colder? We will learn how to do this calculation in the simplest way in Chapter 3, and add sophistication to the calculation in Chapter 4. It turns out that if the atmospheric composition were the same as today's atmosphere throughout Earth's history, then Earth should have been completely frozen over 4 billion years ago, and given that ice reflects sunlight so well, it should still be completely frozen over today. However, we know that incidents of global ice coverage are rare in Earth history, if indeed they happened at all; we know with rather more certainty that the Earth is not solidly frozen over today. This contradiction is generally known as the *Faint Young Sun Paradox*, though of course, like most paradoxes, it is not paradoxical at all once one understands what is going on. Calling it a "paradox" is just a way of starkly bringing home the fact that to account for the basic facts of Earth's climate history, the atmosphere must indeed have undergone massive changes – changes of a sort that could substantially affect climate. How much would we need to increase CO_2 or CH_4 in order to make up for the faintness of the young Sun? This is another of the Big Questions. It will be answered in Chapter 4. A related Big Question is the extent to which CH_4 (or some other long-lived greenhouse gas) substituted for CO_2 in maintaining the Earth's warmth during the Archean. There is scattered evidence from mineral composition of fossil soils that during some periods of the Archean the CO_2 concentration might not have been high enough to offset the Faint Young Sun. This has led some researchers to jump to the conclusion that CH_4 played the decisive role throughout the Archean, but such a viewpoint rests on exceedingly shaky evidence. It matters a great deal whether CO_2 or CH_4 did the trick, since the long-term control of the two gases is governed by very different processes, with very different time scales. The matter, at present, must be considered unresolved.

A corollary to the above resolution of the Faint Young Sun Paradox is that any atmosphere that would be sufficient to keep the Early Earth unfrozen would make it uninhabitably hot with the present solar output. The atmosphere must have somehow changed in lock step with the brightening Sun, in precisely such a manner as to keep the Earth in a habitable temperature range – one where liquid water exists at or near the surface, but where the water never gets hotter than about 100 °C – or indeed, even hotter than about 50 °C if cyanobacteria are to survive. It defies belief that the required co-evolution of atmosphere with solar output could maintain the required temperature range purely by coincidence, so it seems likely that some sort of temperature-regulating mechanism is in operation. In Chapter 8 we will see how the Urey reaction can participate in a geochemical thermostat for the Earth and similar planets. The Snowball episodes represent temporary, and evidently rare, breakdowns of the temperature-regulating mechanism. Whatever the regulating mechanism may be, it must be sufficiently fail-safe to allow for recovery from global or near-global glaciations.

Given what we laid out earlier concerning the myriad processes churning out turmoil in atmospheric composition, the reader should quite rightly feel it a bit silly to make the stipulation "all other things being equal" in the statement of the Faint Young Sun Paradox. Indeed, even if the solar output were constant over time, any number of the aspects of atmospheric evolution we discussed earlier would have been sufficient to freeze or fry the Earth, so the gradual increase in solar output has no special claim on our attention in this regard. To be fair, when the "paradox" was first laid out, much less was known about the ways the Earth's atmosphere evolved, and with much less certainty. From the standpoint of current science, however, the traditional framing of the Faint Young Sun Paradox leaves a lot to be desired. It would be more satisfactory to refer to the Habitability Problem, which could be stated as follows: *How can the temperature of a planet be maintained in the habitable range for billions of years, in the face of geological, geochemical, and biological turmoil in atmospheric greenhouse gas composition, and in the face of gradual increase in solar output?* This is indeed one of the grandest of questions. The material discussed in Chapter 8 provides a plausible solution, but the book is far from closed on the Habitability Problem.

The history of the Faint Young Sun problem reveals something magnificent and deeply inspiring about the nature of discovery in Earth and planetary climate. The basic physics underpinning the energy source of stars was worked out by Hans Bethe by 1939, and the existence of benign conditions on the Early Earth was known (or at least inferred) even earlier. Yet it was not until 1972 that Carl Sagan and George Mullen put the two bits together and inferred that there was indeed a big problem, requiring a profound answer.[4] This insight sparked a revolution in thinking about planetary climate, which was in its own way as earth-shaking as the discovery of DNA was in its own domain. This history highlights a lovely thing about our subject: the most profound new phenomena are often discovered by putting together a few bits of basic physics and chemistry in creative new ways. For the most part, new ideas come from playing with simple models, not from enormous incomprehensible computer simulations that take huge teams to put together. The entire goal of this book is to teach students to think the way Carl Sagan and other innovators did, and to provide the tools needed to build the simple models needed to turn a bright idea into real science.

The problem of maintaining long-term climate stability is not just an issue for Earth. Planets that could support some form of life are naturally of special interest, and while other forms of life than those we know of might prefer quite different conditions than prevail on Earth, the long-term maintenance of climate in a fairly narrow "habitable" range clearly would make it easier for life to evolve and persist. Any potentially life-bearing planet in any solar system anywhere must negotiate a way to maintain long-term habitability in the face of the gradual increase of the brightness of its sun and gradual or not-so-gradual changes in the composition of its atmosphere. Naturally, such considerations apply to the evolution of the climates of other planets in our very own Solar System. Venus and Mars did not manage to maintain their habitability, or perhaps were never in a habitable state. How close did we come to the same fate?

1.4 GOLDILOCKS IN SPACE: EARTH, MARS, AND VENUS

Until well into the 1960s, science fiction stories about Venus generally portrayed it as a steamy jungle planet, but one where intrepid explorers could perhaps survive unprotected on the surface. The idea of a jungle and breathable air was of course unfounded speculation, but the general picture of the climate was not wholly without merit. After all, the dense reflective cloud deck of Venus was readily observable – it is what makes Venus so bright as the "evening star" – and the reflection of sunlight could easily make up for the fact that Venus is closer to the Sun than is Earth. In fact in Chapter 3 we'll see that the reflectivity *more than* makes up for the proximity. In the late 1950s the picture of Venus as a habitable world began to unravel. Recall that the temperature of Earth's Moon was determined by examining infrared radiation from that body. Viewed in the infrared spectrum Venus appeared quite cool, but in the microwave ("radar") spectrum it was far too bright. In fact, seen in the microwave, Venus radiated like a body with a temperature well in excess of 600 K (327 °C). A popular hypothesis at the time was that the anomalous microwave emission arose in the upper portions of the Venusian atmosphere. Another view held that the microwave emission came from the surface of the planet, and that the atmosphere was transparent to microwaves but relatively opaque to infrared. The latter idea suffered from the lack of a plausible mechanism to make the surface of Venus so hot. Then, in 1960, the young Carl Sagan proposed that Venus has a very thick atmosphere rich in greenhouse gases, which would heat up the surface to the required temperature. Little was known about the mass

---

[4] Sagan and Mullen proposed accumulation of atmospheric ammonia as a resolution to the paradox, but later work on atmospheric chemistry showed that sunlight destroys ammonia too rapidly to allow it to build up to the required concentrations. Attention later shifted to CO_2 and CH_4.

of the atmosphere or its composition at the time, but Sagan developed simple models of the greenhouse effect of a thick atmosphere, which showed that the trick could be accomplished with an atmosphere consisting of mostly carbon dioxide with some water vapor mixed in, having a total mass three or four times that of the Earth's atmosphere. Sagan even recognized that since the planet was too hot for water vapor at the hypothesized concentration to condense and reach the surface as liquid, the Urey reaction (which removes carbon dioxide from the atmosphere and turns it into limestone) could not take place. This would make it easy for carbon dioxide to accumulate in the atmosphere, though even Sagan did not envision just how far this would go.

A series of interplanetary probe missions over the next two decades – four US Mariner missions, two US Pioneer missions and sixteen Soviet Venera missions including eight Venera missions that returned data successfully from the surface – refined the estimates of surface temperature and substantially revised the conception of the atmospheric mass and composition. By the late 1970s, it was known that the surface temperature was nearly uniform at 737 K. The atmosphere was found to be much more massive than originally thought, in fact sufficiently massive to raise the surface pressure to 92 times that of Earth's atmosphere. And it was found that the atmosphere consisted almost entirely of carbon dioxide, with only traces of water vapor remaining. The thick clouds that give Venus its high reflectivity were found to be made not of water, but of droplets of sulfur dioxide and concentrated sulfuric acid. It took the better part of another decade before the challenges of dealing with the effect of such an exotic atmosphere on climate were fully mastered and a satisfactory account of the high surface temperature could be given. Still, the initial exploration of the problem was carried out with simple models very like the ones we will introduce in the first half of Chapter 4. The discovery of the true nature of Venus' climate is another illustration of the tenet that big ideas grow from little models.

Mars yielded up its climatic secrets somewhat earlier, because it was not hidden behind the thick atmospheric veil that complicated observation of Venus. Mars was observed in the infrared using the Mt. Wilson telescope during the opposition (time of closest approach) of 1926 and 1927. Like all infrared observations, the interpretation was complicated by interference from the Earth's atmosphere. By 1947, the understanding of the effect of atmospheres on the emission of infrared light had progressed to the point that the mass and composition of the Martian atmosphere could be estimated. Based on these measurements, Kuiper estimated that Mars had an atmosphere that was almost entirely CO_2, with a sufficient mass that the surface pressure on Mars would be only 0.03% of the surface pressure on Earth. This turns out to be an underestimate of the true mass by about a factor of 20, but even so, the picture of Mars as a nearly airless planet was not far wrong. By way of comparison, infrared observations of Venus interpreted in the 1940s using similar techniques suggested that the surface pressure of Venus was a fifth that of Earth – an underestimate by a factor of nearly 500. The Mt. Wilson infrared observations of Mars also indicated that the atmosphere was almost entirely CO_2, with almost no water – so little, in fact, that water vapor wouldn't condense until temperatures fell below $-60\,°C$, and at those temperatures it would condense into frost, not liquid. Infrared observations showed further that the visible polar ice caps of Mars are most probably made of water ice rather than frozen CO_2. Temperature estimates based on the Mt. Wilson infrared observations were less informative. Nicholson and Pettit, in the same paper in which they discussed lunar temperatures, noted a very large day/night cycle of Martian infrared, indicating extreme diurnal contrasts unlike those found on Earth. They concluded that this was due to the lack of water vapor in the Martian atmosphere, but we shall encounter the true reason in Chapter 7. Writing in 1947,

Adel reported quantitative estimates of Martian surface temperature ranging from as low as 236 K to as high as 318 K (or even higher if the surface was assumed to emit infrared inefficiently, as does granite). Some of the variation in reported temperatures may have been due to the fact that these were not whole-disk observations, insofar as Mars exhibits extreme temperature contrasts. The higher end of the estimates based on Mt. Wilson observations turned out to be far greater than the actual maximum ground temperature encountered anywhere on the planet.

Given the thin atmosphere, what does theory lead us to expect about the Martian surface temperature? By 1947, it was a simple exercise to compute the expected temperature of airless bodies like the Moon, using arguments like those we will discuss near the beginning of Chapter 3. Planets with atmosphere present more of a problem. The atmosphere of Mars is thin compared with that of Earth, but how thin does an atmosphere have to be (particularly if it's pure CO_2) in order to have a minimal effect on planetary temperature? We'll learn how to answer that question in the latter portions of Chapter 3. Based on similar reasoning, Kuiper, writing in 1947, inferred correctly that the atmosphere would have only minor effect on temperature, in particular allowing severe night-time temperature drops. By the 1940s Mars was already looking inhospitable – a mostly cold, dry, nearly airless body where (it was still hoped) conceivably lichens might eke out a living but certainly not Thuvia, Maid of Mars.

Ground-based and theoretical estimates of the Martian climate improved gradually over the next decade, but the real breakthrough came with the Mariner flyby of 1965 and the two Viking orbiters and landers of 1976. Spaceborne infrared observations gave the first detailed picture of geographic variations of the ground and air temperature, and Viking provided *in situ* air temperature measurements on the ground. These observations consolidated the picture of Mars as a planet which (as we find it now) has more in common with the airless Moon than with Earth. Even hopes of lichens were dashed, though it is too soon to give up on bacterial life, especially in view of the innovative chemical entrepreneurship shown by non-photosynthetic bacteria on Earth. In discussing Martian temperatures, one must take care to distinguish the air temperature from the temperature of the ground itself, as the two differ considerably. At high noon in the tropics, the ground can indeed briefly get as warm as 300 K, though the temperature drops by 100 K or more at night. The air temperature also shows an extreme day/night cycle, but the peak daytime air temperatures are far less than the peak ground temperature; at the Viking Lander 1 site, in the tropics at 22 °N latitude, the daytime air temperature never exceeds 260 K, and plummets to under 200 K at night. The reason the air temperatures are so much lower than the ground temperatures will become clear in Chapters 4 and 6.

There are considerable seasonal and latitudinal variations in daytime temperature, and the southern hemisphere polar summer is notably warmer than the northern hemisphere polar summer. Night-time temperatures are comparatively uniform both geographically and seasonally; the Mars surface cools so fast that once it is dark, it evidently doesn't matter much whether it has been dark for a few hours or a hundred days. The southern hemisphere winter pole does get notably colder than the rest of the planet, dropping to as low as 160 K. The Viking landers also provided the first clear picture of surface pressure variations on Mars. They showed that the Martian surface pressure varies from a high of about 1% of Earth's surface pressure to a low of about 0.6%, with the lowest values occurring in southern hemisphere winter. Since surface pressure gives a measure of the mass of air in the atmosphere (see Chapter 2), the large variation of pressure indicates that a considerable portion of Mars' CO_2 atmosphere snows out over the north pole in northern winter and the south pole in southern winter, only to sublimate back into the atmosphere as spring approaches. A theory for the seasonal cycle of Martian temperature and pressure will be developed in Chapter 7.

So, Venus, like the porridge tasted by Goldilocks, is too hot. Mars is too cold, and Earth is just right. One could quite reasonably object that this is a view prejudiced by our own status as a form of terrestrial life, and that conditions "too hot" by our standards could well be "just right" for somebody else. However, it appears that there's nobody home on either Venus or Mars (not even a microbial somebody), so if conditions there are indeed "just right" for somebody, it must not be very easy for such a somebody to evolve, given that it didn't happen in the past four billion years.

Presumably, Venus started out with a composition rather similar to Earth. What went wrong? Why did it keep most of its CO_2 in its atmosphere, whereas most of Earth's CO_2 got bound up in carbonate rocks? Where did its water go? The answer came in 1967 with the theory of the *runaway greenhouse*, formulated first by M. Kombayashi and independently rediscovered shortly thereafter by Andrew Ingersoll of Caltech. This theory puts together two simple bits of physics, the first being that water vapor content of a saturated atmosphere increases exponentially with temperature (Chapter 2), and the second being that water vapor is a greenhouse gas (Chapter 4). When the two are put together, it is found that a planet which receives sufficient solar radiation can get into a runaway cycle where the planet warms in response to absorbed sunlight, which causes more water vapor to enter the atmosphere, which causes more greenhouse effect, which leads to further warming in an unstable feedback loop that doesn't end until the entire ocean is evaporated into the atmosphere. At that point, the water vapor in the upper atmosphere breaks down into hydrogen and oxygen under the influence of high-energy solar radiation, and the hydrogen escapes to space while the oxygen reacts with rocks. Without liquid water, the Urey reaction which turns CO_2 into limestone cannot take place, so all the outgassed CO_2 stays in the atmosphere. The runaway greenhouse theory (explored in Chapter 4) gives rather precise predictions of the circumstances under which a runaway can occur, and explains why Earth did not undergo a runaway despite the fact that it has a water ocean. The work of Kombayashi and Ingersoll is another example of the general idea that big ideas come from simple models. Their reasoning was based on simple radiation models of the sort developed in the first half of Chapter 4. This work also illustrates another general principle of planetary climate: profound results can be obtained by combining a few bits of very basic physics in a novel way. The general problem of water loss from Venus is one of the Big Questions.

Radar mapping from the Magellan orbiter of 1990 revealed another remarkable fact about Venus: unlike the Earth, with long-lived continents and a gradually subducted sea floor, Venus has a young-looking uncratered surface, suggesting that the crust may have been engulfed and resurfaced as recently as 500 million years ago. This has important implications for planetary habitability. Evidently, the formation of a planetary crust, as at the end of the Hadean, is not the end of the peril from fires in the deep. For habitability, the crust has to be relatively stable, engulfed slowly in subduction zones (as is the Earth's sea floor) rather than being subject to episodic catastrophic volcanism as seems to have been the case on Venus. However, if the crust is engulfed too slowly, then limestone that forms by the Urey reaction is only sluggishly recycled, leading to a drawdown of atmospheric CO_2 and (under some circumstances) a very cold planet; since photosynthesis uses CO_2 as a feedstock, low

CO_2 impedes habitability even if a planet is in an orbit where it does not need the greenhouse effect of atmospheric CO_2 in order to stay warm. The question of when a planet has *plate tectonics* like Earth or when it has episodic catastrophic resurfacing like Venus is another one of the great questions of planetary science, though one we will not take up at any great length in this book.

Mars may be impoverished in atmosphere compared with Venus, but it has something Venus lacks: a geological record of the distant past preserved in its crust. In fact, the ancient features on the surface of Mars are far better preserved than is the case on Earth. Mars appears to have lost most of its atmosphere quite early on, leading to a near-halt in the rate of erosion of surface features. The first high-resolution data of Martian surface features, returned by the Mariner mission, revealed a startling fact. Evidently, Mars was not always the dry, frigid planet cloaked in a tenuous atmosphere that we see today. The Mariner photographs revealed dry river-like channels for which the only likely explanation is flowing surface water, which would be impossible under the conditions of the present Martian climate. A more recent image of this type of feature is shown in Fig. 1.1. The rate of cratering of a planet goes down with time, so the features can be dated by counting superposed craters; many of the major river-like features date to very early in Mars history, perhaps 4 billion years ago. This led to the concept of a "warm, wet Early Mars." But how could Mars have been so much warmer than it is at present, at a time when the Sun was so much fainter? Mars presents an even more extreme version of the Faint Young Sun Paradox than does the Earth. It will probably come as no surprise that it was Carl Sagan who first pointed out the implications of Martian dry river networks. It is still somewhat disputed whether the surface features really demand that Early Mars be warm and wet, but

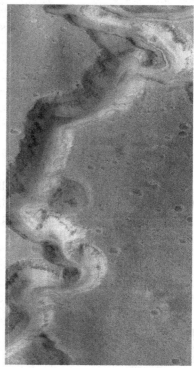

Figure 1.1 Nanedi Vallis on Mars, observed by the Mars Orbiter Camera on the Mars Global Surveyor Mission. The image covers an area 9.8 km wide and 18.5 km long.

adopting the warm-wet view, the resolution of the Faint Young Sun problem, as was the case for Earth, lies in the supposition that the Early Mars atmosphere had a substantially stronger greenhouse effect. What kind of atmosphere could warm Mars to the point that liquid water could flow long distances at the surface? That Big Question is taken up in Chapters 4 and 5.

If Mars started out with such a dense atmosphere, where did it go? Some possible answers are suggested in Chapter 8. Modern high-resolution images suggest other forms of massive climate change on Mars. In particular, there are tropical landform features suggesting that at some time in the past, glaciers formed in the Martian tropics, whereas virtually all the ice is today sequestered at the cold polar regions. The tropical glacier landforms suggest that the equatorial regions of Mars were colder than the poles during parts of Mars' history. How could that be? The answer is provided in Chapter 7.

The fact that Earth maintained habitable conditions while Venus succumbed to a runaway greenhouse and got too hot, and Mars lost its atmosphere and got too cold, raises the question of just how narrowly Earth escaped the fate of Mars and Venus. How much could Earth's orbital distance be changed before it turned into Mars or Venus, and how would the answer to this question change if Earth were more massive (making it easier to hold onto atmosphere) or less massive (making it easier to lose atmosphere)? If Mars were as large as Earth, would it still be habitable today? What if Venus were as small as Mars? Perhaps if the orbits of Mars and Venus were exchanged, our Solar System would have three habitable planets, instead of just one.

The range of orbital distances for which a planet retains Earthlike habitability over billions of years is known as the *habitable zone*. Determining the habitable zone, and how it is affected by planetary size and composition as well as the properties of the parent star, is one of the central problems of planetary climate.

1.5 OTHER SOLAR SYSTEM PLANETS AND SATELLITES

For the *gas giant* planets – Jupiter and Saturn – many of the most striking questions that arise are fluid dynamical in nature. These questions include the origin of the banded multiple-jet structure of the atmospheric flow, and the dynamics of long-lived atmospheric vortices, most famously Jupiter's Great Red Spot. We shall have little to say about such fluid dynamical questions in this book. However, the thermal structure of the atmosphere provides an essential underpinning for any dynamical inquiry. Moreover, the thermal structure determines the rate of heat loss from the planet, and therefore plays a crucial role in the long-term evolution of the gas giants. The thermal structure also affects atmospheric chemistry and the nature of the colorful clouds that allow us to visualize the spectacular fluid patterns on these planets.

The gas giants present an interesting contrast to rocky terrestrial-type planets, because of the lack of a solid surface. Instead of solar radiation having the possibility of penetrating to the ground and being absorbed there, thus heating the atmosphere from below, solar radiation on the gas giants is deposited continuously throughout the upper portion of the atmosphere as the solar beam propagates downward and attenuates. Further, unlike the case of the rocky planets where heat flux from the interior is an insignificant player in climate, the fluid nature of the gas giants allows considerable heat flux from the interior to escape to space. For both Jupiter and Saturn, this heat flux is comparable to the flux of energy received from the Sun. One of our objectives in subsequent chapters will be to learn how the distinct nature of atmospheric driving on the gas giants affects the thermal structure. The gas giants also offer an interesting opportunity to test ideas about how climate is affected by atmospheric composition. These planets are mostly made of H_2, with a lesser amount of He and trace amounts of a range of other substances, including ammonia (NH_3), methane (CH_4), and water, the latter three of which exist in both gaseous and condensed forms. The composition affects the thermodynamics of the atmosphere, as well as the optical properties for both infrared and visible light.

Uranus and Neptune are like the gas giants in that they have no distinct solid surface at any depth that could significantly affect the atmosphere. However, they are usually classified separately as *ice giants* because they contain a much higher proportion of ice-forming substances such as water, ammonia, and methane. The composition of the outer portions of the atmospheres can be determined by spectral observations, and contain a high proportion of hydrogen and helium. The overall density of the planets, however, constrains them to be composed primarily of an *ice mantle*. In the case of Uranus, the ice mantle must make up between 9.3 and 13.4 Earth masses worth of the total mass of the planet, which is 14.5 Earth masses. Similar proportions apply to Neptune. The commonly used term "ice mantle" is somewhat misleading, since the substance is actually a hot, slushy mixture that would be more aptly described as a water–ammonia ocean. Whatever term is used, the very thermal structure that determines the nature of the transition between the ice mantle and the more gaseous outer atmosphere engages all the same issues of atmospheric energy balance as one encounters on other planets. A novel feature of Uranus is its axial tilt. Its axis of rotation is nearly perpendicular to the normal to the plane of the orbit. In other words, the axis lies almost in the plane of the orbit. That means that in the Uranian northern hemisphere summer, the north pole is pointing directly at the Sun and the entire southern hemisphere is in darkness. By way of contrast, the Earth's axis is only tilted by 23.4° relative to the normal at present. The high axial tilt of Uranus potentially gives that planet an extreme seasonal cycle, though it will take a long time to observe it since Uranus' year lasts 84 Earth years. The effect of axial tilt on seasonal cycles of planets are discussed in general terms in Chapter 7. The very low solar radiation received at the distant orbits of Uranus and Neptune leads to extremely cold outer atmospheres, particularly in the case of Neptune. These planets provide an opportunity to examine the novel features of an atmosphere driven by an exceedingly weak trickle of solar energy, supplemented by an equally feeble trickle of heat from the interior. Despite the weak thermal driving, Neptune has by far the strongest winds in the Solar System, as well as a variety of interesting meteorological features. We will not say much about planetary winds, but as in the case of the gas giants, a good understanding of the thermal structure is a prerequisite for any attack on the meteorology.

The gas and ice giants also challenge our notion of *habitable zones* – orbits where a planet has some region where there are Earthlike temperatures allowing for liquid water. The gas and ice giants have no distinct surface, but there is some depth on each of them where the temperature is Earthlike and liquid water can exist. The atmospheres also have plenty of chemical feedstocks for organic molecules, including ammonia and methane. The pressures are no greater than those seen at the bottom of the Earth's ocean. One may have some prejudice in favor of surfaces for life to live on, but it must be recalled that on Earth life first arose in the oceans and indeed stayed there for many billions of years before venturing onto land. The gas and ice giants could just as well be thought of as being "all ocean" rather than "all atmosphere" so it is far from clear that they are inhospitable, at least for chemosynthetic forms of life that do not need much sunlight. Our thinking about habitable zones is overly prejudiced toward life that carries out its existence on a rocky surface.

From the standpoint of planetary climate, one of the most interesting Solar System bodies is not a planet at all, but a satellite. Titan, which orbits Saturn, is a fairly large icy body with a radius that is 76% of that of Mars. Because it is composed of ice rather than rock, the surface gravity is low: $1.35\,m/s^2$, which is actually lower than the Moon's surface gravity, though the Moon is smaller than Titan. What makes Titan interesting, however, is its dense atmosphere. The atmosphere of Titan consists mainly of nitrogen, with a surface pressure about 1.5 times that of Earth. What is even more interesting is that the lower portion of the atmosphere is about 30% methane. At the cold temperatures of Titan (about 95 K) methane can rain out, and it participates in a "hydrological" cycle analogous to that of water on Earth – but operating at a much colder temperature. In subsequent chapters we will develop the physics to examine the similarities and differences between the role of methane on Titan and that of water on Earth. Titan's atmosphere is also a seething organic chemical factory, with complex long-chain hydrocarbon hazes being manufactured from methane in the upper atmosphere. These hazes absorb solar radiation, shade the surface, and are a key player in Titan's climate. Such organic hazes were first discovered on Titan, but there are speculations that similar hazes could have been present in methane-dominated atmospheres of the Early Earth.

A major question about Titan is why it has an atmosphere left at all. Given the low gravity, the N_2 atmosphere would be expected to escape fairly quickly (we will have a look at the relevant physics in Chapter 8). Moreover, the chemical reactions in the atmosphere should gradually convert all the methane into a tarry sludge sequestered at the surface. In some way or other, the atmosphere of Titan must be dynamically maintained by recycling of chemicals deposited on the surface, and by outgassing of nitrogen (probably in the form of ammonia) and CH_4 from the interior. Precisely how this happens is one of the Big Questions of Titan.

Even icy moons without an appreciable atmosphere can manifest features of considerable interest. Jupiter's moon Europa is a case in point. This satellite has a water ice crust between 10 and 50 km thick, but beneath the crust there lies a liquid water ocean. Europa shows an intriguing range of crustal features, including some that suggest melt-through of the ocean. Of course, the existence of the ocean has attracted attention as possible habitat for life. The icy moons challenge the epistemological boundaries of planetary science. At the cold temperatures of Europa's surface, as on Titan, water ice is basically a rock, just as sand can be considered an "ice" of SiO_2 on Earth. The ice-rock forms minerals with ammonia and methane and other compounds, and when warm enough the ice-minerals can flow or melt and lead to cryovolcanism. When studying the crust and interior of Europa or Titan or other icy moons, are we doing geology or oceanography or glaciology? Whatever one wants to call it, these moons are, as has been said, "always icy, never dull."

1.6 FARTHER AFIELD: EXTRASOLAR PLANETS

Until 1988, the Solar System was the only field of play for students of planetary climate, and it provided the only example against which theories of planetary formation could be tested. Revolutionary improvements in detection methods led to the first confirmed detection of a planet orbiting another star in that year, and instrumentation for planetary detection has continued to improve by leaps and bounds. As of the time of writing, over 228 planets orbiting stars within 200 parsecs (652.3 light years) of Earth have been detected, and the rate of detection of new planets is if anything accelerating. Certainly, much of the excitement surrounding the new zoology of planets has revolved around the prospect for detecting a planet that is habitable for life as we know it – or have known it to be in the past few billion years of Earth history. Perfectly aside from the habitability question, though, the rich variety of new planets discovered offers the student of planetary climate stimulus for thinking well outside the box of how the climates of known Solar System planets operate and have evolved over time.

Planets have been detected in orbit about a variety of different kinds of stars, so it is necessary to learn something of how stars are characterized. At the most basic level, stars are classified according to their luminosity (i.e. their net power output) and the temperature of the star at the surface from which the starlight escapes to space. The luminosity is determined by measuring the brightness of the star as seen from Earth's position, and then correcting for the distance between the star and the Earth. For relatively nearby stars, the distance can be measured directly by looking at the tiny shift in angular position of the star as seen from opposite ends of the Earth's orbit, but for stars farther than 500 parsecs (1630.8 light years), more indirect inferences are required. The stellar effective temperature is determined by measuring the spectrum of the starlight – how the brightness changes when the star is observed through different filters. Hotter stars have colors toward the blue end of the spectrum, while cooler stars tend toward the red. Hot stars emit more energy per unit surface area, but a reddish cool star can still have very high luminosity if it is very large, since it then has more surface area that is emitting. The energy sustaining the emission of starlight comes from the fusion of lighter elements into heavier elements. Since hydrogen is by far the most common element in the Universe that can participate in thermonuclear fusion, the overwhelming majority of stars ignite by fusing hydrogen into helium. These stars are known as *Main Sequence* stars, and stellar structure models predict that there is a distinct relation between the luminosity and emission temperature for stars that get their energy in this way. The stellar structure theories imply that the position of a star on the Main Sequence is determined primarily by its mass, with more massive stars being both hotter and more luminous. Moreover, stars spend most of their lifetimes on the main sequence, so that a scatter plot of luminosity vs. temperature of stars – a *Hertzsprung–Russell diagram* – shows most stars to be clustered along the Main Sequence curve. Indeed, the Main Sequence was discovered by plotting catalogs of stars in this way long before the energy source of stars was discovered.

Stars do not evolve *along* the Main Sequence. Rather, they enter it at a certain position when fusion ignites, and remain near the same point for a certain amount of time while gradually brightening. After they have fused a sufficient proportion of their hydrogen fuel into helium, they then leave the Main Sequence for a comparatively short afterlife as a brighter star with a more rapidly evolving spectrum. What happens when a star leaves the Main Sequence depends on the star's mass. The Sun will spend a billion years or so as a red giant, before collapsing into a gradually fading white dwarf. Main Sequence stars are thought to be the best candidates for hosts of habitable planets, since they provide relatively long-term stable stellar environments. Once a star leaves the Main Sequence, the climates of any unfortunate planets the star may have had will have been radically disrupted, if indeed the planets continue to exist at all; there will be relatively little time for any new form of life to establish itself. Nonetheless, planetary systems do exist orbiting stars off the Main Sequence, and these planets, too, have their points of interest. The first confirmed detection of a planet orbiting a Main Sequence star did not occur until 1995.

Just as the luminosity of the Sun increases over time, the luminosity of other Main Sequence stars increases during their time on the Main Sequence. However, the lifetime of a star on the Main Sequence varies greatly with the mass of the star. The mass of the star determines the amount of nuclear fuel available to sustain the star's life on the Main Sequence, while the luminosity gives the rate at which this fuel is consumed. The Main Sequence lifetime of a star with mass $M_\oplus$ and luminosity $\mathcal{L}_\oplus$, scaled to values for the Sun, is estimated by

$$\tau_\oplus = \tau_\odot \frac{M_\oplus}{M_\odot} \frac{\mathcal{L}_\odot}{\mathcal{L}_\oplus}. \tag{1.2}$$

Main Sequence stars have a power law mass–luminosity relationship $\mathcal{L}_\oplus \propto M_\oplus^{3.5}$, so on the Main Sequence the lifetime scales with luminosity according to the law

$$\tau_\oplus = \tau_\odot \left(\frac{\mathcal{L}_\odot}{\mathcal{L}_\oplus}\right)^{1.29}. \tag{1.3}$$

Bright, hot, blue massive stars thus have a much shorter lifetime on the Main Sequence than dim, cool, reddish dwarf stars. The Sun has a Main Sequence lifetime of about 10 billion years, and is nearly halfway through its time on the Main Sequence.

Figure 1.2 shows the Hertzsprung–Russell diagram for stars that are known to host one or more planets. Astronomers designate the color (equivalently surface temperature) of stars according to *spectral classes* given by the letters O, B, A, F, G, K, and M, extending from hottest to coldest, with numbers appended to indicate subdivisions within a spectral class. Our Sun is a class G2 Main Sequence star. The collection of planet-hosting stars in Fig. 1.2 represents a tiny subset of the many millions of stars that have been cataloged. None of the stars hosting planetary systems (so far) have spectral classes hotter than F. The stars

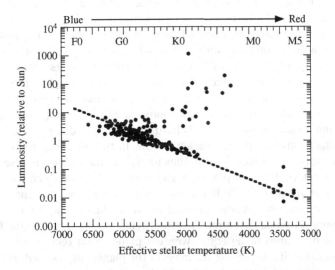

Figure 1.2 Scatter plot of luminosity vs. effective surface temperature for stars about which at least one orbiting planet has been detected as of 2007. Luminosity is given as multiples of luminosity of the Sun. The colder stars are redder while the hotter stars are more blue in color. The letters at the top indicate the standard spectral classification of stars in this temperature range, and the dashed line approximately locates the Main Sequence. The Sun is a class G2 Main Sequence star.

cluster along the Main Sequence simply because there are more stars in general along the Main Sequence than elsewhere. There is also a selection bias in the collection of planet-hosting stars due to the technologies available for detecting planets, which work better for some kinds of stars than for others. Thus, the gap in detections between K and M0 stars may be an artifact of detection bias rather than a reflection of some basic feature of planetary formation.

There is a rich supply of planets orbiting stars between spectral classes G0 and K, as well as a good handful of planets around bright red giant stars off the Main Sequence. The cluster of detections around dim red dwarf stars is particularly interesting, as many of these turn out to represent the most Earthlike planets to date – again a detection bias, because it is easier to detect low-mass Earthlike planets around low-mass stars using present technology. These *M-dwarf* stars are very dim, so a planet has to be in a very close orbit about its star in order to be as warm as Earth. In compensation, these systems have very long lifetimes compared with the Sun and other brighter stars. According to Eq. (1.3), an M5 dwarf spends about 100 times as long on the Main Sequence as the Sun does, and so this kind of star will brighten only very slowly over time. Such a star provides a very stable climate to its planet, and requires much less adjustment of atmospheric conditions than the Earth had to accomplish to resolve the Faint Young Sun problem. In contrast, an F0 star would spend under a billion years on the Main Sequence, and if the history of life on Earth is anything to go by, life around an F0 star would be snuffed out at the prokaryotic stage, before it could even begin to think of making oxygen. Aside from affecting the lifetime, the spectral class of the star affects the degree of absorption of stellar radiation by whatever atmosphere the star's planet may have, and this too provides a lot of novel things to think about when pondering the climates of extrasolar planets. M-dwarf stars are by far the most common types of star in the Universe, and so they have the potential to provide most of the habitable territory for life. It is therefore exceedingly important to come to an understanding of the climate of planets circling M-dwarfs.

That takes care of the stars, but what is known about the extrasolar planets themselves? Here, too, there is a detection bias, since it is much easier to detect massive planets comparable in mass to Jupiter than it is to detect more Earthlike planets. Most planets detected to date are very massive planets which, according to theories of planetary formation, are likely to be gas giants like Jupiter or Saturn, or ice giants like Neptune or Saturn. The full variety of planetary climates offered by the various combinations of planetary mass, orbital characteristics, and stellar characteristics is hard to convey by looking at just a few graphs. We will give a small sampling of this variety below. In the course of the exercises in this book the student will have ample opportunities to explore the wider universe of exoplanets.

One of the key determinants of planetary climate is the rate at which the planet receives energy from its star. This is a function of the luminosity of the star and the distance of the planet from the star, which varies to some extent in the course of the planet's year (as discussed in Chapter 7). Planets that receive more stellar energy flux than the Earth will tend to be hotter, all other things being equal, whereas planets that receive less will tend to be colder. The left panel of Fig. 1.3 shows the mass of the planets discovered so far and plotted against the stellar flux heating the planet at the time of the planet's closest approach to its star. The masses are measured relative to the mass of Jupiter and the fluxes are measured relative to Earth's solar heating. One Earth mass is 0.00315 times the mass of Jupiter. On this diagram, a planet with a relative flux of unity would have an Earthlike temperature if its atmosphere were like Earth's. We see that a great many planets with a mass one-tenth Jupiter mass or greater have been discovered; these are all likely to be gas giants or ice giants

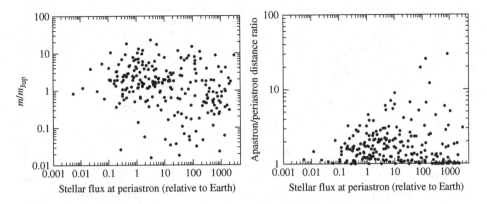

Figure 1.3 Left panel: Scatter plot of mass of extrasolar planets in units of Jupiter masses vs. the flux of stellar energy impinging on the planet at the time of closest approach (the periastron). Right panel: Scatter plot of the ratio of farthest (apastron) to closest (periastron) distance of the planet from its star in the course of the planets orbit vs. the stellar energy at the time of closest approach. The stellar energy flux is given as a multiple of the corresponding flux for the Earth.

made mostly of hydrogen and helium. Though these are in some ways like Jupiters, their climate has no real analog in the Solar System, since most of them are in orbits where the planet receives at least as much stellar energy as the Earth receives from the Sun. These are all "hot Jupiters," and represent climates very different from anything in the Solar System. Could Jupiters receiving Earthlike stellar radiation be habitable for life? That is certainly a Big Question, requiring understanding of the climate of such planets. Even more exotic are the giant planets receiving vastly more stellar flux than the Earth – up to one thousand times as much, in fact. These are "roasters" – very hot gaseous planets in close orbits about their host stars. Planetary formation theory gave no inkling that such things should exist, and indeed the existence of roasters poses real challenges for the theory.

Another way that the new extrasolar planets offer novel climates is in the nature of their orbits. Solar System planets, except for Pluto, have fairly circular orbits. However, most exoplanets have highly elongated orbits with a considerable difference between the distance of closest approach to the star (the *periastron*) and the distance of farthest remove from the star (the *apastron*). The range of orbital elongation is shown in the right panel of Fig. 1.3. Since the stellar energy goes down like the square of the distance from the star, the planets with highly elongated orbits will have novel seasonal cycles unlike any encountered in the Solar System. They would tend to heat up to a great degree at periastron and cool down, perhaps freezing over any ocean, at apastron. Could such a planet be habitable? The answer depends a lot on the thermal response time of the planet's atmosphere and ocean, which could average out the orbital extremes. The relevant physics is discussed in Chapter 7.

The orbital period, or "year," of the extrasolar planets also varies widely, as shown in the left panel of Fig. 1.4. Planets have been discovered with a wide range of orbital periods, ranging from as little as 2 Earth days to as much as 6000 Earth days. Planets in close orbits with short orbital periods are likely to be *tide-locked* and always present the same face to the star, just as the Moon always presents the same face to the Earth. The Super Earths to be discussed shortly are mostly in this category. Tide-locked planets offer novel possibilities for planetary climate. The night side could get very cold, and if the planet has an ocean,

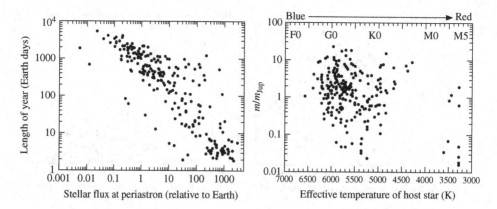

Figure 1.4 Left panel: Scatter plot of orbital period (in Earth days) vs. the flux of stellar energy impinging on the planet at the time of closest approach (the periastron). Right panel: Scatter plot of the mass of the planets (in units of Jupiter masses) vs. the effective radiating temperature of the host star.

it might freeze over completely. The dayside would be hot, but there could be a habitable zone near the ice margin. Further, the transport of moisture and heat from the dayside to the nightside poses interesting questions; the answers are important, since such transports in large measure will determine the nature of the planet's climate.

Low-mass planets are of particular interest because according to theories of planetary formation they have the best chance to have a rocky composition similar to that of Earth, Mars, or Venus. Relatively few planets with a mass of 10 Earth masses (0.03 Jupiter masses) or less have been discovered, but there has been recent progress in this area. The planets discovered so far in this range are all considerably more massive than Earth, and are therefore called *Super Earths*. The left panel of Fig. 1.3 shows that a handful of Super Earths have been discovered with stellar fluxes ranging from 0.25 that of Earth (yielding a too-cold planet) to 1000 times that of Earth (yielding a planet far too hot). The closest to being "just right" is Gliese 581c, with a flux of about three times that of Earth. Would such a planet be habitable, or would it turn into a Venus? The physics needed to answer such questions will be developed in Chapter 4.

The right panel of Fig. 1.4 gives an indication of the masses of planets that have been discovered about stars having various temperatures. The stellar temperature is of interest to climate since it determines the spectrum – the redness or blueness – of the starlight, which in turn affects the absorption of starlight by various atmospheric constituents. Hotter stars also put out more of the energetic ultraviolet radiation, which can have a profound effect on atmospheric chemistry. We see that a few low-mass Super Earths have been found around class G or K stars, but examination of the associated planetary data reveals that these planets are in close orbits and receive several hundred times as much stellar energy as the Earth receives from the Sun. They will unquestionably be extremely hot places, and unlikely to be able to sustain an atmosphere or liquid water. There is, however, a small cluster of Super Earths orbiting very cool stars with temperatures under 3700 K. These stars are Main Sequence *M-dwarfs*. They are very red, very small, and very dim, but by virtue of their dimness planets in close orbits can still have a chance to be habitable. Moreover, dim stars like M-dwarfs have a long lifetime, and therefore provide a stable environment for their planets. Gliese 581 is such a system, but it appears that the two most Earthlike planets still miss being habitable – one likely to turn into a Venus if it has an ocean and the other likely to freeze into a Snowball. Part of the reason for interest in M-dwarfs comes purely from detection technology. It is comparatively easy to detect low-mass planets in close orbit around a low-mass star, and at the same time low-mass dim stars give a planet in a close orbit a chance to be habitable. As time goes on, it is likely that a habitable world around an M-dwarf will be discovered. As detection technology improves, the possibility for discovering Earthlike habitable planets will expand to other spectral classes of stars.

Naturally, one is at this point intensely curious as to the composition of these planets, whether they have water, and what their atmospheres (if any) are made of. Some information can be inferred from planetary formation theory and theories of atmospheric evolution, but there is as yet no great ability to determine atmospheric composition from observations. That will change in the next decade or two, as satellite-borne instruments come online which will be able to determine the spectrum of emission and reflection from extrasolar planets. The anticipated instruments will only return a single spectrum averaged over the whole visible surface of the planet, but a great deal can nonetheless be inferred about atmospheric composition from such data. Learning how to make the most of this single-pixel planetary astronomy, in which the Earth would appear as a pale blue dot (in Carl Sagan's words) opens a whole range of Big Questions. Much of the same radiative transfer used in calculating planetary climate bears on the interpretation of planetary spectrum as well.

1.7 DIGRESSION: ABOUT CLIMATE PROXIES

1.7.1 Overview of proxy data

Instrumental records of climate – that is, records of measurements of temperature and other quantities by scientific instruments – date back at most a few hundred years. The first accurate thermometer was invented in 1654 by Ferdinando II de'Medici, and 200 years passed before anything like a global network of reliable temperature measuring stations began to become available. Written historical records of such events as frost dates, encounters with sea ice and depictions of mountain glacier length provide some information about the climate of the past few millennia, but for the most part one must rely on *climate proxies* for information on what climate was doing a century or more ago. A climate proxy is any measurable thing preserved in the geological record of Earth or other planets, from which some aspect of climate can be inferred; to be useful, a climate proxy must come with a *chronology*, that is, some means of telling what period the proxy dates to. There is a vast and ever-improving array of climate proxies. We have encountered a few already. For example, the existence of river networks on the surface of Mars tells us that at some time in the past the surface of Mars must have been warm enough to support a liquid (probably water) flowing for long distances along the surface; in this case, the chronology at the time of writing comes from counting the number of craters superimposed on the features. Similarly, the existence of marine sedimentary deposits and stromatolites during the Archean provides compelling evidence that the Earth was warm enough to support open ocean water through much of its early history.

Some of the more intuitive kinds of proxies derive from plants and animals that live on land, since physiology places certain constraints on the conditions in which various organisms can grow or thrive. The presence of cold-blooded animals like crocodiles is a sure sign that winters cannot have been much below freezing for extended periods of time. Some kind of plants require tropical conditions to survive, while others require cooler conditions; even the shape of leaves provides information about temperature. A great deal of this evidence is preserved in the fossil record. Where there are trees, the width of tree rings provides a record of annual variations in the temperature of the growing season (though it is sometimes hard to distinguish temperature from rainfall effects). Land-based proxies are only available after the time at which a fairly diverse ecosystem established itself on land. While the first primitive plants colonized land as early as 470 million years ago (in the *Ordovician Period*), and the first plants with stem structures able to conduct water appeared perhaps 430 million years ago (the *Silurian Period*), a diverse land ecosystem did not really get under way until the *Devonian Period*, which began about 416 million years ago. Once plants colonized land, animals in search of a free lunch followed not long afterwards.

One of the richest sources of information about past climates of the Earth comes from material preserved in sea floor sediments. Sediment is laid down in layers like the pages of a book, in which the history of Earth's climate can be read. The Earth is a dynamic planet, and sea floor is constantly being re-created at mid-ocean ridges, and likewise pulled down into the Earth's interior for recycling at subduction zones. For this reason, the deep-ocean marine sediment record is mostly limited to the past 100 million years, and is rather sketchy for the first half of that period. The oldest remaining deep-sea floor is about 180 million years old, and there is precious little of that. Near-shore deposits on continental shelves, on the other hand, can be uplifted and preserved for hundreds of millions, or even billions, of years. Many of the key marine deposits that tell us about the climate of the Neoproterozoic (about 600 million years ago) are now high and dry in Namibia, while others are found in Arctic Canada. These various deposits have a lot of individual personality and are known to geologists by names such as the Acasta Gneiss, the Warrawoona formation, the Akademikerbreen or the Isua Greenstone Belt; we've met many of these already in the preceding sections.

Numerous aspects of the sedimentary record have been used to infer past conditions, and the ingenuity of paleoclimatologists is adding to the list all the time. Some sedimentary proxies involve the physical structure of the sediments, and are independent of chemistry or biology. *Diamictites* are a class of sediments known to be carried by land glaciers discharging into the ocean. They are a sure sign of very cold conditions, since it is only in such cases that a glacier can survive to sea level. *Ice-rafted debris* (IRD) is coarse material that can only be carried offshore by hitching a ride on icebergs or sea ice. *Dropstones* – individual stones the size of pebbles and larger which fall with enough force to deform sediment layers – are a particularly striking form of ice-rafted debris. Inasmuch as stones do not float, they present very convincing evidence for ice. Continental dust in sediments provides an indication of the strength of wind, since larger particles require stronger winds to loft and transport them; the mineral composition of the dust can often mark where the dust came from, and hence the direction of the wind. Surveying the species of fossil algae found in sediments can provide information about the temperature of the layers in which the algae grew, since some organisms require colder temperatures while other require warmer temperatures.

A great deal of information can be gleaned from the chemical composition of the ocean, and of the microscopic creatures which dwell within it. Some chemical proxies do not rely on biology, and others make use of organisms mainly as nearly passive recorders of the composition of the ocean. Still other proxies are more intimately tied to biology, through the effect of temperature on the rate at which an organisms makes use of one element in preference to another. For example, the ratio of magnesium (Mg) to calcium (Ca) in corals and in the shells of certain micro-organisms is dependent on the temperature at which the organisms grew. The strontium–calcium substitution has been used similarly. Organic molecular proxies represent an important new source of data; it has been discovered that certain kinds of micro-organisms produce long-chain organic molecules with a somewhat variable chain length which depends on the temperature at which the organism grows. When these molecules are robust enough to be preserved in the sedimentary record, the ratio of chain lengths can be used to infer past temperatures. Alkenone and TEX-86 biochemical proxies fall into this class. They are useful only over time spans for which organisms producing the molecules exist, and can be presumed to have similar biochemistry to modern analog organisms. Both alkenones and TEX-86 have been used to probe climates lying many tens of millions of years in the past, as well as more recent climates of the past few hundred thousand years.

If the chemical composition of a sedimentary sample has been altered by interaction with ocean water at some time after the sediment was first formed, then the information the sediment provides about past conditions is severely compromised. Post-depositional alteration is known as *diagenesis*, and it is a problem that plagues interpretation of geochemical sedimentary proxies. Paleoceanographers have had to become very clever sedimentary detectives in order to check for the nefarious effects of diagenesis, particularly when dealing with samples more than a few million years old. In some cases the existence of diagenesis has gone undetected for a decade or more, as will be discussed later in connection with the problem of hothouse climates.

1.7.2 Isotopic proxies

The chemical properties of an element are primarily determined by the number of protons in the nucleus (the *atomic number*, which also determines the configuration of the electron cloud. Nuclei also contain neutrons, and atoms having the same atomic number can appear in forms with different numbers of neutrons. These differing forms are known as *isotopes* of the element. Isotopic proxies have proved to be a versatile source of information about past climates. Some isotopes are unstable, and decay into other elements; these can be used as "clocks," to determine when things happened. The original, and most famous, such application is radiocarbon dating, which makes use of the decay of the form of carbon having a molecular weight of 14 (carbon-14, or ^{14}C). Stable isotopes do not decay, and instead provide a tracer of past chemical reactions in which the substance participated. For elements heavier than helium the stable isotopic composition of a planet is determined primarily by the synthesis of the elements in the supernova explosion which gave birth to the material eventually incorporated into the Solar System. The isotopic ratio can in some cases be further changed by the process of planetary formation. For example, oxygen has three stable isotopes: ^{18}O, ^{17}O, and ^{16}O, the last of which is by far the most common. About 1 in 500 atoms of oxygen on Earth are ^{18}O, which is nearly the same ratio as found in the Sun and which presumably represents the composition of the primordial solar nebula. The isotope ^{17}O is less commonly used as a proxy because it is so much rarer than ^{18}O (only 1 in 2500 oxygen atoms on Earth), but variations in its composition are readily detectable with modern measurements, and have important specialized uses. Hydrogen has two stable forms: deuterium (D) which has a proton and a neutron, and normal hydrogen (H) which has only a proton. About one hydrogen molecule in 6500 occurring in Earth's ocean water is D. This differs from the 1:1700 ratio in the outer Sun, because deuterium is destroyed by nuclear fusion in the Sun and in the early solar nebula; curiously, the composition of the outer Sun is nearly the same as Jupiter. The stable isotopes of carbon are ^{13}C and ^{12}C, of which the latter is by far the more common. About 1 in 100 carbon atoms on Earth is ^{13}C. The carbon isotopic system attracts particular attention because carbon is the basis of organic chemistry – the stuff of life as we know it. Stable isotopes of S, N, Ar, and many other elements have proved useful as proxies. In fact, it is hard to think of any measurable isotopic ratio that has not proved useful as a proxy for some aspect of climate or atmospheric composition.

The utility of isotopes as climate proxies derives from what they tell us about processes that sort them out into different reservoirs. Isotopes of an element have *nearly* the same chemical and physical behavior, but not *exactly* the same behavior. The subtle differences come about in part because, at a given temperature, the molecules of the lighter isotopes have a higher typical velocity than those of the heavier isotopes, since all molecules have the same mean kinetic energy at any given temperature. Among other things, this means that lighter isotopic forms of a substance evaporate more readily than heavier ones, so that the vapor phase of a substance is typically depleted in heavy isotopic forms relative to the condensed phase with which it is in contact. We will find this effect particularly useful in the interpretation of water isotopes, but a very similar process applies to the gradual "evaporation" of a planet's atmosphere into outer space, and can be used to constrain the proportion of atmosphere that has been lost in such a fashion.

The rate at which a given element or a molecule containing that element undergoes chemical reactions also depends on the isotope involved. Other things being equal, heavier isotopes would tend to react more slowly, because they have lower speeds and therefore lower collision rates with other reactants than their lighter, nimbler cousins. However, there are more subtle effects involved that can either enhance or retard reaction rates. Specifically, the difference in molecular weight between isotopes can affect the characteristic vibration frequencies of molecule bonds, and this can in turn affect the probability that colliding reactants will stick together. Very often, the degree of preference for one isotopic form over another – the degree of fractionation occurring between reactant pool and reaction product – depends on the temperature at which the reaction is taking place. When this is so, the isotopic composition of the product provides a useful paleothermometer. In Section 1.3 we stated that "certain aspects" of the chemical composition of zircons constrained the temperature of Hadean water with which the zircons were in contact; with this cryptic phrase, we were in fact referring to the ratio of ^{16}O to ^{18}O in the Hadean zircons which, like all silicate minerals, contain oxygen (their chemical formula is $ZrSiO_4$). Similarly, it was the oxygen isotopic ratio of cherts (later confirmed by silicon isotopic ratios) that was used to constrain Archean temperatures in Section 1.3. A problem with all such paleothermometers is that fractionation is *relative to the isotopic composition of the reactant pool,* so that one needs to know the likely composition of the reactant pool in order to infer temperatures from the reaction product (e.g. the cherts or zircons), which are preserved long after the pool of reactants have been dissipated.

Biochemistry also has isotopic preferences. Photosynthetic Earth life (even of the sort that doesn't produce oxygen) prefers the lighter forms of carbon, so that organic material of photosynthetic origin is enriched in ^{12}C and depleted in ^{13}C relative to the inorganic carbon left behind. A record of the isotopic composition of inorganic carbon is preserved in carbonate minerals (e.g. $CaCO_3$) precipitated out and deposited on the ocean floor, with or without the help of shell-forming organisms. Organic material is directly preserved in sedimentary rocks such as shales. For example, the organic material in the Isua shales is presumed to be of biological origin because its carbon is isotopically light – enriched in ^{12}C relative to the average composition of carbon on Earth. Respiration – eating organic matter and combining it with oxygen to release energy – does not fractionate carbon to any significant extent, so the isotopic signature imprinted by photosynthesis is carried over to those of us creatures in the non-photosynthetic organic realm.

The simplest kind of isotopic fractionation to characterize is *equilibrium fractionation*. To keep things concrete, we will illustrate the concept using oxygen isotopes. Consider two physically or chemically distinct substances, each of which contains oxygen atoms. For example, the two substances could consist of water in its vapor phase and water in its liquid phase. Alternately, the two substances might be different chemical compounds containing oxygen, such as calcium carbonate ($CaCO_3$) and water (H_2O), or silica (SiO_2) and water; the latter pair leads to the chert-based paleothermometer mentioned earlier. Each of the two substances will have some initial ratio of ^{18}O to ^{16}O. Now imagine bringing the two substances into contact, whereupon the two substances exchange heavier and lighter forms of oxygen, changing the initial ratios. After a very long time, the isotopic ratios in each substance will reach equilibrium and stop changing. At that point, we can define the equilibrium fractionation factor $f_{1,2}$ via the relation

$$r_1 = f_{1,2} r_2 \qquad (1.4)$$

where r_1 is the ratio of ^{18}O to ^{16}O in the first substance after equilibrium has been reached, and r_2 is the corresponding ratio for the second substance. The fractionation factors differ for different pairs of substances, but typically (though not invariably) exhibit the following characteristics:

- The fractionation factor is typically quite close to unity;
- The fractionation factor deviates most from unity at low temperatures, and approaches unity as temperature increases;
- The deviation of the fractionation factor from unity increases as the contrast between the masses of the two isotopes increases (e.g. more fractionation for ^{18}O vs. ^{16}O than for ^{17}O vs. ^{16}O).

For example, the oxygen in silica (SiO_2) is isotopically heavy in comparison with the oxygen in the water with which it is in equilibrium; the fractionation factor is 1.036 at $20\,°C$ but falls to 1.018 at $100\,°C$. Similarly, oxygen in liquid water is isotopically heavy in comparison with that in the water vapor with which it is in equilibrium; in this case, the fractionation factor is 1.01 at $20\,°C$ and falls to 1.005 at $100\,°C$. It is the temperature dependence of the fractionation that makes it possible to use isotopic ratios as paleothermometers. Some kinds of fractionation appear to operate in nearly the same way whether the reactions happen within organisms or inorganically. This appears to be the case for carbonate precipitation, which fractionates in much the same way regardless of whether it happens inorganically or in shell-forming organisms. Other forms of fractionation, such as the fractionation of carbon isotopes in photosynthesis, are more inherently biologically mediated, though even in such cases the fractionation factors tend to be similar across broad classes of organisms sharing the same biochemical pathways.

Isotopic composition is usually described using δ notation, which is defined as follows. Let r_A be the ratio of the number of atoms of isotope A in a sample to the number of atoms of the dominant isotope. Typically, r_A will be a rather small number. Next let $r_{A,s}$ be the isotopic ratio for a standard reference sample. Isotopic composition is invariably reported relative to a standard because the analytical instruments currently in use cannot measure the absolute composition to very high accuracy, but they can measure the difference relative to a standard very accurately; the standard is an actual physical substance, natural or manufactured, which can be put into the analytical instrument and serve as a basis for comparison. The choice of standard is a matter of convention, and there are various standards typically used in different contexts. For example, the standard for oxygen and hydrogen isotopes in ice or water is usually taken to be VSMOW, which stands for Vienna Standard Mean Ocean Water. The ratio of ^{18}O to ^{16}O in VSMOW is 1/498.7, and of D to H is 1/6420; this approximates the mean present composition of water in the ocean.

Once one agrees upon a standard, the isotopic composition of a sample can be described in terms of the quantity

$$\delta A \equiv \frac{r_A - r_{A,S}}{r_{A,S}}. \tag{1.5}$$

Thus, negative values of δ indicate that the sample is depleted in isotope A relative to the standard, whereas positive values indicate that the sample is enriched. The δ value is usually expressed as a *per mil*, or parts-per-thousand, value. For example, a δ value of 0.001 would usually be expressed as 1 per mil or 1‰. A difference of 1‰ is often equivalent to a minuscule variation in isotopic concentration, requiring high analytical precision to measure. For example, for the case of $\delta^{18}O$, a 1‰ difference is equivalent to changing the ratio of ^{18}O molecules to ^{16}O molecules by $0.001 \cdot \frac{1}{498.7}$, or a hair over 2 parts per million. Deuterium (D) is even tougher. For deuterium, a difference of 1‰ amounts to a change in the D to H ratio of only 0.16 parts per million, though the challenge is offset by the fact that fractionation in the D/H system is considerably stronger than fractionation in the $^{18}O/^{16}O$ system owing to the lesser relative mass contrast in the latter case.

Carbonate minerals (e.g. $CaCO_3$ or $MgCO_3$) are important recorders of the oxygen and carbon isotopic composition of past environments. For carbonates, the isotopic composition is usually reported relative to the PDB standard, named after a mineral powder made from a naturally occurring fossil carbonate (the "Peedee Belemnite"). The physical powder standard no longer exists, so it has been supplanted by a synthetic equivalent, known as VPDB. It is important to keep the standards in mind when interpreting the isotopic literature. Oxygen isotopes occur in both carbonate and water, and so can be reported relative to either the VSMOW or VPDB standard. The conversion between the two is given by

$$\delta^{18}O(VSMOW) = 1.03091\delta^{18}O(VPDB) + 30.91. \tag{1.6}$$

A useful rule of thumb to keep in mind when interpreting oxygen isotopes in marine carbonates is that a carbonate reading zero relative to VPDB would be in equilibrium with water having zero $\delta^{18}O(VSMOW)$, if the carbonate formed at a temperature of around $18\,°C$. This means that a carbonate $\delta^{18}O(PDB)$ of "around" zero goes along with the waters in which they formed having $\delta^{18}O(VSMOW)$ of "around" zero. Later, we'll clarify just what we mean by "around," and how that depends on temperature.

Having introduced the δ notation and the VPDB standard, we are now in a position to get more quantitative about the things to be learned from the stable isotopes of carbon preserved in carbonates. The quantity of interest is $\delta^{13}C$, reported relative to the VPDB reference. Carbon dioxide is cooked out of carbonates in the interior of the Earth, and outgasses from volcanoes, subduction zones, and the mid-ocean ridge with $\delta^{13}C \approx -6‰$. In a steady state, the flux of carbon in this carbon dioxide is balanced by the burial of carbon in the form of inorganic carbonates (e.g. $CaCO_3$) and organic carbon (schematically CH_2O). In the long run, most of this burial is in sea-floor sediments, since whatever forms on land tends to eventually get washed into the ocean. Note that even carbonate precipitated biologically in the form of shells of organisms is considered inorganic material, and acts pretty much (though not exactly) like inorganically precipitated carbonate. There are some kinds of recently evolved plants that use photosynthetic pathways that fractionate carbon very little, but for the most part, photosynthetically produced organic carbon has $\delta^{13}C$ values that are about 25‰ lower than that of the carbon dioxide reservoir from which the photosynthetic organisms make their substance. For example, CO_2 in the atmosphere today has $\delta^{13}C \approx -8‰$, while land plants have $\delta^{13}C$ values running from $-32‰$ to $-25‰$ and marine organic carbon has $\delta^{13}C \approx -25‰$. Fossil fuels, which are made from ancient land plants (in the case of coal) or marine organisms (in the case of oil), have $\delta^{13}C \approx -25‰$.

If f_{org} is the fraction of carbon which is buried in the form of organic carbon, δ_0 is the $\delta^{13}C$ of carbon outgassed from the Earth's interior, δ_{org} is the $\delta^{13}C$ of the organic carbon buried and δ_{carb} is the $\delta^{13}C$ of the carbonate carbon buried, then mass balance implies

$$\delta_0 = f_{org}\delta_{org} + (1 - f_{org})\delta_{carb}. \tag{1.7}$$

If data from both organic and carbonate sediments are available, this formula can be used directly to infer f_{org}. It is instructive, however, to make use of the intrinsic fractionation between inorganic carbon and photosynthetically produced organic carbon to infer the isotopic compositions of the two burial fluxes as a function of f_{org}. Note that because photosynthetic fractionation is relative to the composition of the inorganic carbon pool, it is incorrect to assume that $\delta_{org} \approx -25‰$. It could be considerably heavier, if the inorganic pool has very positive $\delta^{13}C$. Let's consider two limiting cases. Suppose that $f_{org} \approx 1$, so that nearly all of the carbon outgassed from the Earth's interior is intercepted by photosynthesis and buried as organic carbon. In that case, mass balance requires that $\delta_{org} \approx \delta_0 \approx -6‰$. This can only happen if the inorganic carbon pool consists of isotopically heavy carbon with $\delta^{13}C \approx (-6 + 25)‰ = 19‰$. Since the carbonate that precipitates from an inorganic carbon pool tends to be isotopically heavier than the pool itself, the trickle of carbonate precipitated in this situation will have $\delta^{13}C$ in excess of 19‰. The precise value depends on aspects of ocean chemistry we will not pursue here. In the opposite limit, when $f_{org} \approx 0$ and there is little organic carbon burial, then $\delta_{carb} \approx \delta_0 \approx -6‰$. The inorganic carbon pool this carbonate precipitates from is somewhat isotopically light compared with the carbonate itself, and the organic carbon fractionates relative to that value, yielding $\delta_{org} < (-6 - 25)‰ = -31‰$. The typical situation over the past two billion years has been for δ_{carb} to be somewhat positive, between 2‰ and 5‰, while δ_{org} hovers around $-22‰$. There are periods, however, when carbon isotopes undergo considerable excursions from the typical situation. These *carbon isotope excursions* provide an important window into big events in the carbon cycle.

The above picture applies only when the carbon cycle is in a steady state. When any part of the carbon cycle is significantly out of equilibrium – for example, when one is building up a new pool of stored organic carbon in land plants and soils – the simple input–output isotopic calculation no longer works. One must then do a detailed accounting of the flows of carbon between the various reservoirs involved, and the attendant isotopic fractionations. This can be a very intricate process, especially since the fractionations involved are generally temperature-dependent. There are other important aspects of the isotopic carbon cycle we have swept under the rug, such as the important information that can be gained from vertical gradients of $\delta^{13}C$ in carbonates.

There is another biological process that can leave a distinct mark on the carbon isotopic record, namely *methanogenesis*. When there is oxygen around, organic matter generally gets decomposed into CO_2 by respiration. In anaerobic environments, methanogens get the goodies instead, and turn organic matter into CH_4. This is a multistep process, each step of which fractionates carbon. The result is CH_4 which is much lighter isotopically than the organic feedstock from which it was produced. Biologically produced methane today has $\delta^{13}C$ values on the order of $-50‰$. When the atmosphere–ocean system is rich in oxygen, as has been the case for the past half billion years, methanogenesis plays a very minor role in the carbon cycle and usually leaves little imprint in the isotopic record. A possible exception to this general rule may occur as a result of gradual accumulation of large amounts of methane in the form of exotic ices called *clathrates*, which can form in ocean-floor sediments and under arctic permafrost layers. If some event occurs which suddenly releases this stored methane into the atmosphere or ocean, the isotopically light methane quickly oxidizes into CO_2 which works its way into the carbonates. The net result is a negative carbonate carbon isotope excursion, the magnitude of which tells us something about the quantity of methane released relative to the net carbon in the ocean–atmosphere system.

At present, the arsenal of climate proxies is much more limited for other planets than it is for Earth. Biologically mediated proxies are obviously not in the cards for planets that seem to have no biology, but many of the abiological proxies used on Earth would be equally useful on other planets; cherts on Mars would provide much the same kind of information as cherts have provided on Earth. The use of these chemical proxies is limited primarily by the weight and power consumption of analytical instruments needed to carry out some of the analyses that are typically done on Earth materials. The same constraint applies to many of the means of determining chronology based on radioactive decay. Such techniques can be applied to other planets with a preserved geological record (notably Mars), but must await sample return missions. Meanwhile, a considerable amount has been accomplished by remote-sensing from planetary orbiters, and from low-power instrumentation on landers. The landforms of Mars have been imaged in great detail, and constitute a proxy for past climates with regard to occurrence of water and glacial activity. The mineralogy of the surface can be determined by a range of remote-sensing techniques, so a fair amount is known about the occurrence of clay minerals (a signature of water and weathering) in the ancient crust of Mars, and other minerals such as the iron compound hematite also tell us something about the aqueous environment of Early Mars. On all planets, a study of the isotopic composition of atmospheric gases (possible by *in situ* and remote spectroscopic means) provides valuable information on the source of the atmosphere and how much has been lost over time, insofar as lighter isotopes escape to space more readily than heavy isotopes. With sample return missions and improved robotic exploration, the future promises a rich expansion in planetary proxy studies, not least the prospect of drilling the Martian polar glaciers to see what climate mysteries they record.

1.7.3 Hydrogen and oxygen isotopes in sea water and marine sediments

We will turn our attention now to the isotopes of hydrogen and oxygen contained in water and in sediments precipitated from the water column. We will learn what the concentration of these isotopes tells us about the volume of glacier ice and the temperature of various parts of the ocean. "Normal" water is $H_2^{16}O$, but other isotopes of hydrogen or oxygen can substitute for the most prevalent isotopes, leading to various forms of heavy water, notably $HD^{16}O$ and $H_2^{18}O$.

At any given temperature light molecules, on average, travel with higher velocity than heavier molecules. This implies that during evaporation, the lighter isotopic forms of water evaporate more readily than the heavier forms, leading the vapor to be enriched in light water and depleted in heavy water relative to the liquid reservoir. Furthermore, when water vapor condenses into liquid or ice, the heavier forms condense more readily because the

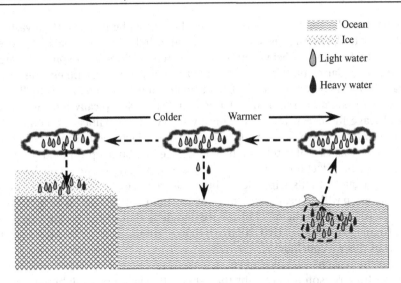

Figure 1.5 Sketch showing how the growth of ice sheets on land affects the isotopic composition of ocean water. The water vapor which evaporates from the ocean is enriched in lighter forms of water, and becomes more isotopically light as the heavy forms of water preferentially rain or snow out before the remainder is deposited on the glacier. This process systematically transfers isotopically light water to the glacier, leaving the ocean isotopically heavy.

slower-moving molecules can more easily stick together without bouncing off each other. In consequence, the rainfall is enriched in heavy forms relative to the vapor in the air, while the vapor left behind in the air is further enriched in the light forms and further depleted in the heavy forms. This is a form of distillation, very similar to the process by which one makes brandy from wine (or moonshine from fermented corn mash). Alcohol is more volatile than water, so the vapor in contact with heated wine is enriched in alcohol relative to the liquid; if part of the vapor cools and condenses, the water condenses out preferentially, leaving a potent essence at the end of the still if the remaining vapor is then condensed separately.

The way the distillation process affects the isotopic composition of sea water is sketched in Fig. 1.5. Let's suppose that the ocean starts with $\delta^{18}O$ of zero relative to VSMOW. Water will evaporate into the air until the air becomes saturated with water vapor (a concept that will be made precise in Chapter 2). Since heavy molecules evaporate less readily than light molecules, the water vapor will be depleted in ^{18}O relative to the ocean – in other words, it will be *isotopically light*. More precisely, the ratio of ^{18}O to ^{16}O for the water vapor in the air will be less than the ratio of the original ocean water, leading to a negative $\delta^{18}O$ for the vapor. The amount of depletion depends weakly on temperature. At 273 K, the water vapor $\delta^{18}O$ is shifted by −11.7 ‰ relative to the ocean. At 290 K the shift is −10.1‰ and at 350 K the shift is −6.0‰.[5] This reduction in isotopic contrast between reservoirs as temperature increases is typical of almost all isotopic fractionation problems. Because the amount

[5] In reality, the isotopic composition of water vapor in the atmosphere just above the ocean deviates somewhat from the equilibrium value, because the water vapor is in a dynamic balance between evaporation from the ocean and mixing of dry air into the layer from aloft. It is only when the air is saturated with water vapor that the equilibrium fractionation applies exactly.

of water stored in the form of water vapor in the atmosphere is utterly dwarfed by the amount of water in the ocean, the selective removal of light isotopes makes the ocean only very, very slightly isotopically heavy. But what if we removed the atmosphere's water vapor, sequestered it in a glacier on land, and repeated the process many times over until a substantial fraction of the ocean water had been transformed into an isotopically light glacier? In that case, the systematic removal of large volumes of isotopically light water from the ocean would leave the ocean water isotopically heavy by a significant amount – it would have a significantly positive $\delta^{18}O$. Thus, the degree to which the ocean is enriched in isotopically heavy forms of water tells us how much ice has built up on land. As ice volume becomes greater, the $\delta^{18}O$ (or similarly, δD) of the remaining ocean water becomes more positive. As an example, let's suppose we build an ice sheet by removing 200 m depth of water from an ocean with a mean depth of 4 km, assuming the glacier to be built from vapor with $\delta^{18}O = -10\permil$. If δ_i is the $\delta^{18}O$ of the ice and δ_0 is that of the ocean water, then conservation of molecules implies that $200\delta_i + (4000 - 200)\delta_0 = 0$, if the ocean started with $\delta^{18}O = 0$. From this we conclude $\delta_0 = 0.526\permil$.

In fact, for the reasons sketched in Fig. 1.5, the water that eventually snows out to form glaciers is much more isotopically light than the $-10\permil$ value one might expect from just looking at the vapor in equilibrium with ocean water. The initially evaporated water vapor may have $\delta^{18}O = -10\permil$, but on the way to the cold polar regions, some of that water will rain out back into the ocean, and the condensed water is isotopically heavy relative to the vapor, since heavy species condense more readily. That means that each time some atmospheric water vapor is lost to rainfall or snowfall back into the ocean, the vapor left behind becomes lighter. The precise extent of the additional lightening by the time the snow eventually falls out on a glacier depends on the amount of water lost on the way, which is in turn a function of the temperature difference between where the water was picked up from the ocean and where it was dropped on the glacier. Over the past 100 000 years, the $\delta^{18}O$ of the snow falling on Greenland has varied from $-42\permil$ in the coldest times to values around $-35\permil$ today. Antarctic ice is somewhat isotopically lighter than Greenland ice, and has $\delta^{18}O$ values ranging from $-40\permil$ to $-55\permil$ depending on location and age of the ice. Because of the additional fractionation on the way to the pole, the formation of the glacier would leave the ocean much more enriched in heavy isotopes than our previous estimate suggested. For glaciers having isotopic compositions comparable to the present ones, removing 200 m of ocean to build glaciers would leave the ocean enriched by about $+2\permil$, rather than a mere $0.526\permil$. The preceding discussion also shows that in order to translate the δ value of the ocean into an ice volume, one needs some estimate of the isotopic composition of the glaciers being formed. For the present glaciers, this can be determined by drilling into the ice, but for past ice that no longer exists one must rely on modeling.

So, if we could go back in time and grab a bucket of sea water, measuring its isotopic composition would tell us the volume of ice on the Earth, and this would tell us much about how cold the planet was. Wouldn't it be awfully nice if there were some way to do that?

1.7.4 Forams to the rescue

As it happens, Nature has provided a handy way of determining the isotopic composition of past ocean waters, via the good works of single-celled shelly amoeba-like organisms known as *foraminifera*, nicknamed *forams* (see Fig. 1.6). These creatures build distinctive calcium carbonate ($CaCO_3$) shells which record the state of the water in which they grew. Because the shells have such diverse and unmistakable shapes, it is easy to recognize and select out

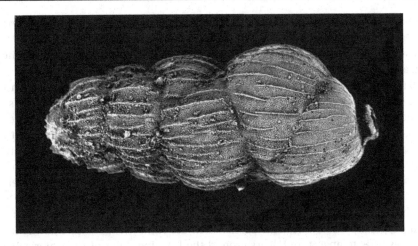

Figure 1.6 The shell of a departed benthic foram (*Uvigerina cushmani*). The specimen is about one half millimeter in length. Image credit: © Dr. E. Rohling, used with permission.

the species which live at the depth level one wishes to investigate. The two principal types of forams are *benthic* which live on the sea floor, and *planktonic* which live near the ocean surface; some planktonic forams require light because they make use of photosynthesis, and these are the ones whose depth habitat is most assured. The shells of both types wind up in tidy layers in the marine sediments, allowing one to read the state of the ocean in both depth and time, from a single sediment core. Benthic forams appear in the fossil record as early as 525 million years ago, but their use as paleoclimate indicators has been primarily restricted to the period over which significant portions of sea floor have survived – approximately the past 70 million years.

Forams are not just passive recorders of the isotopic composition of ocean waters, however. As is the case for any chemically distinct pair of reservoirs in contact, the oxygen in foram carbonate is systematically fractionated relative to the isotopic composition of sea water. There is some dispute as to the extent to which this fractionation can be thought of as equilibrium fractionation, but the fractionation factors behave much like inorganic equilibrium fractionation factors and do not differ greatly amongst species. Carbonate prefers the heavier forms of oxygen, and at a temperature of 18 °C, the ^{18}O to ^{16}O of precipitated carbonate is greater than the ratio for the water in which it precipitates by a factor of about 1.03. In other words, carbonate is about 30‰ enriched in ^{18}O compared with the water with which it is in equilibrium. As is typically the case for equilibrium fractionation, the degree of enrichment increases as temperature decreases. The change in fractionation with temperature is usually expressed as a *paleotemperature equation*. Many paleotemperature equations have been given, based on laboratory-cultured organisms, on field observations of recently living forams, and on laboratory measurements of inorganic carbonate precipitation. All give similar results, though the differences are important if one is interested in high accuracy. A general feel for the numbers is adequately given by the following paleotemperature equation, which applies to the benthic foram *Uvigerina*,

$$T = 17.97 - 4.0 \cdot (\delta_c(\text{VPDB}) - \delta_w(\text{VSMOW})) \tag{1.8}$$

where T is the temperature in degrees C at which the foram grew, δ_c (VPDB) is the measured $\delta^{18}O$ of the foram carbonate reported relative to VPDB, and δ_w(VSMOW) is the $\delta^{18}O$ of the water in which the foram grew, reported relative to VSMOW. Both δ values in the above equation are to be expressed in per mil units.

The temperature dependence of foram fractionation is a two-edged sword. On the one hand, the temperature dependence allows forams to be used as paleothermometers. On the other hand, it means it is hard to disentangle ice volume effects from temperature effects. According to Eq. (1.8), a $\delta^{18}O$ variation in foram carbonates of 2‰ could represent an temperature change of 8 K where the forams grew, or instead a change of 2‰ in the water in which the forams grew – corresponding to an ice volume change equivalent to roughly 200 m of sea level. The use of benthic forams mitigates this ambiguity to some extent, since the deep ocean temperature is much more uniform than surface temperature. This is so because the deep ocean is filled with waters that are created at the coldest parts of the surface ocean, typically located near the poles. When the climate is in a state having ice at one or both poles, this temperature hovers around the freezing point of sea water, whence the benthic oxygen isotopes primarily reflect ice volume rather than temperature – though the temperature effect is still by no means negligible. A 2 K variation in deep ocean temperature, which is not implausible even in icy conditions, leads to about a 0.5‰ variation in the $\delta^{18}O$ of carbonates, which if attributed instead to sea water composition would translate into an ice volume variation equivalent to about 50 m of sea level. At the other extreme, when the climate is in a state without ice at either pole, the isotopic composition of ocean water itself can be considered nearly fixed, and the benthic foram isotopes provide an indication of the polar temperature, regardless of where the sediment core is actually drilled. In this case, a 2‰ increase in benthic foram $\delta^{18}O$ would indicate an 8 K cooling of polar temperature.

Benthic forams thus provide a valuable overall indicator of the state of the climate, giving an indication of polar minimum temperature in ice-free climates, and ice volume in icy climates. Since greater ice volume generally goes with a colder climate, and both high ice volume and low temperature make the $\delta^{18}O$ of foram carbonate more positive, high benthic foram $\delta^{18}O$ indicates a cold climate while low benthic foram $\delta^{18}O$ goes along with a relatively warm climate. In ice-free climates, planktonic forams can be used to estimate surface temperature, but in icy climates the need to subtract out the ice volume effect makes it hard to get accurate surface temperature estimates by this means.

Forams also preserve other chemical signatures that are useful in reconstructing the past state of the climate system. Notably, they provide a record of the $\delta^{13}C$ of inorganic carbon of the ocean. Abiologically precipitated sea-floor carbonates, or carbonates not associated with individual microfossils, also do this, but the additional depth information available by using benthic vs. planktonic forams provides valuable information about the state of the oceanic carbon cycle. The use of fractionation as a paleothermometer is not limited to isotopes. Notably, magnesium (Mg) substitutes for calcium (Ca) in foram shell carbonates, to an extent that depends on temperature; hence the Mg to Ca ratio in foram shells can be used as a paleothermometer.[6] As with oxygen isotopes, the fractionation is relative to the composition of sea water, but the Mg to Ca ratio of ocean water is not affected by formation of glaciers, and hence evolves relatively slowly over geologic time. For this reason, magnesium–calcium paleothermometry has become a crucial tool for probing the past few million years

---

[6] Magnesium–calcium paleothermometry can also be used with cores drilled into corals. The growth zone of corals has a distinct depth preference, which has proved particularly useful in estimating ocean surface temperature.

of climate history, and the tool is being extended farther backward as understanding of the longer-term evolution of the oceanic Mg to Ca ratio improves.

1.8 THE PROTEROZOIC CLIMATE REVISITED: SNOWBALL EARTH

With a few more tools at our disposal, we once more pick up the thread of Earth's climate history, starting with a more detailed look at certain aspects of the Proterozoic – the geological eon that extends from 2.5 billion years ago to 543 million years ago, and is subdivided into the Paleo-, Meso-, and Neoproterozoic, in order of decreasing age. From now on, we will have increasing need to refer to the various subdivisions of the geological time scale by name, so these are summarized in Fig. 1.7.

So far we have not said much about continents and continental drift. Continents consist of light material that has floated to the outer portions of the solid Earth and been incorporated into the crust. This material is too buoyant to be subducted into the interior of the planet to any significant degree, so once continental material had completely segregated, it remained at the surface as a kind of scum resting atop the churning cauldron of Earth's interior fluid motions. The continental material is constantly being pushed around, broken up and rearranged in a process known as *continental drift*. The Earth is the only planet in the Solar System that has continents in this sense, but it is likely that extrasolar rocky planets of a sufficient size to retain the heat that drives internal motions, and having a surface temperature similar to Earth, would also exhibit the dichotomy of drifting continents vs. areas of rapidly recycled mantle material (analogous to Earth's sea floors). It is at present unresolved whether a water ocean is necessary to maintaining this state of affairs. That is indeed one of the Big Questions of planetary science, but it is not one which we will take up in this book.

Continents are important to climate for three main reasons: they are a platform upon which polar glaciers can form; they are the primary sites of the silicate weathering reaction

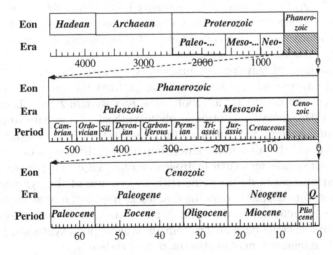

Figure 1.7 The geological time scale. Numbers on the scales represent millions of years before present. The entire span of time before the Phanerozoic is also known as the Precambrian. Q stands for Quaternary.

that governs atmospheric CO_2, and the amount of weathering is strongly affected by the continental configuration; they affect the geometry of ocean basins, and hence the ability of oceans to transport heat from one latitude to another. In addition, they provide distinct habitat for novel forms of life, though they were not colonized to any great degree (if at all) until land plants evolved in the late Ordovician and early Silurian.

There is not enough preserved continental crust to get a clear idea of the distribution of the continents until the very end of the Proterozoic. Geophysical modeling and indirect geochemical evidence has led to a prevailing belief that the total volume of continental material was similar to the present volume throughout the Proterozoic, but this is not a very well settled area of geophysics. By the close of the Proterozoic, however, the picture of the distribution of continents begins to clarify; we will begin showing maps of paleogeography when we come to discuss that time.

The Proterozoic was the Age of Microbes. Indeed, in terms of the functioning of the biogeochemical cycles needed to sustain life, we are all still basically the guests of the prokaryotes, but up until the very end of the Proterozoic, single-celled organisms had the world to themselves, with the more complex eukaryotic form of microbial life making a definite appearance roughly midway through the Proterozoic. The Proterozoic was above all a time of adjustment of biosphere and climate to the massive changes wrought by oxygenation of the atmosphere and ocean. It is a matter of current debate as to whether the oxygenation began in this eon because oxygenic photosynthesis only evolved at this time, or because other factors kept photosynthetic oxygen from accumulating earlier; be that as it may, the oxygenation occurred in fits and starts throughout the Proterozoic, and atmospheric oxygen levels probably reached values comparable to modern ones by the close of the eon. One effect of oxygenation we have already noted is that it is likely to have reduced the importance of CH_4 as a greenhouse gas, with CO_2 (aided by water vapor) becoming increasingly dominant. More speculatively, oxygenation could have changed the abundance of other greenhouse gases that could conceivably have been significant in the anoxic atmosphere, such as N_2O and SO_2. The nature of this greenhouse turnover, and the extent to which it constituted a habitability crisis for our planet, is another of the Big Questions. In Chapter 4 we will learn how to evaluate the relative importance of the various greenhouse gases. Oxygenation would have also sharply limited the possibility for H_2 to accumulate in the atmosphere, which would have consequences for methanogenic ecosystems that used H_2 and CO_2 as a feedstock.

It must be emphasized that "photosynthesis" is not synonymous with "oxygenic photosynthesis." Indeed, oxygenic photosynthesis arose from the combination of two more ancient anoxygenic photosynthetic metabolisms – *photosystem I* and *photosystem II* – in a single organism. The anoxygenic forms of photosynthesis still survive to this day, in the form of green sulfur bacteria and purple sulfur bacteria. These metabolisms are much less efficient than oxygenic photosynthesis, since they harvest the energy of sunlight by making use of hydrogen sulfide, molecular hydrogen, or oxidized iron in various combinations. Oxygenic photosynthesis can do the same trick making use of the hydrogen in ordinary water, which is far more abundant than the feedstocks used by anoxygenic photosynthesis. The carbon isotope fractionation in ancient deposits such as the Isua Greenstone Belt, if it is indeed due to life, could have arisen from anoxygenic photosynthesis, though various anoxic fermentation metabolisms can also fractionate carbon.

Oxygenic photosynthesis turns CO_2 into O_2 and organic carbon, represented schematically as CH_2O. In order for the oxygen to accumulate, the organic carbon produced by photosynthesis must be buried before it can be re-oxidized by other bacteria, which would just take the free oxygen back out of the system and turn it back into CO_2. For this reason, the evolution of oxygen is intimately tied in with the carbon cycle. Since organic carbon is isotopically light compared with the carbon in CO_2 outgassed from the interior of the Earth, the long-term evolution of $\delta^{13}C$ in carbonates gives us a window into the carbon cycle, and a good overview of what is going on in the Proterozoic. As noted earlier, when the carbon cycle is approximately in a steady state, the $\delta^{13}C$ of carbonate is driven more positive as the proportion of organic carbon burial to total carbon burial increases. One must be cautious about this interpretation, however, since the carbon cycle is thrown wildly out of equilibrium in the course of the Snowball episodes that constitute the most dramatic features of Proterozoic climate.

With regard to oxygenation, however, organic carbon burial is not the whole story. Various compounds of sulfur also play a key role in the cycling of oxygen through the Earth system; there is evidence that the role of the sulfur cycle is much more prominent in the Proterozoic than during later times. Bacteria are clever, and have ways of oxidizing organic matter that do not use up free O_2. In a process called *sulfate reduction*, certain bacteria can react the sulfate ion (SO_4^{--}) with organic carbon to produce bicarbonate (HCO_3^-) and the stinky rotten-egg gas hydrogen sulfide (H_2S), which reacts with iron oxides and a little bit of free oxygen to produce water and the mineral pyrite (FeS_2). If the pyrite is then buried without being oxidized further, the net process turns organic matter into mineral carbonate while leaving much of the O_2 liberated by oxygenic photosynthesis in the atmosphere/ocean system. Precisely how much is left in the atmosphere and ocean depends on where the oxygen in sulfate and iron oxides comes from, but the net result is that pyrite burial liberates oxygen in a way that doesn't show up in the carbon isotope record.

The participation of sulfur in a variety of reactions involving oxygen makes the stable isotopes of sulfur – ^{32}S, ^{33}S, ^{34}S, and ^{36}S, of which the first is dominant – a key source of information about past behavior of oxygen. By comparing the degree of fractionation for the three minor isotopes, one can get additional information. One form of fractionation is *mass-independent fractionation* (MIF), which is nearly the same for all three minor isotopes. This kind of fractionation is believed to be produced only in photochemical reactions involving high-energy ultraviolet light, and photochemical models of the atmosphere indicate that mass-independent sulfur fractionation cannot be preserved in the sedimentary record unless the atmospheric oxygen concentration is extremely low – 10^{-5} of the present concentration or less, according to current estimates. It is the sulfur MIF proxy that tells us that Archean oxygen levels are nearly zero; other oxygen proxies only require that Archean oxygen be below 1% of present atmospheric concentration.[7]

The more conventional mass-dependent sulfur fractionation is mediated by sulfate-reducing bacteria. The fractionation nearly disappears when sulfate concentration in ocean water is low, so a strong mass-dependent sulfur fractionation indicates both high sulfate concentration and strong productivity of sulfate-reducing bacteria. An increase in sulfate concentration, in turn, is generally taken as indicative of a rise in atmospheric oxygen, since that permits more oxidation of pyrite into sulfate on land. Once oxygen builds up to the point that at least near-shore bottom waters become oxygenated, a host of additional bacterially mediated sulfur reactions, called *disproportionation reactions*, become possible, and these provide additional means of producing sedimentary sulfides (e.g. pyrite) that are isotopically light in sulfur. The interpretation of mass-dependent sulfur isotope fractionation

---

[7] The atmospheric chemistry models upon which this interpretation of the sulfur MIF is based rest on somewhat shaky assumptions, however.

is an exceedingly complex subject, which is likely to remain in a considerable state of flux for some time to come.

The proxy record shows a great deal of activity toward the beginning of the Proterozoic (during the *Paleoproterozoic*), and also toward the end of the Proterozoic (in the later parts of the *Neoproterozoic*). In between lies a billion-year period that has sometimes been called "the most boring period in Earth history." During this Big Yawn, which stretched from about 1.8 billion years ago to 800 million years ago, carbonate δ^{13}C held steady near 0‰, indicating steady organic carbon burial. Mass-dependent sulfur isotope fractionation suggests oxygen levels of around 10% modern concentrations during this period, though the upper bound on oxygen during this period is not well constrained. There is no indication of any significant glaciation. Eukaryotes appear in the microfossil record towards the beginning of the Big Yawn, but appeared to have little effect on biogeochemical cycling or climate evolution – unless perhaps they were somehow the cause of the long period of climate stability. It would not be surprising if closer study eventually revealed more features of interest in this period, but at this point we'll turn our attention to the more manifestly dramatic doings at the beginning and end of the Proterozoic.

All lines of evidence point to a Great Oxidation Event at the beginning of the Proterozoic. Preservation of mass-independent sulfur fractionation in sediments ceases abruptly between 2.45 and 2.3 billion years ago, and is never seen again throughout the rest of Earth history. A hiatus in banded iron formations begins at about this time. It is estimated that the oxygen content of the atmosphere soared to values well in excess of 1% of the present level but crashed back to lower levels afterwards, as witnessed by a transient reappearance of banded iron formations; the peak value in the event is at present not well constrained. A subsequent increase in mass-dependent sulfur isotope fractionation preserved in the sediments is indicative of an increase in sulfate concentration in the ocean, most likely associated with an increase of oxidation of pyrite on land. Based on this evidence and disappearance of banded iron formations, it is estimated that oxygen levels in the atmosphere recovered to somewhere around 10% of present atmospheric concentrations around 1.7 billion years ago, and stayed there until 700 million years ago when there was a further oxygenation event.

The Paleoproterozoic is characterized by wild swings in the δ^{13}C of carbonates. Around the time of the Great Oxygenation event, δ^{13}C has a major positive excursion, reaching values as high as 10‰ before eventually subsiding to the lower values characteristic of the middle Proterozoic. This indicates a major transition in the carbon cycle, most likely an increase in the proportion of organic carbon burial. It is very suggestive of a take-off of oxygenic photosynthesis, but whether the cause is evolutionary, ecological, or a matter of factors that allow better burial of carbon is a matter of dispute. There are several major glaciations within the Paleoproterozoic, of which one – the Makganyene alluded to earlier – was a Snowball event in which ice reached tropical latitudes. The Makganyene Snowball occurred within the interval between 2.32 and 2.22 billion years ago, and fine-scale examination of arguably synchronous glacial deposits in the Duitschland formation (Transvaal, South Africa) indicates extreme carbon isotope excursions associated with major Paleoproterozoic glaciations: Carbonate $\delta^{13}C$ was around 5‰ before the glaciation, then dropped to zero or even negative values as the glaciation progressed, recovering slowly afterwards. We'll be able to probe similar features in more detail in connection with the Neoproterozoic Snowballs. The Paleoproterozoic presents us with a puzzle whose pieces include oxygen, the effect of oxygen on greenhouse gases, the carbon cycle, and glaciation. Figuring out how these pieces fit together is one of the Big Questions.

The Neoproterozoic has many features in common with the Paleoproterozoic. The extreme carbonate carbon isotope excursions which had been dormant for so long resumed in the Neoproterozoic. There are several major glaciations during the Neoproterozoic, and two of these were Snowball events in which ice reached tropical latitudes. The more recent of the two Snowballs is the Marinoan event, which occurred about 640 million years ago; the older is the Sturtian, centered on 710 million years ago. Neoproterozoic Snowball-related geological formations exhibit a distinctive sequence of events. The scene starts with high carbonate δ^{13}C, up to 5‰, which is in fact higher than the modern value and indicative of a greater proportion of organic carbon burial than is the case at present. Then, the δ^{13}C drops, falling to zero or even negative values. At some point in this decline, one sees diamictites and other glacial deposits. The δ^{13}C continues to drop, and above the glacial deposits one finds *cap carbonates* – very unusual carbonate features that are believed to require very high deposition rates from waters highly supersaturated in carbonate. In the carbonates overlying the glacial deposit, the δ^{13}C becomes negative, typically around −5‰ which is about the value for abiotic carbon outgassing from the Earth's interior. The δ^{13}C gradually recovers to positive values over a long (but somewhat unconstrained) period of time afterwards. A particularly clear depiction of this sequence of events is given in the review by Hoffman and Schrag listed in the Further Reading for this chapter.

However, not all major carbon isotope excursions are associated with Snowball events. In fact, the greatest carbon isotope excursion in Earth history – the *Shuram excursion* – sets in gradually after a conventional glaciation which is thought to reach only to midlatitudes (the *Gaskiers* glaciation). The Shuram excursion brings the carbonate δ^{13}C all the way down to −12‰. There is no known process that could bring the carbonate δ^{13}C so far below the mantle outgassing value if the carbon cycle is in equilibrium. Indeed, the δ^{13}C is so implausibly low in the Shuram that it was long thought to be an artifact of diagenetic alteration. The Shuram is an enigmatic event – indeed one of the Big Questions. Current thinking has it that the Shuram is associated with a transient reorganization of the carbon cycle, in which a large isotopically light pool of suspended organic carbon in the ocean is oxidized and deposited as carbonate.

In fact, a lot is going on with oxygen across the Neoproterozoic, though it is a bit hard to determine what, where, and when. What is certain is that oxygen must have been high – even near present levels – right down to the ocean bottom by the end of the Neoproterozoic, since bottom-dwelling animals appear in the fossil record by this time, and it is unquestionable that such creatures require a great deal of oxygen. At the other side of the Neoproterozoic, around 700 million years ago, there is further evidence of oxygenation, in that a sharp rise in mass-dependent fractionation in sedimentary sulfides indicates the expression of sulfur disproportionation reactions, which indicates an oxygenation of at least some of the bottom waters. Another important clue as to what is going on is the reappearance of banded iron formations in connection with the Neoproterozoic Snowball events, suggesting that the ocean once more went anoxic, most plausibly as a result of global ice cover shutting down photosynthesis.

A very Big Question is why all this excitement suddenly resumed after nearly a billion years of stasis.

The Snowball events of the early and late Proterozoic are some of the most dramatic events of Earth history. We have used the term "Snowball" to refer to any glaciation where there is evidence of glaciation at tropical latitudes, but it is a matter of considerable debate whether the oceans were indeed nearly completely frozen over all the way to the Equator during these events. Sometimes the term *Hard Snowball* is used to refer specifically to a state with near-total ice cover. A Big Question of climate physics is whether it is possible to cool down the planet enough to yield land-based ice sheets discharging into the tropics, without also freezing over the tropical ocean completely. This requires an understanding of ice-albedo feedback, which will be developed at several places throughout the book. However, it also involves ocean heat transports (which are good at melting ice) and glacier dynamics, which for the most part are subjects that will be left for another time and place.

The Snowball phenomenon is pregnant with Big Questions, the most obvious of which are: How do you get in? How do you get out? And if your planet does succumb to a global Snowball, how long does it take to get out again? Is it a matter of centuries, millions of years, or billions of years? On Earth, the upper limit set by the geological record for the duration of Neoproterozoic Snowballs is about 20 million years, and the duration could well have been shorter. However, without a clear understanding of the nature of the event, it is hard to determine whether we just got lucky or whether the event could have lasted much longer.

Most theories for the entry into a Snowball involve the drawdown of whatever greenhouse gas had previously been maintaining the planet's warmth – usually CO_2 and CH_4 in some combination. Various hypotheses include methane destruction by oxygen, weathering enhancement triggered by catastrophic methane release from sediments, weathering enhancement due to continental configuration or production of weatherable rock by massive volcanic eruptions, and (more speculatively) drawdown of CO_2 through runaway photosynthesis and oxygenation. Whatever the mechanism, a key requirement is that the mechanism be compatible with the observed reduction in $\delta^{13}C$ *before* the onset of glaciation. Not all of the relevant biogeochemistry will be treated in this book, but in order to evaluate the hypotheses, it is certainly necessary that one have the tools to assess how low any given greenhouse gas has to go in order to trigger a global glaciation. These tools will be provided in subsequent chapters.

Assuming for the moment that the cooling process caused a Hard Snowball, the next question is how to deglaciate the planet. We will see in Chapter 3 that one would have to wait a billion years or more to exit from a Snowball if the exit were due to increase in solar output alone. Based on rather simple reasoning of the sort that will be covered in the remainder of this book, Kirschvink proposed that once the Earth freezes over, the weathering of silicate to carbonate (which requires liquid water washing over weatherable rocks) ceases, so that CO_2 outgassed from the Earth's interior accumulates in the atmosphere until it reaches concentrations sufficient to cause a deglaciation. This is another illustration of the principle that *Big Ideas come from Simple Models.* A Big Question (treated in subsequent chapters) is: how high does CO_2 have to go in order to trigger deglaciation of a globally ice-covered planet? Much hangs on the answer.

Both cap carbonates and the persistent negative carbon isotope excursion following the Snowball events are consistent with a massive buildup of CO_2 in the atmosphere during the frozen-over period. Once the planet gets warm enough to deglaciate, the powerful precipitation in the ensuing hothouse world would wash great quantities of land carbonates into the ocean, where they would precipitate to form cap carbonates. Further, if photosynthesis nearly shuts off during the glaciated phase, the inorganic carbon that accumulates in the ocean–atmosphere reservoir would have $\delta^{13}C$ comparable to the mantle outgassing value of about −6‰. As this reservoir is gradually transformed by silicate weathering into carbonate sediments, the isotopically light carbon works its way into the carbonates. If the reservoir is big enough at the termination of the Snowball, it can keep the sedimentary carbonate $\delta^{13}C$ light even in the face of a resumption of photosynthesis. When interpreting the carbon isotope excursions in the course of a Snowball, it is essential to keep in mind that the carbon cycle is likely to be far out of equilibrium in the course of these events. A full understanding of the connection between the carbon isotope evolution and the sequence of events surrounding the Snowball requires a detailed accounting of flows of carbon between the many different carbon reservoirs in the Earth system – land carbonate rocks, marine carbonate sediments, atmospheric carbon dioxide, various species of dissolved inorganic carbon, and organic carbon.

Assuming that the exit from a Snowball state does indeed proceed from accumulation of a great deal of CO_2 in the atmosphere, several Big Questions arise in connection with the post-glacial climate. Is there a risk of triggering a runaway greenhouse? If not, how hot does the climate get in the tropics and polar regions? Would it be hot enough to sterilize the planet to any great degree? How long is the post-Snowball recovery? In other words, how long does it take for weathering processes to draw the CO_2 back down to more normal levels?

Other Big Questions include: Why were there no Snowball events during the Big Yawn period of the middle Proterozoic? Why did Snowball events cease at the beginning of the Phanerozoic? Could they happen again, or is this particular threat behind us?

But let us not forget that another Big Question is whether there is a climate state with significant amounts of open water in the tropics, which is nevertheless consistent with the full range of geological data accompanying the Marinoan and Sturtian events. One could pose the same question for the Makganyene event, but there is less data to constrain the answer. The early and late Proterozoic manifests a lot of climate "weirdness" that is not seen elsewhere in Earth history, and such striking signatures would seem to call for an equally dramatic cause, rather than just a minor variation on the theme of ice ages. The global Snowball seems to fit the bill, but it remains to be seen whether other explanations are possible. The climate physics developed throughout this book will give the reader the underpinning needed to assess hypotheses as they develop, and even perhaps to formulate new ones.

Regardless of whether a true global Snowball glaciation ever happened on Earth, the Snowball certainly represents a state a water-covered planet could fall into if the right stellar and atmospheric conditions were encountered. Once a planet falls into a Snowball state, it is likely to stay there for a long time, and the consequences for existing life and evolution of new forms of life are profound. As such, Snowball states are a potential habitability crisis that extrasolar planets need to avoid or surmount. It is therefore worth understanding, in general terms, the physics of entry into, exit from, and duration of Snowball states, as well as the nature of the climate at various stages of the sequence and the effect of the sequence on life. This constitutes another Big Question, about which we will have much to say.

1.9 THE HOTHOUSE/ICEHOUSE DICHOTOMY

The present climate has ice at both poles, and the ice volume has fluctuated episodically between the present amount and a considerably larger extent for the past two million years (about which more anon). However, the present icy climate is not at all typical of Earth history. A careful study of the climate evolution over the past 70 million years illustrates a transition between climate states archetypical of a theme that has been played over with variations during the 543 million years since the close of the Proterozoic. With this latest eon, known as the *Phanerozoic*, we complete the repertoire of the major divisions of geologic time, as summarized in Fig. 1.7. Though the very first preserved multicellular organisms appear at the close of the Proterozoic, the Phanerozoic is the eon in which multicellular organisms of a generally modern form become abundant and diversify, first in the ocean with colonization of land coming towards the middle of the eon. Though the Phanerozoic was not subjected to extreme variations of climate and atmospheric composition rivaling the Snowball or oxygenation transitions of earlier eons, the events of the Phanerozoic are by no means inconsequential.

1.9.1 The past 70 million years

Figure 1.8 shows the paleogeography at the end of the Cretaceous, 65 million years ago. The continent of Antarctica has approached the south pole, and will continue to drift over the next 40 million years or so until it is more nearly centered on the pole. There is open water at the north pole in the late Cretaceous, and the open Arctic Ocean continues throughout the subsequent time through the present. The modern continents of North and South America, Eurasia, and Africa are still early in their separation, leaving a narrow Atlantic ocean and a very broad Pacific. The continents will continue to drift apart as they approach their modern configuration; the Atlantic widens steadily throughout the span of time under discussion.

The record of benthic foram $\delta^{18}O$ in Fig. 1.9 provides a good overview of the climate evolution for the past 70 million years. Towards the beginning of the period, $\delta^{18}O$ is considerably lower than it is in the modern ocean. It reaches a minimum value of $-0.1‰$ around 51 million years ago. Independent geological evidence shows that there was no significant amount of ice sequestered in land glaciers until 36 million years ago, so the variations in $\delta^{18}O$ before this time can be interpreted as polar ocean temperature changes. Melting the present ice in Antarctica and Greenland would leave the ocean with a $\delta^{18}O$ of around $-0.7‰$ relative to VSMOW. Plugging this into the paleotemperature equation, we estimate a high-latitude ocean temperature of $15.5\,°C$. This estimate applies to the coldest seasonal temperature attained either in the Arctic ocean or in the waters surrounding Antarctica.

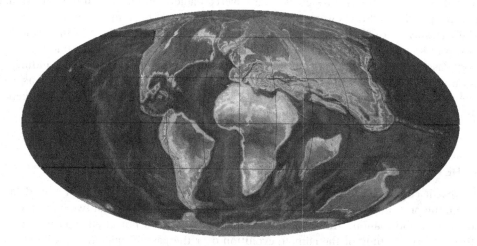

Figure 1.8 Position of the continents at the end of the Cretaceous, 65 million years ago. Continents are shown on a Mollweide map projection, with the north pole at the top and the south pole at the bottom. Light areas are continents while dark gray areas are ocean. Original color version of map © Ron Blakey, used with permission.

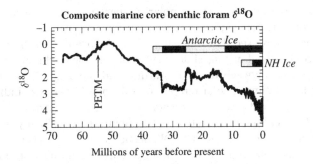

Figure 1.9 A composite record of benthic foram $\delta^{18}O$ (vs. PDB) over the past 70 million years, based on the average of several marine sediment cores. Note that the axis has been reversed so that upward δ excursions represent warmer times. Data from Zachos *et al.* 2001.

This is well above the freezing point of sea water, and so we conclude that the oceans were ice-free year round, even in the Arctic and Antarctic regions which are very cold in the modern climate. We'll refer to this kind of climate state as a *hothouse climate*. The late Cretaceous polar temperatures were about 4 °C cooler than those at the peak warmth, but still warm enough to guarantee ice-free conditions.

Other lines of evidence also support warm northern high-latitude conditions and above-freezing winter conditions. In the early twenty-first century, the first useful deep-time Arctic marine cores were recovered, and TEX-86 proxies applied to these cores indicated Arctic ocean up to 22 °C during the time of the spike marked PETM in Fig. 1.9, with temperatures in the range of 17–18 °C before and after. Fossil vegetation from Arctic land also supports temperatures in this range. Moreover, the abundant evidence that lemurs and crocodiles were able to survive in high northern latitudes points toward mild winters, since these creatures cannot survive sub-freezing temperatures for any significant length of time.

While evidence for warm, ice-free polar conditions in the Eocene and late Cretaceous is unambiguous, the nature of the tropical climate is somewhat problematic. Up until the year 2001, most paleoceanographers would have said, based on planktonic foraminiferal ^{18}O, that the Cretaceous tropical sea surface was no more than two or three degrees warmer than present, and the Eocene tropical sea surface was no warmer than today, and might even have been cooler. This posed the paradox of the "low gradient" climate – how to warm up the planet enough to prevent polar ice, without frying the tropics. In 2001, it was discovered that most of the evidence for a cool tropics was spurious, having been affected by diagenetic alteration of sediments. The surviving non-altered data indicated a warmer tropics, but there was precious little data left after the diagenetically contaminated data was discarded. Gradually, new sediments and new proxies have come to the fore, and the story continues to develop. TEX-86 proxies and Mg/Ca proxies now indicate tropical temperatures of up to 34 °C or even 37 °C in places, in the Eocene warm interval. Tropical surface temperatures were two or three degrees cooler in the late Cretaceous. This still gives a considerable reduction in the pole–Equator temperature gradient as compared with the modern climate, but the problem is not as severe as it was. It is quite certain that the tropical sea surface temperatures could not have been as high as 40 °C, since the planktonic forams that are seen in the sedimentary record could not have survived at such high temperatures. A striking feature of the data is that tropical surface temperatures seem to remain fairly constant throughout the Eocene, though polar temperatures (as indicated by the benthic forams) decrease towards the Oligocene.

Returning to Fig. 1.9, we see that following the peak Eocene warmth of 51 million years ago, the climate commenced a long slide toward the icehouse climate characterizing the latter part of the record. Between the peak warmth and the beginning of the Oligocene 34 million years ago, the minimum polar temperature dropped by 8 °C as indicated by benthic $\delta^{18}O$. At this point, small ephemeral ice sheets began to form on Antarctica, culminating in a more substantial glaciation of Antarctica that lasted until 26 million years ago, somewhat before the beginning of the Miocene. The Oligocene glaciation is visible as a pronounced ditch in the $\delta^{18}O$. This first attempt at glaciating Antarctica didn't last, however, since the climate recovered and returned to a period of generally cold Antarctic conditions with sea ice but with land ice sheets having volume below 50% of the present volume. This situation lasted until 15 million years ago, when the slide toward icehouse conditions resumed. Antarctic ice sheets grew again, but the northern hemisphere land glaciation had not yet been initiated. The first abundant evidence of sea ice, based on ice-rafted debris in polar sediment cores, appears at 14 million years ago. The increase in $\delta^{18}O$ in the next several million years is due to a combination of continued cooling and Antarctic ice sheet growth, culminating in the initiation of the major northern hemisphere ice sheets around 6.5 million years ago, as we enter the Pliocene period. The oxygen isotopes begin to show substantial fluctuations at this time, which grow in amplitude as one enters the Pleistocene. These fluctuations are due to waxing and waning of ice sheets, predominantly in the northern hemisphere – in other words, the coming and going of "ice ages." The nature of the fluctuations will be examined more closely in Section 1.10.

What accounts for the nature of the hothouse climate state, and for the subsequent hothouse/icehouse transition? This is another of the Big Questions. There is no support in astrophysics for solar variability of a magnitude and type that could explain the transition, so attention has settled primarily on long-term fluctuations in the greenhouse gas content of the atmosphere. In an oxygenated atmosphere like that of the Phanerozoic, CO_2 is the only known long-lived greenhouse gas that can build up to concentrations sufficient to cause climate changes of a magnitude comparable to those seen over the Phanerozoic; to get fluctuations of the requisite magnitude requires amplification of the direct CO_2 effect by water vapor feedbacks, and cloud feedbacks can also substantially modify the response. Another thing that makes CO_2 a good suspect for the role of primary agent in Phanerozoic climate evolution is the fact that it is a central participant in all aspects of the inorganic and organic carbon cycle, which offers many possible mechanisms whereby CO_2 could evolve over the long term. There are a great many unresolved issues regarding the CO_2 theory of Phanerozoic climate evolution, but a central part of testing the theory is to understand the way CO_2 and water vapor act in concert to determine the temperature of a planet; the necessary estimates will be given in Chapter 4. One needs to understand how much CO_2 it would take to account for Eocene warmth, before one can decide whether there are plausible geochemical mechanisms that could lead to the required concentration.

The greatest impediment to testing the CO_2 theory of the hothouse/icehouse dichotomy is the difficulty of estimating past CO_2 levels. There are various geochemical and fossil proxies that can be brought to bear on the problem. For example, algae preferentially take up ^{13}C at a rate that depends somewhat on the CO_2 concentration in the water in which they grow. Carbon isotopes in fossil soil carbonate also preserves information about past CO_2, as does the density of pores (*stomata*) in fossil leaves. All estimates known to date are subject to considerable uncertainty. Nonetheless there is support for the idea that CO_2 concentrations around 70 million years ago could have been 6–10 times modern pre-industrial values. The evidence points to a general decline of CO_2 since that time, but there are also some indications that CO_2 may have already attained quite low levels during some periods well before the Pliocene icehouse climate set in. This is a rapidly evolving subject, however, and nothing definitive can be said at present. What is certain is that there are known geochemical mechanisms associated with the Urey reaction and silicate weathering, which have the potential for causing changes in atmospheric CO_2 of the required magnitude, and on a time scale consistent with the observations. These mechanisms will be discussed in Chapter 8. As with any climate problem, on Earth or elsewhere, uncertainties regarding cloud feedbacks complicate the problem of testing theories of climate response. It is not out of the question that part of the answer lies in modulation of cloud albedo by, say, biologically produced sulfur compounds that seed cloud formation. Further, if data should ultimately support the low-gradient picture of the Cretaceous and Eocene hothouse climates, some mechanism will be needed to keep the tropics from overheating while the poles are warmed by elevation of CO_2. This, too, may involve clouds, or it may involve changes in ocean circulation. It has even been suggested, with considerable physical support, that increases in hurricane intensity in a warmer world could provoke ocean circulation changes of the sort required to provoke a low-gradient climate.

Figure 1.9 exhibits a dramatic climate event of considerable importance. The spike in $\delta^{18}O$ at the Paleocene/Eocene boundary (marked "PETM" in the figure, for *Paleocene/Eocene Thermal Maximum*) is not a glitch in the data. It represents a real, abrupt and massive transient warm event. The spike looks small in comparison to the range of isotopic variation over the past 70 million years, but in fact it represents the planet accomplishing two million years' worth of warming in a warm spike that (on closer examination) sets in within 10 000 years and has a duration of around 200 000 years. This isotopic excursion corresponds to a global warming of about 4 °C; other proxy records show that the warming had similar magnitude in the Arctic and at the Equator, and that it extended to the deep ocean. This climate event triggered a mass extinction of benthic species, probably due to a combination of warming, oxygen depletion, and ocean acidification. An important clue as to the cause of the warming is that the record of $\delta^{13}C$ from the same core (not shown) exhibits a major negative excursion at the same time, going from values of about +1.2‰ down to about zero at the bottom of the excursion. This indicates a catastrophic release of large quantities of isotopically light carbon into the climate system, which presumably increased the atmospheric greenhouse effect and led to warming. One possibility is that the release came in the form of methane from destabilized clathrate ices in the ocean sediments; another is that the isotopically light carbon came from oxidation of suddenly exposed organic carbon pools on land, releasing large quantities of CO_2. Based on analysis of the carbon isotope record, it has been estimated that 4000–6000 gigatonnes of carbon were released into the ocean–atmosphere system, if the release were from organic matter. This compares with 700 gigatonnes of carbon in the form of CO_2 in the modern atmosphere. This would considerably enhance the atmospheric greenhouse effect, though the effect would largely wear off after a thousand years or so, over which time about 80% of the released carbon would have worked its way into the ocean. It is far from clear that one can account for the observed magnitude and duration of PETM warming with the amount of carbon one has at one's disposal. This is one of the Big Questions. We shall not answer it in this book, but the reader will be provided with the tools needed to assess the warming caused by various amounts of CO_2 or methane, and also (in Chapter 8) a bit of insight about the partitioning of carbon between atmosphere and ocean. These are tools one must have at hand in order to evaluate any theory of the PETM.

The Cretaceous is closed by the impact of a large asteroid or comet (known generically as a *bolide*). This is known as the *K–T impact event* (for "Cretaceous/Tertiary," Tertiary being an obsolete term for the period following the Cretaceous). This event has little or no expression in the isotope record shown in Fig. 1.9, and is instead identified by the global presence of a layer of the element iridium. The impact crater has also been identified, which allows an estimate of the energy of the impactor. The K–T impactor had effects of extreme consequence despite the lack of an expression in the oxygen isotope record. Notably, it was the dinosaur killer – though many other species went extinct at the same time. Examination of the carbon isotope record shows also that the ocean carbon cycle remained highly perturbed for millions of years. There are many Big Questions associated with the consequences of a bolide impact. What is the mechanism by which the impact causes extinction? Is it direct blast and heat, or some longer-lasting change in the climate? It has been estimated that the impactor released about $5 \cdot 10^{23}$ joules of energy. How does this energy compare to other energy sources in the climate system, and what effects should it have on the atmosphere and ocean? What are the broader effects of a bolide impact on climate, and how long do they last? Does the impact cause a warming (perhaps through release of greenhouse gases) or a cooling (through lofting of a dust and soot cloud)? Some of the climate questions will be taken up in Chapter 4.

The K–T impact event is the archetype of impact events, which have been episodically important throughout Earth history. Similarly the Earth has experienced many other mass extinctions beside that at the K–T boundary, not all of which are clearly associated with a bolide impact. All mass extinctions lead to Big Questions.

1.9.2 Hothouse and icehouse climates over the Phanerozoic

The Cretaceous hothouse climate and the Pleistocene icehouse climate represent opposite extremes of the Earth's typical climate state of the past half billion years. Going back further in time, the Snowball Earth represents an ultra-extreme on the cold end; going further afield in space, the runaway greenhouse represents an ultra-extreme on the hot end, though one that evidently never occurred on Earth. The Earth has experienced many individual hothouse and icehouse episodes in the past half billion years. For times earlier than 70 million years ago one cannot rely on $\delta^{18}O$ records from sea-floor sediments to provide a cross-check on ice volume. Instead, one must look for evidence of glaciation or temperate polar climate preserved as glacial features or fossil animal and plant occurrences on land and in near-coastal environments. This record is far less complete and well-preserved. An accurate long-term sea level record would make it possible to estimate ice volume in the distant past, but recovering and interpreting sea level has proved difficult. It is certainly possible to detect the existence of major glaciations, but good estimates of ice volume are not available for pre-Mesozoic glaciations. The periods between major glaciations – tentatively labeled as hothouse climates below – could well have undetected episodes of polar ice embedded within them. The known episodes of major and minor land glaciation are summarized in Fig. 1.10.

The Cretaceous hothouse conditions extended back throughout the entire Mesozoic, and well into the late Permian. There is evidence that some of these periods, notably the early Jurassic, may have been even warmer than the Eocene. To find another era of glaciation to rival that of the Pliocene and Pleistocene icehouse, one needs to go back to late Carboniferous and early Permian. The 60-million-year period centered on the Permo-Carboniferous boundary 300 million years ago was a time of very extensive glaciation, reaching to lower

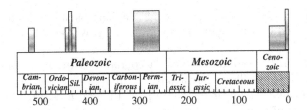

Figure 1.10 Known occurrences of land glaciation during the Phanerozoic. Tall shaded bars indicate periods of major glaciation in which ice reaches latitudes of 50° N or 50° S, while short shaded bars indicate periods of more minor glaciation. Numbers on the scale represent time in millions of years before present.

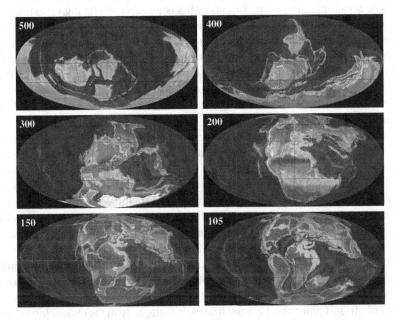

Figure 1.11 Evolution of geography over the Phanerozoic. Map projection and shading as for Fig. 1.8. The numbers give the time in millions of years ago. Original color version of maps © Ron Blakey, used with permission.

latitudes even than the Pleistocene ice ages, though not attaining Snowball proportions. This period is a crucial one for the CO_2 theory of Phanerozoic climate, since it is a time when there is quite strong evidence for low CO_2. The earlier Phanerozoic exhibits comparatively minor glaciations at the end of the Devonian, in the mid-Silurian and for a brief period in the mid-Cambrian, but so far as the Phanerozoic goes, the Permo-Carboniferous glaciation is the big one to explain.

The changing paleogeography, shown in Fig. 1.11, is likely to have influenced climate. In particular, it is bound to be easier for a glacier to accumulate on land if there is land at or near one or both of the poles. Certainly there was land at the south pole during the Carboniferous glaciation and the current glaciation that began in the mid-Cenozoic. However, this is clearly not the whole story, as there was plenty of land at the south pole already 400 million years ago, but the Carboniferous glaciation didn't set in until nearly a hundred million years later. Likewise, Antarctica was already near the south pole during the Cretaceous, but Antarctic glaciation only took off in the mid-Cenozoic. Most likely, fluctuations in CO_2 – probably itself affected by continental configuration – play a crucial role in the timing of glaciations.

A general theme in the evolution of paleogeography is the assembly and breakup of *supercontinents*. From 500 million years ago to 400 million years ago one can see the southern supercontinent Gondwana near the south pole, though there are a few leftover bits of land that are not part of Gondwana. By 300 million years ago, Gondwana has merged with these bits to form the global supercontinent Pangea, which then breaks up into the present continents over the course of the rest of the Phanerozoic. The interiors of supercontinents are isolated from the moderating effects of the oceans on climate, and so could be expected to experience harsh seasonal swings in temperature. Do we expect supercontinent interiors to be deserts or steaming, moist fern forests? This is another of the Big Questions.

Is there a clear dominance of hothouse or icehouse conditions over the past half billion years? The record of the past hundred million years certainly supports the notion that the largely ice-free hothouse is the preferred state of the Earth's climate, but going further back in time it is harder to say whether the apparent dominance of hothouse conditions is an artifact of poor preservation of the polar deposits where glaciers are most likely to have occurred. Some of the episodes we think of as hothouse climates could well have had significant amounts of ice.

In any event, the delineation of the circumstances which favor icehouse or hothouse climates, and the factors governing the transition between the two, constitutes one of the Big Questions of climate science. It seems likely that if the hothouse/icehouse transition of the past 70 million years can be understood, similar mechanisms could be applied to the rest of the Phanerozoic. Variations on the theme would include a greater range of different continental configurations – notably the breakup of supercontinents – as well as biological innovations such as the colonization of land and the evolution of deepwater carbonate shell-forming micro-organisms, both of which can affect the global carbon cycle.

During the Phanerozoic, life on Earth went through a number of mass extinctions rivaling or exceeding the end-Cretaceous event. The biggest mass extinction of all occurred at the end of the Permian, wiping out 96% of all marine species, 70% of all land vertebrates, and a large fraction of all land plants and invertebrates. It is the only mass extinction that included insects to any great extent. There is no clear evidence for a bolide impact at this time, though it remains possible that an impact occurred but failed to leave a trace in the fossil record. In any event, the cause of the end-Permian mass extinction ranks as one of the Big Questions.

1.10 PLEISTOCENE GLACIAL-INTERGLACIAL CYCLES

Now we will take a closer look at what's been going on in the past five million years. The earlier portion of this time is known as the *Pliocene* epoch, and the latter portion, beginning around 1.8 million years ago, is the *Pleistocene*. The choice of the Pliocene–Pleistocene boundary is based on an obsolete notion of the time when northern hemisphere ice sheets attained nearly their modern extent. At the level of geological periods, the Neogene-Quaternary more closely approximates this distinction. The Quaternary period extends all the way to the present, but the Pleistocene is terminated somewhat arbitrarily at the end of the last major ice sheet retreat around 10 000 years ago. The remainder of time up to the present is the *Holocene* epoch. This is a very human-centric division, since the time we are experiencing now is a fairly ordinary Pleistocene-type interglacial period – or at least it was up until the commencement of the industrial revolution. A more rational division would extend the Pleistocene up to about the year 1700, whereafter we would enter something with a name like "Anthropocene" (for reasons to be discussed in Section 1.12).

The Pliocene and Pleistocene are a time of establishment of the great northern hemisphere ice sheets. During this time, the ice sheets settled into a rhythm of expansion and retreat – the rhythm of the coming and going of *ice ages*. The notion of an "ice age" is distinct from that of the "icehouse climate state" introduced earlier. The latter term refers to a span of time (usually several million years) during which there is permanent ice at one or both poles. Within the time embraced by an icehouse climate state, ice volume is not constant, but fluctuates episodically with variations in ice volume that can be on the order of a factor of two (judging from the Pleistocene). Individual episodes of large ice volume within an icehouse climate are referred to as *ice ages*, with the warmer periods in between referred to as *interglacials*, though the ice doesn't come close to disappearing completely. In the Pleistocene, the fluctuation in ice volume is dominated by changes in northern hemisphere ice sheets, but as ice ages come and go, the entire globe becomes colder and warmer.

1.10.1 The marine sediment record

Figure 1.12 shows the $\delta^{18}O$ of benthic forams in a tropical Pacific core over the past 4 million years. The short-period fluctuation represents fluctuations in both ice volume and benthic temperature, and in addition there is a downward trend in temperature and increasing trend in ice volume over the earlier 2 million years of the period. By 2 million years ago, the fluctuations have settled into a fairly regular pattern, with a dominant period of about 40 000 years. The period may be crudely estimated by counting peaks in the isotope record. This says that major ice advances in the early Pleistocene occur roughly every 40 000 years. About 800 000 years ago, there is a major transition in which the amplitude of the glacial–interglacial cycle becomes markedly larger, and the periodicity lengthens to about 100 000 years. During this period, an asymmetry between glaciation and deglaciation becomes readily apparent: the climate cools and ice builds up over long periods of time, but deglaciation occurs rather precipitously.

The periodicity of the ice ages, and the reason for the transition to a dominant 100 000-year cycle later in the Pleistocene, is another of the Big Questions of climate science. We will learn in Chapter 7 that the periodicities are almost certainly connected with the quasiperiodic variations of the Earth's orbital characteristics – namely the tilt of its rotation axis and the departure of the orbit from circularity. The cycles are known as *Milankovic cycles*, after the scientist who first formulated a detailed theory connecting orbital parameters with the coming and going of ice ages. Milankovic's theory was largely ignored for decades, because not enough was known about the pattern of ice ages to give the theory a fair test. It was only revived in the 1970s, when data of the sort given in Fig. 1.12 first became available. Even today, the means by which cycles in orbital characteristics are expressed in the climate record are far from clear, and remain a subject of active research.

This is another case in which the lesson learned from Earth carries over to other planets, as the orbital parameters of Mars also show Milankovic cycles. Compared with Earth, however, the study of how these cycles are expressed in the paleoclimate record of Mars is in its very infancy.

1.10.2 Ice-core records

The ice at the base of the Antarctic ice sheet is about a million years old, so one can also retrieve a record of past climate by drilling cores into the ice. Many aspects of climate are

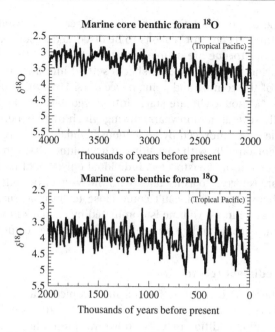

Figure 1.12 Benthic foram oxygen isotopes from the Ocean Drilling Program Core ODP 849 (see Mix *et al.* in the Further Reading for this chapter). Values are reported relative to the PDB standard. The core is located in the tropical Pacific, but the benthic data is representative of the global climate state. Note that the vertical axis has been reversed, so that upward excursions represent warmer and less icy times.

recorded in the ice, but the ones that will concern us here are stable isotopes of water (δD and $\delta^{18}O$), recorded in the ice itself, and composition of past air preserved in bubbles within the ice. The stable isotopes are essentially recorders of temperature; the ice becomes more isotopically light as the polar temperature becomes colder, since more of the heavy isotopes have been distilled out in that case.

The upper panel of Fig. 1.13 shows the δD time series from the Vostok and EPICA ice cores in the Antarctic. The Vostok data is systematically below the EPICA data because Vostok is further inland, higher, and colder, but otherwise tracks the EPICA data. This record only covers the period within which the glacial–interglacial cycles have already settled into a 100 000-year cycle; older ice is too distorted to yield a useful record. This record confirms the 100 000-year cycle seen in the marine cores, and also confirms the asymmetry between slow buildup of glaciers and rapid deglaciation. It also shows that there is a strong Antarctic warming and cooling that occurs in association with the glacial–interglacial cycles.

The CO_2 data is shown in the lower panel of Fig. 1.13. In the course of the glacial–interglacial cycles, CO_2 fluctuates between 180 parts per million (by count of molecules) and 300 parts per million. Moreover, the fluctuation is very nearly synchronous with the warming and cooling. The correlation between CO_2 and temperature does not determine which causes which (or whether they mutually reinforce each other), but since CO_2 is a greenhouse gas, it is certain that the rise in CO_2 warms the planet and reinforces the termination of ice ages, whereas conversely the fall of CO_2 enhances cooling and reinforces the onset of glaciation. The origin of the glacial–interglacial CO_2 cycle is another of the Big Questions. Our understanding of glacial–interglacial cycles cannot be complete without resolving this question.

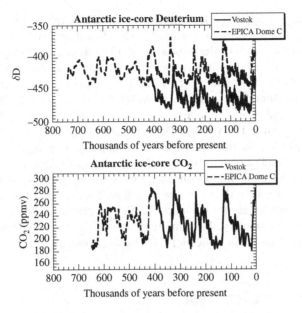

Figure 1.13 Data from the Vostok and EPICA Antarctic ice cores. The upper panel shows the variation in deuterium depletion of the ice, which is a proxy for temperature. Higher (less negative) values indicate warmer conditions. The lower panel gives the variation in the CO_2 concentration in air bubbles trapped in the ice. Concentrations are stated as molecules of CO_2 per million molecules of air.

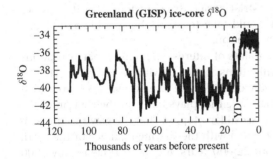

Figure 1.14 Oxygen isotope data from the GISP-2 Greenland ice core. Larger (less negative) values correspond to warmer temperatures.

Greenland ice also records past climate, as seen in Fig. 1.14. The base of the Greenland ice cap is not as old as the Antarctic ice, so one can only go back in time about 100 000 years here. By way of compensation, though, the rate of snow accumulation in Greenland is much higher than in Antarctica, so one can see the past with much higher time resolution.

The Greenland record records a sharp deglaciation leading to the modern era, consistently with the Antarctic record. However, we can see in Greenland that there are many high-frequency temperature fluctuations embedded within the glacial period – especially the period from about 60 000 years ago to 10 000 years ago. These don't have a strict periodicity, but have a time scale on the order of a thousand years, and hence are collectively referred to as *millennial variability*. The expressions of millennial variability seen in Greenland isotopes are called *Dansgaard–Oeschger events* after their discoverers. There are many other climate proxies that reflect millennial variability of the sort seen in Greenland isotopes, and these often represent precipitous switches in the climate state – referred to as *abrupt climate change*. One of the most striking of these abrupt change events occurs as the planet is coming out of the most recent ice age – the *Last Glacial Maximum* (or LGM). The spike marked "B" in the figure is the *Bølling* warm period, and represents a recovery of the climate to full interglacial warmth. However, in the wake of the Bølling the climate abruptly reverted to full glacial temperatures, in an event known as the *Younger Dryas* (marked "YD" in the figure). The Younger Dryas is believed to have been triggered by a massive draining of a glacial lake into the ocean, but generally speaking the mechanisms both of the Younger Dryas and of millennial variability remain as Big Questions that are yet to be resolved. This is an especially important question because the flat, quiescent period at the end of the Greenland record – the *Holocene* period we have lived in for all of the history of civilization – represents an abrupt cessation of high-amplitude millennial variability. What accounts for the uncommonly stable climate of the Holocene, and what would it take to break this situation? This is a Big Question with considerable import indeed.

1.11 HOLOCENE CLIMATE VARIATION

The climate variations of the Holocene have been (at least so far) more subtle than the massive variations we have discussed previously. In part, this is simply because the Holocene is a short period of time, and there hasn't been enough time for something really dramatic to happen, given the relatively slow pace of many of the geological processes that modify climate. But the limited span of the Holocene is not the whole story. The Holocene has not witnessed extreme millennial-scale variability of the sort seen in the preceding glacial time. Given the massive assault on climate by industrial society (see Section 1.12) it becomes all the more pressing to understand what it would take to break the equable and steady Holocene climate enjoyed during the rise of civilization.

Still, the Holocene has not been without its points of interest so far as climate variations go, the more so because there were human civilizations around to witness and be affected by these variations. A key driver of Holocene climate change is the *precessional cycle*, to be discussed in Chapter 7. It is the same precessional cycle that plays a role in the rhythm of the Pleistocene ice ages. The Earth's spin axis precesses like a top, so that the way the "tilt seasons" line up with the "distance seasons" associated with the varying distance of the Earth from the Sun goes through a cycle lasting about 22 000 years. A quarter cycle is only 5500 years, which brings these cycles within the span of recorded history. At present, the Earth is farthest from the Sun when the northern hemisphere points toward the Sun (i.e. during northern summer), and is closest to the Sun when the northern hemisphere points away (i.e. during northern winter). This gives us relatively warm winters and relatively cool summers. Eleven thousand years ago the situation was reversed, and northern hemisphere summers received considerably more sunlight than they do today, particularly at high latitudes. This should have made polar regions considerably warmer than today, but for reasons that are only partly understood, the time of warmest summers was delayed several thousand years, to a time called rather tendentiously the *Climatic Optimum* about 7000 years ago. It is not entirely clear what the climate was "optimal" for, and this is in any event a northern-centric view as southern hemisphere continents if anything experienced a weak cooling at this time (as one would expect from the nature of the precessional cycle). The alternate term *Altithermal* is preferred, as being less value-laden. In any event, at this time the northern polar regions were getting up to 10% more solar radiation in summer than they are today, leading to warmer summers. The warming is expected to be greatest over land, since oceans average out the warm summers and cold winters. Tree ring records indicate that high-latitude land masses were between 2 °C and 4 °C warmer in the summer at the time of the Altithermal. These estimates are corroborated by an expansion of the tree line to higher-altitude land in the European Arctic. Cold-tolerant trees are primarily sensitive to growing-season temperatures, and are little affected by a moderate decrease in winter temperatures; in fact, more severe winters can in some cases be favorable to tree growth, since cold winters interrupt the life cycle of various insect pests. While the Altithermal is certainly connected with the precessional cycle, the magnitude of the warming, the cause of the delay in warming (partly associated with leftover glaciers from the last ice age), and the role of ocean circulations and vegetation changes in the Altithermal, are all active subjects of research.

The precessional cycle has also had a profound effect on the distribution of precipitation in low latitudes. The Sahara is a desert today, but the dry river features known as *wadis* were flowing with water 6000 years ago, at which time the desert was a savannah grassland. This wet period commenced about 14 500 years ago and the Sahara abruptly reverted to desert about 4700 years ago. There are also intriguing indications that the time of initiation of present tropical mountain glaciers follows a precessional cycle. For example, the Peruvian Andean glaciers of the southern hemisphere date back to the last ice age, while the Kilimanjaro ice fields were laid down during the African Humid Period about 10 000 years ago, and Himalayan glaciers of the northern subtropics tend to be even younger. The connection between the seasonal cycle of solar radiation and the tropical precipitation distribution involves atmospheric circulations – monsoons and the Hadley circulation – that cannot be treated without a full understanding of atmospheric fluid dynamics. We will therefore have only limited opportunities to pursue the precessional precipitation cycles in the course of this book, though the treatment of the precessional cycle in solar forcing will provide the student with the necessary background for further study.

The *Little Ice Age* is another Holocene climate fluctuation of considerable interest. This term refers to a period of generally cool northern hemisphere extratropical land temperatures extending from approximately 1500 to 1800. Tree ring estimates suggest the northern hemisphere mean temperature dropped by something over 0.5 °C between the year 1400 and 1600. The cooling is corroborated by records of advances of mountain glaciers, sailors' observations of sea ice, and agricultural records. The Little Ice Age is too short and too recent to have anything to do with the precessional cycle, and while it is possible that fluctuations in the ocean circulation could have produced the cooling, the prime candidate for an explanation of the Little Ice Age is a temporary slight dimming of the Sun. Sunspot observations do indicate a cessation in the normal solar sunspot cycle – called the *Maunder Minimum* – at about the time of the Little Ice Age. However, most estimates of the associated solar output change are far too small to yield a significant cooling. Various mechanisms are under investigation which could amplify the response to the small solar fluctuation, but in the grand scheme of things the Little Ice Age is a rather subtle event, and accordingly hard to understand, particularly in terms of simple models.

To put the Holocene in perspective, it is salutary to note that the "abrupt" PETM event discussed in Section 1.9.1 lasted nearly 200 000 years, and set in over a period of around 10 000 years – as long as the full length of the Holocene. There really has not been much time for things to happen in the Holocene, and the time span of Fig. 1.9 no doubt contains many 10 000-year periods as quiescent as the Holocene has been up until recently. Short as the Holocene is, we will see next that human activities have been able to cause some dramatic changes in the composition of the atmosphere and consequently in the Earth's climate. One wonders what it would take to trigger a hyperthermal event such as the PETM, which is over in the wink of an eye by the standards of Fig. 1.9, but has a duration 20 times as long as the span of human civilization to date.

1.12 BACK TO HOME: GLOBAL WARMING

We have seen that CO_2 has been a major factor in determining climate throughout Earth's history, and that life, in turn, has greatly shaped the carbon cycle. Life is in the midst of disrupting the carbon cycle once more, but this time it is technological life that has provided the necessary innovations. Over billions of years, a great deal of organic carbon has been sequestered in the Earth's crust without oxidizing. Most of this carbon is in very dilute forms which cannot easily be tapped to provide economically useful amounts of energy. However, a very small portion winds up in nearly pure forms that are moreover chemically altered in a fashion that makes them especially convenient as fuels. These are the *fossil fuels* – coal formed from land plants and oil formed in marine environments. Natural gas can be produced by thermal alteration of either coal or oil. Fossil fuels represent concentrated solar energy stored in the form of organic carbon, which has been accumulating over hundreds of millions of years. This pool of readily oxidizable carbon exists precisely because it is in geological formations that have kept it apart from oxygen over the ages. It is only the evolution of technical civilization that is making it possible to dig up and oxidize hundreds of millions of years' worth of stored fossil carbon within a few centuries.

In the year 2005, over 8 gigatonnes ($8 \cdot 10^{12}$ kg) of carbon were released by fossil fuel burning, and annual emissions continue to grow rapidly. There are several ways to see that this is a very big number – a major upset to the natural carbon cycle. First, the pre-industrial atmosphere contained about 600 gigatonnes of carbon, so the 2005 annual emission is fully 1.3% of the undisturbed atmospheric content. If the same amount were released into the atmosphere each year, it would take only 75 years to double the atmospheric CO_2 content, provided all the released CO_2 stayed in the atmosphere. Alternately, one could compare the fossil fuel emissions to the volcanic outgassing which in the long term balances silicate weathering and sustains the carbon cycle. Precise estimates of volcanic outgassing are hard to come by, but generally are on the order of 0.1 gigatonnes of carbon per year or less. Thus, *fossil fuel carbon emissions are 80 times larger than background volcanic outgassing.* In fact, the very largest carbon flux number involved in the whole carbon cycle is the net CO_2 carbon fixed into organic carbon each year by worldwide photosynthesis, and fossil fuel emissions even look impressive when compared to this number. Based on satellite chlorophyll observations, it has been estimated that photosynthesis fixes 100 gigatonnes of carbon each year, about half on land and half in the oceans. The fossil fuel emissions in the year 2005 were fully 8% of this number. In other words, worldwide photosynthetic productivity would have to increase by 8% to take up the fossil fuel CO_2 *and 100% of that carbon would have to be buried as organic matter without being recycled by respiration.* That, of course, would be a completely absurd situation, as virtually all of the photosynthetically fixed carbon is quickly respired back into the atmosphere, largely by bacteria who have had several billion years to become proficient at making use of organic carbon wherever they find it. As an example, land photosynthesis fixes about 50 gigatonnes of carbon each year, but the flux of organic carbon to the oceans in all the world's rivers is a mere 0.4 gigatonnes per year (one-twentieth of fossil fuel carbon emissions). And there is no evidence that much of the remainder of the photosynthetically fixed carbon is remaining on land as soil organic carbon. To say that humans have become a force of geological proportions vastly understates the case, for by this measure human influences on the carbon cycle overwhelmingly dominate the natural sources.

The result of all our busy digging and burning has been a steady increase in atmospheric CO_2. Figure 1.15 shows the time series of atmospheric CO_2 concentration since 1750. Carbon dioxide has a very long atmospheric lifetime, so it is well-mixed. In consequence, one finds nearly the same CO_2 concentration wherever one measures it, so long as the measurement is not in the immediate vicinity of major sources or sinks. The part of the record since 1950 comes from direct analyses of air samples at the Mauna Loa observatory, whereas the earlier part of the record comes from air trapped in bubbles in the ice of the Siple Dome site, Antarctica, but the two records match up well where they meet. At the dawn of the industrial era CO_2 concentrations are near 280 molecules per million (ppmv for short), right where they were left at the end of the most recent ice age. After 1750 the concentrations begin to rise, and by 2007 the concentrations have exceeded 380 ppmv – fully 35% above the pre-industrial value. Most of the increase has happened since the mid twentieth century, and the rate of increase seems to be accelerating along with population and economic growth.

Not all of the carbon released by fossil fuel burning has remained in the atmosphere. Estimates based on careful historical inventories suggest that only about half of the total carbon released to date remains in the atmosphere as carbon dioxide. Most of the remainder has slowly infiltrated the ocean, with a smaller amount having been taken up by the terrestrial ecosystem (net of deforestation). In fact, it has been shown that the rate at which the ocean can take up the excess CO_2 is limited by the mixing between the upper ocean and the deep ocean. This is a slow process, and if all fossil fuel burning were suddenly to cease, it would take in excess of 600 years for 80% of the excess CO_2 to be taken out of the atmosphere. The remainder would stay in the atmosphere for millennia longer, owing to certain chemical processes (discussed briefly in Chapter 8) which limit the ability of the ocean to take up CO_2. The slow net removal rate of CO_2 allows fossil fuel emissions to accumulate in the atmosphere. Another consequence of the long lifetime of CO_2 in the atmosphere is that the climatic effects of elevated CO_2 will persist for centuries to millennia, even after any

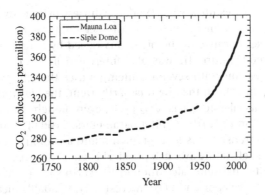

Figure 1.15 Annual mean CO_2 concentration from 1750 to the time of writing. The earlier part of the record is from air trapped in bubbles in the Siple Dome Antarctic ice core. The more recent part of the record is from instrumental measurements at the Mauna Loa observatory. The units are molecules of CO_2 per million molecules of air.

(much to be hoped-for) dramatic restriction of fossil fuel burning. Allowing for uptake by the ocean, there are enough fossil fuel reserves – primarily in the form of coal – to ultimately increase the atmospheric CO_2 concentration to at least six times the pre-industrial value. The number could go much higher if the ocean sink were to become less efficient, or if land ecosystems were to turn around and become a CO_2 source rather than a sink.

This all leads us to a series of very Big Questions: if the rise in CO_2 is allowed to con-tinue to a doubling of the pre-industrial value, how much will the Earth warm? How will the warming be distributed? How much will sea level rise as a result of melting land ice and thermal expansion of ocean water? What will happen to precipitation patterns? How will all of this affect human societies and natural ecosystems? The basic physics needed to treat these questions is identical to what is used to account for the influence of CO_2 and other long-lived greenhouse gases on past climates. The problem in this instance has more imme-diacy as many generations of our descendants will be living with the consequences of our fossil fuel emissions in the next several decades. In order to understand what kind of planet we are leaving these descendants, there is a demand for greater detail in the understanding of the climate changes to be wrought by these rapid increases in atmospheric CO_2.

Interest in the effect of CO_2 changes on climate long predates the kind of data shown in Fig. 1.15 which showed that CO_2 was on the increase, and in fact predates the realization that human activities really could cause CO_2 to increase appreciably. Likewise, global warming was a concern long before it was confirmed that the Earth really was warming in response to increases of CO_2. These things were all anticipated theoretically a century or more before global warming burst onto the scene as an issue of political consequence, and the driving force was basic curiosity about the physics governing planetary temperature. It's a line of inquiry that extends right back to Fourier's pathbreaking inquiry into how an atmosphere affects the energy budget of a planet, and hence its temperature. The discovery of global warming is a great triumph of two centuries of developments in fundamental physics and chemistry. It is not a matter of people having noticed that both CO_2 and temperature were going up, and concluding that the first must be somehow causing the second. Both the rise of CO_2 as a consequence of fossil fuel burning, and the consequent rise in temperature as a response to the Earth's perturbed energy balance, were anticipated long before either was observed.

After Fourier, the tale resumes with Tyndall, whose work on the infrared absorption of CO_2 and water vapor was mentioned near the beginning of this chapter. Tyndall was interested in these gases because of the questions raised by Fourier regarding the factors governing planetary temperature. He was also interested in the recently discovered phe-nomenon of the ice ages, and with several contemporaries thought perhaps ice ages could arise from a reduction in CO_2. In that, he was partly right; the Pleistocene ice ages are cold partly because of the glacial–interglacial CO_2 cycle, even though the ultimate pacemaker of the ice ages is the rhythm of Earth's orbital parameters. Tyndall died, however, before he ever had the chance to translate his measurements into a computation of the Earth's tem-perature. That task was left to the Swedish physical chemist Svante Arrhenius, who in 1896 performed the first self-consistent calculation of the Earth's temperature incorporating the greenhouse effect of water vapor and CO_2. Interestingly, Tyndall's measurements were not sufficient to provide the information about weak absorption over long path lengths, so for the absorption data he needed he turned to Langley's observations of infrared emitted by the Moon. It was a felicitous re-use of data intended originally for determination of the Moon's temperature, and indeed was a more correct use of the data than Langley was able to accomplish. This shows the benefit of curiosity-driven science: measurements taken to satisfy curiosity about lunar temperature wound up being instrumental in permitting an evaluation of the effect of the Earth's atmosphere on the Earth's temperature. Astronomers initiated the study of infrared as an observational technique, but the radiative transfer work stimulated by their needs soon provided the crucial tool needed to understand planetary climate. Arrhenius not only estimated the Earth's then-current temperature, but also estimated how much it would warm if the amount of CO_2 in the atmosphere were to double. Using clever scaling analyses from Langley's data, he was able to do this without a firm knowledge of just what the atmosphere's CO_2 content actually was. Not long afterwards, he realized that industrial burning of coal was dumping CO_2 into the atmosphere, and could eventually bring about a doubling; he described this process as "evaporating our coal mines into the atmosphere." At then-current rates of consumption, it appeared that a doubling would take up to a millennium, and Arrhenius would no doubt have been surprised to know that his own great-grandchildren could well live to witness the doubling. This takes our story to about 1900. What happened then?

A long hiatus. In part, there was little sense of urgency, because a failure to anticipate the explosive growth of fossil fuel use looming in the coming decades led to a belief that any problem lay in the far-distant future. Besides that, two unfortunate turns of events held back the study of global warming for decades. The first was a highly touted experimental study published in 1900 by the prominent physicist Knut Ångström, which purported to show that the radiative effects of CO_2 are "saturated," i.e. that the gas already absorbs as much as it can at the atmosphere's then-present concentration, so increases would have no effect. A concomitant and closely associated (and equally wrong) idea was that the strong absorption of water vapor would completely swamp any effect CO_2 might have. The experiment turned out to be wrong, but such was Ångström's reputation and such was the resistance to the idea that humans could change climate that it was decades before anybody definitively checked the result. Moreover, it turns out that even if Ångström had been right, it would not have negated the greenhouse effect; this misunderstanding hinged on the poorly developed understanding of radiative transfer in a temperature-stratified atmosphere. The "gray gases" we will study in the first half of Chapter 4 are "saturated" in the sense of Ångström, but nonetheless allow for an increase in the greenhouse effect as more greenhouse gas is added to the atmosphere. The second barrier to progress was the belief that the huge carbon content of the ocean would buffer the atmosphere, overwhelming anything human industry could have thrown at it. The carbonate chemistry needed to defeat this idea was largely worked out by the 1930s; indeed, it requires nothing more than is taught routinely in high-school chemistry courses today. However, it had not been assimilated into a coherent and widely appreciated picture of the uptake rate of CO_2 by the oceans, in part because of lack of knowledge of the rate of mixing between the upper ocean and the deep ocean. A paper published by Revelle and Suess in 1957 is widely credited with having broken the logjam, but in fact mentions the essential carbonate buffering mechanism (fully worked out by earlier researchers, and cited as such) almost as an afterthought. The attempt of Revelle and Suess no doubt helped to revive interest in the question of oceanic CO_2 uptake rates, but in fact the paper came to exactly the wrong conclusion – that fossil fuel emissions were unlikely to lead to any significant increase in atmospheric CO_2 concentration. The true implications of the carbonate buffer for CO_2 increase due to fossil fuel burning was finally brought out clearly in a paper two years later, by Bert Bolin and Erik Eriksson.

Despite Revelle and Suess's conclusion, the idea gained hold that somebody should actually systematically check and see what atmospheric CO_2 was doing. This program was initiated by Charles Keeling while at Caltech, and was subsequently encouraged by Revelle.

Keeling's work culminated in the Mauna Loa data shown in Fig. 1.15. The techniques for recovering past CO_2 from air bubbles trapped in ice were not to be developed until the 1970s, so Keeling had to wait a decade or so before it was clear that CO_2 was really rising, and a bit more time after that before there was a clear idea of just how high CO_2 already was relative to the pre-industrial value. This work, together with developments in infrared radiative transfer stimulated by astronomical observation and military interest in infrared target detection, led to new breakthroughs in the formulation of radiative transfer. The work culminated in 1967 with the calculation by Manabe and Wetherald of the Earth's temperature using modern radiative physics. They were also able to calculate the warming due to a doubling of CO_2, allowing for expected changes in water vapor content as the planet warmed. This was not the end of the story, which indeed continues today, since there was much to be done in terms of embedding the radiative transfer in a fully consistent computation incorporating the fluid dynamics of the atmosphere – a *general circulation model*. It was, however, the beginning of the modern chapter of the study of global warming. With the publication of the Charney report by the US National Academy of Sciences in 1979, global warming began to be perceived as a real threat. The powers that be were slow to awaken to the magnitude of the problem, and several more years were to pass before the creation of the Intergovernmental Panel on Climate Change in 1988, which initiated regular, comprehensive surveys of the state of the science surrounding global warming. At the time of writing, the world still awaits substantive action to curb fossil fuel emissions.

All aspects of the essential chemistry, radiative physics and thermodynamics underlying the prediction of human-caused global warming have been verified in numerous laboratory experiments or observations of the Earth and other planets. Other aspects of the effect of increasing greenhouse gases rely on complex collective behavior of the interacting parts of the climate system; this includes behavior of clouds and water vapor, sea ice and snow, and redistribution of heat by atmospheric winds and ocean currents. Such things are impossible to test in laboratory experiments. To some extent, aspects of our theories of the collective behavior have been tested against the seasonal cycle of Earth, interannual variability, and past climates, as well as attempts to simulate other planetary climates. The ultimate test of the theory, though, is to verify it against the uncontrolled and inadvertent experiment we are conducting on Earth's own climate. Can we see the predicted warming in data? This is not an easy task. For one thing, the atmospheric CO_2 increase is only a small part of the way towards doubling, and the climate has not even fully adjusted to the effect of this amount of extra radiative forcing: oceans take time to warm up, and delay the effect for many years (for reasons to be discussed in Chapter 7). Thus, so far the signal of the human imprint on climate is fairly small. Set against that is a fair amount of noise complicating the detection of the signal. Climate, even unperturbed by human influence, is not steady from year to year, but is subject to a certain amount of natural variability. This can be due to volcanic eruptions and subtle variations in the brightness of the Sun. There are also various natural cycles in the ocean–atmosphere system that cause the planet to be a bit warmer or colder from one year to the next. Chief among these is the El Niño phenomenon of the tropical Pacific. During El Niño years, the coupled dynamics of the tropical ocean and atmosphere causes warm water to spread throughout the Pacific, leading to a warming of mean surface temperatures both in the tropics and further afield. La Niña years represent a bunching up of the warm water, and an accentuated upwelling of cold water, leading to cold years. The two phases alternate erratically, with a typical time scale of three to five years.

The fact that the signal is hard to detect does not mean that global warming is of little consequence. The difficulty arises precisely because we are trying to detect the signal before it becomes so overwhelmingly large as to be obvious. Given the long lifetime of CO_2 in the atmosphere, it would be highly desirable to keep the signal from ever getting that large, as if it ever does it will take many centuries to subside. Let's now take a look at some of the data, and see if there are any signs that the theoretically anticipated warming is really taking place.

Figure 1.16 shows a time series of estimated global mean surface temperature, based on recorded temperatures measured with thermometers. There is a lot of arduous statistics and data archaeology behind this simple little curve. Particularly for the data going into the early part of the curve, there has been a need to standardize measurements to allow for the various different ways of taking a temperature reading. Most of the oceanic measurements, for example, were taken by commercial or military ships of one sort or another, and some of the entries in ships' logs record things like "bait tank temperature" or "engine inlet temperature." There has also been a need to screen out stations that have been strongly affected by local land use changes such as urbanization, and to avoid spurious trends due to changes in the spatial distribution of temperature measurements (e.g. fewer Antarctic readings once Antarctic whaling essentially ceased). To help correct for biases in individual classes of temperature measurements, the long-term trends are presented as *anomalies* relative to the average of a station's long-term average standardized to a fixed base period (e.g. 1951–80 for the data in Fig. 1.16).

There is little temperature trend between 1880 and 1920, but between 1920 and 2005 the temperature has risen by nearly 1 °C. The rise has not been steady and uninterrupted, however. It takes the form of an early rise between 1920 and 1940, followed by a 30-year period when temperatures remained fairly flat, whereafter the temperature rise resumed and has continued to the present. Given that CO_2 has been rising at an ever-increasing rate over the industrial period, why was the warming interrupted between 1940 and 1970? The answer lies largely in another effect of human activities on climate. Burning of fossil fuels, and especially coal, releases sulfur compounds into the atmosphere which form tiny highly reflective droplets known as *sulfate aerosols*. By 1995,the effect was finally quantified with sufficient accuracy to permit reasonable estimates of the effect, and it began to appear that most of the evolution of climate of the twentieth and twenty-first centuries could be accounted for by a combination of rising greenhouse gases (mainly CO_2) due to human activity, with an offsetting cooling effect of sulfate aerosols. The reason small particles are so good at scattering light back to space is discussed in Chapter 5, where the optical properties of sulfate aerosols will be discussed in detail. By the year 2000, the greenhouse warming signal had unquestionably risen above both the noise of natural variability and the offsetting effect of aerosol cooling. Sulfur is an active element in many actual and hypothetical planetary atmospheres, and so the study of sulfate aerosols on Earth informs other planetary problems, including the clouds of Venus and Venus-like extrasolar planets.

The Earth's emissions in the microwave spectrum have been monitored continuously by satellite-borne instruments since 1979, and these observations make it possible in principle to obtain reconstructions of atmospheric temperature trends which are independent of the somewhat inhomogeneous surface station network. Processing the microwave data accurately enough to obtain reliable temperature trends proved very difficult, and there were many false steps along the way. Nonetheless, the main problems were resolved by early in the twenty-first century. The microwave temperature retrievals give the temperature of the atmosphere averaged over fairly deep layers, in contrast to the surface stations which measure near-surface air temperature. The left panel of Fig. 1.17 shows the satellite

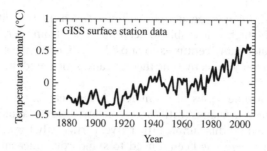

Figure 1.16 Global average annual mean surface temperature since 1870, estimated from surface temperature observations. Data source is the NASA GISS surface station analysis. The temperature is given as an anomaly relative to the mean temperature for the years 1951 to 1980. To turn these into actual global mean temperatures in degrees Celsius, add 14 °C to the anomaly.

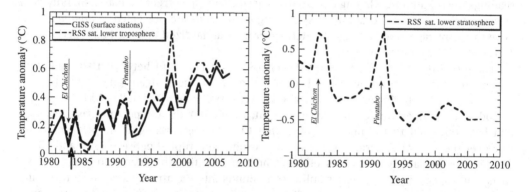

Figure 1.17 Atmospheric temperature time series derived from analysis of microwave satellite data. Left panel: Mean temperature anomaly for the layer of the atmosphere below about 5 km, compared with the GISS instrumental record. Temperature is given as an anomaly relative to the same base value as used in the GISS instrumental record. Right panel: Temperature anomaly for the lower stratosphere (layer from about 15 km to 25 km). In both figures, the El Chichon and Pinatubo volcanic eruptions are marked. In the left panel, major El Niño events are indicated by upward open-headed arrows.

retrieval of temperature in the layer of the atmosphere known as the *lower troposphere* – extending from sea level to roughly 5 km in altitude. The satellite record tracks the GISS surface station record very closely, with the exception of the very strong 1997 El Niño, during which the satellite indicates that the lower tropospheric layer warmed considerably more than the near-surface air. Both satellite and GISS records reproduce the cooling caused by the El Chichon eruption (which overwhelmed the 1982 El Niño) and the Pinatubo eruption (which accentuated the La Niña cooling following the 1991 El Niño, leading to a very cold year). The substantial agreement between the satellite and surface station record proves beyond doubt that the warming observed in recent times is not an artifact of any supposed inadequacies of the surface station record.

The situation looks quite different higher up in the atmosphere. The right panel of Fig. 1.17 shows the temperature trend in a portion of the atmosphere called the *lower stratosphere*, extending from about 15 to 25 km in altitude. Here, volcanic eruptions produce a pronounced warming, as opposed to the cooling seen at lower layers. This suggests that volcanic aerosols heat the upper atmosphere. The vertical profile of warming and cooling produced by high-altitude infrared or solar-absorbing layers will be discussed in Chapter 4, and the reflective effect of aerosols will be brought into the picture in Chapter 5. Leaving out the warm spikes associated with volcanic eruptions, the lower stratosphere appears to have undergone a pronounced cooling over the span of the satellite record. Is stratospheric cooling compatible with CO_2-induced warming in the lower troposphere? This is a Big Question that is resolved in Chapter 4. Ozone destruction also cools the stratosphere, since ozone absorbs sunlight. That portion of the cooling should go away as ozone recovers as a consequence of the Montreal Protocol banning ozone-destroying chlorofluorocarbons.

The Big Question of how much the Earth will warm upon a doubling or quadrupling of CO_2, and how fast it will do so, engages a number of associated questions. Insofar as water vapor is itself a powerful greenhouse gas, any tendency for water vapor content to increase with temperature will amplify the warming caused by CO_2. This is known as *water vapor feedback*. This feedback is now considered to be on quite secure ground, but the study of the behavior of water vapor in the atmosphere offers many challenges, and is a problem of considerable subtlety. In subsequent chapters, we will provide the underpinnings needed for a study of this host of questions. Clouds present an entirely greater order of difficulty, as they warm the planet through their effect on outgoing infrared radiation, but cool the planet through their reflection of solar radiation. The net effect depends on the complex processes determining cloud height, cloud distribution, cloud particle size, and cloud water or ice content. The infrared effects of clouds will be discussed in Chapter 4 and the reflective effects of clouds on sunlight will be discussed in Chapter 5. Uncertainties about the behavior of clouds are the main reason we do not know precisely how much warmer the planet will ultimately get if we double the CO_2 concentration. Typical predictions of equilibrium global average warming for a doubling of CO_2 range from a low of around $2\,°C$ to a high of around $6\,°C$, with some potential for even greater warming with a low (but currently unquantifiable) probability. Because of other uncertainties in the system (particularly the magnitude of the aerosol effect and especially the indirect aerosol effect on cloud brightness) simulations with a range of different cloud behaviors can all match the historical climate record so far, but nonetheless yield widely different forecasts for the future. There is no analysis at present that excludes the possibility of the higher end of the forecast range, for which the effects would likely be catastrophic. There are other feedbacks in the climate system that complicate the forecast. These include feedbacks from melting snow and ice, and from the dynamics of glaciers on land. They also include changes in vegetation, and changes in the ocean circulation which can affect the delay due to burial of heat in the deep ocean.

Global warming – perhaps more aptly called "global climate disruption" – is an event of geological proportions, but one which is caused by human activities. The natural range of CO_2 for the past 800 000 years, and almost certainly for the entire two million years of the Pleistocene, has been 180 to 280 molecules per million. Owing to human activities, the CO_2 concentration is already far above the top of the natural range that has prevailed for the entire lifetime of the human species, and without action will become much higher still. The human species and the natural ecosystems we share the Earth with have adapted over the Pliocene and Pleistocene to glacial–interglacial cycles, but a world with doubled CO_2

will subject them in the course of two centuries or less to a temperature jump to levels far warmer than the top of the range to which societies and organisms have adapted. Even if climate sensitivity is at the low end of the predicted range and if human societies hold the line at a doubling of CO_2, the resulting $2\,^\circ C$ warming represents a substantial climate change; it takes a great deal to change the mean temperature of the entire globe, and a $2\,^\circ C$ global mean increase is a summary statistic that masks much higher regional changes and potentially quite massive effects on sea ice, glaciers, and ecosystems. If climate sensitivity turns out to be at the high end, the warming could be $4\,^\circ C$ or more, and if that is compounded by an increase to four times pre-industrial CO_2 the global mean increase could reach $8\,^\circ C$. That is twice the degree of warming in the PETM, and though the PETM looks abrupt, it is very likely to have set in on a longer time scale than it would take human industrial society to burn the remaining reserves of fossil fuels. If this is allowed to happen, it will take thousands of years for the climate to recover to a normal state. Could global warming disrupt the natural glacial–interglacial cycle? What would the consequences of that be? Those are indeed Big Questions.

As seen by paleoclimatologists 10 million years in the future, whatever species they may be, the present era of catastrophic release of fossil fuel carbon will appear as an enigmatic event which will have a name of its own, much as paleoclimatologists and paleobiologists refer today to the PETM or the K–T boundary event. The fossil carbon release event will show up in ^{13}C proxies of the carbon cycle, in dissolution of ocean carbonates through acidification of the ocean, through mass extinctions arising from rapid warming, and through the moraine record left by retreating mountain glaciers and land-based ice sheets. As an event, it is unlikely to permanently destroy the habitability of our planet, any more than did the K–T event or the PETM. Still, a hundred generations or more of our descendants will be condemned to live in a planetary climate far different from that which nurtured humanity, and in the company of a greatly impoverished biodiversity. Biodiversity does recover over the course of millions of years, but that is a very long time to wait, if indeed there are any of our species left around at the time to do the waiting. Extinction may not be precisely forever, but it is close enough.

1.13 THE FATE OF THE EARTH AND THE LIFETIME OF BIOSPHERES

Even if a planet enters a habitable phase at some stage in its life, it will not remain habitable forever; various kinds of crises can bring its habitability to an abrupt or gradual end. This brings us to the Big Question of lifetime of biospheres; the answer has implications for how likely it is that complex or intelligent life will have had time to evolve elsewhere in the Universe.

Certainly, the Earth's habitability will end when the Sun leaves the Main Sequence and expands into a red giant. Perhaps some of the outer planets or their satellites will enter a brief habitable phase at that time, but it will not be long-lasting. That particular crisis is about four billion years in Earth's future, but other habitability crises are likely to set in long before then. In particular, as the Sun continues to brighten, at some point the brightness will outstrip the ability of the silicate weathering process to compensate by drawing down CO_2. At that point the Earth would succumb to a runaway greenhouse, become lethally hot, and eventually lose its water to space. When will that happen? That is a Big Question, and some current estimates put the remaining natural lifetime of Earth's biosphere at as little as a half billion years. Given that it took four billion years of Earth history before intelligent life emerged, that makes our existence look like quite a close call. Even before the runaway stage, silicate weathering will draw down CO_2 to the point where most forms of photosynthesis will no longer be able to operate. Can more efficient forms of photosynthesis fill in the gap? That's a Big Question as well, but one of a primarily biological nature that we will not attempt to answer.

As the Sun's luminosity increases, Earth may become uninhabitable, but other planets in the Solar System may become more hospitable; in any event they will go through interesting transformations. Mars will warm up, but given that it has little or no active tectonics to generate a new atmosphere, it is unlikely to become Earthlike unless some artificial means is found to give it a more substantial atmosphere. Could Europa melt and become a waterworld? What will happen to Titan as the Sun gets brighter?

Alternately the end could come by ice rather than fire. Earth's life and climate are ultimately maintained by a brew consisting of solar energy and the CO_2 outgassing from the interior. The CO_2 has a warming effect of its own, which can be modified by organisms that intercept it and transform it into oxygen, methane, or other compounds. If the tectonic release of CO_2 ceases, as it will once the Earth exhausts its interior heat sources, all that will come to an end. There will be nothing to offset silicate weathering, and CO_2 will draw down until the Earth turns into a Snowball – unless the runaway greenhouse from a brightening Sun gets us first.

This class of questions naturally generalizes to the question of how the time scales that limit the biosphere's lifetime would be different for planetary systems around other stars. We have already mentioned that hotter stars have a shorter life on the Main Sequence, while cooler stars last longer; the former will have planets with short-lived biospheres compared with the latter. The question of how long the silicate weathering thermostat can cope with changing stellar luminosity, and how long outgassing can sustain the climate, is far subtler, and will have interesting dependences on the planet's size, composition, and orbit. There could well be chemical cycles other than silicate weathering and CO_2 outgassing that could provide climate regulation; the search for such possibilities is still in its infancy. There could also be novel habitability crises, associated with long-term evolution of planetary systems with highly eccentric orbits or systems perturbed by binary star companions.

All this climate catastrophe presupposes no intervention by the inhabitants. In fact, there are quite realistic possibilities for technologically adept inhabitants to stave off the catastrophe at least until their star leaves the Main Sequence. A runaway greenhouse could be prevented by simply reducing the effective stellar brightness, through orbital sunshades or injection of reflecting aerosols into the upper atmosphere. Indeed, such *geoengineering* fixes have been proposed to offset the global warming effect of anthropogenic CO_2 increases. They are a rather desperate and alarming prospect as a solution to global warming, since they offset a climate forcing lasting a thousand years or more with a fix requiring more or less annual maintenance if catastrophe is not to strike; far better to keep CO_2 from getting dangerously high in the first place. However, if the alternative a half billion years out is a runaway greenhouse, the risk of maintaining sunshades will no doubt seem quite acceptable. The loss of CO_2 outgassing as Earth's tectonic cycle ceases also has a relatively easy technical fix. Inhabitants could use a small portion of the energy received from the star to cook CO_2 back out of carbonates, in a process nearly identical to that by which cement is manufactured. Given the slow rate of silicate weathering, only modest quantities of carbonate would have to be processed. All this can be done, but it would appear to require long-term planning and intelligent intervention. A good understanding of the principles of planetary climate will be needed by any beings contemplating such interventions. This book, we hope, will be a good place to start.

1.14 WORKBOOK

1.14.1 Your computational toolkit

The following problems introduce the computer and numerical analysis skills needed to do the problems in the Workbook sections of each chapter. There are a few specialized techniques which will be introduced in later chapters, but these basic skills will see the student a very long way indeed, if thoroughly mastered.

The numerical algorithms are introduced below by name, but no attempt has been made to describe or motivate the algorithm here. For that, the student should consult the book *Numerical Recipes.* Should you write your own routines, or use ready-made routines from a library of some sort? Ideally, one should do a bit of both – writing a basic implementation to see what is going on, and then eventually going over to a full-featured professionally written implementation for heavy-duty work. If time is short, it is reasonable to cut corners by going directly to the packaged routines, so long as one has some idea about what is going on inside the black box. In any event, the intent is that after completion of the problems in this section, the reader will be able to carry out all the necessary numerical operations.

Data referred to here and elsewhere in the Workbook sections of this book is provided as part of the online supplement. It is presumed that the instructor has downloaded the data and put it in a place that is available to everybody who needs it. The reader who is doing the problems on his or her own can simply download the data and put it wherever convenient.

Python tips: First, of course, you should learn some basic Python programming. This can be done through the tutorial material provided as part of the online supplement, or through information found at the website python.org. The Python implementations are for the most part done in an object-oriented fashion. This may sound fancy and high-tech, but it's really quite simple. A kind of object is defined by a class, which allows one to create objects that perform some kind of task. This could be integration of a differential equation, evaluation of a definite integral or just about anything. If one provides the needed specifications, an object can be called like a function, indexed like an array, or even have arithmetic operations defined on it; an object can store data as well as means to manipulate the data. Objects make it easy to substitute alternate algorithms that perform the same function, without having to do major surgery on the rest of the program. Object-oriented programming is a very versatile technique, which is implemented in all modern programming languages, including even MATLAB and FORTRAN 90. Python provides far and away the gentlest introduction to object-oriented programming, however. This style of programming has not yet pervaded scientific simulation, but the reader would do well to gain some facility with it, regardless of which programming language is the language of choice.

For the Python user, the basic courseware tools are implemented in the module ClimateUtilities.py. This module contains utilities for data input and output, graphics, and numerical analysis. Since the objects and functions defined in this module are so extensively used in solving the Workbook problems, you should normally import it into the Python interpreter using

```
from ClimateUtilities import *
```

Pay particularly close attention to the Curve class defined in ClimateUtilities. It allows you to create Curve objects that make it easy to do input and output of tabular data without dealing with low-level details, and also facilitates graphics and data manipulation. As with all Python objects, you can find out what features the Curve class has by typing dir(Curve), and you can access built-in documentation by typing help(Curve).

The other main courseware modules are phys.py, and planets.py. The first of these, phys.py, provides universal physical constants, thermodynamic properties, and frequently used functions for computing various physical quantities. The second, planets.py, contains data on selected astronomical bodies. The use of these two modules will be introduced in the Workbook sections of later chapters.

The online supplement provides Python example scripts showing how to use the tools in ClimateUtilities to solve problems of the type introduced below. It also provides a number of numerical analysis tutorials which introduce the basic ideas behind the key algorithms. These can be used to supplement the deeper discussion provided in *Numerical Recipes*.

Problem 1.1 *Plotting a function*

Make a graph of Eq. (1.1) showing the evolution of solar luminosity over time. As the vertical axis, you may use the luminosity relative to its present value. Show values out to twice the current age of the Sun. On the same plot show a straight-line approximation to the function, which yields the same values as Eq. (1.1) at $t = 0$ and $t = t_\odot$.

There are two different ways to go about making the plot. One alternative is to generate the data to be plotted in the same software package you use for plotting, and then carry out the operations (commands, menu choices, etc.) needed to do the plot. The second alternative is to generate the data in one programming environment, write it out to a text file, and then read the file into the plotting function of your choice. This can be handy if you want to use commercial plotting software that produces professional-looking figures.

Python tips: Using the courseware in Python, you do this by making a list of arguments to the function, making a list of corresponding function values, installing each list in a Curve object, and then plotting. If the curve object is MyCurve, you plot using plot(MyCurve). Check the examples in the supplementary material to see how to set plot options such as axis labels, titles, etc. As an alternative to plotting within Python, you can use the dump method of the Curve object to write out the data to a text file, which you can then read into the plotting software of your choice. For example, MyCurve.dump("out.txt") writes the data in MyCurve to a text file called out.txt.

Problem 1.2 *Reading and writing tabular data*

Tabular data is a very common form of scientific dataset. A tabular dataset consists of a number of columns of identical length. Each column represents a quantity of interest, and each row represents a set of the quantities pertaining to a given "sample." The "sample" could represent some individual chunk of rock, but more commonly, it will represent some point in space or time. For example, we could have a three-column dataset in which the first column is time, the second is air temperature, and the third is the moisture content of the air. Or, the first column could be height above the surface of a planet while the second column was air pressure and the third air density.

When the amount of data involved is not huge, it is convenient to store tabular data in the form of plain human-readable text files. Since these files can be examined in any text editor, it is easy to determine their structure and how to read the data into just about any software. There are still a few variations on the theme that one must keep in mind. First, there must always be some scheme to separate one column from another. This is usually done using a delimiter character, which is often a tab character, but could also be an arbitrary number of blank spaces or a comma. Second, columns may have text labels at the top, giving the name of the variable. Third, there may be header information before the data, which gives information about the dataset; less commonly, there can also be documentation at the end of the data. Finally, there may be missing data, and there are various ways to indicate missing data. Often it's done with a numeric code which is easily recognizable as being not real data. Sometimes, though, missing data is indicated by a character or group of characters (e.g. "MISSING," or "NA"). That can make problems with software that is expecting a number and tries to treat the entry as a number. All these things must be taken into account when reading in tabular data. The datasets to be analyzed in the Workbook sections of this book will all use numeric codes to indicate missing data, but one can always use a text editor to globally replace a text code with a numeric code.

This problem will be built around a tabular set of data from the Vostok ice core. The dataset consists of four columns: the depth in the core, an estimate of the age of the ice, the deuterium composition of the ice relative to a standard (which is a measure of temperature at the time the ice was deposited), and a deuterium-based estimate of the past Antarctic temperature relative to the present. This file, and all the other ones you will need for this problem, are in the `Chapter1Data/iceCores` subdirectory of the Workbook datasets directory. First, open the file in a text editor and take a look at it to see what's there. Most text editors will not show the differences between tabs and spaces by default, though the better ones will have an option to make this difference visible. Read the header information, to get an idea of what the dataset means and where it came from.

Let's now try reading in some data from the Vostok ice core. The full file, which contains detailed documentation in a header plus column labels, with tab-delimited data following, is in the file you just looked at in the text editor. Let's start more simply, though. The file `vostokT.DataOnlyTabDelim.txt` contains only the data, with no header information whatever, in tab-delimited form. Read it in and take a look at some values. The description of the columns is given at the top of the file you looked at previously in your text editor. Next try reading in the data in space-delimited form, from file `vostokT.DataOnlySpaceDelim.txt`, to get an idea of how to handle that kind of data in your software. Then, try loading in tab-delimited data with column labels, from `vostokT.ColHeadTab.txt`. Finally, see if you can load in the data and column header part of the full file, `vostokT.txt`. Documentation is often provided at the top of tabular data, so if your software doesn't provide a procedure for skipping the header and extracting the data, you may need to use a text editor to do this by hand before loading data into your software for further analysis.

Once you have read in the data, make a graph of the temperature vs. ice age. Also, make a graph of the age vs. depth.

In addition to being able to read tabular data, you should be able to write tabular data to a file for later use. To practice this skill, take any of the Vostok ice core datasets you read in, and write a new text file which contains only the ice-age and temperature anomaly columns (the second and fourth columns of the original dataset). Read it back in and compare to the original data to make sure everything has worked properly.

Python tips: Data in a `Curve` object can be written to a text file using the `dump` method. In the courseware, the function `readTable` takes a filename as its argument, reads the data in the text file, and returns a `Curve` object containing all the data. This object can then be used for plotting, or for further data processing. The data must be in tabular form, with columns delimited by some consistent character (usually a tab). readTable is usually smart enough to strip off header information and spot where the data begins and identify column headings. The dataset you read in typically has more information in it than you will want to plot on a single graph. To plot just selected columns, use the extract method of your Curve object. e.g. if c is your Curve, then c.extract(["t","Z"]) will return a new Curve containing only the columns labeled t and Z.

Problem 1.3 *Arithmetic on columns*
One of the most basic forms of computation is the performance of arithmetic on rows or columns of tabular data. This kind of operation is very familiar from spreadsheets and commercial graphing software. All modern computer languages allow the performance of such operations on *arrays* or *vectors* without the need to write a loop dealing with each entry separately. An array or vector can be thought of as a row or column of data.

To exercise this skill, make up a column of data corresponding to pressure levels in an atmosphere. The pressure will be measured in millibars (mb), but the units are immaterial for this problem. The top of the atmosphere should start at 10 mb, and the data should extend to 1000 mb, which is a typical surface pressure. Next compute a column giving temperature on the *dry adiabat*, which you will learn about in Chapter 2. The dry adiabat for Earth air is defined by the formula $T(p) = T_g \cdot (p/p_s)^{2/7}$, where T_g is the ground temperature and p_s is the surface pressure. Use $T_g = 300$ K. Use the data in this column to make a third column which gives the corresponding temperature in degrees Fahrenheit (just hold your nose and pretend you are making a plot to be used in a television weather show).

To do a somewhat more interesting application of these skills, first read in the instrumental surface temperature record in the file GISSTemp.txt, located in the GlobalWarming subdirectory of the Workbook datasets directory for this chapter. These are estimates of global mean temperature. The first column gives the year, and the remaining columns give the temperature for each month, and for selected annual and seasonal averages. The temperatures are given as deviations from a baseline temperature, measured in hundredths of a degree C. Divide the annual average column (labeled "J-D") by 100 and add 14 °C, to convert to actual temperature. Plot this curve and take a look at it to make sure it is reasonable (apart from the fact that it is going up at an alarming rate).

Now take the difference of the June–August column ("JJA") and the December–February column ("DJF"). These correspond to northern hemisphere summer, but remember that these are global means, so that northern summer is being averaged with southern winter. Global mean JJA still tends to be warmer than global mean DJF, because the seasonal cycle is strongest in the northern hemisphere. However, note that the columns give the *deviation* of the temperature from the baseline average for these months, so the difference you computed gives the *change* in the strength of the global mean seasonal cycle, not its absolute magnitude. Divide your result by 100 to convert the deviation to degrees. Do you notice any interannual variability in the strength of the seasonal cycle? What is the typical time scale? Do you notice any trend?

Python tips: The Curve object supports arithmetic on columns. Specifically, if "x" is a data column in the curve object c, then c["x"] returns an array on which arithmetic can be done just as if it were a regular number. For example, 2*c["x"] is an array in which the entries are all doubled, c["x"]**4 returns an array consisting of the fourth powers of entries, etc. You can even do things like c["T"] = c["T"] + 273.15 to convert a temperature from Celsius to Kelvin in-place, and can do similar things to create new columns that do not initially exist. You can compute the average of a column using constructions like `sum(c["T"])/len(c["T"])`. Many other array arithmetic operations can be done using the array functions available through `numpy` or its older version `Numeric`.

Problem 1.4 *Interpolation*

It commonly happens that one has data from observations or a model tabulated on a discrete set of points, but needs to be able to determine the value at an arbitrary point within the range covered by the data. The technique required to deal with this situation is called *interpolation*. An interpolation routine takes an array of values of the independent variable $[X_0, X_1, ..., X_n]$ and corresponding data values $[Y_0, Y_1, ..., Y_n]$, and on this basis provides a function $f(x)$ which estimates the value of the independent variable at an arbitrary value x of the dependent variable. In general, the list of independent variables should be sorted in monotonically increasing or decreasing order before carrying out interpolation.

The simplest form of interpolation is *linear interpolation*, where one determines the intermediate values by drawing a straight line connecting the neighboring data points. *Polynomial interpolation* is similar, except that one finds the m nearest points to x and uses them to determine an $(m-1)^{th}$ order polynomial that exactly goes through all the data points. m is usually taken to be a fairly small number, perhaps between 2 and 10. The case $m = 2$ corresponds to linear interpolation. There are more sophisticated interpolation techniques, which you can read about in *Numerical Recipes*, but polynomial interpolation will be adequate for our purposes.

Write a basic general purpose linear interpolation routine. Try it out on the data `X=[1,2,3,4]`, `Y=[1,8,27,64]`, subsampled from $y = x^3$, and plot the results at high resolution to make sure everything is working. Then, write or learn to use a polynomial interpolation routine. Do the same, and verify that the procedure recovers the cubic behavior exactly. Try subsampling the function $y = \sin(x)$ with various numbers of points, and see how well polynomial interpolation of various orders reproduces the original function.

Next, read in the Vostok temperature dataset discussed in Problem 1.2. The first column is depth, and this dataset is already averaged down to a uniform 1 m depth interval for each data point. However, this layer depth corresponds to a longer time interval for old ice than for the young ice which has not been as stretched out; snow accumulation rate also affects the depth–time relation. The second column is the ice age, and this scale has non-uniform intervals. Suppose you need time series data on regular time grid with uniform 100-year intervals? Create such a series of deuterium composition (third column, a measure of the past temperature) by doing a polynomial interpolation using the ice-age data as the independent variable array and the deuterium data as the dependent variable array. Try including different numbers of points in the polynomial interpolation to see how that affects the results.

Python tips: Polynomial interpolation from a table of values is implemented in the courseware by the function `polint`. This function evaluates a polynomial that goes through *all* the points in the table. In practice, this works best if the fit is done based just on a few points in the table nearest to the interpolation target. The full procedure, including identification of the n nearest points, is implemented by the `interp` class, which creates a function (strictly speaking, a callable object) that returns the interpolated value of the dependent variable given the desired dependent variable value as its argument.

Problem 1.5 *Root finding*

Find the solution of $x - \exp(-x) = 0$. First do it by the bisection method on the interval $[0, 1]$. Then do it using Newton's method.

Python tips: In the courseware, Newton's method is implemented via the class `newtSolve`, which creates an object that finds the zeroes of a given function given an initial guess.

Problem 1.6 *Quadrature*

Evaluate the definite integral $\int_0^a \exp(-x^2)dx$ for $a = 2$ using the trapezoidal rule approximation. Then, use multiple trapezoidal rule evaluations together with Romberg extrapolation to obtain a more accurate result with less work. Next use the numerical evaluation to show how the integral varies as a function of a. What happens at very large a?

Python tips: In the courseware, quadrature using the trapezoidal rule and Romberg extrapolation is implemented via the class `romberg`, which creates an object that carries out the integral. Note that if you want to try writing your own Romberg extrapolation quadrature routine, the function `polint` can be used to do the needed polynomial extrapolation, as can the class `interp`.

Problem 1.7 *Ordinary differential equation integration in one dimension*

Numerically solve the logistic equation

$$\frac{dY}{dt} = aY \cdot (1 - Y) \tag{1.9}$$

subject to the initial condition $Y(0) = 0.01$. Compare results using the midpoint method with results using Runge–Kutta, and discuss the dependence on the time step you use.

Python tips: In the courseware, the Runge–Kutta integration is implemented via the class `integrator`, which creates an object which carries out the integration. You should try implementing your own midpoint method as a loop in Python.

Problem 1.8 *Ordinary differential equation integration in two dimensions*

Numerically compute the solution of the system

$$\frac{dY}{dt} = -\frac{Z}{Y^2 + Z^2}, \frac{dZ}{dt} = \frac{Y}{Y^2 + Z^2} \tag{1.10}$$

subject to the initial condition $Y = 1$, $Z = 0$. Compare results for the midpoint and Runge–Kutta methods, and discuss the dependence on the time step you use.

Python tips: This can be done using the `integrator` class, which works for an arbitrary number of dependent variables.

1.14.2 Basic physics and chemistry

Problem 1.9 The acceleration of gravity on Earth is $10 \, \text{m/s}^2$, in round numbers. How much force (in N, i.e. newtons) does a 1 kg object exert when at rest? How much energy (in J, i.e. joules) does it take to lift a 1 kg object to a height of 100 m?

On the average about 200 J of solar energy falls on each square meter of Earth's surface each second – that's $200 \, \text{W/m}^2$ (1 W, or watt, is 1 J/s). If all this energy could be converted to mechanical work, how many kg/s can be lifted to 100 m?

There are about 10 000 kg of air over each square meter of Earth's surface. How long would it take to lift all the Earth's air to an altitude of 200 km, if all the incoming solar energy could be used for this purpose? (You may assume the acceleration of gravity to be constant over this range of altitudes.) Why doesn't this happen to Earth's atmosphere?

Problem 1.10 A small comet with a mass of $4 \cdot 10^9$ kg strikes the Earth with a speed of 10 km/s. How much energy is released? Express the answer in joules (J).

Problem 1.11 The surface gravitational acceleration on Titan is $1.35\,\text{m/s}^2$. The radius of Titan is $2575\,\text{km}$. Use Newton's law of gravitation to determine Titan's mass. (The gravitational constant G is $6.674 \cdot 10^{-11}\,\text{m}^3/\text{kg} \cdot \text{s}^2$.)

Problem 1.12 Some aggressive space aliens halt the orbital velocity of the Moon, so that it falls like a rock upon the Earth. How much kinetic energy does the Moon have when it hits the surface? (*Hint:* Use conservation of kinetic plus potential energy.) What is its speed at the time of impact, ignoring air resistance? Does the speed depend on the mass of the Moon?

Problem 1.13 An impactor with the mass of the Moon hits the Earth with a speed of $15\,\text{km/s}$. The object is made of silicate rock. Is the energy released enough to vaporize the object? Does your answer depend on the mass of the object? (*Note:* It takes about $2 \cdot 10^7\,\text{J}$ to vaporize $1\,\text{kg}$ of silicate rock.)

Problem 1.14 *Bolometric magnitude*
What is relevant for the climate of a planet is the power output of the planet's host star – its luminosity. What astronomers *observe* is the flux of energy from the star which reaches a telescope on Earth or in orbit. Suppose that the energy flux measured at the telescope is F. Astronomers generally transform the flux into *magnitudes*, defined by the expression

$$M = C - 2.5 \log_{10}(F) \qquad (1.11)$$

where C is a constant chosen by convention which depends on the zero-magnitude reference. This magnitude is called the *apparent magnitude* because it corresponds to how bright the star seems as seen from Earth. The expression has its roots in the perception of relative brightness by human vision.

The magnitude defined above is called the apparent *bolometric magnitude* if it is based on the total flux over all wavelengths. In this case, the zero-magnitude convention is $C = -18.98$ if F is measured in W/m^2. With this convention, the star Vega has approximately zero apparent magnitude. The Sun has an apparent bolometric magnitude of -26.82 as seen from Earth's mean orbit.

Assume that the star is at a distance d from the telescope and that its luminosity is $\mathcal{L}_\odot$. The radiation from the star spreads out uniformly over a sphere of radius d, and there is no absorption by the intervening interstellar medium. Find an expression for $\mathcal{L}_\odot$ in terms of the apparent bolometric magnitude and d/r_0, where r_0 is the radius of the Earth's mean orbit. *Hint:* Use the apparent magnitude of the Sun, given above.

The red dwarf star Gliese 581, which hosts some interesting planets, has an apparent bolometric magnitude of 8.12, and is located at a distance of 6.27 parsecs (1 parsec = $3.086 \cdot 10^{16}\,\text{m}$). What is its luminosity relative to that of the Sun? If a planet were in orbit about Gliese 581, how close would it have to be to the star in order for the stellar flux incident upon the planet to equal the solar flux incident upon Earth?

Absolute magnitude is the magnitude as seen from a standard distance of 10 parsecs. What is the absolute bolometric magnitude of the Sun? Of Gliese 581?

There's actually a bit more to learn about magnitudes. Astronomers typically report the magnitude as seen through various standard filters, for example "V-magnitudes" for yellowish light. These must be converted to bolometric magnitudes using a bolometric correction factor. This issue will be explored later, in Problem 3.11.

Problem 1.15 Luminosity is the net power output of a star, which we will measure in W. The proton process fuses four protons (hydrogen nuclei) into helium, yielding $4.17 \cdot 10^{-12}$ J of energy. Assuming constant luminosity, find a relation between the mass of the star and its lifetime on the Main Sequence, assuming the star to be initially hydrogen-dominated. How does your answer change if you use Eq. (1.1) to take into account the increase in luminosity over time? In this application of the formula, the parameter $t_\odot$ should assumed to represent half the star's lifetime on the Main Sequence, and the formula should be considered valid only out to times $2\, t_\odot$.

If T is the surface temperature of the star, then the radiation flux in W/m^2 out of the surface is $5.67 \cdot 10^{-8} T^4$. You will learn more about this formula in Chapter 3. Use this to estimate the surface temperature of the star in terms of its radius. This also gives you the spectral class of the star, since cooler stars are redder. Then, if you have a relation between the mass of the star and its radius, you have a relation between lifetime and spectral class. Use this to show that the bluer stars are more short-lived, assuming that the density is independent of mass. A more accurate calculation would need to take into account the compressibility of the star's substance.

Problem 1.16 The Sun puts out energy at a rate of $3.84 \cdot 10^{26}$ W. A resting human consumes (and puts out) energy at a rate of about 100 W. Look up the mass of the Sun, and determine which of these two bodies requires more energy production per unit of mass. Using the data given in Problem 1.15, determine how much mass of hydrogen would constitute a lifetime supply of food if humans could eat hydrogen and fuse it into helium.

Problem 1.17 In 2009, the world fossil fuel emissions of carbon in the form of CO_2 were about 9 Gt (1 Gt = 1 gigatonne = 10^9 metric tonne = 10^{12} kg). If all of this carbon were turned into calcite ($CaCO_3$), how many Gt of calcite would you have? If it were spread evenly over the state of Kansas (surface area 213 000 km^2), how deep would the layer be? The density of calcite is about 2700 kg/m^3.

Problem 1.18 How many molecules of CO_2 are produced by the annual emission of 9 Gt carbon? Earth's atmosphere contains about 10^{44} molecules in all, most of which are N_2 and O_2. If all the CO_2 from fossil fuel emissions stayed in the atmosphere, how much would the atmospheric CO_2 concentration increase each year? Express your answer in terms of the number of CO_2 molecules per million air molecules (i.e. ppm molar, also called ppmv).

1.14.3 Stable isotope calculations

Problem 1.19 Some ice that you dig up in Greenland has a $\delta^{18}O$ of $-37‰$. What is the ratio of ^{18}O to ^{16}O in this ice? Recent ice in Antarctica has $\delta D = -438‰$. What is the ratio of D to H in this ice?

Problem 1.20 Assuming the mean δD for Antarctic ice to be $-420‰$, how much does the removal of 300 m depth of ocean water to form an Antarctic glacier shift the δD of the remaining ocean water?

Problem 1.21 Suppose that we write the equilibrium fractionation factor for phase 2 to phase 1 as $f_{1,2} = 1 + \epsilon$, where ϵ generally has small magnitude. Show that the shift in δ values between the two phases is

$$\delta_1 - \delta_2 = (1 + \delta_2)\epsilon. \tag{1.12}$$

In the typical case, where δ_2 is small, this formula leads to the convenient approximation that the shift in δ between the phases is constant. This approximation breaks down if $|\delta_2|$ is not small. The precise statement of the physics at constant temperature is that the fractionation factor remains constant as mass is transferred from one reservoir to another. The approximation of a constant shift in δ is adequate for most common purposes, though.

Suppose that at 290 K the $\delta^{18}O$ of water vapor in equilibrium with liquid having $\delta^{18}O = 0$ is $-10‰$. What is the value of ϵ?

Problem 1.22 A mass of liquid water with initial $\delta^{18}O$ of zero is brought into contact with dry air in a closed box. The mass evaporates until the air is saturated (i.e. holds as much water as it can). The isotopes are in equilibrium between liquid and vapor phases. The box is big enough that once equilibrium is reached, 1% of the mass of the water has evaporated. What is the $\delta^{18}O$ of the vapor at that point? What is the $\delta^{18}O$ of the remaining liquid? Then, new dry air is brought into the system, flushing out the moist air, and the process is repeated, allowing another 1% of the remaining water to evaporate. What are the $\delta^{18}O$ values now? Continue in this fashion, and make a graph of the time series of the $\delta^{18}O$ of the vapor exhausted from the box and the $\delta^{18}O$ of the remaining water. You may assume that the entire experiment takes place at a fixed temperature of 290 K. (See Problem 1.21 for useful information.)

Problem 1.23 *A tale of two lakes*
A full lake is separated from an empty lake basin by a tall mountain range. The full lake initially is filled with water having $\delta^{18}O = 0‰$. A wind starts to blow from the full lake towards the empty basin, picking up water from the lake as it goes. The water that it picks up is in equilibrium with the lake water at all times. As the moisture-laden air blows up the mountain slope, moisture condenses out bit by bit. Each time moisture condenses out, it flows back into the originally full lake. The water that condenses is always in equilibrium with the vapor from which it condenses. As the moist air climbs the mountain, first 1% condenses out. Then another 1% of the remaining condenses out and so on. This happens 50 times by the time the top of the mountain is reached. What is the $\delta^{18}O$ of the vapor that reaches the top of the mountain? You may ignore the effect of temperature on the fractionation factor, and assume it remains constant. (See Problem 1.21 for useful information.)

The air that crosses the top of the mountain mixes with some very frigid air, and all of the remaining vapor condenses out, falls into the initially empty lake and begins to fill it. Eventually the first lake will empty and all the water will have been transferred to the second lake. Describe the time evolution of the $\delta^{18}O$ of each lake, shown as a function of the fraction of the initial lake volume that has been transferred from one lake to the other.

Now, suppose that instead of all the water being transferred, the second lake basin is shallow and there is a river channel connecting the top back to the first lake. When the second lake fills to one-quarter the initial volume of the first, water flows back into the first lake at the same rate as water rains out into the smaller lake. The water running back mixes uniformly with the water in the big lake. What is the $\delta^{18}O$ in each lake once this process reaches equilibrium? What would the $\delta^{18}O$ be if the mountain were made taller, so that 75 precipitation steps of 1% each were made as the air crossed the mountain?

Say why this process is analogous to the isotopic effect of the transfer of sea water into an accumulating land glacier which eventually equilibrates by melting at a rate equal to the accumulation of new snow. State why making the mountain taller has an effect on the isotopic composition of the second lake which is analogous to the effect of a polar temperature decrease on the isotopic composition of the Antarctic or Greenland glacier.

Problem 1.24 During the PETM, a large quantity of isotopically light carbon was released into the atmosphere-ocean system. What does the change in measured $\delta^{13}C$ say about the amount of carbon released? This problem deals with that question in a simplified setting.

Suppose that 700 Gt of carbon was released into the atmosphere in the form of CO_2, from oxidation of organic carbon having $\delta^{13}C = -25‰$. Suppose all this CO_2 stays in the atmosphere. How much does the atmospheric $\delta^{13}C$ shift if the atmosphere initially contains 700 Gt of carbon in the form of CO_2 (roughly equal to present Earth levels)? Does the answer depend on what the initial atmospheric $\delta^{13}C$ is? What happens if the atmosphere initially contains 1400 Gt of carbon? What would happen if the carbon instead were released in the form of biogenic methane, with $\delta^{13}C = -50‰$, which was subsequently oxidized into CO_2 in the atmosphere?

The organic carbon release used in this problem is unrealistically low because on the real Earth, most of the CO_2 added to the atmosphere will be taken up by the ocean over a few thousand years, and the $\delta^{13}C$ in the atmospheric carbon will come into equilibrium with the $\delta^{13}C$ in the oceanic carbon. As a very crude attempt at incorporating this effect, assume that 7000 Gt was released, and that the ^{13}C is shared equally amongst a total atmosphere-ocean carbon pool which is about 100 times the mass of carbon in the atmosphere alone. Assume further that the $\delta^{13}C$ of carbonates deposited in the ocean mirrors the shift in the $\delta^{13}C$ of the total carbon pool. Compare your result to the 1.2‰ lightening observed in PETM carbonates. Note that one needs to know quite a bit of oceanic carbonate chemistry to do this problem properly, accounting for the isotopic exchanges of carbon amongst all the reservoirs in the system. Note further that the size of the carbon pool the atmosphere exchanges with is a quite separate question from how much of the added CO_2 remains in the atmosphere after various lengths of time.

Problem 1.25 Consider a planet with no photosynthesis. The planet has no ocean, though there is enough water around to keep the soil damp and sustain some forms of life. Carbon dioxide outgassing from the interior creates a pure CO_2 atmosphere containing a mass M of carbon in the form of CO_2. Then outgassing ceases, and the atmosphere stops accumulating. The $\delta^{13}C$ of the carbon in the atmosphere is $-6‰$ at this point. The surface mineralogy on this planet does not permit the formation of carbonates.

At this point, oxygenic photosynthesis evolves, and starts converting the atmosphere into O_2, which stays in the air, and organic carbon, which accumulates in a layer on the ground. The photosynthetic organisms fractionate carbon by $-25‰$ *relative to whatever value the air has at the time they grow*. All their carbon comes from the air. In the first year, they convert 1% of the CO_2 into organic carbon, which accumulates in a layer. Then they die, and the next year a new crop grows, which converts 1% of the remaining CO_2. Graph the time series of the $\delta^{13}C$ in the air, resulting from this process. If, after a long time, one were to drill a core into the organic carbon layer, what would the profile of $\delta^{13}C$ vs. depth look like? You may assume that none of the carbon in the organic layer ever gets oxidized back into CO_2.

Now assume instead that the carbon cycle is in equilibrium between volcanic outgassing and organic burial. Each year, 1% of the carbon in the atmosphere is buried in the organic layer, and each year that same amount is replaced by outgassing at $-6‰$. Describe how the $\delta^{13}C$ in the organic layer and atmosphere behave in that case. Does the behavior depend on the equilibrium mass M of carbon in the atmosphere? What happens to the mass of O_2 in the atmosphere as time progresses? Why doesn't this happen on Earth?

Problem 1.26 The file `BenthicTropPacificB.txt`, located in the `marineCores` subdirectory of the data directory for this chapter, contains $\delta^{13}C$ and $\delta^{18}O$ data from benthic forams, recovered from a tropical Pacific ocean sediment core. The data covers the past 5 million years. Read in the data. What is the typical variation in $\delta^{13}C$ over this time period? If the changes were entirely due to changes in organic carbon burial rate, which times would have high burial rate and which would have low burial rate? Try to estimate the implied changes in burial rate. (*Note:* In reality $\delta^{13}C$ is subject to a number of other influences, including the amount of CO_2 available in the ocean).

Now look at the $\delta^{18}O$ data. Convert the data to ice volume, assuming the entire variation is due to ice. Convert the data to temperature using the paleotemperature equation, assuming the entire variation is due to changes in the deep ocean temperature. Which limiting case do you think is closer to the truth?

1.15 FOR FURTHER READING

Many of the problems and explorations in the Workbook sections of this book require the use of some basic numerical analysis. The necessary algorithms are enumerated and exercised in the Workbook section of this chapter. The essential reference for the derivation and implementation of the algorithms is the *Numerical Recipes* series. The current edition is:

- Press WH, Teukolsky SA, Vetterling WT and Flannery BP 2007: *Numerical Recipes* 3rd Edition: *The Art of Scientific Computing*. Cambridge University Press.

The algorithm description is independent of programming language, but the implementations given in this particular edition are based on c++. Earlier editions are available for other programming languages. The c language implementations in the 1992 edition *Numerical Recipes in c* provide a clean basis for re-implementation in other programming languages, particularly convenient for the reader who wishes to avoid some of the intricacies of c++. The reader will need only a very small part of the material covered the *Numerical Recipes* opus; the relevant sections are pointed out in the Workbook for this chapter.

Extrasolar planet observations grow by leaps and bounds. New planets are added to the catalog on a practically monthly basis, and there are frequent updates to the database of planetary characteristics. For the most current data, consult the online *Extrasolar Planets Encyclopedia*, at `http://exoplanet.eu`. The data shown in this chapter is from the paper

- Butler RP *et al.* 2006: Catalog of nearby exoplanets, *Astrophys. J.* **646**.

Conditions during the earliest period of Earth's history, the time required for the crust to form and cool, and the time evolution of heat flux from the interior of the Earth to the surface are discussed in:

- Sleep NH, Zahnle K and Neuhoff PS 2001: Initiation of clement surface conditions on the earliest Earth, *Proc. Nat. Acad. Sci.* **98**, 3666–3672.
- Turcotte DL 1980: On the thermal evolution of the Earth, *Earth Planet. Sci. Lett.*, **48**, 53–58.

The long-term evolution of the brightness of the Sun and similar stars is discussed in:

- Gough DO 1981: Solar interior structure and luminosity variations, *Solar Phys.* **74**, 21–34.
- Sackmann I-J, Boothroyd AI and Kraemer KE 1993: Our Sun. III. Present and future, *Astrophys. J.* **418**, 457–468.

For a very engaging introduction to what we know about life on the Early Earth, the reader is directed to the book

- Knoll A 2004: *Life on a Young Planet.* Princeton University Press.

The original paper on the Faint Young Sun problem is

- Sagan C and Mullen G 1972: Earth and Mars: Evolution of atmospheres and surface temperatures, *Science* **177**, 52–56.

A good review of the history of oxygen on Earth and the proxy methods used to infer this history can be found in

- Canfield DE 2005: The early history of atmospheric oxygen: homage to Robert M. Garrels, *Annu. Rev. Earth Planet. Sci.* **33**, 1–36. doi:10.1146/annurev.earth.33.092203.122711.

For a general introduction to the Snowball Earth problem, see

- Hoffman PF and Schrag DP 2002: The snowball Earth hypothesis: testing the limits of global change, *Terra Nova* **14**, 129–155.

The discussion of late Cretaceous and Cenozoic paleoclimate drew largely on the following papers:

- Sluijs A *et al.* 2006: Subtropical Arctic Ocean temperatures during the Palaeocene/Eocene thermal maximum, *Nature* **441**, 610–613, doi:10.1038/nature04668.
- Moran K *et al.* 2006: The Cenozoic palaeoenvironment of the Arctic Ocean, *Nature* **441**, 601–605, doi:10.1038/nature04800.
- Pearson PN *et al.* 2007: Stable warm tropical climate through the Eocene epoch, *Geology* **35**, 211–214, doi:10.1130/G23175A.
- Forster A *et al.* 2007: Mid-Cretaceous (Albian–Santonian) sea surface temperature record of the tropical Atlantic Ocean, *Geology* **35**, 919–922, doi:10.1130/G23874A.
- Mix AC *et al.* 1995. Benthic foraminifera stable isotope record form Site 849, 0–5 Ma: Local and global climate changes. Pages 371–412 in NG Pisias *et al.*, editors, *Proceedings of the Ocean Drilling Program*, Scientific Results 138, College Station, Texas, USA.

Evolution of Phanerozoic climate, occurrence of glaciations, and evolution of CO_2 content of the atmosphere are discussed in

- Crowley TJ and Berner RA 2001: CO_2 and climate change, *Science* **292**, 870–872, doi:10.1126/science.1061664.
- Zachos J *et al.* 2001: trends, rhythms and aberrations in global climate 65 Ma to present, *Science*, **292**, 686–693, doi:10.1126/science.1059412.

Veizer's long-term fossil ^{18}O tropical temperature record, discussed in Crowley and Berner (2001), is not generally considered reliable.

The GISP, Vostok, and EPICA ice-core records are described in

- Grootes PM and Stuiver M 1997: Oxygen 18/16 variability in Greenland snow and ice with 10^3 to 10^5-year time resolution. *J. Geophys. Res.* **102**, 26455–26470.
- Petit JR *et al.* 2001: Vostok ice core data for 420,000 years, *IGBP PAGES/World Data Center for Paleoclimatology Data Contribution Series No. 2001-076*. NOAA/NGDC Paleoclimatology Program, Boulder CO, USA.
- Petit JR *et al.* 1999: Climate and atmospheric history of the past 420,000 years from the Vostok ice core, Antarctica, *Nature* **399**, 429–436.
- Siegenthaler TF *et al.* 2005: Stable carbon cycle–climate relationship during the late Pleistocene. *Science* **310**, 1313–1317.

The intellectual history surrounding anthropogenic climate change ("global warming") is surveyed in the following book and accompanying website:

- Weart S 2008: *The Discovery of Global Warming*. Harvard University Press.
- http://www.aip.org/history/climate/index.html

Additional information can be found in *The Warming Papers*, a set of critical readings with essays by D. Archer and the author, forthcoming from Wiley/Blackwell circa 2010.

CHAPTER TWO

Thermodynamics in a nutshell

2.1 OVERVIEW

The atmospheres which are our principal objects of study are made of compressible gases. The compressibility has a profound effect on the vertical profile of temperature in these atmospheres. As things progress it will become clear that the vertical temperature variation in turn strongly influences the planet's climate. To deal with these effects it will be necessary to know some thermodynamics – though just a little. This chapter does not purport to be a complete course in thermodynamics. It can only provide a summary of the key thermodynamic concepts and formulae needed to treat the basic problems of planetary climate. It is assumed that the student has obtained (or will obtain) a more fundamental understanding of the general subject of thermodynamics elsewhere.

2.2 A FEW OBSERVATIONS

The temperature profile in Fig. 2.1, measured in the Earth's tropics, introduces most of the features that are of interest in the study of general planetary atmospheres. It was obtained by releasing an instrumented balloon (radiosonde) which floats upward from the ground, and sends back data on temperature and pressure as it rises. Pressure goes down monotonically with height, so the lower pressures represent greater altitudes. The units of pressure used in the figure are *millibars* (mb). One bar is very nearly the mean sea-level pressure on Earth, and there are 1000 mb in a bar.

Pressure is a very natural vertical coordinate to use. Many devices for measuring atmospheric profiles directly report pressure rather than altitude, since the former is generally easier to measure. More importantly, most problems in the physics of climate require knowledge only of the variation of temperature and other quantities with pressure; there are relatively few cases for which it is necessary to know the actual height corresponding to a

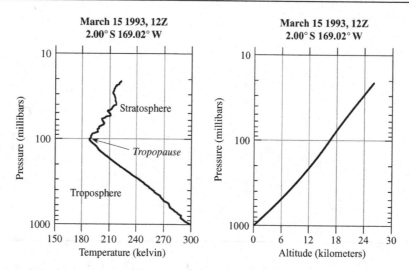

Figure 2.1 Left panel: Temperature profile measured at a point in the tropical Pacific. Right panel: The corresponding altitude. The measurements were obtained from a radiosonde ("weather balloon") launched at 12Z (an abbreviation for Greenwich Mean Time) on March 15, 1993.

given pressure. Pressure is also important because it is one of the fundamental thermodynamic variables determining the state of the gas making up the atmosphere. Atmospheres in essence present us with a thermodynamic diagram conveniently unfolded in height. Throughout, we will use pressure (or its logarithm) as our fundamental vertical coordinate.

However, for various reasons one might nevertheless want to know at what altitude a given pressure level lies. By altitude tracking of the balloon, or using the methods to be described in Section 2.3, the height of the measurement can be obtained in terms of the pressure. The right panel of Fig. 2.1 shows the relation between altitude and pressure for the sounding shown in Fig. 2.1. One can see that the height is very nearly linearly related to the log of the pressure. This is the reason it is often convenient to plot quantities vs. pressure on a log plot. If p_0 is representative of the largest pressure of interest, then $-\ln(p/p_0)$ is a nice height-like coordinate, since it is positive and increases with height.

We can now return to our discussion of the critical aspects of the temperature profile. The most striking feature of the temperature sounding is that the temperature goes down with altitude. This is a phenomenon familiar to those who have experienced weather in high mountains, but the sounding shows that the temperature drop continues to altitudes much higher than sampled at any mountain peak. This sounding was taken over the Pacific Ocean, so it also shows that the temperature drop has nothing to do with the presence of a mountain surface. The temperature drop continues until a critical height, known as the *tropopause*, and above that height (100 mb, or 16 km in this sounding) begins to increase with height.[1] The portion of the atmosphere below the tropopause is known as

---

[1] Though the temperature minimum in vertical profiles of the Earth's atmosphere is the most obvious indicator of a transition between distinct layers, we will eventually be able to provide a more dynamically based definition of the tropopause. In the definition we'll ultimately settle on, the tropopause need not be marked by a temperature minimum, though on Earth the temperature minimum lies only somewhat above the dynamically defined tropopause, and therefore serves as an approximate indicator of the tropopause location.

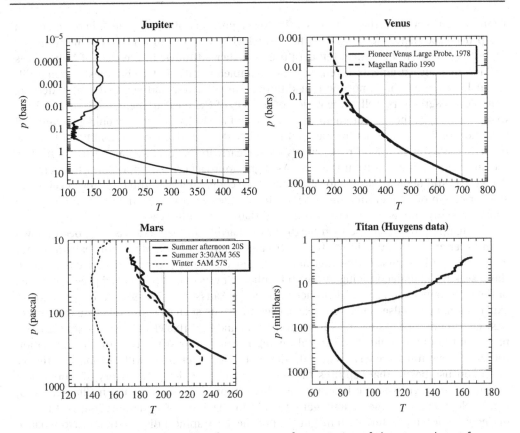

Figure 2.2 The vertical profile of temperature for a portion of the atmosphere of Jupiter (upper left panel), for the atmosphere of Venus (upper right panel), and for the atmosphere of Mars (lower left panel) and that of Titan (lower right panel). Titan data is from the Huygens probe, and Jupiter data is from the Galileo probe. The Venus Magellan and Mars data derive from observations of radio transmission through the atmosphere, taken by the Magellan (late 1980s) and Global Surveyor orbiters, respectively. The information on the lower portion of the Venus atmosphere comes from one of the four 1978 Pioneer Venus probes (the others show a similar pattern). The Jupiter data derives from *in situ* deceleration measurements of the Galileo probe. The full Mars profile dataset reveals considerable seasonal and geographical variation. The profiles shown here were taken in the southern hemisphere by the Mars Global Surveyor. The warmest one is in the late afternoon of 1998 in the summer subtropics while the next warmest is at night-time under otherwise similar conditions. The coldest Mars sounding shown in the plot is from the winter south polar region. 100 Pascal = 1 mb.

the *troposphere*, whereas the portion immediately above is the *stratosphere*. "Tropo" comes from the Greek root for "turning" (as in "turning over"), while "Strato" refers to stratification. The reasons for this terminology will become clear shortly. The stratosphere was discovered in 1900 by Léon Philippe Teisserenc de Bort, the French pioneer of instrumented balloon flights.

The sounding we have shown is typical. In fact, a similar pattern is encountered in the atmospheres of many other planets, as indicated in Fig. 2.2 for Venus, Mars, Jupiter, and

Titan. In common with the Earth case, the lower portions of these atmospheres exhibit a sharp decrease of temperature with height, which gives way to a region of more gently decreasing, or even increasing, temperature at higher altitudes. In the case of Venus, it is striking that measurements taken with two completely different techniques, probing different locations of the atmosphere at different times of day and separated by a decade, nonetheless agree very well in the region of overlap of the measurements. This attests both to the accuracy of the measurement techniques, and the lack of what we would generally call "weather" on Venus, at least insofar as it is reflected in temperature variability. In the case of Mars, the temperature decrease shows most clearly in the local summer afternoon when the surface is still warm from the Sun. As night approaches, the upper-level temperature decrease is still notable, but the lower atmosphere cools rapidly leading to a low-level *inversion*, or region of temperature increase. In the Martian polar winter, the whole atmosphere cools markedly, and is much more isothermal than in the other cases.

The temperature decrease with height in the Earth's atmosphere has long been known from experience of mountain weather. It became a target of quantitative investigation not long after the invention of the thermometer, and was early recognized as a challenge to those seeking an understanding of the atmosphere. It was one of the central preoccupations of the mountaineer and scientist Horace Bénédict de Saussure (1740–99). In the quest for an explanation, many false steps were taken, even by greats such as Fourier, before the correct answer was unveiled. As will be shown in the remainder of this chapter, some simple ideas based on thermodynamics and vertical mixing provide at least the core of an explanation for the temperature decrease. Towards the end of Chapter 3 we will introduce a theory of tropopause height that captures the essence of the problem; the theory of tropopause height will be revisited with increasing sophistication at various points in Chapters 4, 5, and 6. Nonetheless, some serious gaps remain in the state of understanding of the rate of decrease of temperature with height, and of the geographical distribution of tropopause height. In Chapters 3 and 4 we will see that the energy budget of a planet is crucially affected by the vertical structure of temperature; therefore, a thorough understanding of this feature is central to any theory of planetary climate.

2.3 DRY THERMODYNAMICS OF AN IDEAL GAS

2.3.1 The equation of state for an ideal gas

The three thermodynamic variables with which we will mainly be concerned are: temperature (denoted by T), pressure (denoted by p), and density (denoted by ρ). Temperature is proportional to the average amount of kinetic energy per molecule in the molecules making up the gas. We will always measure temperature in kelvin, the units of which are the same size as degrees Celsius (or centigrade), except offset so that absolute zero – the temperature at which molecular motion ceases – occurs at 0 K. In Celsius degrees, absolute zero occurs at about $-273.15\,°C$. In essence, a system is in *thermodynamic equilibrium* precisely when it can be characterized by a single temperature, in that all subcomponents of the system exchange energy until they all have the same temperature. Equilibrium is achieved by heat flowing from hotter subsystems to colder subsystems. Even for macroscopic objects such as atmospheres, within which temperature varies from place to place, thermodynamic equilibrium remains a useful concept since one can usually subdivide the object into small parcels within which temperature is nearly uniform, but which nonetheless contain a huge number of molecules.

Pressure is defined as the force per unit area exerted on a surface in contact with the gas, in the direction perpendicular to the surface.[2] It is independent of the orientation of the surface, and can be defined at a given location by making the surface increasingly small. In the SI units we employ throughout this book, pressure is measured in pascals (Pa); 1 pascal is 1 newton of force per square meter of area, or equivalently $1 kg/(ms^2)$. For historical reasons, atmospheric pressures are often measured in "bars" or "millibars." One bar, or equivalently 1000 millibars (mb) is approximately the mean sea-level pressure of the Earth's current atmosphere. We will often lapse into using mb as units of pressure, because the unit sounds comfortable to atmospheric scientists. For calculations, though, it is important to convert millibars to pascals. This is easy, because 1 mb = 100 Pa. Hence, we should all learn to say "hectopascal" in place of "millibar." It may take some time. When pressures are quoted in millibars or bars, one must make sure to convert them to pascals before using the values in any thermodynamic calculations.

Density is simply the mass of the gas contained in a unit of volume. In SI units, it is measured in kg/m^3.

For a perfect gas, the three thermodynamic variables are related by the perfect gas *equation of state*, which can be written

$$p = knT \tag{2.1}$$

where p is the pressure, n is the number of molecules per unit volume (which is proportional to density), and T is the temperature. k is the *Boltzmann thermodynamic constant*, a universal constant having dimensions of energy per unit temperature. Its value depends only on the units in which the thermodynamic quantities are measured. n is the *particle number density* of the gas. To relate n to mass density, we multiply it by the mass of a single molecule of the gas. Almost all of this mass comes from the protons and neutrons in the molecule, since electrons weigh next to nothing in comparison. Moreover, the mass of a neutron differs very little from the mass of a proton, so for our purposes the mass of the molecule can be taken to be $M \cdot \mu$ where μ is the mass of a proton and M is the *molecular weight* – an integer giving the count of neutrons and protons in the molecule. (The equivalent count for an individual atom of an element is the *atomic weight*.) The density is thus $\rho = n \cdot M \cdot \mu$. If we define the *universal gas constant* as $R^* \equiv k/\mu$ the perfect gas equation of state can be re-written

$$p = \frac{R^*}{M}\rho T. \tag{2.2}$$

In SI units, $R^* = 8314.5 \, (m/s)^2/K$. We can also define a gas constant $R = R^*/M$ particular to the gas in question. For example, dry Earth air has a mean molecular weight of 28.97, so $R_{dryair} = 287 \, (m/s)^2/K$, in SI units.

If μ is measured in kilograms, then $1/\mu$ is the number of protons needed to make up a kilogram. This large number is known as a *Mole*, and is commonly used as a unit of measurement of numbers of molecules, just as one commonly counts eggs by the dozen. For any substance, a quantity of that substance whose mass in kilograms is equal to the molecular weight of the substance will contain one Mole of molecules. For example, 2 kg of

---

[2] Pressure can equivalently be defined as the amount of momentum per unit area per unit time which passes *in both directions* through a small hoop placed in the gas. This definition is equivalent to the exerted-force definition because when a molecule with velocity v and mass m bounces elastically off a surface, the momentum change is $2mv$, but only half of the molecules are moving toward the surface at any given time. The momentum flux definition, in contrast, counts molecules going through the hoop in both directions.

H_2 is a Mole of hydrogen molecules, while $32\,kg$ of the most common form of O_2 is a Mole of molecular oxygen. If n were measured in Moles/m^3 instead of molecules per m^3, then density would be $\rho = n \cdot M$. One can also define the *gram-mole* (or *mole* for short), which is the number of protons needed to make a gram; this number is known as *Avogadro's number*, and is approximately $6.022 \cdot 10^{23}$.

Generally speaking, a gas obeys the perfect gas law when it is tenuous enough that the energy stored in forces between the molecules making up the gas is negligible. Deviations from the perfect gas law can be important for the dense atmosphere of Venus, but for the purposes of the current atmosphere of Earth or Mars, or the upper part of the Jovian or Venusian atmosphere, the perfect gas law can be regarded as an accurate model of the thermodynamics.

An extension of the concept of a perfect gas is the *law of partial pressures*. This states that, in a mixture of gases in a given volume, each component gas behaves just as it would if it occupied the volume alone. The pressure due to one component gas is called the *partial pressure* of that gas. Consider a gas which is a mixture of substance A (with molecular weight M_A) and substance B (with molecular weight M_B). The partial pressures of the two gases are

$$p_A = kn_A T, \quad p_B = kn_B T \tag{2.3}$$

or equivalently,

$$p_A = R_A \rho_A T, \quad p_B = R_B \rho_B T \tag{2.4}$$

where $R_A = R^*/M_A$ and $R_B = R^*/M_B$. The same temperature appears in both equations, since thermodynamic equilibrium dictates that all components of the system have the same temperature. The ratio of partial pressures of any two components of a gas is a convenient way to describe the composition of the gas. From Eq. (2.3), $p_A/p_B = n_A/n_B$, so the ratio of partial pressure of A to that of B is also the ratio of number of molecules of A to the number of molecules of B. This ratio is called the *molar mixing ratio*. When we refer to a mixing ratio without qualification, we will generally mean the molar mixing ratio. Alternately, one can describe the composition in terms of the ratio of partial pressure of one component to total pressure of the gas ($p_A/(p_A + p_B)$ in the two-component example). Summing the two partial pressure equations in Eq. (2.3), we see that this is also the ratio of number of molecules of A to total number of molecules; hence we will use the term *molar concentration* for this ratio.[3] If ξ_A is the molar mixing ratio of A to B, then the molar concentration is $\xi_A/(1 + \xi_A)$, from which we see that the molar concentration and molar mixing ratio are nearly the same for substances which are very dilute (i.e. $\xi_A \ll 1$). We will use the notation η_A throughout the book to denote the molar concentration of substance A.

Exercise 2.1 Show that a mixture of gases with molar concentrations $\eta_A = n_A/(n_A + n_B)$ and $\eta_B = n_B/(n_A + n_B)$ behaves like a perfect gas with mean molecular weight $M = \eta_A M_A + \eta_B M_B$

---

[3] The term *volumetric* mixing ratio or concentration is often used interchangeably with the term *molar*, as in "ppmv" for "parts per million by volume." The reason for this nomenclature is that the volume occupied by a given quantity of gas at a fixed temperature and pressure is proportional to the number of molecules of the gas contained in that quantity. To see this, write $n = N/V$, where N is the number of molecules and V is the volume they occupy. Then, the ideal gas law can be written in the alternate form $V = (kT/p)N$. Hence the ratio of standardized volumes is equal to the molar mixing ratio, and so forth. Abbreviations like "ppmv" for molar mixing ratios are common and convenient, because the "v" can unambiguously remind us that we are talking about a volumetric (i.e. molar) mixing ratio or concentration, whereas in an abbreviation like "ppmm" one is left wondering whether the second "m" means "mass" or "molar."

(i.e. derive the expression relating total pressure $p_A + p_B$ to total density $\rho_A + \rho_B$ and identify the effective gas constant). Compute the mean molecular weight of dry Earth air. (Dry Earth air consists primarily of 78.084% N_2, 20.947% O_2, and 0.934% Ar, by count of molecules.)

The *mass mixing ratio* is the ratio of the mass of substance A to that of substance B in a given parcel of gas, i.e. ρ_A / ρ_B. From Eq. (2.4) it is related to the molar mixing ratio by

$$\frac{\rho_A}{\rho_B} = \frac{M_A}{M_B}\frac{p_A}{p_B}. \tag{2.5}$$

Throughout this book, we will use the symbol r to denote mass mixing ratios, with subscripts added as necessary to distinguish the species involved. Yet another measure of composition is *specific concentration*, defined as the ratio of the mass of a given substance to the total mass of the parcel (e.g. $\rho_A / (\rho_A + \rho_B)$ in the two-component case). We will use the symbol q, with subscripts as necessary, to denote the specific concentration of a substance. The term *specific humidity* generally refers to the specific concentration of water vapor, though by extension we will use it to refer to the specific concentration of other condensable substances. Using the law of partial pressures, the specific concentration of substance A in a mixture is related to the molar concentration by

$$q_A \equiv \frac{\rho_A}{\rho_{\text{tot}}} = \frac{M_A}{\bar{M}}\frac{p_A}{p_{\text{tot}}} = \frac{M_A}{\bar{M}}\eta_A \tag{2.6}$$

where $\bar{M}$ is the mean molecular weight of the mixture, with the mean computed using weighting according to molar concentrations of the species, as in Exercise 2.1.

All of the ratios we have just defined are convenient to use because, unlike densities, they remain unchanged as a parcel of air expands or contracts, provided the constituents under consideration do not undergo condensation, chemical reaction or other forms of internal sources or sinks. Hence, for a compressible gas, two components A and B are well mixed relative to each other if the mixing ratio between them is independent of position.

Constituents will tend to become well mixed over a great depth of the atmosphere if they are created or destroyed slowly, if at all, relative to the characteristic time required for mixing. In the Earth's atmosphere, the mixing ratio of oxygen to nitrogen is virtually constant up to about 80 km above the surface. The mixing ratio of carbon dioxide in air can vary considerably in the vicinity of sources at the surface, such as urban areas where much fuel is burned, or under forest canopies when photosynthesis is active. Away from the surface, however, the carbon dioxide mixing ratio varies little. Variations of a few parts per million can be detected in the relatively slowly mixed stratosphere, associated with the industrial-era upward trend in fossil fuel carbon dioxide emissions. Small seasonal and interhemispheric fluctuations in the tropospheric mixing ratio, associated with variations in the surface sources, can also be detected. For most purposes, though, carbon dioxide can be regarded as well mixed throughout the atmosphere. In contrast, water vapor has a strong internal sink in Earth's atmosphere, because it is condensable there; hence its mixing ratio shows considerable vertical and horizontal variations. Carbon dioxide, methane, and ammonia are not condensable on Earth at present, but their condensation can become significant in colder planetary atmospheres.

Exercise 2.2 (a) In the year 2000, the concentration of CO_2 in the atmosphere was about 370 parts per million molar. What is the ratio p_{CO_2}/p_{tot}? Estimate p_{CO_2} in mb at sea level. Does the molar concentration differ significantly from the molar mixing ratio? What is the mass mixing ratio of CO_2 in air? What is the mass mixing ratio of *carbon* (in the form of

CO_2) in air – i.e. how many kilograms of carbon would have to be burned into CO_2 in order to produce the CO_2 in 1 kg of air? (*Note:* The mean molecular weight of air is about 29.) (b) The molar concentration of O_2 in Earth air is about 20%. How many grams of O_2 does a 1 liter breath of air contain at sea level (1000 mb)? At the top of Qomolangma (a.k.a. "Mt. Everest," about 300 mb)? Does the temperature of the air (within reasonable limits) affect your answer much?

2.3.2 Specific heat and conservation of energy

Conservation of energy is one of the three great pillars upon which the edifice of thermodynamics rests. When expressed in terms of changes in the state of matter, it is known as the *First Law of Thermodynamics*. When a gas expands or contracts, it does work by pushing against the environment as its boundaries move. Since pressure is force per unit area, and work is force times distance, the work done in the course of an expansion of volume dV is pdV. This is the amount of energy that must be added to the parcel of gas to allow the increase in volume to take place. For atmospheric purposes, it is more convenient to write all thermodynamic relations on a per unit mass basis. Dividing V by the mass contained in the volume yields ρ^{-1}, whence the work per unit mass is $pd(\rho^{-1})$. This is not the end of energy accounting. Changing the temperature of a unit mass of the substance while holding volume fixed changes the energy stored in the various motions of the molecules by an amount $c_v dT$, where c_v is a proportionality factor known as the *specific heat at constant volume*. For example, it takes about 720 joules of energy to raise the temperature of 1 kg of air by 1 K while holding the volume fixed. For ideal gases, the specific heat can depend on temperature, though the dependence is typically weak. For non-ideal gases, specific heat can depend on pressure as well.

Combining the two contributions to energy change we find the expression for the amount of energy that must be added per unit mass in order to accomplish a change of both temperature and volume:

$$\delta Q = c_v dT + pd(\rho^{-1}). \tag{2.7}$$

Using the perfect gas law, the heat balance can be re-written in the form

$$\delta Q = c_v dT + d(p\rho^{-1}) - \rho^{-1}dp = (c_v + R)dT - \rho^{-1}dp. \tag{2.8}$$

From this relation, we can identify the *specific heat at constant pressure*, $c_p \equiv c_v + R$, which is the amount of energy needed to warm a unit mass by 1 K while allowing it to expand enough to keep pressure constant.

The units in which we measure temperature are an artifact of the marks one researcher or other once decided to put on some device that responded to heat and cold. Since temperature is proportional to the energy per molecule of a substance, it would make sense to set the proportionality constant to unity and simply use energy as the measure of temperature. This not being common practice, one has occasion to make use of the *Boltzmann thermodynamic constant*, k, which expresses the proportionality between temperature and energy. More precisely, each degree of freedom in a system with temperature T has a mean energy $\frac{1}{2}kT$. For example, a gas made of rigid spherical atoms has three degrees of freedom per atom (one for each direction it can move), and therefore each atom has energy $\frac{3}{2}kT$ on average; a molecule which could store energy in the form of rotation or vibration would have more degrees of freedom, and therefore each molecule would have more energy at any given temperature. The energy–temperature relation is made possible by an important thermodynamic principle, the *equipartition principle*, which states that in equilibrium, each degree of freedom accessible to a system gets an equal share of the total energy of the system. This is, in essence, the definition of thermodynamic equilibrium, and for equilibrium to be achieved it is necessary for the parts of the system to interact strongly enough for a sufficiently long time that energy is exchanged and randomized amongst all parts of the system. In contrast to physical constants like the speed of light, the Boltzmann constant should not be considered a fundamental constant of the Universe. It is just a unit conversion factor.

Exercise 2.3 There are 20 students and one professor in a well-insulated classroom measuring 20 meters by 20 meters by 3 meters. Each person in the classroom puts out energy at a rate of 100 watts (1 watt = 1 joule/second). The classroom is dark, except for a computer and LCD projector which together consume power at a rate of 200 watts. The classroom is filled with air at a pressure of 1000 mb (no extra charge). The room is sealed so no air can enter or leave, and has an initial temperature of 290 K. How much does the temperature of the classroom rise during the course of a 1-hour lecture?

2.3.3 Entropy, reversibility, and potential temperature; the Second Law

One cannot use Eq. (2.8) to define a "heat content" Q of a state (p, T) relative to a reference state (p_0, T_0), because the amount of heat needed to go from one state to another depends on the path in pressure–temperature space taken to get there; the right hand side of Eq. (2.8) is not an exact differential. However, it can be made into an exact differential by dividing the equation by T and using the perfect gas law as follows:

$$ds \equiv \frac{\delta Q}{T} = c_p \frac{dT}{T} - R\frac{dp}{p} = c_p d \ln\left(Tp^{-R/c_p}\right) \tag{2.9}$$

assuming c_p to be constant. This equation defines the *entropy*, $s \equiv c_p \ln\left(Tp^{-R/c_p}\right)$. Entropy is a nice quantity to work with because it is a *state variable* – its change between two states is independent of the path taken to get from one to the other. A process affecting a parcel of matter is said to be *adiabatic* if it occurs without addition or loss of heat from the parcel. By definition, $\delta Q = 0$ for adiabatic processes. In consequence, adiabatic processes leave entropy unchanged, *provided that the changes in state of the system are slow enough that the system remains close to thermodynamic equilibrium at all times*. The latter condition is satisfied for all atmospheric phenomena that will concern us, but could be violated, for example, in the case of explosive adiabatic expansion of a formerly confined gas into a vacuum. Entropy can also be defined for gases whose specific heat depends on temperature and pressure, for inhomogeneous mixtures of gases, and for non-ideal gases.

The *Second Law of Thermodynamics* states that entropy never decreases for energetically closed systems – systems to which energy is neither added nor subtracted in the course of their evolution. The formal derivation of the law from the microscopic properties of molecular interactions is in many ways an unfinished work of science, but the tendency toward an increase in entropy – an increase in disorder – seems to be a nearly universal property of systems consisting of a great many interacting components. A process during which the entropy remains constant is *reversible*, since it can be run both ways.

The Second Law is perhaps more intuitive when restated in the following way: *In an energetically closed system, heat flows from a hotter part of the system to a colder part of the system, causing the system to evolve toward a state of uniform temperature.* To see that this statement is equivalent to the entropy-increase principle, consider a thermally insulated box of gas having uniform pressure, but within which the left half of the mass is at temperature T_1 and the right half of the mass is at temperature $T_2 < T_1$. Now suppose that we transfer an amount of heat δQ from the left half of the box to the right half. This transfer leaves the net energy unchanged, but it changes the entropy. Specifically, according to Eq. (2.9), the entropy change summed over the two halves of the gas is $ds = \left(\frac{1}{T_2} - \frac{1}{T_1}\right)\delta Q$. Since $T_2 < T_1$, this change is positive only if $\delta Q > 0$, representing a transfer from the hotter to the colder portion of the gas. Entropy can be increased by further heat transfers until $T_1 = T_2$, at which point the maximum entropy state has been attained.

The Second Law endows the Universe with an arrow of time. If one watches a movie of a closed system and sees that the system starts with large fluctuations of temperature (low entropy) and proceeds to a state of uniform temperature (high entropy), one knows that time is running forward. If one sees a thermally homogeneous object spontaneously generate large temperature inhomogeneities, then one knows that the movie is being run backward. Note that the Second Law applies only to closed systems. The entropy of a subcomponent can decrease, if it exchanges energy with the outside world and increases the entropy of the rest of the Universe. This is how a refrigerator works.

Entropy is a very general concept, of which we have seen only the most basic instance. For a homogeneous ideal gas near thermodynamic equilibrium, the notions of reversible (i.e. isentropic) and adiabatic processes are equivalent, but caution must be exercised when extending this picture to more complex systems involving mixtures of gases. For example, if a box of gas at uniform temperature T contains pure N_2 in its left half and pure O_2 in its right half, then entropy will increase when the two gases spontaneously mix, even if no energy is let into the box. The entropy can still be defined in terms of changes in $\delta Q/T$, but it requires careful attention to what precisely is meant by δQ and to which subsystems the heat changes are being applied. The references given in the Further Reading section of this chapter provide a deeper and more general understanding of the use of entropy in solving thermodynamic problems.

Now let's get back to basics. Entropy can be used to determine how the temperature of an air parcel changes when it is compressed or expanded adiabatically. This is important because it tells us what happens to temperature as a bit of the atmosphere is lifted from low altitudes (where the pressure is high) to higher altitudes (where the pressure is lower), provided the lifting occurs so fast that the air parcel has little time to exchange heat with its surroundings. If the initial temperature and pressure are (T, p), then conservation of entropy tells us that the temperature T_0 found upon adiabatically compressing or expanding to pressure p_0 is given by $Tp^{-R/c_p} = T_0 p_0^{-R/c_p}$ This leads us to define the *potential temperature*

$$\theta = T\left(\frac{p}{p_0}\right)^{-R/c_p} \tag{2.10}$$

which is simply the temperature an air parcel would have if reduced adiabatically to a reference pressure p_0. Like entropy, potential temperature is conserved for adiabatic processes.

To understand why the presence of cold air above warm air in the sounding of Fig. 2.1 does not succumb immediately to instability, we need only look at the corresponding profile of potential temperature, shown in Fig. 2.3. This figure shows that potential temperature increases monotonically with height. This profile tells us that the air aloft is cold, but that

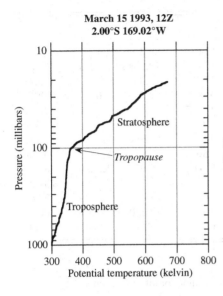

March 15 1993, 12Z
2.00°S 169.02°W

Figure 2.3 The dry potential temperature profile for the sounding in Fig. 2.1.

if it were pushed down to lower altitudes, compression would warm it to the point that it is warmer than the surrounding air, and thus being positively buoyant, will tend to float back up to its original level rather than continuing its descent. We see also where the stratosphere gets its name: potential temperature increases very strongly with height there, so air parcels are very resistant to vertical displacement. This part of the atmosphere is therefore strongly stratified.

The troposphere is stable, but has much weaker gradients of θ. In a compressible atmosphere, a well-stirred layer would have constant θ rather than constant T, since it is the former that is conserved for adiabatic processes such as would be caused by rapid vertical displacements. This is the essence of the explanation for why temperature decreases with height: turbulent stirring relaxes the troposphere towards constant θ, yielding the *dry adiabat*

$$T(p) = \theta \cdot \left(\frac{p}{p_0} \right)^{R/c_p}. \tag{2.11}$$

In this formula, θ has the constant value $T(p_0)$.

On the dry adiabat, the slope $d\ln T/d\ln p$ has the value R/c_p. From the first equality in Eq. (2.9), it can be seen that this result remains valid even if R/c_p depends on temperature and pressure. In general, the slope $d\ln T/d\ln p$ provides a convenient measure of how sharply temperature decreases in the vertical; positive values correspond to decrease of temperature with height, since pressure decreases with altitude. The *dry adiabatic slope* is then R/c_p. In atmospheric science, it is common to characterize the temperature structure by $-dT/dz$ – the *lapse rate* – but there are very few circumstances in which the altitude z is a convenient coordinate for climate calculations. Unless noted otherwise, we will (somewhat unconventionally) use the term *lapse rate* to refer to the slope $d\ln T/d\ln p$. With this definition, the dry adiabatic lapse rate would be R/c_p. (See Problem 2.29 for an exploration of the more conventional use of the term.)

It is evident from Fig. 2.3 that something prevents θ from becoming completely well mixed. An equivalent way of seeing this is to compare the observed temperature profile

	H_2O	CH_4	CO_2	N_2	O_2	H_2	He	NH_3
Crit. point T	647.1	190.44	304.2	126.2	154.54	33.2	5.1	405.5
Crit. point p	221.e5	45.96e5	73.825e5	34.0e5	50.43e5	12.98e5	2.28e5	112.8
Triple point T	273.16	90.67	216.54	63.14	54.3	13.95	2.17	195.4
Triple point p	611.	0.117e5	5.185e5	0.1253e5	0.0015e5	0.072e5	0.0507e5	0.061e5
L vap(b.p.)	22.55e5	5.1e5	–	1.98e5	2.13e5	4.54e5	0.203e5	13.71e5
L vap(t.p.)	24.93e5	5.36e5	3.97e5	2.18e5	2.42e5	??	??	16.58e5
L fusion	3.34e5	0.5868e5	1.96e5	0.2573e5	0.139e5	0.582e5	??	3.314e5
L sublimation	28.4e5	5.95e5	5.93e5	2.437e5	2.56e5	??	??	19.89e5
ρ liq(b.p.)	958.4	450.2	1032.	808.6	1141.	70.97	124.96	682.
ρ liq(t.p.)	999.87	??	1110.	??	1307.	??	??	734.2
ρ solid	917.	509.3	1562.	1026.	1351.	88.	200.	822.6
$c_p(0\,°C/1\,bar)$	1847.	2195.	820.	1037.	916.	14 230.	5196.	2060.
$\gamma(c_p/c_v)$	1.331	1.305	1.294	1.403	1.393	1.384	1.664	1.309

Table 2.1 Thermodynamic properties of selected gases.
Latent heats of vaporization are given at both the boiling point (the point where saturation vapor pressure reaches 1 bar) and the triple point. Liquid densities are given at the boiling point and the triple point. For CO_2 the "boiling point" is undefined, so the liquid density is given at 253 K/20 bar instead. Note that the maximum density of liquid water is 1000.00 kg/m^3 and occurs at −4 °C. Densities of solids are given at or near the triple point. All units are SI, so pressures are quoted as Pa with the appropriate exponent. Thus, 1 bar is written as 1e5 in the table. Missing data is indicated by ??

with the dry adiabat. For example, if the air at 1000 mb in Fig. 2.3, having temperature 298 K, were lifted dry-adiabatically to the tropopause, where the pressure is 100 mb, then the temperature would be $298 \cdot \left(\frac{100}{1000}\right)^{2/7}$, i.e. 154.3 K (using the value $R/c_p = 2/7$ for Earth air). This is much colder than the observed temperature, which is 188 K. We will see shortly that in the Earth's atmosphere, condensation of water vapor is one of the factors in play, though it is not the only one affecting the tropospheric temperature profile. The question of what determines the tropospheric θ gradient is at present still largely unsettled, particularly outside the tropics.

It is no accident that the value of R/c_p for air lies close to the ratio of two small integers. It is a consequence of the equipartition principle. Using methods of statistical thermodynamics, it can be shown that a gas made up of molecules with n degrees of freedom has $R/c_p = 2/(n + 2)$. Using the expression for the gas constant in terms of the specific heats, the adiabatic coefficient can also be written as $R/c_p = 1 - 1/\gamma$, where $\gamma = c_p/c_v$; for exact equipartition, $\gamma = 1 + 2/n$. The measured values of γ for a few common atmospheric gases are shown in Table 2.1. Helium comes close to the theoretical value for a molecule with no internal degrees of freedom, underscoring that excitation of electron motions plays little role in heat storage for typical planetary temperatures. The diatomic molecules have values closest to the theoretical value for $n = 5$, one short of what one would expect from adding two rotational and one vibrational internal degrees of freedom to the three directional degrees of freedom. Among the triatomic molecules, water acts roughly as if it had $n = 6$ while carbon dioxide is closer to $n = 7$. The two most complex molecules, methane and ammonia, are also characterized by $n = 7$. The failure of thermodynamics to access all the degrees of freedom classically available to a molecule is a consequence of quantum theory. Since the energy stored in states of motion of a molecule in fact comes in discrete-sized chunks, or "quanta," one can have a situation where a molecule hardly ever gets enough energy from a collision to excite even a single vibrational degree of freedom, for example, leading to the phenomenon of partial excitation or even non-excitation of certain classical degrees of freedom. This is one of many ways that the quantum theory, operating on exceedingly tiny spatial scales, exerts a crucial control over macroscopic properties of matter that can affect the very habitability of the Universe. Generally speaking, the higher the temperature gets, the more easy it is to excite internal degrees of freedom, leading to a decrease in γ. This quantum effect is the chief reason that specific heats vary somewhat with temperature.

Exercise 2.4 (a) A commercial jet airliner cruises at an altitude of 300 mb. The air outside has a temperature of 240 K. To enable the passengers to breathe, the ambient air is compressed to a cabin pressure of 1000 mb. What would the cabin temperature be if the air were compressed adiabatically? How do you think airlines deal with this problem? (b) Discuss whether the lower portion of the Venus temperature profile shown in Fig. 2.2 is on the dry CO_2 adiabat. Do the same for the summer afternoon Mars sounding. (c) Assume that the Jupiter sounding is on a dry adiabat, and estimate the value of R/c_p for the atmosphere. Based on your result, what is the dominant constituent of the Jovian atmosphere likely to be? What other gas might be mixed with the dominant one?

2.4 STATIC STABILITY OF INHOMOGENEOUS MIXTURES

An atmosphere is *statically unstable* if an air parcel displaced from its original position tends to continue rising or sinking instead of returning to its original position. Such a state will tend to mix itself until it becomes stable. Static stability is important to planetary climate because it affects the vertical mixing which creates a planet's troposphere.

For a well-mixed atmosphere, the potential temperature profile tells the whole story about static stability, since, according to the ideal gas law, the density of an air parcel with potential temperature θ_0 will be $\rho_0 = p_1/(R\theta_0 \cdot (p_1/p_0)^{R/c_p})$ upon being elevated to an altitude with pressure $p_1 < p_0$. The ambient density there is $\rho_1 = p_1/(R\theta_1 \cdot (p_1/p_0)^{R/c_p})$. The displaced parcel will be negatively buoyant and return toward its original position if $\rho_0 > \rho_1$, which is true if and only if $\theta_0 < \theta_1$, i.e. if the potential temperature increases with height.

For an inhomogeneous atmosphere, this is no longer the case, since the gas constant R depends on the mean molecular weight of the mixture, which varies from place to place. As an example, we may consider an atmosphere which has uniform θ computed on the basis of a reference pressure p_0, but which consists of pure N_2 for $p > p_0$ and pure CO_2 for $p < p_0$. One immediately notes that the system has an unstable density jump at p_0, since the density is $p_0/R_{N_2}\theta$ just below the interface and $p_0/R_{CO_2}\theta$ just above the interface. Since N_2 has lower molecular weight (28) than CO_2 (44), the gas constant for N_2 is considerably greater than the gas constant for CO_2. Given that $R_{N_2} > R_{CO_2}$, the density is greater just above the interface than it is just below the interface. This is an unstable situation, and the N_2 layer will tend to mix itself into the CO_2, despite the constancy of θ.

The phenomenon is very familiar: it is why helium balloons rise in air, even when they are at the same temperature as their surroundings. The low molecular weight of helium makes it lighter (i.e. lower density) than air having the same temperature and pressure.

Exercise 2.5 Make sense of the following statement: "For the Earth's atmosphere, moist air is lighter than dry air." Would this still be true for a planet whose atmosphere is mainly H_2?

Suppose now that a parcel of N_2 is lifted from just below the interface to an altitude where the pressure is some lower pressure p_1. Let the density of the parcel when it arrives at its destination be ρ_{lift} and the density of the ambient CO_2 be ρ_{amb}. The density difference is then

$$\rho_{lift} - \rho_{amb} = \frac{p_1}{\theta} \left(\frac{1}{R_{N_2}(p_1/p_0)^{-(R/c_p)_{N_2}}} - \frac{1}{R_{CO_2}(p_1/p_0)^{-(R/c_p)_{CO_2}}} \right). \tag{2.12}$$

The biggest effect in this equation comes from the fact that $R_{N_2} > R_{CO_2}$, which assures that the N_2 retains its buoyancy as it is lifted. A much weaker effect comes from the fact that R/c_p is slightly greater for N_2 (0.286) than for CO_2 (0.230). This modulates the density difference as the parcel is lifted, but the effect is slight. Even lifting a great distance, so that $p_1/p_0 = 0.01$, the pressure factor in the first term is only modestly greater than the pressure factor in the second term (3.74 vs. 2.89), yielding a modest reduction in the buoyancy of the lifted parcel. For other pairs of gases, the difference in R/c_p could be more significant, and in principle it could also go in the opposite direction and increase buoyancy rather than reducing it.

For an arbitrary profile of composition and temperature, one can define a *potential density*, which is the density an air parcel would have if compressed or expanded adiabatically to a standard reference pressure p_0. Using the gas law, and the fact that mixing ratios are conserved (whence R/c_p is conserved on adiabatic compression or expansion of the parcel), the potential density relative to reference level p_0 is

$$\tilde{\rho}(p|p_0) = \frac{p_0}{R\theta}$$
$$= \frac{p_0}{RT} \left(\frac{p}{p_0} \right)^{R/c_p} = \rho(p) \left(\frac{p}{p_0} \right)^{R/c_p - 1}. \tag{2.13}$$

The R and c_p in this equation must be taken from the values prevailing at pressure p, since these values are determined by composition, which in this calculation is presumed to remain fixed as the parcel is displaced to the reference pressure p_0. When $\tilde{\rho}(p|p_0)$ increases with p, then the system will be stable in the sense that a parcel from $p < p_0$ will be positively buoyant when pushed down to p_0 and thus tend to return to its original level; similarly a parcel lifted to p_0 from some pressure $p > p_0$ will be negatively buoyant and also tend to return to where it came from. When the atmosphere is homogeneous, R/c_p is constant and one can determine whether a parcel at pressure p_1 will be buoyant when displaced adiabatically to p_2 by adiabatically reducing both to the standard pressure p_0 and comparing the densities there. In other words, for a homogeneous atmosphere, one can tell immediately from a single plot of potential density relative to p_0 which regions are stable and which regions will tend to overturn. When the atmosphere is inhomogeneous, this is no longer the case. When a parcel with density $\rho(p_1)$ is displaced to p_2, it will be positively buoyant if

$$\rho(p_1) \left(\frac{p_1}{p_2} \right)^{R(p_1)/c_p(p_1) - 1} < \rho(p_2). \tag{2.14}$$

However, if both parcels are reduced to a common pressure p_0, then the first parcel would be buoyant relative to the second if

$$\rho(p_1)\left(\frac{p_1}{p_0}\right)^{R(p_1)/c_p(p_1)-1} < \rho(p_2)\left(\frac{p_2}{p_0}\right)^{R(p_2)/c_p(p_2)-1} \tag{2.15}$$

i.e.

$$\rho(p_1)\left(\frac{p_1}{p_0}\right)^{R(p_1)/c_p(p_1)-1}\left(\frac{p_0}{p_2}\right)^{R(p_2)/c_p(p_2)-1} < \rho(p_2). \tag{2.16}$$

This yields the same criterion as the correct criterion given in Eq. (2.14) only if R/c_p is constant.

The lack of a globally valid potential density complicates the precise analysis of the static stability of inhomogeneous atmospheres. Strictly speaking, one needs to examine potential density profiles for a range of different p_0 covering the atmosphere. In practice, the potential density based on a single p_0 can often provide a useful indication of static stability within a reasonably thick layer of the atmosphere because the variations of R/c_p are usually slight unless the compositional variations are extreme.

An alternate approach to dealing with the effect of composition on buoyancy is to define a modified potential temperature based on a *virtual temperature*. The virtual temperature is the temperature at which the gas law for a standard composition (e.g. dry air) would yield the same density as the true gas law taking into account the effect of the actual composition on density (e.g. moist air). This approach is a common way of dealing with the effect of water vapor on buoyancy in Earth's atmosphere. It is equivalent to the potential density approach and shares the same limitations. Making use of a virtual temperature can be convenient for some purposes, but it can also be confusing in that one needs to keep track of the contexts in which one uses the virtual temperature and the ones in which one must use the actual temperature.

In planetary atmospheres, compositional variations often arise as a result of condensation, evaporation, or sublimation of one of the atmospheric components. For example, the Martian atmosphere contains a moderate amount of Ar and N_2, and when CO_2 condenses at the winter pole, it leaves an Ar and N_2 enriched layer near the surface. Because these gases have lower molecular weight than CO_2, this layer tends to be buoyant and convect upward. When CO_2 sublimates from the polar glacier in the spring, the pure CO_2 layer resulting is statically stable relative to the Ar and N_2 rich atmosphere above. The evaporation of light CH_4 from the surface of Titan makes the low-level air positively buoyant with respect to the overlying N_2, favoring upward mixing. Water vapor has a small but significant effect on buoyancy on the present Earth; the importance of the effect would increase sharply under warmer conditions in which the water vapor content of the atmosphere is greater. Condensation can also lead to compositional variations in the interior of atmospheres, and in some circumstances it may be necessary to take these into account when determining where the vertical mixing that creates the troposphere takes place. In principle, chemical reactions could also create compositional gradients strong enough to create buoyancy, though there do not at present appear to be any atmospheres where this is known to be a significant effect.

2.5 THE HYDROSTATIC RELATION

The hydrostatic relation relates pressure to altitude and the mass distribution of the atmosphere, and provides the chief reason that pressure is the most natural vertical coordinate to use in most atmospheric problems. Consider a column of any substance at rest, and suppose that the density of the substance as a function of height z is given by $\rho(z)$. Suppose further that the range of altitudes being considered is small enough that the acceleration of gravity is essentially constant; the magnitude of this acceleration will be called g, and the force of gravity is taken to point along the direction of decreasing z. Now, consider a slice of the column with vertical thickness dz, having cross-sectional area A in the horizontal direction. Since pressure is simply force per unit area, then the change in pressure from the base of this slice to the top of this slice is just the force exerted by the mass. By Newton's second law, then, we have

$$Adp = -Agdm = -Ag\rho dz \qquad (2.17)$$

where dm is the increment of mass in the column per unit area. An immediate consequence of this relation is that

$$dm = -\frac{dp}{g} \qquad (2.18)$$

which states that the amount of mass in a slab of atmosphere is proportional to the thickness of that slab, measured in pressure coordinates. A further consequence, upon dividing by dz, is the relation

$$\frac{dp}{dz} = -\rho g. \qquad (2.19)$$

This differential equation expresses the *hydrostatic relation*. It is exact if the substance is at rest (hence the "static"), but if the material of the column is in motion, the relation is still approximately satisfied provided the acceleration is sufficiently small, compared with the acceleration of gravity. In practice, the hydrostatic relation is very accurate for most problems involving large-scale motions in planetary atmospheres. It would not be a good approximation within small-scale intense updrafts or downdrafts where the acceleration of the fluid may be large. Derivation of the precise conditions under which the hydrostatic approximation holds requires consideration of the equations of fluid motion.

An important consequence of the hydrostatic relation is that it enables us to determine the total mass of an atmosphere through measurements of pressure taken at the surface alone. Integrating Eq. (2.18) from the surface ($p = p_s$) to space ($p = 0$) yields the relation

$$m = \frac{p_s}{g} \qquad (2.20)$$

where m is the total mass of the atmosphere located over a unit area of the planet's surface. Note that this relation presumes that the depth of the layer containing almost all the mass of the atmosphere is sufficiently shallow that gravity can be considered constant throughout the layer. Given that gravity decays inversely with the square of distance from the planet's center, this is equivalent to saying that the atmosphere must be shallow compared with the radius of the planet. For a well-mixed substance A with mass-specific concentration q_A relative to the whole atmosphere, the mass of substance A per square meter of the planet's surface is just mq_A.

Using the perfect gas law to eliminate ρ from Eq. (2.19) yields

$$\frac{dp}{dz} = -\frac{g}{RT}p \tag{2.21}$$

where R is the gas constant for the mixture making up the atmosphere. This has the solution

$$p(z) = p_s \exp\left(-\frac{g}{R\bar{T}}z\right), \quad \bar{T}(z) = \left(\frac{1}{z}\int_0^z T^{-1}dz\right)^{-1}. \tag{2.22}$$

Here, $\bar{T}(z)$ is the harmonic mean of temperature in the layer between the ground and altitude z. If temperature is constant, then pressure decays exponentially with *scale height* RT/g. Because temperature is measured relative to absolute zero, the mean temperature $\bar{T}(z)$ can be relatively constant despite fairly large variations of temperature within the layer. In consequence, pressure typically decays roughly exponentially with height even when temperature is altitude-dependent.

Exercise 2.6 Compute the mass of the Earth's atmosphere, assuming a mean surface pressure of 1000 mb. (The Earth's radius is 6378 km, and the acceleration of its gravity is 9.8 m/s². Compute the mass of the Martian atmosphere, assuming a mean surface pressure of 6 mb. (Mars' radius is 3390 km, and the acceleration of its gravity is 3.7 m/s².)

Note that the hydrostatic relation applies only to the total pressure of all constituents; it does not apply to partial pressures individually. However, in the special case in which the gases are well mixed, the total mass of each well-mixed component can still be determined from surface data alone. One simply multiplies the total mass obtained from surface pressure, by the appropriate (constant) mass-specific concentration. It is important to recognize, however, that even for a well-mixed gas the mass per unit area of a constituent with partial pressure p_A is *not* simply p_A/g; it is $(M_A/\overline{M})(p_A/g)$, where M_A is the molecular weight of substance A and $\overline{M}$ is the mean molecular weight of the mixture, which varies according to how much of A is added to the mixture. This becomes especially important to keep straight when dealing with situations in which a certain amount of, say, CO_2 is added to an atmosphere. Sometimes when one talks about adding "1 bar of CO_2 to an atmosphere" it means adding an amount of CO_2 that would increase the partial pressure of CO_2 by 1 bar; however, a phrase like this is often used to mean adding an amount of CO_2 that would have a surface pressure of 1 bar if the CO_2 were alone in the atmosphere. The two things are not the same, as illustrated by the following exercise, which also makes the point that the partial pressure of the rest of the atmosphere does not remain fixed as one adds or takes away CO_2, or any other gas with a different molecular weight than the mean.

Exercise 2.7 Suppose an atmosphere initially consists of 1 bar of pure N_2. One then adds an amount of CO_2 to the atmosphere that would yield a surface pressure of 1 bar if it were all alone in an atmosphere. Show that at the end of this process the surface pressure is 2 bar, the partial pressure of CO_2 is 0.76 bar, the partial pressure of N_2 is 1.24 bar, and the molar concentration of CO_2 is 38%.

In the study of atmospheric dynamics, the hydrostatic equation is used to compute the pressure gradients which drive the great atmospheric circulations. Outside of dynamics, there are rather few problems in physics of climate that require one to know the altitude corresponding to a given pressure level. Our main use of the hydrostatic relation in this book will be in the form of Eq. (2.18), which tells us the mass between two pressure surfaces.

The hydrostatic relation also allows us to derive a useful alternate form of the heat budget, by re-writing the heat balance equation as follows:

$$\delta Q = c_p dT - \rho^{-1} dp = c_p dT - \rho^{-1} \frac{dp}{dz} dz = d(c_p T + gz) \tag{2.23}$$

assuming c_p to be constant. The quantity $c_p T + gz$ is known as the *dry static energy*. Dry static energy provides a more convenient basis for atmospheric energy budgets than entropy, since changes in dry static energy following an air parcel are equal to the net energy added to or removed from the parcel by heat sources such as solar radiation. For example, if there are no horizontal transports and if there is no net flux of energy between the atmosphere and the underlying planetary surface, then the rate of change of the net dry static energy in a (dry) atmospheric column is the difference between the rate at which solar energy flows into the top of the atmosphere and the rate at which infrared radiation leaves the top of the atmosphere; one needs to know nothing about how the heat is deposited within the atmosphere in order to determine how the net dry static energy changes. This is not the case for entropy.

Note that the dry static energy as defined above is actually the energy *per unit mass* of atmosphere. Thus, the total energy in a column of atmosphere, per unit surface area, is $\int (c_p T + gz) \rho \, dz$, which by the hydrostatic relation is equal to $\int (c_p T + gz) dp/g$ if the pressure integral is taken in the direction of increasing pressure.

2.6 THERMODYNAMICS OF PHASE CHANGE

When a substance changes from one form to another (e.g. water vapor condensing into liquid water or gaseous carbon dioxide condensing into dry ice) energy is released or absorbed even if the temperature of the mass is unchanged after the transformation has taken place. This happens because the amount of energy stored in the form of intermolecular interactions is generally different from one form, or *phase*, to another. The change of phase from one form to another is called a *phase transition*. The amount of energy released when a unit of mass of a substance changes from one phase to another, holding temperature constant, is known as the *latent heat* associated with that phase change. By convention, latent heats are stated as positive numbers, with the phase change going in the direction that releases energy. Phase transitions are *reversible*. If one kilogram of matter releases L joules of energy in going from phase A to phase B, it will take the same L joules of energy to turn the mass back into phase A. The units of latent heat are energy per unit mass (joules per kilogram in SI units).

Condensable substances play a central role in the atmospheres of many planets and satellites. On Earth, it is water that condenses, both into liquid water and ice. On Mars, CO_2 condenses into dry ice in clouds and in the form of frost at the surface. On Jupiter and Saturn, not only water but ammonia (NH_3) and a number of other substances condense. The thick clouds of Venus are composed of condensed sulfuric acid. On Titan it is methane, and on Neptune's moon Triton nitrogen itself condenses. Table 2.1 lists the latent heats for the liquid–vapor (evaporation), liquid–solid (fusion) and solid–vapor (sublimation) phase transitions for a number of common constituents of planetary atmospheres. Water has an unusually large latent heat; the condensation of 1 kg of water vapor into ice releases nearly five times as much energy as the condensation of 1 kg of carbon dioxide gas into dry ice. This is why the relatively small amount of water vapor in Earth's present atmosphere can nonetheless have a great effect on atmospheric structure and dynamics. Ammonia also has an unusually large latent heat, though not so much as water. In both cases, the anomalous latent heat arises from the considerable energy needed to break hydrogen bonds in the condensed phase.

Like most thermodynamic properties, latent heat varies somewhat with temperature. For example, the latent heat of vaporization of water is $2.5 \cdot 10^6$ J/kg at $0\,°$C, but only $2.26 \cdot 10^6$ J/kg at $100\,°$C. For precise calculations, the variation of latent heat must be taken into account, but nonetheless for many purposes it will be sufficient to assume latent heat to be constant over fairly broad temperature ranges.

The three main phases of interest are solid, liquid, and gas (also called vapor), though other phases can be important in exotic circumstances. There is generally a *triple point* in temperature–pressure space where all three phases can co-exist. Above the triple point temperature, the substance undergoes a vapor–liquid phase transition as temperature is decreased or pressure is increased; below the triple point temperature, vapor condenses directly into solid, once thermodynamic equilibrium has been attained. For water, the triple point occurs at a temperature of 273.14 K and pressure of 6.11 mb (see Table 2.1 for other gases). Generally, the triple point temperature can also be taken as an approximation to the "freezing point" – the temperature at which a liquid becomes solid – because the freezing temperature varies only weakly with pressure until very large pressures are reached. Though we will generally take the freezing point to be identical to the triple point in our discussions, the effect of pressure on freezing of liquid can nonetheless be of great importance at the base of glaciers and in the interior of icy planets or moons, and perhaps also in very dense, cold atmospheres.

Typically, the solid phase is more dense than the liquid phase, but water again is exceptional. Water ice floats on liquid water, whereas carbon dioxide ice would sink in an ocean of liquid carbon dioxide, and methane ice would sink in a methane lake on Titan. This has profound consequences for the climates of planets with a water ocean such as Earth has, since ice formed in winter remains near the surface where it can be more readily melted when summer arrives.

Exercise 2.8 Per square meter, how many joules of energy would be required to evaporate a puddle of methane on Titan, having a depth of 20 m?

Atmospheres can transport energy from one place to another by heating an air parcel by an amount δT, moving the parcel vertically or horizontally, and then cooling it down to its original temperature. This process moves an amount of heat $c_p \delta T$ per unit mass of the parcel. Latent heat provides an alternate way to transport energy, since energy can be used to evaporate liquid into an air parcel until its mixing ratio increases by δr, moving it, and then condensing the substance until the mixing ratio returns to its original value. This process transports an amount of heat $L\delta r$ per unit mass of the planet's incondensable air, and can be much more effective at transporting heat than inducing temperature fluctuations, especially when the latent heat is large. "Ordinary" heat – the kind that feels hot when you touch it, and that is stored in the form of the temperature increase of a substance – is known in atmospheric circles as "sensible" heat.

All gases are condensable at low enough temperatures or high enough pressures. On Earth (in the present climate) CO_2 is not a condensable substance, but on Mars it is. The ability of a gas to condense is characterized by the *saturation vapor pressure*, p_{sat} of that gas, which may be a function of any number of thermodynamic variables. When the partial pressure p_A of gas A is below $p_{sat,A}$, more of the gas can be added, raising the partial pressure, without causing condensation. However, once the partial pressure reaches $p_{sat,A}$, any further addition of A will condense out. The state $p_A = p_{sat,A}$ is referred to as "saturated" with regard to substance A. Each condensed state (e.g. liquid or solid) will have its own distinct saturation vapor pressure. Rather remarkably, for a mixture of perfect gases, the saturation vapor pressure of each component is independent of the presence of the other gases. Water vapor mixed with 1000 mb worth of dry air at a temperature of 300 K will condense when it reaches a partial pressure of 38 mb; a box of pure water vapor at 300 K condenses at precisely the same 38 mb. If a substance "A" has partial pressure p_A that is below the saturation vapor pressure, it is said to be "subsaturated," or "unsaturated." The degree of subsaturation is measured by the *saturation ratio* $p_A/p_{sat,A}$, which is often stated as a percent. Applied to water vapor, this ratio is called the *relative humidity*, and one often speaks of the relative humidity of other substances, e.g. "methane relative humidity" instead of saturation ratios. Note that the relative humidity is also equal to the ratio of the *mixing ratio* of the substance A in a given mixture to the *mixing ratio* the substance would have if the substance were saturated. This is different from the ratio of *specific humidity* to *saturation specific humidity*, or the ratio of *molar concentration* to *saturation molar concentration* except when the mixing ratio is small.

It is intuitively plausible that the saturation vapor pressure should increase with increasing temperature, as molecules move faster at higher temperatures, making it harder for them to stick together to form a condensate. The temperature dependence of saturation vapor pressure is expressed by a remarkable thermodynamic relation known as the *Clausius–Clapeyron equation.* It is derived from very general thermodynamic principles, via a detailed accounting of the work done in an reversible expansion–contraction cycle crossing the condensation threshold, and requires neither approximation nor detailed knowledge of the nature of the substance condensing. The relation reads

$$\frac{dp_{sat}}{dT} = \frac{1}{T}\frac{L}{\rho_v^{-1} - \rho_c^{-1}} \tag{2.24}$$

where ρ_v is the density of the less condensed phase, ρ_c is the density of the more condensed phase, and L is the latent heat associated with the transformation to the more condensed phase. For vapor to liquid or solid transitions, $\rho_c \gg \rho_v$, enabling one to ignore the second term in the denominator of Eq. (2.24). Further, upon substituting for density from the perfect gas law, one obtains the simplified form

$$\frac{dp_{sat}}{dT} = \frac{L}{R_A T^2} p_{sat} \tag{2.25}$$

where R_A is the gas constant for the substance which is condensing. If we make the approximation that L is constant, then Eq. (2.24) can be integrated analytically, resulting in

$$p_{sat}(T) = p_{sat}(T_0)e^{-\frac{L}{R_A}\left(\frac{1}{T} - \frac{1}{T_0}\right)} \tag{2.26}$$

where T_0 is some reference temperature. This equation shows that saturation water vapor content is very sensitive to temperature, decaying rapidly to zero as temperature is reduced and increasing rapidly as temperature is increased. The rate at which the change occurs is determined by the characteristic temperature $\frac{L}{R_A}$ appearing in the exponential. For the transition of water vapor to liquid, it has the value 5420 K at temperatures near 300 K. For CO_2 gas to dry ice, it is 3138 K, and for methane gas to liquid methane it is 1031 K. Equation (2.26) seems to imply that the p_{sat} asymptotes to a constant value when $T \gg L/R_A$. This is a spurious limit, though, since the assumption of constant L invariably breaks down over such large temperature ranges. In fact, L typically approaches zero at some *critical*

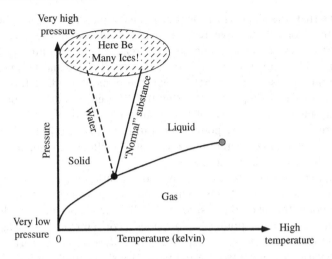

Figure 2.4 The general form of a phase diagram showing the regions of temperature-pressure space where a substance exists in solid, liquid, or gaseous forms. The triple point is marked with a black circle while the critical point is marked with a gray circle. The solid–liquid phase boundary for a "normal" substance (whose solid phase is denser than its liquid phase) is shown as a solid curve, whereas the phase boundary for water (ice less dense than liquid) is shown as a dashed curve. The critical point pressure is typically several orders of magnitude above the triple point pressure, while the critical point temperature is generally only a factor of two or three above the triple point temperature. Therefore, the pressure axis on this diagram should be thought of as logarithmic, while the temperature axis should be thought of as linear. This choice of axes also reflects the fact that the pressure must typically be changed by an order of magnitude or more to cause a significant change in the temperature of the solid–liquid phase transition.

temperature, where the distinction between the two phases disappears. For water vapor, this *critical point* occurs at a temperature and pressure of 647.1 K and 221 bars. For carbon dioxide, the critical point occurs for the vapor-liquid transition, at 304.2 K and 73.825 bars. Critical points for other atmospheric gases are shown in Table 2.1. At high pressures, the solid-liquid phase boundary does not typically terminate in a critical point, but instead gives way to a bewildering variety of distinct solid phases distinguished primarily by crystal structure.

Exercise 2.9 Show that the slope $d \ln p_{sat}/dT$ becomes infinite as $T \to 0$. Show that it decreases monotonically with T provided the latent heat decreases or stays constant as T increases. Show that the curve $p_{sat}(T)$ is infinitely flat near $T = 0$, in the sense that all the derivatives $d^n p_{sat}/dT^n$ vanish there. In Fig. 2.4 why is the curvature of the phase boundary sketched the way it is at low temperature?

Figure 2.4 summarizes the features of a typical phase diagram. Over ranges of a few bars of pressure, the solid-liquid boundary can be considered nearly vertical. In fact the exact form of the Clausius–Clapeyron relation (Eq. (2.24)) tells us why the boundary is nearly vertical and how it deviates from verticality. Because the difference in density between solid and liquid is typically quite small while the latent heat of fusion is comparatively large,

Eq. (2.24) implies that the slope dp/dT is very large (i.e. nearly vertical). The equation also tells us that in the "normal" case where ice is denser than liquid, the phase boundary tilts to the right, and so the freezing temperature increases with pressure; at fixed pressure, one can cause a cold liquid to freeze by squeezing it. The unusual lightness of water ice relative to the liquid phase implies that instead the phase boundary tilts to the left; one can melt solid ice by squeezing it. Substituting the difference in density between water ice and liquid water, and the latent heat of fusion, into Eq. (2.24), we estimate that 100 bars of pressure decreases the freezing point temperature by about 0.74 K. This is roughly the pressure caused by about a kilometer of ice on Earth. The effect is small, but can nonetheless be significant at the base of thick glaciers.

Below the triple point temperature, the favored transition is gas–solid, and so the appropriate latent heat to use in the Clausius–Clapeyron relation is the latent heat of sublimation. Above the triple point, the favored transition is gas–liquid, whence one should use the latent heat of vaporization. The triple point (T, p) provides a convenient base for use with the simplified Clausius–Clapeyron solution in Eq. (2.26), or indeed for a numerical integration of the relation with variable L. Results for water vapor are shown in Fig. 2.5. These results were computed using the constant-L approximation for sublimation and vaporization, but in fact a plot of the empirical results on a logarithmic plot of this type would not be distinguishable from the curves shown. The more exact result does differ from the constant-L idealization by a few percent, which can be important in some applications. Be that as it may, the figure reveals the extreme sensitivity of vapor pressure to temperature. The vapor pressure ranges from about 0.1 pascals at 200 K (the tropical tropopause temperature) to 35 mb at a typical

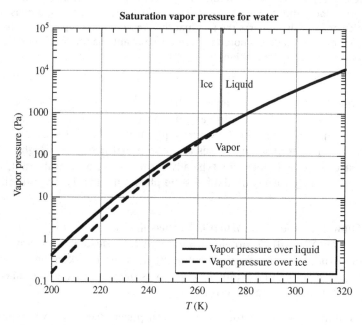

Figure 2.5 Saturation vapor pressure for water, based on the constant-L form of the Clausius–Clapeyron relation. Curves are shown for vapor pressure based on the latent heat of vaporization, and (below freezing) for latent heat of sublimation. The latter is the appropriate curve for sub-freezing temperatures.

tropical surface temperature of 300 K, rising further to 100 mb at 320 K. Over this span of temperatures, water ranges from a trace gas to a major constituent; at temperatures much above 320 K, it rapidly becomes the dominant constituent of the atmosphere. Note also that the distinction between the ice and liquid phase transitions has a marked effect on the vapor pressure. Because the latent heat of sublimation is larger than the latent heat of vaporization, the vapor pressure over ice is lower than the vapor pressure over liquid would be, at sub-freezing temperatures. At 200 K, the ratio is nearly a factor of three.

Exercise 2.10 Let's consider once more the case of the airliner cruising at an altitude of 300 mb, discussed in an earlier exercise. Suppose that the ambient air at flight level has 100% relative humidity. What is the relative humidity once the air has been brought into the cabin, compressed to 1000 mb, and chilled to a room temperature of 290 K?

Once the saturation vapor pressure is known, one can compute the molar or mass mixing ratios with respect to the background non-condensable gas, if any, just as for any other pair of gases. The saturation vapor pressure is used in this calculation just like any other partial pressure. For example, the molar mixing ratio is just p_{sat}/p_a, if p_a is the partial pressure of the non-condensable background. Note that, while the saturation vapor pressure is independent of the pressure of the gas with which the condensable substance is mixed, the saturation mixing ratio is not.

Exercise 2.11 What is the saturation molar mixing ratio of water vapor in air at the ground in tropical conditions (1000 mb and 300 K)? What is the mass mixing ratio? What is the mass-specific humidity? What is the molar mixing ratio (in ppm) of water vapor in air at the tropical tropopause (100 mb and 200 K)?

For liquids, one defines the *boiling point* as the temperature at which the vapor pressure equals the ambient pressure in the liquid. For an unconfined liquid at sea level on the modern Earth, this pressure is 1 bar at the liquid surface, which defines the sea-level boiling point. The significance of the boiling point is that when the temperature exceeds this value, the vapor pressure is great enough that bubbles of the vapor phase can form in the interior of the liquid, allowing the liquid–vapor phase transition to occur over a much greater surface area than is the case if one had to rely just on evaporation into a subsaturated environment at the liquid surface.

2.7 THE MOIST ADIABAT

When air is lifted, it cools by adiabatic expansion, and if it gets cold enough that one of the components of the atmosphere begins to condense, latent heat is released. This makes the lifted air parcel warmer than the dry adiabat would predict. Less commonly, condensation may occur as a result of subsidence and compression, since the increase of partial pressure of one of the compressed atmospheric components may overwhelm the increase in saturation vapor pressure resulting from adiabatic warming. Whichever direction leads to condensation, if we assume further that the condensation is efficient enough that it keeps the system at saturation, the resulting temperature profile will be referred to as the *moist adiabat*, regardless of whether the condensing substance is water vapor (as on Earth) or something else (CO_2 on Mars or methane on Titan). We now proceed to make this quantitative.

2.7.1 One-component condensable atmospheres

The simplest case to consider is that of a single-component atmosphere, which can attain cold enough temperatures to reach saturation and condense. This case is relevant to present Mars, which has an almost pure CO_2 atmosphere that can condense in the cold winter hemisphere and at upper levels at any time of year. A pure CO_2 atmosphere with a surface pressure on the order of two or three bars is a commonly used model of the atmosphere of Early Mars, though the true atmospheric composition in that instance is largely a matter of speculation. Another important application of a single-component condensable atmosphere is the pure steam (water vapor) atmosphere, which occurs when a planet with an ocean gets warm enough that the mass of water which evaporates into the atmosphere dominates the other gases that may be present. This case figures prominently in the *runaway greenhouse* effect that will be studied in Chapter 4.

For a single-component atmosphere, the partial pressure of the condensable substance is in fact the total atmospheric pressure. Therefore, at saturation, the pressure is related to the temperature by the Clausius–Clapeyron relation. To find the saturated moist adiabat, we simply solve for T in terms of p_{sat} in the Clausius–Clapeyron relation, and recall that $p = p_{sat}$ because we are assuming the atmosphere to be saturated everywhere. Using the simplified form of Clausius–Clapeyron given in Eq. (2.26), the saturated moist adiabat is thus

$$T(p) = \frac{T_0}{1 - \frac{RT_0}{L} \ln \frac{p}{p_{sat}(T_0)}} \qquad (2.27)$$

where R is the gas constant for the substance making up the atmosphere. This equation is really just an alternate form of Clausius–Clapeyron. It can be thought of as a formula for the "dew-point" or "frost-point" temperature corresponding to pressure p: given a box of gas at fixed pressure p, condensation will occur when the temperature is made lower than the temperature $T(p)$ given by Eq. (2.27).

Without loss of generality, we may suppose that T_0 is taken to be the surface temperature, so that $p_{sat}(T_0)$ is the surface pressure p_s. Since the logarithm is negative, the temperature decreases with altitude (recalling that lower pressure corresponds to higher altitude). Further, the factor multiplying the logarithm is the ratio of the surface temperature to the characteristic temperature L/R. Since the characteristic temperature is large, the prefactor is small, and as a result the temperature of saturated adiabat for a one-component atmosphere varies very little over a great range of pressures. For example, in the case of the CO_2 vapor–ice transition, an atmospheric surface pressure of 7 mb (similar to that of present Mars) would be in equilibrium with a surface dry-ice glacier at a temperature of 149 K; at 0.07 mb – one one-hundredth of the surface pressure – the temperature on the saturated adiabat would only fall to 122 K.

Exercise 2.12 In the above example, what would the temperature aloft have been if there were no condensation and the parcel were lifted along the dry adiabat?

The criterion determining whether condensation occurs on ascent or descent for an arbitrary one-component atmosphere is derived in Problem 2.50. Note that for a single-component saturated atmosphere, the supposition that the atmosphere is saturated is sufficient to determine the temperature profile, regardless of the means by which the saturation is maintained. Thus, it does not actually matter to the result whether the saturation is

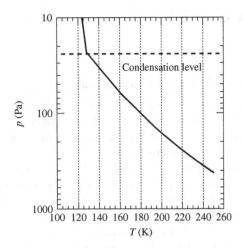

Figure 2.6 The adiabatic profile for a pure CO_2 atmosphere with a surface pressure of 450 Pa (4.5 mb) and a surface temperature of 255 K. The conditions are similar to those encountered on the warmer portions of present-day Mars.

maintained by condensation due to vertical displacement of air parcels, or some other physical mechanism such as radiative cooling of the atmosphere to the point that condensation occurs.

Unless there is a reservoir of condensate at the surface to maintain saturation, it would be rare for an atmosphere to be saturated all the way to the ground. Suppose now that a one-component atmosphere has warm enough surface temperature that the surface pressure is lower than the saturation vapor pressure computed at the surface temperature. In this case, when a parcel is lifted by convection, its temperature will follow the *dry* or *non-condensing* adiabat, until the temperature falls so much that the gas becomes saturated. The level at which this occurs is called the *lifted condensation level*. Above the lifted condensation level, ascent causes condensation and the parcel follows the saturated adiabat. Since the temperature curve along the saturated adiabat falls with altitude so much less steeply than the dry adiabat, it is very easy for the two curves to intersect provided the surface temperature is not exceedingly large. An example for present summer Martian conditions (specifically, like the warmest sounding in Fig. 2.2) is shown in Fig. 2.6. A comparison with the Martian profiles in Fig. 2.2 indicates that something interesting is going on in the Martian atmosphere. For the warm sounding, whose surface temperature is close to 255 K, the entire atmosphere aloft is considerably warmer than the adiabat, and the temperature nowhere comes close to the condensation threshold, though the very lowest portion of the observed atmosphere, below the 200 Pa level, does follow the dry adiabat quite closely. Clearly, something we haven't taken into account is warming up the atmosphere. A likely candidate for the missing piece is the absorption of solar energy by CO_2 and dust.

Although results like Fig. 2.6 show a region of weak temperature dependence aloft which bears a superficial resemblance to the stratosphere seen in Earth soundings (and also at the top of the Venus, Jupiter, and Titan soundings), one should not jump to the conclusion that the stratosphere is caused by condensation. This is not generally the case, and there are other explanations for the upper atmospheric temperature structure, which will be taken up in the next few chapters.

2.7.2 Mixtures of condensable with non-condensable gases

As a final step up on the ladder of generality, let us consider a mixture of a condensable substance with a substance that doesn't condense under the range of temperatures encountered in the atmosphere under consideration. This might be a mixture of condensable methane on Titan with non-condensable nitrogen, or condensable carbon dioxide on Early Mars with non-condensable nitrogen, or water vapor on Earth with a non-condensable mixture of oxygen and nitrogen. Whatever the substance, we distinguish the properties of the condensable substance with the subscript "c," and those of the non-condensable substance by the subscript "a" (for "air"). We now need to do the energy budget for a parcel of the mixture, assuming that it is initially saturated, and that it is displaced in such a way that condensation (releasing latent heat) must occur in order to keep the parcel from becoming supersaturated. We further introduce the assumption that essentially all of the condensate is immediately removed from the system, so that the heat storage in whatever mass of condensate is left in suspension may be neglected. This is a reasonable approximation for water or ice clouds on Earth, but even in that case the slight effect of the mass of retained condensate on buoyancy can be significant in some circumstances. In other planetary atmospheres the effect of retained condensate could be of greater importance. The temperature profile obtained by assuming condensate is removed from the system is called a *pseudoadiabat*, because the process is not truly reversible. One cannot return to the original saturated state, because the condensate is lost. At the opposite extreme, if all condensate is retained, it can be re-evaporated when the parcel is compressed, allowing for true reversibility.

Let the partial pressure, density, molecular weight, gas constant, and specific heat of the non-condensable substance be p_a, ρ_a, M_a, R_a, and c_{pa}, and similarly for the condensable substance. Further, let L be the latent heat of the phase transition between the vapor and condensed phase of the condensable substance, and let $p_{c,sat}(T)$ be the saturation vapor pressure of this substance, as determined by the Clausius–Clapeyron relation. The assumption of saturation amounts to saying that $p_c = p_{c,sat}(T)$; if the parcel were not at saturation, there would be no condensation and we could simply use the dry adiabat based on a non-condensing mixture of substance "a" and "c."

Now consider a parcel consisting of a mass m_a of non-condensable gas with an initial mass m_c of condensable gas occupying a volume V. Suppose that the temperature is changed by an amount dT, the partial pressure of non-condensable gas is changed by an amount dp_a, and the partial pressure of the condensable gas is changed by an amount dp_c. The mechanical work term for the mixture is $Vd(p_a + p_c)$, but by the definition of density $V = m_a/\rho_a = m_c/\rho_c$. Thus, the total heat budget of the parcel can be written in the form

$$(m_a + m_c)\delta Q = m_a c_{pa} dT - \frac{m_a}{\rho_a} dp_a + m_c c_{pc} dT - \frac{m_c}{\rho_c} dp_c + L dm_c \qquad (2.28)$$

where m_c is the mass of condensable substance *in the vapor phase*. Thus, when some vapor is condensed out, dm_c is negative, yielding a negative contribution to δQ when L is positive. Since condensation releases latent heat, this sign may seem counter-intuitive. Recall, however, that condensation puts the system in a *lower energy state*, which is why there is heat available to be "released." If the system is energetically closed ($\delta Q = 0$), then the negative contribution of the latent heat term is offset by an increase in the remaining terms – e.g. an increase in temperature.

Exercise 2.13 Show that when there is no condensation ($dm_c = 0$) the adiabat $T(p)$ obtained by setting $\delta Q = 0$ has the same form as the usual dry adiabat, but with R and c_p taking on the appropriate mean values for the mixture.

The heat budget does not contain any term corresponding to heat storage in the condensed phase, since it is assumed that all condensate disappears from the parcel by precipitation. The usual way to change dp_a would be by lifting, causing expansion and reduction of pressure. Now, we divide by $m_c T$, make use of the perfect gas law to substitute for ρ_a and ρ_c, and make use of the fact that $m_c/m_a = (M_c/M_a)(p_c/p_a)$, since m_c/m_a is just the mass mixing ratio, denoted henceforth by r_c. This yields

$$(1 + r_c)\frac{\delta Q}{T} = \left[c_{pa}\frac{dT}{T} - R_a\frac{dp_a}{p_a} \right] + \left[c_{pc} r_c \frac{dT}{T} - r_c R_c \frac{dp_c}{p_c} \right] + \frac{L}{T} dr_c. \tag{2.29}$$

The first two bracketed groups of terms on the right hand side can be recognized as the contribution of the two substances to the non-condensing entropy of the mixture, weighted according to the relative abundance of each species. If there is no condensation, the mixing ratio is conserved as the parcel is displaced to a new pressure, $dr_c = 0$, and the expression reduces to the equivalent of Eq. (2.9), leading to the dry adiabat for a mixture. At this point, we introduce the saturation assumption, which actually consists of two parts: First, we assume that the air parcel is initially saturated, so that before being displaced, $p_c = p_{c,sat}(T)$ and $r_c = r_{sat} = \epsilon p_{c,sat}(T)/p_a$, where ϵ is the ratio of molecular weights M_c/M_a and $p_{c,sat}(T)$ is determined by the Clausius–Clapeyron relation. Second, we assume that a displacement conserving r_c would cause supersaturation, so that condensation would occur and bring the partial pressure p_c back to the saturation vapor pressure corresponding to the new value of T. Usually, this would occur as a result of ascent and cooling, since cooling strongly decreases the saturation vapor pressure. Typically (though not inevitably), the effect of compressional warming on saturation vapor pressure dominates the effect of increasing partial pressure, so that subsidence of initially saturated air follows the dry adiabat.

Assuming that the displacement causes condensation, we may replace p_c by $p_{c,sat}(T)$ and r_c by r_{sat} everywhere in Eq. (2.29). Next, we use Clausius–Clapeyron to re-write $dp_{c,sat}$, observing that

$$\frac{dp_{c,sat}}{p_{c,sat}} = d\ln p_{c,sat} = \frac{d\ln p_{c,sat}}{dT} dT \tag{2.30}$$

and

$$dr_{sat} = \epsilon d\frac{p_{c,sat}}{p_a} = \epsilon \frac{p_c}{p_a} d\ln\frac{p_c}{p_a} = r_{sat} \cdot (d\ln p_{c,sat} - d\ln p_a). \tag{2.31}$$

Upon substituting into Eq. (2.29) and collecting terms in $d\ln T$ and $d\ln p_a$ we find

$$(1 + r_{sat})\frac{\delta Q}{T} = \left(c_{pa} + \left(c_{pc} + \left(\frac{L}{R_c T} - 1 \right)\frac{L}{T} \right) r_{sat} \right) d\ln T - \left(1 + \frac{L}{R_a T} r_{sat} \right) R_a d\ln p_a.$$

$$\tag{2.32}$$

To obtain the adiabat, we set $\delta Q = 0$, which leads to the following differential equation defining $\ln T$ as a function of $\ln p_a$:

$$\frac{d \ln T}{d \ln p_a} = \frac{R_a}{c_{pa}} \frac{1 + \frac{L}{R_a T} r_{sat}}{1 + \left(\frac{c_{pc}}{c_{pa}} + \left(\frac{L}{R_c T} - 1 \right) \frac{L}{c_{pa} T} \right) r_{sat}}. \tag{2.33}$$

Note that this expression reduces to the dry adiabat, as it should, when $r_{sat} \to 0$.

Exercise 2.14 What would the slope $d \ln T / d \ln p_a$ be for a non-condensing mixture of the two gases? (*Hint:* $\ln p = \ln p_a + const$ in this case.) Why doesn't Eq. (2.33) reduce to this value as $L \to 0$? (*Hint:* Think about the way Clausius–Clapeyron has been used in deriving the moist adiabat, and what it implies for variations of p_c.)

A displaced parcel of atmosphere will follow the moist adiabat if the condensable substance condenses in the course of the displacement, but it will follow the dry adiabat if the displacement causes the condensable substance to become subsaturated. Does condensation occur on ascent (which lowers the total pressure) or descent (which raises the total pressure)? To answer this question, we must compare the moist adiabatic slope $d \ln T / d \ln p$ computed from Eq. (2.33) with the dry adiabatic slope R/c_p, with R and c_p computed as the appropriate weighted average for the mixture. When the moist adiabatic slope is lower than R/c_p, then lifting a parcel adiabatically creates enough cooling that the parcel becomes supersaturated, and condensation occurs. Conversely, if the moist adiabatic slope is less than R/c_p, condensation occurs on descent instead.

The only problem with applying this criterion is that Eq. (2.33) gives us $d \ln T / d \ln p_a$ whereas we need $d \ln T / d \ln p$. Here, we will restrict attention to the dilute case, where there is little enough condensable substance present that $p \approx p_a$. In this case $d \ln T / d \ln p \approx d \ln T / d \ln p_a$ and $R/c_p \approx R_a/c_{p,a}$. The opposite limit of the condensable-dominated atmosphere is explored in Problem 2.50, and the general case in Problem 2.51.

An examination of the properties of gases indicates that c_{pc}/c_{pa} is typically of order unity, whereas $L/(R_c T)$ is typically very large, so long as the temperature is not exceedingly great. If one drops the smaller terms from the denominator of Eq. (2.33), one finds that the temperature gradient along the moist adiabat is weaker than that along the dry adiabat provided $\epsilon L/c_{pa} T > 1$, which is typically the case. Here, condensation occurs on ascent and warms the ascending parcel through the release of latent heat. This behavior can fail when the latent heat is weak or the non-condensable specific heat is very large, whereupon the heat added by condensation has little effect on temperature. It is in this regime that condensation happens on *descent* rather than ascent. Given the thermodynamic properties of common atmospheric constituents, it is not an easy regime to get into. Even the case of H_2O condensation in an H_2 Jovian or Saturnian atmosphere yields condensation on ascent, despite the high specific heat of H_2. In that case, the high specific heat of H_2 is canceled out by the low value of ϵ, which is typical behavior since specific heat tends to scale inversely with molecular weight – substances with low molecular weight have more degrees of freedom per kilogram. The quantity $\epsilon L/c_{pa} T$ decreases with temperature, especially since L approaches zero as the critical point is approached, and this suggests that condensation might occur during descent at very high temperatures. A proper evaluation of the possibility would require taking into account non-ideal gas effects, the temperature dependence of specific heat, the neglected condensate density term in Clausius–Clapeyron, and probably also non-dilute effects.

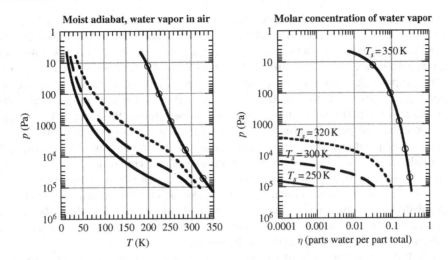

Figure 2.7 The moist adiabat for saturated water vapor mixed with Earth air having a partial pressure of 1 bar at the surface. Results are shown for various values of surface temperature, ranging from 250 K to 350 K. The left panel shows the temperature profile, while the right shows the profile of molar concentration of water vapor. A concentration value of 0.1 would mean that 1 molecule in 10 of the atmosphere is water vapor.

Everything on the right hand side of Eq. (2.33) is either a thermodynamic constant, or can be computed in terms of $\ln T$ and $\ln p_a$. Therefore, the equation defines a first order ordinary differential equation which can be integrated (usually numerically) to obtain T as a function of p_a. Usually one wants the temperature as a function of total pressure, rather than partial pressure of the non-condensable substance. This is no problem. Once $T(p_a)$ is known, the corresponding total pressure at the same point is obtained by computing $p = p_a + p_{c,sat}(T(p_a))$. To make a plot, or a table, one treats the problem parametrically: computing both T and p as functions of p_a. When the condensable substance is dilute, then $p_{c,sat} << p_a$, and $p \approx p_a$, so Eq. (2.33) gives the desired result directly.

Figure 2.7 shows a family of solutions to Eq. (2.33), for the case of water vapor in Earth air. When the surface temperature T_s is 250 K, there is so little moisture in the atmosphere that the profile looks like the dry adiabat right to the ground. As temperature is increased, a region of weak gradients appears near the ground, representing the effect of latent heat on temperature. This layer gets progressively deeper as temperature increases and the moisture content of the atmosphere increases. When the surface temperature is 350 K, so much moisture has entered the atmosphere that the surface pressure has actually increased to over 1300 mb. Moreover, the moisture-dominated region extends all the way to 10 Pa (0.1 mb), and even at 100 Pa (1 mb) the atmosphere is 10% water by volume. Thus, for moderate surface temperatures, there is little water high up in the atmosphere. When the surface temperature approaches or exceeds 350 K, though, the "cold trap" is lost, and a great deal of water is found aloft, where it is exposed to the destructive ultraviolet light of the Sun and the possibility of thermal escape to space. In subsequent chapters, it will be seen that this phenomenon plays a major role in the life cycle of planets, and probably accounts for the present hot, dry state of Venus.

The mixing ratio of the condensable component varies along the moist adiabat, so the reader may wonder whether these variations can lead to compositionally induced buoyancy along the lines discussed in Section 2.4. Suppose a parcel starts on the moist adiabat at pressure p, with temperature $T(p)$, and that the parcel is saturated. For the sake of definiteness, let's suppose also that condensation occurs on ascent, so ascending displacements are saturated and remain on the moist adiabat. Since there is a unique moist adiabat going through the starting point, and since saturation also determines the composition uniquely once temperature and pressure are known, then when the parcel is lifted to pressure p_1 it has the same temperature, composition, and density as the ambient air there. Composition does not induce buoyancy, because the variations of R/c_p are constrained by the saturation assumption, and have been taken into account in the computation of the moist adiabat. In this sense, the moist adiabat is neutrally stable against condensing ascent. On the other hand, a subsiding air parcel will become subsaturated, and descending along the dry adiabat will become warmer than the surrounding air, causing it to be positively buoyant. In this sense, the moist adiabat is stable against subsidence.[4]

Saturated air can go up without expenditure of energy, but it takes energy to push it down. However, when an air parcel is initially *subsaturated* and must be lifted some distance before it becomes saturated and follows the moist adiabat, then the full range of issues discussed in Section 2.4 come into play. In that case, the compositional variations in the subsaturated layer can either favor or inhibit the triggering of vertical mixing. On a related note, suppose that the temperature profile $T(p)$ happens to follow the moist adiabat, but that the air is unsaturated; this is in fact the case over most of the Earth's tropics, for dynamical reasons that are beyond the scope of this book. In that case, if a saturated air parcel from the ocean surface is lifted through the dry surroundings, it will follow the moist adiabat but because its composition will be different from the unsaturated surroundings it will in fact be positively buoyant if the condensable has lower molecular weight than the non-condensable background, as is the case for water vapor in Earth air. If the condensable has higher molecular weight, the parcel will be negatively buoyant and vertical mixing will be choked off. Real atmospheres consist of a mix of saturated and unsaturated regions, and the representation of vertical mixing in these cases poses a considerable challenge. Compositional effects on Earth at its present temperature are slight, but for atmospheres in which the condensable is non-dilute, these effects become more and more important. With regard to the modeling of convection, this represents largely unexplored territory. It is probably important for the case of methane on Titan, but would also be important on Earthlike planets having temperatures some tens of degrees or more warmer than the present Earth.

The mass of retained condensate – cloud mass – can also affect the buoyancy of lifted or subsiding parcels, since condensate alters the density of an air parcel. In such cases one also needs to consider cooling due to evaporation of condensate when the air parcel subsides, warms, and becomes subsaturated. In quiescent conditions retained condensate mass is expected to be a small fraction of the total mass of an air parcel, since a large condensate loading usually leads to coalescence of cloud droplets and subsequent removal by precipitation. The effects of condensate loading are small, but significant, to convection in

---

[4] Compositional effects will alter the buoyancy of subsiding parcels, and in principle could make subsiding parcels negatively buoyant in extreme cases. For typical atmospheric gases the compositional effect is too small to overwhelm the positive buoyancy due to compressional heating, but if one could find a situation where composition rendered subsiding parcels unstable to descent, the resulting convection would be very interesting indeed to study.

the present Earth's atmosphere, but the possibility should be kept in mind that condensate loading may play a more prominent role in planetary atmospheres that have not been as extensively studied as Earth's.

2.7.3 Moist static energy

When the condensable substance is dilute, it is easy to define a *moist static energy* which is a generalization of the dry static energy defined in Eq. (2.23). This is accomplished by multiplying Eq. (2.29) by T to obtain the heat budget, dropping the terms proportional to r_c (which are small in the dilute limit), and making use of $R_a T/p_a = 1/\rho_a \approx 1/\rho$. The hydrostatic relation is used to re-write the pressure work term in precisely the same way as was done for dry static energy. Then, if we further assume that the temperature range of interest is small enough that L may be regarded as constant, we obtain

$$\delta Q = d(c_{pa}T + gz + Lr_c). \tag{2.34}$$

Note that this relation does not assume that the condensable mixing ratio r_c is saturated. For adiabatic flow in the dilute case, this moist static energy will be conserved following an air parcel whether or not condensation occurs. If energy is added to or taken away from the parcel, then the change in moist static energy is equal to the net amount of energy added or taken away, per unit mass of the parcel.

When the condensate is non-dilute, things are a bit more complicated. In this case significant amounts of mass can be lost from the atmosphere in the course of condensation, and in essence the precipitation of condensate can take away significant amounts of energy with it. In order to deal with the heat storage in condensate, one must make use of the specific heat of the condensed phase, which we will refer to as $c_{pc\ell}$ (regardless of whether the condensate is liquid or solid); the behavior of this specific heat is inextricably linked to the changes in latent heat with temperature through the thermodynamic relation

$$\frac{dL}{dT} = c_{pc} - c_{pc\ell} \tag{2.35}$$

which is valid when the condensate density is much greater than the vapor density. This relation is essentially a consequence of energy conservation. Given this link, we will take into account the change of latent heat with temperature in carrying out the following analysis.

First let's analyze the energy budget per unit mass of the non-condensable substance. We'll write the pressure work term in Eq. (2.28) in the alternate form $(m_a/\rho_a)dp$. On dividing the equation by m_a and using Eq. (2.35) and the hydrostatic relation we obtain

$$(1 + r_c)\delta Q = d[(c_{pa} + r_c c_{pc\ell})T + (1 + r_c)gz + L(T)r_c] - (c_{pc\ell}T + gz)dr_c. \tag{2.36}$$

The quantity in square brackets is thus identified as the moist static energy per unit mass of non-condensable substance; it will be denoted by the symbol $\mathfrak{M}$. The second term on the right hand side, involving dr_c, represents the sink of moist static energy due to the heat and potential energy carried away by the condensate.

This form of moist static energy can be inconvenient to use, because the $(1 + r_c)$ weighting on the left hand side makes it hard to do the energy budget of a column of air knowing only the net input of energy into the column. The expression also becomes inconvenient when the atmosphere becomes dominated by the condensable substance, leading to very large values of r_c. We can formulate the moist static energy per unit *total* mass of the gas by dividing Eq. (2.36) by $(1 + r_c)$. After carrying out a few basic manipulations, we find

$$\delta Q = d \left[\frac{\mathfrak{M}}{1 + r_c} \right] - \left(c_{pc\ell} T + gz - \frac{\mathfrak{M}}{1 + r_c} \right) d\ln(1 + r_c). \qquad (2.37)$$

The term $\mathfrak{M}/(1 + r_c)$ is the desired moist static energy per unit of total gaseous mass, and the second term is the corresponding sink due to precipitation.

Exercise 2.15 Re-write the expression $\mathfrak{M}/(1 + r_c)$ in terms of the mass specific concentration q. What is the form of this expression when $q \approx 1$? What happens to the precipitation sink term in this limit?

An alternate approach to dealing with moist static energy in the non-dilute case is to write an energy budget per unit total mass (condensate included) for an air parcel that retains its condensate. To allow for precipitation, one then deals explicitly with the energy loss occurring when some of the retained condensate is removed from the air parcel. This approach is especially useful in situations when the mass of retained condensate can be appreciable. It can be carried out using the moist entropy expressions derived in Emanuel (1994) (see Further Reading). Further modifications to the expression for moist static energy are required if both ice and liquid phases are present in the atmosphere, in order to account for the latent heat of the solid/liquid phase transition.

Because the change in moist static energy of an atmospheric column can be determined from the energy fluxes into the top of the column and the energy fluxes between the bottom of the column and the underlying surface, moist static energy provides a convenient basis for diagnostics of atmospheric energy flows. For the same reason, it provides a convenient basis for the formulation of simplified vertically averaged energy balance models (EBM) of climate. Such models are taken up briefly in Chapter 9. Moist static energy is also important to the formulation of simplified representations of buoyancy-driven vertical mixing in a statically unstable layer of the atmosphere. In such applications, when radiation or some other process creates a region of static instability, it is presumed that mixing will proceed until some layer of the atmosphere is reset to a state of neutral stability. The constraint that the moist static energy following mixing should be the same as that of the initial state (within a suitably chosen layer of atmosphere) plays a crucial role in determining what the temperature profile is following mixing.

2.8 WORKBOOK

Python tips: Many of the following problems require thermodynamic data such as specific heats, latent heats, etc. Most of the necessary data is given in the thermodynamic table for this chapter, and the rest can be found easily from standard print and online references. Python users can get the data directly from the `phys` courseware module, which defines gas objects for a number of common gases. Each gas object contains the data for the corresponding gas. For example `phys.N2.cp` is the specific heat (at standard temperature) for N_2. Type `help(phys)` and `help(gas)` for more documentation, or look directly at `phys.py`. The `phys` module also contains general constants such as the Boltzmann thermodynamic constant (`phys.k`) and many others, as well as a number of handy functions and routines which will be introduced below as needed. To use the `phys` module, type `import phys` from inside the Python interpreter.

2.8.1 Basic dry thermodynamics

Problem 2.1 A spherical spaceship with radius 10 meters is pressurized to an interior pressure of 1000 mb. The exterior pressure can be regarded as zero. What is the net force (in newtons) exerted on the walls of the spaceship?

Problem 2.2 A hollow metal sphere of radius 1 meter is cut in half and then fitted back together with an airtight rubber gasket. All the air is then removed from the inside using a vacuum pump. The external air pressure is 700 mb. What is the total force (in newtons) exerted by air on each hemisphere? What is the component of this force in the direction parallel to a diameter perpendicular to the plane along which the two hemispheres are joined? How much force would have to be exerted in order to pull the two hemispheres apart?

Problem 2.3 What is the density of Earth air at a temperature of 0 °C (centigrade) and pressure of 1000 mb? At the same pressure, what is the density of air in a hot air balloon in which the temperature has been increased to 50 °C? If the balloon is approximately spherical and has a radius of 3 meters, how much mass could it lift in Earth's gravity, if the ambient temperature is 0 °C? (*Recall:* Archimedes' law states that the buoyancy force is equal to the weight of the fluid displaced. The opposing force on the balloon is the force of gravity on the air in the balloon itself, which must be subtracted from the buoyancy force.)

Problem 2.4 This is an elementary computer problem which exercises the skills needed to write and use a function.

Write a function rho(p,T) which returns the density of a gas given the pressure and the temperature. Check your function by recomputing the density of air at 0 °C and 1000 mb. The value of the universal gas constant can be hard-wired into the function, but you will need to find some way to specify the molecular weight of the gas in question. This can be hard-wired into the function (requiring you to edit the function when you want to change gases) or made into an additional argument of the function (requiring you to remember to pass a value which you probably aren't changing too often). A better solution is to put the data in a shared space that can be accessed by all functions. The particular way one does this depends on which computer language one is using.

Python tips: Treat the molecular weight of the gas as a global, so that you can use the function for different gases by just re-defining the value of the global before calling the function.

Problem 2.5 A bicycle tire with a mass of 0.1 kg when empty, and a volume of 3 liters is pumped up with Earth air to a pressure of 4 bars. It is pumped up slowly, so that its temperature remains at the ambient air temperature of 290 K. What is the mass of the tire after it has been pumped up? On Earth, what does the tire *weigh* (in newtons)? What would the mass of the tire be if it were filled to the same pressure with CO_2 instead of air? With He?

Problem 2.6 Compute the density of CO_2 at the surface of Mars, where the pressure is 6 mb and the temperature is 220 K. Compute the density of N_2 at the surface of Titan, where the pressure is 1.5 bar and the temperature is 95 K. Compute the density of a pure CO_2 atmosphere at the surface of Venus, where the pressure is 92 bars and the temperature is 737 K.

Problem 2.7 Assuming the Earth's atmosphere to contain 20% O_2 by count of molecules and 80% N_2, what is the mass mixing ratio of oxygen? What is the mass-specific concentration?

Problem 2.8 The air in a room with dimensions 3 m by 20 m by 20 m has a CO_2 concentration of 300 ppmv. How many kilograms of *carbon* would have to be burned into the air (or exhaled by students) in order to double the CO_2 concentration in the room? If the O_2 concentration is initially 20% (molar), what is the concentration after the burning (or respiration) has taken place?

Problem 2.9 *Making the Earth's oxygen*
Photosynthesis performs a kind of solar-powered electrolysis, taking water (H_2O) and carbon dioxide (CO_2) and using them to make organic matter – where the carbon and hydrogen and some of the oxygen go – and releasing the rest of the oxygen in the process. For the purposes of this problem, you can consider an organic molecule to be CH_2O, though real organics are typically more complex. The Earth's atmosphere contains about 20% O_2 (molar), and the oxygen can be considered well-mixed out to a great altitude.

Consider a patch of ocean of some surface area A. How deep a layer of water needs to be converted by photosynthesis to create the oxygen in the air above that patch? How many moles of organic matter will be made in the process? How many kilograms?

Problem 2.10 *First Law for a non-ideal gas*
Beginning from the First Law in the form in Eq. (2.7), show that for a gas with a general equation of state $\rho(p, T)$ the First Law can be re-written

$$\delta Q = c_p dT - \frac{p}{\rho^2} \frac{\partial \rho}{\partial p}\Big|_T dp \tag{2.38}$$

and find an expression for c_p in terms of c_v. Show that when ρ is given by the ideal gas equation of state, your results reduce to the expressions derived in the text for the ideal case.

Problem 2.11 Suppose that sunlight heats a patch of the Earth's surface at a rate of 300 W/m^2, and that turbulence instantly distributes this heat over a column of boundary layer of depth 500 m. This means that 300 joules of energy go into a column with base of 1 square meter and height of 500 m each second. Assuming the boundary layer to be filled with Earth air with density 1 kg/m^3, at what rate does the temperature of the column increase, if there is no heat loss from the column? Re-do the calculation for the present Martian climate, with a pure CO_2 atmosphere having surface pressure of 6 mb, assuming the density to be constant over 500 m, the temperature to be constant at 240 K, and the solar radiation to be 150 W/m^2.

Problem 2.12 Use the Boltzmann constant to calculate the mean kinetic energy of a helium atom in a gas with a temperature of 300 K. What is the typical speed with which the atom moves? Recall that kinetic energy is $mv^2/2$, and that each direction of motion counts as a separate degree of freedom. (The mass of a helium atom is 6.642×10^{-27} kg.)

Problem 2.13 *Temperature-dependent specific heat*
This is a computer problem involving temperature-dependent specific heats. To do the problem, you will need to know the following programming techniques: (1) Defining and

Gas	A	B	C	D	E
N_2	931.857	293.529	−70.576	5.688	1.587
CO_2	568.122	1254.249	−765.713	180.645	−3.105
NH_3	1176.213	2927.717	−904.470	113.009	11.128

Table 2.2 Shomate coefficients for selected gases, valid from 300 K to 1300 K for CO_2 and NH_3, and 300 K to 6000 K for N_2.

evaluating a function; (2) tabulating a set of values and either plotting them or writing them out for plotting with a separate program; (3) summing up a set of values in a loop.

The *Shomate equation* is an empirical formula for the dependence of specific heat on temperature. It works well for a broad range of gases. The formula reads:

$$c_p = A + B * (T/1000) + C * (T/1000)^2 + D * (T/1000)^3 + E * (T/1000)^{-2} \qquad (2.39)$$

where T is the temperature in kelvin and $A, ..., E$ are gas-dependent constants. Some coefficient sets are given in Table 2.2.

Write a function to compute cp(T). Explore the behavior for the three gases, taking note of the degree to which the specific heat varies. Make a plot of the temperature dependence. From a physical standpoint, why do the specific heats increase with temperature? Measurements show that the specific heat of He is very nearly temperature-independent. Why?

Compute the energy it would take to raise the temperature of 1 kg of N_2 from 300 K to 700 K, in steps of 0.1 K (assuming c_p to be constant within each step). Compare this to the value you would have gotten if c_p were assumed constant at its value in the middle of the temperature range. (*Hint:* Write a loop to sum the energies in each step.)

Python tips: It is convenient to store the Shomate coefficients for each gas in a list, e.g.

```
N2_coeffs = [931.857,293.529,-70.576,5.688,1.587]
```

If you do this, you can set all the coefficients at once with a statement like A,B,C,D,E = N2_coeffs. A good trick is to define the lists as globals outside the function definition for cp(T), and to have the function refer to the list of coefficients as, e.g., ShomateCoeffs. Then, to switch from one gas to another, all you have to do is set ShomateCoeffs to the list you want before you call cp(T), as in ShomateCoeffs = N2_coeffs.

2.8.2 Potential temperature, dry adiabats and inhomogeneous atmospheres

Problem 2.14 The aforementioned bicycle tire with a volume of 3 liters is again pumped up with Earth air to a pressure of 4 bars. This time, it is pumped up so rapidly that there is not enough time for it to lose any heat to the surroundings. Assuming the ambient air has a temperature of 290 K, what is the temperature inside the tire immediately after it has been pumped up? What is the pressure after the tire has had time to cool down to the ambient temperature?

Problem 2.15 The atmosphere of Jupiter is predominantly H_2. A space probe determines that the temperature is 150 K at a pressure of 10^5 Pa. At what pressure does the temperature reach 300 K if the atmosphere is well mixed and there is no condensation of any substance? Compare your results with the observed temperature profile for Jupiter given in the text.

Problem 2.16 The current atmosphere of Earth consists of about 20% O_2 by moles, and 80% N_2. Compute the dry adiabat (i.e. temperature vs. pressure with fixed θ) for the pre-biotic atmosphere without any oxygen, and compare with the dry adiabat for an atmosphere with 20% oxygen and 80% nitrogen.

Problem 2.17 You live on a planet with a pure CO_2 atmosphere with surface pressure of 1 bar. The acceleration of gravity is $10 \, \mathrm{m/s^2}$. Design a "nitrogen balloon" (filled with N_2) to lift a mass of 100 kg, under conditions when the low-level temperature is 280 K.

Problem 2.18 A wind of pure N_2 with a temperature of 100 K blows over a methane (CH_4) lake on Titan. The lake has a temperature T, and a thin layer of atmosphere in contact with the lake exchanges heat so that its temperature is also T. Enough methane evaporates from the lake to bring the molar concentration of methane in this layer to a value η. The surface pressure is 1.5 bar. Since we are assuming the atmosphere is pure N_2 above a thin layer in direct contact with the lake, the methane evaporation has essentially no impact on surface pressure. For what range of T and η will the layer of atmosphere in contact with the lake be positively buoyant relative to the pure N_2 just above it?

Problem 2.19 Referring to the Venus temperature profile in Fig. 2.2, discuss the extent to which the Magellan observations of the temperature of the Venus atmosphere are consistent with a CO_2 dry adiabat. Extrapolate the temperature from the highest observed pressure to the 92 bar surface of the planet, and compare the result to the observed surface temperature of 737 K. (Note that part of the mismatch is due to the assumption of constant c_p and the inaccuracy of the perfect gas law at high pressures.) For CO_2 gas, $R/c_p \approx 0.2304$.

Problem 2.20 In this computer problem, you will compute the dry adiabat $T(p)$ for an ideal gas whose specific heat depends on temperature, in accord with the Shomate equation (see Problem 2.13). In addition to basic skills such as defining functions and loops, you will need to know how to write programs that find approximate solutions to an ordinary differential equation of the form $dY/dx = f(x, Y)$.

First, use the First Law of Thermodynamics to derive a differential equation for $d \ln T / d \ln p$ assuming $\delta Q = 0$. This defines the dry adiabat. Note that since c_p is a function of T, you can no longer treat it as a constant in doing the integral.

Write a program that tabulates approximate solutions to the differential equation. Note that your dependent variable is $Y \equiv \ln T$ whereas the right hand side of the differential equation involves T. This is not a problem, since you can write $T = \exp(\ln T)$. In writing your program, assume that $c_p(T)$ is defined by the Shomate equation.

Apply your program to obtain an improved approximation to the dry adiabat in a pure CO_2 Venusian atmosphere, which you originally computed in Problem 2.19. Start your computation at the ground ($p_s = 92$ bars) with the observed mean surface temperature of Venus (737 K). Integrate up to the 100 mb level, and compare the temperatures you get with those in the Magellan observations shown in Fig. 2.2 of the text. Make a plot comparing your calculations with the dry adiabat obtained by keeping c_p constant at 820 J/kg.

2.8.3 Data Lab: Analysis of temperature soundings

Problem 2.21 *Analysis of tropical Earth soundings*
In this Data Lab we'll look at a set of balloon soundings of the atmosphere over the tropical Pacific, taken from shipboard during the CEPEX field campaign. The data is located in the subdirectory cepex_sondes. The files are named according to the date/time of release, in the form yymmddhh.txt.

To get started, let's read in a profile and take a look at it. First read the data from 93031512.txt into your software. Note that the pressure is in millibars. The data files also contain relative humidity data, but we will not be using that for now. Make a plot of $T(p)$, and make sure it looks like the plot given in the text. To follow the usual conventions in a sounding plot, you should use a logarithmic pressure axis, and reverse the axis so that high pressure appears at the bottom.

We are now ready to take a more systematic look at tropopause height. The temperature minimum may be taken as a rough guide to the boundary of the troposphere, though it is known that there are some inadequacies in defining the tropopause this way. The general procedure is to look at the temperature array (or "column") T which has been read into your software, and find the index i which corresponds to the minimum temperature. Then the corresponding element i of the pressure array p is the tropopause pressure you want. The precise procedure for doing this depends on the language or software you are using. Try out the procedure on a few files in the dataset, then write a function that takes the temperature and pressure array and returns the tropopause pressure.

Next, write a loop that loops over all the files in the dataset and finds the tropopause pressure for each. (This may be easy or difficult, depending on what kind of programming language you are using.) Summarize the results you find. How much does tropopause pressure vary among these soundings? All of the data in this dataset has been interpolated to a common set of pressure levels. This makes it easy to compute statistics over the sounding. Write a loop to compute the average temperature profile over the whole dataset. The loop will look rather like the one you used for analyzing tropopause height, but instead of just computing the tropopause pressure, the body of the loop will accumulate an array whose ith element is the cumulative sum of the temperatures at the ith pressure level. You would also keep track of the number of soundings read, then divide by this at the end to get the average. Try this out. Then improve your loop so it also computes the standard deviation of the temperature at each level about its mean at that level. At what heights does temperature vary the most? At what heights is it most weakly variable?

Next, pick a sounding, and use the ideal gas law to compute the density as a function of pressure. Remember that you need to convert the pressure from mb to pascals when carrying out the calculation. Plot the computed density.

Now, write a function that computes the dry potential temperature profile $\theta(p)$ from data in a specified file. Look at a few plots. Which parts of the atmosphere are very statically stable? What part is closest to a dry adiabat? Process the entire dataset (or as much as you can), and examine the values of $\theta(600\,\text{mb}) - \theta(1000\,\text{mb})$ and $\theta(200\,\text{mb}) - \theta(600\,\text{mb})$. Discuss your results.

Python tips: Recall that you can use the function readtable in ClimateUtilities to read each sounding into a Curve() object. For example if datapath is the string containing the directory where the sounding data files are located, then

```
sounding = readTable(datapath+"93031512.txt")
sounding.YlogAxis = sounding.reverseY = sounding.switchXY =True
```

will read the data into a `Curve()` object called `sounding`. The second line sets the correct plotting options for a sounding.

There are a few functions available to `Python` which can help you find the temperature minimum and hence tropopause pressure. For any indexed object `data`, such as a list, the function `min` will find the minimum value in the list, and the list method `index` will find the index where that value occurs. Thus, if T is your temperature array, `i = T.index(min(T))` finds the index where the minimum occurs, then `p[i]` is the corresponding pressure. If you are using the numpy or an equivalent array module, then `i = numpy.argmin(T)` will directly find the index of the minimum temperature.

The `glob` function in the `glob` module, which is part of most standard Python distributions, can generate lists of filenames in a directory, which you can then loop over in doing your analysis. `glob.glob()` accepts wildcard characters in its argument, the most important of which is the asterisk. Thus, `glob.glob("*.txt")` returns a list of all the files in the current directory whose names end in ".txt". Try it out on a few of your directories, then try the following:

```
from glob import glob
from ClimateUtilities import *
datapath = <Your Path Here>
files = glob(datapath+"93*.txt") #Makes a list of files matching pattern
for file in files:
    sounding = readTable(file)
    i = numpy.argmin(sounding["T"]) #Finds index at which minimum T occurs
    print file,sounding["p"][i],T[i]
```

In doing manipulations with the data, you can make use of the fact that a `Curve()` object returns arrays on which arithmetic can be done without the need for a loop over elements. For example, to compute an average temperature, you first create an array of the correct length initialized to zero, using `Tbar = numpy.zeros (len(sounding["T"]),numpy.float)`. Then, inside the loop over files, you only need a statement like `Tbar = Tbar + c["T"]`. Similarly, you can compute the dry potential temperature array from your temperature and pressure array using a single-line statement.

Problem 2.22 *Analysis of midlatitude and polar Earth soundings*
To see how the vertical structure of the atmosphere changes at mid and high latitudes, we turn to data from a global historical network of regular sounding stations. This dataset is rather sparse in the tropics, and even more so over the tropical oceans, which is why we used a special dataset to investigate the tropical structure. The dataset provided here is a very tiny subset of the vast CARDS sounding dataset, and is located in CARDS_sondes. The files are named according to the station number, date, and time. A list of stations and their locations is given in the file `stations.txt`.

Pick a few midlatitude stations, and look at the temperature profiles. How well defined is the tropopause? How does the tropopause height vary throughout the year? How do the values of $\Delta\theta/\Delta p$ in the lower and upper troposphere compare with those you found in the tropical case?

Now focus on northern hemisphere winter, and try to discover something about how tropopause height varies with latitude.

Finally, pick one of the Antarctic soundings, and take a look at the seasonal cycle of temperature. How high is the tropopause in summer? Can a tropopause be identified in the dead of winter? What is the temperature structure near the ice surface? What do you think is happening there to cause that temperature structure?

Problem 2.23 *Analysis of planetary soundings*

First, we offer two very small datasets, which represent nearly all that is known about the vertical structure of the atmospheres of Jupiter and Venus. `Chapter2Data/ GalileoProfiles/` contains one brave sounding of the Jovian atmosphere, conducted by the Galileo probe. Note that the pressure is reported in bars. Take a look at this sounding, and compare the temperature profile with the dry adiabat for various mixtures of H_2 and He, ranging from all hydrogen to all helium. From the appearance of Jupiter it is clear that the atmosphere is full of clouds, and hence condensation is widespread. Why do you think the dry adiabat works so well for this sounding?

The next small dataset consists of a set of temperature profiles for Venus. The first group of data is from a set of four probes dropped into the Venus atmosphere by the Pioneer Venus mission. These soundings are contained in `VenusPioneerProfiles`. The second group of data was taken by the Magellan probe, whose main mission was actually radar mapping of the surface of the planet. This data was obtained by examining the transmission of radio waves through the Venusian atmosphere. This data set only covers roughly the upper 7 bars of the atmosphere, whose surface pressure is 90 bars. The soundings are contained in `Chapter2Data/VenusMagellanProfiles`.

Make plots of the soundings. Do you see any structure resembling a stratosphere? Do the temperatures differ much from one sounding to another? Compare selected Magellan soundings with the Pioneer Venus soundings.

In what part of the atmosphere do you think convection is occurring? Compare the temperature profile in this region with the dry adiabat for CO_2.

Now we turn to a bigger dataset, with more scope for exploration. The Mars Global Surveyor mission measured thousands and thousands of temperature profiles of Mars using the same radio occultation technique as the Magellan dataset. The data extends nearly to the ground. The main limitation of this dataset is that, owing to orbital geometry, it covers a limited range of latitudes and times of day. In particular, one is limited essentially to late afternoon and night-time data. Hence, none of these soundings reflect times of most active convection through a strongly heated surface. The dataset is located in `Chapter2Data/MarsRadioOccProfiles`. What is provided here is only a tiny sample of the full dataset. The small subset of the data included in the Workbook datasets focus on the southern hemisphere, because the orbital geometry allows a better examination of the warmer part of the day than is the case for the northern hemisphere data. Some relatively warm summer midlatitude evening soundings are given in the subdirectory `SummerSHevening`, while corresponding night-time and near-dawn soundings are in the `SummerSHnight` subdirectory. Some winter data in the southern hemisphere polar region is given in the `WinterSHPolarNight` subdirectory. The description line at the top of each file contains the geographic location of the sounding, and also local solar time and the angle the Sun makes with the local vertical (the solar zenith angle). The information tag also contains the *subsolar latitude*, also known sometimes as the *latitude of the sun*. This is the latitude at which the Sun is directly overhead at noon, and provides a useful measure of where we are in the seasonal cycle. For example, when the subsolar latitude is 0 degrees, we are at an equinox. When it is −20 degrees, it is near the southern hemisphere summer solstice, and when it is +20 degrees it is near northern hemisphere summer solstice. Note that not all of the soundings go all the way to the surface.

First take a look at the midlatitude early evening soundings. Do they generally contain a region where temperature decreases with height? Compute and analyze the profile of potential temperature for these soundings, assuming a pure CO_2 atmosphere with constant c_p. Are any of these soundings on the dry adiabat? What is the typical static stability (i.e. $d\theta/dp$)? The very high potential temperatures in the very upper part of the atmosphere may make it hard to see the structure in the lower portions, so you may want to try eliminating these from the graph. As an equivalent way of addressing the same question, compare the temperature profiles with the $T(p)$ corresponding to the dry adiabat based on the observed temperature nearest the ground.

Next take a look at the night-time midlatitude soundings. Where do these differ most from the early evening soundings? Compute and plot $\theta(p)$ for these soundings and discuss the results.

Now take a look at the polar night soundings. The main difference in the physical context of these soundings is that the midlatitude atmosphere has experienced the warming effect of the sun within the past day, whereas the polar night soundings have not been touched by sunlight for months. How does the low-level temperature compare with that of the midlatitude soundings? How about the upper-level temperature? Discuss the vertical profile. Are these soundings typically more or less statically stable than the midlatitude soundings?

Python tips: The description line at the top of a text data file is read into the returned `Curve()` object c as `c.description` when you read in the data using `readTable`.

You can use `Python`'s array cross section notation to subset the levels when computing potential temperature and creating a new curve object for plotting. For example if c is the `Curve()` object you read in from the dataset using `readTable`, `c["p"][0:10]` will give you the first 10 pressure levels, starting from the first pressure in the sounding (which is the highest pressure available, in this dataset).

2.8.4 The hydrostatic relation

Problem 2.24 Venus has a surface pressure of 92 bar, and a surface gravity of $8.87\,m/s^2$. 3.5% of the atmosphere (by mole fraction) consists of N_2. Compute the mass of N_2 per unit surface area of Venus, and compare with the corresponding number for Earth's atmosphere.

Problem 2.25 Suppose that the CO_2 concentration in the Earth's atmosphere is 300 ppmv, and that it is well mixed. Compute the mass of *carbon* located over each square meter of the Earth's surface. Compute the total mass of carbon in the Earth's atmosphere. How does the carbon per unit area on Earth compare with the value for the present Martian atmosphere, consisting of pure CO_2 with a surface pressure of about 6 mb?

Problem 2.26 Titan has a mostly nitrogen atmosphere with a surface pressure of about 1.5 bar. The gravitational acceleration at the surface of Titan is $1.35\,m/s^2$. Titan's radius is about 2600 km. How much mass of atmosphere is there above each square meter of surface? How does this compare to the mass of air over each square meter of Earth's surface? What is the total mass of Titan's atmosphere?

Problem 2.27 The temperature of Titan's troposphere, which contains most of the mass of the atmosphere, is about 100 K. The atmosphere is mostly N_2. Estimate the density scale height of Titan's atmosphere and compare it to Titan's radius. How does the situation on Titan differ from that of Earth? For which body is the atmosphere more accurately described as a "thin spherical shell"?

Problem 2.28 *An atmosphere for the Moon*
At some time in the future the surface of the Earth becomes uninhabitable, and the few remaining residents decide to move to the Moon. To make life more congenial, they decide to give the Moon an N_2/O_2 atmosphere with a surface pressure of 1 bar. They also add in enough carbon dioxide to bring the mean surface temperature up to 280 K. How much total mass needs to be brought in to create this atmosphere? You may assume the required mass of carbon dioxide to be negligible. If the mass comes in the form of comets, each 10 km in radius, how many are needed? Estimate the scale height of the atmosphere, and compare it to the radius of the Moon.

    Python tips: The radius of the moon is `planets.Moon.a`, and the acceleration of gravity at the surface of the Moon is `planets.Moon.g`.

Problem 2.29 Use the hydrostatic relation to show that $dT/dz = -g/c_p$ assuming the dry potential temperature to be constant. Do this first by differentiating the dry adiabatic temperature profile $T(p)$ and using the chain rule and hydrostatic relation. Then show that you get the same answer directly by assuming the dry static energy to be independent of height.

    The quantity g/c_p is called the *dry adiabatic lapse rate*. Compute the typical value for dry air on Earth, CO_2 on Mars, CO_2 on Venus, and N_2 on Titan (all based on surface gravity).

Problem 2.30 In this computer problem, you will find numerical solutions to the hydrostatic equation for an ideal gas atmosphere with general $T(p)$ (alternately $T(z)$). The solution is done in two different ways. The techniques you will need are: (1) Numerical quadrature (i.e. evaluation of a definite integral); and (2) numerical solution of an ordinary differential equation.

    As a prelude, first integrate the hydrostatic relation analytically, to find $p(z)$ for an atmosphere whose temperature profile is given by the ideal gas dry adiabat, $T(p) = T_s(p/p_s)^{R/c_p}$. From the answer to Problem 2.29, this is equivalent to $T(z) = T_s - (g/c_p)z$, where g is the acceleration of gravity. This solution will serve as a check on your numerical solution.

    One approach to evaluating the hydrostatic relation is to use Eq. (2.22), which gives you $p(z)$ directly once you know $\bar{T}(z)$. The temperature $\bar{T}$ can be found for any z by carrying out the definite integral given in Eq. (2.22). This works as long as you know T as a function of z. Write a program to find $p(z)$ using this method. To keep your program general, let the temperature profile be specified by an arbitrary function `T(z)`. Apply your program to the dry adiabat for Earth air, with surface pressure 1 bar and surface temperature 300 K. Compare with the analytical result.

    The above method does not work if you know $T(p)$ rather than $T(z)$. Since it is easy to measure pressure and temperature but relatively hard to measure altitude, balloon and planetary probe soundings often provide $T(p)$. There is an easy fix for this problem, since Eq. (2.21) can be re-written as $dz/d\ln p = -RT(p)/g$, whence one can get $z(\ln p)$ by carrying out a definite integral with respect to $\ln p$. Write a program to do this, and test it on the dry adiabat. Note that if you want a graph of $p(z)$, you can just make a graph of $z(p)$ using this routine, and switch the axes (or equivalently, turn it on its side).

The second approach is to solve the hydrostatic relation by numerically integrating the ordinary differential equation in the form given by Eq. (2.21). This method can actually be faster, since it doesn't require the evaluation of exponentials, and has the virtue of very easily giving results for a whole range of altitudes, without having to re-do the calculation from scratch for each new altitude. Since T is a function of p, Eq. (2.21) is an ordinary differential equation of the form $dY/dx = f(x, Y)$ with $Y \equiv p$ and $x \equiv z$. Write a program to solve the program in this form. Apply it to the dry adiabatic profile defined above, and check against the analytic result.

2.8.5 Latent heat and Clausius–Clapeyron

Problem 2.31 A spherical comet 1 km in radius and with a density twice that of water hits the ocean with a velocity of 20 km/s. If all the energy is converted to heat and used to vaporize water, how many kilograms of steam result?

Problem 2.32 Consider a glacier of water ice that is 1 km thick and covers an area of $10^8 \, \text{km}^2$ (about a fifth of the area of the Earth). Suppose that the ice absorbs $50 \, \text{W/m}^2$ of solar radiation, and all of this is used to either sublimate (into gas) or melt (into water) the ice. How long does it take for the glacier to disappear in each of these cases? Give your answer in years. The density of ice is about $930 \, \text{kg/m}^3$. What determines whether sublimation or melting occurs?

Problem 2.33 Titan receives somewhat less than $5 \, \text{W/m}^2$ of sunlight at its surface. If all of this energy were used to evaporate methane, about how long would it take to evaporate a lake of methane with a depth of 1 meter?

Python tips: The latent heat of evaporation of methane is given in `phys.CH4.L_vaporization`.

Problem 2.34 This computer exercise requires that you learn how to write a function in a programming language. It applies this skill to evaluating the Clausius–Clapeyron relation.

The simplified constant L form of the Clausius–Clapeyron relation is adequate for many purposes. To evaluate it, you need to specify a reference temperature and vapor pressure, the latent heat of the phase transition under consideration, and the gas constant (or equivalently the molecular weight). Using your favorite programming language, write a function `psat(T)` which computes the saturation vapor pressure as a function of temperature. The constants needed to evaluate the function can be treated as global constants, hard-wired into the function, or added to the argument list.

Put in the constants appropriate to CO_2 gas in equilibrium with a solid. It is convenient to use the triple point pressure and temperature for the reference pressure and temperature. Use the function to estimate how low temperature would have to be for CO_2 to condense at the Martian surface, assuming the 7 mb surface pressure typical of the present planet.

Python tips: Commonly, one makes a function because one wants to evaluate it many times for the same gas but different values of T. If you add the thermodynamic constants to the argument list, you are stuck typing in a long argument list when all you really wanted to change is T. Making the constants into globals solves this problem, but then makes it hard to deal with saturation vapor pressure functions for several different gases in the same program. `phys` provides a better way to deal with the problem, making use of the fact that objects can be made callable. Namely, it provides the class `satvps_function` which takes the needed gas parameters as arguments and creates a function-like object that stores the constants it needs. The triple point temperature and pressure provide a convenient set of base conditions for use in the formula. Thus, to create a constant-L saturation vapor pressure function for CO_2 sublimation, one would write

```
gas = phys.CO2 # get the gas object you want
pCO2 = phys.satvps_function(gas.TriplePointT,
  gas.TriplePointP,gas.MolecularWeight, gas.L_sublimation)
```

This creates a callable object pCO2, whereafter you can evaluate the vapor pressure by simply writing pCO2(T). This is a simple but very useful technique. Take a look at the class definition in phys to see how it is done, since you may want to use the technique yourself sometime.

Actually, it is pretty tedious to have to enter all the thermodynamic constants in the argument list. Why not just give all the constants at once in the form of a gas object? In fact, through the power of *polymorphism*, the `satvps_function` class recognizes what kind of thing it's given as an argument, and acts accordingly. Therefore, you could get the same result as above using phys.satvps_function(phys.CO2). By default, this will create a function that uses the latent heat of sublimation for temperatures below the triple point, and the latent heat of vaporization for temperatures above the triple point. But what if you want to force the function to use one latent heat or the other regardless of temperature? For this purpose, the class allows an optional second argument. If its value is the string `ice´, the latent heat of sublimation is assumed. If it is `liquid´ the latent heat of vaporization is assumed.

Problem 2.35 Titan has a surface temperature of about 95 K. Suppose that there is a lake of liquid methane (CH_4) on the surface, and that the air just above the lake is in equilibrium with the lake, and therefore saturated with respect to methane. Use the simplified form of Clausius–Clapeyron to determine the partial pressure of methane in the air near the surface. What would the partial pressure be if the temperature rose to 120 K? Based on an N_2 partial pressure of 1.5 bar at the surface, what would the molar mixing ratio of methane in the atmosphere be? What would the mass mixing ratio, the molar concentration, and the mass specific concentration be? What would the molar mixing ratio be if the temperature rose to 120 K? What would the mass mixing ratio, the molar concentration, and the mass specific concentration be?

Problem 2.36 Transition from liquid to vapor (evaporation) can happen at any temperature. It doesn't require a threshold temperature to be exceeded before it can happen. What then is the "boiling point," which certainly represents a kind of vapor–liquid transition? In fact, boiling is not much different from evaporation. What is special about boiling is that the saturation vapor pressure becomes greater than the *total* atmospheric pressure at the liquid surface. This means that the vapor pressure can push aside fluid and form bubbles in the interior of the fluid, which can then escape by rising. It is a much faster process than evaporation, because the transition can happen throughout the fluid, and not just at the gas–liquid interface.

Compute the boiling point of water at an altitude where the pressure is 300 mb.

Suppose a river with a temperature of 300 K erupted from the interior of present Mars, where the surface pressure is about 7 mb. Would the river boil? How high would the surface pressure have to be to prevent boiling at this temperature?

Now, think of Glurg the Titanian, who would like to boil up liquid methane to make his tea. The surface pressure of Titan is about 2 bars (mostly nitrogen). How hot does his stove have to get?

Problem 2.37 *Water on the Hadean Earth*
Ancient zircon crystals tell us that liquid water existed somewhere on the surface of the Earth as early as 4.2 billion years ago. What does this tell us about the surface temperature? Does it imply that the temperatures had to be below 100 °C (the "boiling point")? To answer this question, compute the saturation partial pressure of water vapor at an ocean surface as a function of surface temperature. Do this three ways. First, use the exponential form of the Clausius–Clapeyron relation with constant latent heat, based on triple point data. Next, use the exponential form based on "boiling point" data (i.e. that the vapor pressure is 1 bar at 373.15 K. Finally, use the empirical Antoine equation:

$$p_{sat}(T) = 10^{A-B/(T+C)} \tag{2.40}$$

where A, B, and C are empirical coefficients determined by fits to experimental data. In the range 373–647 K, the coefficients are $(A, B, C) = (5.2594, 1810.94, -28.665)$, if the pressure is given in bars and temperature in K. This fit gets the critical point pressure slightly wrong, but it is adequate for the purposes of the present problem.

Using your results, answer the following. At 100 °C (373.15 K) what is the partial pressure of water vapor at the surface? What fraction of the molecules of the atmosphere are water at the surface (assuming that the non-condensing part of the atmosphere has partial pressure of 1 bar)? Does the ocean boil under these conditions? What if the surface temperature is 200 °C instead? At this temperature, what proportion of the mass of water on the planet is in vapor form in the atmosphere, assuming the total water mass to be the same as that of the present ocean ($1.4 \cdot 10^{21}$ kg)? How hot would the surface have to be in order for all the inventory of water to go into the atmosphere? Specifically, show that at the critical point temperature for water, almost all of the Earth's oceanic inventory of water has gone into the atmosphere. Do you think that it is a coincidence that the equivalent pressure of the Earth's ocean is close to the critical point pressure?

Hint: At temperatures of 400 K and above, it is a reasonable approximation that water vapor dominates the mass of the atmosphere, so that by the hydrostatic law, the mass of water vapor in the atmosphere can be approximated by $4\pi a^2 \cdot p_{sat}/g$, where p_{sat} is the saturation water vapor pressure at the surface (in Pa) and a is the radius of the Earth. Problem 2.54 will show you how to accurately compute the total mass of water in cooler conditions when water vapor does not dominate the atmosphere.

Remark: Although you will find that the existence of liquid water does not itself limit the surface temperature to below 100 °C, the oxygen isotopic content of the zircons does tend to argue for moderate surface temperatures.

Problem 2.38 Suppose that the latent heat of vaporization varies linearly with temperature, according to the law $L = a - bT$ with a and b positive, up to the temperature where latent heat vanishes. Use the Clausius-Clapeyron relation to find an expression for $p_{sat}(T)$, given its value at some T_0. Make a sketch illustrating how the saturation vapor pressure differs from the constant-L case.

Problem 2.39 In this computer problem you will implement functions which provide empirical fits to the vapor pressure for water, taking into account the variations of latent heat with temperature.

It has been found that the function satvpw(T) defined below provides an accurate fit to the measured vapor pressure over liquid water, for temperatures between the freezing point and 400 K.

$$\text{satvpw}(T) = e_0 \cdot 10^{a+b+c+d} \tag{2.41}$$

where

$$a = -7.90298 \cdot (T_0/T - 1), b = 5.02808 \log_{10}(T_0/T),$$
$$c = -1.3816 \cdot 10^{-7} \cdot 10^{11.344(1-T/T_0)-1}, \tag{2.42}$$
$$d = 8.1328 \cdot 10^{-3} \cdot 10^{-3.49149(T_0/T-1)-1}$$

where $T_0 = 373.15$ K and $e_0 = 101\,324.6$ Pa. For ice, the corresponding function satvpi(T) has the same form but with $T_0 = 273.15$ K, $e_0 = 610.71$ Pa,

$$a = -9.09718(T_0/T - 1), b = -3.56654 \log_{10}(T_0/T), c = 0.876793(1 - T/T_0) \tag{2.43}$$

and $d = 0$. The fit for ice is valid from the freezing point down to about 100 K.

As it turns out, there's a further complication: in real atmospheres, liquid water can exist in a supercooled state well below the triple point temperature. Typically, the mix of supercooled water and ice shifts from all water to all ice as the temperature is reduced below the triple point. The following function (taken from a typical climate model) uses the empirical function for ice for temperatures below $-20\,°C$, the empirical function for liquid for temperatures above $0\,°C$, and shades linearly between ice and water saturation as temperature is increased from $-20\,°C$ to $0\,°C$, specifically:

$$\text{satvpg}(T) = 0.05(273.15 - T) \cdot \text{satvpi}(T) + 0.05(T - 253.15) \cdot \text{satvpw}(T) \tag{2.44}$$

where satvpi and satvpw are the vapor pressures over ice and liquid water, defined previously.

Write functions to compute satvpw(T), satvpi(T), and satvpg(T). Compare satvpw(T) with the idealized saturation vapor pressure function for water based on a constant latent heat of vaporization, using the triple point temperature and pressure as the base values. Can you see the difference between the two when you plot them both with a logarithmic pressure axis? To get a more accurate view of the error, compute a table of (satvp1(T) – satvpw(T))/satvpw(T), where satvp1 is the idealized function. Try using the boiling point pressure and temperature (1 bar at 373.15 K) as the reference value for satvp1 and see how the comparison differs.

Compare the empirical and idealized vapor pressures for ice in the range 150 K to 273 K.

Compare satvpw(T) with a simplified function consisting of the constant L function for ice below the triple point and the constant L function for liquid water above the triple point. In the simplified function, you should use the triple point temperature and pressure to provide the reference value. Do the comparison for the range 250 K to 290 K.

Python tips: The functions described above are provided for you as `phys.satvpw`, `phys.satvpi`, and `phys.satvpg`

Problem 2.40 Suppose that you are given a function $e(T)$ which gives the saturation vapor pressure of a substance as a function of temperature, and you need to find the latent heat of the phase transition, as a function of T. This situation might arise if the vapor pressure function you are using arises from an empirical fit to data. This is the case for the function satvpg(T) in Problem 2.39.

Use the Clausius-Clapeyron relation to derive a formula for L in terms of $d \ln e / dT$. You can approximate the derivative using the formula

$$\frac{d \ln e}{dT} \approx \frac{\ln e(T + dT) - \ln e(T - dT)}{2 dT} \tag{2.45}$$

where dT is some suitably small number. Writing the derivative function this way allows you to pass any saturation vapor pressure function to the procedure, and get an answer without re-coding. Now write a function using the approximate derivative which returns the latent heat for any argument T. It's a good idea to make dT an argument of the function, so that you can try out different values easily. Note that in most computer languages the function name e can also be passed to your latent heat function as an argument. If you do this, you can use your latent heat function for several different choices of $e(T)$ without recoding. The function needs the gas constant R for the material in question. You could add the gas constant to the argument list of the function, but it makes more sense to leave it as a global to be set in your main program before you use the function.

First, check your latent heat function by passing it a saturation vapor pressure function made using the constant L approximation, and make sure that you get the right value of latent heat back. What happens if you make dT too small? If you make it too large?

Now apply your procedure to satvpg. Plot the results in the range from 200 K to 400 K, and discuss. Compare the computed latent heat with the computed latent heats for satvpw and satvpi. Why do you think the effective latent heat for satvpg has a spike near $-20\,^\circ$C?

Problem 2.41 Apply the function you wrote in Problem 2.40 to determine the implied latent heat as a function of temperature in the Antoine vapor pressure formula used in Problem 2.37.

Problem 2.42 This problem is inspired by Fritz Leiber's story, "A Pail of Air." The Earth has a close encounter with a large interplanetary object, and is flung off into the cold night of space far from the Sun. As the Earth gradually cools down the atmosphere begins to condense, until essentially all of the atmosphere is deposited in a condensed layer at the surface of the planet. For the purposes of this problem, you may assume that dry Earth air consists of a mixture of 80% N_2 and just under 20% O_2 (by mole fraction), with the balance consisting of CO_2 with a mole fraction of 380 ppmv. To keep things simple, we'll ignore atmospheric water vapor, which would quickly snow out and add to the existing surface layer of water ice and snow. The Earth's initial surface pressure is 10^5 Pa, and the acceleration of gravity is 9.8 m/s^2.

Describe the sequence of events that occurs as the atmosphere cools. Will the atmosphere condense into solid or liquid forms? Will all the components of the atmosphere be mixed together at the surface, or will they occur in layers? Why? How thick would the layer of each substance be, and in what order would the layers occur? For the purposes of this problem, you may assume that any snow (of any substance) that forms becomes so compacted that the density is the same as that of the pure solid.

Problem 2.43 *Water content of Earth's atmosphere*
Consider an atmosphere with the temperature profile:

$$T(p) = \begin{cases} T_s \cdot \left(\frac{p}{p_s}\right)^a & \text{for } p > p_t, \\ T_s \cdot \left(\frac{p_t}{p_s}\right)^a & \text{otherwise.} \end{cases} \tag{2.46}$$

This provides an adequate model of the tropical temperature profile with $p_t = 100\,\text{mb}$ and $a = 0.176$. Find the profile of the saturation specific humidity for water, q_s, associated with this profile. What happens in the isothermal stratosphere if you insist that the water vapor remain saturated? Suppose now that there is no water in the stratosphere ($p < p_t$). From the definition of specific humidity, the total mass of water per square meter in a column of the atmosphere is then:

$$m_w = \int_{p_s}^{p_t} q_s \frac{dp}{g}. \tag{2.47}$$

Compute this integral approximately, by dividing the column up into many thin layers and summing the water in each. Discuss the dependence on surface temperature. The *liquid equivalent depth* is the depth of the layer of liquid water that would be produced if all the vapor were condensed out of the column and spread uniformly over its base. It is m_w/ρ_{liq}, where ρ_{liq} is the density of liquid water (about $1000\,\text{kg/m}^3$). What is the equivalent liquid depth of water in a saturated tropical atmosphere?

Problem 2.44 Examine some of the polar winter soundings from the Mars Global Surveyor Radio Occultation dataset introduced in Problem 2.23, and discuss whether there are any parts of the atmosphere where condensation of CO_2 into CO_2 ice is likely to have occurred. To do this, you need to use the Clausius–Clapeyron relation to compute the saturation vapor pressure at each level in the sounding, using the temperature data, and then compare the saturation pressure with the actual pressure. Alternately, you can solve the simplified Clausius–Clapeyron formula for $T(p)$ and compare this directly with the temperature sounding.

Problem 2.45 According to some interpretations of geologic data from the Neoproterozoic (about 600 million years ago), the Earth went through one or more episodes of global glaciation. The phenomenon is known as "Snowball Earth." The general thinking has it that the Earth would exit from a Snowball state after sufficient CO_2 has built up in the atmosphere to warm the planet to the point where it deglaciates.

In the Snowball state, the planet becomes very cold, because an ice surface reflects a great deal of solar radiation. Simulations show that the surface temperature at the winter pole can drop to 160 K, though it warms to about 240 K in the balmy polar summer. Under these circumstances, how high would the CO_2 partial pressure have to rise before CO_2 condensation sets in at the winter polar surface? Based on a 1 bar surface pressure for the non-CO_2 part of the air, what would the molar mixing ratio and molar concentration of CO_2 have to be to cause condensation? How do your answers change if the surface temperature is increased to 200 K? What do you think condensation would do to the global CO_2 concentration?

2.8.6 Moist adiabats

Problem 2.46 *Springtime for Europa*

Something is about to happen. Something wonderful. To promote life on Jupiter's moon Europa, which currently is composed of a liquid water ocean covered by a very thick water ice crust, the alien race which built Tycho Magnetic Anomaly 1 ignites thermonuclear fusion on Jupiter, heating Europa to the point that its icy crust melts, leaving it with a globally ocean covered surface having a temperature of 280 K. Water vapor is the only source of atmosphere for this planet. Describe what the atmosphere would be like, and calculate $T(p)$ for this atmosphere. Give a rough estimate of the depth (in km) of the layer containing most of the mass of the atmosphere. (The gravitational acceleration at Europa's surface is $1.3\,\mathrm{m/s^2}$.)

Problem 2.47 *Martian polar winter temperature structure*

During the winter near each of the Martian poles, CO_2 condenses out from the atmosphere to form a layer of CO_2 snow which accumulates at the surface. As springtime approaches, this snow layer sublimates into the atmosphere until it is completely gone, leaving only the underlying water ice behind. What is the surface pressure when the surface temperature is 145 K, 150 K, 151 K, and 152 K, assuming that some CO_2 ice remains at the surface in these conditions? Plot the temperature profile $T(p)$ you expect for these cases, and justify your reasoning. In each case, show results for p ranging down to $\frac{1}{1000}$ of the surface pressure. Do you think any CO_2 snow is likely to be left at the surface of Mars when the polar temperature reaches 160 K?

For the purposes of this problem, you may assume that Mars has a pure CO_2 atmosphere.

Hint: To make it easier to compare the profiles, you may want to use $-\ln(p/p_s)$ as your vertical coordinate.

Problem 2.48 Consider a planet with a methane–ammonia atmosphere, both of which are assumed in saturation so that lifting will cause both methane and ammonia to condense. Make a table and plot of the temperature vs. pressure for this atmosphere, and of the mixing ratio of methane to ammonia as a function of pressure.

Hint: The partial pressures of the two individual components are given simply by their respective Clausius–Clapeyron relations. How do you then get the total pressure from the partial pressures? Note that finding an expression for $T(p)$ is difficult, but computing $p(T)$ is easy. You can use the latter to make graphs and tables just as easily as the former.

Problem 2.49 Using the formula for the slope of the saturated moist adiabat for a condensable/non-condensable mixture, compute $d\ln(T)/d\ln(p_a)$ for a mixture of water vapor and Earth air at a pressure of 1000 mb and (a) a temperature of 300 K, (b) a temperature of 250 K. List the values of the individual terms in the numerator and denominator, so that you can get a feel for which are big and which are small. Compare the values with the slope of the dry adiabat. Note that $p_a \approx p$ in this temperature range, which simplifies the calculation.

Problem 2.50 Consider an atmosphere on the saturated one-component moist adiabat $T_{ad}(p)$ given by Eq. (2.27). Suppose that a parcel of atmosphere is lifted from a pressure p to a slightly lower pressure $p - \delta p$ (with $\delta p > 0$). Suppose that the parcel is lifted without condensation, i.e. along the dry adiabat. By comparing the resulting temperature with $T_{ad}(p - \delta p)$, show that condensation occurs on ascent provided the slope $d\ln T_{ad}/d\ln p$ is less than the dry adiabatic slope R/c_p. Using Clausius–Clapeyron to compute the moist adiabatic slope, show that this criterion is equivalent to $c_p T/L < 1$. Show that if the criterion is not satisfied, then condensation occurs on descent instead.

Another way to state the criterion is to define the characteristic temperature L/c_p, and to say that condensation occurs on ascent when T is below the critical temperature. Compute L/c_p for CO_2, H_2O, N_2, CH_4, and NH_3, using both the latent heat of sublimation and the latent heat of vaporization in each case. For the purposes of this problem you may use the latent heats and specific heats given in Table 2.1. Show that in each of these cases the criterion for condensation on ascent is automatically satisfied as long as the temperature is below the critical point temperature. Note that to determine what happens as the critical point is approached, one must use a more general form of Clausius–Clapeyron and take into account the temperature variations of L and c_p, as well as the departures from the ideal gas equation of state.

Problem 2.51 In this problem you will compare the moist adiabatic slope with the dry adiabatic slope for an arbitrary mixture of a saturated condensable gas with a non-condensable gas. Though the calculation can be done manually, it is much easier to write a small program to carry out the procedure outlined below, given temperature and gas properties as input. For the purposes of this problem you may assume the latent heat to be independent of temperature.

Divide up the problem as follows. First, using the saturation vapor pressure, compute R/c_p for the mixture using the appropriately weighted gas constants and specific heats for the two components. This is the dry adiabatic slope. Next, using Eq. (2.33), compute $d\ln T/d\ln p_a$. Finally, to convert this to $d\ln T/d\ln p$ you need to multiply by $d\ln p_a/d\ln p$, which is equal to $(p/p_a)(dp_a/dp)$. Compute this quantity by noting $p = p_a + p_c$, $d(p_a + p_c)/dp_a = 1 + dp_c/dp_a$, and $dp_c/dp_a = (dp_c/dT)/(dT/dp_a)$ You can directly evaluate (dp_c/dT) using Clausius–Clapeyron, since the condensable substance is assumed saturated. You can evaluate (dT/dp_a) using the value of $d\ln T/d\ln p_a$ which you computed already.

Carrying out this procedure allows you to evaluate the moist adiabatic slope with a minimum of tedious algebra. For a range of temperatures extending from temperatures where the condensable is dilute to temperatures where the condensable is dominant, compare the moist adiabatic slope to the dry adiabatic slope for each of the following mixtures: (1) Condensable H_2O in N_2, (2) condensable H_2O in H_2, (3) condensable CH_4 in H_2, (4) condensable CO_2 in N_2, (5) condensable CH_4 in N_2.

Problem 2.52 *Numerical computation of dilute moist adiabat*
For the dilute case, in which the mixing ratio of the condensable substance is small, the slope of the moist adiabat is given approximately by $d\ln T/d\ln p = (R_a/c_{p,a})(1+A)/(1+A\cdot B)$, where $A \equiv Lq_{sat}/R_a T$ and $B \equiv (R_a/c_{p,a})(L/R_c T)$. q_{sat} is the saturation mass concentration of the condensable substance, which is approximately the same as the saturation mass mixing ratio in the dilute case. What approximations do you have to make in order to recover this formula from the full non-dilute formula given in the text? (*Hint:* Lq/RT can be large even if q itself is small.)

Note that both A and B can be written as functions of $\ln T$ and $\ln p$, so if you define y to be the former and x to be the latter, we have a differential equation of the form $dy/dx = F(x,y)$. Write a computer program that solves this differential equation using the Euler and Midpoint methods. (Optionally, you may want to try Runge–Kutta.) Show solutions for the case of water vapor in Earth air, with surface temperature set to 250 K, 300 K, and 320 K. Compare with the dry adiabat. Examine the values of q in your solutions and comment on the validity of the dilute approximation. Try this out for some other combination of gases (e.g. CH_4 in N_2 with surface pressures of around 1.5 bar and temperatures of 80 K to 100 K, approximating Titan conditions).

You may assume that L is independent of T, with the exception that you should use the latent heat for gas to liquid or gas to solid according to whether the temperature is above or below the freezing point. Do your results change much if you implement this refinement in the case of water vapor?

In this problem you should not use any canned numerical analysis package; you should cook up your own from scratch.

Problem 2.53 *Computing the moist adiabat*

In this computer problem you will numerically integrate the equation determining the moist adiabat for a mixture of a non-condensable and a condensable component, use your routine to reproduce Fig. 2.7 in the text, and then go on to further explorations. Besides basic programming skills, you will need to know how to find numerical solutions to an ordinary differential equation (ODE) on the computer.

In Problem 2.49 you already wrote a function which computes the slope $d\ln T/d\ln p_a$ in terms of arguments $\ln T$ and $\ln p_a$ and the thermodynamic constants. For generality, you should allow for an arbitrary choice of non-condensable and condensable substances, rather than hard-wiring in the thermodynamic constants. You should assume that L is constant. Test out your slope function in some critical limits to make sure it is behaving properly.

Using the slope function, write a program to determine the moist adiabat for the general non-dilute case, using ODE integration. Recall that, as described in the text, you find the profile $T(p)$ using a two-step process: first you compute $T(p_a)$ by using an ODE integrator, then you use the result to compute the total pressure from the formula $p = p_a + p_{c,sat}(T(p_a))$. Using this procedure, you essentially make a list of values $T(p_a)$ and the corresponding $p(p_a)$, which can then be plotted or analyzed. Verify that this procedure automatically reduces to the one-component moist adiabat for hot conditions where $p_a \ll p_{c,sat}(T)$.

Now, use your system to reproduce the graph of the moist adiabat given in the text (Fig. 2.7). Extend the results to higher temperature, and compare with what you would have gotten with a pure steam atmosphere.

Python tips: The integration to find the moist adiabat is implemented in the Chapter Script `MoistAdiabat.py`. It makes use of `gas` objects to allow for generality in the choice of condensable and non-condensable components. The same computation is implemented as a callable object `MoistAdiabatComp` in the `phys` courseware module. The object implementation allows interpolation to a fixed pressure grid, which makes intercomparison of calculations for different parameters easier, and also facilitates comparison with data. You should try to implement the moist adiabat computation yourself before taking a look at the courseware implementations.

Problem 2.54 Consider an atmosphere consisting of a saturated mixture of condensable water vapor with non-condensable Earth air. The way we have been setting up the calculation, the partial pressure of non-condensable air, p_a is specified at the ground, and the total pressure p (including water vapor) is computed. From the hydrostatic relation, the total mass of the atmosphere per square meter is p/g, but the mass of air is not p_a/g and the mass of water is not p_w/g because water vapor is not well mixed, i.e. $p_w/(p_w + p_a)$ is not constant and because the molecular weight of water differs to that of air.

Write a function to compute the total mass of water vapor and of air in Earth's atmosphere as a function of p_a at the surface and the surface temperature. Explore the behavior with p_a fixed at 1 bar, as the temperature ranges from 270 K to 400 K. How much does the mass of non-condensing air vary with surface temperature? Why does it vary? Compare the mass of air with p_a/g, and give an argument explaining the sign of the discrepancy. How does the mass of water in the atmosphere compare with the estimate provided by p_w/g, which is accurate when water vapor dominates the atmosphere?

Hint: If $q(p)$ is the mass concentration of a substance, then the mass of the substance per unit area is $\int_{p_s}^0 q(p)dp/g$. The integral can be approximated by a sum over layers. In order to compute $q(p)$ from saturation vapor pressure, you will need to make use of the numerical evaluation of the moist adiabat $T(p)$ which you carried out in one or more of the previous problems.

Python tips: You can do this problem by modifying the Chapter Script `MoistAdiabat.py`, or by using the `phys.MoistAdiabat` class to make a function to compute the moist adiabat and mass concentration, then summing up the output.

Problem 2.55 *Comparison of tropical Earth soundings with the moist adiabat*
Pick a few tropical Earth soundings from the datasets analyzed in Problem 2.21 and see how well they match the moist adiabat starting at some pressure p_1 off the ground. You will need to experiment a bit to find the best value of p_1; loosely speaking, it corresponds to a boundary layer height, or in some cases a lifted condensation level. At what pressure level do the soundings start to become warmer than the moist adiabat? This level provides an estimate of the level reached by convection. The convection level, estimated in this way, will always be below the level at which the temperature minimum occurs. Compare the pressures of the two levels in a few soundings selected from the tropical dataset.

Look at the relative humidity data in the soundings. Do the observations follow the moist adiabat even when the atmosphere is substantially undersaturated? The soundings on March 24, 25, 26, 27, which were taken in the subtropics, provide a particularly clear indication of this behavior.

Finally, examine some midlatitude soundings from the datasets analyzed in Problem 2.22. Where is the temperature minimum? How well can you fit the data below the temperature minimum with a moist adiabat?

Problem 2.56 *Moist adiabat on Titan*
The atmosphere of Titan consists of a mixture of N_2 and CH_4 (methane). The methane is condensable at Titan temperatures but for the purposes of this problem the nitrogen can be regarded as non-condensable. (In reality, it does condense a little bit in Titan's upper troposphere.) Assume that there is sufficient methane ice or liquid at the surface to maintain methane saturation.

Compute and plot the moist adiabatic $T(p)$ for Titan assuming 1.5 bar of N_2 at the ground. You may also assume that methane is saturated throughout the atmosphere (requiring a reservoir of methane at the ground). Plot and discuss results for the mixing ratio as a function of p as well. Discuss results for three different cases: (a) $T = 95$ K at the surface, as for Titan at present, (b) a hot Titan, with $T = 120$ K, and (c) a cold Titan, with $T = 80$ K.

Discuss how the adiabat compares with the dry adiabat and the single-component methane condensing adiabat.

Compare the lower portion of the Huygens Titan sounding to the dry N_2 adiabat and to the saturated moist adiabat based on the observed low-level temperature and pressure. Which adiabat provides the better fit to the data? How accurate must the temperature measurement be to distinguish the two? (The data you need is in the `Titan` subdirectory of the Workbook datasets for this chapter.)

Python tips: You can modify the Chapter Script `MoistAdiabat.py` for use with a mixture as defined above. Gas data for for N_2 and CH_4 are both in the `phys` module, under the names `N2` and `CH4`. Alternately you can use the `MoistAdiabat` object in the `phys` module.

Problem 2.57 Find the moist adiabat for a mixture of 1 bar of N_2 (measured by surface pressure) with CO_2 in saturation, in the surface temperature range 200 K to 300 K. How does the surface pressure depend on temperature?

Problem 2.58 As a challenging variant of Problem 2.57, consider the case where there are 2 bars partial pressure of CO_2 at the surface and 1 bar of N_2, and where the surface temperature is warm enough that the CO_2 is unsaturated at low levels. Find the lifted condensation level, and patch the result at this level to the numerically determined moist adiabat further aloft. This is considered by many to be a model of the atmosphere of Early Mars.

Problem 2.59 *Moist adiabat on gas giants*
The atmospheres of the gas giants are mostly H_2, mixed with small quantities of various condensable substances. Explore the behavior of the moist adiabat for H_2 mixed with each of the following condensables taken one at a time: H_2O, CH_4, NH_3. For what range of temperatures and pressures does each of these substances appreciably alter the dry H_2 adiabat? For each case, restrict attention to the regime in which the molar concentration of the condensable remains under 10%.

For more accuracy, you should take into account that the non-condensing part of gas giant atmospheres is actually a mixture of H_2 and He, just as present Earth air is mostly a mixture of N_2 and O_2. For Jupiter, He makes up about 10% of the non-condensable total, whereas for Saturn it is about 3%. What do you need to modify in your calculation in order to incorporate the effect of He? Discuss how much difference this makes in the CH_4 case.

2.9 FOR FURTHER READING

For a deeper discussion of thermodynamics, see

- Feynman RP, Leighton RB and Sands M 2005: *The Feynman Lectures on Physics*, Vol. 1. Addison Wesley.

A physically based derivation of Clausius–Clapeyron is found in Section 45-3, and a good discussion of entropy is found in Section 44-6.

A very thorough discussion of moist thermodynamics and of the representation of the effects of moist convection on the Earth's atmosphere can be found in

- Emanuel KA 1994: *Atmospheric Convection.* Oxford University Press.

An example of the computation of the dilute moist adiabat for a non-ideal non-condensable background gas can be found in the Appendix of

- Kasting JF 1991: CO_2 condensation and the climate of Early Mars. *Icarus* **94**, 1–13.

The extension to the non-dilute case would be straightforward were it not for the fact that the saturation vapor pressure can no longer be computed independently of the non-condensable gas state. This subject is under investigation for the N_2/CH_4 system on Titan, and may also be relevant for non-ideal mixed CO_2/H_2O mixtures on hot, wet planets such as Early Venus. The planetary implications of non-ideal non-dilute behavior are still at a very early stage of development.

A very nice analysis of the way the mixing of inhomogeneities increases entropy is given in

- Pauluis O and Held IM 2002: Entropy budget of an atmosphere in radiative-convective equilibrium. Part II: Latent heat transport and moist processes. *J. Atmos. Sci.* **59**, 140–149.

This particular analysis applies to the mixing of moist and dry air, and shows that the mixing is a big part of what makes moist convection irreversible.

For thermodynamic properties of gases and associated phase transitions, the NIST Chemistry WebBook and the Air Liquide Gas Encyclopedia found at

- http://webbook.nist.gov/
- http://encyclopedia.airliquide.com/encyclopedia.asp

are very useful. Properties for more exotic conditions can often be found in the *Journal of Physical and Chemical Reference Data*, and results pertinent to planetary atmospheres are often reported in the journal *Icarus*.

CHAPTER THREE

Elementary models of radiation balance

3.1 OVERVIEW

Our objective is to understand the factors governing the climate of a planet. In this chapter we will be concerned with energy balance and planetary temperature. Certainly, there is more to climate than temperature, but equally certainly temperature is a major part of what is meant by "climate," and greatly affects most of the other processes which come under that heading.

From the preceding chapter, we know that the temperature of a chunk of matter provides a measure of its energy content. Suppose that the planet receives energy at a certain rate. If uncompensated by loss, energy will accumulate and the temperature of some part of the planet will increase without bound. Now suppose that the planet loses energy at a rate that increases with temperature. Then, the temperature will increase until the rate of energy loss equals the rate of gain. It is this principle of energy balance that determines a planet's temperature. To quantify the functional dependence of the two rates, one must know the nature of both energy loss and energy gain.

The most familiar source of energy warming a planet is the absorption of light from the planet's star. This is the dominant mechanism for rocky planets like Venus, Earth, and Mars. It is also possible for energy to be supplied to the surface by heat transport from the deep interior, fed by radioactive decay, tidal dissipation, or high temperature material left over from the formation of the planet. Heat flux from the interior is a major player in the climates of some gas giant planets, notably Jupiter and Saturn, because fluid motions can easily transport heat from the deep interior to the outer envelope of the planet. The sluggish motion of molten rock, and even more sluggish diffusion of heat through solid rock, prevent internal heating from being a significant part of the energy balance of rocky planets. Early in the history of a planet, when collisions are more common, the kinetic energy brought to the planet in the course of impacts with asteroids and planetesimals can be a significant part of the planet's energy budget.

There are many ways a planet can gain energy, but essentially only one way a planet can lose energy. Since a planet sits in the hard vacuum of outer space, and its atmosphere is rather tightly bound by gravity, not much energy can be lost through heated matter streaming away from the planet. The only significant energy loss occurs through emission of electromagnetic radiation, most typically in the infrared spectrum. The quantification of this rate, and the way it is affected by a planet's atmosphere, leads us to the subject of *blackbody radiation.*

3.2 BLACKBODY RADIATION

It is a matter of familiar experience that a sufficiently hot body emits light – hence terms like "red hot" or "white hot." Once it is recognized that light is just one form of electromagnetic radiation, it becomes a natural inference that a body with any temperature at all should emit some form of electromagnetic radiation, though not necessarily visible light. Thermodynamics provides the proper tool for addressing this question.

Imagine a gas consisting of two kinds of molecules, labeled A and B. Suppose that the two species interact strongly with each other, so that they come into thermodynamic equilibrium and their statistical properties are characterized by the same temperature T. Now suppose that the molecules A are ordinary matter, but that the "molecules" B are particles of electromagnetic radiation (*photons*) or, equivalently, electromagnetic waves. If they interact strongly with the A molecules, whose energy distribution is characterized by their temperature T in accord with classical thermodynamics, the energy distribution of the electromagnetic radiation should also be characterized by the same temperature T. In particular, for any T there should be a unique distribution of energy amongst the various frequencies of the waves. This spectrum can be observed by examining the electromagnetic radiation leaving a body whose temperature is uniform. The radiation in question is known as *blackbody radiation* because of the assumption that radiation interacts strongly with the matter; any radiation impinging on the body will not travel far before it is absorbed, and in this sense the body is called "black" even though, like the Sun, it may be emitting light. Nineteenth century physicists found it natural to seek a theoretical explanation of the observed properties of blackbody radiation by applying well-established thermodynamical principles to electromagnetic radiation as described by Maxwell's classical equations. The attempt to solve this seemingly innocuous problem led to the discovery of quantum theory, and a revolution in the fundamental conception of reality.

Radiation is characterized by direction of propagation and frequency (and also polarization, which will not concern us). For electromagnetic radiation, the frequency ν and wavelength λ are related by the *dispersion relation* $\nu\lambda = c$, where c is a constant with the dimensions of velocity. Because visible light is a familiar form of electromagnetic radiation, c is usually called "the speed of light." The wavenumber, defined by $n = \lambda^{-1} = \nu/c$, is often used in preference to frequency or wavelength. The wavenumber can be viewed as the frequency measured in alternate units, and so we will often refer to wavenumber and frequency interchangeably. Although SI units are preferred throughout this book, we follow spectroscopic convention and make an exception for wavenumber when dealing with infrared radiation, which will usually be measured in cm^{-1} since it yields comfortable and familiar ranges of numbers. Wavelengths themselves will sometimes be measured in μm (microns, or 10^{-6} m). Figure 3.1 gives the approximate regions of the electromagnetic spectrum corresponding to common names such as "radio waves" and so forth.

If a field of radiation consists of a mixture of different frequencies and directions, the mixture is characterized by a *spectrum*, which is a function describing the proportions of each type of radiation making up the blend. A spectrum is a *density* describing the amount of electromagnetic energy contained in a unit volume of the space (three-dimensional position, frequency, direction) needed to characterize the radiation, just as the mass density of a three-dimensional object describes the distribution of mass in three-dimensional space.

Before proceeding, we must pause and talk a bit about how the "size" of collections of directions is measured in three dimensions. For collections of directions on the plane, the measure of the "size" of the set of directions between two directions is just the angle between those directions. The angle is typically measured in radians; the measure of the angle in radians is the length of the arc of a unit circle whose opening angle is the angle we are measuring. The set of all angles in two dimensions is then 2π radians, for example. A collection of directions in three-dimensional space is called a *solid angle*. A solid angle can sweep out an object more complicated than a simple arc, but the "size" or measure of the solid angle can be defined through a generalization of the radian, known as the *steradian*. The measure in steradians of a solid angle made by a collection of rays emanating from a point P is defined as the area of the patch of the unit sphere centered on P which the rays intersect. For example, a set of directions tracing out a hemisphere has measure 2π steradians, while a set of directions tracing out the entire sphere (i.e. all possible directions) has measure 4π steradians. If we choose some specific direction (e.g. the vertical) as a reference direction, then a direction in three-dimensional space can be specified in terms of two angles, θ and ϕ, where θ is the angle between the reference direction and the direction we are specifying, and ϕ is the angle along a circle centered on the reference direction. These angles define a spherical polar coordinate system with the reference direction as axis; $0 \geq \theta \leq \pi$ and $0 \geq \phi \leq 2\pi$. In terms of the two direction angles, the differential of solid angle Ω is $d\Omega = \sin\theta d\theta \, d\phi = -(d\cos\theta)(d\phi)$. Generally, when writing the expression for $d\Omega$ in the latter form we drop the minus sign and just remember to flip the direction of integration to make the solid angle turn out positive. We recover the area of the unit sphere by integrating $d\Omega$ over $\cos\theta = -1$ to $\cos\theta = 1$ and $\phi = 0$ to $\phi = 2\pi$. A similar integration shows that the set of directions contained within a cone with vertex angle $\Delta\theta$ measured relative to the altitude of the cone has measure $2\pi(1 - \cos\Delta\theta)$ steradians. A narrow cone with $\Delta\theta \ll 1$ has measure $\pi(\Delta\theta)^2$ steradians.

We wish to characterize the energy in the vicinity of a point $\vec{r}$ in three-dimensional space, with frequency near ν and direction near that given by a unit vector $\hat{n}$. The energy spectrum $\Sigma(\vec{r}, \nu, \hat{n})$ at this point is defined such that the energy contained in a finite but small sized neighborhood of the point $(\vec{r}, \nu, \hat{n})$ is $\Sigma dV d\nu \, d\Omega$, where dV is a small volume of space, $d\nu$ is the width of the frequency band we wish to include, and $d\Omega$ measures the range of solid angles we wish to include.

Since electromagnetic waves in a vacuum move with constant speed c, the energy *flux* through a flat patch perpendicular to $\hat{n}$ with area dA is simply $c\Sigma dA d\nu \, d\Omega$, which defines the flux spectrum $c\Sigma$. In SI units, the flux spectrum has units of (watts/m^2)/(Hz · steradian), where the hertz (Hz) is the unit of frequency, equal to one cycle per second. The flux spectrum defined in this way is usually called the *spectral irradiance*; integrated over all frequencies, it is called the *irradiance*.

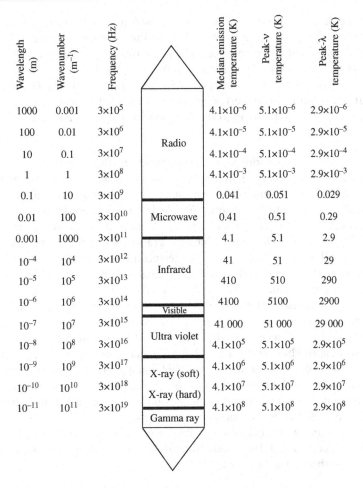

Wavelength (m)	Wavenumber (m⁻¹)	Frequency (Hz)		Median emission temperature (K)	Peak-ν temperature (K)	Peak-λ temperature (K)
1000	0.001	$3{\times}10^5$		$4.1{\times}10^{-6}$	$5.1{\times}10^{-6}$	$2.9{\times}10^{-6}$
100	0.01	$3{\times}10^6$	Radio	$4.1{\times}10^{-5}$	$5.1{\times}10^{-5}$	$2.9{\times}10^{-5}$
10	0.1	$3{\times}10^7$		$4.1{\times}10^{-4}$	$5.1{\times}10^{-4}$	$2.9{\times}10^{-4}$
1	1	$3{\times}10^8$		$4.1{\times}10^{-3}$	$5.1{\times}10^{-3}$	$2.9{\times}10^{-3}$
0.1	10	$3{\times}10^9$		0.041	0.051	0.029
0.01	100	$3{\times}10^{10}$	Microwave	0.41	0.51	0.29
0.001	1000	$3{\times}10^{11}$		4.1	5.1	2.9
10^{-4}	10^4	$3{\times}10^{12}$		41	51	29
10^{-5}	10^5	$3{\times}10^{13}$	Infrared	410	510	290
10^{-6}	10^6	$3{\times}10^{14}$		4100	5100	2900
			Visible			
10^{-7}	10^7	$3{\times}10^{15}$		41 000	51 000	29 000
10^{-8}	10^8	$3{\times}10^{16}$	Ultra violet	$4.1{\times}10^5$	$5.1{\times}10^5$	$2.9{\times}10^5$
10^{-9}	10^9	$3{\times}10^{17}$	X-ray (soft)	$4.1{\times}10^6$	$5.1{\times}10^6$	$2.9{\times}10^6$
10^{-10}	10^{10}	$3{\times}10^{18}$	X-ray (hard)	$4.1{\times}10^7$	$5.1{\times}10^7$	$2.9{\times}10^7$
10^{-11}	10^{11}	$3{\times}10^{19}$	Gamma ray	$4.1{\times}10^8$	$5.1{\times}10^8$	$2.9{\times}10^8$

Figure 3.1 The electromagnetic spectrum. The median emission temperature is the temperature of a blackbody for which half of the emitted power is below the given frequency (or equivalently, wavelength or wavenumber). The peak-ν temperature is the temperature of a blackbody for which the peak of the Planck density in frequency space is at the stated frequency. The peak-λ temperature is the temperature of a blackbody for which the peak of the Planck density in wavelength space is at the stated wavelength. The infrared (or *IR*) spectrum is subdivided into *near infrared* having wavelengths shorter than 5 μm and *thermal infrared* having wavelengths between 5 and 200 μm. The exact boundaries of these subdivisions are not rigidly defined. Bodies with Earthlike temperatures radiate primarily in the thermal infrared, whereas all Main Sequence stars have significant output in the near infrared, as well as visible and ultraviolet. The ultraviolet spectrum is subdivided in various ways, depending on the application. *UVA* is 0.315 to 0.4 μm, *UVB* is 0.28 to 0.315 μm and *UVC* is 0.1 to 0.28 μm. *Far ultraviolet (FUV)* is 0.122 to 0.2 μm while *extreme ultraviolet (EUV)* is 0.01 to 0.122 μm.

Exercise 3.1 The SI unit of energy is the *joule*, J, which is 1 newton · meter. A watt (W) is 1 J/s. A typical resting human in not-too-cold weather requires about 2000 calories/day. (A Calorie is the amount of energy needed to increase the temperature of 1 kg of pure water by 1 K.) Convert this to a power consumption in W, using the fact that 1 Calorie = 4184 J.

On the average, the flux of solar energy reaching the Earth's surface is about 240 W/m². Assuming that food plants can convert solar energy to usable food calories with an efficiency of 1%, what is the maximum population the Earth could support? (The radius of the Earth is about 6371 km.)

The bold assumption introduced by Max Planck is that electromagnetic energy is exchanged only in amounts that are multiples of discrete *quanta*, whose size depends on the frequency of the radiation, in much the same sense that a penny is the quantum of US currency. Specifically, the quantum of energy for electromagnetic radiation having frequency ν is $\Delta E = h\nu$, where h is now known as *Planck's constant*. It is (so far as currently known) a constant of the universe, which determines the granularity of reality. Planck's constant is an exceedingly small number ($6.626 \cdot 10^{-34}$ joule-seconds), so quantization of energy is not directly manifest as discreteness in the energy changes of everyday objects. A 1 watt blue nightlight (wavelength 0.48 μm, or frequency $6.24 \cdot 10^{14}$ Hz), emits $2.4 \cdot 10^{18}$ photons each second, so it is no surprise that the light appears to be a continuous stream. If a bicycle were hooked to an electrical brake that dissipated energy by driving a blue light, emitting photons, the bike would indeed slow down in discontinuous increments, but the velocity increment, assuming the bike and rider to have a mass of 80 kg, would be only 10^{-10} m/s; if one divides a 1 m/s decrease of speed into 10^{10} equal parts, the deceleration will appear entirely continuous to the rider. Nonetheless, the aggregate effect of microscopic graininess of energy transitions exerts a profound influence on the macroscopic properties of everyday objects. Blackbody radiation is a prime example of this.

Once the quantum assumption was introduced, Planck was able to compute the irradiance (flux spectrum) of blackbody radiation with temperature T using standard thermodynamic methods. The answer is

$$B(\nu, T) = \frac{2h\nu^3}{c^2} \frac{1}{e^{h\nu/kT} - 1} \tag{3.1}$$

where k is the Boltzmann thermodynamic constant defined in Chapter 2. $B(\nu, T)$ is known as the *Planck function*. Note that the Planck function is independent of the direction of the radiation; this is because blackbody radiation is *isotropic*, i.e. equally intense in all directions. In a typical application of the Planck function, we wish to know the flux of energy exiting the surface of a blackbody through a small nearly flat patch with area dA, over a frequency band of width $d\nu$. Since energy exits through this patch at all angles, we must integrate over all directions. However, energy exiting in a direction which makes an angle θ to the normal to the patch contributes a flux $(B\,dA\,d\nu\,d\Omega)\cos\theta$ through the patch, since the component of flux parallel to the patch carries no energy through it. Further, using the definition of a steradian, $d\Omega = 2\pi d\cos\theta$ for the set of all rays making an angle θ relative to the normal to the patch. Integrating $B\cos\theta\,d\Omega$ from $\theta = 0$ to $\theta = \pi/2$, and using the fact that B is independent of direction, we then find that the flux through the patch is $\pi B\,dA\,d\nu$. This is also the amount of electromagnetic energy in a frequency band of width $d\nu$ that would pass each second through a hoop enclosing area dA (from one chosen side to the other), placed in the interior of an ideal blackbody; an equal amount passes through the hoop in the opposite sense.

The way the angular distribution of the radiation is described by the Planck function can be rather confusing, and requires a certain amount of practice to get used to. The following exercise will test the readers' comprehension of this matter.

Exercise 3.2 A radiation detector flies on an airplane a distance H above an infinite flat plain with uniform temperature T. The detector is connected to a watt-meter which reports the total radiant power captured by the detector. The detector is sensitive to rays coming in at angles $\leq \delta\theta$ relative to the direction in which the detector is pointed. The area of the aperture of the detector is δA. The detector is sensitive to frequencies within a small range $\delta\nu$ centered on ν_0.

If the detector is pointed straight down, what is the power received by the detector? What is the size of the "footprint" on the plain to which the detector is sensitive? How much power is emitted by this footprint in the detector's frequency band? Why is this power different from the power received by the detector?

How do your answers change if the detector is pointed at an angle of $45°$ relative to the vertical, rather than straight down?

The Planck function depends on frequency only through the dimensionless variable $u = h\nu/(kT)$. Recalling that each degree of freedom has energy $\frac{1}{2}kT$ in the average, we see that u is half the ratio of the quantum of energy at frequency ν to the typical energy in a degree of freedom of the matter with which the electromagnetic energy is in equilibrium. When u is large, the typical energy in a degree of freedom cannot create even a single photon of frequency ν, and such photons can be emitted only by those rare molecules with energy far above the mean. This is the essence of the way quantization affects the blackbody distribution – through inhibition of emission of high-frequency photons. On the other hand, when u is small, the typical energy in a degree of freedom can make many photons of frequency ν, and quantization imposes less of a constraint on emission. The characteristic frequency kT/h defines the crossover between the classical world and the quantum world. Much lower frequencies are little affected by quantization, whereas much higher frequencies are strongly affected. At 300 K, the crossover frequency is 6240 gigahertz, corresponding to a wavenumber of $20\,814\,\mathrm{m}^{-1}$, or a wavelength of 48 μm; this is in the far infrared range.

In terms of u, the Planck function can be re-written

$$B(\nu, T) = \frac{2k^3 T^3}{h^2 c^2} \frac{u^3}{e^u - 1}. \tag{3.2}$$

In the classical limit, $u \ll 1$, and $u^3/(\exp(u) - 1) \approx u^2$. Hence, $B \approx 2kT\nu^2/c^2$, which is independent of h. In a classical world, where $h = 0$, this form of the spectrum would be valid for all frequencies, and the emission would increase quadratically with frequency without bound; a body with any non-zero temperature would emit infrared at a greater rate than microwaves, visible light at a greater rate than infrared, ultraviolet at a greater rate than visible, X-rays at a greater rate than ultraviolet, and so forth. Bodies in equilibrium would cool to absolute zero almost instantaneously through emission of a burst of gamma rays, cosmic rays, and even higher-frequency radiation. This is clearly at odds with observations, not least the existence of the Universe. We are saved from this catastrophe by the fact that h is non-zero, which limits the range of validity of the classical form of B. At frequencies high enough to make $u \gg 1$, then $u^3/(\exp(u)-1) \approx u^3\exp(-u)$ and the spectrum decays somewhat more slowly than exponentially as frequency is increased. The peak of B occurs at $u \approx 2.821$, implying that the frequency of maximum emission is $\nu \approx (2.821k/h)T \approx 58.78 \cdot 10^9 T$. The peak of the frequency spectrum increases linearly with temperature. This behavior, first deduced empirically long before it was explained by quantum theory, is known as the *Wien Displacement Law.*

Because the emission decays only quadratically on the low-frequency side of the peak, but decays exponentially on the high-frequency side, bodies emit appreciable energy at frequencies much lower than the peak emission, but very little at frequencies much higher. For example, at one-tenth the peak frequency, a body emits at a rate of 4.8% of the maximum value. However, at ten times the peak frequency, the body emits at a rate of only $8.9 \cdot 10^{-9}$ of the peak emission. The microwave emission from a portion of the Earth's atmosphere with temperature 250 K (having peak emission in the infrared) is readily detectable by satellites, whereas the emission of visible light is not.

Since B is a density, one cannot obtain the corresponding distribution in wavenumber or wavelength space by simply substituting for ν in terms of wavenumber or wavelength in the formula for B. One must also take into account the transformation of $d\nu$. For example, to get the flux density in wavenumber space (call it B_n) we use $B(\nu, T)d\nu = B(n \cdot c, T)d(n \cdot c) = cB(n \cdot c, T)dn$, whence $B_n(n, T) = cB(n \cdot c, T)$. Thus, transforming to wavenumber space changes the amplitude but not the shape of the flux spectrum. The Planck density in wavenumber space is shown for various temperatures in Fig. 3.2. Because the transformation of the density from frequency to wavenumber space only changes the labeling of the vertical axis of the graph, one can obtain the wavenumber of maximum emission in terms of the frequency of maximum emission using $n_{max} = \nu_{max}/c$. An important property of the Planck function, readily verified by a simple calculation, is that $dB/dT > 0$ for all wavenumbers. This means that the Planck function for a large temperature is strictly above one for a lower temperature, or equivalently, that increasing temperature increases the emission at each individual wavenumber.

If one transforms to wavelength space, however,

$$B(\nu, T)d\nu = B(c/\lambda, T)d(c/\lambda) = -\frac{c}{\lambda^2}B(c/\lambda, T)d\lambda = \frac{2k^5 T^5}{h^4 c^3}\frac{u^5}{e^u - 1}d\lambda = B_\lambda d\lambda \qquad (3.3)$$

where $u = h\nu/kT = hc/\lambda kT$, as before. Transforming to wavelength space changes the shape of the flux spectrum. B_λ has its maximum at $u \approx 4.965$, which is nearly twice as large as the value for the wavenumber or frequency spectrum.

Since the location of the peak of the flux spectrum depends on the coordinate used to measure position within the electromagnetic spectrum, this quantity has no intrinsic physical meaning, apart from being a way to characterize the shape of the curve coming out of some particular kind of measuring apparatus. A more meaningful quantity can be derived from the *cumulative flux spectrum*, whose value at a given point in the spectrum is the same regardless of whether we use wavenumber, wavelength, $\log \lambda$, or any other coordinate to describe the position within the spectrum. The cumulative flux spectrum is defined as

$$F_{cum}(\nu, T) = \int_0^\nu \pi B(\nu', T)d\nu' = \int_\infty^\lambda \pi B_\lambda(\lambda', T)d\lambda'. \qquad (3.4)$$

Note that in defining the cumulative emission we have included the factor π which results from integrating over all angles of emission in a hemisphere. $F_{cum}(\nu, T)$ thus gives the power emitted per square meter for all frequencies less than ν, or equivalently, for all wavelengths greater than c/ν. This function is shown for various temperatures in the lower panel of Fig. 3.2, where it is plotted as a function of wavenumber. The value of ν for which $F_{cum}(\nu, T)$ reaches half the net emission $F_{cum}(\infty, T)$ provides a natural characterization of the spectrum.

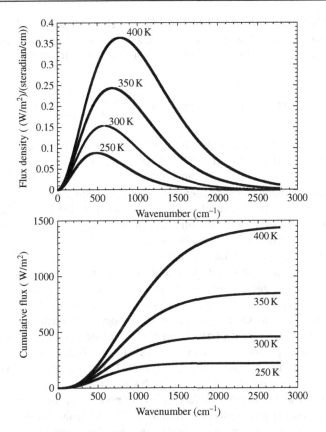

Figure 3.2 The spectrum of blackbody radiation for the various temperatures indicated on the curves. Upper panel: The Planck density in wavenumber space. Lower panel: The cumulative emission as a function of wavenumber. Note that the density has been transformed such that the density times *dn* is the power per unit solid angle per unit area radiated in a wavenumber interval of width *dn*.

We will refer to this characteristic frequency as the *median emission frequency*. The median emission wavelength and wavenumber is defined analogously. Whether one uses frequency, wavelength or some other measure, the median emission is attained at $u \approx 3.503$. For any given coordinate used to describe the spectrum, the (angle-integrated) Planck density in that coordinate is the derivative of the cumulative emission with respect to the coordinate. Hence the peak in the Planck density just gives the point at which the cumulative emission function has its maximum slope. This depends on the coordinate used, unlike the point of median emission. Figure 3.1 shows the portion of the spectrum in which blackbodies with various temperatures dominantly radiate. For example, a body with a temperature of around 4 K radiates in the microwave region; this is the famous "Cosmic Microwave Background Radiation" left over from the Big Bang.[1] A body with a temperature of 300 K radiates in the

---

[1] What is remarkable about this observed cosmic radiation is not so much that it is in the microwave region, but that it has a blackbody spectrum, which says much about the interaction of radiation with matter in the early moments of the Universe.

infrared, one with a temperature of a few thousand kelvin radiates in the visible, and one with a temperature of some tens of thousands of kelvin would radiate in the ultraviolet.

Next, we evaluate $F_{cum}(\infty, T)$, to obtain the total power F exiting from each unit area of the surface of a blackbody:

$$F = \int_0^\infty \pi B(\nu, T) d\nu = \int_0^\infty \pi B(u, T) \frac{kT}{h} du = \left[\frac{2\pi k^4}{h^3 c^2} \int_0^\infty \frac{u^3}{e^u - 1} du \right] T^4 = \sigma T^4 \qquad (3.5)$$

where[2] $\sigma = 2\pi^5 k^4 / (15 c^2 h^3) \approx 5.67 \cdot 10^{-8} \, \mathrm{W/m^2 K^4}$. The constant σ is known as the *Stefan-Boltzmann constant*, and the law $F = \sigma T^4$ is the *Stefan-Boltzmann law*. This law was originally deduced from observations, and Boltzmann was able to derive the fourth-power scaling in temperature using classical thermodynamic reasoning. However, classical physics yields an infinite value for the constant σ. The formula for σ clearly reveals the importance of quantum effects in determining this constant, since σ diverges like $1/h^3$ if we try to pass to the classical limit by making h approach zero.

An important property of an ideal blackbody is that the radiation leaving its surface depends only on the temperature of the body. If a blackbody is interposed between an observer and some other object, all properties of the object will be hidden from the observer, who will see only blackbody radiation corresponding to the temperature of the blackbody. This remark allows us to make use of blackbody theory to determine the emission from objects whose temperature varies greatly from place to place, even though blackbody theory applies, strictly speaking, only to extensive bodies with uniform temperature. For example, the temperature of the core of the Earth is about 6000 K, but we need not know this in order to determine the radiation emitted from the Earth's surface; the outermost few millimeters of rock, ice, or water at the Earth's surface contain enough matter to act like a blackbody to a very good approximation. Hence, the radiation emitted from the surface depends only on the temperature of this outer skin of the planet. Similarly, the temperature of the core of the Sun is about 16 000 000 K and even at a distance from the center equal to 90% of the visible radius, the temperature is above 600 000 K. However, the Sun is encased in a layer a few hundred kilometers thick which is sufficiently dense to act like a blackbody, and which has a temperature of about 5780 K. This layer is known as the *photosphere*, because it is the source of most light exiting the Sun. Layers farther out from the center of the Sun can be considerably hotter than the photosphere, but they have a minimal effect on solar radiation because they are so tenuous. In Chapter 4 we will develop more precise methods for dealing with tenuous objects, such as atmospheres, which peter out gradually without having a sharply defined boundary.

An ideal blackbody would be opaque at all wavelengths, but it is a common situation that a material acts as a blackbody only in a limited range of wavelengths. Consider the case of window glass: It is transparent to visible light, but if you could see it in the infrared it would look as opaque as stone. Because it interacts strongly with infrared light, window glass emits blackbody radiation in the infrared range. At temperatures below a few hundred K, there is little blackbody emission at wavelengths shorter than the infrared, so at such temperatures the net power per unit area emitted by a pane of glass with temperature T is very nearly σT^4, even though it doesn't act like a blackbody in the visible range. Liquid water, and water ice, behave similarly. Crystalline table salt, and carbon dioxide ice, are nearly transparent in

---

[2] The definite integral $\int_0^\infty (u^3/(e^u - 1)) du$ was determined by Euler, as a special case of his study of the behavior of the Riemann zeta function at even integers. It is equal to $6\zeta(4) = \pi^4/15$.

the infrared as well as in the visible, and in consequence emit radiation at a much lower rate than expected from the blackbody formula. (They would make fine windows for creatures having infrared vision.) There is, in fact, a deep and important relation between absorption and emission of radiation, which will be discussed in Section 3.5.

3.3 RADIATION BALANCE OF PLANETS

As a first step in our study of the temperature of planets, let's consider the following idealized case:

- The only source of energy heating the planet is absorption of light from the planet's host star.
- The *planetary albedo*, or proportion of sunlight reflected by the planet as a whole including its atmosphere, is spatially uniform.
- The planet is spherical, and has a distinct solid or liquid surface which radiates like a perfect blackbody.
- The planet's temperature is uniform over its entire surface.
- The planet's atmosphere is perfectly transparent to the electromagnetic energy emitted by the surface.

The uniform temperature assumption presumes that the planet has an atmosphere or ocean which is so well stirred that it is able to mix heat rapidly from one place to another, smoothing out the effects of geographical fluctuations in the energy balance. The Earth conforms fairly well to this approximation. The equatorial annual mean temperature is only 4% above the global mean temperature of 286 K, while the north polar temperature is only 10% below the mean. The most extreme deviation occurs on the high Antarctic plateau, where the annual mean south polar temperature is 21% below the global mean. The surface temperature of Venus is even more uniform than that of Earth. That of Mars, which in our era has a thin atmosphere and no ocean, is less uniform. Airless, rocky bodies like the Moon and Mercury do not conform at all well to the uniform temperature approximation.

Light leaving the upper layers of the Sun and most other stars takes the form of blackbody radiation. It is isotropic, and its flux and flux spectrum conform to the blackbody law corresponding to the temperature of the photosphere, from which the light escapes. Once the light leaves the surface of the star, however, it expands through space and does not interact significantly with matter except where it is intercepted by a planet. Therefore, it is no longer blackbody radiation, though it retains the blackbody spectrum. In the typical case of interest, the planet orbits its star at a distance that is much greater than the radius of the star, and itself has a radius that is considerably smaller than the star and is hence yet smaller than the orbital distance. In this circumstance, all the rays of light which intersect the planet are very nearly parallel to the line joining the center of the planet to the center of its star; the sunlight comes in as a nearly *parallel beam*, rather than being isotropic, as would be the case for true blackbody radiation. The parallel-beam approximation is equivalent to saying that, as seen from the planet, the Sun occupies only a small portion of the sky, and as seen from the Sun the planet also occupies only a small portion of the sky. Even for Mercury, with a mean orbital distance of 58 000 000 km, the Sun (whose radius is 695 000 km) occupies an angular width in the sky of only about $2 \cdot (695\,000/58\,000\,000)$ radians, or 1.4°.

The solar flux impinging on the planet is also reduced, as compared with the solar flux leaving the photosphere of the star. The total energy per unit frequency leaving the star is $4\pi r_\odot^2 (\pi B(\nu, T_\odot))$, where $r_\odot$ is the radius of the star and $T_\odot$ is the temperature of its photosphere. At a distance r from the star, the energy has spread uniformly over a sphere whose surface area is $4\pi r^2$; hence at this distance, the energy flux per unit frequency is $\pi B r_\odot^2 / r^2$, and the total flux is $\sigma T_\odot^4 r_\odot^2 / r^2$. The latter is the flux seen by a planet at orbital distance r, in the form of a beam of parallel rays. It is known as the solar "constant," and will be denoted by $L_\odot$, or sometimes simply L where there is no risk of confusion with latent heat. The solar (or stellar) "constant" depends on a planet's distance from its star, but the *luminosity* of the star is an intrinsic property of the star at any given stage of the star's life. The stellar luminosity is the net power output of a star, and if the star's emission can be represented as blackbody radiation, the luminosity is given by $\mathcal{L}_\odot = 4\pi r_\odot^2 \sigma T_\odot^4$. In this book the terms "solar constant," "stellar constant," "solar flux," and "stellar flux" are all to be considered synonymous, and to refer to the parallel-beam flux from the star measured at any given position in a planet's orbit. Most of the time, the terms will be used to refer to the average flux over the planet's year, though in Chapter 7 we will be concerned with the seasonal variations of the fluxes for planets in non-circular orbits. The term "solar constant" as used here must be distinguished from the number $L_\odot$ which sometimes goes by the same name, and which is an actual constant approximately equal to the Sun's flux at the Earth's mean orbit, averaged over the 11-year solar cycle.

We are now equipped to compute the energy balance of the planet, subject to the preceding simplifying assumptions. Let a be the planet's radius. Since the cross-sectional area of the planet is πa^2 and the solar radiation arrives in the form of a nearly parallel beam with flux $L_\odot$, the energy per unit time impinging on the planet's surface is $\pi a^2 L_\odot$; the rate of energy absorption is $(1 - \alpha)\pi a^2 L_\odot$, where α is the albedo. The planet loses energy by radiating from its entire surface, which has area $4\pi a^2$. Hence the rate of energy loss is $4\pi a^2 \sigma T^4$, where T is the temperature of the planet's surface. In equilibrium the rate of energy loss and gain must be equal. After cancelling a few terms, this yields

$$\sigma T^4 = \frac{1}{4}(1 - \alpha)L_\odot. \tag{3.6}$$

Note that this is independent of the radius of the planet. The factor $\frac{1}{4}$ comes from the ratio of the planet's cross-sectional area to its surface area, and reflects the fact that the planet intercepts only a disk of the incident solar beam, but radiates over its entire spherical surface. This equation can be readily solved for T. If we substitute for $L_\odot$ in terms of the photospheric temperature, the result is

$$T = \frac{1}{\sqrt{2}}(1 - \alpha)^{1/4} \sqrt{\frac{r_\odot}{r}} T_\odot. \tag{3.7}$$

Formula 3.7 shows that the blackbody temperature of a planet is much less than that of the photosphere, so long as the orbital distance is large compared with the stellar radius. From the displacement law, it follows that the planet loses energy through emission at a distinctly lower wavenumber than that at which it receives energy from its star. This situation is illustrated in Fig. 3.3. For example, the energy received from our Sun has a median wavenumber of about $15\,000\,\mathrm{cm}^{-1}$, equivalent to a wavelength of about $0.7\,\mu\mathrm{m}$. An isothermal planet at Mercury's orbit would radiate to space with a median emission

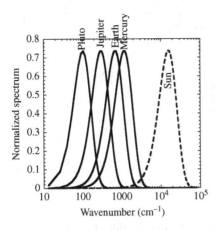

Figure 3.3 The Planck density of radiation emitted by the Sun and selected planets in radiative equilibrium with absorbed solar radiation (based on the observed short-wave albedo of the planets). The Planck densities are transformed to a logarithmic spectral coordinate, and all are normalized to unit total emission.

wavenumber of $1100\,\mathrm{cm}^{-1}$, corresponding to a wavelength of $9\,\mu\mathrm{m}$. An isothermal planet at the orbit of Mars would radiate with a median wavenumber of $550\,\mathrm{cm}^{-1}$, corresponding to a wavelength of $18\,\mu\mathrm{m}$.

Exercise 3.3 A planet with zero albedo is in orbit around an exotic hot star having a photospheric temperature of $100\,000\,\mathrm{K}$. The ratio of the planet's orbit to the radius of the star is the same as for Earth (about 215). What is the median emission wavenumber of the star? In what part of the electromagnetic spectrum does this lie? What is the temperature of the planet? In what part of the electromagnetic spectrum does the planet radiate? Do the same if the planet is instead in orbit around a brown dwarf star with a photospheric temperature of $600\,\mathrm{K}$.

The separation between absorption and emission wavenumber will prove very important when we bring a radiatively active atmosphere into the picture, since it allows the atmosphere to have a different effect on incoming vs. outgoing radiation. Since the outgoing radiation has longer wavelength than the incoming radiation, the flux of emitted outgoing radiation is often referred to as *outgoing longwave radiation*, and denoted by *OLR*. For a non-isothermal planet, the *OLR* is a function of position (e.g. latitude and longitude on an imaginary sphere tightly enclosing the planet and its atmosphere). We will also use the term to refer to the outgoing flux averaged over the surface of the sphere, even when the planet is not isothermal. As for the other major term in the planet's energy budget, we will refer to the electromagnetic energy received from the planet's star as the *shortwave*, *solar*, or *stellar* flux. Most stars, our Sun not excepted, have their primary output in the visible, ultraviolet, and near-infrared part of the spectrum, all of which are shorter in wavelength than the thermal infrared – the *OLR* – by which planets lose energy to space. (See Fig. 3.1 for the definitions of ultraviolet, near infrared, and thermal infrared.)

Formula 3.7 is plotted in Fig. 3.4 for a hypothetical isothermal planet with zero albedo. Because of the square-root dependence on orbital distance, the temperature varies only weakly with distance, except very near the star. Neglecting albedo and atmospheric effects, Earth would have a mean surface temperature of about $280\,\mathrm{K}$. Venus would be only $50\,\mathrm{K}$ warmer than the Earth and Mars only $53\,\mathrm{K}$ colder. At the distant orbit of Jupiter, the blackbody equilibrium temperature falls to $122\,\mathrm{K}$, but even at the vastly more distant orbit of Neptune the temperature is still as high as $50\,\mathrm{K}$. The emission from all of these planets lies

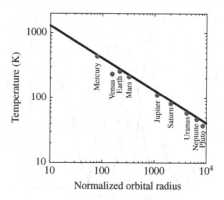

Figure 3.4 The equilibrium blackbody temperature of an isothermal spherical zero-albedo planet, as a function of distance from a Sun having a photospheric temperature of 5800 K. The orbital distance is normalized by the radius of the Sun. Dots show the equilibrium blackbody temperature of the Solar System planets, based on their actual observed albedos.

in the infrared range, though the colder planets radiate in the deeper (lower wavenumber) infrared. An exception to the strong separation between stellar and planetary temperature is provided by the "roasters" – a recently discovered class of extrasolar giant planets with $\frac{r}{r_\odot}$ as low as 5. Such planets can have equilibrium blackbody temperatures as much as a third that of the photosphere of the parent star. For these planets, the distinction between the behavior of incoming and outgoing radiation is less sharp.

It is instructive to compare the ideal blackbody temperature with observed surface temperature for the three Solar System bodies which have both a distinct surface and a thick enough atmosphere to enforce a roughly uniform surface temperature: Venus, Earth, and Saturn's moon Titan. For this comparison, we calculate the blackbody temperature using the observed planetary albedos, instead of assuming a hypothetical zero albedo planet as in Fig. 3.4. Venus is covered by thick, highly reflective clouds, which raise its albedo to 0.75. The corresponding isothermal blackbody temperature is only 232 K (as compared with 330 K in the zero albedo case). This is far less than the observed surface temperature of 740 K. Clearly, the atmosphere of Venus exerts a profound warming effect on the surface. The warming arises from the influence of the atmosphere on the infrared emission of the planet, which we have not yet taken into account. Earth's albedo is on the order of 0.3, leading to a blackbody temperature of 255 K. The observed mean surface temperature is about 285 K. Earth's atmosphere has a considerably weaker warming effect than that of Venus, but it is nonetheless a very important warming, since it brings the planet from subfreezing temperatures where the oceans would almost certainly become ice-covered, to temperatures where liquid water can exist over most of the planet. The albedo of Titan is 0.21, and using the solar constant at Saturn's orbit we find a blackbody temperature of 85 K. The observed surface temperature is about 95 K, whence we conclude that the infrared effects of Titan's atmosphere moderately warm the surface.

The way energy balance determines surface temperature is illustrated graphically in Fig. 3.5. One first determines the way in which the mean infrared emission per unit area depends on the mean surface temperature T_s; for the isothermal blackbody calculation, this curve is simply σT_s^4. The equilibrium temperature is determined by the point at which the *OLR* curve intersects the curve giving the absorbed solar radiation (a horizontal line in the present calculation). In some sense, the whole subject of climate comes down to an ever-more sophisticated hierarchy of calculations of the curve $OLR(T_s)$; our attention will soon turn to the task of determining how the *OLR* curve is affected by an atmosphere. With increasing sophistication, we will also allow the solar absorption to vary with T_s, owing to changing clouds, ice cover, vegetation cover, and other characteristics.

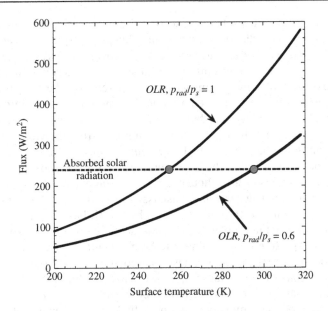

Figure 3.5 Determination of a planet's temperature by balancing absorbed solar energy against emitted longwave radiation. The horizontal line gives the absorbed solar energy per unit surface area, based on an albedo of 0.3 and a solar constant of $1370\,\mathrm{W/m^2}$. The *OLR* is given as a function of surface temperature. The upper curve assumes the atmosphere has no greenhouse effect ($p_{rad} = p_s$), while the lower *OLR* curve assumes $p_{rad}/p_s = 0.6$, a value appropriate to the present Earth.

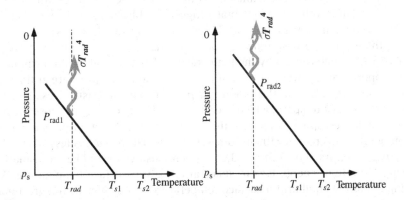

Figure 3.6 Sketch illustrating how the greenhouse effect increases the surface temperature. In equilibrium, the outgoing radiation must remain equal to the absorbed solar radiation, so T_{rad} stays constant. However, as more greenhouse gas is added to the atmosphere, p_{rad} is reduced, so one must extrapolate temperature further along the adiabat to reach the surface.

We will now consider an idealized thought experiment which illustrates the essence of the way an atmosphere affects *OLR*. Suppose that the atmosphere has a temperature profile $T(p)$ which decreases with altitude, according to the dry or moist adiabat. Let p_s be the surface pressure, and suppose that the ground is strongly thermally coupled to the atmosphere by turbulent heat exchanges, so that the ground temperature cannot deviate much from that of the immediately overlying air. Thus, $T_s = T(p_s)$. If the atmosphere were transparent to infrared, as is very nearly the case for nitrogen or oxygen, the *OLR* would be σT_s^4. Now, let's stir an additional gas into the atmosphere, and assume that it is well mixed with uniform mass concentration q. This gas is transparent to solar radiation but interacts strongly enough with infrared that when a sufficient amount is mixed into a parcel of air, it turns that parcel into an ideal blackbody. Such a gas, which is fairly transparent to the incoming shortwave stellar radiation but which interacts strongly with the outgoing (generally infrared) emitted radiation, is called a *greenhouse gas*, and the corresponding effect on planetary temperature is called the *greenhouse effect*. Carbon dioxide, water vapor, and methane are some examples of greenhouse gases, and the molecular properties that make a substance a good greenhouse gas will be discussed in Chapter 4. The mass of greenhouse gas that must be mixed into a column of atmosphere with base of $1\,\text{m}^2$ in order to make that column act begin to act like a blackbody is characterized by the *absorption coefficient* κ, whose units are m^2/kg. Here we'll assume κ to be independent of frequency, temperature, and pressure, though for real greenhouse gases, κ depends on all of these. Since the mass of greenhouse gas in a column of thickness Δp in pressure coordinates is $q\Delta p/g$, then the definition of κ implies that the slab acts like a blackbody when $\kappa q\Delta p/g > 1$. When $\kappa q p_s/g < 1$ then the entire mass of the atmosphere is not sufficient to act like a blackbody and the atmosphere is said to be *optically thin*. For optically thin atmospheres, infrared radiation can escape from the surface directly to space, and is only mildly attenuated by atmospheric absorption. When $\kappa q p_s/g \gg 1$, the atmosphere is said to be *optically thick*.

If the atmosphere is optically thick, we can slice the atmosphere up into a stack of slabs with thickness Δp_1 such that $\kappa q\Delta p_1/g = 1$. Each of these slabs radiates like an ideal blackbody with temperature approximately equal to the mean temperature of the slab. Recall, however, that another fundamental property of blackbodies is that they are perfect *absorbers* (though if they are only blackbodies in the infrared, they will only be perfect absorbers in the infrared). Hence *infrared radiation escapes to space only from the topmost slab*. The *OLR* will be determined by the temperature of this slab alone, and will be insensitive to the temperature of lower portions of the atmosphere. The pressure at the bottom of the topmost slab is Δp_1. We can thus identify Δp_1 as the characteristic pressure level from which radiation escapes to space, which therefore will be called p_{rad} in subsequent discussions. The radiation escaping to space – the *OLR* – will then be approximately $\sigma T(p_{rad})^4$. Because temperature decreases with altitude on the adiabat the *OLR* is less than σT_s^4 to the extent that $p_{rad} < p_s$. As shown in Fig. 3.6, a greenhouse gas acts like an insulating blanket, reducing the rate of energy loss to space at any given surface temperature. All other things being equal the equilibrium surface temperature of a planet with a greenhouse gas in its atmosphere must be greater than that of a planet without a greenhouse gas, in order to radiate away energy at a sufficient rate to balance the absorbed solar radiation. The key insight to be taken from this discussion is that *the greenhouse effect only works to the extent that the atmosphere is colder at the radiating level than it is at the ground*.

For real greenhouse gases, the absorption coefficient varies greatly with frequency. Such gases act on the *OLR* by making the atmosphere very optically thick at some frequencies, less optically thick at others, and perhaps even optically thin at still other frequencies. In portions of the spectrum where the atmosphere is more optically thick, the emission to space originates in higher (and generally colder) parts of the atmosphere. In reality, then, the infrared escaping to space is a blend of radiation emitted from a range of atmospheric levels, with some admixture of radiation from the planet's surface as well. The concept of an *effective radiating level* nonetheless has merit for real greenhouse gases. It does not represent a distinct physical layer of the atmosphere, but rather characterizes the mean depth from which infrared photons escape to space. As more greenhouse gas is added to an atmosphere, more of the lower parts of the atmosphere become opaque to infrared, preventing the escape of infrared radiation from those regions. This increases the altitude of the effective radiating level (i.e. decreases p_{rad}). Some of the implications of a frequency-dependent absorption coefficient are explored in Problem 3.29, and the subject will be taken up at great length in Chapter 4.

From an observation of the actual *OLR* emitted by a planet, one can determine an equivalent blackbody radiating temperature T_{rad} from the expression $\sigma T_{rad}^4 = OLR$. This temperature is the infrared equivalent of the Sun's photospheric temperature; it is a kind of mean temperature of the regions from which infrared photons escape, and p_{rad} represents a mean pressure of these layers. For planets for which absorbed solar radiation is the only significant energy source, T_{rad} is equal to the ideal blackbody temperature given by Eq. (3.7). The arduous task of relating the effective radiating level to specified concentrations of real greenhouse gases is treated in Chapter 4.

Figure 3.7 illustrates the reduction of infrared emission caused by the Earth's atmosphere. At every latitude, the observed *OLR* is much less than it would be if the planet radiated to space at its observed surface temperature. At the Equator the observed *OLR* is $238 \, W/m^2$, corresponding to a radiating temperature of $255 \, K$. This is much less than the observed surface temperature of $298K$, which would radiate at a rate of $446 \, W/m^2$ if the atmosphere did not intervene. It is interesting that the gap between observed *OLR* and the computed surface emission is less in the cold polar regions, and especially small at the winter pole. This happens partly because, at low temperatures, there is simply less infrared emission for the atmosphere to trap. However, differences in the water content of the atmosphere, and differences in the temperature profile, can also play a role. These effects will be explored in Chapter 4.

Gases are not the only atmospheric constituents which affect *OLR*. Clouds consist of particles of condensed substance small enough to stay suspended for a long time. They can profoundly influence *OLR*. Gram for gram, condensed water interacts much more strongly with infrared than does water vapor. In fact, a mere 20 grams of water in the form of liquid

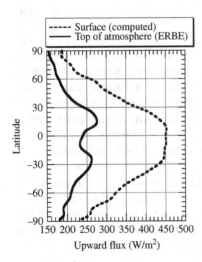

Figure 3.7 The Earth's observed zonal-mean *OLR* for January, 1986. The observations were taken by satellite instruments during the Earth Radiation Budget Experiment (ERBE), and are averaged along latitude circles. The figure also shows the radiation that would be emitted to space by the surface (σT_s^4) if the atmosphere were transparent to infrared radiation.

droplets of a typical size is sufficient to turn a column of air 500 m thick by 1 m square into a very nearly ideal blackbody. To a much greater extent than for greenhouse gases, a water cloud layer in an otherwise infrared-transparent atmosphere really can be thought of as a discrete radiating layer. The prevalence of clouds in the high, cold regions of the tropical atmosphere accounts for the dip in *OLR* near the Equator, seen in Fig. 3.7. Clouds are unlike greenhouse gases, though, since they also strongly reflect the incoming solar radiation. It's the tendency of these two large effects to partly cancel that makes the problem of the influence of clouds on climate so challenging. Not all condensed substances absorb infrared as well as water does. Liquid methane (important on Titan) and CO_2 ice (important on present and Early Mars) are comparatively poor infrared absorbers. They affect *OLR* in a fundamentally different way, through reflection instead of absorption and emission. This will be discussed in Chapter 5.

In a nutshell, then, here is how the greenhouse effect works. From the requirement of energy balance, the absorbed solar radiation determines the effective blackbody radiating temperature T_{rad}. This is not the surface temperature; it is instead the temperature encountered at some pressure level in the atmosphere p_{rad}, which characterizes the infrared opacity of the atmosphere, specifically the typical altitude from which infrared photons escape to space. The pressure p_{rad} is determined by the greenhouse gas concentration of the atmosphere. The surface temperature is determined by starting at the fixed temperature T_{rad} and extrapolating from p_{rad} to the surface pressure p_s using the atmosphere's lapse rate, which is approximately governed by the appropriate adiabat. Since temperature decreases with altitude over much of the depth of a typical atmosphere, the surface temperature so obtained is typically greater than T_{rad}, as illustrated in Fig. 3.6. Increasing the concentration of a greenhouse gas decreases p_{rad}, and therefore increases the surface temperature because temperature is extrapolated from T_{rad} over a greater pressure range. It is very important to recognize that greenhouse warming relies on the decrease of atmospheric temperature with height, which is generally due to the adiabatic profile established by convection. The greenhouse effect works by allowing a planet to radiate at a temperature colder than the surface, but for this to be possible, there must be some cold air aloft for the greenhouse gas to work with.

For an atmosphere whose temperature profile is given by the dry adiabat, the surface temperature is

$$T_s = (p_s/p_{rad})^{R/c_p} T_{rad}. \tag{3.8}$$

With this formula, the Earth's present surface temperature can be explained by taking $p_{rad}/p_s = 0.67$, whence $p_{rad} \approx 670$ mb. Earth's actual radiating pressure is somewhat lower than this estimate, because the atmospheric temperature decays less strongly with height than the dry adiabat. The high surface temperature of Venus can be accounted for by taking $p_{rad}/p_s = 0.0095$, assuming that the temperature profile is given by the non-condensing adiabat for a pure CO_2 atmosphere. Given Venus' 93 bar surface pressure, the radiating level is 880 mb which, interestingly, is only slightly less than Earth's surface pressure. Earth radiates to space from regions quite close to its surface, whereas Venus radiates only from a thin shell near the top of the atmosphere. Note that from the observed Venusian temperature profile in Fig. 2.2, the radiating temperature (253 K) is encountered at $p = 250$ mb rather than the higher pressure we estimated. As for the Earth, our estimate of the precise value p_{rad} for Venus is off because the ideal-gas non-condensing adiabat is not a precise model of the actual temperature profile. In the case of Venus, the problem most likely comes from the ideal-gas assumption and neglect of variations in c_p, rather than condensation.

	Obs. OLR (W/m^2)	Abs. solar (W/m^2)	T_{rad} (actual)	T_{rad} (solar)
Jupiter	14.3	12.7	126 K	110 K
Saturn	4.6	3.8	95 K	81 K
Uranus	0.52	0.93	55 K	58 K
Neptune	0.61	0.38	57 K	47 K

Table 3.1 The energy balance of the gas giant planets, with inferred radiating temperature. The solar-only value of T_{rad} given in the final column is the radiating temperature that would balance the observed absorbed solar energy, in the absence of any internal heat source.

The concept of radiating level and radiating temperature also enables us to make sense of the way energy balance constrains the climates of gas giants like Jupiter and Saturn, which have no distinct surface. The essence of the calculation we have already done for rocky planets is to use the top-of-atmosphere energy budget to determine the parameters of the adiabat, and then extrapolate temperature to the surface along the adiabat. For a non-condensing adiabat, the atmospheric profile compatible with energy balance is $T(p) = T_{rad}(p/p_{rad})^{R/c_p}$. This remains the appropriate temperature profile for a (non-condensing) convecting outer layer of a gas giant, and the only difference with the previous case is that, for a gas giant, there is no surface to act as a natural lower boundary for the adiabatic region. At some depth, convection will give out and the adiabat must be matched to some other temperature model in order to determine the base of the convecting region, and to determine the temperature of deeper regions. There is no longer any distinct surface to be warmed by the greenhouse effect, but the greenhouse gas concentration of the atmosphere nonetheless affects $T(p)$ through p_{rad}. For example, adding some additional greenhouse gas to the convecting outer region of Jupiter's atmosphere would decrease p_{rad}, and therefore increase the temperature encountered at, say, the 1 bar pressure level.

The energy balance suffices to uniquely determine the temperature profile because the non-condensing adiabat is a one-parameter family of temperature profiles. The saturated adiabat for a mixture of condensing and non-condensing gases is also a one-parameter family, defined by Eq. (2.33), and can therefore be treated similarly. If the appropriate adiabat for the planet had more than one free parameter, additional information beyond the energy budget would be needed to close the problem. On the other hand, a single-component condensing atmosphere such as described by Eq. (2.27) yields a temperature profile with no free parameters that can be adjusted so as to satisfy the energy budget. The consequences of this quandary will be taken up as part of our discussion of the runaway greenhouse phenomenon, in Chapter 4.

Using infrared telescopes on Earth and in space, one can directly measure the OLR of the planets in our Solar System. In the case of the gas giants, the radiated energy is substantially in excess of the absorbed solar radiation. Table 3.1 compares the observed OLR to the absorbed solar flux for the gas giants. With the exception of Uranus, the gas giants appear to have a substantial internal energy source, which raises the radiating temperature to values considerably in excess of what it would be if the planet were heated by solar absorption alone. Uranus is anomalous, in that it appears to be emitting less energy than it receives from the Sun. Uncertainties in the observed OLR for Uranus would actually allow the emission to be in balance with solar absorption, but would still appear to preclude any significant internal energy source. This may indicate a profound difference in the internal dynamics of Uranus. On the other hand, the unusually large tilt of Uranus' rotation axis means that Uranus has an unusually strong seasonal variation of solar heating, and it may be that the hemisphere that has been observed so far has not yet had time to come into equilibrium, which would throw off the energy balance estimate.

Because it is the home planet, Earth's radiation budget has been very closely monitored by satellites. Indirect inferences based on the rate of ocean heat uptake indicate that the top-of-atmosphere radiation budget is currently out of balance, the Earth receiving about $1 \, W/m^2$ more from solar absorption than it emits to space as infrared.[3] This is opposite from the imbalance that would be caused by an internal heating. It is a direct consequence of the rapid rise of CO_2 and other greenhouse gases, caused by the bustling activities of Earth's human inhabitants. The rapid greenhouse gas increase has cut down the *OLR*, but because of the time required to warm up the oceans and melt ice, the Earth's temperature has not yet risen enough to restore the energy balance.

Exercise 3.4 A typical well-fed human in a resting state consumes energy in the form of food at a rate of 100W, essentially all of which is put back into the surroundings in the form of heat. An astronaut is in a spherical escape pod of radius r, far beyond the orbit of Pluto, so that it receives essentially no energy from sunlight. The air in the escape pod is isothermal. The skin of the escape pod is a good conductor of heat, so that the surface temperature of the sphere is identical to the interior temperature. The surface radiates like an ideal blackbody.

Find an expression for the temperature in terms of r, and evaluate it for a few reasonable values. Is it better to have a bigger pod or a smaller pod? In designing such an escape pod, should you include an additional source of heat if you want to keep the astronaut comfortable?

How would your answer change if the pod were cylindrical instead of spherical? If the pod were cubical?

Bodies such as Mercury or the Moon represent the opposite extreme to the uniform-temperature limit. Having no atmosphere or ocean to transport heat, and a rocky surface through which heat is conducted exceedingly slowly, each bit of the planet is, to a good approximation, thermally isolated from the rest. Moreover, the rocky surface takes very little time to reach its equilibrium temperature, so the surface temperature at each point is very nearly in equilibrium with the instantaneous absorbed solar radiation, with very little day–night or seasonal averaging. In this case, averaging the energy budget over the planet's surface gives a poor estimate of the temperature, and it would be more accurate to compute the instantaneous equilibrium temperature for each patch of the planet's surface in isolation. For example, consider a point on the planet where the Sun is directly overhead at some particular instant of time. At that time, the rays of sunlight come in perpendicularly to a small patch of the ground, and the absorbed solar radiation per unit area is simply $(1 - \alpha)L_\odot$; the energy balance determining the ground temperature is then $\sigma T^4 = (1 - \alpha)L_\odot$, without the factor of $\frac{1}{4}$ we had when the energy budget was averaged over the entire surface of an isothermal planet. For Mercury, this yields a temperature of 622 K, based on the mean

---

[3] At the time of writing, top-of-atmosphere satellite measurements are not sufficiently accurate to permit direct observation of this imbalance.

orbital distance and an albedo of 0.1. This is similar to the observed maximum temperature on Mercury, which is about 700 K (somewhat larger than the theoretical calculation because Mercury's highly elliptical orbit brings it considerably closer to the Sun than the mean orbital position). The Moon, which is essentially in the same orbit as Earth and shares its solar constant, has a predicted maximum temperature of 384 K, which is very close to the observed maximum. In contrast, the maximum surface temperature on Earth stays well short of 384 K, even at the hottest time of day in the hottest places. The atmosphere of Mars in the present epoch is thin enough that this planet behaves more like the no-atmosphere limit than the uniform-temperature limit. Based on a mean albedo of 0.25, the local maximum temperature should be 297 K, which is quite close to the observed maximum temperature.

More generally speaking, when doing energy balance calculations the temperature we have in mind is the temperature averaged over an appropriate portion of the planet and over an appropriate time interval, where what is "appropriate" depends on the response time and the efficiency of the heat transporting mechanisms of the planet under consideration. Correspondingly, the appropriate incident solar flux to use is the incident solar flux per unit of radiating surface, averaged consistently with temperature. We will denote this mean solar flux by the symbol S. The term *insolation* will be used to refer to an incident solar flux of this type, sometimes with additional qualifiers as in "surface insolation" to distinguish the flux reaching the ground from that incident at the top of the atmosphere. For an isothermal planet, $S = \frac{1}{4}L_\odot$, while at the opposite extreme $S = L_\odot$ for the instantaneous response at the *subsolar point* – the point on the planet at which the sun is directly overhead. In other circumstances it might be appropriate to average along a latitude circle, or over a hemisphere. A more complete treatment of geographical, seasonal, and diurnal temperature variations will be given in Chapter 7.

Exercise 3.5 Consider a planet which is tide-locked to its sun, so that it always shows the same face to the sun as it proceeds in its orbit (just as the Moon always shows the same face to the Earth). Estimate the mean temperature of the dayside of the planet, assuming the illuminated face to be isothermal, but assuming that no heat leaks to the nightside.

3.4 ICE-ALBEDO FEEDBACK

Albedo is not a static quantity determined once and for all time when a planet forms. In large measure, albedo is determined by processes in the atmosphere and at the surface which are highly sensitive to the state of the climate. Clouds consist of suspended tiny particles of the liquid or solid phase of some atmospheric constituent; such particles are very effective reflectors of visible and ultraviolet light, almost regardless of what they are made of. Clouds almost entirely control the albedos of Venus, Titan, and all the gas giant planets, and also play a major role in Earth's albedo. In addition, the nature of a planet's surface can evolve over time, and many of the surface characteristics are strongly affected by the climate. Table 3.2 gives the albedo of some common surface types encountered on Earth. The proportions of the Earth covered by sea ice, snow, glaciers, desert sands, or vegetation of various types are determined by temperature and precipitation patterns. As climate changes, the surface characteristics change too, and the resulting albedo changes feed back on the state of the climate. It is not a "chicken and egg" question of whether climate causes albedo or albedo causes climate; rather it is a matter of finding a consistent state compatible with the physics of the way climate affects albedo and the way albedo

Table 3.2 Typical values of albedo for various surface types.

Surface type	Albedo
Clean new H_2O snow	0.85
Bare sea ice	0.5
Clean H_2O glacier ice	0.6
Deep water	0.1
Sahara Desert sand	0.35
Martian sand	0.15
Basalt (any planet)	0.07
Granite	0.3
Limestone	0.36
Grassland	0.2
Deciduous forest	0.14
Conifer forest	0.09
Tundra	0.2

These are only representative values. Albedo can vary considerably as a function of detailed conditions. For example, the ocean albedo depends on the angle of the solar radiation striking the surface (the value given in the table is for near-normal incidence), and the albedo of bare sea ice depends on the density of air bubbles.

affects climate. In this sense, albedo changes lead to a form of climate *feedback*. We will encounter many other kinds of feedback loops in the climate system.

Among all the albedo feedbacks, that associated with the cover of the surface by highly reflective snow or ice plays a distinguished role in thinking about the evolution of the Earth's climate. Let's consider how albedo might vary with temperature for a planet entirely covered by a water ocean – a reasonable approximation to Earth, which is $\frac{2}{3}$ ocean. We will characterize the climate by the global mean surface temperature T_s, but suppose that, like Earth, the temperature is somewhat colder than T_s at the poles and somewhat warmer than T_s at the Equator. When T_s is very large, say greater than some threshold temperature T_0, the temperature is above freezing everywhere and there is no ice. In this temperature range, the planetary albedo reduces to the relatively low value (call it α_0) characteristic of sea water. At the other extreme, when T_s is very, very low, the whole planet is below freezing, the ocean will become ice-covered everywhere, and the albedo reduces to that of sea ice, which we shall call α_i. We suppose that this occurs for $T_s < T_i$, where T_i is the threshold temperature for a globally frozen ocean. In general T_i must be rather lower than the freezing temperature of the ocean, since when the mean temperature $T_s = T_{freeze}$ the equatorial portions of the planet will still be above freezing. Between T_i and T_0 it is reasonable to interpolate the albedo by assuming the ice cover to decrease smoothly and monotonically from 100% to zero. The phenomena we will emphasize are not particularly sensitive to the detailed form of the interpolation, but the quadratic interpolation

$$\alpha(T) = \begin{cases} \alpha_i & \text{for } T \leq T_i, \\ \alpha_0 + (\alpha_i - \alpha_0)\frac{(T-T_0)^2}{(T_i-T_0)^2} & \text{for } T_i < T < T_0 \\ \alpha_0 & \text{for } T \geq T_0 \end{cases} \qquad (3.9)$$

qualitatively reproduces the shape of the albedo curve which is found in detailed calculations. In particular, the slope of albedo vs. temperature is large when the temperature is low and the planet is nearly ice-covered, because there is more area near the Equator, where ice melts first. Conversely, the slope reduces to zero as the temperature threshold for an ice-free planet is approached, because there is little area near the poles where the last ice survives; moreover, the poles receive relatively little sunlight in the course of the year, so the albedo there contributes less to the global mean than does the albedo at lower latitudes. Note that this description assumes an Earthlike planet, which on average is warmest near the Equator. As will be discussed in Chapter 7, other orbital configurations could lead to the poles being warmer, and this would call for a different shape of albedo curve.

Ice-albedo feedback of a similar sort could arise on a planet with land, through snow accumulation and glacier formation on the continents. The albedo could have a similar temperature dependence, in that glaciers are unlikely to survive where temperatures are very much above freezing, but can accumulate readily near places that are below freezing – *provided there is enough precipitation.* It is the latter requirement that makes land-based snow/ice-albedo feedback much more complicated than the oceanic case. Precipitation is determined by complex atmospheric circulation patterns that are not solely determined by local temperature. A region with no precipitation will not form glaciers no matter how cold it is made. The present state of Mars provides a good example: its small polar glaciers do not advance to the Equator, even though the daily average equatorial temperature is well below freezing. Still, for a planet like Earth with a widespread ocean to act as a source for precipitation, it may be reasonable to assume that most continental areas will eventually become ice-covered if they are located at sufficiently cold latitudes. In fairness, we should point out that even the formation of sea ice is considerably more complex than we have made it out to be, particularly since it is affected by the mixing of deep unfrozen water with surface waters which are trying to freeze.

Earth is the only known planet that has an evident ice/snow-albedo feedback, but it is reasonable to inquire as to whether a planet without Earth's water-dominated climate could behave analogously. Snow is always "white" more or less regardless of the substance it is made of, since its reflectivity is due to the refractive index discontinuity between snow crystals and the ambient gas or vacuum. Therefore, a snow-albedo feedback could operate with substances other than water (e.g. nitrogen or methane). Titan presents an exotic possibility, in that its surface is bathed in a rain of tarry hydrocarbon sludge, raising the speculative possibility of "dark glacier" albedo feedbacks. Sea ice forming on Earth's ocean gets its high albedo from trapped air bubbles, which act like snowflakes in reverse. The same could happen for ices of other substances, but sea ice-albedo feedback is likely to require a water ocean. The reason is that water, alone among likely planetary materials, floats when it freezes. Ice forming on, say, a carbon dioxide or methane ocean would sink as soon as it formed, preventing it from having much effect on surface albedo.

Returning attention to an Earthlike waterworld, we write down the energy budget

$$(1 - \alpha(T_s))\frac{L_\odot}{4} = OLR(T_s). \tag{3.10}$$

This determines T_s as before, with the important difference that the solar absorption on the left hand side is now a function of T_s instead of being a constant. Analogously to Fig. 3.5, the equilibrium surface temperature can be found by plotting the absorbed solar radiation and the OLR vs. T_s on the same graph. This is done in Fig. 3.8, for four different choices of $L_\odot$. In

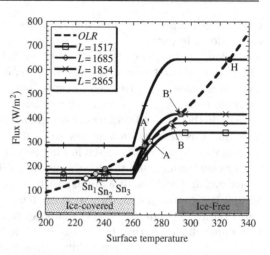

Figure 3.8 Graphical determination of the possible equilibrium states of a planet whose albedo depends on temperature in accordance with Eq. (3.9). The *OLR* is computed assuming the atmosphere has no greenhouse effect, and the albedo parameters are $\alpha_0 = 0.1$, $\alpha_i = 0.6$, $T_i = 260\,\mathrm{K}$, and $T_0 = 290\,\mathrm{K}$. The solar constant for the various solar absorption curves is indicated in the key.

this plot, we have taken $OLR = \sigma T^4$, which assumes no greenhouse effect.[4] In contrast with the fixed-albedo case, the ice-albedo feedback allows the climate system to have *multiple equilibria*: there can be more than one climate compatible with a given solar constant, and additional information is required to determine which state the planet actually settles into. The nature of the equilibria depends on $L_\odot$. When $L_\odot$ is sufficiently small (as in the case $L_\odot = 1517\,\mathrm{W/m^2}$ in Fig. 3.8) there is only one solution, which is a very cold globally ice-covered Snowball state, marked Sn_1 on the graph. Note that the solar constant that produces a unique Snowball state exceeds the present solar constant at Earth's orbit. Thus, were it not for the greenhouse effect, Earth would be in such a state, and would have been for its entire history. When $L_\odot$ is sufficiently large (as in the case $L_\odot = 2865\,\mathrm{W/m^2}$ in Fig. 3.8) there is again a unique solution, which is a very hot globally ice-free state, marked H on the graph. However, for a wide range of intermediate $L_\odot$, there are three solutions: a Snowball state (Sn_2), a partially ice covered state with a relatively large ice sheet (e.g. A), and a warmer state (e.g. B) which may have a small ice sheet or be ice-free, depending on the precise value of $L_\odot$. In the intermediate range of solar constant, the warmest state is suggestive of the present or Pleistocene climate when there is a small ice cap, and suggestive of Cretaceous-type hothouse climates when it is ice-free. In either case, the frigid Snowball state is available as an alternate possibility.

As the parameter $L_\odot$ is increased smoothly from low values, the temperature of the Snowball state increases smoothly but at some point an additional solution discontinuously comes into being at a temperature far from the previous equilibrium, and splits into a pair as $L_\odot$ is further increased. As $L_\odot$ is increased further, at some point, the intermediate temperature state merges with the Snowball state, and disappears. This sort of behavior, in which the behavior of a system changes discontinuously as some control parameter is continuously varied, is an example of a *bifurcation*.

Finding the equilibria tells only part of the story. A system placed exactly at an equilibrium point will stay there forever, but what if it is made a little warmer than the equilibrium? Will it heat up yet more, perhaps aided by melting of ice, and ultimately wander far from the

[4] Of course, this is an unrealistic assumption, since a waterworld would inevitably have at least water vapor – a good greenhouse gas – in its atmosphere.

equilibrium? Or will it cool down and move back toward the equilibrium? Similar questions apply if the state is made initially slightly cooler than an equilibrium. This leads us to the question of *stability*. In order to address stability, we must first write down an equation describing the time evolution of the system. To this end, we suppose that the mean energy storage per unit area of the planet's surface can be written as a function of the mean temperature; let's call this function $E(T_s)$. Changes in the energy storage could represent the energy required to heat up or cool down a layer of water of some characteristic depth, and could also include the energy needed to melt ice, or released by the freezing of sea water. For our purposes, all we need to know is that E is a monotonically increasing function of T_s. The energy balance for a time-varying system can then be written

$$\frac{dE(T_s)}{dt} = \frac{dE}{dT_s}\frac{dT_s}{dt} = G(T_s) \tag{3.11}$$

where $G = \frac{1}{4}(1-\alpha(T_s))L_\odot - OLR(T_s)$. We can define the generalized heat capacity $\mu(T) = dE/dT$, which is positive by assumption. Thus,

$$\frac{dT_s}{dt} = \frac{G(T_s)}{\mu(T_s)}. \tag{3.12}$$

By definition, $G = 0$ at an equilibrium point T_{eq}. Suppose that the slope of G is well-defined near T_{eq} – in formal mathematical language, we say that G is continuously differentiable at T_{eq}, meaning that the derivative of G exists and is a continuous function for T_s in some neighborhood of T_{eq}. Then, if $dG/dT_s < 0$ at T_s, it will also be negative for some finite distance to the right and left of T_s. This is the case for points a and c in the net flux curve sketched in Fig. 3.9. If the temperature is made a little warmer than T_{eq} in this case, $G(T_s)$ and hence $\frac{dT_s}{dt}$ will become negative and the solution will move back toward the equilibrium. If the temperature is made a little colder than T_{eq}, $G(T_s)$ and hence $\frac{dT_s}{dt}$ will become positive, and the solution will again move back toward the equilibrium. In contrast, if $dG/dT_s > 0$ near the equilibrium, as for point b in the sketch, a temperature placed near the equilibrium moves away from it, rather than toward it. Such equilibria are *unstable*. If the slope happens to be exactly zero at an equilibrium, one must look to higher derivatives to determine stability. These are "rare" cases, which will be encountered only for very special settings of the parameters. If the d^2G/dT^2 is non-zero at the equilibrium, the curve takes the form

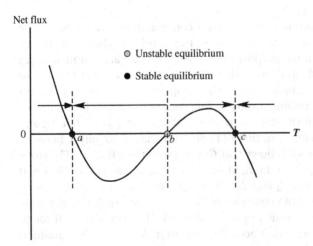

Net flux

○ Unstable equilibrium

● Stable equilibrium

Figure 3.9 Sketch illustrating stable vs. unstable equilibrium temperatures.

of a parabola tangent to the axis at the equilibrium. If the parabola opens upwards, then the equilibrium is stable to displacements to the left of the equilibrium, but unstable to displacements to the right. If the parabola opens downwards, the equilibrium is unstable to displacements to the left but stable to displacements to the right. Similar reasoning applies to the case in which the first non-vanishing derivative is higher order, but such cases are hardly ever encountered.

Exercise 3.6 Draw a sketch illustrating the behavior near marginal equilibria with $d^2G/dT^2 > 0$ and $d^2G/dT^2 < 0$. Do the same for equilibria with $d^2G/dT^2 = 0$, having $d^3G/dT^3 > 0$ and $d^3G/dT^3 < 0$.

It is rare that one can completely characterize the behavior of a nonlinear system, but one-dimensional problems of the sort we are dealing with are exceptional. In the situation depicted in Fig. 3.9, G is positive and dT/dt is positive throughout the interval between b and c. Hence, a temperature placed anywhere in this interval will eventually approach the solution c arbitrarily closely – it will be *attracted* to that stable solution. Similarly, if T is initially between a and b, the solution will be attracted to the stable equilibrium a. The unstable equilibrium b forms the boundary between the *basins of attraction* of a and c. No matter where we start the system within the interval between a and c (and somewhat beyond, depending on the shape of the curve further out), it will wind up approaching one of the two stable equilibrium states. In mathematical terms, we are able to characterize the *global behavior* of this system, as opposed to just the *local behavior* near equilibria.

At an equilibrium point, the curve of solar absorption crosses the *OLR* curve, and the stability criterion is equivalent to stating that the equilibrium is stable if the slope of the solar curve is less than that of the *OLR* curve where the two curves intersect. Using this criterion, we see that the intermediate-temperature large ice-sheet states, labeled A and A' in Fig. 3.8, are unstable. If the temperature is made a little bit warmer then the equilibrium the climate will continue to warm until it settles into the warm state (B or B') which has a small or non-existent ice sheet. If the temperature is made a little bit colder than the equilibrium, the system will collapse into the Snowball state (Sn_2 or Sn_3). The unstable state thus defines the boundary separating the basin of attraction of the warm state from that of the Snowball state.

Moreover, if the net flux $G(T)$ is continuous and has a continuous derivative (i.e. if the curve has no "kinks" in it), then the sequence of consecutive equilibria always alternates between stable and unstable states. For the purpose of this theorem, the rare marginal states with $dG/dT = 0$ should be considered "wildcards" that can substitute for either a stable or unstable state. The basic geometrical idea leading to this property is more or less evident from Fig. 3.9, but a more formalized argument runs as follows: Let T_a and T_b be equilibria, so that $G(T_a) = G(T_b) = 0$. Suppose that the first of these is stable, so $dG/dT < 0$ at T_a, and also that the two solutions are consecutive, so that $G(T)$ does not vanish for any T between T_a and T_b. Now if $dG/dT < 0$ at T_b, then it follows that $G > 0$ just to the left of T_b. The slope near T_a similarly implies that $G < 0$ just to the right of T_a. Since G is continuous, it would follow that $G(T) = 0$ somewhere between T_a and T_b. This would contradict our assumption that the two solutions are consecutive. In consequence, $dG/dT \geq 0$ at T_b. Thus, the state T_b is either stable or marginally stable, which proves our result. The proof goes through similarly if T_a is unstable. Note that we didn't actually need to make use of the condition that dG/dT be continuous everywhere: it's enough that it be continuous near the equilibria, so we can actually tolerate a few kinks in the curve.

A consequence of this result is that, if the shape of $G(T)$ is controlled continuously by some parameter like $L_\odot$, then new solutions are born in the form of a single marginal state which upon further change of $L_\odot$ splits into a stable/unstable or unstable/stable pair. The first member of the pair will be unstable if there is a pre-existing stable solution immediately on the cold side of the new one, as is the case for the Snowball states Sn in Fig. 3.8. The first member will be stable if there is a pre-existing unstable state on cold side, or a pre-existing stable state on the warm side (e.g. the state H in Fig. 3.8). What we have just encountered is a very small taste of the very large and powerful subject of *bifurcation theory*.

3.4.1 Faint Young Sun, Snowball Earth and hysteresis

We now have enough basic theoretical equipment to take a first quantitative look at the Faint Young Sun problem. To allow for the greenhouse effect of the Earth's atmosphere, we take $p_{rad} = 670$ mb, which gives the correct surface temperature with the observed current albedo $\alpha = 0.3$. How much colder does the Earth get if we ratchet the solar constant down to 960 W/m^2, as it was 4.7 billion years ago when the Earth was new? As a first estimate, we can compute the new temperature from Eq. (3.8) holding p_{rad} and the albedo fixed at their present values. This yields 261 K. This is substantially colder than the present Earth. The fixed albedo assumption is unrealistic, however, since the albedo would increase for a colder and more ice-covered Earth, leading to a substantially colder temperature than we have estimated. In addition, the strength of the atmospheric greenhouse effect could have been different for the Early Earth, owing to changes in the composition of the atmosphere.

An attempt at incorporating the ice-albedo feedback can be made by using the energy balance Eq. (3.10) with the albedo parameterization given by Eq. (3.9). For this calculation, we choose constants in the albedo formula that give a somewhat more realistic Earthlike climate than those used in Fig. 3.8. Specifically, we set $\alpha_0 = 0.28$ to allow for the albedo of clouds and land, and $T_0 = 295$ K to allow a slightly bigger polar ice sheet. The position of the equilibria can be determined by drawing a graph like Fig. 3.8, or by applying a root-finding algorithm like Newton's method to Eq. (3.10). The resulting equilibria are shown as a function of $L_\odot$ in Fig. 3.10, with p_{rad} held fixed at 670 mb. Some techniques for generating diagrams of this type are developed in Problem 3.34. For the modern solar constant, and $p_{rad} = 670$ mb, the system has a stable equilibrium at $T_s = 286$ K, close to the observed modern surface temperature, and is partially ice-covered. However, the system has a second stable equilibrium, which is a globally ice-covered Snowball state having $T_s = 249$ K. Even today, the Earth would stay in a Snowball state if it were somehow put there. The two stable equilibria are separated by an unstable equilibrium at $T_s = 270$ K, which defines the boundary between the set of initial conditions that go to the "modern" type state, and the set that go to a Snowball state. The attractor boundary for the modern open-ocean state is comfortably far from the present temperature, so it would not be easy to succumb to a Snowball.

Now we turn down the solar constant, and re-do the calculation. For $L_\odot = 960$ W/m^2, there is only a single equilibrium point if we keep $p_{rad} = 670$ mb. This is a stable Snowball state with $T_s = 228$ K. Thus, if the Early Earth had the same atmospheric composition as today, leading to a greenhouse effect no stronger than the present one, the Earth would have inevitably been in a Snowball state. The open-ocean state only comes into being when

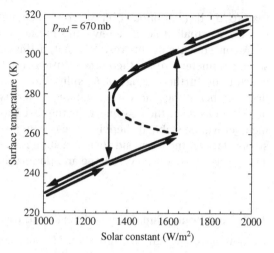

Figure 3.10 Hysteresis diagram obtained by varying $L_\odot$ with p_{rad}/p_s fixed at 0.67. Arrows indicate path followed by the system as $L_\odot$ is first increased, then decreased. The unstable solution branch is indicated by a dashed curve.

$L_\odot$ is increased to $1330\,W/m^2$, which was not attained until the relatively recent past. This contradicts the abundant geological evidence for prevalent open water throughout several billion years of Earth's history. Even worse, if the Earth were initially in a stable Snowball state four billion years ago, it would stay in that state until $L_\odot$ increased to $1640\,W/m^2$, at which point the stable Snowball state would disappear and the Earth would deglaciate. Since this far exceeds the present solar constant, the Earth would be globally glaciated today. This even more obviously contradicts the data.

The currently favored resolution to the paradox of the Faint Young Sun is the supposition that the atmospheric composition of the Early Earth must have resulted in a stronger greenhouse effect than the modern atmosphere produces. The prime candidate gases for mediating this change are CO_2 and CH_4. The radiative basis of the idea will be elaborated further in Chapter 4, and some ideas about why the atmosphere might have adjusted over time so as to maintain an equable climate despite the brightening Sun are introduced in Chapter 8. Figure 3.11 shows how the equilibria depend on p_{rad}, with $L_\odot$ fixed at $960\,W/m^2$. Whichever greenhouse gas is the Earth's savior, if it is present in sufficient quantities to reduce p_{rad} to $500\,mb$ or less, then a warm state with an open ocean exists (the upper branch in Fig. 3.11). However, for $420\,mb < p_{rad} < 500\,mb$ a stable Snowball state also exists, meaning that the climate that is actually selected depends on earlier history. If the planet had already fallen into a Snowball state for some reason, the early Earth would stay in a Snowball unless the greenhouse gases built up sufficiently to reduce p_{rad} below $420\,mb$ at some point.

Figures 3.10 and 3.11 illustrate an important phenomenon known as *hysteresis*: the state in which a system finds itself depends not just on the value of some parameter of the system, but the history of variation of that parameter. This is possible only for systems that have multiple stable states. For example, in Fig. 3.10 suppose we start with $L_\odot = 1000\,W/m^2$, where the system is inevitably in a Snowball state with $T = 230\,K$. Let's now gradually increase $L_\odot$. When $L_\odot$ reaches $1500\,W/m^2$ the system is still in a Snowball state, having $T = 254\,K$, since we have been following a stable solution branch the whole way. However, when $L_\odot$ reaches $1640\,W/m^2$, the Snowball solution disappears, and the system makes a sudden transition from a Snowball state with $T = 260\,K$ to the only available stable solution, which is an ice-free state having $T = 301\,K$. As $L_\odot$ increases further to $2000\,W/m^2$, we follow the warm, ice-free state and the temperature rises to $316\,K$. Now suppose we begin to gradually dim

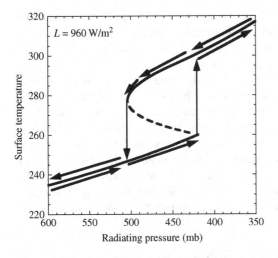

Figure 3.11 As in Fig. 3.10, but varying p_{rad} with $L_\odot = 960\,\mathrm{W/m^2}$.

the Sun, perhaps by making the Solar System pass through a galactic dust cloud. Now, we follow the upper, stable branch as $L_\odot$ decreases, so that when we find ourselves once more at $L_\odot = 1500\,\mathrm{W/m^2}$ the temperature is 294 K and the system is in a warm, ice-free state rather than in the Snowball state we enjoyed the last time we were there. As $L_\odot$ is decreased further, the warm branch disappears at $L_\odot = 1330\,\mathrm{W/m^2}$ and the system drops suddenly from a temperature of 277 K into a Snowball state with a temperature of 246 K, whereafter the Snowball branch is again followed as $L_\odot$ is reduced further. The trajectory of the system as $L_\odot$ is increased then decreased back to its original value takes the form of an open loop, depicted in Fig. 3.10.

The thought experiment of varying $L_\odot$ in a hysteresis loop is rather fanciful, but many atmospheric processes could act to either increase or decrease the greenhouse effect over time. For the very young Earth, with $L_\odot = 960\,\mathrm{W/m^2}$, the planet falls into a Snowball when p_{rad} exceeds 500 mb, and thereafter would not deglaciate until p_{rad} was reduced to 420 mb or less (see Fig. 3.11). The boundaries of the hysteresis loop, which are the critical thresholds for entering and leaving the Snowball, depend on the solar constant. For the modern solar constant, the hysteresis loop operates between $p_{rad} = 690\,\mathrm{mb}$ and $p_{rad} = 570\,\mathrm{mb}$. It takes less greenhouse effect to keep out of the Snowball now than it did when the Sun was fainter, but the threshold for initiating a Snowball in modern conditions is disconcertingly close to the value of p_{rad} which reproduces the present climate.

The fact that the freeze–thaw cycle can exhibit hysteresis as atmospheric composition changes is at the heart of the Snowball Earth phenomenon. An initially warm state can fall into a globally glaciated Snowball if the atmospheric composition changes in such a way as to sufficiently weaken the greenhouse effect. Once the threshold is reached, the planet can fall into a Snowball relatively quickly – in a matter of a thousand years or less – since sea ice can form quickly. However, to deglaciate the Snowball, the greenhouse effect must be increased far beyond the threshold value at which the planet originally entered the Snowball state. Atmospheric composition must change drastically in order to achieve such a great increase, and this typically takes many millions of years. When deglaciation finally occurs, it leaves the atmosphere in a hyper-warm state, which only gradually returns to normal as the atmospheric composition evolves in such a way as to reduce the greenhouse effect. As discussed in Chapter 1, there are two periods in Earth's past when geological evidence suggests that one or more Snowball freeze–thaw cycles may have occurred. The first is in the Paleoproterozoic, around 2 billion years ago. At this time, $L_\odot \approx 1170\,W/m^2$, and the thresholds for initiating and deglaciating a Snowball are $p_{rad} = 600\,mb$ and $p_{rad} = 500\,mb$ in our simple model. For the Neoproterozoic, about 700 million years ago, $L_\odot \approx 1290\,W/m^2$ and the thresholds are at $p_{rad} = 650\,mb$ and $p_{rad} = 540\,mb$.

The boundaries of the hysteresis loop shift as the solar constant increases, but there is nothing obvious in the numbers to suggest why a Snowball state should have occurred in the Paleoproterozoic and Neoproterozoic but not at other times. Hysteresis associated with ice-albedo feedback has been a feature of the Earth's climate system throughout the entire history of the planet. Hysteresis will remain a possibility until the solar constant increases sufficiently to render the Snowball state impossible even in the absence of any greenhouse effect (i.e. with $p_{rad} = 1000\,mb$). Could a Snowball episode happen again in the future, or is that peril safely behind us? These issues require an understanding of the processes governing the evolution of Earth's atmosphere, a subject that will be taken up in Chapter 8.

Exercise 3.7 Assuming an ice albedo of 0.6, how high does $L_\odot$ have to become to eliminate the possibility of a Snowball state? Will this happen within the next five billion years? What if you assume there is enough greenhouse gas in the atmosphere to make $p_{rad}/p_s = 0.5$?

Note: The evolution of the solar constant over time is approximately $L_\odot(t) = L_{\odot p} \cdot (0.7 + (t/22.975) + (t/14.563)^2)$, where t is the age of the Sun in billions of years ($t = 4.6$ being the current age) and $L_{\odot p}$ is the present solar constant. This fit is reasonably good for the first 10 billion years of solar evolution.

The "cold start" problem is a habitability crisis that applies to waterworlds in general. If a planet falls into a Snowball state early in its history, it could take billions of years to get out if one needs to wait for the Sun to brighten. The time to get out of a Snowball could be shortened if greenhouse gases built up in the atmosphere, reducing p_{rad}. How much greenhouse gas must build up to deglaciate a Snowball? How long would that take? What could cause greenhouse gases to accumulate on a Snowball planet? These important questions will be taken up in subsequent chapters.

Another general lesson to be drawn from the preceding discussion is that the state with a stable, small ice cap is very fragile, in the sense that the planetary conditions must be tuned rather precisely for the state to exist at all. For example, with the present solar constant, the stable small ice cap solution first appears when p_{rad} falls below 690 mb. However, the ice cap shrinks to zero as p_{rad} is reduced somewhat more, to 615 mb. Hence, a moderate strengthening in the greenhouse effect would, according to the simple energy balance model, eliminate the polar ice entirely and throw the Earth into an ice-free Cretaceous hothouse state. The transition to an ice-free state of this sort is continuous in the parameter being varied; unlike the collapse into a Snowball state or the recovery from a Snowball, it does not result from a bifurcation. In light of its fragility, it is a little surprising that the Earth's present small-icecap state has persisted for the past two million years, and that similar states have occurred at several other times in the past half billion years. Does the simple energy balance model exaggerate the fragility of the stable small-icecap state? Does some additional feedback process adjust the greenhouse effect so as to favor such a state while resisting the peril of the Snowball? These are largely unresolved questions. Attacks on the first question require comprehensive dynamical models of the general circulation, which we will not encounter in the present volume. We will take up, though not resolve, the second question in Chapter 8. It is worth noting that small-icecap states like those of the past two million years appear to be relatively uncommon in the most recent half billion years of Earth's history, for which data is good enough to render a judgement about ice cover. The typical state appears to be more like the warm relatively ice-free states of the Cretaceous, and perhaps this reflects the fragility of the small-icecap state.

The simple models used above are too crude to produce very precise hysteresis boundaries. Among the many important effects left out of the story are water vapor radiative feedbacks, cloud feedbacks, the factors governing albedo of sea ice, ocean heat transports, and variations in atmospheric heat transport. The phenomena uncovered in this exposition are general, however, and can be revisited across a hierarchy of models. Indeed, the re-examination of this subject provides an unending source of amusement and enlightenment to climate scientists.

3.4.2 Climate sensitivity, radiative forcing and feedback

The simple model we have been studying affords us the opportunity to introduce the concepts of *radiative forcing, climate sensitivity coefficient*, and *feedback factor*. These diagnostics can be applied across the whole spectrum of climate models, from the simplest to the most comprehensive.

Suppose that the mean surface temperature depends on some parameter Λ, and we wish to know how sensitive T is to changes in that parameter. For example, this parameter might be the solar constant, or the radiating pressure. It could be some other parameter controlling the strength of the greenhouse effect, such as CO_2 concentration. Near a given Λ, the sensitivity is characterized by $dT/d\Lambda$.

Let G be the net top-of-atmosphere flux, such as used in Eq. (3.11). To allow for the fact that the terms making up the net flux depend on the parameter Λ, we write $G = G(T, \Lambda)$. If we take the derivative of the energy balance requirement $G = 0$ with respect to Λ, we find

$$0 = \frac{\partial G}{\partial T}\frac{dT}{d\Lambda} + \frac{\partial G}{\partial \Lambda} \tag{3.13}$$

so that

$$\frac{dT}{d\Lambda} = -\frac{\frac{\partial G}{\partial \Lambda}}{\frac{\partial G}{\partial T}}. \tag{3.14}$$

The numerator in this expression is a measure of the *radiative forcing* associated with changes in Λ. Specifically, changing Λ by an amount $\delta\Lambda$ will perturb the top-of-atmosphere radiative budget by $\frac{\partial G}{\partial \Lambda}\delta\Lambda$, requiring that the temperature change so as to bring the energy budget back into balance. For example, if Λ is the solar constant L, then $\frac{\partial G}{\partial \Lambda} = \frac{1}{4}(1 - \alpha)$. If Λ is the radiating pressure p_{rad}, then $\frac{\partial G}{\partial \Lambda} = -\frac{\partial OLR}{\partial p_{rad}}$. Since OLR goes down as p_{rad} is reduced, a reduction in p_{rad} yields a positive radiative forcing. This is a warming influence.

Radiative forcing is often quoted in terms of the change in flux caused by a standard change in the parameter, in place of the slope $\frac{\partial G}{\partial \Lambda}$ itself. For example, the radiative forcing due to CO_2 is typically described by the change in flux caused by doubling CO_2 from its pre-industrial value, with temperature and everything else held fixed. This is practically the same thing as $\frac{\partial G}{\partial \Lambda}$ if we take $\Lambda = \log_2 pCO_2$, where pCO_2 is the partial pressure of CO_2. Similarly, the climate sensitivity is often described in terms of the temperature change caused by the standard forcing change, rather than the slope $\frac{dT}{d\Lambda}$. For example, the notation ΔT_{2x} would refer to the amount by which temperature changes when CO_2 is doubled.

The denominator of Eq. (3.14) determines how much the equilibrium temperature changes in response to a given radiative forcing. For any given magnitude of the forcing, the response will be greater if the denominator is smaller. Thus, the denominator measures the *climate sensitivity*. An analysis of ice-albedo feedback illustrates how a feedback process affects the climate sensitivity. If we assume that albedo is a function of temperature, as in Eq. (3.9), then

$$\frac{\partial G}{\partial T} = -\frac{1}{4} L \frac{\partial \alpha}{\partial T} - \frac{\partial OLR}{\partial T}. \tag{3.15}$$

With this expression, Eq. (3.14) can be re-written

$$\frac{dT}{d\Lambda} = -\frac{1}{1+\Phi} \left[\frac{\frac{\partial G}{\partial \Lambda}}{\frac{\partial OLR}{\partial T}} \right] \tag{3.16}$$

where

$$\Phi = \frac{1}{4} L \frac{\frac{\partial \alpha}{\partial T}}{\frac{\partial OLR}{\partial T}}. \tag{3.17}$$

In writing this equation we primarily have ice-albedo feedback in mind, but the equation is valid for arbitrary $\alpha(T)$ so it could as well describe a variety of other processes. The factor in square brackets in Eq. (3.16) is the sensitivity the system would have if the response were unmodified by the change of albedo with temperature. The first factor determines how the sensitivity is increased or decreased by the feedback of temperature on albedo. If $-1 < \Phi < 0$ then the feedback increases the sensitivity – the same radiative forcing produces a bigger temperature change than it would in the absence of the feedback. When $\Phi = -\frac{1}{2}$, for example, the response to the forcing is twice what it would have been in the absence of the feedback. The sensitivity becomes infinite as $\Phi \to -1$, and for $-2 < \Phi < -1$ the feedback is so strong that it actually reverses the sign of the response as well as increasing its magnitude. On the other hand, if $\Phi > 0$, the feedback reduces the sensitivity. In this case it is a *stabilizing feedback*. The larger Φ gets, the more the response is reduced. For example, when $\Phi = 1$ the response is half what it would have been in the absence of feedback. Note that the feedback term is the same regardless of whether the radiative forcing is due to changing L, p_{rad}, or anything else.

As an example, let's compute the feedback parameter Φ for the albedo–temperature relation given by Eq. (3.9), under the conditions shown in Fig. 3.10. Consider in particular the upper solution branch, which represents a stable partially ice-covered climate like that of the present Earth. At the point $L_{\odot} = 1400 \, \text{W/m}^2$, $T = 288 \, \text{K}$ on this branch, we find $\Phi = -0.333$. Thus, at this point the ice-albedo feedback increases the sensitivity of the climate by a factor of about 1.5. At the bifurcation point $L_{\odot} \approx 1330 \, \text{W/m}^2$, $T \approx 277 \, \text{K}$, $\Phi \to -1$ and the sensitivity becomes infinite. This divergence merely reflects the fact that the temperature curve is vertical at the bifurcation point. Near such points, the temperature change is no longer linear in radiative forcing. It can easily be shown that the temperature varies as the square root of radiative forcing near a bifurcation point, as suggested by the plot.

The ice-albedo feedback increases the climate sensitivity, but other feedbacks could be stabilizing. In fact Eq. (3.17) is valid whatever the form of $\alpha(T)$, and shows that the albedo feedback becomes a stabilizing influence if albedo increases with temperature. This could conceivably happen as a result of vegetation feedback, or perhaps dissipation of low clouds. The somewhat fanciful Daisyworld example in the Workbook section at the end of this chapter provides an example of such a stabilizing feedback.

The definition of the feedback parameter can be generalized as follows. Suppose that the energy balance function G depends not only on the control parameter Λ, but also on some other parameter $\aleph$ which varies systematically with temperature. In the previous example, $\aleph(T)$ is the temperature-dependent albedo. We write $G = G(T, \aleph(T), \Lambda)$. Following the same line of reasoning as we did for the analysis of ice-albedo feedback, we find

$$\Phi = \frac{\frac{\partial G}{\partial \aleph}\frac{\partial \aleph}{\partial T}}{\frac{\partial G}{\partial T}}. \tag{3.18}$$

For example, if $\aleph$ represents the concentration of water vapor on Earth, or of methane on Titan, and if $\aleph$ varies as a function of temperature, then the feedback would influence G through the *OLR*. Writing $OLR = OLR(T, R(T), \Lambda)$, then the feedback parameter is

$$\Phi = \frac{\frac{\partial OLR}{\partial \aleph}\frac{\partial \aleph}{\partial T}}{\frac{\partial OLR}{\partial T}} \tag{3.19}$$

assuming the albedo to be independent of temperature in this case. Now, since *OLR* increases with T and *OLR* decreases with $\aleph$, the feedback will be destabilizing ($\Phi < 0$) if $\aleph$ increases with T. (One might expect $\aleph$ to increase with T because Clausius–Clapeyron implies that the saturation vapor pressure increases sharply with T, making it harder to remove water vapor by condensation, all other things being equal.) Note that in this case the water vapor feedback does not lead to a runaway, with more water leading to higher temperatures leading to more water in a never-ending cycle; the system still attains an equilibrium, though the sensitivity of the equilibrium temperature to changes in a control parameter is increased.

3.5 PARTIALLY ABSORBING ATMOSPHERES

The assumption underpinning the blackbody radiation formula is that radiation interacts so strongly with matter that it achieves thermodynamic equilibrium at the same temperature as the matter. It stands to reason, then, that if a box of gas contains too few molecules to offer much opportunity to intercept a photon, the emission will deviate from the blackbody law. Weak interaction with radiation can also arise from aspects of the structure of a material which inhibit interaction, such as the crystal structure of table salt or carbon dioxide ice. In either event, the deviation of emission from the Planck distribution is characterized by the *emissivity*. Suppose that $I(\nu, \hat{n})$ is the observed flux of radiation at frequency ν emerging from a body in the direction $\hat{n}$. Then the emissivity $e(\nu, \hat{n})$ is defined by the expression

$$I(\nu, \hat{n}) = e(\nu, \hat{n})B(\nu, T) \tag{3.20}$$

where T is the temperature of the collection of matter we are observing. Note that in assigning a temperature T to the body, we are assuming that the matter itself is in a state of thermodynamic equilibrium. The emissivity may also be a function of temperature and pressure. We can also define a mean emissivity over frequencies, and all rays emerging from a body. The mean emissivity is

$$\bar{e} = \frac{\int_{\nu, \Omega} e(\nu, \hat{n})B(\nu, T) \cos\theta \, d\nu \, d\Omega}{\sigma T^4} \tag{3.21}$$

where θ is the angle of the ray to the normal to the body's surface and the angular integration is taken over the hemisphere of rays leaving the surface of the body. With this definition, the net flux emerging from any patch of the body's surface is $F = \bar{e}\sigma T^4$. Even if e does not depend explicitly on temperature, $\bar{e}$ will be temperature-dependent if e is frequency-dependent, since the relative weighting of different frequencies, determined by $B(v, T)$, changes with temperature.

A blackbody has unit emissivity at all frequencies and directions. A blackbody also has unit absorptivity, which is just a restatement of the condition that blackbodies interact strongly with the radiation field. For a non-blackbody, we can define the absorptivity $a(v, \hat{n})$ by shining light at a given frequency and direction at the body and measuring how much is reflected and how much comes out the other side. Specifically, suppose that we shine a beam of electromagnetic energy with direction $\hat{n}$, frequency v, and flux F_{inc} at the test object. Then we measure the *additional* energy flux coming out of the object once this beam is turned on. This outgoing flux may come out in many different directions, because of scattering of the incident beam; in exotic cases, even the frequency could differ from the incident radiation. Let T and R be the transmitted and reflected energy flux, integrated over all angles and frequencies. Then, the absorptivity is defined by taking the ratio of the flux of energy left behind in the body to the incident flux. Thus,

$$a(v, \hat{n}) = \frac{F_{inc} - (T + R)}{F_{inc}}. \tag{3.22}$$

The Planck function is unambiguously the natural choice of a weighting function for defining the mean emissivity $\bar{e}$ for an object with temperature T. There is no such unique choice for defining the mean absorptivity over all frequencies and directions. The appropriate weighting function is determined by the frequency and directional spectrum of the incident radiation which requires a detailed knowledge of its source. If the incident radiation is a blackbody with temperature T_{source} then $\bar{a}$ should be defined with a formula like Eq. (3.20), using $B(v, T_{source})$ as the weighting function. Note that the weighting function is defined by the temperature of the *source* rather than by the temperature of the object doing the absorbing. As was the case for mean emissivity, the temperature dependence of the weighting function implies that $\bar{a}$ will vary with T_{source} even if $a = a(v)$ and is not explicitly dependent on temperature.

Absorptivity and emissivity might appear to be independent characteristics of an object, but observations and theoretical arguments reveal an intimate relation between the two. This relation, expressed by *Kirchhoff's law of radiation*, is a profound property of the interaction of radiation with matter that lies at the heart of all radiative transfer theory. Kirchhoff's law states that the emissivity of a substance at any given frequency equals the absorptivity measured at the same frequency. It was first inferred experimentally. The hard-working spectroscopists of the late nineteenth centuries employed their new techniques to measure the emission spectrum $I(v, \hat{n}, T)$ and absorptivity $a(v, \hat{n}, T)$ of a wide variety of objects at various temperatures. Kirchhoff found that, with the exception of a few phosphorescent materials whose emission was not linked to temperature, all the experimental data collapsed onto a single universal curve, independent of the material, once the observed emission was normalized by the observed absorptivity. In other words, virtually all materials fit the relation $I(v, \hat{n}, T)/a(v, \hat{n}, T) = f(v, T)$ with the same function f. If we take the limit of a perfect absorber – a perfectly "black" body – then $a = 1$ and we find that f is in fact what we have been calling the *Planck* function $B(v, T)$. In fact, it was this extrapolation to a perfect absorber that originally led to the formulation of the notion of blackbody radiation. Since $f = B$ and $I = eB$, we recover the statement of Kirchhoff's law in the form $e/a = 1$.

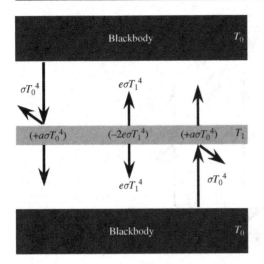

Figure 3.12 Sketch illustrating thought experiment for demonstrating Kirchhoff's law in the mean over all wavenumbers. In the annotations on the sketch, $a = \bar{a}(T_0)$ and $e = \bar{e}(T_1)$.

The thought experiment sketched in Fig. 3.12 allows us to deduce Kirchhoff's law for the mean absorptivity and emissivity from the requirements of the Second Law of Thermodynamics. We consider two infinite slabs of a blackbody material with temperature T_0, separated by a gap. Into the gap, we introduce a slab of partially transparent material with mean absorptivity $\bar{a}(T_1)$ and mean emissivity $\bar{e}(T_1)$, where T_1 is the temperature of the test material. Note that this system is energetically closed. We next require that the radiative transfer between the blackbody material and the test object cause the system to evolve toward an isothermal state. In other words we are *postulating* that radiative heat transfers satisfy the Second Law. A *necessary* condition for radiative transfer to force the system to evolve toward an isothermal state is that the isothermal state $T_0 = T_1$ be an equilibrium state of the system; if it were not, an initially isothermal state would spontaneously generate temperature inhomogeneities. Energy balance requires that $2\bar{a}(T_0)\sigma T_0^4 = 2\bar{e}(T_1)\sigma T_1^4$. Kirchhoff's law then follows immediately by setting $T_0 = T_1$ in the energy balance, which then implies $\bar{a}(T_0) = \bar{e}(T_0)$. Note that the mean absorptivity in this statement is defined using the Planck function at the common temperature of the two materials as the weighting function.

A modification of the preceding argument allows us to show that in fact the emissivity and absorptivity should be equal at each individual frequency, and not just in the mean. To simplify the argument, we will assume that e and a are independent of direction. The thought experiment we employ is similar to that used to justify Kirchhoff's law in the mean, except that this time we interpose frequency-selective mirrors between the test object and the blackbody material, as shown in Fig. 3.13. The mirrors allow the test object to exchange radiant energy with the blackbody only in a narrow frequency band $\Delta\nu$ around a specified frequency ν. The energy budget for the test object now reads $2e(\nu)B(\nu, T_1)\Delta\nu = 2a(\nu)B(\nu, T_0)\Delta\nu$. Setting $T_1 = T_0$ so that the isothermal state is an equilibrium, we find that $e(\nu) = a(\nu)$.

The preceding argument, presented in the form originally given by Kirchhoff, is the justification commonly given for Kirchhoff's law. It is ultimately unsatisfying, as it applies equilibrium thermodynamic reasoning to a system in which the radiation field is manifestly out of equilibrium with matter; in the frequency-dependent form, it invokes the existence of mirrors with hypothetical material properties; worse, it takes as its starting point that

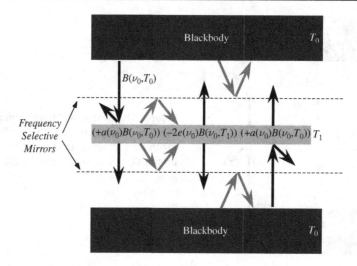

Figure 3.13 Sketch illustrating thought experiment for demonstrating Kirchhoff's law for a narrow band of radiation near frequency ν_0. The thin dashed lines represent ideal frequency-selective mirrors, which pass frequencies close to ν_0, but reflect all others without loss.

radiative heat transfer will act like other heat transfers to equalize temperature, whereas we really ought to be able to demonstrate such a property from first principles of the interaction of radiation with molecules. The great mathematician David Hilbert was among many who recognized these difficulties; in 1912 he presented a formal justification that eliminated the involvement of hypothetical ideal selective mirrors. The physical content of Hilbert's proof is that one doesn't need an ideal mirror, if one requires that a sufficient variety of materials with different absorbing and emitting properties will all come into an isothermal state at equilibrium. Hilbert's derivation nonetheless relied on an assumption that radiation would come into equilibrium with matter at each individual wavelength considered separately. While Kirchhoff did the trick with mirrors, Hilbert, in essence, did the trick with axioms instead, leaving the microscopic justification of Kirchhoff's law equally obscure. It is in fact quite difficult to provide a precise and concise statement of the circumstances in which a material will comply with Kirchhoff's law. Violations are quite commonplace in nature and in engineered materials, since it is quite possible for a material to store absorbed electromagnetic energy and emit it later, perhaps at a quite different frequency. A few examples that come to mind are phosphorescent ("glow in the dark") materials, fluorescence (e.g. paints that glow when exposed to ultraviolet, or "black" light), frequency-doubling materials (used in making green laser pointers), and lasers themselves. In Nature, such phenomena involve insignificant amounts of energy, and are of no known importance in determining the energy balance of planets. We will content ourselves here with the statement that all known liquid and solid planetary materials, as well as the gases making up atmospheres, conform very well to Kirchhoff's law, except perhaps in the most tenuous outer reaches of atmospheres where the gas itself is not in thermodynamic equilibrium.

When applying Kirchhoff's law in the mean, careful attention must be paid to the weighting function used to define the mean absorptivity. For example, based on the incident solar spectrum, the Earth has a mean albedo of about 0.3, and hence a mean absorptivity of 0.7. Does this imply that the mean emissivity of the Earth must be 0.7 as well? In fact, no such implication can be drawn, because Kirchhoff's law only requires that the mean emissivity and absorptivity are the same when averaged over identical frequency weighting functions. Most of the Earth's thermal emission is in the infrared, not the visible. Kirchhoff's law indeed requires that the *visible* wavelength emissivity is 0.7, but the net thermal emission of the Earth in this band is tiny compared with the infrared, and contributes almost nothing to the Earth's net emission. Specifically, the Planck function implies that at 255 K the emission in visible wavelengths is smaller than the emission in infrared wavelengths by a factor of about 10^{-19}. Thus, if the infrared emission from some region were $100 \, W/m^2$, the visible emission would be only $10^{-17} W/m^2$. Using $\Delta E = h\nu$ to estimate the energy of a photon of visible light, we find that this amounts to an emission of only 50 visible light photons each second, from each square meter of radiating surface. This tiny outgoing *thermal* emission of visible light should not be confused with the much larger outgoing flux of *reflected* solar radiation.

It is a corollary of Kirchhoff's law that $e \leq 1$. If the emissivity were greater than unity, then by Kirchhoff's law, the absorptivity would also have to be greater than unity. In consequence, the amount of energy absorbed by the body per unit time would be greater than the amount delivered to it by the incident radiation. By conservation of energy, that would imply the existence of an internal energy source. However, any internal energy source would ultimately be exhausted, violating the assumption that the system is in a state of equilibrium which can be maintained indefinitely.

3.6 OPTICALLY THIN ATMOSPHERES: THE SKIN TEMPERATURE

Since the density of an atmosphere always approaches zero with height, in accordance with the hydrostatic law, one can always define an outer layer of the atmosphere that has so few molecules in it that it will have low infrared emissivity. We will call this the *skin layer*. What is the temperature of this layer? Suppose for the moment that it is transparent to solar radiation, and that atmospheric motions do not transport any heat into the layer; thus, it is heated only by infrared upwelling from below. Because the emissivity of the skin layer is assumed small, little of the upwelling infrared will be absorbed, and so the upwelling infrared is very nearly the same as the *OLR*. The energy balance is between absorption and emission of infrared. Since the skin layer radiates from both its top and bottom, the energy balance reads

$$2 e_{ir} \sigma T_{skin}^4 = e_{ir} OLR. \tag{3.23}$$

Hence,

$$T_{skin} = \frac{1}{2^{\frac{1}{4}}} \left(\frac{OLR}{\sigma} \right)^{\frac{1}{4}} = \frac{1}{2^{\frac{1}{4}}} T_{rad} \tag{3.24}$$

where T_{rad} is defined as before. Thus, the skin temperature is colder than the blackbody radiating temperature by a factor of $2^{-\frac{1}{4}}$. The skin temperature is the natural temperature the outer regions of an atmosphere would have in the absence of *in situ* heating by solar absorption or other means. Note that the skin layer does not need any interior heat transfer mechanism to keep it isothermal, since the argument we have applied to determine T_{skin} applies equally well to any sublayer of the skin layer.

A layer that has low emissivity, and hence low absorptivity, in some given wavelength band is referred to as being *optically thin* in this band. A layer could well be optically thick in the infrared, but optically thin in the visible, which is in fact the case for strong greenhouse gases.

Now let's suppose that the entire atmosphere is optically thin, right down to the ground, and compute the pure radiative equilibrium in this system in the absence of heat transfer by convection. We will also assume that the atmosphere is completely transparent to the incident solar radiation. Let S be the incident solar flux per unit surface area, appropriate to the problem under consideration (e.g. $\frac{1}{4}L_\odot$ for the global mean or $L_\odot$ for temperature at the subsolar point on a planet like modern Mars). Since the atmosphere has low emissivity, the heating of the ground by absorption of downwelling infrared emission coming from the atmosphere can be neglected to lowest order. Since the ground is heated only by absorbed solar radiation, its temperature is determined by $\sigma T_s^4 = (1 - \alpha)S$, just as if there were no atmosphere at all. In other words, $p_{rad} = p_s$ because the atmosphere is optically thin, so that the atmosphere does not affect the surface temperature no matter what its temperature structure turns out to be. Next we determine the atmospheric temperature. The whole atmosphere has small *but non-zero* emissivity so that the skin layer in this case extends right to the ground. The atmosphere is then isothermal, and its temperature T_a is just the skin temperature $2^{-1/4}T_s$.

The surface is thus considerably warmer than the air with which it is in immediate contact. There would be nothing unstable about this situation if radiative transfer were truly the only heat transfer mechanism coupling the atmosphere to the surface. In reality, the air molecules in contact with the surface will acquire the temperature of the surface by heat conduction, and turbulent air currents will carry the warmed air away from the surface, forming a heated, buoyant layer of air. This will trigger convection, mixing a deep layer of the atmosphere within which the temperature profile will follow the adiabat. The layer will grow in depth until the temperature at the top of the mixed layer matches the skin temperature, eliminating the instability. This situation is depicted in Fig. 3.14. The isothermal, stably stratified region above the mixed region is the stratosphere in this atmosphere, and the lower, adiabatic region is the troposphere; the boundary between the two is the tropopause. We have just formulated a theory of tropopause height for optically thin atmospheres. To make it quantitative, we need only require that the adiabat starting at the surface temperature match to the skin temperature at the tropopause. Let p_s be the surface pressure and p_{trop} be the tropopause pressure. For the dry adiabat, the requirement is then $T_s(p_{trop}/p_s)^{R/c_p} = T_{skin}$.

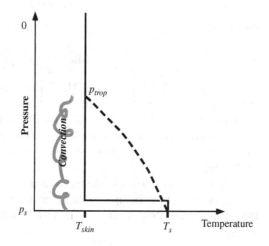

Figure 3.14 The unstable pure radiative equilibrium for an optically thin atmosphere (solid line) and the result of adjustment to the adiabat by convection (dashed line). The adjustment of the temperature profile leaves the surface temperature unchanged in this case, because the atmosphere is optically thin and has essentially no effect on the OLR.

Since $T_s = 2^{1/4} T_{skin}$, the result is

$$\frac{p_{trop}}{p_s} = 2^{-\frac{c_p}{4R}}. \tag{3.25}$$

Note that the tropopause pressure is affected by R/c_p, but is independent of the solar flux S.

The stratosphere in the preceding calculation differs from the observed stratosphere of Earth in that it is isothermal rather than warming with altitude. The factor we have left out is that real stratospheres often contain constituents that absorb solar radiation. To rectify this shortcoming, let's consider the effect of solar absorption on the temperature of the skin layer. Let e_{ir} be the infrared emissivity, which is still assumed small, and a_{sw} be the short-wave (mostly visible) absorptivity, which will also be assumed small. Note that Kirchhoff's law does not require $e_{ir} = a_{sw}$, as the emissivity and absorptivity are at different wavelengths. The solar absorption of incident radiation is $a_{sw}S$. We will assume that the portion of the solar spectrum which is absorbed by the atmosphere is absorbed so strongly that it is completely absorbed before reaching the ground. This is in fact the typical situation for solar near-infrared and ultraviolet. In this case, one need not take into account absorption of the upwelling solar radiation reflected from the surface.

Exercise 3.8 Show that if the atmosphere absorbs uniformly throughout the solar spectrum, then the total absorption in the skin layer is $(1 + (1 - a_{sw})\alpha_g)a_{sw}S$, where α_g is the solar albedo of the ground. Show that the *planetary albedo* – i.e. the albedo observed at the top of the atmosphere – is $(1 - a_{sw})^2 \alpha_g$.

The energy balance for the skin layer now reads

$$2e_{ir}\sigma T^4 = e_{ir}OLR + a_{sw}S. \tag{3.26}$$

Hence,

$$T = T_{skin}\left(1 + \frac{a_{sw}}{e_{ir}}\frac{S}{OLR}\right)^{\frac{1}{4}} \tag{3.27}$$

where T_{skin} is the skin temperature in the absence of solar absorption. The formula shows that solar absorption always increases the temperature of the skin layer. The temperature increases as the ratio of shortwave absorption to infrared emissivity is made larger. So long as the temperature remains less than the solar blackbody temperature, the system does not violate the Second Law of Thermodynamics, since the radiative transfer is still acting to close the gap between the cold atmospheric temperature and the hot solar temperature. As the atmospheric temperature approaches that of the Sun, however, it would no longer be appropriate to use the infrared emissivity, since the atmosphere would then be radiating in the shortwave range. Kirchhoff's law would come into play, requiring $a/e = 1$. This would prevent the atmospheric temperature from approaching the photospheric temperature.

If the shortwave absorptivity is small, the skin layer can be divided into any number of sublayers, and the argument applies to determine the temperature of each one individually. This is so because the small absorptivity of the upper layers does not take much away from the solar beam feeding absorption in the lower layers. We can then infer that the temperature of an absorbing stratosphere will increase with height if the absorption increases with height, making a_{sw}/e_{ir} increase with height.

Armed with our new understanding of the optically thin outer portions of planetary atmospheres, let's take another look at a few soundings. The skin temperature, defined

Table 3.3 Computed skin temperatures of selected planets.

Planet	Skin temperature (K)
Venus	213
Earth	214
Mars (255 K sfc)	214
Jupiter	106
Titan	72

The Mars case was done assuming a 255 K local surface temperature.

in Eq. (3.24), provides a point of reference. It is shown for selected planets in Table 3.3. Except for the Martian case, these values were computed from the global mean *OLR*, either observed directly (for Jupiter) or inferred from the absorbed solar radiation. In the case of present Mars, the fast thermal response of the atmosphere and surface makes the global mean irrelevant. Hence, assuming the atmosphere to be optically thin, we compute the skin temperature based on the upwelling infrared from a typical daytime summer surface temperature corresponding to the Martian soundings of Fig. 2.2. The tropical Earth atmosphere sounding shown in Fig. 2.1 shows that the temperature increases sharply with height above the tropopause. This suggests that solar absorption is important in the Earth's stratosphere. For Earth, the requisite solar absorption is provided by ozone, which strongly absorbs solar ultraviolet. This is the famous "ozone layer," which shields life on the surface from the sterilizing effects of deadly solar ultraviolet rays. However, it is striking and puzzling that virtually the entire stratosphere is substantially colder than the skin temperature based on the global mean radiation budget. The minimum temperature in the sounding is 188 K, which is fully 26 K below the skin temperature. If anything, one might have expected the tropical temperatures to exceed the global mean skin temperatures, because the local tropospheric temperatures are warmer than the global mean. A reasonable conjecture about what is going on is that high, thick tropical clouds reduce the local *OLR*, thus reducing the skin temperature. However, the measured tropical *OLR* in Fig. 3.7 shows that at best clouds reduce the tropical *OLR* to 240 W/m^2, which yields the same 214 K skin temperature computed from the global mean budget. Apart from possible effects of dynamical heat transports, the only way the temperature can fall below the skin temperature is if the infrared emissivity becomes greater than the infrared absorptivity. This is possible, without violating Kirchhoff's law, if the spectrum of upwelling infrared is significantly different from the spectrum of infrared emitted by the skin layer. We will explore this possibility in the next chapter.

Referring to Fig. 2.2 we see that the temperature of the Martian upper atmosphere declines steadily with height, unlike Earth; this is consistent with Mars' CO_2 atmosphere, which has only relatively weak absorption in the solar near infrared spectrum. The Martian upper atmosphere presents the same quandary as Earth's though, in that the temperatures fall well below the skin temperature estimates. Just above the top of the Venusian troposphere, there is an isothermal layer with temperature 232 K, just slightly higher than the computed skin temperature. However, at higher altitudes, the temperature falls well below the skin temperature, as for Mars.

Between 500 mb and 100 mb, just above Titan's troposphere, Titan has an isothermal layer with temperature 75 K, which is very close to the skin temperature. Above 100 mb, the atmosphere warms markedly with height, reaching 160 K at 10 mb. The solar absorption in

Titan's stratosphere is provided mostly by organic haze clouds. Jupiter, like Titan, has an isothermal layer just above the troposphere, whose temperature is very close to the skin temperature. Jupiter's atmosphere also shows warming with height; its upper atmosphere becomes nearly isothermal at 150 K, which is 44 K warmer than the skin temperature. This indicates the presence of solar absorbers in Jupiter's atmosphere as well, though the solar absorption is evidently more uniformly spread over height on Jupiter than it is on Earth or Titan.

We have been using the term "stratosphere" rather loosely, without having attempted a precise definition. It is commonly said, drawing on experience with Earth's atmosphere, that a stratosphere is an atmospheric layer within which temperature increases with height. This would be an overly restrictive and Earth-centric definition. The dynamically important thing about a stratosphere is that it is much more stably stratified than the troposphere, i.e. that its temperature goes down less steeply than the adiabat appropriate to the planet under consideration. The stable stratification of a layer indicates that convection and other dynamical stirring mechanisms are ineffective or absent in that layer, since otherwise the potential temperature would become well mixed and the temperature profile would become adiabatic. An isothermal layer is stably stratified, because its potential temperature increases with height; even a layer like that of Mars' upper atmosphere, whose temperature decreases gently with height, can be stably stratified. We have shown that an optically thin stratosphere is isothermal in the absence of solar absorption. Indeed, this is often taken as a back-of-the-envelope model of stratospheres in general, in simple calculations. In the next chapter, we will determine the temperature profile of stratospheres that are not optically thin.

In a region that is well mixed in the vertical, for example by convection, temperature will decrease with height. Dynamically speaking, such a mixed layer constitutes the troposphere. By contrast the stratosphere may be defined as the layer above this, within which vertical mixing plays a much reduced role. Note, however, that the temperature minimum in a profile need not be coincident with the maximum height reached by convection; as will be discussed in Chapter 4, radiative effects can cause the temperature to continue decreasing with height above the top of the convectively mixed layer. Yet a further complication is that, in midlatitudes, large-scale winds associated with storms are probably more important than convection in carrying out the stirring which establishes the tropopause.

We conclude this chapter with a few comparisons of observed tropopause heights with the predictions of the optically thin limit. We'll leave Venus out of this comparison, since its atmosphere is about as far from the optically thin limit as one could get. On Mars, using the dry adiabat for CO_2 and a 5 mb surface pressure puts the tropopause at 2.4 mb, which is consistent with the top of the region of steep temperature decline seen in the daytime Martian sounding in Fig. 2.2. For Titan, we use the dry adiabat for N_2 and predict that the tropopause should be at 816 mb, which is again consistent with the sounding. If we use the methane/nitrogen moist adiabat instead of the nitrogen dry adiabat, we put the tropopause distinctly higher, at about 440 mb. Because the moist adiabatic temperature decreases less rapidly with height than the dry adiabat, one must go to greater elevations to hit the skin temperature (as in Fig. 3.14). The tropopause height based on the saturated moist adiabat is distinctly higher than seems compatible with the sounding, from which we infer that the low levels of Titan must be undersaturated with respect to methane. Using $R/c_p = \frac{2}{7}$ for Earth air and 1000 mb for the surface pressure, we find that the Earth's tropopause would be at 545 mb in the optically thin, dry limit. This is somewhat higher in pressure (lower in altitude) that the actual midlatitude tropopause, and very much higher in pressure than the tropical tropopause. Earth's real atmosphere is not optically thin, and the lapse rate is less steep than the dry adiabat owing to the effects of moisture. The effects of optical thickness will be treated in detail in Chapter 4, but we can already estimate the effect of using the moist adiabat. Using the computation of the water-vapor/air moist adiabat described in Chapter 2, the tropopause rises to 157 mb, based on a typical tropical surface temperature of 300 K and the skin temperature estimated in Table 3.3. This is much closer to the observed tropopause (defined as the temperature minimum in the sounding), with the remaining mismatch being accounted for by the fact that the minimum temperature is appreciably colder than the skin temperature.

3.7 WORKBOOK

Python tips: The phys module contains Planck's constant (phys.h), the speed of light (phys.c), and the Planck function (invoked as phys.B(nu,T)).

3.7.1 Basic concepts

Problem 3.1 A comet hits the Earth with a speed of 15 km/s. What mass does the comet have to have in order for the energy delivered by the impact to equal the solar energy received by the Earth in one day? (Assume the albedo is 30%.) If the comet has the same density as water, what would its radius be?

Problem 3.2 With the present solar luminosity, the Earth absorbs solar energy at a rate of about 240 W/m^2, averaged over the planet's surface. Suppose that the Earth had no way of losing energy. Assuming the absorbed energy to be uniformly distributed over the entire mass of the planet, how much would the Earth heat up after one million years? After one billion years? For the purposes of this problem, you may assume the Earth to have a mean specific heat of 840 J/kg K, which is the value for basalt rock at surface pressure. The mass of the Earth is $6 \cdot 10^{24}$ kg.

Problem 3.3 Estimate how bright it would look in the daytime on the surface of Neptune's moon Triton. Do this by comparing the flux received from the Sun at Triton with the illumination provided by a candle carried into a dark room. A candle puts out about 1 W of visible light. Assume the candle light bounces off a light-colored wall at a distance of d meters, and compute the flux from the candle which hits the wall. How far would the wall have to be in order to look about as bright as the surface of Triton?

Problem 3.4 The V-filter used by astronomers corresponds to the peak sensitivity of human vision. It has a central wavelength of 0.55 μm and a bandwidth of about 0.1 μm. If the flux density in wavenumber space passing through this band is measured in W/(m^2 cm^{-1}) then the constant C used in defining zero magnitude for this band is -29.836. (See Problem 1.14 for the definition of magnitude.) Note than when defining filtered magnitudes, flux density is used in the formula rather than flux itself. This makes the definition less sensitive to the filter bandwidth.

A telescope with a collector area of 1 m^2 observes a star with apparent V-magnitude M_V. Make a table showing how many V-band photons strike the collector per second, for $M_V = 0, ..., 20$.

Hint: Make sure to convert to frequency in Hz when using Planck's constant, and to convert the bandwidth to units compatible with the flux density. This problem involves a mishmash of different units, but that's just like real life; at least you have been spared learning what a jansky is!

Problem 3.5 In the text we derived the $1/r^2$ law for solar radiation by dividing the net power output of a star by the area of a sphere of radius r with the star at its center. In this problem, you will rederive the same result by summing up the blackbody radiation coming from each patch of the star's visible surface, making use of the angular distribution of blackbody radiation. This problem tests your understanding of the Planck density and the way the angular distribution is represented. It also provides a way to compute the solar constant in cases where the brightness is not uniform across the star's surface.

Consider an observer at a distance $r \gg r_\odot$. You need to compute the power incident on a small flat patch P with area dA, oriented perpendicular to the line joining the observer to the center of the star. Each patch $P_\odot$ of the visible surface of the star, having area $dA_\odot$, contributes a flux (per unit frequency) $B(\nu, T_\odot) dA_\odot \, d\Omega \cos\theta$ to the net flux going through P. In this formula, $d\Omega$ is the (small) solid angle made up by the set of rays from $P_\odot$ which go through P, and θ is the angle between the normal to the star's surface at $P_\odot$ and the line joining the center of $P_\odot$ to the center of P. Complete the derivation by responding to the following:

- Why do we need the factor $\cos\theta$?
- Find an expression for $d\Omega$ in terms of dA and r.
- Find an expression for $\cos\theta$ in terms of the position on the star's surface.
- Integrate $B d\Omega \cos\theta$ with respect to $dA_\odot$ over the visible surface of the star. Note that since $r \gg r_\odot$, the visible surface (i.e. the part of the surface from which some rays pass through the patch P) is approximately a hemisphere.

If you did everything properly, you should obtain the same answer as we got using the previous method, namely $\pi B r_\odot^2 / r^2$.

3.7.2 The Planck function

Problem 3.6 Consider a patch of the photosphere of the Sun with an area of $10 \, \text{m}^2$. The patch is small enough that it can be considered flat. Assume that the temperature of the photosphere is 6000 K and that it radiates like an ideal blackbody. Now consider all the light with wavenumbers between 10 000 and 10 100 cm^{-1}, which leaves the patch in directions making angles between 45 degrees and 50 degrees with the normal to the patch. How much power (in W) is contained in the light meeting these requirements? What is the rate (in W) at which energy is carried *through* the patch by such light? For the purposes of this problem, you may assume the Planck function to be constant within the specified wavenumber band.

Problem 3.7 What is the total power radiated by a blackbody sphere of radius 1 m having uniform temperature 300 K, in the wavenumber range between 500 cm^{-1} and 750 cm^{-1}. You may assume the Planck density to be approximately constant over this range. Compare this to the total power radiated over all wavenumbers. *Note:* Blackbody radiation is emitted with equal intensity in all directions, so you must remember to take this into account in computing the total energy flux coming from the surface.

Problem 3.8 Compute the *wavenumber* of maximum emission of an object with temperature 200 K. What is the rate of energy emission from each square meter of the object's surface, in the wavenumber band extending 50 cm^{-1} on either side of the maximum? You may assume for the purposes of this calculation that the Planck function is constant over this range of wavenumbers. How good is this assumption?

Problem 3.9 Write down the Planck density corresponding to the case in which position in the electromagnetic spectrum is measured by $\ln \lambda$, where λ is the wavelength. Estimate the value of $u = h\nu/(kT)$ for which this density has its peak. (You will probably need to do a small numerical calculation to do this.) How does your result change if you use the log of the wavenumber or frequency as the coordinate instead?

Problem 3.10 *Color index and stellar temperature*
Suppose we observe the flux density from a star at two different frequencies ν_1 and ν_2, with $\nu_2 > \nu_1$. Let the measured flux densities be F_1 and F_2. Suppose further that the star radiates like an ideal blackbody with temperature $T_\odot$. The two flux densities will then be proportional to the Planck function at the corresponding frequencies, multiplied by the inverse-square attenuation of the starlight. Write the flux densities as astronomical magnitudes using $M_j = C_j - 2.5 \log_{10} F_j$, where C_j is the constant defining zero magnitude for each band. Show that $M_2 - M_1$ is independent of the distance from the star. This difference is called a *color index*. Plot the color index as a function of $T_\odot$ for ν_2 corresponding to the astronomer's blue ("B") band at 0.44 μm and ν_1 corresponding to the visual ("V") band at 0.55 μm, and show that the magnitude difference is a monotonically decreasing function of temperature. Find analytic expressions in the low-frequency and high-frequency limit, and use these to get an approximate expression for $T_\odot$ as a function of the color index. Given an observation of the color index alone, how would you know which, if either, of these two approximations is applicable? *Hint:* In the low-frequency case you will need to retain one term beyond the lowest order expansion in frequency in order to get a dependence of the color index on $T_\odot$.

Astronomers define color indices for specific standard filter bands. The B–V color index is very commonly reported as a proxy for the temperature of an object. Various empirical fits are available, which take into account the actual standard filter characteristics. For example, for temperatures between 3000 K and 9000 K, the empirical relation is $M_B - M_V = -3.684 \log_{10} T + 14.551$. Compare this fit to the results you got in your calculation above, which in essence assume infinitely narrow filters. In making this comparison, choose $C_1 - C_2$ so that the agreement is exact at 6000 K.

Problem 3.11 *Bolometric correction*
Problem 1.14 introduced the concept of bolometric magnitude. Earth-based astronomers rarely measure the bolometric magnitude directly, because the atmosphere absorbs much of the spectrum and it is hard to take accurate all-wave flux measurements. Direct measurement of bolometric magnitude is becoming more common with the advent of space-based telescopes, but it is still common for astronomers to report V-magnitudes and add a temperature-dependent correction to convert to bolometric magnitudes assuming a blackbody spectrum. This term is called the *bolometric correction*, denoted by BC.

Suppose that the observed flux density near frequency ν_0 is F_0. Compute the total flux F_{tot} integrated over all wavenumbers, assuming that the spectrum is proportional to a Planck spectrum with temperature $T_\odot$. Show that for any given ν_0, F_{tot}/F_0 depends only on $T_\odot$. Using the definition of magnitude, this ratio is converted to a bolometric correction using

$BC = C_{bol} - C_0 - 2.5\log_{10}(F_{tot}/F_0)$, where C_{bol} and C_0 are the constants used to define the zero-magnitude point for the bolometric and filtered fluxes, respectively. For V-magnitudes this difference is about 10.86. Plot the bolometric correction as a function of $T_\odot$, for ν_0 corresponding to the V-band at 0.55 µm.

The following empirical fit gives the bolometric correction taking into account the actual filter characteristics of the standard V-filter:

$$BC = -8.499x^4 + 13.421x^3 - 8.131x^2 - 3.901x - 0.438, \quad x \equiv \log_{10} T_\odot - 4. \qquad (3.28)$$

Compare this with the narrow-filter calculation you did above.

Problem 3.12 Show that if both ν_1 and ν_2 are in the frequency range where $h\nu/kT \ll 1$, then the total flux per steradian emitted between ν_1 and ν_2 is $\frac{2}{3}(kT/c^2)\left(\nu_2^3 - \nu_1^3\right)$.

Problem 3.13 *Microwave brightness of Venus*
The thick CO_2 atmosphere of Venus is nearly opaque in the infrared, so observations of Venus in the infrared spectrum provide no direct information about the temperature of the ground. However, the atmosphere is nearly transparent to microwave radiation, so the microwave emission of the planet can be used to infer its surface temperature. Indeed, this is how it was inferred in the 1960s that the surface of the planet was hotter than a pizza oven, rather than the steamy jungle world earlier science fiction authors had envisioned.

Assuming the ground to radiate like a blackbody in the microwave, compute the net power (in W) radiated by Venus in the wavelength band from 1 millimeter to 100 millimeters, if the surface temperature is uniform at 737 K. What would the radiated power be if the surface temperature were 300 K instead? *Hint:* You can use the result from Problem 3.12. Why is this valid?

At its closest approach, Venus is about 41 million km distant from Earth. Using the $1/r^2$ law, what microwave energy flux (in W/m²) would be seen from Earth orbit by a microwave antenna directed toward Venus? How much microwave power would be collected by an antenna with area 100 m²?

Problem 3.14 Compute the total power radiated by a person with a normal body temperature of 37 °C. Why is this so much greater than the typical daily energy consumed by a person in the form of food (equivalent to about 100 W)? Next, compute the power radiated by the person in the visible wavelength band (about 0.5 to 1 µ m). Approximately how many visible light photons per second are radiated? About how long would you have to wait before the person emitted a *single* ultraviolet photon (about 0.1 µm wavelength)?

For the purposes of estimating the surface area needed in this problem, you may assume that the person is shaped approximately like a rectangular prism, with height 1.5 m, width 0.5 m, and depth 0.25 m.

Problem 3.15 Derive an approximate form of the cumulative blackbody emission spectrum, valid for large values of $h\nu/(kT)$. *Hint:* Write the cumulative spectrum as $\sigma T^4 - I_1(h\nu/(kT))$, and use the fact that the dominant contribution to the integral I_1 comes from the vicinity of the lower limit of integration, $h\nu/(kT)$.

Problem 3.16 Estimate the total power radiated by the Earth at wavelengths equal to or shorter than visible light (about 0.5 µm. For the purposes of this problem, assume that the Earth has a uniform radiating temperature of 255 K.

Problem 3.17 In the text we needed to evaluate the integral $\int_0^\infty (u^3/(e^u - 1))du$ in order to determine the Stefan–Boltzmann constant. Using a Taylor series expansion of the integrand, show that the integral is equal to $6\zeta(4)$, where

$$\zeta(n) = \sum_{j=1}^{\infty} \frac{1}{j^n}. \tag{3.29}$$

Hint: Multiply the numerator and denominator of the integrand by e^{-u} and then Taylor expand in e^{-u}. Equivalently, the expansion can be done as the geometric series $1 + a + a^2 + \cdots$ where $a = e^{-u}$. Note that the series is convergent for all $u > 0$ because $e^{-u} < 1$.

3.7.3 Basic energy balance

Problem 3.18 You are designing a spherical planet to be placed at the orbit of Mercury. The planet will have a nitrogen atmosphere that has no effect on the infrared radiated by the planet, but is dense enough and mixes heat rapidly enough that the entire planetary surface is isothermal. What albedo should the planet have in order for its surface temperature to be a comfortable 300 K?

Problem 3.19 A cylindrical space station with length h and radius r is in an orbit about the Sun at a distance where the solar constant is $L_\odot$. The space station has zero albedo in the shortwave range and radiates as a perfect blackbody in the longwave (infrared) range. The flow of air inside keeps the entire station at the same temperature, and the skin is a good conductor of heat, so that its temperature is the same as that of the interior. The orientation of the station is such that the axis of the cylinder is always perpendicular to the line joining the center of the station to the center of the Sun. Find an expression for the temperature of the station. Put in numbers corresponding to the mean solar constant at Earth's orbit, assuming $r = h$.

Now suppose that the equipment in the interior of the space station consumes 1 megawatt of solar-generated electrical power, which is dissipated as heat. How much warmer would this make the station once the equipment was turned on? To get rid of this excessive heat, you are to design a radiator, which is a large, thin flat plate heated by pumped water from the space station so that its temperature is the same as the interior of the space station. The radiator is perfectly reflective in the shortwave range, but acts as a perfect blackbody in the infrared range. How large should the radiator plate be in order to get rid of the excess heat? For the purposes of this part of the problem you may assume $r = h = 50$ m.

Problem 3.20 Compute the equilibrium temperature at the subsolar point of Europa, which is in orbit around Jupiter and therefore has the same solar constant as that planet. The greenhouse effect of Europa's tenuous atmosphere can be neglected. For the purposes of this problem, you may assume that the albedo of Europa is 0.67. Assuming Europa to have a water-ice surface, what would be the saturation vapor pressure of the water vapor atmosphere immediately above the subsolar point? Suppose that there is some methane, carbon dioxide, and ammonia mixed in with the ice. What would be the partial pressures of these gases?

Problem 3.21 Consider a planet covered in water ice with a uniform albedo of 0.7. The planet is tide-locked, so that the same face always points to the Sun; the other side is in perpetual night. The atmosphere has negligible greenhouse effect. Compute the solar constant needed to begin melting the ice under the following three alternate scenarios: (a) the atmosphere is so efficient at transporting heat that the entire surface of the planet (dayside and nightside) has the same temperature; (b) the atmosphere is only moderately efficient, so that the dayside temperature is uniform but essentially no energy is carried away to the nightside; (c) there is no atmosphere, so that each bit of the planet's surface is in equilibrium with the solar radiation it absorbs.

Problem 3.22 The physicist Freeman Dyson has speculated that an advanced civilization might enclose its star within a spherical shell of material, and live on the inner surface. This gives them more living space, and avoids wasting all that starlight that would ordinarily escape to space. Consider a red dwarf star with photospheric temperature 4000 K and radius 200 000 km. What should the radius of the Dyson sphere be in order to have an inner surface temperature of 300 K? Assume that the sphere material is a perfect blackbody with zero albedo at all wavelengths, and that it is a good conductor of heat so that the inner surface and outer surface temperature are identical. You may also assume that the temperature of the photosphere remains unchanged after the Dyson sphere is built. If you were an astronomer looking for such a Dyson sphere, in what part of the electromagnetic spectrum should you look? What radius should the Dyson sphere have if it is instead made of a good thermal insulator, so that the outer surface is significantly colder than the inner surface? In the latter case, assume that the heat flux through the shell is proportional to the temperature difference between the inner and outer surface, and that this flux must match the radiation to space from the outer surface.

Challenge question: In either case, is it actually reasonable to assume the photosphere temperature stays unchanged? If you think it should change, what value should it take on assuming: (a) the fusion energy output of the star remains unchanged, and (b) the radius of the Dyson sphere is adjusted so that the inner temperature is 300 K as before?

Problem 3.23 The discovery of a planetary system orbiting the red dwarf star Gliese 581 was announced in 2007. This system is of particular interest, because two of the planets (581c and 581d) have masses just a few times that of Earth, and so are presumably solid (ice or rocky) bodies rather than gas giants. The star has a luminosity 0.013 times that of the Earth's Sun. Gliese 581c orbits at a mere 0.073 A.U. (Astronomical Units) from its star, and 581d orbits at 0.25 A.U. The temperature of the photosphere of Gliese 581 is 3480 K.

Compute the equilibrium temperatures of these two planets, assuming that they are isothermal spherical bodies with an albedo of 0.3, and assuming that the atmosphere has no greenhouse effect. What is the gap between the median-emission wavenumber of the star and that of each of the planets? Compare this to the situation of Mercury in our own Solar System. Based on the data given, estimate the radius of the star in the Gliese system.

Problem 3.24 *Freeze-out time of magma ocean*
This problem is an idealization of the calculation of the freeze-out time of a magma ocean such as might have occurred on Earth following the Moon-forming impact. Earthlike magma freezes at 2036 K at near-surface pressures. We will assume the entire planet is liquid and convecting vigorously, and that the entire planet is isothermal at this temperature until enough energy has been removed to freeze the whole planet solid. The latent heat required to freeze the magma is on the order of $3 \cdot 10^5$ J/kg. Under these assumptions, compute the time needed for the magma ocean to freeze, assuming the radius of the planet to be a,

the density to be ρ and the atmosphere to be transparent to infrared. The latter assumption implies that the planet radiates at a rate of σT^4 per unit surface area, where T is the surface temperature (held constant at the melting point). Estimate the time, in years, for this to happen on the Earth, assuming a density of $5500\,\text{kg/m}^3$, and verify the statement in Chapter 1 that this is a matter of a few hundred years. In doing this calculation, does it matter much how much solar radiation the Earth absorbs? How does the answer change if your planet is bigger?

Note that this calculation produces an underestimate of the time for the entire planet to freeze-out, and an overestimate of the time required for a crust to first form. The reason is that in reality heat will be removed first from an upper layer of the magma ocean, allowing a crust to form at some point less than the freeze-out time for an isothermal magma ocean. Once the crust forms, the insulating properties of solid rock will greatly reduce the heat flow from the interior, extending the time needed for the rest of the planet to solidify.

Problem 3.25 *Energy balance of satellites*
A spherical satellite with radius a_1 is in a circular orbit at a distance r_1 from its planet, which is also spherical. The planet has radius a_0 and is located a distance r_0 from its star; $r_1 \gg a_0$ and $a_0 \gg a_1$. Both the planet and the satellite are perfectly spherical, both are isothermal. The planet has zero albedo in the shortwave (solar) spectrum, but the satellite has a shortwave albedo of unity so that it is not directly heated by the star. The satellite absorbs perfectly in the infrared spectrum.

Derive an expression for the temperature of the satellite, and compare this to the temperature the satellite would have if it had zero shortwave albedo and were heated directly by the Sun. Based on these results, do you think one needs to include the infrared emission of Saturn when computing the temperature of Titan (which orbits at a distance of about one million km from Saturn)?

How do your results change if you relax the assumption $r_1 \gg a_0$, i.e. you allow the orbit to be so close that the planet "fills the sky"? (This regime is unlikely for natural satellites, but is quite common for artificial satellites in low orbit about the Earth or other planets.)

Problem 3.26 A sphere rests on an infinite horizontal plane with uniform temperature T. Both the sphere and the plane are perfect blackbodies. The sphere does not absorb any sunlight. There is no atmosphere, so the only transfer of heat is by radiation. What is the temperature distribution of the surface of the sphere if it is a perfect insulator so no heat is transferred from one place to another within the sphere? What is the temperature of the sphere if it is filled with a well-stirred fluid which keeps the surface isothermal?

3.7.4 The greenhouse effect

Problem 3.27 For Jupiter, the observed *OLR* is $14.3\,\text{W/m}^2$. Compute the effective radiating temperature. Referring to the Jupiter temperature profile given in Chapter 2, estimate the effective radiating pressure of Jupiter. Is there more than one possible value? Which of these do you consider more likely?

Problem 3.28 *Early Mars climate*
The surface of Mars has river-like channel networks that suggest that at sometime in the past Mars was warm and wet. Some of these features date back to a time several billion

years ago when the solar luminosity (and hence the solar constant at the orbit of Mars) was about 25% less than its current value. One school of thought states that the mean surface temperature of Mars would have to have reached values at the freezing point (273 K) or above to account for such features.

Assume that Early Mars had a dense atmosphere that kept its surface temperature approximately uniform, and that the planet had an albedo similar to the present one (about 0.25). Compute the greenhouse effect – as measured by p_{rad}/p_s – needed to bring the surface temperature of Early Mars up to the freezing point under the following alternate assumptions regarding the vertical profile of temperature:

- (a) A pure CO_2 atmosphere on the dry adiabat;
- (b) A mixture of equal parts (molar), of CO_2 and N_2 on the dry adiabat.

Note that if κ is the gray absorption coefficient for the gas, then $p_{rad}/p_s = g/\kappa p_s$, so you are in essence computing the surface pressure of CO_2 needed to make Mars warm and wet. In Chapter 4 this problem will be done with real gas radiation physics, and it will seen in that case that the effects of CO_2 condensation on the profile can significantly inhibit warming. Why might CO_2 condensation aloft affect the warming?

For the purposes of this problem, you may assume that the ground temperature is equal to the low-level air temperature.

Problem 3.29 Consider an atmosphere whose temperature profile is on the dry adiabat with $R/c_p = \frac{2}{7}$, all the way down to zero pressure (infinite altitude). The surface temperature is $300 K$ and the surface pressure is $10^5 Pa$. The gravity is the same as Earth's gravity, $9.8 m/s^2$. We mix in a hypothetical greenhouse gas with mass concentration q. The greenhouse gas has the property that its absorption coefficient κ is $1 m^2/kg$ for wavelengths between $15 \mu m$ and $18 \mu m$, and is zero elsewhere. Mixing in the greenhouse gas does not change the temperature profile, and you may assume that q is small enough that the change in surface pressure may be neglected. The radiating pressure for any given κ is estimated using $\kappa q p_{rad}/g = 1$, and the radiating temperature is the temperature at this pressure level. Using this estimate, answer the following questions.

The radiating temperature, or *brightness temperature*, for any give frequency v is the temperature T_{rad} such the actual emission to space is equal to the blackbody emission $\pi B(v, T_{rad})$. Sketch a graph of the brightness temperature vs. frequency, showing curves for a range of q from zero to 0.1. Compute a graph of the way the spectrum of outgoing radiation (as observed from space) depends on frequency, for the same range of q. Finally, graph the behavior of the *net* outgoing radiation (i.e. the integral of the curves in the previous question) as a function of q.

Hint: You can answer the last question using the fact that the integral of $\pi B(v, T)$ over all v is σT^4, together with a numerical evaluation of the integral over the 15 to 18 μm band. When the frequency interval is small enough that B is essentially constant over the frequency range, then the numerical integral should be approximately $\pi B(v_0, T)\Delta v$, where v_0 is the frequency at the center of the band. How well does this approximation match the numerically computed value?

Problem 3.30 Consider an atmosphere in which, for some reason, the temperature *increases* with height, according to the formula $T(p) = T_s + a \cdot (p_s - p)$ with a a positive constant. Suppose that the addition of a greenhouse gas elevates the radiating level to $p_{rad} < p_s$.

Find an expression for the dependence of the surface temperature on p_{rad}, assuming that the atmosphere is in equilibrium with an absorbed solar flux $S \cdot (1 - \alpha)$ per unit surface area of the planet.

Problem 3.31 Taking into account its observed albedo, Titan absorbs $2.94 \, \text{W/m}^2$ of solar radiation (averaged over its entire surface). The observed surface temperature is 95 K. If you assume that the temperature profile of the atmosphere is given by the dry adiabat for pure N_2 having a surface pressure of 1.5 bar, what would the radiating pressure for Titan have to be in order to account for the observed surface temperature?

Problem 3.32 *Analysis of the Earth's observed radiation budget*
To do this Data Lab, you will need to know how to read tabular data from a text file using whatever software you are using for data analysis, and to perform simple arithmetic manipulations of the data. This Data Lab can be done using most common spreadsheets and many kind of plotting software, as well as in programming languages such as `Python` or `Matlab`.

The course data directory contains a number of plain text files containing data for the zonal mean (i.e. mean along latitude circles) of quantities measured by the Earth Radiation Budget Experiment, and also surface temperature data. These are contained in the subdirectory `Chapter3Data/ERBE/`.

For July 1988, carry out the following analyses:

* From the *OLR* data, compute the effective radiating temperature T_{rad} as a function of latitude.
* Plot it on the same graph as the actual surface temperature. Also, make a plot of the difference between T_{rad} and T_s. Where is the difference large? Where is it small? Discuss your results, taking note of where it is summer and where it is winter.
* Estimate the radiating pressure p_{rad} assuming the surface pressure to be 1000 mb and assuming the atmosphere to be on the dry adiabat. Plot p_{rad} as a function of latitude. Note that this will be an overestimate of p_{rad} because the real atmosphere is closer to the moist adiabat.

Python tips: The Chapter Script `ERBEplot.py` gives an example of how to read the data, plot it, and perform simple manipulations of the data.

Problem 3.33 *The atmosphere as heat engine*
A *heat engine* is a device that makes use of the temperature difference between two reservoirs in order to do mechanical work. It transfers energy from the hot reservoir to the cold reservoir; the difference between the energy drawn from the hot reservoir and the energy deposited in the cold reservoir is the mechanical work that can be done by the device. For a cycle that starts and ends at the same (p, T) state, Eq. (2.8) implies that $W \equiv -\oint \rho^{-1} dp = \oint T ds$. The final step proceeds from the fact that $\delta Q = T ds$, where s is the entropy, and the fact that $\oint c_p dT = 0$ around a closed loop if c_p is constant. (If c_p is not constant the same result applies, but must be re-phrased in terms of internal energy, defined by $dU = c_p dT$.)

A planetary atmosphere works like a heat engine working between the temperature of the star's photosphere (6000 K for our Sun) and the temperature of space (essentially zero). Only a small part of this temperature difference can be tapped to do mechanical work, however. Consider a dry atmosphere consisting of an ideal gas which is transparent to the incoming solar radiation. All the incoming energy is absorbed at the planet's surface, whereafter it is transferred by heat conduction to the air in contact with the ground. Using the above thermodynamic relation, show that if the atmosphere is transparent to infrared as well, so that the only energy loss to space is by radiation from the ground, then the system cannot do any mechanical work. *Hint:* Think about the surface heating up by a certain amount in time Δt owing to solar absorption, and then cooling down by the same amount owing to radiation of infrared.

Now suppose that the atmosphere is not transparent to infrared, and that the radiation escapes to space from a pressure $p_{rad} < p_s$, where p_s is the surface pressure. Consider a cycle where the atmosphere is maintained on a dry adiabat with surface air temperature equal to ground temperature by the following process. Solar absorption increases the ground temperature and low-level air temperature by an amount ΔT in a time Δt. The air, whose temperature is now $T_s + \Delta T$, becomes buoyant and rises *adiabatically* (i.e. at constant s) to the pressure p_{rad}, where it cools down by radiation until its temperature is equal to the ambient adiabat $T_s \cdot (p_{rad}/p_s)^{R/c_p}$ After that it sinks adiabatically back to the ground, where the process starts all over again. Both the heating and cooling occur at constant pressure. Sketch this cycle on thermodynamic diagrams in $(p, 1/\rho)$ and (T, s) space.

Compute the work done in the course of the preceding cycle, and describe how it depends on p_{rad}/p_s. This is the work available to drive the atmospheric circulation. The solar energy per unit mass put into the parcel in time Δt is $c_p \Delta T$. Compute the efficiency of the atmospheric heat engine by dividing W (the work done per unit mass) by this quantity. Show that the thermodynamic efficiency increases as the greenhouse effect is made stronger.

3.7.5 Ice-albedo feedback, hysteresis and bifurcation

Problem 3.34 *Computing the ice-albedo hysteresis diagram*
Calculations with a complete real gas radiation simulation indicate that, for a CO_2 concentration of 300 ppmv, and with an atmosphere on the moist adiabat, a reasonable fit to the actual *OLR* curve in the range of 220 K to 310 K is the linear fit $OLR(T) = a + b \cdot (T - 220)$ where $a = 113\,\mathrm{W/m^2}$ and $b = 2.177\,\mathrm{W/m^2K}$. Compute the ice-albedo hysteresis diagram giving the set of equilibrium temperatures as a function of the solar constant $L_\odot$. Note that there is a simple trick for getting the bifurcation plot. The equation determining the equilibrium is $\frac{1}{4}L_\odot(1 - \alpha(T)) = OLR(T)$. Instead of specifying $L_\odot$ and finding the T that satisfy the equation, we can re-write the equation as

$$L_\odot = 4\frac{OLR(T)}{1 - \alpha(T)}. \tag{3.30}$$

Now, if we call the right hand side $G(T)$, then $G(T)$ gives the unique value of $L_\odot$ which supports the temperature T. Hence, to get the bifurcation diagram, you can just plot $G(T)$ and then turn it sideways.

Use the same albedo–temperature function defined in Chapter 3. Assume that the albedo for an ice-free Earth is 0.2 and for an ice-covered Earth is 0.6.

Based on your calculations, if CO_2 were held constant how much would $L_\odot$ have to be reduced from its modern value before Earth was forced to fall into an inevitable Snowball state? Using the inverse-square law and assuming a circular orbit, compute how far out from the Sun the Earth would have to be displaced (relative to its present orbit) to achieve this solar constant. Conversely, how close to the Sun would you have to place the Earth before a Snowball state became impossible? *Note:* The assumption of fixed CO_2 is very unrealistic, since tectonically active planets with water have a way of adjusting CO_2 in response to changes in the solar constant. This will be discussed in Chapter 8.

Python tips: The script `IceAlbedoZeroD.py`, found in the Chapter Scripts collection for this chapter, calculates and plots hysteresis diagrams using this method, and also makes various other plots useful in understanding bifurcations due to ice-albedo feedback. You can easily modify this script to solve this and similar problems, by redefining the `OLR(T)` function, and (in other cases) also the albedo function.

Problem 3.35 *Daisyworld*

This is a variant on a theme due to Lovelock. Daisyworld is a black planet (zero albedo) upon which white daisies (albedo α_d) can grow. The daisies are temperature-sensitive – if the planet is too cold, the daisies die out and decompose, leaving the planet black. The same happens if the planet is too hot. In between, the daisy coverage varies such that $\alpha(T) = \alpha_d \cdot 4(T - T_1)(T_2 - T)/(T_1 - T_2)^2$, where T_1 and T_2 are the temperatures determining the survival range for the daisies.

Assuming $OLR(T) = \sigma T^4$, discuss the possible equilibrium states of the climate and the way they vary as the solar constant is changed. Which solutions are stable? Which are unstable? Does this system exhibit hysteresis as the solar constant increases and decreases?

For simplicity you may assume $T_1 = 200\,\mathrm{K}$ and $T_2 = 300\,\mathrm{K}$ if you want, but you should discuss how the qualitative behavior of the system changes as you vary α_d between zero and unity.

3.7.6 Kirchhoff's law, emissivity, absorptivity

Problem 3.36 Consider a sandwich made of three slabs of material The two outer slabs are blackbodies, with temperatures T_1 and T_2 respectively. Between these two is a slab which is a gray body having emissivity e. What is the temperature of the gray-body slab?

Problem 3.37 Suppose that the ground has temperature T_s, which is a given. The ground has unit infrared emissivity. There is one isothermal layer above the ground with emissivity e_1, and another one above that with emissivity e_2. Neither layer absorbs any solar radiation, and neither layer is affected by any heat transfer mechanism other than radiation.

Find an expression for the temperatures of the two layers. Do not assume that the emissivities are small.

Problem 3.38 *Solar-absorbing stratospheric cloud*

Consider an atmosphere which is transparent to solar and infrared radiation, except for a high stratospheric cloud which has emissivity e in the infrared and absorptivity a in the solar spectrum. The cloud is above the region reached by convection, so it is in equilibrium with the absorbed energy from upwelling infrared coming from the planet's surface, the solar radiation which it absorbs directly, and the infrared cooling due to the cloud's emission; it receives no heat at all from fluid dynamical transports. The surface of the planet has constant albedo α, and has unit emissivity in the infrared. The whole system is in equilibrium with an incident solar flux S per unit area of the planet's surface.

Compute the surface temperature and the temperature of the cloud. Under what circumstances does the cloud increase the surface temperature? Under what circumstances does it cool the surface? *Note:* this "cloud" is somewhat like the Earth's ozone layer, which absorbs ultraviolet from the Sun, but also has significant infrared emissivity.

Hint: First do the problem with $\alpha = 0$ to get a feel for what is going on. Then add in the terms in the budget that come from reflection of solar radiation off the ground.

Problem 3.39 Consider an optically thin atmosphere which has a non-zero emissivity e within a very narrow frequency band of width $\Delta \nu$ centered on frequency ν_0. This atmosphere is above a surface having a known temperature T_g, which radiates like a blackbody. What is the temperature of the atmosphere if no convection is allowed to take place? Explore how the temperature varies as a function of ν_0. Does the result depend on $\Delta \nu$? On the emissivity? Put in numbers corresponding to Martian afternoon conditions: $T_g = 230\,\mathrm{K}$, and ν_0 corresponding to a wavenumber of $650\,\mathrm{cm}^{-1}$ (approximately the center of the principal absorption band of CO_2).

3.7.7 Second-law challenges

These rather open-ended questions provide some food for thought.

Problem 3.40 *Do burning glasses violate the Second Law?*
With a magnifying glass on a clear, sunny day, you can focus sunlight to a small dot and achieve extremely high temperatures. The same could be done with a parabolic mirror. Suppose you are conducting this experiment in the vacuum of space, so you needn't be concerned with losses due to atmospheric absorption or reflection. Based on the Second Law, what would you expect the maximum achievable temperature to be at the focus? Do magnifying glasses and mirrors indeed conform to this limit? What processes, if any, keep these devices from violating the limit? *Hint:* Think about the implications of the Sun's finite angular size, as seen from any given orbit.

Problem 3.41 *Do solar-electric furnaces violate the Second Law?*
One could set up a large array of solar cells in the desert, and then use the electricity generated to power an electric furnace in which the electric energy is dissipated in a small enough volume to achieve an arbitrarily high temperature. Do such devices violate the Second Law?

Problem 3.42 *Frequency doublers*
Green laser pointers make use of a nonlinear frequency-doubling device, which takes near-infrared laser input and doubles the frequency to produce green light. This is not a filter, but a transformer of frequency. If such a frequency doubler could be made lossless so that it didn't consume any energy, could you use it to make a device that violated the Second Law (i.e. an energetically closed device that would spontaneously generate temperature gradients from an isothermal state)? *Hint:* Think about what you could do if such a frequency doubler were used in place of the selective mirrors in Kirchhoff's argument for Kirchhoff's law.

3.8 FOR FURTHER READING

For more information on electromagnetic waves and electromagnetic radiation, I recommend

- Jackson JD 1998: *Classical Electrodynamics.* Wiley.
- Feynman RP, Leighton RB and Sands M 2005: *The Feynman Lectures on Physics*, Vol 2. Addison Wesley.

An engaging and accessible intellectual history of the quantum theory can be found in

- Pais A 1991: *Niels Bohr's Times.* Oxford University Press.

For a derivation of the Planck distribution, see Chapter 1 of

- Rybicki GB and Lightman AP 2004: *Radiative Processes in Astrophysics.* Wiley-VCH.

The reader seeking a comprehensive introduction to non-relativistic quantum theory will find it in Volume 1 of

- Messiah A 1999: *Quantum Mechanics.* Dover.

In the Dover edition, this book is a bargain, and repays a lifetime of close study.

FOUR

Radiative transfer in temperature-stratified atmospheres

4.1 OVERVIEW

Our objective in this chapter is to treat the computation of a planet's energy loss by infrared emission in sufficient detail that the energy loss can be quantitatively linked to the actual concentration of specific greenhouse gases in the atmosphere. Unlike the simple model of the greenhouse effect described in the preceding chapter, the infrared radiation in a real atmosphere does not all come from a single level; rather, a bit of emission is contributed from each level (each having its own temperature), and a bit of this is absorbed at each intervening level of the atmosphere. The radiation comes out in all directions, and the rate of emission and absorption is strongly dependent on frequency. Dealing with all these complexities may seem daunting, but in fact it can all be boiled down to a conceptually simple set of equations which suffice for a vast range of problems in planetary climate.

It was shown in Chapter 3 that there is almost invariably an order of magnitude separation in wavelengths between the shortwave spectrum at which a planet receives stellar radiation and the longwave (generally infrared) spectrum at which energy is radiated to space. This is true throughout the Solar System, for cold bodies like Titan and hot bodies like Venus, as well as for bodies like Earth that are habitable for creatures like ourselves. The separation calls for distinct sets of approximations in dealing with the two kinds of radiation. Infrared is both absorbed and emitted by an atmosphere, at typical planetary temperatures. However, the long infrared wavelengths are not appreciably scattered by molecules or water clouds, so scattering can be neglected in many circumstances. One of the particular challenges of infrared radiative transfer is the intricate dependence of absorption and emission on wavelength. The character of this dependence is linked to the quantum transitions in molecules whose energy corresponds to infrared photons; it requires an infrared-specific description.

In contrast, planets do not emit significant amounts of radiation in the shortwave spectrum, though shortwave scattering by molecules and clouds is invariably significant; absorption of shortwave radiation arises from quite different molecular processes than those involved in infrared absorption, and its wavelength dependence has a correspondingly different character. Moreover, solar radiation generally reaches the planet in the form of a nearly parallel beam, whereas infrared from thermal emission by the planet and its atmosphere is more nearly isotropic. The approximations pertinent to shortwave radiation will be taken up in Chapter 5, where we will also consider the effects of scattering on thermal infrared.

We'll begin with a general formulation of the equations of plane-parallel radiative transfer without scattering, in Section 4.2. Although we will be able to derive certain general properties of the solutions of these equations, the equations are not very useful in themselves because of the problem of wavelength dependence. To gain further insight, a detailed examination of an idealized model with wavelength-independent infrared emissivity will be presented in Section 4.3. A characterization of the wavelength dependence of the absorption of real gases, and methods for dealing with that dependence, will be given in Sections 4.4 and 4.5.

4.2 BASIC FORMULATION OF PLANE-PARALLEL RADIATIVE TRANSFER

We will suppose that the properties of the radiation field and the properties of the medium through which it travels are functions of a single coordinate, which we will take to be the pressure in a hydrostatically balanced atmosphere. (Recall that in such an atmosphere there is a one-to-one correspondence between pressure and altitude.) This is the *plane-parallel* assumption. Although the properties of planetary atmospheres vary geographically with horizontal position within the spherical shell making up the atmosphere, in most cases it suffices to divide up the sphere into patches of atmosphere which are much larger in the horizontal than they are deep, and over which the properties can be considered horizontally uniform. In this case, vertical radiative transfer is much more important than horizontal transfer, and the atmosphere can be divided up into a large number of columns that act independently, insofar as radiative transfer is concerned.

In this section, we will develop an approximate form of the equations of plane-parallel radiative transfer. The errors introduced in this approximation are small enough that the resulting equations are sufficiently accurate to form the basis of the infrared radiative transfer component of virtually all large-scale climate models. These equations will certainly be good enough for addressing the broad-brush climate questions that are our principal concern.

4.2.1 Optical thickness and the Schwarzschild equations

Although the radiation field varies in space only as a function of pressure, p, its intensity depends also on direction. Let $I(p, \hat{n}, \nu)$ be the flux density of electromagnetic radiation propagating in direction $\hat{n}$, measured at point p. This density is just like the Planck function $B(\nu, T)$, except that we allow it to depend on direction and position. The technical term for this flux density is *spectral irradiance*. Now we suppose that the radiation propagates through a thin layer of atmosphere of thickness δp as measured by pressure. The absorption of energy at frequency ν is proportional to the number of molecules of absorber encountered; assuming the mixing ratio of the absorber to be constant within the layer for small δp, the number of molecules encountered will be proportional to δp, in accord with the hydrostatic law. By Kirchhoff's law, the absorptivity and emissivity of the layer are the same; we'll call the value $\delta\tau_\nu$, and keep in mind that in general it will be a function of ν. Let θ be the angle between the direction of propagation $\hat{n}$ and the vertical, as shown in Fig. 4.1. Now, let $\delta\tau_\nu^*$ be the emissivity (and absorptivity) of the layer for radiation propagating in the direction $\theta = 0$. We may define the proportionality between emissivity and pressure through the relation $\delta\tau_\nu^* = -\kappa\delta p/g$ where g is the acceleration of gravity and κ is an absorption coefficient. It has units of area per unit mass, and can be thought of as an absorption cross-section per unit mass – in essence, the area taken out of the incident beam by the absorbers contained in a unit mass of atmosphere. In general κ is a function of frequency, pressure, temperature, and the mixing ratios of the various greenhouse gases in the atmosphere. Passing to the limit of small δp, we can define an *optical thickness* coordinate through the differential equation

$$\frac{d\tau_\nu^*}{dp} = -\frac{1}{g}\kappa. \tag{4.1}$$

Since pressure decreases with altitude, τ_ν^* increases with altitude. Radiation propagating at an angle θ relative to the vertical acts just like vertically propagating radiation, except that the thickness of each layer through which the beam propagates, and hence the number of absorbing molecules encountered, is increased by a factor of $1/\cos\theta$. Hence, the optical thickness for radiation propagating with angle θ is simply $\tau_\nu = \tau_\nu^*/\cos\theta$. The equations of radiative transfer can be simplified by using either τ_ν or τ_ν^* in place of pressure as the vertical coordinate.

The specific absorption cross-section κ depends on the number of molecules of each greenhouse gas encountered by the beam and the absorption properties characteristic to each kind of greenhouse gas molecule. Letting q_i be the mass-specific concentration of greenhouse gas i, we may write

$$\kappa(\nu, p, T) = \sum_{i=0}^{n} \kappa_i(\nu, p, T)q_i(p). \tag{4.2}$$

The specific concentrations q_i depend on p because we are using pressure as the vertical coordinate, and the concentration of the gas may vary with height; a well-mixed greenhouse gas would have constant q_i. The dependence of the coefficients κ_i on p and T arises from certain aspects of the physics of molecular absorption, to be discussed in Section 4.4.

Equation (4.1) defines the optical thickness $\tau_\nu^*(p_1, p_2)$ for the layer between pressures p_1 and p_2. Unless κ is constant, it is not proportional to $|p_1 - p_2|$, but it is a general consequence of the definition that $\tau_\nu^*(p_1, p_2) = \tau_\nu^*(p_1, p') + \tau_\nu^*(p', p_2)$ if p' is between p_1 and p_2. Consider an atmosphere with a single greenhouse gas having concentration $q(p)$. Then, if κ_G is independent of p the optical thickness can be expressed as $\tau_\nu^*(p_1, p_2) = \kappa_G\ell$ where ℓ is the *path*, defined by

$$\ell(p_1, p_2) = -\int_{p_1}^{p_2} q(p)\frac{dp}{g}. \tag{4.3}$$

The boundaries of the layer are generally chosen so as to make the path and optical thickness positive. The path is the mass of greenhouse gas in the layer, per square meter of the planet's surface, and in SI units has units of kg/m^2. If the greenhouse gas is well mixed then $\ell = q \cdot (p_1 - p_2)/g$. Now, it often happens that κ_G increases linearly with pressure – a phenomenon known as *pressure broadening* (alternatively *collisional broadening* for reasons

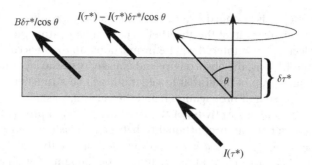

Figure 4.1 Sketch of the radiative energy balance for a slab of atmosphere illuminated by incident radiation from below.

that will eventually become clear). If we write $\kappa_G(p) = \kappa_G(p_0) \cdot (p/p_0)$, then we can define an *equivalent path*

$$\ell_e = -\int_{p_1}^{p_2} q(p) \frac{p}{p_0} \frac{dp}{g} \tag{4.4}$$

such that $\tau_\nu^*(p_1, p_2) = \kappa_G(p_0)\ell_e$ much as before. The equivalent path still has units of mass per unit area, but because of the pressure weighting will differ from the actual path. For example, if the greenhouse gas is well mixed then

$$\ell_e = \frac{1}{g} q \frac{1}{2p_0} \left(p_1^2 - p_2^2 \right) = q \frac{p_1 - p_2}{g} \frac{p_1 + p_2}{2p_0}. \tag{4.5}$$

The equivalent path is thus the actual path weighted by the ratio of the mean pressure to the reference pressure.

Consider now the situation illustrated in Fig. 4.1, in which radiation at a given frequency and angle is incident on a slab of atmosphere from below. In general, part of the incident radiation is scattered into other directions. However, for infrared and longer wave radiation interacting with gases, such scattering is negligible; scattering is also negligible for infrared interacting with condensed cloud particles made of substances such as water, which are strong absorbers. Here, we shall neglect scattering, though it will be brought back into the picture in Chapter 5. The radiation at the same angle which comes out the top of the slab is then the incident flux minus the small amount absorbed in the slab, plus the small amount emitted. Thus

$$I\left(\tau_\nu^* + \delta\tau_\nu^*, \hat{n}, \nu\right) = \left(1 - \frac{\delta\tau_\nu^*}{\cos\theta}\right) I\left(\tau_\nu^*, \hat{n}, \nu\right) + B\left(\nu, T\left(\tau_\nu^*\right)\right) \frac{\delta\tau_\nu^*}{\cos\theta} \tag{4.6}$$

or, passing to the limit of small $\delta\tau_\nu^*$,

$$\frac{d}{d\tau_\nu^*} I\left(\tau_\nu^*, \hat{n}, \nu\right) = -\frac{1}{\cos\theta} \left[I\left(\tau_\nu^*, \hat{n}, \nu\right) - B\left(\nu, T\left(\tau_\nu^*\right)\right)\right]. \tag{4.7}$$

This is the general form of the *Schwarzschild equation* for radiative transfer without scattering. For a precise solution, one needs to solve this equation separately for each θ and then integrate over angles to get the net upward and downward fluxes. The angular distribution of radiation changes with distance from the source, since radiation propagating near the direction $\theta = 0$ or $\theta = \pi$ decays more gradually than radiation with θ nearer to $\pi/2$. Hence, radiation that starts out isotropic at the source (as is the blackbody emission) tends to become more forward-peaked as it propagates. For some specialized problems, it is indeed necessary to solve for the angular distribution explicitly in this fashion, which is rather computationally demanding. Fortunately, the isotropy of the blackbody source term tends to keep longwave radiation isotropic enough to allow one to make do with a much more economical approximate set of equations.

We can derive an equation for the net upward flux per unit frequency, I_+, by multiplying Eq. (4.7) by $\cos\theta$ and integrating over all solid angles in the upward-pointing hemisphere. Integrating over the downward hemisphere yields the net downward flux I_-. However, because of the factor $1/\cos\theta$ on the right hand side of Eq. (4.7), the hemispherically averaged intensity appearing on the right hand side is not I_+. Instead, it is $\int I(\tau^*, \hat{n}, \nu)d\Omega$, or equivalently $\int_0^{\pi/2} 2\pi I(\tau^*, \theta, \nu)\sin\theta d\theta$. One cannot proceed further without some assumption about the angular distribution. If we assume that the distribution remains approximately isotropic, by virtue of the isotropic source B, then $I(\tau^*, \theta, \nu)$ is independent of θ, and hence the problematic integral becomes $2\pi I \int_0^{\pi/2} \sin\theta d\theta$ which is equal to $2I_+$ under the assumption of isotropy. This result yields a closed equation for I_+. It states that, if the radiation field remains approximately isotropic, the decay rate is the same as for unidirectional radiation propagating with an angle $\bar{\theta}$ such that $\cos\bar{\theta} = \frac{1}{2}$, i.e. $\bar{\theta} = 60°$. From now on we will deal only with this approximate angle-averaged form of the equations, and use $\tau_\nu = \tau^*/\cos\bar{\theta}$ as our vertical coordinate. The choice $\cos\bar{\theta} = \frac{1}{2}$ is by no means a unique consequence of the assumption of isotropy. The fact is that an isotropic distribution is not an exact solution of Eq. (4.7) except in a few very special limits, so that the choice we make is between different errors of roughly the same magnitude. If we had calculated $\cos\bar{\theta}$ by multiplying Eq. (4.7) by $(\cos\theta)^2$ instead of $\cos\theta$ before averaging over angles, we would have concluded $\cos\bar{\theta} = \frac{2}{3}$ and this would be an equally valid choice within the limitations of the isotropic approximation. Sometimes, a judicious choice of $\cos\bar{\theta}$ is used to maximize the fit to an angle-resolved calculation in some regime of particular interest. For the most part we will simply use $\cos\bar{\theta} = \frac{1}{2}$ in our calculations unless there is a compelling reason to adopt a different value.

In terms of τ_ν, the equations for the upward and downward flux are

$$\frac{d}{d\tau_\nu}I_+ = -I_+ + \pi B(\nu, T(\tau_\nu))$$
$$\frac{d}{d\tau_\nu}I_- = I_- - \pi B(\nu, T(\tau_\nu)).$$

(4.8)

These are known as the *two-stream equations*, and will serve as the basis for all subsequent discussion of radiative transfer in this book, save that we will incorporate the neglected scattering term in Chapter 5. The two-stream equations without scattering will generally be referred to simply as the *Schwarzschild equations*, after the physicist Karl Schwarzschild who did much to develop their use within astrophysics.

Because of the neglect of scattering, the equations for I_+ and I_- are uncoupled, and each consists of a linear, inhomogeneous first order differential equation. The solution can be obtained by substituting $I_+ = A(\tau_\nu)\exp(-\tau_\nu)$, and similarly for I_-, which reduces the problem to evaluation of a definite integral for A. The result is

$$I_+(\tau_\nu, \nu) = I_+(0)e^{-\tau_\nu} + \int_0^{\tau_\nu} \pi B(\nu, T(\tau_\nu'))\, e^{-(\tau_\nu - \tau_\nu')}d\tau_\nu'$$
$$I_-(\tau_\nu, \nu) = I_-(\tau_\infty)e^{-(\tau_\infty - \tau_\nu)} + \int_{\tau_\nu}^{\tau_\infty} \pi B(\nu, T(\tau_\nu'))\, e^{-(\tau_\nu' - \tau_\nu)}d\tau_\nu'$$

(4.9)

where τ'_v is a dummy variable and τ_∞ is the optical thickness of the entire atmosphere, i.e. $\tau_v^*(p_s, 0)/\cos\bar{\theta}$. Note that τ_∞ depends on v in general, though we have suppressed the subscript for the sake of readability. The top of the atmosphere ($p = 0$) is at τ_∞. The physical content of these equations is simple: $I_+(\tau_v, v)$ consists of two parts. The first is the portion of the emission from the ground which is transmitted by the atmosphere (the first term in the expression for I_+). The second is the radiation emitted by the atmosphere itself, which appears as an exponentially weighted average (the second term in the expression for I_+) of the emission from all layers below τ_v, with more distant layers given progressively smaller weights. Similarly, $I_-(\tau_v, v)$ is an exponentially weighted average of the emission from all layers above τ_v, plus the transmission of incident downward flux. The atmospheric emission to space will be most sensitive to temperatures near the top of the atmosphere. This emission will dominate the *OLR* when the atmosphere is fairly opaque to the radiation emitted from the ground, whereas the transmitted ground emission will dominate when the atmosphere is fairly transparent. The downward radiation into the ground will be most sensitive to temperatures nearest the ground.

In the long run, it will save us some confusion if we introduce special notation for temperatures and fluxes at the boundaries; this will prove especially important when there is occasion to switch back and forth between pressure and optical thickness as a vertical coordinate. The temperature at the top of the atmosphere ($p = 0$ or $\tau = \tau_\infty$) will be denoted by T_∞, and the temperature of the air at the bottom of the atmosphere ($p = p_s$ or $\tau = 0$) will be called T_{sa}. For planets with a solid or liquid surface this is the temperature of the gas in immediate contact with the surface. For such planets, one must distinguish the temperature of the air from the temperature of the surface (the "ground") itself, which will be called T_g. The outgoing and incoming fluxes at the top of the atmosphere will be called $I_{+,\infty}(v)$ and $I_{-,\infty}(v)$ respectively, while the upward and downward fluxes at the bottom of the atmosphere will be called $I_{+,s}(v)$ and $I_{-,s}(v)$.

For planets with a liquid or solid surface, we require that $I_{+,s}(v)$ be equal to the upward flux emitted by the ground, which is $\pi e(v)B(v, T_g)$, where $e(v)$ is the emissivity of the ground. Continuity of the fluxes is required because, the air being in immediate contact with the ground, there is no medium between the two which could absorb or emit radiation, nor is there any space where radiation "in transit" could temporarily reside. We generally assume that there is no infrared radiation incident on the top of the atmosphere, so that the upper boundary condition is $I_{-,\infty} = 0$. The incident solar radiation does contain some near infrared, but this is usually treated separately as part of the shortwave radiation calculation (see Chapter 5). For planets orbiting stars with cool photospheres, such as red giant or dwarf stars, it might make sense to allow $I_{-,\infty}$ to be non-zero and treat the incoming infrared simultaneously with the internally generated thermal infrared. Since the radiative transfer equations are linear in the intensities, it is a matter of taste whether to treat the incoming stellar infrared in this way, or as part of the calculation dealing with the shorter wave part of the incoming stellar spectrum.

For gas or ice giant planets, which have no distinct solid or liquid surface, we do not usually try to model the whole thermal structure of the planet all the way to its center. It typically suffices to specify the temperature and convective heat flux from the interior at some level which is sufficiently deep that the density has increased to the point that the fluid making up the atmosphere can be essentially considered a blackbody. Once the optically thick regime is reached, one doesn't need to know the temperature deeper down in order to do the radiation calculation, any more than one needs to know the temperature of the Earth's core to do radiation on Earth.

The weighting function appearing in the integrands in Eq. (4.9) is the *transmission function*. Written as a function of pressure, it is

$$\mathcal{T}_\nu(p_1, p_2) = e^{-|\tau_\nu(p_1) - \tau_\nu(p_2)|}. \tag{4.10}$$

$\mathcal{T}_\nu(p_1, p_2)$ is the proportion of incident energy flux at frequency ν which is transmitted through a layer of atmosphere extending from p_1 to p_2; whatever is not transmitted is absorbed in the layer. Note that $\mathcal{T}_\nu(p, p')d\tau'_\nu = d\mathcal{T}_\nu$ (with p held constant), if $p < p'$, and $\mathcal{T}_\nu(p, p')d\tau'_\nu = -d\mathcal{T}_\nu$ if $p > p'$. Using this result Eq. (4.9) can be re-written

$$I_+(\tau_\nu, \nu) = I_{+,s}(\nu)\mathcal{T}_\nu(p, p_s) - \int_{p'=p}^{p_s} \pi B(\nu, T(p'))d\mathcal{T}_\nu(p, p')$$

$$I_-(\tau_\nu, \nu) = I_{-,\infty}(\nu)\mathcal{T}_\nu(0, p) + \int_{p'=0}^{p} \pi B(\nu, T(p'))d\mathcal{T}_\nu(p, p'). \tag{4.11}$$

In the integrals above, the differential of $\mathcal{T}_\nu$ is meant to be taken with p held fixed. Integration by parts then yields the following alternate form of the solution to the two-stream equations:

$$I_+(p, \nu) = \pi B(\nu, T(p)) + (I_{+,s}(\nu) - \pi B(\nu, T_{sa}))\mathcal{T}_\nu(p, p_s) + \int_p^{p_s} \pi \mathcal{T}_\nu(p, p')dB(\nu, T(p'))$$

$$I_-(p, \nu) = \pi B(\nu, T(p)) + (I_{-,\infty}(\nu) - \pi B(\nu, T_\infty))\mathcal{T}_\nu(0, p) - \int_0^p \pi \mathcal{T}_\nu(p, p')dB(\nu, T(p')). \tag{4.12}$$

Neither of these forms of the solution is particularly convenient for analytic work, but either one can be used to good advantage when carrying out approximate integrations via the trapezoidal rule (see Section 4.4.6). For analytical work, and some kinds of numerical integration, it helps to re-write the integrand using $dB = (dB/dT)(dT/dp')dp'$. The result is

$$I_+(p, \nu) = \pi B(\nu, T(p)) + (I_{+,s}(\nu) - \pi B(\nu, T_{sa}))\mathcal{T}_\nu(p, p_s) + \int_p^{p_s} \pi \mathcal{T}_\nu(p, p')\frac{dB}{dT}|_{T(p')}\frac{dT}{dp'}dp'$$

$$I_-(p, \nu) = \pi B(\nu, T(p)) + (I_{-,\infty}(\nu) - \pi B(\nu, T_\infty))\mathcal{T}_\nu(0, p) - \int_0^p \pi \mathcal{T}_\nu(p, p')\frac{dB}{dT}|_{T(p')}\frac{dT}{dp'}dp'. \tag{4.13}$$

A considerable advantage of any of the forms in Eq. (4.11), (4.12), or (4.13) is that the integration variable p' is no longer dependent on frequency. This will prove particularly useful when we come to consider real gases, for which the optical thickness has an intricate dependence on frequency. The first two terms in the expression for the fluxes in either Eq. (4.12) or (4.13) give the exact result for an isothermal atmosphere; in each case, the first of the two terms represents the contribution of the local blackbody radiation, whereas the second accounts for the modifying effect of the boundaries. The boundary terms vanish at points far from the boundary, where $\mathcal{T}$ is small. Note that the boundary term for I_+ vanishes identically if the upward flux at the boundary has the form of blackbody radiation with temperature equal to the surface air temperature. For a planet with a solid or liquid surface, this would be the case if the ground temperature equals the surface air temperature and the ground has unit emissivity.

The main reason for dealing with radiative transfer in the atmosphere is that one needs to know the amount of energy deposited in or withdrawn from a layer of atmosphere by radiation. This is the radiative heating rate (with negative heating representing a cooling). It is obtained by taking the derivative of the net flux, which gives the difference between the energy entering and leaving a thin layer. The heating rate per unit optical thickness, per unit frequency, is thus

$$\mathfrak{H}_v = -\frac{d}{d\tau_v}(I_+(\tau_v, v) - I_-(\tau_v, v)). \tag{4.14}$$

This must be integrated over all frequencies to yield the net heating rate. For making inferences about climate, one ordinarily requires the heating rate per unit mass rather than the heating rate per unit optical depth. This is easily obtained using the definition of optical depth, specifically,

$$H_v = g\frac{d}{dp}(I_+ - I_-) = g\frac{d\tau_v}{dp}\frac{d}{d\tau_v}(I_+ - I_-) = \frac{\kappa}{\cos\bar{\theta}}\mathfrak{H}_v. \tag{4.15}$$

When integrated over frequency this heating rate has units W/kg. One can convert into a temperature tendency K/s by dividing this value by the specific heat c_p.

4.2.2 Some special solutions of the two-stream equations

Beer's law

Suppose that the atmosphere is too cold to radiate significantly at the frequency under consideration. In that case, $B(v, T) \approx 0$ and the internal source vanishes. This would be the case, for example, if v is in the visible light range and the temperature of the atmosphere is Earthlike. In this case, the solutions are simply $I_+ = I_+(0)\exp(-\tau_v)$ and $I_- = I_-(\tau_\infty)\exp(\tau_v - \tau_\infty)$. The exponential attenuation of radiation is known as *Beer's law*. Here we've neglected scattering, but in Chapter 5 we'll see that a form of Beer's law still applies even if scattering is taken into account.

Infinite isothermal medium

Consider next an unbounded isothermal medium. In this case, it is readily verified that $I_+ = I_- = \pi B(v, T)$ is an exact solution to (4.9). The right hand sides of the equations vanish, but the derivatives on the left hand sides vanish also, because T is independent of τ_v. Hence, in an unbounded isothermal medium, the radiation field reduces to uniform blackbody radiation.

Since the fluxes are independent of τ_v, the radiative heating rate vanishes, from which we recover the fact that blackbody radiation is in equilibrium with an extended body of isothermal matter.

Exercise 4.1 Derive this result from Eq. (4.9); from Eq. (4.13).

Finite-thickness isothermal slab

Now let's consider an isothermal layer of finite thickness, embedded in an atmosphere which is completely transparent to radiation at frequency v. We suppose further that there is no radiation at this frequency incident on the layer from either above or below. We are free to define $\tau_v = 0$ at the center of the layer, so that $\tau_v = \frac{1}{2}\tau_\infty$ at the top of the layer and $\tau_v = -\frac{1}{2}\tau_\infty$ at the bottom. The boundary conditions corresponding to no incident flux are $I_- = 0$ at the top of the layer and $I_+ = 0$ at the bottom of the layer. The solution $I_+ = I_- = \pi B$ is still a *particular* solution within the layer, since the layer is isothermal, but it does not satisfy the boundary conditions. A homogeneous solution must be added to each flux in order to satisfy the boundary conditions. The homogeneous solutions are just exponentials, and so we easily find that the full solution within the layer is

$$I_+(\tau_\nu, \nu) = \left[1 - \exp\left(-\left(\tau + \frac{\tau_\infty}{2}\right)\right)\right]\pi B$$
$$I_-(\tau_\nu, \nu) = \left[1 - \exp\left(+\left(\tau - \frac{\tau_\infty}{2}\right)\right)\right]\pi B. \tag{4.16}$$

Exercise 4.2 Derive this result from Eq. (4.13).

The radiation emitted out of the top of the layer is $I_+(\frac{1}{2}\tau_\infty, \nu)$, or $(1 - \exp(-\tau_\infty))\pi B$, which reduces to the blackbody value πB when the layer is optically thick for the frequency in question, i.e. $\tau_\infty(\nu) \gg 1$. The same applies for the emission out of the bottom of the layer, *mutatis mutandum*. Note that in the optically thick limit, $I_+ = I_- = \pi B$ through most of the layer, and the inward-directed intensities only fall to zero in the two relatively thin skin layers near the top and bottom of the slab.

In the opposite extreme, when the slab is optically thin, both τ and τ_∞ are small. Using the first order Taylor expansion of the exponentials, we find that the emission out of the top of the layer is $\tau_\infty \pi B$, and similarly for the bottom of the layer. Hence, τ_∞ in this case is just the bulk emissivity of the layer. This is consistent with the way we constructed the Schwarzschild equations, which can be viewed as a matter of stacking a great number of individually optically thin slabs upon each other.

Substituting into Eq. (4.14), the heating rate for this solution is

$$\mathfrak{H}_\nu = -[\exp(-\tau) + \exp(\tau)]\exp\left(-\frac{\tau_\infty}{2}\right)\pi B. \tag{4.17}$$

In the optically thick case, the heating rate is nearly zero in the interior of the slab, but there is strong radiative cooling within about a unit optical depth of each surface. In this case the radiation drains heat out of a thin skin layer near each surface, causing intense cooling there. In the optically thin limit, the cooling is distributed uniformly throughout the slab.

It turns out that condensed water is a much better infrared absorber than the same mass of water vapor. Hence, an isolated absorbing layer such as we have just considered can be thought of as a very idealized model of a cloud. The following slight extension makes the connection with low lying stratus clouds, such as commonly found over the oceans, more apparent.

Exercise 4.3 Instead of being suspended in an infinite transparent medium, suppose that the cloud is in contact with the ground, and that the ground has the same temperature as the cloud. We still assume that the air above the cloud is transparent to radiation at the frequency under consideration. Compute the upward and downward fluxes, and the radiative heating rate, in this case.

This exercise shows that convection in boundary layer stratus clouds can be driven by cooling at the top, rather than heating from below. This is rather important, since the reflection of sunlight by the cloud makes it hard to warm-up the surface. Entrainment of dry air due to top-driven convection is one of the main mechanisms for dissipating such clouds.

Optically thick limit

We now depart from the assumption of constant temperature. While allowing T to vary in the vertical, we assume the atmosphere to be *optically thick* at frequency ν. This means that a small change in pressure p amounts to a large change in the optical thickness coordinate τ_ν. Referring to Eq. (4.1), we see that the assumption of optical thickness is equivalent to the assumption that $\kappa \delta p / g \gg 1$, where δp is the typical amount by which one has to change the pressure in order for the temperature to change by an amount comparable to its mean value. For most atmospheres, it suffices to take δp to be the depth of the whole atmosphere (measured in pressure coordinates), namely p_s, so that the optical thickness assumption becomes $\tau_\infty = \kappa p_s / g \gg 1$. Since κ depends on frequency, an atmosphere may be optically thick near one frequency, but optically thin near another.

The approximate form of the fluxes in the optically thick limit can be most easily derived from the integral expression in the form given in Eq. (4.9). Consider first the expression for I_+. Away from the immediate vicinity of the bottom boundary, the boundary term proportional to $\mathcal{T}_\nu(p, p_s)$ is exponentially small and can be dropped. To simplify the integral, we note that $\mathcal{T}_\nu(p, p')$ is very small unless p' is close to p. Therefore, as long as the temperature gradient is a continuous function of p', it varies little over the range of p' for which the integrand contributes significantly to the integral. Hence dT/dp' can be replaced by its value at p, which can then be taken outside the integral. Likewise, dB/dT can be evaluated at $T(p)$, so that this term can also be taken outside the integral. Finally, if one is not too close to the bottom boundary,

$$\int_p^{p_s} \mathcal{T}_\nu(p', p)\,dp' \approx \int_p^\infty \mathcal{T}_\nu(p', p)\,dp' = \frac{g\cos\bar{\theta}}{\kappa}\int_\tau^\infty \mathcal{T}_\nu(\tau', \tau)\,d\tau' = \frac{g\cos\bar{\theta}}{\kappa},\qquad(4.18)$$

whence the upward flux in the optically thick limit becomes

$$I_+(\nu, p) = \pi B(\nu, T(p)) + \pi \frac{g\cos\bar{\theta}}{\kappa}\frac{dB}{dT}\Big|_{T(p)}\frac{dT}{dp}.\qquad(4.19)$$

Near the bottom boundary, the neglected boundary term would have to be added to this expression. In addition, Eq. (4.18) would need to be corrected to allow for the fact that there is not room for $\int \mathcal{T}$ to integrate out to its asymptotic value for an infinitely thick layer.

Using identical reasoning, the downward flux becomes

$$I_-(\nu, p) = \pi B(\nu, T(p)) - \pi \frac{g\cos\bar{\theta}}{\kappa}\frac{dB}{dT}\Big|_{T(p)}\frac{dT}{dp}\qquad(4.20)$$

so long as one is not too near the top of the atmosphere. Near the top of the atmosphere, the neglected boundary term becomes significant.

In both expressions the second term, proportional to the temperature gradient, becomes progressively smaller as κ is made larger and the atmosphere becomes more optically thick. To lowest order, then, the upward and downward fluxes are both equal to the blackbody radiation flux at the local temperature. In this sense, the optically thick limit looks "locally isothermal." The term proportional to the temperature gradient represents a small correction to the locally isothermal behavior. In the expression for I_+, for example, if $dT/dp > 0$ the correction term makes the upward flux somewhat greater than the local blackbody value. This makes sense, because a small portion of the upwelling radiation comes from lower layers where the temperature is warmer than the local temperature. Note that the correction term depends on ν through the frequency dependence of κ, as well as through the frequency dependence of dB/dT.

The radiation exiting the top of the atmosphere ($I_{+,\infty}$) is of particular interest, because it determines the rate at which the planet loses energy. In the optically thick approximation, we find that as long as dT/dp is finite at $p = 0$, $I_{+,\infty}$ becomes close to $\pi B(\nu, T_\infty)$ as the atmosphere is made more optically thick. Hence, at frequencies where the atmosphere is optically thick, the planet radiates to space like a blackbody with temperature equal to that of the upper regions of the atmosphere – the regions "closest" to outer space.

Similarly, the downward radiation ($I_{-,s}(\nu)$) from the atmosphere into the ground – sometimes called the *back-radiation* – is of interest because it characterizes the radiative effect of the atmosphere on the surface energy budget. In the optically thick limit, $I_{-,s}(\nu) = \pi B(\nu, T_{sa})$ to lowest order, so that the atmosphere radiates to the ground like a blackbody with temperature equal to the low-level air temperature. If $dT/dp > 0$ at the ground, as is typically the case, then the correction term slightly reduces the downward radiation, because some of the radiation into the ground comes from higher altitudes where the air is colder. Suppose now that the surface temperature T_g is equal to the air temperature T_{sa}, and that the surface has unit emissivity at the frequency under consideration. In that case, the net radiative heating of the ground is

$$I_{-,s}(\nu) - B(\nu, T_g) = I_{-,s}(\nu) - B(\nu, T_{sa}) = -\pi \frac{g \cos \bar{\theta}}{\kappa} \frac{dB}{dT}|_{T_{sa}} \frac{dT}{dp}|_{p_s} \tag{4.21}$$

at frequencies where the solar flux is negligible. This is negative when $dT/dp > 0$, representing a radiative cooling of the ground. The radiative cooling vanishes in the limit of large κ. In the optically thick limit, then, the surface cannot get rid of heat by radiation unless the ground temperature becomes larger than the low-level air temperature. Remember, though, that the radiative heating of the ground is but one term in the surface energy budget coupling the surface to the atmosphere. Turbulent fluxes of moisture and heat also exchange energy between the surface and the atmosphere, and these become dominant when the radiative term is weak.

In the optically thick limit, the net flux is

$$I_+ - I_- = 2\pi \frac{g \cos \bar{\theta}}{\kappa} \frac{dB}{dT} \frac{dT}{dp} \tag{4.22}$$

whence the radiative heating rate is

$$H_\nu = \frac{d}{dp}\left[D(\nu, p) \frac{dT}{dp} \right] \tag{4.23}$$

where

$$D(\nu, p) = 2\pi \frac{g^2 \cos \bar{\theta}}{\kappa} \frac{dB}{dT}|_{T(p)}. \tag{4.24}$$

Hence, in the optically thick limit, the heating and cooling caused by radiative transfer acts just like a thermal diffusion in pressure coordinates, with the diffusivity given by $D(\nu, p)$. Since $dB/dT > 0$, the radiative diffusivity is always positive. It becomes weak as κ becomes large. Note that the diffusive approximation to the heating is only valid when one is not too close to the top or bottom of the atmosphere. Near the boundaries, the neglected boundary terms contribute an additional heating which is exponentially trapped near the top and bottom of the atmosphere. The effect of the boundary terms is explored in Problem 4.4.

Consider an atmosphere which is transparent to solar radiation, and within which heat is redistributed only by infrared radiative transfer. Equation (4.15) then requires that the net upward flux $I_+ - I_-$ must be independent of altitude when integrated over all wavenumbers. This constant flux is non-zero, since the infrared flux through the system is set by the rate at which infrared escapes from the top of the atmosphere – namely, the OLR. Integrating Eq. (4.22) over the infrared yields an expression for dT/dp in terms of the OLR and the frequency-integrated diffusivity; because both OLR and diffusivity are positive, it follows that $dT/dp > 0$ for an optically thick atmosphere in pure infrared radiative equilibrium – that is, the temperature decreases with altitude. The more optically thick the atmosphere becomes, the smaller is D, and hence the stronger is the temperature variation in equilibrium. Pure radiative equilibrium will be discussed in detail in Sections 4.3.4 and 4.7, and the optically thick limit is explored in Problem 4.7.

Optically thin limit

The *optically thin* limit is defined by $\tau_\infty \ll 1$. Since $\tau_\nu \leq \tau_\infty$ and $\tau'_\nu \leq \tau$ in Eq. (4.9), all the exponentials in the expression for the fluxes are close to unity. Moreover, the integral is carried out over the small interval $[0, \tau_\nu]$, and hence is already of order τ_∞ or less. It is thus a small correction to the first term, and we may set the exponentials in the integrand to unity and still have an expression that is accurate to order τ_∞. The boundary terms are not integrated, though, so we must retain the first two terms in the Taylor series expansion of the exponential to achieve the same accuracy. With these approximations, the fluxes become

$$I_+(\tau_\nu) = (1 - \tau_\nu)I_{+,s} + \int_0^{\tau_\nu} \pi B\left(\nu, T\left(\tau'_\nu\right)\right) d\tau'_\nu$$

$$I_-(\tau_\nu) = (1 - (\tau_\infty - \tau_\nu))I_{-,\infty} + \int_{\tau_\nu}^{\tau_\infty} \pi B\left(\nu, T\left(\tau'_\nu\right)\right) d\tau'_\nu. \tag{4.25}$$

In this case, the upward flux is the sum of the upward flux from the boundary (diminished by the slight atmospheric absorption on the way up) with the sum of the unmodified blackbody emission from all the layers below the point in question. The downward flux is interpreted similarly.

In order to discuss the radiation escaping the top of the atmosphere and the back-radiation into the ground, we introduce the mean emission temperature $\bar{T}_\nu$, defined by solving the relation

$$B(\bar{T}_\nu, \nu) = \frac{1}{\tau_\infty} \int_0^{\tau_\infty} B\left(\nu, T\left(\tau'_\nu\right)\right) d\tau'_\nu. \tag{4.26}$$

With this definition, the boundary fluxes are

$$I_{+,\infty} = (1 - \tau_\infty)I_{+,s} + \tau_\infty \pi B(\bar{T}_\nu, \nu)$$

$$I_{-,s} = (1 - \tau_\infty)I_{-,\infty} + \tau_\infty \pi B(\bar{T}_\nu, \nu). \tag{4.27}$$

According to this expression, an optically thin atmosphere acts precisely like an isothermal slab with temperature $\bar{T}_\nu$ and (small) emissivity τ_∞. It is only in the optically thin limit that the radiative effect of the atmosphere mimics that of an isothermal slab.

Substituting the approximate form of the fluxes into the expression for radiative heating rate, we find

$$H_\nu = \frac{\kappa}{\cos \bar{\theta}} \cdot [(I_{+,s} + I_{-,\infty}) - 2\pi B(\nu, T(p))]. \tag{4.28}$$

This is small, because κ is small in the optically thin limit. The first pair of terms are always positive, and represent heating due to the proportion of incident fluxes which are absorbed in the atmosphere. The second term is always negative, and represents cooling by blackbody emission of the layer of air at pressure p. In contrast to the general case or the optically thick case, the cooling term is purely a function of the local temperature; radiation emitted by each layer escapes directly to space or to the ground, without being significantly captured and re-emitted by any other layer.

Typical greenhouse gases are optically thin in some spectral regions and optically thick in others. We have seen that the infrared heating rate becomes small in both limits. From this result, we deduce the following general principle: *the infrared heating rate of an atmosphere is dominated by the spectral regions where the optical thickness is order unity.* If an atmosphere is optically thick throughout the spectrum, the heating is dominated by the least thick regions; if it is optically thin throughout, it is dominated by the least thin regions.

4.3 THE GRAY GAS MODEL

We will see in Section 4.4 that for most atmospheric gases κ, and hence the optical thickness, has an intricate dependence on wavenumber. This considerably complicates the solution of the radiative transfer equations, since the fluxes must be solved for individually on a very dense grid of wavenumbers, and then the results integrated to yield the net atmospheric heating, which is the quantity of primary interest. The development of shortcuts that can improve on a brute-force integration is an involved business, which in some respects is as much art as science, and leads to equations whose behavior can be difficult to fathom. The radiative transfer equations become much simpler if the optical thickness is independent of wavenumber. This is known as the *gray gas approximation.* For gray gases, the Schwarzschild equations can be integrated over wavenumber, yielding a single differential equation for the net upward and downward flux. More specifically, we shall assume only that the optical thickness is independent of wavenumber within the infrared spectrum, and that the temperature of the planet and its atmosphere is such that essentially all the emission of radiation lies in the infrared. Instead of integrating over all wavenumbers, we integrate only over the infrared range, thus obtaining a set of equations for the net infrared flux. Because of the assumption regarding the emission spectrum, the integrals of the Planck function $\pi B(\nu, T)$ over the infrared range can be well approximated by σT^4.

With the exception of clouds of strongly absorbing condensed substances like water, the gray gas model yields a poor representation of radiative transfer in real atmospheres, for which the absorption is typically strongly dependent on wavenumber. Nonetheless, a thorough understanding of the gray gas model provides the starting point for any deeper inquiry into atmospheric radiation. Here, we can find many of the fundamental phenomena laid bare, because one can get much farther before resorting to detailed numerical computations. Further, gray gas radiation has proved valuable as a placeholder radiation scheme in theoretical studies involving the coupling of radiation to fluid dynamics, when one wants to focus on dynamical phenomena without the complexity and computational expense of real gas radiative transfer. Sometimes, a simple scheme which is easy to understand is better than an accurate scheme which defies comprehension.

The gray gas versions of the two-stream Schwarzschild equations are obtained by making τ_ν independent of frequency and integrating the resulting equations over all frequencies.

The result is

$$\frac{d}{d\tau}I_+ = -I_+ + \sigma T(\tau)^4$$
$$\frac{d}{d\tau}I_- = I_- - \sigma T(\tau)^4. \tag{4.29}$$

Gray gas versions of the solutions given in the previous section can similarly be obtained by integrating the relations over all frequencies, taking into account that τ is now independent of ν. The expressions have precisely the same form as before, except that I_+ and I_- now represent total flux integrated over all longwave frequencies, and every occurrence of πB is replaced by σT^4. To avoid unnecessary proliferation of notation, when the context allows little possibility of confusion we will use the same symbols I_+ and I_- to represent the longwave-integrated flux as we used earlier to represent the frequency-dependent flux spectrum. When we need to emphasize that a flux is a frequency-dependent spectrum, we will include the dependence explicitly (as in "$I_+(\nu)$" or "$I_+(\nu, p)$"; when we need to emphasize that a flux represents the longwave-integrated net flux, we will use an overbar (as in $\bar{I}_+$).

4.3.1 *OLR* and back-radiation for an optically thin gray atmosphere

The *OLR* and surface back-radiation for an optically thin gray atmosphere are obtained by integrating Eq. (4.27) over all frequencies. The result is

$$I_{+,\infty} = (1 - \tau_\infty)I_{+,s} + \tau_\infty \sigma \bar{T}^4$$
$$I_{-,s} = (1 - \tau_\infty)I_{-,\infty} + \tau_\infty \sigma \bar{T}^4 \tag{4.30}$$

where the mean atmospheric emission temperature is given by

$$\bar{T}^4 = \frac{1}{\tau_\infty} \int_0^{\tau_\infty} T^4 d\tau. \tag{4.31}$$

The first term in the expression for $I_{+,\infty}$ is the proportion of upward radiation from the ground which escapes without absorption by the intervening atmosphere, while the second is the emission to space added by the atmosphere. In the expression for $I_{-,s}$ the first term is the proportion of incoming infrared flux which reaches the surface without absorption, while the second is the downward emission from the atmosphere. Note that for an optically thin atmosphere the atmospheric emission to space is identical to the atmospheric emission to the ground; in this regard the atmosphere radiates like an isothermal slab with temperature $\bar{T}$. According to Eq. (4.27), a non-gray atmosphere behaves similarly, if it is optically thin *for all frequencies*.

4.3.2 Radiative properties of an all-troposphere dry atmosphere

Let's consider an atmosphere for which the convection is so deep that it establishes a dry adiabat throughout the depth of the atmosphere. Thus, $T(p) = T_s(p/p_s)^{R/c_p}$ all the way to $p = 0$. We wish to compute the *OLR* for this atmosphere, which is done by substituting this $T(p)$ into the gray gas form of Eq. (4.9) and evaluating the integral for I_+ at $\tau = \tau_\infty$, i.e. the top of the atmosphere. Since the temperature is expressed as a function of pressure, it is necessary to substitute for pressure in terms of optical thickness in order to carry out the integral. We'll suppose that κ is a constant, so that $\tau_\infty - \tau = \tau_\infty p/p_s$. Using this to eliminate pressure from $T(p)$, the integral for *OLR* becomes

$$OLR = I_{+,s}e^{-\tau_\infty} + \int_0^{\tau_\infty} \sigma T_s^4 \left(\frac{\tau_\infty - \tau'}{\tau_\infty}\right)^{4R/c_p} e^{-(\tau_\infty - \tau')}d\tau'$$

$$= I_{+,s}e^{-\tau_\infty} + \int_0^{\tau_\infty} \sigma T_s^4 \left(\frac{\tau_1}{\tau_\infty}\right)^{4R/c_p} e^{-\tau_1}d\tau_1 \qquad (4.32)$$

$$= I_{+,s}e^{-\tau_\infty} + \sigma T_s^4 \tau_\infty^{-4R/c_p} \int_0^{\tau_\infty} \tau_1^{4R/c_p} e^{-\tau_1}d\tau_1.$$

The second line is derived by introducing a new dummy variable $\tau_1 = \tau_\infty - \tau'$. This is the optical depth measured relative to the top of the atmosphere, and the re-expressed integral is computed by integrating from the top down, rather than from the ground up. The first term on the right hand side of Eq. (4.32) represents the proportion of the upward surface radiation which survives absorption by the atmosphere and reaches space. The second term is the net emission from the atmosphere. In the optically thin limit, the integral becomes small and the exponential in the first term approaches unity; thus, the OLR approaches the emission from the ground, $I_{+,s}$. As the atmosphere is made more optically thick, the boundary term becomes exponentially small, and the integral becomes more and more dominated by the emission from the upper reaches of the atmosphere. However, to obtain the optically thick limit, we cannot use the gray gas form of Eq. (4.19), since dT/dp becomes infinite at $p = 0$ when the dry adiabat extends all the way to the top of the atmosphere.

In the optically thick limit, $\tau_\infty \gg 1$, the first term becomes exponentially small and the upper limit of the integral can be replaced by ∞, yielding the expression

$$OLR = \sigma T_s^4 \tau_\infty^{-4R/c_p} \Gamma\left(1 + \frac{4R}{c_p}\right) \qquad (4.33)$$

where Γ is the Gamma function, defined by $\Gamma(s) \equiv \int_0^\infty \zeta^{s-1} exp(-\zeta)d\zeta$. Using integration by parts, $\Gamma(s) = (s-1)\Gamma(s-1)$, while $\Gamma(1) = 1$, so $\Gamma(n) = (n-1)!$. For Earth air, $4R/c_p = \frac{8}{7}$ so $\Gamma(1 + 4R/c_p)$ will be close to $\Gamma(2)$, which is unity; in fact it is approximately 1.06. For any of the gases commonly found in planetary atmospheres, $\Gamma(1 + 4R/c_p)$ will be an order unity constant. As the atmosphere is made more optically thick, the OLR goes down algebraically like τ_∞^{-4R/c_p}, becoming much less than the value σT_s^4 prevailing for a transparent atmosphere. The OLR approaches zero as τ_∞ is made large because the temperature vanishes at the top of the atmosphere, and as the atmosphere is made more optically thick, the OLR is progressively more dominated by the emission from the cold upper reaches of the atmosphere.

The calculation can be related to the conceptual greenhouse effect model introduced in the previous chapter by computing the effective radiating pressure p_{rad}. Recall that $\sigma T_{rad}^4 = OLR$, so

$$\sigma T_s^4 \left(\frac{p_{rad}}{p_s}\right)^{4R/c_p} = OLR = \sigma T_s^4 \tau_\infty^{-4R/c_p} \Gamma\left(1 + \frac{4R}{c_p}\right) \qquad (4.34)$$

whence $(p_{rad}/p_s) = \tau_\infty^{-1}(\Gamma(1 + 4R/c_p))^{c_p/4R}$. This formula implies that the radiation to space comes essentially from the top unit optical depth of the atmosphere. If an atmosphere has optical depth $\tau_\infty = 100$, then it is only the layer between roughly the top of the atmosphere ($\tau = 100$) and $\tau = 99$ which dominates the OLR. For the all-troposphere model, the maximum temperature of the top unit optical depth approaches zero as the atmosphere is made more optically thick, because this entire layer corresponds to pressures approaching zero ever more closely as κ is made larger.

If $(1 - \alpha)S$ is the absorbed solar radiation per unit area of the planet's surface, then the surface temperature in balance with this radiation is obtained by setting the OLR equal to $(1 - \alpha)S$. Solving for the surface temperature, we find that in the optically thick all-troposphere limit, the surface temperature is

$$T_S = \left[((1 - \alpha)S/\sigma)^{\frac{1}{4}}\right] \Gamma \left(1 + \frac{4R}{c_p}\right)^{-\frac{1}{4}} \cdot \tau_\infty^{R/c_p}. \tag{4.35}$$

The term in square brackets is the temperature the planet would have in the absence of any atmosphere. As τ_∞ increases, the surface becomes warmer without bound. This constitutes our simplest quantitative model of the greenhouse effect for a temperature-stratified atmosphere. Note that the greenhouse warming depends on the lapse rate. For an isothermal atmosphere ($R/c_p = 0$) there is no greenhouse warming. For fixed optical depth, the greenhouse warming becomes larger as the R/c_p, and hence the lapse rate, becomes large. For Venus, the absorbed solar radiation is approximately $163\,W/m^2$, owing to the high albedo of the planet. For a pure CO_2 atmosphere $R/c_p \approx 0.2304$, for which $\Gamma \approx 0.969$. Then, the $737\,K$ surface temperature of Venus can be accounted for if $\tau_\infty = 156$, which is a very optically thick atmosphere. This is essentially the calculation used by Carl Sagan to infer that the dense CO_2 atmosphere of Venus could give it a high enough surface temperature to account for the then-mysterious anomalously high microwave radiation emitted by the planet (microwaves being directly emitted to space by the hot surface without significant absorption by the atmosphere).

Exercise 4.4 This exercise illustrates the fact that if the Earth's atmosphere acted like a gray gas, then a doubling of CO_2 would make us toast! Using Eq. (4.35), find the τ_∞ that yields a surface temperature of $285\,K$ for the Earth's absorbed solar radiation (about $270\,W/m^2$ allowing a crude correction for net cloud effects). Now suppose we double the greenhouse gas content of the atmosphere. If the Earth's greenhouse gases were gray gases, this would imply doubling the value of τ_∞ from the value you just obtained. What would the resulting temperature be? Note that this rather alarming temperature doesn't even fully take into account the amplifying effect of water vapor feedback.

An examination of the radiative heating rate profile for the all-troposphere case provides much insight into the processes which determine where the troposphere leaves off and where a stratosphere will form. We'll assume that $I_{-,\infty} = 0$ and that the turbulent heat transfer at the ground is efficient enough that $T_{sa} = T_g$. Consider first the optically thin limit, for which the gray gas version of Eq. (4.28) is

$$H = \frac{\kappa}{\cos\bar{\theta}} \cdot \left[\sigma T_g^4 - 2\sigma T(p)^4\right] \tag{4.36}$$

assuming the stated boundary conditions. Since the radiative heating rate is non-zero, the temperature profile will not be in a steady state unless some other source of heating and cooling is provided to cancel the radiative heating. According to Eq. (4.36), the atmosphere is cooling at low altitudes, where $T > T_g/2^{\frac{1}{4}}$, i.e. where the local temperature is greater than the skin temperature. The cooling will make the atmosphere's potential temperature lower than the ground temperature, which allows the air in contact with the ground to be positively buoyant. The resulting convection brings heat to the radiatively cooled layer, allowing a steady state to be maintained if the convection is vigorous enough. However, in the upper atmosphere, where $T < T_g/2^{\frac{1}{4}}$, the atmosphere is being *heated* by upwelling infrared radiation, and there is no obvious way that convection could provide the cooling needed to make this region a steady state. Instead, the atmosphere in this region is expected to warm until a stratosphere in pure radiative equilibrium forms. Indeed, the tropopause as estimated by the boundary between the region of net heating and net cooling is located at the point where $T(p)$ equals the skin temperature; this is precisely the same result as we obtained in the steady state model of the tropopause for an optically thin atmosphere, as discussed in Section 3.6.

In the optically thick limit it is easiest to infer the infrared heating profile from an examination of the expression for net infrared flux, which becomes

$$I_+ - I_- = 2\frac{g\cos\bar{\theta}}{\kappa}(4\sigma T^3)\frac{dT}{dp} = 8\sigma T_g^4\frac{R}{c_p}\frac{g\cos\bar{\theta}}{\kappa p_s}\left(\frac{p}{p_s}\right)^{4R/c_p-1} \tag{4.37}$$

in the all-troposphere gray gas case. Recall that this expression breaks down in thin layers within roughly a unit optical depth of the bottom and top boundaries. The formula shows that whether the bulk of an optically thick atmosphere is heating or cooling depends on the lapse rate. The formula is valid even if κ depends on pressure and temperature. For constant κ, if $4R/c_p > 1$ the optically thick net flux decreases with height, and most of the atmosphere is *heated* by infrared radiation, and hence we expect a deep stratosphere and shallow troposphere. If $4R/c_p < 1$, corresponding to a weaker temperature lapse rate, most of the atmosphere instead experiences infrared radiative cooling, so we expect a deep troposphere. For constant κ, most gases would produce a deep stratosphere in the optically thick limit. This conclusion is greatly altered by pressure broadening. If $\kappa(p) = (p/p_s)\kappa(p_s)$, then the appearance of the pressure factor in the denominator alters the flux profile. Specifically, we now only require that $4R/c_p < 2$ in order for the flux to increase with height, yielding radiative cooling throughout the depth where the approximation is valid. Most atmospheric gases satisfy this criterion, and therefore most gases ought to yield a deep troposphere in the optically thick limit. Real gases are typically optically thick at some wavenumbers but optically thin at others, and we shall see in Section 4.8 how the competition plays out

Figure 4.2 shows numerically computed profiles of net infrared flux ($I_+ - I_-$) without pressure broadening, for a range of optical thicknesses, with $R/c_p = \frac{2}{7}$. In this case, $4R/c_p > 1$, and we expect deep heating in the optically thick limit. For $\tau_\infty = 50$ the profile does follow the optically thick approximate form over most of the atmosphere, and exhibits a decrease in flux with height, implying deep heating. There is a thin layer of cooling near the ground, where the optically thick formula breaks down. When $\tau_\infty = 10$, the flux only conforms to the optically thick limit near the center of the atmosphere; there is a region of infrared cooling that extends from the ground nearly to 70% of the surface pressure. The case $\tau_\infty = 1$ looks quite like the optically thin limit, with the lower half of the atmosphere cooling and the upper half heating. Numerical computations incorporating pressure broadening confirm the predictions of the optically thick formula. Specifically, the boundary between cooling and heating is unchanged for optically thin atmospheres, but rises to $p/p_s = 0.24$ for $\tau_\infty = 10$ and to $p/p_s = 0.11$ for $\tau_\infty = 50$. The profiles are not shown here, but are explored in Problem 4.3.

The troposphere is defined as the layer stirred by convection, and since hot air rises, buoyancy-driven convection transports heat upward where it is balanced by radiative cooling. Therefore, at least the upper region of a troposphere invariably experiences radiative cooling. In the calculation discussed above, the layer with cooling, fated to become the

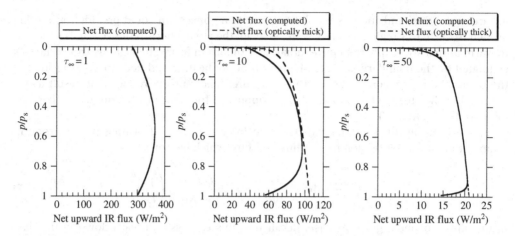

Figure 4.2 Net infrared flux ($I_+ - I_-$) for the all-troposphere gray gas model, for $\tau_\infty =$ 1, 10, and 50. In the latter two cases, the dashed line gives the result of the optically thick approximation. The surface temperature is fixed at 300 K, and the temperature profile is the dry adiabat with $R/c_p = \frac{2}{7}$.

troposphere, occurs in the lower portion of the atmosphere. In Fig. 4.2 one notices that the radiative cooling decreases as the atmosphere is made more optically thick, suggesting that tropospheric convection becomes more sluggish in an optically thick atmosphere, there being less radiative cooling to be offset by convective heating. However, one should note also that this sequence of calculations is done with fixed surface temperature, and that the *OLR* decreases as optical thickness is made larger. Hence, in the optically thick cases, it takes less absorbed solar radiation to maintain the surface temperature of the planet. There is less flux of energy through the system, and correspondingly less convection. Mars, at a more distant orbit than Earth, receives less solar energy; if Mars were given an atmosphere with enough greenhouse effect to warm it up to Earthlike temperatures, one would expect the radiative cooling in its troposphere to be less than Earth's, and one would expect the convection to be more sluggish.

The presence of a stratosphere causes the *OLR* to exceed the values implied by the all-troposphere calculation, since the upper portions of an atmosphere with a stratosphere are warmer than the all-troposphere model would predict. If the stratosphere is optically thin, it has a minor effect on the *OLR*; in essence, the all-troposphere *OLR* formula provides a good estimate if the effective radiating level is below the tropopause. If the stratosphere becomes optically thick, then the *OLR* is in fact determined by the stratospheric structure. Problem 4.2 explores some aspects of the effect of an optically thick stratosphere on *OLR*. Puzzling out the effect of the stratosphere on *OLR* is rather tricky, because the tropopause height itself depends on the optical thickness of the atmosphere. An optically thin atmosphere obviously cannot have an optically thick stratosphere, but an optically thick atmosphere can nevertheless have an optically thin stratosphere if the tropopause height increases rapidly enough with τ_∞. The gray gas radiative cooling profiles discussed above suggest that the stratosphere becomes optically thick when $4R/c_p > 1$. In contrast, for $4R/c_p < 1$ the radiatively cooled layer extends toward the top of the atmosphere in the optically thick limit, and hence the stratosphere could remain optically thin.

4.3.3 A first look at the runaway greenhouse

We have seen in Chapter 2 that the mass of an atmosphere in equilibrium with a reservoir of condensed substance (e.g. a water ocean) is not fixed. It increases with temperature in accordance with the dictates of the Clausius–Clapeyron relation. If the condensable substance is a greenhouse gas, then the optical thickness τ_∞ increases with temperature. This tends to reduce the *OLR*, offsetting or even reversing the tendency of rising temperature to increase the *OLR*. What are the implications of this for the dependence of *OLR* on surface temperature, and for planetary energy balance? The resulting phenomena are most commonly thought about in connection with the effects of a water ocean on evolution of a planet's climate, but the concept generalizes to any condensable greenhouse gas in equilibrium with a large condensed reservoir. We'll take a first look at this problem here, in the context of the gray gas model.

In the general case, we would like to consider an atmosphere in which the condensable greenhouse gas is mixed with a non-condensable background of fixed mass (which may also have a greenhouse effect of its own). This is the case for water vapor in the Earth's atmosphere, for methane on Titan, and probably also for water vapor in the early atmosphere of Venus. It could also have been the case for mixed nitrogen–CO_2 atmospheres on Early Mars, with CO_2 playing the role of the condensable component. We will eventually take up such atmospheres, but the difficulty in computing the moist adiabat for a two-component atmosphere introduces some distractions which get in the way of grasping the key phenomena. Hence, we'll start with the simpler case in which the atmosphere consists of a pure condensable component in equilibrium with a reservoir (an "ocean," or perhaps a glacier). In this case, the saturated moist adiabat is given by the simple analytic formula Eq. (2.27), obtained by solving the simplified form of the Clausius–Clapeyron relation for temperature in terms of pressure. We have already seen in Chapter 2 that a mixed atmosphere is dominated by the condensable component at high temperatures, so if we are primarily interested in the high-temperature behavior, the use of the one-component condensable atmosphere is not at all a bad approximation.

We write $T(p) = T_0 / \left(1 - \frac{RT_0}{L} \ln \frac{p}{p_0}\right)$, where (p_0, T_0) are a *fixed* reference temperature and pressure on the saturation curve, such as the triple point temperature and pressure. If the surface pressure is p_s, then the surface temperature is $T_s = T(p_s)$. Hence, specifying surface pressure is equivalent to specifying surface temperature in this problem. To keep the algebra simple, we'll assume a constant specific absorption κ, and absorb $\cos \bar{\theta}$ into the definition of κ. Then $\tau_\infty = \kappa p_s/g$, which increases as T_s is made larger. Further, for constant specific absorption, $p/p_0 = (\tau_\infty - \tau')/\tau_0$ where $\tau_0 = \kappa p_0/g$. Now, the choice of the reference temperature and pressure (p_0, T_0) is perfectly arbitrary, and we'll get the same answer no matter what choice we make (within the accuracy of the approximate form of Clausius–Clapeyron we are using). Hence, we are free to set $p_0 = g/\kappa$ so that $\tau_0 = 1$. T_0 then implicitly depends on κ, and becomes larger as κ gets smaller. T_0 is the temperature at the level of the atmosphere where the optical depth measured relative to the top of the atmosphere is unity.

Substituting the one-component $T(p)$ into the integral giving the solution to the Schwarzschild equation, and substituting for pressure in terms of optical thickness, we find

$$
\begin{aligned}
OLR &= I_+(0)e^{-\tau_\infty} + \int_0^{\tau_\infty} \sigma \frac{T_0^4}{\left(1 - \frac{RT_0}{L} \ln \frac{p}{p_0}\right)^4} e^{-(\tau_\infty - \tau')} d\tau' \\
&= I_+(0)e^{-\tau_\infty} + \sigma T_0^4 \int_0^{\tau_\infty} \frac{1}{\left(1 - \frac{RT_0}{L} \ln \tau_1\right)^4} e^{-\tau_1} d\tau_1
\end{aligned}
$$

(4.38)

where we have in the second line defined a new dummy variable $\tau_1 = \tau_\infty - \tau'$ as before. The surface temperature enters the expression for *OLR* only through τ_∞, which is proportional to surface pressure. In the optically thin limit, the integral on the right hand side of the expression is small (because τ_∞ is small). This happens at low surface temperatures, because p_s is small when the surface temperature is small. The *OLR* then reduces to the first term, which is approximately $I_+(0)$, i.e. the unmodified upward radiation from the surface. In the optically thick limit, which occurs for high surface temperatures, the term proportional to $I_+(0)$ is negligible, and the second term dominates. This term consists of the flux σT_0^4 multiplied by a non-dimensional integral. Recall that T_0 is a constant dependent on the thermodynamic and infrared optical properties of the gas making up the atmosphere; it does not change with surface temperature. Because of the decaying exponential in the integrand, the integral is dominated by the contribution from the vicinity of $\tau_1 = 0$, and will therefore become independent of τ_∞ for large τ_∞.[1] In the optically thick (high temperature) limit, then, the integral is a function of RT_0/L alone. From this we conclude that the *OLR* becomes independent of surface temperature in the limit of large surface temperature (and hence large τ_∞). This limiting *OLR* is known as the *Kombayashi–Ingersoll limit*. It was originally studied in connection with the long-term history of water on Venus, using a somewhat different argument than we have presented here. We shall use the term to refer to a limiting *OLR* arising from the evaporation of any volatile greenhouse gas reservoir, whether computed using a gray gas model or a more realistic radiation model.

It is readily verified that the integral multiplying σT_0^4 approaches unity as RT_0/L approaches zero. In fact, for typical atmospheric gases L/R is a very large temperature, on the order of several thousand kelvins. Hence, unless the specific absorption is exceedingly small, RT_0/L tends to be small, typically on the order of 0.1 or less. For $RT_0/L = 0.1$, the integral has the value of 0.905. Thus, the limiting *OLR* is essentially σT_0^4. Recalling that T_0 is the temperature of the moist adiabat at one optical depth unit down from the top of the atmosphere, we see that the limiting *OLR* behaves very nearly as if all the longwave radiation were emitted from a layer one optical depth unit from the top of the atmosphere.

Figure 4.3 shows some results from a numerical evaluation of the integral in Eq. (4.38). For small surface temperatures, there is little atmosphere, and the *OLR* increases like σT_s^4. As the surface temperature is made larger, the atmosphere becomes thicker and the *OLR* eventually asymptotes to a limiting value, as predicted. In accordance with the argument given above, the limiting *OLR* should be slightly less than the blackbody flux corresponding to the temperature T_0 found one optical depth down from the top of the atmosphere. This temperature depends on g/κ, which is the pressure one optical depth down from the top. For $g/\kappa = 100\,\mathrm{Pa}$, solving the simplified Clausius–Clapeyron relation for T at $100\,\mathrm{Pa}$ yields $T_0 = 250.3\,\mathrm{K}$, whence $\sigma T_0^4 = 222.6\,\mathrm{W/m^2}$; for $g/\kappa = 1000\,\mathrm{Pa}$, $T_0 = 280.1\,\mathrm{K}$ and $\sigma T_0^4 = 349.0\,\mathrm{W/m^2}$. These values are consistent with the numerical results shown in the graph.

Note that for a given atmospheric composition (which determines κ) the Kombayashi–Ingersoll limit depends on the acceleration of gravity. A planet with stronger surface gravity will have a higher Kombayashi–Ingersoll limit than one with weaker gravity. An explicit formula for the dependence on g/κ is obtained by substituting for T_0 using the formula for the single-component saturated adiabat, Eq. (2.27). Thus, the limiting *OLR* can be written in the form

---

[1] Technically, the integral diverges at extremely large τ_∞, because the denominator of the integrand can vanish. This is an artifact of assuming a constant latent heat and has no physical significance.

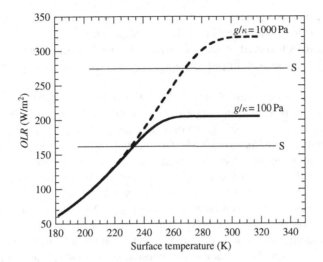

Figure 4.3 *OLR* vs. surface temperature for a one-component gray gas condensable atmosphere in equilibrium with a reservoir. Calculations were done for thermodynamic parameters L and R corresponding to water vapor. Results are shown for two different values of g/κ, where κ is the specific cross-section of the gas and g is the acceleration of gravity. The two horizontal lines marked S indicate two different values of absorbed solar radiation, and their intersections with the *OLR* curves yield the corresponding equilibrium temperatures.

$$OLR_\infty = A(L/R)\sigma T_0^4 = A(L/R)\frac{\sigma(L/R)^4}{\ln(\kappa p^*/g)^4} \tag{4.39}$$

where $A(L/R)$ is the order unity constant discussed previously and $p^* = p_{ref}\exp(L/RT_{ref})$. In the formula for p^*, (p_{ref}, T_{ref}) is any point on the saturation vapor pressure curve, for example the triple point temperature and pressure. p^* is an enormous pressure ($2.3 \cdot 10^{11}$ Pa for water vapor), so Eq. (4.39) predicts that the Kombayashi–Ingersoll limit increases as surface gravity increases, since increasing g makes the logarithm in the denominator smaller. The apparent singularity when $\kappa p^*/g = 1$ is spurious, as the approximations we have made break down long before that value is reached.

We are now prepared to describe the runaway greenhouse phenomenon. Let $(1 - \alpha)S$ be the absorbed solar radiation per unit surface area of the planet, and let the limiting *OLR* computed above be OLR_{max}. If $(1 - \alpha)S < OLR_{max}$ the planet will come to equilibrium in the usual way, warming up until it loses energy by infrared radiation at the same rate as it receives energy from its star. But what happens if $(1 - \alpha)S > OLR_{max}$? In this case, as long as there is still an ocean or other condensed reservoir to feed mass into the atmosphere, the planet cannot get rid of all the solar energy it receives no matter how much it warms up; hence the planet continues to warm until the surface temperature becomes so large that the entire ocean has evaporated into the atmosphere. The temperature at this point depends on the mass and composition of the volatile reservoir. For example, the Earth's oceans contain enough mass to raise the surface pressure to about 100 bars if dumped into the atmosphere in the form of water vapor. The ocean has been exhausted when the saturation vapor pressure reaches this value. Using the simplified exponential form of the Clausius–Clapeyron relation to extrapolate the vapor pressure from the sea-level boiling point (1 bar at 373.15 K), we estimate that this vapor pressure is attained at a surface temperature of about 550 K. This estimate is inaccurate, because the latent heat of vaporization varies appreciably over the range of temperatures involved. A more exact value based on measurements of properties of steam is 584 K, but the grim implications for survival or emergence of life as we know it are largely the same.

At temperatures larger than that at which the ocean is depleted, the mass of the atmosphere becomes fixed and no longer increases with temperature. The greenhouse gas content of the atmosphere – which in the present case is the entire atmosphere – no longer increases with temperature. As a result, the OLR is once more free to increase as the surface becomes warmer, and the planet will warm-up until it reaches an equilibrium at a temperature warmer than that at which the ocean is depleted. The additional warming required depends on the gap between the Kombayashi–Ingersoll limit and the absorbed solar radiation. This situation is depicted schematically in Fig. 4.4. Once the ocean is gone, the lower atmosphere is unsaturated and air can be lifted some distance before condensation occurs. The resulting atmospheric profile is on the dry adiabat in the lower atmosphere, transitioning to the moist adiabat at the altitude where condensation starts. The situation is identical to that depicted for CO_2 in Fig. 2.6. Rain will still form in the condensing layer. Much of it will evaporate in the lower non-condensing layer; some of it may reach the ground, but the resulting puddles will tend to evaporate rapidly back into the highly undersaturated lower atmosphere. As surface temperature is made larger, the altitude where condensation sets in moves higher, until at very large temperatures the atmosphere behaves like a non-condensing dry system (albeit one where the entire atmosphere may consist of water vapor).

The runaway greenhouse phenomenon may explain how Venus wound up with such a radically different climate from Earth, despite having started out in a rather similar state. The standard story goes something like this: Venus started with an ocean, and with most of its CO_2 bound up in rocks as is the case for Earth. However, it was just closer enough to the Sun to trigger a runaway greenhouse. Once the entire ocean had evaporated into the atmosphere, there was so much water vapor in the upper atmosphere that it could be

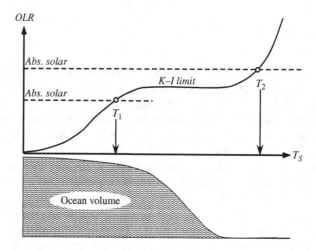

Figure 4.4 Schematic picture of the termination of a runaway greenhouse upon depletion of the volatile reservoir.

broken apart by energetic solar ultraviolet rays, whereafter the light hydrogen could escape to space. The highly reactive oxygen left behind would react to form minerals at the surface. Once there was no more liquid water in play, the reactions that bind up carbon dioxide in rocks could no longer take place (as will be explained in Chapter 8), so all the planet's CO_2 outgassed from volcanism and stayed in the atmosphere, leading to the hot, dry super-dense atmosphere of modern Venus.

Assuming habitability to require a reservoir of liquid water, the Kombayashi–Ingersoll limit for water determines the inner orbital limit for habitability, since if the solar constant exceeds the limiting flux a runaway will ensue and any initial ocean will not persist. It also determines how long it takes before the planet's Sun gets bright enough to trigger a runaway, and thus sets the lifetime of a water-dependent biosphere (Earth's included). Accurate calculations of the Kombayashi–Ingersoll limit are therefore of critical importance to understanding the limits of habitability both in time and orbital position. The gray gas model is not good enough to determine the value of p_0 appropriate to a given gas, and so cannot be used for accurate evaluations of the runaway greenhouse threshold. We can at least say that, all other things being equal, a planet with larger surface gravity will be less susceptible to the runaway greenhouse. This is so because $p_0 = g/\kappa$, whence larger g implies larger p_0, which implies in turn larger $T(p_0)$ and hence a larger limiting OLR. This observation may be relevant to the class of extrasolar planets known as "Super Earths."

We will revisit the runaway greenhouse using more realistic radiation physics in Section 4.6. Some consequences of the effects of the stratosphere and of clouds will be brought into the picture in Chapter 5.

The runaway greenhouse phenomenon is usually thought of in conjunction with water vapor, but the concept applies equally well to any situation where there is a volatile reservoir of greenhouse gas, whether it be in solid or liquid form. For example, one could have a runaway greenhouse in association with the sublimation of a large CO_2 ice cap, or in association with the evaporation of a methane or ammonia ocean. In fact, the Kombayashi–Ingersoll limit determines whether a planet would develop a reservoir of condensed substance at its surface (a glacier or ocean), given sustained outgassing of that substance in the absence of any chemical sink. As the gas builds up in the atmosphere, the pressure increases and it would eventually tend to condense at the surface, preventing any further gaseous accumulation. However, the greenhouse effect of the gas warms the surface, which increases the saturation vapor pressure. The Kombayashi–Ingersoll limit tells us which effect wins out as surface pressure increases. Earth is below the threshold for water, so we have a water ocean. Venus is above the threshold for water and CO_2, so both accumulate as gases in the atmosphere (apart from possible escape to space). When we revisit the problem with real gas radiation, we will be able to say whether CO_2 would form a condensed reservoir on Earth or Mars, or CH_4 on Titan, given sustained outgassing in the absence of a chemical sink.

4.3.4 Pure radiative equilibrium for a gray gas atmosphere

For the temperature profiles discussed in Sections 4.3.2 and 4.3.3, the net infrared radiative heating computed from Eq. (4.14) is non-zero at virtually all altitudes; generally the imbalance acts to cool the lower atmosphere and warm the upper atmosphere. In using such solutions to compute OLR and back-radiation, we are presuming that convective heat fluxes will balance the cooling and keep the troposphere in a steady state. The upper atmosphere will continue to heat, and ultimately reach equilibrium creating a stratosphere, but in the all-troposphere idealization we presume that the stratosphere is optically thin enough that it doesn't much affect the *OLR*.

Now, we'll investigate solutions for which, in contrast, the net radiative heating vanishes individually at each altitude. Such solutions are in *pure radiative equilibrium*, as opposed to *radiative-convective equilibrium*. First we'll consider the case in which the only radiative heating is supplied by infrared; later we'll bring heating by atmospheric solar absorption into the picture. Pure radiative equilibrium is the opposite extreme from the all-troposphere idealization, and tells us much about the nature of the stratosphere, and the factors governing the tropopause height.

Assuming the atmosphere to be transparent to solar radiation, pure radiative equilibrium requires that the frequency-integrated longwave radiative heating $\mathcal{D}$ vanish for all τ. From the gray gas version of Eq. (4.14), we then conclude that $I_+ - I_-$ is independent of τ. Applying the upper boundary condition, we find that this constant is $I_+(\tau_\infty)$, which is the *OLR*. Now, by taking the difference between the equations for I_+ and I_- we find

$$0 = \frac{d}{d\tau}(I_+ - I_-) = -(I_+ + I_-) + 2\sigma T^4 \tag{4.40}$$

which gives us the temperature in terms of $(I_+ + I_-)$. Next, taking the sum of the equations for I_+ and I_- yields

$$\frac{d}{d\tau}(I_+ + I_-) = -(I_+ - I_-). \tag{4.41}$$

This is easily solved by noting that $-(I_+ - I_-) = const = -OLR$. In consequence,

$$2\sigma T^4 = (I_+ + I_-) = (1 + \tau_\infty - \tau)OLR \tag{4.42}$$

where we have again used the boundary condition at τ_∞. This expression gives us the pure radiative equilibrium temperature profile $T(\tau)$. In pure radiative equilibrium, the temperature always approaches the skin temperature at the top of the atmosphere, where $\tau = \tau_\infty$. This recovers the result obtained in the previous chapter, in Section 3.6. When the atmosphere is optically thin, $\tau_\infty - \tau$ is small throughout the atmosphere, and the entire atmosphere becomes isothermal with temperature equal to the skin temperature. When the atmosphere is not optically thin, the temperature decreases gently with height, approaching the skin temperature as the top of the atmosphere is approached.

Equation (4.42) also gives us the upward and downward fluxes, since we now know both $I_+ - I_-$ and $I_+ + I_-$ at each τ. In particular, the downward flux into the ground is

$$I_-(0) = \frac{1}{2}((I_+ + I_-) - (I_+ - I_-)) = \frac{1}{2}((1 + \tau_\infty)OLR - OLR) = \frac{1}{2}\tau_\infty OLR. \tag{4.43}$$

For an optically thin atmosphere, the longwave radiation returned to the ground by the atmosphere is only a small fraction of that emitted to space. As the atmosphere becomes optically thick, the radiation returned to the ground becomes much greater than that emitted to space, because the radiative equilibrium temperature near the ground becomes large and the optical thickness implies that the radiation into the ground is determined primarily by the low-level temperature. If we assume the planet to be in radiative equilibrium with the absorbed solar radiation $(1 - \alpha)S$, where α is the albedo of the ground, then $OLR = (1 - \alpha)S$ and the radiative energy budget of the ground is

$$\sigma T_g^4 = (1 - \alpha)S + I_-(0) = (1 - \alpha)S \cdot \left(1 + \frac{1}{2}\tau_\infty\right) \tag{4.44}$$

where T_g is the surface temperature. This, together with the temperature profile determined by Eq. (4.42), determines what the thermal state of the system would be in the absence of heat transport mechanisms other than radiation. For an optically thin atmosphere, the surface temperature is only slightly greater than the no-atmosphere value. As τ_∞ becomes large, the surface temperature increases without bound. Note that, while this formula yields a greenhouse warming of the surface, the relation between surface temperature and τ_∞ is different from that given by the all-troposphere radiative-convective calculation in Eq. (4.35), because the pure radiative equilibrium temperature profile is different from the adiabat which would be established by convection.

Let's now compare the surface temperature with the temperature of the air in immediate contact with the surface. From Eq. (4.42) we find that the low-level air temperature is determined by $\sigma T(0)^4 = (1 - \alpha)S \cdot \left(\frac{1}{2} + \frac{1}{2}\tau_\infty\right)$. Taking the ratio,

$$\frac{T(0)}{T_g} = \left(\frac{\frac{1}{2} + \frac{1}{2}\tau_\infty}{1 + \frac{1}{2}\tau_\infty}\right)^{1/4}. \tag{4.45}$$

Thus, the surface is always warmer than the overlying air in immediate contact with it. In the previous chapter, we saw that this was the case for pure radiative equilibrium in an optically thin atmosphere, but now we have generalized it to arbitrary optical thickness. In the optically thin limit, the formula reduces to our earlier result, $T(0) = 2^{-1/4}T_g$. In the optically thick limit, $T_g - T(0) = \frac{1}{4}T_g/\tau_\infty$, whence the temperature jump (relative to surface temperature) falls to zero as the atmosphere is made more optically thick. As we already discussed in Section 3.6, cold air immediately above a warmer surface constitutes a very unstable situation. Under the action of diffusive or turbulent heat transfer between the surface and the nearby air, a layer of air near the surface will heat up to the temperature of the surface, whereafter it will be warmer than the air above it. Being buoyant, it will rise and lead to convection, which will stir up some depth of the atmosphere and establish an adiabat – creating a troposphere.

In pure radiative equilibrium, the surface heating inevitably gives rise to convection. However, it is also possible that the temperature profile in the interior of the atmosphere may become unstable to convection, even without the benefit of a surface. This is a particularly important possibility to consider for gas or ice giant planets, which have no distinct surface to absorb solar radiation and stimulate convection. To determine stability, we must compute the lapse rate dT/dp in radiative equilibrium, and see if it is steeper than that of the adiabat (moist or dry) appropriate to the atmosphere. Taking the derivative of (4.42) with respect to optical thickness, we find

$$8\sigma T^3 \frac{dT}{d\tau} = -OLR \tag{4.46}$$

whence, using $d/dp = (d\tau/dp)(d/d\tau)$, we find

$$\frac{d\ln T}{d\ln p} = -\frac{1}{4}\frac{1}{(1 + \tau_\infty - \tau)}p\frac{d\tau}{dp}. \tag{4.47}$$

Stability is determined by comparing the slope of the adiabat to the radiative-equilibrium slope we have just computed. For the dry adiabat, the atmosphere is stable where

$$\frac{R}{c_p} \geq -\frac{1}{4}p\frac{d\tau}{dp}\frac{1}{(1 + \tau_\infty - \tau)}. \tag{4.48}$$

The factor p appearing on the right hand side of this equation guarantees that the upper portion of the atmosphere will always be stable, unless $d\tau/dp$ blows up like $1/p$ or faster as $p \to 0$. Moreover, optically thin atmospheres are always stable throughout the depth of the atmosphere. This is so because the denominator is close to unity in the optically thin limit, while $-p \, d\tau/dp = \kappa p/(g \cos \bar{\theta}) < \tau_\infty \ll 1$. Since optically thin atmospheres are nearly isothermal in pure radiative equilibrium, it is hardly surprising that they are statically stable.

In the case of constant absorption coefficient κ, we have $p \, d\tau/dp = -\kappa p/g \cos \bar{\theta}$, which is just $\tau - \tau_\infty$. Thus, the stability condition becomes

$$\frac{R}{c_p} \geq \frac{1}{4} \frac{\tau_\infty - \tau}{(1 + \tau_\infty - \tau)}. \tag{4.49}$$

The right hand side has its maximum at the ground $\tau = 0$, and the maximum value is $\frac{1}{4} \tau_\infty/(1 + \tau_\infty)$. The more optically thick the atmosphere is, the more unstable it is near the ground. For large optical thickness, the stability criterion becomes the remarkably simple statement $4R/c_p \geq 1$. Dry Earth air, with $R/c_p = 2/7$, just misses being unstable by this criterion, and pure non-condensing water vapor is almost precisely on the boundary. Pure non-condensing CO_2, NH_3, and CH_4 just barely satisfy the condition for instability near the ground when the atmosphere is optically thick.

Typical atmospheres, however, will be more unstable than the constant κ calculation suggests. As will be explained in Section 4.4, *collisional broadening* typically causes the absorption coefficient to increase linearly with pressure, out to pressures of several bars. With collisional broadening, $\kappa(p) = \kappa(p_s) \cdot (p/p_s)$ for a combination of well-mixed greenhouse gases with pressure-independent concentrations. In this case $-p \, d\tau/dp = 2\tau_\infty$ at the surface, so that the dry adiabatic stability condition in the optically thick limit becomes $2R/c_p \geq 1$. The extra factor of 2 compared with the case without pressure broadening destabilizes all well-mixed atmospheres, provided they are sufficiently optically thick near the ground. The maximum value of the stability parameter occurs for mono-atomic gases like helium, which has $2R/c_p = 0.8$ and easily meets the criterion for instability.

Other processes can destabilize the atmosphere as well. For example, on the moist adiabat the slope $-d \ln T/d \ln p$ is always less than the dry adiabatic slope, which deepens the layer within which the radiative-equilibrium atmosphere is unstable. In addition, a sharp decrease of optical thickness with height tends to destabilize the atmosphere, particularly if it occurs in a place where the atmosphere is optically thick. This happens whenever there is a layer where the concentration of absorbers decreases strongly with height. Because of Clausius–Clapeyron, this situation often happens in the lower portions of atmospheres in equilibrium with a reservoir of condensable greenhouse gas (water vapor on Earth or methane on Titan, for example). In this case, the condensable substance has a destabilizing effect through its influence on the radiative equilibrium, as well as through its effect on the adiabat. Condensed water is a good infrared absorber, so radiative equilibrium drives cloud tops to be unstable. In contrast, the atmosphere is stabilized where the infrared absorber concentration increases strongly with height; this situation is less typical, but can happen at the bottom of a water cloud.

4.3.5 Effect of atmospheric solar absorption on pure radiative equilibrium

Now we will examine how the absorption of solar radiation within an atmosphere affects the temperature structure of the atmosphere in radiative equilibrium. The prime application of this calculation is to understand the thermal structure of stratospheres. Under what circumstances does the temperature of a stratosphere increase with height? The effect of solar absorption on gas giant planets like Jupiter is even more crucial. There being no distinct surface to absorb sunlight, *all* solar driving of the atmosphere for gas giants comes from deposition of solar energy within the atmosphere. In this case, the profile of absorption determines in large measure where, if anywhere, the radiative equilibrium atmosphere is unstable to convection, and therefore where a troposphere will tend to form. The answer determines whether convection on gas giants is driven in part by solar heating as opposed to ascent of buoyant plumes carrying heat from deep in the interior of the planet.

In the Earth's stratosphere, solar absorption is largely due to the absorption of ultraviolet by ozone. On Earth as well as other planets having appreciable water in their atmospheres, absorption of solar near-infrared by water vapor and water clouds is important. CO_2 also has significant near-infrared absorption, which is relatively unimportant at present-day CO_2 concentrations on Earth, but significant on the Early Earth when CO_2 concentrations were much higher; solar near-infrared absorption by CO_2 is of course important in the CO_2-dominated atmospheres of Mars (present and past) and Venus. Solar absorption by dust is important to the Martian thermal structure throughout the depth of the atmosphere. On Titan, it is solar absorption by organic haze clouds that control the thermal structure of the upper atmosphere. Solar absorption is also crucial to the understanding of the influence of greenhouse gases like CH_4 and SO_2, which strongly absorb sunlight in addition to being radiatively active in the thermal infrared. Strong solar absorption also would occur in the high-altitude dust and soot cloud that would be lofted in the wake of a global thermonuclear war or asteroid impact (the "nuclear winter" problem).

Since the Schwarzschild equations in this chapter are used to describe the infrared flux alone, the addition of solar heating does not change these equations. The heating due to solar absorption only alters the condition for local equilibrium, which now involves the deposition of solar as well as infrared flux. We write the solar heating rate per unit optical depth in the form $Q_\odot = dF_\odot/d\tau$, where $F_\odot$ is the net downward solar flux as a function of infrared optical depth. At the top of the atmosphere, $F_\odot = (1 - \alpha)S$, where α is the planetary albedo – that is, the albedo measured at the top of the atmosphere. Since atmospheres at typical planetary temperatures do not emit significantly in the solar spectrum, there is no internal source of solar flux and therefore $F_\odot$ must decrease monotonically going from the top of the atmosphere to the ground.

The net radiative heating at a given position is now the sum of the infrared and solar term, i.e.

$$-\frac{d}{d\tau}(I_+ - I_-) + \frac{d}{d\tau}F_\odot = 0. \tag{4.50}$$

Integrating this equation and requiring that the top-of-atmosphere energy budget be in balance with the local absorbed solar radiation, we find

$$(I_+ - I_-) - F_\odot = 0. \tag{4.51}$$

At the top of the atmosphere, this reduces to $OLR - (1 - \alpha)S = 0$, which is the requirement for top-of-atmosphere energy balance. Because the solar absorption does not change the infrared Schwarzschild equations, Eq. (4.41) is unchanged from the case of pure radiative equilibrium without solar absorption. Substituting Eq. (4.51) and integrating, we obtain

$$I_+ + I_- = \int_\tau^{\tau_\infty} F_\odot(\tau')d\tau' + (1 - \alpha)S. \tag{4.52}$$

In writing this expression we have made use of the boundary condition $I_+ - I_- = OLR = (1 - \alpha)S$ at the top of the atmosphere. The heat balance equation (4.40) needs to be slightly modified, since the infrared cooling now balances the solar heating, instead of being set to zero. Thus,

$$\frac{d}{d\tau} F_\odot = \frac{d}{d\tau}(I_+ - I_-) = -(I_+ + I_-) + 2\sigma T^4 \qquad (4.53)$$

from which we infer

$$2\sigma T^4 = \frac{d}{d\tau} F_\odot + \int_\tau^{\tau_\infty} F_\odot(\tau')d\tau' + (1 - \alpha)S. \qquad (4.54)$$

This gives the vertical profile of temperature in terms of the vertical profile of the solar flux; the previous case (without solar absorption) can be recovered by setting $F_\odot = const = (1 - \alpha)S$. At the top of the atmosphere, the integral in Eq. (4.54) vanishes, and the temperature becomes identical to the temperature of a skin layer heated by solar absorption, derived in Chapter 3 (Eq. (3.27)).

Taking the derivative with respect to τ yields

$$\frac{d}{d\tau} 2\sigma T^4 = \frac{d^2}{d\tau^2} F_\odot - F_\odot. \qquad (4.55)$$

This equation provides a simple criterion determining when the solar absorption causes the temperature to increase with height. When there is no absorption, $F_\odot$ is a constant and since it is positive the temperature decreases with height. The quantity $\sqrt{\left| F_\odot^{-1} d^2 F_\odot / d\tau^2 \right|}$ is the local solar flux decay rate, expressed in units of infrared optical depth. Where the local solar decay rate is less than unity – meaning solar flux is attenuated at a lower rate than infrared – the radiative equilibrium temperature decreases with height. Where the local solar decay rate is greater than unity, the temperature increases with height. Note that it is the solar extinction rate *relative to the infrared extinction rate* that counts. One can make the temperature increase with height either by increasing the solar opacity or decreasing the infrared opacity (generally by decreasing the greenhouse gas concentration). The profile of solar absorption is also sensitive to the vertical distribution of solar absorbers. Where solar absorbers increases sharply with height, as is the case for ozone on Earth or organic haze on Titan, the stratospheric temperature increases with height.

By way of illustration, let's suppose that the net downward solar flux decays exponentially as it penetrates the atmosphere. Specifically, let $F_\odot = (1 - \alpha)S \exp(-(\tau_\infty - \tau)/\tau_S)$, where τ_S is the decay rate of solar radiation, measured in infrared optical depth units. When τ_S is large, solar absorption is weak compared with infrared absorption, and one must go a great distance before the solar beam is appreciably attenuated. Conversely, when τ_S is small, solar absorption is strong and the solar beam decays to zero over a distance so short that infrared is hardly attenuated at all. With the assumed form of the solar flux, the temperature profile is given by

$$2\frac{\sigma T^4}{(1 - \alpha)S} = 1 + \tau_S + \left(\frac{1}{\tau_S} - \tau_S\right) e^{-(\tau_\infty - \tau)/\tau_S}. \qquad (4.56)$$

If $\tau_S > 1$ the temperature decreases with height, and if $\tau_S < 1$ the temperature increases with height. Defining the skin temperature as $T_{skin} \equiv \left(\frac{1}{2\sigma}(1 - \alpha)S\right)^{1/4}$, the temperature at the top of the atmosphere is $(1 + 1/\tau_S)^{1/4} T_{skin}$, which reduces to the skin temperature when τ_S is large and becomes much greater than the skin temperature when τ_S is small. If the atmosphere is deep enough that essentially all solar radiation is absorbed before reaching the ground, then the exponential term vanishes in the deep atmosphere and the deep atmosphere becomes isothermal with temperature $(1 + \tau_S)^{1/4} T_{skin}$. Thus, when τ_S is small, all the solar radiation is absorbed within a thin layer near the top of the atmosphere. The temperature increases rapidly with height in this layer, but the bulk of the atmosphere below is approximately isothermal at the skin temperature. The strong solar absorption causes the deep atmosphere, and the ground (if there is one), to be colder than it would have been in the absence of an atmosphere. This *anti-greenhouse effect* arises because the deep atmosphere is heated only by downwelling infrared emitted by the solar-absorbing layer. This limit is relevant to the *nuclear winter* phenomenon, in which energetic explosions and fires loft a global or hemispheric solar-absorbing soot and dust cloud to high altitudes. The same situation would occur in the aftermath of a large bolide (asteroid or comet) impact. In either case, the atmosphere below the soot layer would become as frigid as the depth of winter, but moreover the relaxation to a uniform temperature state would shut off the convection which in large measure drives the hydrological cycle.

It can happen that the atmosphere is deep enough to absorb all solar radiation before it reaches the ground, even if the rate of solar absorption is weak and $\tau_S \gg 1$. This would happen if the atmosphere was so optically thick in the infrared that $\tau_\infty / \tau_S \gg 1$ despite τ_S being large. In this case, the deep atmosphere is still isothermal, but it becomes much hotter than the skin temperature – indeed it becomes hotter without bound as τ_S is made larger. In this case, it is the top of the atmosphere which equilibrates at the relatively cold skin temperature, while the deep atmosphere exhibits a strong greenhouse effect. Because the deep atmosphere is isothermal, it is stable and will not generate a troposphere. This is a possible model for the internal state of a gas or ice giant with little internal heat flux, whose atmosphere is optically thick in the infrared but for which there is only weak solar absorption in the deep atmosphere.

The solution given in Eq. (4.56) shows that one can account for the temperature increase in the Earth's stratosphere if the upper atmosphere strongly absorbs solar radiation. As a model of the Earth's atmosphere, however, it has the shortcoming that if one makes the stratospheric absorption strong enough to yield a temperature increase with height, essentially all the solar beam is depleted in the stratosphere, leaving an isothermal lower atmosphere that will not convect and create a troposphere. What happens in Earth's real stratosphere is that ozone is a good absorber only in the ultraviolet part of the solar spectrum. Once the ultraviolet is depleted, the remaining flux making its way into the lower atmosphere is only weakly absorbed by the atmosphere. In terms of Eq. (4.55), the solar decay rate $\sqrt{\left| F_\odot^{-1} d^2 F_\odot / d\tau^2 \right|}$ is large in the upper atmosphere, where there is still plenty of ultraviolet to absorb; in consequence, the temperature increases with height there. In the lower atmosphere the solar decay rate becomes small, so that the radiative equilibrium temperature decreases with height. There is also plenty of solar radiation left to heat the ground, destabilize the atmosphere, and create a troposphere.

The solution in Eq. (4.56) also explains why human-caused increases in CO_2 over the past century have led to tropospheric warming but stratospheric cooling, as illustrated in Fig. 1.17. Increasing the greenhouse gas concentration is equivalent to increasing τ_∞. If one plots temperature as a function of pressure for a sequence of increasing τ_∞, the phenomenon is immediately apparent in cases where the upper-level solar absorption is sufficiently strong. The behavior is explored in Problems 4.12. Without solar absorption, increasing τ_∞ warms the atmosphere at every level, though the amount of warming decreases with height as the temperature asymptotes to the skin temperature. With solar absorption, however, the increased infrared cooling of the upper atmosphere offsets more and more of the warming due to solar absorption, leading to a cooling there. In the real atmosphere, convection modifies the temperature profile in the lower atmosphere. Further, one must take into account real gas infrared and solar absorption in order to quantitatively account for the observed temperature trends. Nonetheless, the gray gas pure radiative equilibrium calculation captures the essence of the mechanism.

The general lesson to take away from this discussion is that solar absorption near the top of the atmosphere stabilizes the atmosphere, reduces the greenhouse effect, and cools the lower portion of the atmosphere and also the ground. This is important in limiting the effectiveness of greenhouse gases like CH_4 and SO_2, which significantly absorb solar radiation when their concentration becomes very high. It is also the way high-altitude solar-absorbing haze clouds on Titan and perhaps Early Earth act to cool the troposphere. The soot and dust clouds lofted by an asteroid impact act similarly. In contrast, solar absorption concentrated near the ground has an effect which is not much different from simply reducing the albedo of the ground itself.

4.4 REAL GAS RADIATION: BASIC PRINCIPLES

4.4.1 Overview: *OLR* through thick and thin

It would be exceedingly bad news for planetary habitability if real greenhouse gases were gray gases (see Exercise 4.4). Greenhouse gas concentrations would have to be tuned exceedingly accurately to maintain a planet in a habitable temperature range, and there would be little margin for error. Thus, it is of central importance that, for real gases, *OLR* varies much more gradually with greenhouse gas concentration than it would for an idealized gray gas.[2] This is another area in which the quantum nature of the Universe directly intervenes in macroscopic phenomena governing planetary climate.

Infrared radiative transfer is a very deep and complex subject, and mastery of the material in this section will still not leave the reader prepared to write state-of-the-art radiation codes. Nor will we cover the myriad engineering tricks large and small which are needed to make a radiation code fast enough to embed in a general circulation model, where it will need to be invoked a dozen times per model day at each of several thousand grid points. We do aspire to provide enough of the basic physics to allow the student to understand why *OLR* is less sensitive to the concentration of a typical real gas than to a gray gas, and to help the student develop some intuition about the full possible range of behaviors of greenhouse gases on Earth and other planets, now and in the distant past or future. Such an understanding should extend even to greenhouse gases that are not at present commonly considered in the context of climate, or implemented in standard "off the shelf" radiation models. What would you do, for example, if you found yourself wondering whether SO_2 or H_2S significantly affected the climate of Early Earth or Mars? The gray gas model does not

---

[2] Lest there be any misunderstanding, we must emphasize at this point that "less sensitive" does not mean "insensitive." If CO_2 were a gray gas, then doubling its concentration, as we are poised to do within the century, would be unquestionably lethal. Because CO_2 is not in fact a gray gas, the results may be merely catastrophic.

provide an adequate first attack on such problems. We thus aspire to provide enough of the basic algorithmic equipment to allow the student to build simplified radiative models from scratch, which get the *OLR* and infrared heating profiles roughly correct.

Even though we will have recourse to a "professionally" written radiation code in Section 4.5, we would at least like to draw back the curtain a little bit, so that the reader will not be left with the all-too-common notion that radiation routines are black boxes, the internal workings of which can only be understood by the high priesthood of radiative transfer. Hopefully, this will also open the door to entice more people into innovative work on the subject.

Since the main point is to understand how the wavenumber dependence of absorptivity affects the sensitivity of *OLR* to greenhouse gas concentration, we'll begin with a discussion of the spectrum of outgoing longwave radiation in an idealized case. Let's consider a planet whose surface radiates like an ideal blackbody in the infrared, having an atmosphere whose air temperature at the surface T_{sa} is equal to the ground temperature T_g. The temperature $T(p)$ is monotonically decreasing with height in the troposphere, and is patched continuously to an isothermal stratosphere having temperature T_{strat}. The atmosphere consists mostly of infrared-transparent N_2 and O_2 with a surface pressure of 10^5 Pa, like Earth. Unlike Earth, the only greenhouse gas is a mythical substance (call it Oobleck), which is a bit like CO_2, but much simpler to think about. It has the same molecular weight as CO_2, but its absorption coefficient $\kappa_{Oob}(\nu)$ has an absorption band centered on wavenumber $\nu_0 = 600\,\text{cm}^{-1}$. Within $100\,\text{cm}^{-1}$ of ν_0, κ_{Oob} has the constant value κ_0. Outside this limited range of wavenumbers, Oobleck is transparent to infrared, i.e. $\kappa_{Oob} = 0$. To make life even simpler for the atmospheric physicists of this planet, κ_{Oob} is independent of both temperature and pressure. Like real CO_2, the specific concentration of Oobleck (q_{Oob}) is constant throughout the depth of the atmosphere.

What does the spectrum of *OLR* look like for this planet? The answer is shown in the left panel of Fig. 4.5. In this figure, we have assumed that the Oobleck molecule has an absorptivity of $1\,\text{m}^2/\text{kg}$. Then, with a molar concentration of 300 ppmv (like CO_2 in the 1960s), the specific concentration is $4.6 \cdot 10^{-4}$ and the optical thickness $\kappa_0 q_{Oob} p_s / g \cos \bar{\theta}$ is 9.4 within the absorption band. Since the atmosphere is optically thick in this wavenumber region, infrared radiation in this part of the spectrum exits the atmosphere with the temperature of the stratosphere. This is exactly what we see in the graph. Outside the absorption band, the atmosphere is transparent, and hence infrared leaves the top of the atmosphere at the much higher temperature of the ground. The overall appearance of the *OLR* spectrum is that the greenhouse gas has "dug a ditch" in the spectrum of *OLR*, or perhaps "taken a bite" out of it. The ditch in the spectrum reduces the total *OLR* of the planet, but not so much so as if the absorption were strong throughout the spectrum, as would be the case for a gray gas. This is the typical way that real greenhouse gases work: they make the atmosphere optically thick in a limited part of the spectrum, while leaving it fairly transparent elsewhere. The strength of the greenhouse effect is not so much a matter of how deep the ditch is, but how wide.

Oobleck is a very contrived substance, but the above exercise gives us a fair idea of what to look for when interpreting real observations of the spectrum of *OLR*. Figure 4.6, giving the *OLR* spectrum of Mars observed at two different seasons by the TES instrument on Global Surveyor, is a case in point. Mars has an essentially pure CO_2 atmosphere complicated only by optically thin ice clouds and dust clouds (which can be very thin between major dust storms). The planet thus provides perhaps the purest illustration of the CO_2 radiative effect available in the Solar System. In Fig. 4.6 a CO_2 "ditch" centered on $670\,\text{cm}^{-1}$ is evident

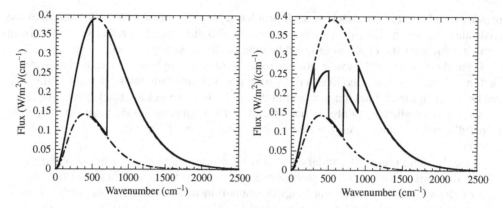

Figure 4.5 The *OLR* spectrum for a hypothetical gas which has a piecewise constant absorption coefficient. The dashed-dotted lower curve is the blackbody spectrum corresponding to the stratospheric temperature T_{strat} while the dashed upper curve is the blackbody spectrum corresponding to the surface temperature T_g. The calculation was carried out for $T_g = 280$ K and $T_{strat} = 200$ K, and with a greenhouse gas concentration sufficient to make the optical thickness ≈ 10 in the central absorption band. Left panel: The gas has an absorption coefficient of $1\,m^2/kg$ within a single absorption band extending from 500 to $700\,cm^{-1}$. Right panel: The gas has additional weak absorption bands from 300 to $500\,cm^{-1}$ and 700 to $900\,cm^{-1}$, within which the absorption coefficient is $0.125\,m^2/kg$.

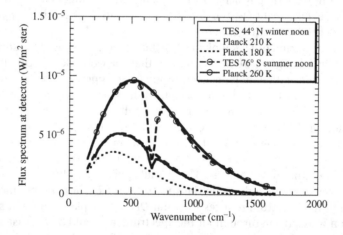

Figure 4.6 Some representative *OLR* spectra for Mars, observed by the Thermal Emission Spectrometer (TES) on Mars Global Surveyor. A northern hemisphere winter case and southern hemisphere summer case are shown. These observations were taken under nearly cloud-free and dust-free conditions.

both in the afternoon and sunset spectra. At the trough of this ditch, the radiation exits the atmosphere with a radiating temperature of about 180 K in the winter case and 200 K in the summer case. These temperatures are similar to the coldest temperatures encountered in the upper atmosphere of Mars (see Fig. 2.2), and are compatible with the strong decrease of temperature with height seen in the soundings. Away from the CO_2 ditch, the atmosphere appears transparent, and the emission resembles the blackbody emission from a land surface having temperature 260 K in the summer case and 210 K in the winter case. These numbers are compatible with the observed range of ground temperature on Mars, cross-checked by near-surface data from landers.

In a situation like that shown in the left panel of Fig. 4.5, the *OLR* is as low as it is going to get, provided the stratospheric temperature is held fixed. Increasing the greenhouse gas concentration q_G cannot lower the *OLR* further since in the spectral region where the gas is radiatively active the atmosphere is already radiating at the coldest available temperature.[3] Suppose, however, that instead of the gas being transparent outside the central absorption band, there is a set of weaker absorption bands waiting in the wings on either side of the primary band – a gas we may call "Two-Band Oobleck." In this case, illustrated in the right panel of Fig. 4.5, the effect of the weaker bands on *OLR* is not yet saturated, and increases in q_G will cause the *OLR* to go down until these bands, too, are saturated. But what if there are yet weaker absorption bands waiting a bit farther out? Then further increases of q_G will yield additional decreases in the *OLR*. One can imagine making the process continuous by making the width of the bands smaller, and the jump in absorption coefficients between adjacent bands smaller. Real greenhouse gases act very much like this, as they almost invariably have absorption whose overall strength decays strongly with distance in wavenumber from a central peak. The rate at which the absorption decays with distance determines the rate at which the *OLR* decreases as the greenhouse gas concentration is made larger.

From Eq. (4.11), (4.12), or (4.13), if we know the transmission function, we can carry out the integral needed to obtain the radiative fluxes. As we shall see shortly, in most cases the dependence of κ on wavenumber is so intricate that solving the problem by doing a brute-force integral over wavenumber is prohibitive if one aims to do the calculation enough times to gain some insight from modeling a climate (even in a single dimension). In any event, doing the calculation with enough spectral resolution to directly resolve all the wiggles in $\kappa(\nu)$ provides much more information about spectral variability than is needed in most cases. What we really want is to understand something about the properties of the transmission function averaged over a finite-sized spectral region of width Δ, centered on a given frequency ν. Specifically, let's choose Δ to be small enough that the Planck function B and its derivative dB/dT are both approximately constant over the spectral interval of width Δ. In that case, when the solution for the flux given in Eq. (4.12) or its alternate forms is averaged over Δ, B can be treated as nearly independent of ν and taken outside the average. In consequence, the resulting band-averaged equations have precisely the same form as the original ones, save that the fluxes are replaced by average fluxes like

$$\bar{I}_+(\nu, p) = \frac{1}{\Delta} \int_{\nu-\Delta/2}^{\nu+\Delta/2} I_+(\nu', p) d\nu' \qquad (4.57)$$

and the transmission function is replaced by

$$\bar{\mathcal{T}}_\nu(p, p') = \frac{1}{\Delta} \int_{\nu-\Delta/2}^{\nu+\Delta/2} \mathcal{T}_{\nu'}(p, p') d\nu'. \qquad (4.58)$$

---

[3] This example is somewhat contrived, since increasing the concentration of a greenhouse gas generally cools the stratosphere. However, it serves to illustrate the way additional weak absorption bands influence the *OLR*. The additional *OLR* reduction from cooling of the stratosphere as q_G increases is a secondary effect. Since the temperature there is already so low, it would not throw off the result very much to simply replace the *OLR* at the depths of the ditch to zero.

We need to learn how to derive properties of $\bar{\mathcal{T}}_\nu(p, p')$. The essential challenge is that the nonlinear exponential function stands between the statistics of κ_ν and the statistics of $\mathcal{T}_\nu$.

The transmission function satisfies the *multiplicative property*, that

$$\mathcal{T}_\nu(p_1, p_2) = \mathcal{T}_\nu(p_1, p')\mathcal{T}_\nu(p', p_2) \tag{4.59}$$

if p' is between p_1 and p_2. The multiplicative property means that the transmission along a path through the atmosphere can be obtained by taking the product of the transmissions along any number of constituent parts of the path. The band-average transmission loses this valuable property, because for two general functions f and g, $\int f(\nu)g(\nu)d\nu \neq (\int f(\nu)d\nu)(\int g(\nu)d\nu)$. The equality holds only if the two functions are uncorrelated, which is not generally the case for the transmission in two successive parts of a path. In the first part of the path, the strongly absorbed frequencies are used up first, and are no longer available for absorption in the second part of the path. The system has memory, and one can think of the light as becoming "tired," or depleted more and more in the easily absorbed frequencies the longer it travels, with the result that the absorption in the latter parts of the path are weaker than they would be if fresh light were being absorbed.

4.4.2 The absorption spectrum of real gases

We will now take a close look at the absorption properties of CO_2, in order to introduce some general ideas about the nature of the absorption of infrared radiation by molecules in a gas. Continuing to use CO_2 as an example, these ideas will be developed in Sections 4.4.3, 4.4.4, and 4.4.6 into a computationally efficient means of calculating infrared fluxes in a real gas atmosphere. A survey of the spectral characteristics of selected other greenhouse gases will be given in Sections 4.4.7 and 4.4.8. We will often refer to the graph of the absorption coefficient of a substance vs. wavenumber as its *absorption spectrum*, even though the graph in this case represents an ordinary function and, unlike the flux or energy spectra referred to earlier, does not represent the density of anything.

Figure 4.7 shows the absorption coefficient of CO_2 as a function of wavenumber, for pure CO_2 gas at a pressure of 1 bar and a temperature of 293 K. In some spectral regions, e.g. 1100–1800 cm^{-1}, CO_2 at this temperature and pressure is essentially transparent. This is a *window region* through which infrared can easily escape to space if no other greenhouse gas intervenes. For a 285 K blackbody, 60 W/m^2 can be lost through this window. There are two major bands in which absorption occurs. For Earthlike temperatures, the lower wavenumber band, from about 450 to 1100 cm^{-1}, is by far the most important. At 285 K the blackbody emission in this band is 218 W/m^2 out of a total of 374 W/m^2, so the absorption in this band is well tuned to intercept terrestrial infrared and thus to reduce *OLR*. The blackbody emission in the higher wavenumber band, from 1800 to 2500 cm^{-1}, is only 6 W/m^2. This band has a minor effect on *OLR* for Earth, but it can become important for much hotter planets like Venus, and even for Earth is important for the absorption of solar near-infrared. Within either band, the absorption coefficient varies by more than eight orders of magnitude.

The absorption does not vary randomly. It is arranged around six peaks (three in each major band), with the overall envelope of the absorption declining approximately exponentially with distance from the peak. However, there is a great deal of fine-scale variation within the overall envelope. Zooming in on a typical region in the inset to Fig. 4.7 we see that the absorption can vary by an order of magnitude over a wavenumber range of only a

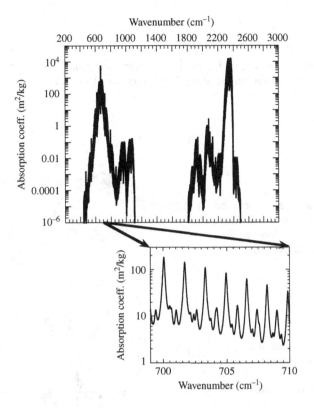

Figure 4.7 The absorption coefficient vs. wavenumber for pure CO_2 at a temperature of 293 K and pressure of 10^5 Pa. This graph is not the result of a measurement by a single instrument, but is synthesized from absorption data from a large number of laboratory measurements of spectral features, supplemented by theoretical calculations. The inset shows the detailed wavenumber dependence in a selected spectral region.

few tenths of a cm^{-1}. Most significantly, the absorption peaks sharply at a discrete set of frequencies, known as *spectral lines.*

Why does the absorption peak at preferred frequencies? In essence, molecules are like little radio receivers, tuned to listen to light only at certain specific frequencies. Since energy is conserved, the absorption or emission of a photon must be accompanied by a change in the internal energy state of the molecule. It is a consequence of quantum mechanics that the internal energy of a molecule can only take on values drawn from a finite set of possible energy states, the distribution of which is determined by the structure of the molecule. If there are N states, there are $N(N-1)/2$ possible transitions, and each one leads to a possible absorption/emission line as illustrated in Fig. 4.8. Transitions between different energy states of a molecule's electron configuration do not significantly contribute to infrared absorption in most planetary situations. The energy states involved in infrared absorption and emission are connected with displacement of the nuclei in the molecule, and take the form of vibrations or rotations. Every molecule has an equilibrium configuration, in which each nucleus is placed so that the electromagnetic forces from the other nuclei and from the electron cloud sum up to zero. A displacement of the nuclear positions will result

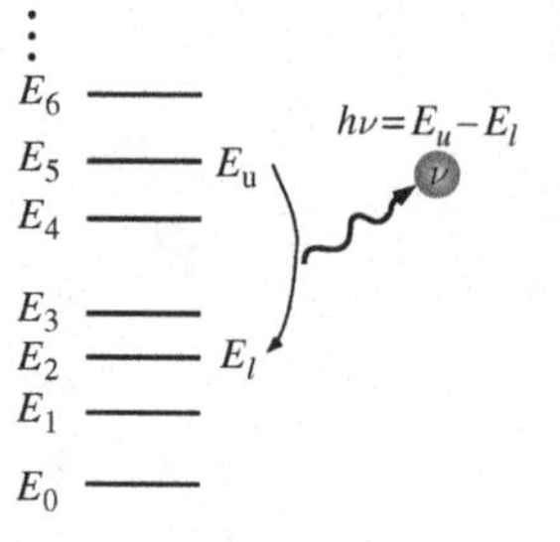

Figure 4.8 Schematic of emission of a photon by transition from a higher energy state to a lower energy state. Subscripts u and l stand for upper, lower.

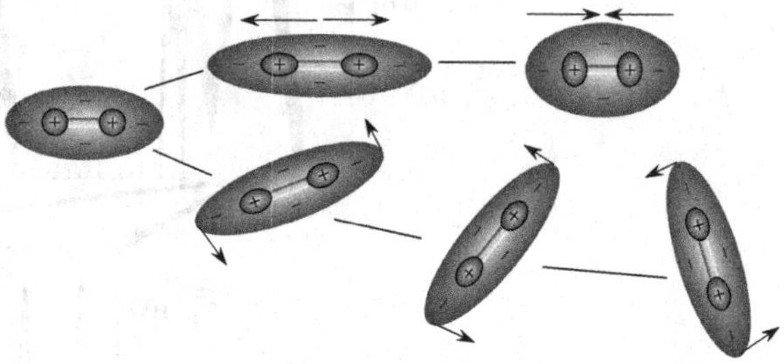

Figure 4.9 Vibration and rotation modes of a diatomic molecule made of a pair of identical atoms, with associated charge distributions.

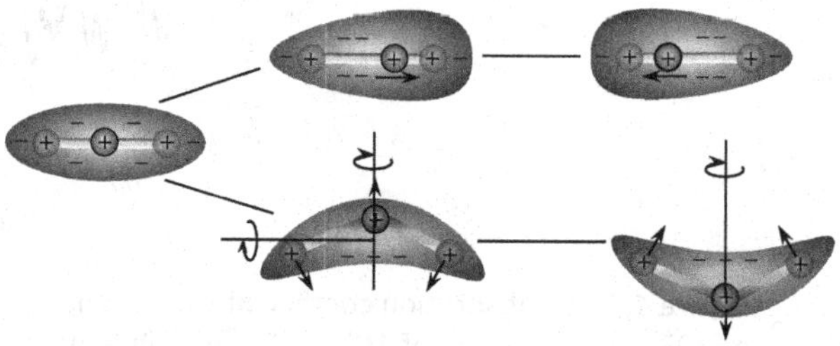

Figure 4.10 Vibration and rotation modes of a linear symmetric triatomic molecule (like CO_2), with associated charge distributions.

in a restoring force that brings the system back toward equilibrium, leading to vibrations. The nuclei can be thought of as being connected with quantum-mechanical springs (one between each pair of nuclei) of different spring constants, and the vibrations can be thought of as arising from a set of coupled quantum-mechanical oscillators. Rigid molecules, held together by rigid rods rather than springs, would have rotational states but not vibrational states. The fact that molecules are not rigid causes the rotational states to couple to the vibrational states, through the Coriolis and centrifugal forces.

Noble gases (He, Ar, etc.) are monatomic, have only electron transitions, and are not active in the infrared. A diatomic molecule (Fig. 4.9) has a set of energy levels associated with the oscillation caused by pulling the nuclei apart and allowing them to spring back; it also has a set of energy levels associated with rotation about either of the axes perpendicular to the line joining the nuclei. Centrifugal force couples the stretching to the rotation. Triatomic molecules (Fig. 4.10) have an even richer set of vibrations and rotations, especially if their equilibrium state is bent rather than linear (Fig. 4.11). Polyatomic molecules like CH_4, NH_3, SF_6, and the chlorofluorocarbons (e.g. CFC-12, which is CCl_2F_2) have yet more complex modes of vibration and rotation. As the set of energy states becomes richer and more complex, the set of differences between states fills in more and more of the spectrum, making the molecule a better infrared absorber.

For a molecule to be a good infrared absorber and emitter, it is not enough that it have transitions whose energy corresponds to the infrared spectrum. In order for a photon to

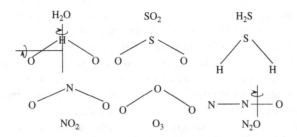

Figure 4.11 Some polar triatomic molecules. Two different modes of rotation are indicated for the H_2O molecule. There is a third mode of rotation about an axis perpendicular to the page.

be absorbed or emitted, the associated molecular motions must also couple strongly to the electromagnetic field. Although the quantum nature of radiation is crucial for many purposes, when it comes to the interaction of infrared or longer-wavelength radiation with molecules, one can productively think of the interaction in semiclassical terms. The reason is that the wavelength of infrared is on the order of $10\,\mu m$, which is two to three orders of magnitude larger than the size of the molecules we will be considering. Thus, one can think of the infrared light as providing a large-scale fluctuating electric and magnetic field which alters the environment in which the molecule finds itself, and exerts a force on the constituent parts of the molecule. This force displaces the nuclei and electron cloud, and excites vibration or rotation. Conversely, a vibrating or rotating molecule creates a moving charge distribution, which classically radiates an electromagnetic wave. While one must fully take into account quantum effects in describing molecular motion, one need not for our purposes confront the much harder problem of quantizing the electromagnetic field as well (the problem of "quantum field theory"). The only way in which we make use of the quantum nature of the electromagnetic field is in converting the energy difference $E_u - E_\ell$ into a frequency of light, via $\Delta E = h\nu$.

The strongest interaction is between an electromagnetic field and a particle with a net charge. A charged particle will experience a net force when subjected to an electric field, which will cause the particle to accelerate. However, ions are extremely rare throughout most of a typical planetary atmosphere. The molecules involved in determining a planet's energy balance are almost invariably electrically neutral. The next best thing to having a net charge is to have a disproportionate part of the molecule's negatively charged electron cloud bunched up on one side of the molecule, while a compensating excess of positive charged nuclei are at the other side. This creates a *dipole moment*, which experiences a net torque when placed in an electric field, causing the dipole axis to try to align with the field. Interactions associated with higher order moments than the dipole lead to absorption many orders of magnitude weaker than the dipole absorption, and can be ignored for most planetary climate purposes.

Many common atmospheric molecules have no dipole moment in their unperturbed equilibrium state. Such *non-polar* molecules can nonetheless couple strongly to the electromagnetic field. They do so because vibration and rotation can lead to a dipole moment through distortion of the equilibrium positions of the electron cloud and the nuclei. As illustrated in Fig. 4.9, diatomic molecules made of two identical atoms do not acquire a dipole moment under the action of either rotation or stretching. Symmetric diatomic molecules, such as N_2, O_2, and H_2, in fact have plenty of rotational and vibrational transitions that are in the infrared range. Because the associated molecular distortions have no dipole moment, however, these gases are essentially transparent to infrared unless they are strongly perturbed by frequent collisions. This is why the most common gases in Earth's atmosphere – N_2 and O_2 – do not contribute to Earth's greenhouse effect. However, it is important to recognize that situations in which diatomic molecules become good greenhouse gases are in fact quite common in planetary atmospheres. When there are frequent collisions, such as happen in the high-density atmospheres of Titan and on all the giant planets, diatomic molecules acquire enough of a dipole moment during the time collisions are taking place that the electromagnetic field can indeed interact with their transitions quite strongly. This makes N_2 and H_2 the most important greenhouse gases on Titan, and H_2 a very important greenhouse gas on all the gas giant planets. In terms of volume of atmosphere affected, hydrogen is by far the most important greenhouse gas in the entire Solar System. Collision-induced absorption of this type forms a *continuum* in which the absorption is a very smooth function of wavenumber, without any significant line structure. Polyatomic molecules can also have significant continua, existing alongside the line spectra. Continuum absorption will be discussed in Section 4.4.8.

Carbon dioxide is a linear molecule with the two oxygens symmetrically disposed about the central carbon, as illustrated in Fig. 4.10. A uniform stretch of such a molecule does not create a dipole moment, but a vibrational mode which displaces the central atom from one side to the other does. In addition, bending modes of CO_2 have a fluctuating dipole moment, which can in turn be further influenced by rotation. Both these modes are illustrated schematically in Fig. 4.10. Modes of this sort make CO_2 a very good greenhouse gas – the more so because the typical energies of the transitions involved happen to correspond to frequencies near the peak of the Planck function for Earthlike temperatures.

Some molecules – called *polar* – have a dipole moment even in their undisturbed state. Most common diatomic gases made of two different elements – notably HF and HCl – are polar, and their vibrational and rotational modes cause fluctuations in the dipole which make them quite good infrared absorbers. They are not commonly thought of as greenhouse gases, because they are highly chemically reactive and do not appear in radiatively significant quantities in any known planetary atmosphere. However, one must keep an open mind about such things. Most triatomic atmospheric gases (H_2O, SO_2, O_3, NO_2, and H_2S, among others) are polar. Carbon dioxide, a symmetric linear molecule with the carbon at the center, is a notable exception. Ammonia (NH_3) is also polar, having its three hydrogens sticking out on one side like legs of a tripod attached to the nitrogen atom at the other side. Polar molecules couple strongly to the electromagnetic field, and their asymmetry also gives them a rich set of coupled rotation and vibration modes with many opportunities for transitions corresponding to the infrared spectrum. The spectrum is enriched because rotation about the axis with the largest moment of inertia (shown as the vertical axis for the water molecule in Fig. 4.11) causes the wing molecules to fling outwards, changing the bond angle and the dipole moment. The molecule can also rotate about an axis perpendicular to the plane of the figure, leading to a distinct set of energy levels. Further, energy can be stored in rotations about the axis with minimum moment of inertia (shown as horizonal in the figure). For a linear molecule like CO_2, rotation about the corresponding axis has essentially no energy.

Let's return now to the matter of how the *OLR* varies as a function of the greenhouse gas content of an atmosphere. Essentially, we revisit the discussion of Fig. 4.5, but this time in the context of the actual absorption spectrum of CO_2 instead of the hypothetical gas discussed earlier. We use the same temperature profile as in Fig. 4.5, but the *OLR* is computed using the fully resolved absorption coefficient as a function of wavenumber, reproduced in the lower panel of Fig. 4.12. The horizontal lines in this panel indicate the spectral regions that are optically thick for CO_2 paths of $\frac{1}{10}$, 1, and 1000 kg/m². The upper panel shows the corresponding *OLR* for the same three paths. The *OLR* was computed at the full spectral resolution of the lower panel, but was smoothed over bands of width 10 cm⁻¹ to make the pattern easier to see. The smoothing is done in such a way that the integral of the smoothed *OLR* curve over wavenumber yields the same net value as the integral over the original unsmoothed curve. This calculation of *OLR* is still not completely realistic, since, to keep things simple, it is carried out as if the absorption coefficient were uniform throughout the depth of the atmosphere. In reality, κ_{CO2} varies with both temperature and pressure, though we'll see eventually that to a good approximation this variation can be handled through the introduction of an *equivalent path*, which is generally somewhat more than half the actual full-atmosphere path, if the reference pressure at which absorption coefficients are stated is taken to be the surface pressure.

Figure 4.12 explains why the *OLR* reduction is approximately logarithmic in greenhouse gas concentration for CO_2 and similar greenhouse gases. The key thing to note is that the absorption coefficient in the principal band centered on 675 cm⁻¹ decays exponentially with distance from the center. Hence, as the CO_2 path is increased by a factor of 10, from $\frac{1}{10}$

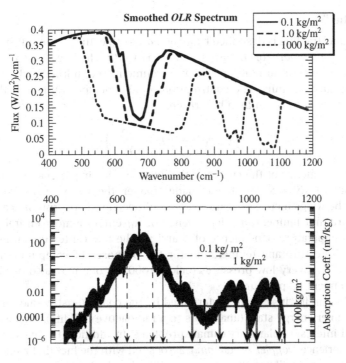

Figure 4.12 Lower panel: The absorption coefficient for CO_2 at 1 bar and 300 K, in the wavenumber range of interest for Earthlike and Marslike planets. The horizontal lines show the wavenumber range within which the optical thickness exceeds unity for CO_2 paths of $\frac{1}{10}$, 1, and 1000 kg/m². Upper panel: The corresponding *OLR* for the three path values, computed for the same temperature profiles as in Fig. 4.5. The *OLR* has been averaged over bands of width 10 cm⁻¹.

to $1 \, kg/m^2$, the width of the ditch within which the radiating temperature is reduced to cold stratospheric values increases only like the logarithm of the ratio of paths. This is true for paths as small as $0.01 \, kg/m^2$ and as large as $100 \, kg/m^2$. However, when the path gets as large as $1000 \, kg/m^2$, the weak absorption bands on the shoulder, near 950 and $1050 \, cm^{-1}$, start to become important, and enhance the optical thickness beyond what one would expect on the basis of the central absorption peak. A path of $1000 \, kg/m^2$ corresponds to a partial pressure of CO_2 of about 66 mb for Earth's gravity, or equivalently a molar mixing ratio of about 6.6% for Earth's current surface pressure. This is far in excess of any CO_2 concentration on Earth likely to have been attained in the past 300 million years, but is well within the range of what has been contemplated at the end of a Neoproterozoic Snowball episode, or earlier during the Faint Young Sun period. Many greenhouse gases also have a central absorption peak with exponential skirts, and these will also exhibit a nearly logarithmic dependence of *OLR* on the concentration of the corresponding greenhouse gas.

With the notable exception of the collision-induced continuum discussed in Section 4.4.8, the absorption spectrum of a gas is built by summing up the contributions of the thousands of spectral lines from each of the radiatively active constituents of the gas. To proceed further, then, we must look more deeply into the nature of the lines and how they are affected by pressure and temperature.

4.4.3 I walk the line

An individual spectral line is described by a line *position* (i.e. the wavenumber at the center), a line *shape*, a line *strength* (or *intensity*), and a line *width*. The line shape is described by a non-dimensional function of non-dimensional argument, $f(x)$, normalized so that the total area under the curve is unity. The contribution of a single spectral line to the absorption coefficient for substance G can then be written

$$\kappa_G(\nu, p, T) = \frac{S}{\gamma} f\left(\frac{\nu - \nu_c}{\gamma}\right) \tag{4.60}$$

where ν_c is the frequency of the center of the line, S is the line intensity, and γ is the line width. Note that $\int \kappa_G d\nu = S$. As a line is made broader, the area remains fixed, so that the absorption in the wings increases at the expense of decreased absorption near the center.

The pressure and temperature dependence of κ_G enters almost entirely through the pressure and temperature dependence of S and γ. The line center ν_c can be regarded as independent of pressure and temperature for the purposes of computation of planetary radiation balance. At very low pressures (below 1000 Pa), one may also need to make the line shape dependent on pressure.

Every line has an intrinsic width determined by the characteristic time for spontaneous decay of the higher energy state (analogous to a radioactive half-life). This width is far too narrow to be of interest in planetary climate problems. In addition, the lines of a molecule in motion will experience *Doppler broadening*, associated with the fact that a molecule moving towards a light source will see the frequency shifted to higher values, and conversely for a molecule moving away. For molecules in thermodynamic equilibrium, the velocities have a Gaussian distribution, and so the line shape becomes $f(x) = \exp(-x^2)/\sqrt{\pi}$. The width is $\gamma = \gamma(T) = \nu_c \frac{\nu}{c}$, where $\nu = \sqrt{2RT}$, R being the gas constant for the molecule in question. ν is a velocity, which is essentially the typical speed of a molecule at temperature T. For CO_2 at 250 K, the Doppler line width for a line with center $600 \, cm^{-1}$ is only about $0.0006 \, cm^{-1}$.

The type of line broadening of primary interest in planetary climate problems is *collisional broadening*, alternatively called *pressure broadening*. Collisional broadening arises because the kinetic energy of a molecule is not quantized, and therefore if a molecule has experienced a collision sufficiently recently, energy can be borrowed from the kinetic energy in order to make up the difference between the photon's energy and the energy needed to jump one full quantum level. The theory of this process is exceedingly complex, and in many regards incomplete. There is a simple semi-classical theory that predicts that collision-broadened lines should have the *Lorentz line shape* $f(x) = 1/(\pi \cdot (1 + x^2))$, and this shape seems to be supported by observations, at least within a hundred widths or so of the line center. For the Lorentz shape, absorption decays rather slowly with distance from the center; 10 half-widths γ from the center, the Lorentz absorption has decayed to only $\frac{1}{101}$ of its peak value, whereas the Gaussian Doppler-broadened line has decayed to less than 10^{-43} of its peak. There are both theoretical and observational reasons to believe that the very far tails of collision-broadened lines die off faster than predicted by the Lorentz shape. A full discussion of this somewhat unsettled topic is beyond the level of sophistication which we aspire to here, but the shape of far tails has some important consequences for the continuum absorption, which will be taken up briefly in Section 4.4.8.

In the simplest theories leading to the Lorentz line shape, the width of a collision-broadened line is proportional to the mean collision frequency, i.e. the reciprocal of the time between collisions. The Lorentz shape is valid in the limit of infinitesimal duration of collisions; it is the finite time colliding molecules spend in proximity to each other that leads to deviations from the Lorentz shape in the far tails, but there is at present no general theory for the far-tail shape. For many common planetary gases the line width is on the order of a tenth of a cm^{-1} when the pressure is 1 bar and the temperature is around 300 K. For fixed temperature, the collision frequency is directly proportional to pressure, and laboratory experiment shows that the implied proportionality of line width to pressure is essentially exact. Holding pressure fixed, the density goes down in inverse proportion to temperature while the mean molecular velocity goes up like the square root of temperature. This should lead to a collision frequency and line width that scales like $1/\sqrt{T}$. Various effects connected with the way the collision energy affects the partial excitation of the molecule lead to the measured temperature exponent differing somewhat from its ideal value of $\frac{1}{2}$. Putting both effects together, if the width is known at a standard state (p_0, T_0), then it can be extrapolated to other states using

$$y(p, T) = y(p_0, T_0)\frac{p}{p_0}\left(\frac{T_0}{T}\right)^n \tag{4.61}$$

where n is a line-dependent exponent derived from quantum mechanical calculations and laboratory measurements. It is tabulated along with standard-state line widths in spectral line databases. One must typically go to very low pressures before Doppler broadening starts to become important. For example, for a collision-broadened line with width 0.1 cm^{-1} at 1 bar, the width does not drop to values comparable to the Doppler width until the pressure falls to 6 mb – comparable to the middle stratosphere of Earth or the surface pressure of Mars. Even then, the collision broadening dominates the absorption when one is not too close to the line center, because the Lorentz shape tails fall off so much more gradually than the Gaussian.

Another complication is that the collision-broadened line width depends on the molecules doing the colliding. Broadening by collision between molecules of like type is called *self-broadening*, while that due to dissimilar molecules is called *foreign broadening*.

The simple Lorentz theory would suggest a proportionality to collision rate, which is a simple function of the ratio of molecular weights. Not only does the actual ratio of self- to foreign-broadened width deviate from what would be expected by this ratio, but the ratio actually varies considerably from one absorption line to another. For CO_2, for example, some self-broadened lines have essentially the same width as for the air-broadened case,[4] whereas others can have widths nearly half again as large. For water vapor, the disparity is even more marked. Evidently, some kinds of collisions are better at partially exciting energy levels than others. There is no good theory at present that enables one to anticipate such effects. Standard spectral databases tabulate the self-broadened and air-broadened widths at standard temperature and pressure, but if one were interested in broadening of water vapor by collisions with CO_2 (important for Early Mars) or broadening of NH_3 by collisions with H_2 (on Jupiter or Saturn), one would have to either find specialized laboratory experiments or extrapolate based on molecular weights and hope for the best.

The line intensities are independent of pressure, but they do increase with temperature. For temperatures of interest in most planetary atmospheres, the temperature dependence of the line intensity is well described by

$$S(T) = S(T_0) \left(\frac{T}{T_0} \right)^n \exp \left(-\frac{h\nu_\ell}{k} \left(\frac{1}{T} - \frac{1}{T_0} \right) \right) \tag{4.62}$$

where n is the line-width exponent defined above and $h\nu_\ell$ is the energy of the lower energy state in the transition that gives rise to the line. This energy is tabulated in standard spectroscopic databases, and is usually stated as the frequency ν_ℓ. Determination of the lower state energy is a formidable task, since it means that one must assign an observed spectral line to a specific transition. When such an assignment cannot be made, one cannot determine the temperature dependence of the strength of the corresponding line.

Now let's compute the average transmission function associated with a single collision-broadened spectral line in a band of wavenumbers of width Δ. We'll assume that the line is narrow compared with Δ, so that the absorption coefficient can be regarded as essentially zero at the edges of the band. Without loss of generality, we can then situate the line at the center of the band. The mean transmission function is

$$\bar{\mathcal{T}}(p_1, p_2) = \frac{1}{\Delta} \int_{-\Delta/2}^{\Delta/2} \left[\exp \left(-\frac{1}{g\pi \cos\bar{\theta}} \int_{p_1}^{p_2} \frac{S(T)\gamma q}{\nu'^2 + \gamma^2} dp \right) \right] d\nu' \tag{4.63}$$

where $\nu' = \nu - \nu_c$. The argument of the exponential is just the optical thickness of the layer between p_1 and p_2, and to keep the notation simple we will assume the integral to be taken in the sense that makes it positive. The double integral and the nonlinearity of the exponential make this a hard beast to work with, but there are two limits in which the result becomes simple. When the layer of atmosphere between p_1 and p_2 is optically thin even at the center of the line, where absorption is strongest, the line is said to be in the *weak line regime*. All lines are in this regime in the limit $p_2 \to p_1$, though if the line is very narrow or the intensity is very large, the atmospheric layer might have to be made exceedingly small before the weak line limit is approached. For weak lines the exponential can be approximated as $\exp(-\delta\tau) \approx 1 - \delta\tau$, whence

---

[4] "Air-broadening" refers to broadening by collisions with dry Earth air, though sometimes the measurements are performed with an idealized N_2/O_2 mixture, or even pure N_2.

$$\bar{\mathcal{T}}(p_1, p_2) \approx 1 - \frac{1}{\Delta} \int_{-\Delta/2}^{\Delta/2} \frac{1}{g\pi \cos\bar{\theta}} \int_{p_1}^{p_2} \frac{S(T)\gamma q}{\nu'^2 + \gamma^2} dp d\nu'$$

$$= 1 - \frac{1}{\Delta} \frac{1}{g\cos\bar{\theta}} \int_{p_1}^{p_2} S(T)q dp \qquad (4.64)$$

$$= 1 - \frac{1}{\Delta} S(T_0)\ell_w$$

where T_0 is a constant reference temperature and the *weighted path* for weak lines is

$$\ell_w(p_1, p_2) \equiv \frac{1}{g\cos\bar{\theta}} \int_{p_1}^{p_2} \frac{S(T(p))}{S(T_0)} q(p) dp. \qquad (4.65)$$

Note that for weak lines, the averaged transmission is independent of the line width. From the expression for $\bar{\mathcal{T}}$ we can define the *equivalent width* of the line, $W \equiv S(T_0)\ell_w$. To understand the meaning of the equivalent width, imagine that absorption takes *all* of the energy out of the incident beam within a range of wavenumbers of width W, leaving the rest of the spectrum undisturbed. The equivalent width W is defined such that the amount of energy thus removed is equal to the amount removed by the actual absorption, which takes just a little bit of energy out of each wavenumber throughout the spectrum.

When the layer of atmosphere between p_1 and p_2 is optically thick at the line center, the transmission is reduced to nearly zero there. This defines the *strong line* limit. For strong lines, there is essentially no transmission near the line center; all the transmission occurs out on the wings of the lines. Since essentially nothing gets through near the line centers anyway, there is little loss of accuracy in replacing the line shape by its far-tail form, $\pi^{-1}S\gamma/\nu'^2$. With this approximation to the line shape, the band-averaged transmission may be written:

$$\bar{\mathcal{T}}(p_1, p_2) \approx \frac{1}{\Delta} \int_{-\Delta/2}^{\Delta/2} \left[\exp\left(-\frac{1}{\nu'^2} \frac{1}{g\pi \cos\bar{\theta}} \int_{p_1}^{p_2} S(T)\gamma q dp \right) \right] d\nu'$$

$$= \frac{1}{\Delta} \int_{-\Delta/2}^{\Delta/2} \exp\left(-\frac{X}{\nu'^2} \right) d\nu' \qquad (4.66)$$

$$= \frac{1}{2\zeta_m} \int_{-\zeta_m}^{\zeta_m} \exp\left(-\frac{1}{\zeta^2} \right) d\zeta$$

where $X \equiv S(T_0)\gamma(p_0, T_0)\ell_s/\pi$, and the weighted path for strong lines is

$$\ell_s \equiv \frac{1}{g\cos\bar{\theta}} \int_{p_1}^{p_2} \frac{S(T(p))}{S(T_0)} \left(\frac{T_0}{T} \right)^n \frac{p}{p_0} q(p) dp. \qquad (4.67)$$

The third line in the expression for $\bar{\mathcal{T}}$ comes from introducing the rescaled dummy variable $\zeta \equiv \nu'/\sqrt{X}$; the limit of integration then becomes $\zeta_m = \Delta/\left(2\sqrt{X}\right)$. Unless the path is enormous, ζ_m will be very large, because the averaging interval Δ is invariably taken to be much larger than the typical line width (otherwise there would be little point in averaging). For $\zeta_m \gg 1$, the integral in the last line can be evaluated analytically, and is

$$\int_0^{\zeta_m} \exp\left(-\frac{1}{\zeta^2} \right) d\zeta \approx \zeta_m - \sqrt{\pi} \qquad (4.68)$$

(see Problem 4.13). Therefore, substituting for X, the expression for $\bar{\mathcal{T}}$ in the strong limit becomes

$$\bar{\mathcal{T}}(p_1, p_2) \approx 1 - \frac{1}{\Delta} 2\sqrt{S(T_0)\gamma(p_0)\ell_s}. \qquad (4.69)$$

For strong lines the equivalent width is $W \equiv 2\sqrt{S(T_0)\gamma(p_0)\ell_s}$. In this case, the width of the chunk taken out of the spectrum increases like the square root of the path because the absorption coefficient decreases like $1/\nu'^2$ with distance from the line center, implying that the width of the spectral region within which the atmosphere is optically thick scales like the square root of the path. Unlike weak lines, strong lines really do take almost all of the energy out of a limited segment of the spectrum. The multiplicative property for transmission is equivalent to an additive property for equivalent width. The nonlinearity of the square root linking path to equivalent width in the strong line case thus means that the band-averaged transmission has lost the multiplicative property. As in our earlier general discussion of this property, the loss stems from the progressive depletion of energy in parts of the spectrum near the line center.

The pressure-weighting of the strong line path reflects the fact that, away from the line centers, the atmosphere becomes more optically thick as pressure is increased and the absorption is spread over a greater distance around each line. Note that if we choose as the reference pressure p_0 any pressure that remains between p_1 and p_2, then $\ell_s \rightarrow \ell_w$ as $p_1 \rightarrow p_2$. In this case, one can use the strong line path ℓ_s regardless of the pressure range, since the strong line path reduces to the correct weak line path for thin layers where the weak line approximation becomes valid. A common choice for the reference pressure is the average $(p_1 + p_2)/2$, but one could just as well choose one of the endpoints of the interval instead. In the case of a well-mixed greenhouse gas (constant q_G) for a nearly isothermal layer, the equivalent path becomes $1/\cos\bar{\theta}$ times $\frac{1}{2g}q_G\left(p_2^2 - p_1^2\right)/p_0$, which reduces to the actual mass path $q_G(p_2 - p_1)/g$ if p_0 is taken to be the average. In this case, one gets the correct transmission by using the conventional mass path with absorption coefficients computed for the average pressure of the layer. This is known as the *Curtiss–Godson* approximation.

In solving radiative transfer problems related to planetary climate, one typically takes the bandwidth Δ large enough that the band contains a great many lines. For example, there are about 600 CO_2 lines in the band between 600 and $625\,\mathrm{cm}^{-1}$. In the weak line limit the transmission is linear in the absorption coefficient, so one can simply sum the equivalent widths of all the lines in the band to obtain the total equivalent width $W = \sum W_i$. For strong lines, the situation is a bit more complicated, because of the nonlinearity of the exponential function. For the same reasons one loses the multiplicative property of transmission upon band averaging, one generally loses the additive property of equivalent widths. There is one important case in which additivity of equivalent widths is retained, however. If the lines are *non-overlapping*, in the sense that they are far apart compared with the width over which each one causes significant absorption, then the absorption from each line behaves almost as if the line were acting in isolation. In this case, each line essentially takes a distinct chunk out of the spectrum, and the equivalent widths can be summed up to yield the net transmission.

The additivity of strong-line equivalent widths breaks down at large paths. Since each W_i increases like the square root of the path, eventually the sum exceeds Δ, leading to the absurdity of a negative transmission. What is going wrong is that, as the equivalent widths become large, the absorption regions associated with each line start to overlap. One is trying to take away the same chunk of the spectrum more than once. This doesn't work for spectra any more than it works for ten hungry people trying to eat an eighth of a pizza each. One approach which has met with considerable success is to assume that the lines are randomly placed, so that the transmission functions due to each line are uncorrelated.

This is *Goody's Random Overlap Approximation*. For uncorrelated transmission functions, the band-averaged transmissions can be multiplied, yielding

$$\bar{\mathcal{T}} \approx \left(1 - \frac{W_1}{\Delta}\right)\left(1 - \frac{W_2}{\Delta}\right)\cdots\left(1 - \frac{W_N}{\Delta}\right)$$

$$= \exp\sum\ln\left(1 - \frac{W_i}{\Delta}\right) \qquad\qquad (4.70)$$

$$\approx \exp\left(-\frac{1}{\Delta}\sum W_i\right).$$

The last step follows from the assumption that each individual equivalent width is small compared with Δ. Note that when the sum of the equivalent widths is small compared with Δ, this expression reduces to the previous expression given for individual or non-overlapping lines.

The line parameters laid out above – position, width, strength, and temperature scaling – lie at the heart of most real gas radiative transfer calculations. There being thousands of spectral lines for dozens of substances of interest in planetary climate, teasing out the data one needs from the original literature is a daunting task. Fortunately, a small but dedicated group of spectroscopists[5] have taken on the task of validating, cross-checking, and assembling the available line data into a convenient database known as HITRAN. It is suitable for most planetary calculations, though it must sometimes be supplemented with information on absorption that is not associated with spectral lines (the *continuum* absorption), and with additional data on weak lines which are important in the extremely hot, dense atmosphere of Venus. Instructions for obtaining the HITRAN database, along with sources for additional spectral data of use on Venus, Titan, and Jupiter, are given in the references section at the end of this chapter, and the software supplement to this book provides a simple set of routines for reading and performing calculations with the HITRAN database.

4.4.4 Behavior of the band-averaged transmission function

Although the absorption spectrum has very complex behavior, the band-averaged transmission function averages out most of the complexity. The definition of the transmission guarantees that it decays monotonically as $|p_1 - p_2|$ increases and the path increases, but in addition the decay is invariably found to be smooth, proceeding without erratic jumps, kinks, or other complex behavior. This smoothness is what makes computationally economical radiative transfer solutions possible, and the various schemes for carrying out the calculation of fluxes amount to different ways of exploiting the smoothness of the band-averaged transmission function.

By way of example, the band-averaged transmission function for CO_2 is shown for three different bands in Fig. 4.13. The calculation of $\bar{\mathcal{T}}_\nu(p_1, p_2)$ was carried out using a straightforward – and very time-consuming – integration of the transmission over frequency; at each frequency in the integrand, one must do an integral of $\kappa_{CO2}(\nu, p)$ over pressure, and each of those κ must be evaluated as a sum over the contributions of up to several hundred lines. Temperature was held constant at 296 K and a constant mass-specific concentration of 0.0005 (330 ppmv) of CO_2 mixed with air was assumed. The pressure p_1 was held fixed at 100 mb, while p_2 was varied from 100 mb to 1000 mb. This plot thus gives an indication of the upward flux transmitted from each layer of the atmosphere, as seen looking down

---

[5] May they live forever!

from the Earth's tropical tropopause. The results are plotted as a function of the pressure-weighted strong line path, which for constant q and T is $q \cdot \left(p_2^2 - p_1^2 \right) / (2 g p_0 \cos \bar{\theta})$, where the reference pressure p_0 is taken to be 10^5 Pa. Plotting the results this way makes it easier to compare them with theoretical expectations, and also makes it easier to generalize the results to transmission between different pairs of pressure levels, which will have different amounts of pressure broadening. The rationale for using the strong line path is that the lines are narrow enough that almost all parts of the spectrum are far from the line centers in comparison to the width, and in such cases the collision-broadened absorption coefficient increases linearly with pressure almost everywhere. This behavior is incorrect near the line centers, but the error in the transmission introduced by this shortcoming is minimal, since the absorption is so strong there that the contribution to the transmission is essentially zero anyway. This reasoning – based directly on what we have learned from the strong line limit – is at the basis of most representations of pressure-broadening effects in radiative calculations. Here, we only are using it as a graphical device, since the transmission itself is computed without approximation. Note that the strong line path becomes proportional to the (weak line) mass path $q \cdot (p_2 - p_1)/(g \cos \bar{\theta})$ when $p_2 \rightarrow p_1$, with proportionality constant p_1 / p_0. In the present calculation, when p_2 is at its limit of 1000mb, the path is about 5kg/m^2, which is about half the unweighted mass path over the layer. This reflects the fact that the lower pressure over most of the layer weakens the absorption relative to the reference value at $p = p_0$.

Apart from noticing that the transmission function is indeed smooth, we immediately remark that the transmission first declines sharply, as portions of the spectrum with the highest absorption coefficient are absorbed. At larger paths, the spectrum becomes progressively more depleted in easily absorbed wavenumbers, and the decay becomes slower. For the two strongly absorbing bands in the left panel, the transmission curve becomes nearly vertical at small paths, as suggested by the square-root behavior of the strong line limit. There is guaranteed to be a weak-line region at sufficiently small paths, where the slope becomes finite, but in these bands the region is so tiny it is invisible. In fact, the strong line transmission function in Eq. (4.69) fits the calculated transmission in the 575–600 cm^{-1} band almost exactly throughout the range of paths displayed, when used with the random-overlap modification in Eq. (4.70). For the more strongly absorbing 600–625 cm^{-1} band the fit is very good out to paths of 1.5 kg/m^2, but thereafter the actual transmission decays considerably more rapidly than the strong line form. This mismatch occurs because the derivation of the strong line transmission function assumes that the absorption coefficients within the band approach zero arbitrarily closely: as more and more radiation is absorbed, there is always some region where the absorption coefficient is arbitrarily close to zero, which leads to ever-slower decay. In reality, overlap between the skirts of the lines leads to finite-depth valleys between the peaks (see the inset of Fig. 4.7), and the absorption is bounded below by a finite positive value. The decay of the transmission at large paths is determined by the local minima in the valleys, and will tend toward exponential decay, rather than the slower decay predicted by the strong line approximation.

For the weakly absorbing band shown in the right panel of Fig. 4.13, a hint of weak-line behavior can be seen at small values of the path, with the result that the behavior diverges noticeably from the best strong line fit. The representation of the transmission can be improved by adopting a two-parameter fit tailored to give the right answer in both the weak and strong limits. The *Malkmus model* is a handy and widely used example of this approach. It is defined by

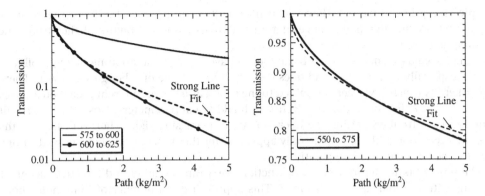

Figure 4.13 The band-averaged transmission as a function of path, for the three different bands, as indicated. In each case, the transmission is computed between a fixed pressure $p_1 = 100$ mb and a higher pressure p_2 ranging from 100 mb to 1000 mb. Calculations were carried out assuming the temperature to be constant at 296 K, with a constant CO_2 specific concentration of $q = 0.0005$, and assuming a mean propagation angle $\cos \bar{\theta} = \frac{1}{2}$. Results are plotted as a function of the pressure-weighted path for strong lines, $q \cdot \left(p_2^2 - p_1^2 \right) /(2 g p_0 \cos \bar{\theta})$, where $p_0 = 1000$ mb. In the left panel, the best fit to the strong line transmission function is shown as a dashed curve; the fit is essentially exact for the 575–600 cm^{-1} band, so the fitted curve isn't visible. For the weaker absorption band in the right panel, fits are shown both for the strong line and the Malkmus transmission function, but the Malkmus fit is essentially exact and cannot be distinguished.

$$\sum W_i = 2 \frac{R^2}{S} \frac{p_1}{p_0} \left(\sqrt{1 + \frac{S^2}{R^2} \left(\frac{p_0}{p_1} \right)^2 \ell_s} - 1 \right) \qquad (4.71)$$

where R and S are the parameters of the fit.[6] The parameters can be identified with characteristics of the absorption spectrum in the band by looking at the weak line (small ℓ_s) and strong line (large ℓ_s) limits. For small ℓ_s, the sum of the equivalent widths is $S \cdot (p_0/p_1)\ell_s = S\ell_w$, so by comparing with Eq. (4.64) we identify S as the sum of the line intensities. For large ℓ, the sum is $2\sqrt{R^2 \ell}$, whence on comparison with Eq. (4.69) we identify R^2 as the sum of $\gamma_i(p_0)S_i$ for all the lines in the band. The parameters R and S can thus be determined directly from the database of line intensities and widths, though in some circumstances it can be advantageous to do a direct fit to the results of a line-by-line calculation like that in 4.13 instead. One uses the Malkmus equivalent-width formula with the random-overlap transformation given in Eq. (4.70), so as to retain validity at large paths. With the Malkmus model, the transmission function in the weakly absorbing 550–575 cm^{-1} band can be fit almost exactly. Since the Malkmus model reduces to the strong line form at large paths, it fits the transmission functions in the left panel of Fig. 4.13 at least as well

---

[6] The factor p_1/p_0 deals with the difference between the strong line and weak line paths, and is necessary so that the limits work out properly for small and large path. There is some flexibility in defining this factor. It is common to use $\frac{1}{2}(p_1 + p_2)/p_0$ to make things look more symmetric in p_1 and p_2. This slightly changes the way the function interpolates between the weak and strong limits, without changing the endpoint behavior.

as the strong line curve did. However, it does nothing to improve the fit of the strongly absorbing case at large paths, since that mismatch arises from a failure of the strong line assumption itself.

The Malkmus model is a good basic tool to have in one's radiation modeling toolkit. It works especially well for CO_2, and does quite well for a range of other gases as well. There are other fits which have been optimized to the characteristics of different greenhouse gases (e.g. *Fels–Goody* for water vapor), and fits with additional parameters. Most of the curve-fit families have troubles getting the decay of the transmission right when very large paths are involved, though if the trouble only appears after the transmission has decayed to tiny values, the errors are inconsequential.

Empirical fits to the transmission function are a time-honored and effective means of dealing with infrared radiative transfer. This approach has a number of limitations, however. We have already seen some inadequacies in the Malkmus model when the path gets large; patching up these problems leads to fits with more parameters, and finding fits that are well-tailored to the characteristics of some new greenhouse gas one wants to investigate can be quite involved. It also complicates the implementation of the algorithm to have to use different classes of fits for different gases, and maybe even according to the band being considered. A more systematic and general approach is called for. The one we shall pursue now, known as *exponential sums*, has the additional advantage that it can be easily generalized to allow for the effects of scattering, which is not possible with band-averaged fits like the Malkmus model. As a gentle introduction to the subject, let's consider the behavior of the integral

$$\bar{\mathcal{T}}(\ell) = \frac{1}{\Delta} \int_{\nu_0 - \Delta/2}^{\nu_0 + \Delta/2} e^{-\kappa_G(\nu)\ell} d\nu \tag{4.72}$$

where κ_G is the absorption coefficient for a greenhouse gas G, and ℓ is a mass path. This would in fact be the exact expression for the band-averaged transmission for a simplified greenhouse gas whose absorption coefficient is independent of pressure and temperature. In this case, the path ℓ between pressure p_1 and p_2 is simply the unweighted mass path $|\int q dp|/(g \cos \bar{\theta})$, which reduces to $q|p_1 - p_2|/(g \cos \bar{\theta})$ if the concentration q is constant.

The problem we are faced with is the evaluation of the integral of a function $f(x)$ which is very rapidly varying as a function of x. The ordinary way to approximate the integral is as a Riemann–Stieltjes sum, dividing the interval up into N sub-intervals $[x_j, x_{j+1}]$ and summing the areas of the rectangles, i.e.

$$\int_0^1 f(x) dx \approx \sum_1^N f\left(\frac{x_j + x_{j+1}}{2}\right)(x_{j+1} - x_j). \tag{4.73}$$

The problem with this approach is that a great many rectangles are needed to represent the complex area under the curve $f(x)$. Instead, we may define the function $H(a)$, which is the sum of the lengths of the intervals for which $f(x) \leq a$, as illustrated in Fig. 4.14. Now, the integral can be approximated instead by the sum

$$\int_0^1 f(x) dx = \int_{f=f_{min}}^{f_{max}} f dH(f) \approx \sum_1^M \frac{f_j + f_{j+1}}{2}(H(f_{j+1}) - H(f_j)) \tag{4.74}$$

where we have divided the range of the function f (i.e. $[f_1, f_2]$) into M partitions. This representation can be very advantageous if $H(f)$ is a much more smoothly varying function than $f(x)$. To mathematicians, this form of the approximation of an integral by a sum is the

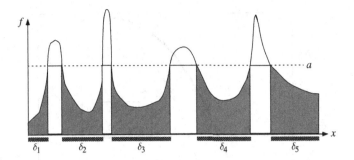

Figure 4.14 Evaluation of the area under a curve by Lebesgue integration.

first step in the magnificent apparatus of *Lebesgue integration*, leading onwards to what is known as measure theory, which forms the basis of rigorous real analysis.

The idea is to apply the Lebesgue integration technique to the transmission function defined in Eq. (4.72), with the absorption coefficient κ_G playing the role of f and the frequency ν playing the role of x. Thus, if $H(a)$ is the sum of the lengths of the frequency intervals in the band for which $\kappa_G \leq a$, then $H(a) = 0$ when a is less than the minimum of κ_G, and $H(a)$ approaches the bandwidth Δ when a approaches the maximum of κ_G. The transmission function can then be written approximately as

$$\bar{\mathcal{T}}(\ell) = \int_{\kappa_{min}}^{\kappa_{max}} e^{-\kappa_G \ell} dH(\kappa_G) \approx \sum_{1}^{M} e^{-(\kappa_{j+1} + \kappa_j)\ell/2}(H(\kappa_{j+1}) - H(\kappa_j)). \tag{4.75}$$

This is the exponential sum formula. It can be regarded as an M term fit to the transmission function, much as the Malkmus model is a two-parameter fit. The Lebesgue integration technique amounts to a simple reshuffling of the terms in the integrand: we collect together all wavenumbers with approximately the same κ, compute the transmission for that value, and then weight the result according to "how many" such wavenumbers there are.

Because the absorption coefficient varies over such an enormous range, it is more convenient to work with $H(\ln \kappa)$ rather than $H(\kappa)$. A typical result for CO_2 is shown in Fig. 4.15, computed for two bands at a pressure of 100 mb. The function is quite smooth, and can be reasonably well characterized by ten points or fewer. In contrast, given that the typical line width at 100 mb is only 0.01 cm^{-1}, evaluation of the transmission integral in the Riemann form, Eq. (4.73), would require at least 25 000 points in a band of width 25 cm^{-1}. Thus, the exponential sum approach is vastly more economical of computer time than a direct line-by-line integration would be.

The decay of the transmission with path length described by Eq. (4.75) is exactly analogous to the decay in time of the concentration of a mixture of radioactive substances with different half-lives. The short-lived things go first, leading to rapid initial decay of concentration; as time goes on, the mixture is increasingly dominated by the long-lived substances, and the decay rate is correspondingly slower. The way the transmission function converges as additional terms are included in the exponential sum formula is illustrated in Fig. 4.16. Specifically, we divide the range of absorption coefficients into 20 bins equally spaced in $\log \kappa_G$, and then truncate H so as to keep only the N largest absorption coefficients, with N ranging from 1 (retaining only the strongest absorption) to 20 (retaining all absorption coefficients including the weakest). When only the strongest absorptions are included, the

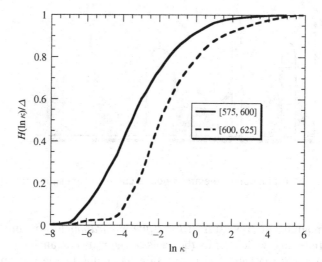

Figure 4.15 Cumulative probability function of the natural log of the absorption coefficient for CO_2. Results are given for the 600–625 cm^{-1} and 575–600 cm^{-1} bands, and were computed at a pressure of 100 mb and a temperature of 296 K.

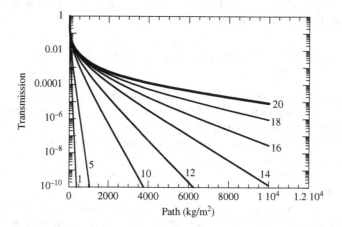

Figure 4.16 Convergence of exponential sum representation of the band-averaged transmission function as the number of terms is increased. The calculation is for CO_2 in a wavenumber band near the peak absorption.

steep decay of transmission for small paths is correctly represented, but the transmission function decays too strongly at large paths. As more of the weaker absorption terms are included, the weaker decay of the transmission is well represented out to larger and larger paths. The ability to represent the decay rate of transmission over a very large range of paths is one of the two advantages of exponential sums over the Malkmus approach, the other advantage being the ability to incorporate scattering effects. An analytical example exploring related features of the exponential integral in Eq. (4.75) is given in Problem 4.14.

If it were not for the dependence of absorption coefficient on pressure and temperature, the exponential sum representation would be exact in the limit of sufficiently many terms.

The computational economy of exponential sums comes at a cost, however, which is scrambling the information about which absorption coefficient corresponds to which frequency. This is not a problem if one is dealing with a layer with essentially uniform pressure and temperature, but it becomes a cause for concern in the typical atmospheric case where one is computing the transmission over a layer spanning considerable variations in temperature and pressure. The problem is that changing pressure or temperature changes the *shape* of the distribution $H(\kappa)$, and there is no rigorously correct way to deal with this within the exponential sum framework. In the discussion of line shapes, for example, we learned that increasing p reduces the peak absorption, but increases absorption between peaks. In terms of the probability distribution of κ, this means that the largest and smallest values of κ become less prevalent at the same time that the intermediate values become more prevalent. At very large values of pressure where the lines become extremely broad, κ becomes a smooth function of frequency within each band and the probability distribution becomes concentrated on a single mean value of κ. The effect of temperature on the shape of $H(\kappa)$ can be even more complex, since the temperature-dependence coefficients of line strength can differ greatly even for neighboring lines.

All these problems notwithstanding, experience has shown that one can obtain a reasonably accurate approximation to the band-averaged transmission function by assuming that all the absorption coefficients within a band have separable scaling of the form

$$\kappa_G(p, T) = \kappa_G(p_0, T_0)F(p/p_0, T/T_0). \tag{4.76}$$

Given scaling of this sort, one can compute the transmission for a path through an inhomogeneous atmosphere using Eq. (4.75), by defining a suitable equivalent path. For example, if the specific concentration of the greenhouse gas is q_G, the absorption scaling is linear in pressure, and the temperature dependence scales according to a function $S(T/T_0)$, then the equivalent path for the layer between p_1 and p_2 is

$$\ell = \frac{1}{g} \int_{p_1}^{p_2} q_G(p)\frac{p}{p_0}S\left(\frac{T(p)}{T_0}\right) dp. \tag{4.77}$$

The temperature-scaling function S would be computed separately for each band. The use of linear scaling in pressure is justified by the fact that most of the spectrum is far from line centers, so absorption scales like the strong line approximation as long as the pressure is not so large that line widths become comparable to the width of the band under consideration. The use of a single temperature-scaling function is harder to justify theoretically, but seems to be supported by numerical experiment.

When q_G is not small, there is one additional complication to take into account when defining the equivalent path. Namely, the absorption coefficient for self-broadened collisions is generally different from that for foreign-broadened collisions, so one must scale the absorption coefficient according to the proportion of self vs. foreign collisions. Like the temperature scaling factor, the ratio of self to foreign broadening can vary considerably even amongst nearby lines. Typically, though, one simply chooses a representative ratio of self to foreign broadening which is assumed constant within each band over which the distribution H is computed. Let's call this ratio a_{self}. The molar concentration of the greenhouse gas is $(\bar{M}/M_G)q_G$, where $\bar{M}$ is the mean molecular weight of the mixture; hence the proportion of collisions which are self collisions is $(\bar{M}/M_G)q_G$ while the proportion of foreign collisions is $1 - (\bar{M}/M_G)q_G$. Then, if H is computed using the foreign-broadened absorption at the standard pressure, the appropriate equivalent path to use in computing the transmission is

$$\ell = \frac{1}{g} \int_{p_1}^{p_2} \left(1 + \frac{\bar{M}}{M_G} q_G(p)(a_{self} - 1)\right) q_G(p) \frac{p}{p_0} S\left(\frac{T(p)}{T_0}\right) dp. \tag{4.78}$$

For a pure one-component atmosphere, $(\bar{M}/M_G)q_G = 1$ and one simply uses the self-broadened absorption coefficients in preparing the distribution $H(\kappa)$, rather than going through the intermediary of defining a_{self} for each band.

Because the scaling of absorption coefficient with pressure and temperature is only approximate, it is important to compute H for a reference pressure and temperature that is characteristic of the general range of interest for the atmosphere under consideration, so as to minimize the amount of scaling needed. Typically, one might use a half or a tenth of the surface pressure as the reference pressure, and a mid-tropospheric temperature as the reference temperature; if one were primarily interested in stratospheric phenomena, or if one were computing OLR on a planet like Venus where most of the OLR comes from only the uppermost part of the atmosphere, pressures and temperatures characteristic of a higher part of the atmosphere would be more appropriate.

Modern professionally written radiative transfer codes attempt to get around the inaccuracies of temperature and pressure scaling by using an extension of the exponential sums method known as *correlated-k*. The basic idea behind this method is to explicitly compute a database of absorption distribution functions H covering the range of pressure and temperature values encountered in the atmosphere, rather than generating them from rescaling of a single distribution function. The mathematical justification of the way these distribution functions are used to compute the transmission is poor, but the method reduces to exponential sums when scaling is valid. It is thus guaranteed to be no worse than exponential sums, and comparisons with detailed line-by-line calculations indicate that it commonly performs well, though it is hard to say when the method should work and when it shouldn't. The exponential sums approach suffices to give the reader an understanding of the basic principles of real gas radiative effects, so we will use that method as the basis of most of our further discussion of the subject. The reader wishing to learn how to make use of the correlated-k method is directed to the reference given in the Further Reading section of this chapter.

4.4.5 Dealing with multiple greenhouse gases

We now know how to efficiently compute the band-averaged transmission function for a single greenhouse gas acting alone. It is commonly the case, however, that two or more greenhouse gases are simultaneously present – CO_2, CH_4, and water vapor in Earth's case, for example. How do we compute the averaged transmission function in this situation? The issues are closely related to those discussed in Section 4.4.1 in connection with the loss of the multiplicative property in band-averaged transmission functions. Similar reasoning shows that the average transmission function for two greenhouse gases acting together generally differs from the product of the averaged transmission functions of each of the individual gases taken alone. Fortunately, however, the special circumstances under which the multiplicative property holds for multiple gases are expected to be fairly common.

By way of illustration, let's consider an idealized greenhouse gas A with transmission function $\mathcal{T}_A(\nu)$ in a wavenumber band of width Δ, with the property that $\mathcal{T}_A = a$ on a set of wavenumber sub-intervals with length adding up to $r \cdot \Delta$, but with $\mathcal{T}_A = 1$ elsewhere; naturally, we require $a < 1$ and $r \leq 1$. Consider a second greenhouse gas B with $\mathcal{T}_B = b$ on a set of wavenumber sub-intervals with length adding up to $s \cdot \Delta$, and with $\mathcal{T}_B = 1$ elsewhere.

The band-averaged transmission for gas A in isolation is $r \cdot a + (1 - r)$, and for B in isolation is $s \cdot b + (1 - s)$. What is the transmission when both gases act in combination?

The answer depends on how the regions of absorption of gas A - spectral regions where $\mathcal{T}_A < 1$ - line up with those of gas B. We can distinguish three limiting cases. First, the regions of absorption of the two gases may be perfectly correlated, in which case $r = s$, $\mathcal{T}_A(\nu) = a$ precisely where $\mathcal{T}_B(\nu) = b$, and

$$
\begin{aligned}
\bar{\mathcal{T}}(\nu) &= \frac{1}{\Delta} \int_{\nu-\Delta/2}^{\nu+\Delta/2} \mathcal{T}_A(\nu')\mathcal{T}_B(\nu') d\nu' \\
&= rab + (1 - r) \\
&= \bar{\mathcal{T}}_A\bar{\mathcal{T}}_B - r(1 - r)(a + b - ab - 1).
\end{aligned}
\tag{4.79}
$$

Further, $a + b - ab = a + (1 - a)b$, and so this expression has a value lying between a and b, both of which are less then unity. Hence $(a+b-ab-1) < 0$ and $r(1-r)(a+b-ab-1) \leq 0$, given that $0 \leq r \leq 1$. We conclude that the mean transmission function for the two gases acting in concert is always greater than or equal to the product of the individual transmission functions, the equality applying only when $r = 0$ or $r = 1$, i.e. when there is no absorption or completely uniform absorption. For example, if we take $r = \frac{1}{2}$ and $a = b = \frac{1}{10}$, the mean transmission function for the two gases acting together is $\frac{101}{200}$, or just a bit over a half, whereas the individual transmission functions are $\frac{11}{20}$ each, multiplying out to $\frac{121}{400}$ or just over a quarter. When the absorption regions of the two gases coincide, the gases acting together transmit considerably more radiation than one would infer by allowing each gas to act independently in sequence. This happens because one of the gases uses up some of the frequencies that the other gas would like to absorb.

Exercise 4.5 Derive the final equality in Eq. (4.79). Sketch graphs of transmission functions vs. frequency for the two gases for *two different* cases illustrating perfectly correlated absorption regions. Evaluate the mismatch between $\bar{\mathcal{T}}$ and $\bar{\mathcal{T}}_A\bar{\mathcal{T}}_B$ for $a = b = r = \frac{1}{2}$. Allowing $a, b,$ and r to vary over all possible values, what is the greatest possible mismatch?

At the other extreme, the absorption regions of the two gases may be completely *disjoint*, so that $\mathcal{T}_A = 1$ wherever $\mathcal{T}_B = b < 1$ and $\mathcal{T}_B = 1$ wherever $\mathcal{T}_A = a < 1$. For this to be possible, we require $r + s \leq 1$. In the disjoint case,

$$
\begin{aligned}
\bar{\mathcal{T}}(\nu) &= \frac{1}{\Delta} \int_{\nu-\Delta/2}^{\nu+\Delta/2} \mathcal{T}_A(\nu')\mathcal{T}_B(\nu') d\nu' \\
&= ra + sb + (1 - (r + s)) \\
&= \bar{\mathcal{T}}_A\bar{\mathcal{T}}_B - rs(1 - a)(1 - b).
\end{aligned}
\tag{4.80}
$$

In the disjoint case, then, the transmission for the two gases acting together is always *less than* the compounded transmission of the two gases acting independently.

Exercise 4.6 Derive the final equality in Eq. (4.80). Sketch some transmission functions illustrating the disjoint case. Put in a few numerical values for $a, b, r,$ and s to show the size of the mismatch between $\bar{\mathcal{T}}$ and $\bar{\mathcal{T}}_A\bar{\mathcal{T}}_B$. What is the greatest possible mismatch in the disjoint case?

As the final limiting case, suppose that the absorption of the two gases is *uncorrelated*, so that at any given frequency the probability that $\mathcal{T}_A = a$ is r regardless of the value of $\mathcal{T}_B$

there, and the probability that $\mathcal{T}_B = b$ is s regardless of the value of $\mathcal{T}_A$ there. This situation is also known as the *random overlap* case. In this case

$$
\begin{aligned}
\bar{\mathcal{T}}(\nu) &= \frac{1}{\Delta} \int_{\nu-\Delta/2}^{\nu+\Delta/2} \mathcal{T}_A(\nu')\mathcal{T}_B(\nu')d\nu' \\
&= r(1-s)a + s(1-r)b + rsab + (1-r)(1-s) \\
&= \bar{\mathcal{T}}_A\bar{\mathcal{T}}_B.
\end{aligned}
\tag{4.81}
$$

The reasoning behind the second line is that $r(1-s)$ is the probability that only the first gas is absorbing, $s(1-r)$ is the probability that only the second gas is absorbing, rs is the probability that both gases are absorbing, and $(1-r)(1-s)$ is the probability that neither gas is absorbing. Multiplying out the terms in the product of $\mathcal{T}_A$ and $\mathcal{T}_B$ we find that in the random overlap case, the mean transmission of the two gases acting together is precisely the same as the compounded transmission of the two gases acting independently.

The properties illustrated by the three cases just discussed can be generalized to an arbitrary set of transmission functions. Let $\mathcal{T}_A$ and $\mathcal{T}_B$ be any two transmission functions, and define the fluctuation

$$
\mathcal{T}'_A = \mathcal{T}_A - \bar{\mathcal{T}}_A
\tag{4.82}
$$

and similarly for $\mathcal{T}_B$. Then,

$$
\overline{\mathcal{T}_A\mathcal{T}_B} = \bar{\mathcal{T}}_A\bar{\mathcal{T}}_B + \overline{\mathcal{T}'_A\mathcal{T}'_B}.
\tag{4.83}
$$

From this we conclude that the transmission of the two gases acting in concert is greater than the product of the individual transmissions if the two transmissions are positively correlated, less than the product of the individual transmissions if the two transmissions are negatively correlated, and equal to the product if the two transmissions are uncorrelated.

In fact, by exercising just a little more mathematical sophistication, it is possible to go further and put an upper bound on the amount by which the mean transmission function for joint action by the two gases deviates from the product of the individual transmission functions. The key is to use a handy and powerful relation known as the *Schwartz Inequality*, which states that for any two functions $f(x)$ and $g(x)$, $(\overline{fg})^2 \leq \overline{f}^2\overline{g}^2$, where an overbar indicates an average over x. The equality applies only when $f(x)$ is proportional to $g(x)$. Applying the Schwartz Inequality to Eq. (4.83), we find that the deviation satisfies the inequality

$$
\left|\overline{\mathcal{T}'_A\mathcal{T}'_B}\right| \leq \sqrt{\overline{(\mathcal{T}'_A)^2}}\sqrt{\overline{(\mathcal{T}'_B)^2}}.
\tag{4.84}
$$

In other words, in the worst case the deviation can become as large as the product of the standard deviations of the two individual transmission functions. Since $0 \leq \mathcal{T}'_{A,B} \leq 1$, the maximum standard deviation is $\frac{1}{2}$, occurring when each transmission function is zero for half of the frequencies in the band and unity for the other half; the error in random overlap in this case is $+\frac{1}{4}$ if the transmissions are perfectly correlated and $-\frac{1}{4}$ if the transmissions are perfectly anticorrelated. These errors should be compared to the random-overlap value for the limiting case, which is $\frac{1}{4}$. The effects of non-random overlap are potentially severe.

Fortunately, the situation is rarely as bad as the worst case suggests. The positions of spectral lines are a sensitive function of molecular structure, so it is a reasonable guess that the absorption spectra of dissimilar molecules should be fairly uncorrelated. Thus, in most circumstances one can get a reasonable approximation to the joint transmission function by computing the individual transmission functions for each gas and taking the product of the individual transmission functions. It is fairly easy, if computationally expensive, to test the accuracy of the random-overlap assumption in a given band by computing the correlations of the full frequency-dependent transmission functions in the band. However, finding a general characterization of the correction to random-overlap, and the way the correction depends on the concentrations of the individual gases, is an intricate art which we will not pursue here.

4.4.6 A homebrew radiation model

We have now laid out all the ingredients that go into a real gas radiation model, and are ready to begin assembling them. The ingredients are:

- A means of computing the band-averaged transmission over a specified wavenumber range;
- The band-averaged integral (Eq. (4.11), (4.12), or (4.13)) giving the band-averaged solution to the Schwarzschild equation in terms of the preceding transmission functions;

and the recipe is:

- Divide the spectrum into bands of a suitable width;
- Prepare in advance: Malkmus coefficients or exponential sum coefficients $H(\ln \kappa)$ for each band, for each greenhouse gas present in significant quantities in the atmosphere;
- Program up a function to compute the band-averaged transmission in each band, using the coefficients prepared in the previous step;
- If there are multiple greenhouse gases, do the preceding for each individual greenhouse gas and combine the resulting transmission functions, allowing suitably for the nature of the overlap between absorption bands of the competing gases (for advanced chefs only!);
- Use the resulting transmission in a numerical approximation to the integral in Eq. (4.11), (4.12), or (4.13) in each band to get the band-averaged fluxes;
- Sum up the fluxes in each band to get the total flux;
- Serve up the fluxes to the rest of the climate model and enjoy.

In the typical climate simulation application, one is given a list of values of temperature and greenhouse gas concentrations tabulated on a finite grid of pressure levels p_j for $j = 0, ..., N$, and one must compute the fluxes based on this information. Either of Eq. (4.11) or (4.12) provides a suitable basis for numerical evaluation when one is working from atmospheric data tabulated on a grid. In writing down the approximate expressions for the flux, we will adopt the convention that $j = 0$ at the top of the atmosphere and that $j = N$ represents the ground. We shall use the superscript (k) to refer to quantities averaged or integrated over the band k, centered on frequency $\nu^{(k)}$ and having width $\Delta^{(k)}$. Let's define the gridded quantities:

$$
\begin{aligned}
B_j &\equiv B\left(\nu^{(k)}, \frac{1}{2}(T_j + T_{j+1})\right) \Delta^{(k)} \\
\bar{\mathcal{T}}_{ij}^{(k)} &\equiv \bar{\mathcal{T}}^{(k)}(p_i, p_j) \\
e_{ij} &\equiv \bar{\mathcal{T}}_{ij}^{(k)} - \bar{\mathcal{T}}_{i(j+1)}^{(k)}.
\end{aligned}
\tag{4.85}
$$

The trapezoidal-rule approximation to the expression for upward flux in band (k), based on Eq. (4.11), is then simply

$$I_+^{(k)}(p_i) = I_{+,s}^{(k)} \bar{\mathbb{T}}_{iN}^{(k)} + \sum_{j=i}^{N} B_j e_{ij}. \tag{4.86}$$

The expression for the downward flux follows a similar form. B_j is the blackbody emission from layer j, and the flux at a given level is a weighted sum of the emissions from each layer below (for upward flux) or above (for downward flux) layer i. The weighting coefficient e_{ij} characterizes the joint effects of the emissivity at layer j and the absorptivity by all layers between i and j.

Exercise 4.7 Write down the analogous trapezoidal-rule approximation to I_-.

Exercise 4.8 Write down analogous trapezoidal-rule approximations to I_+ and I_- based on the form of the solution given in Eq. (4.12). What would be the advantages of using this form of solution?

To implement Eq. (4.86) and its variants as a computer algorithm, one generally writes a function which computes the transmission between levels p_i and p_j. The rest of the algorithm is independent of the form this function takes, and so one can easily switch from one representation to another (e.g. Malkmus to exponential sums) by simply switching functions. One can equally easily use different representations for different bands. The transmission function requires as arguments the transmission parameters for the band under consideration (e.g. R and S parameters for Malkmus, or the H distribution for exponential sums), as well as enough information to compute the equivalent path. For a well-mixed greenhouse gas, if we are ignoring temperature scaling effects the equivalent path is simply $q_G \frac{1}{2} \left| \left(p_1^2 - p_2^2 \right) \right| / (p_0 g \cos \bar{\theta})$, and one can simply make the concentration q_G and the pressures p_1 and p_2 arguments of the transmission function. For an inhomogeneous path, arising when q_G varies with height or one needs to take into account temperature scaling which also varies in height, the path is determined by an integral. In this case, it is inefficient to recompute the path from scratch each time. Since the equivalent path can be computed incrementally using $\ell(p_1, p_2 + \delta p) = \ell(p_1, p_2) + \ell(p_2, p_2 + \delta p)$, it is better to use the equivalent path as an argument to the transmission function, and compute the path from layer i to each layer j iteratively in the same loop in Eq. (4.86) where the weighted emission is computed.

In the preceding algorithm, we have used exponential sums to represent the transmission function appearing in the integral form of the solution to the Schwarzschild equations. However, because the equations are linear in the fluxes, and because the exponential sum method is a weighted sum of calculations for a number of different absorption coefficients, exactly the same results can be obtained by organizing the calculation in a quite different way. Namely, instead of working from the integral form of the solution, we can work directly with a set of independent Schwarzschild equations (Eq. (4.8)) – one for each κ going into the exponential sum for a given band; as usual, the band would be chosen narrow enough that $B(\nu, T)$ could be assumed independent of frequency within the band, so we wouldn't need to know anything about which set of wavenumbers each κ corresponded to. With a 10-term exponential sum, for example, we would solve the Schwarzschild equation for each of the 10 values of κ, then form a weighted sum of the 10 resulting fluxes. This alternate formulation is not available with band-averaged transmission function models such as the Malkmus model. The weighted differential-equation approach offers a number of advantages over the transmission-function approach. A single solution gives the fluxes at all levels, making optimal use of calculations for preceding levels. This is useful when computing heating rate profiles. Moreover, it is easy to use a high-order integrator to obtain high accuracy with fewer levels. These two advantages make the method computationally attractive, but there is a further advantage that is even more compelling for our purposes: it is straightforward to extend this method to incorporate scattering, whereas it is essentially impossible to do so with band-averaged transmission approaches. We will carry out this program in Chapter 5. One could well ask why the weighted differential-equation approach hasn't completely taken over the business of radiation modeling. There is some evidence that, when scattering is unimportant, pressure and temperature scaling can be done more accurately in band-averaged transmission models, but in large measure the transmission models are a holdover from an earlier day when many radiation calculations were done on paper, and when slow computers required highly tuned special approximations in order to speed up the calculation. It does seem that the exponential sum (and its close cousin the correlated-k) approach are gradually taking over. We have nonetheless chosen to introduce the transmission function approach first, because it corresponds better to what is going on in most existing radiation models, and because the form of the solution gives considerable direct insight into the factors governing the fluxes at a given level.

4.4.7 Spectroscopic properties of selected greenhouse gases

Now we will provide a survey of the infrared absorption properties of a few common greenhouse gases, with particular emphasis on the "Big Three" that determine much of climate evolution of Earth, Mars, and Venus from the distant past through the distant future: CO_2, H_2O, and CH_4. In each case, the spectral results shown are for the dominant isotopic form of the gas, e.g. $^{12}C^{16}O_2$ for the case of carbon dioxide. Other isotopic forms (*isotopologs*) can have significantly different spectra, particularly when the substitution of a heavier or lighter isotope changes the symmetry of the molecule, as in HDO or $^{12}C^{16}O^{18}O$. The reader should keep these isotopic effects in mind, since the asymmetric molecules can have strong absorption in parts of the spectrum where the symmetric molecule has essentially no absorption lines at all. In such cases, the asymmetric isotopologs can be radiatively important even if they are present only in small quantities. When such is the case, one needs to know the isotopic composition of an atmosphere before one can accurately compute the climate. This is a considerable challenge for atmospheres that cannot be directly observed.

This section will deal only with that part of absorption that can be identified as being caused by nearby spectral lines associated with energy transitions of the molecule in question; the *continuum absorption* which is not so directly associated with spectral lines will be discussed separately. Although CH_4 is a greenhouse gas on Titan, most of its contribution there comes from continuum absorption rather than its line spectrum.

Though we focus on the "Big Three" greenhouse gases, there are many other greenhouse gases of known or potential importance on Earth and other planets. These include N_2O, SF_6, chlorofluorocarbons, and hydrofluorocarbons. For that matter, ozone itself is a very potent greenhouse gas, though its greenhouse effect is typically offset by the surface cooling effect arising from its strong absorption of ultraviolet. Moreover, the reader is encouraged to keep an open mind with regard to what might be a greenhouse gas. At present, NO_2 and SO_2 do not exist in sufficient quantities on any known planet to be important as a greenhouse gas, but with different atmospheric chemistries occurring in the past or on as-yet undiscovered planets, the situation could well be different. For that matter, things like

SiO_2 that we consider rocks on Earth could be gases and clouds on "roasters" – extrasolar gas giants in near orbits – and there one ought to give some thoughts to their effect on thermal infrared.

When interpreting the absorption spectra to be presented below, it is useful to keep the Planck function in mind. Absorption is not very important where there is little flux to absorb, so the relevant part of the absorption curve varies with the temperature of the planet under consideration. For Titan at 100 K, three-quarters of the emission is at wavenumbers below 350 cm^{-1}. For Earth at 280 K, three-quarters of the emission occurs below 1000 cm^{-1}. Mars is slightly colder, and the threshold wavenumber is therefore a bit less. For Venus at 737 K, three-quarters of the emission is below 2550 cm^{-1}, but moreover the flux beyond this wavenumber amounts to over 4000 W/m^2. This near-infrared flux is vastly in excess of the mere 170 W/m^2 of absorbed solar radiation which maintains the Venusian climate. For Venus one needs to consider the absorption out to higher wavenumbers than for Earth or Mars, and given the large fluxes involved and the huge mass of CO_2 present, exquisite attention to detail and to the effect of weak lines that can normally be ignored is required. For the aforementioned roaster planets, the "thermal" emission extends practically out to the visible range.

An ideal greenhouse gas absorbs well in the thermal infrared part of the spectrum but is transparent to incoming solar radiation. The following survey concentrates on thermal infrared – the part of the spectrum that is involved in *OLR* – but it should be remembered that some of the gases discussed, such as methane, absorb quite strongly in the solar near-infrared spectrum. This can compromise their effectiveness as greenhouse gases when they reach concentrations high enough that the solar absorption becomes significant. Even the weaker solar absorption from CO_2 and H_2O can affect the thermal structure of atmospheres in significant ways. The implications of real gas solar absorption will be taken up in more detail in Chapters 5.

Carbon dioxide

We have already introduced a fair amount of information about the absorption properties of CO_2 but here we will present the data in a more systematic and general way, which will make it easier to compare CO_2 with other greenhouse gases. In a line graph of a wildly varying function like the absorption, such as shown in Fig. 4.7, one only sees the maximum and minimum defining the envelope of the absorption. There is no useful information about the relative probabilities of the values in between. To get around these problems, we divide up the spectrum into slices of a fixed width (50 cm^{-1} in the results presented through-out this section), and then compute the minimum, maximum, median, and the 25th and 75th percentiles of the log of the absorption in each band. By plotting these statistics vs. wavenumber, it is possible to present a fairly complete picture of the probability distribution of the absorption. To make the absorption data easier to interpret, we exponentiate the median, quartiles, min and max, and plot the resulting values on a logarithmic vertical axis. Plots of this sort for CO_2 at a temperature of 260 K are shown in Fig. 4.17. The left panel is computed at a pressure of 100 mb while the right panel is computed at 1000 mb. Both results are for air-broadened lines. As for the other absorption spectra we shall discuss, these results are for the dominant isotopic form of CO_2, i.e. $^{12}C^{16}O_2$. Other isotopic forms can have substantially different properties, particularly the asymmetric forms involving one heavy oxygen and one lighter oxygen.

For Earthlike and Marslike temperatures, only the lower frequency absorption region, from about 400–1100 cm^{-1}, affects the *OLR*, since there is little emission in the range of

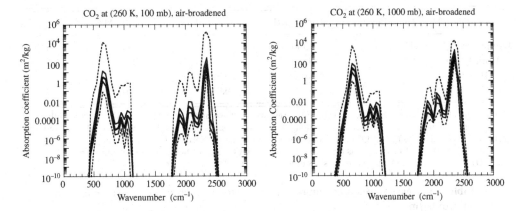

Figure 4.17 The minimum, 25th percentile, median, 75th percentile, and maximum absorption coefficients in bands of width $50 \, cm^{-1}$, computed for CO_2 in air from spectral line data in the HITRAN database. The left panel shows results at an air pressure of $100 \, mb$, whereas the right panel gives results at 1 bar. Both are calculated at a temperature of $260 \, K$.

the higher frequency band. For Earth, the higher frequency band can contribute to solar absorption, since the solar radiation reaching the Earth in that part of the spectrum amounts to $7.2 \, W/m^2$ (or about $3 \, W/m^2$ at the orbit of Mars). However, most of the absorption of solar near-infrared, whether by water vapor or CO_2, occurs at higher frequencies. Venus, however, has significant surface emission in the range of the higher frequency group, which therefore has a significant effect on that planet's *OLR*.

The absorption spectrum depends on temperature and pressure, and on whether the lines are broadened by collision with air or by collisions with molecules of the same type (e.g. CO_2 with CO_2 in the present case); rather than present page upon page of graphs, the strategy is to show the results for a single standard temperature and pressure, and make use of appropriate temperature- and pressure-weighted equivalent paths to apply the standard absorption to a wider variety of atmospheric circumstances. We will adopt $T = 260 \, K$, $p = 100 \, mb$ and air-broadening as our standard conditions when discussing other greenhouse gases in this section. The comparison of the two panels of Fig. 4.17 illustrates the nature of the pressure scaling. Increasing the pressure by a factor of 10 shifts the minimum, median, and quartiles upward by a similar factor. The main exception to the scaling is the maximum absorption in each band, which is reduced and becomes more tightly clustered to the median. This is the expected behavior, since increased pressure reduces the absorption near the centers of absorption lines. Since this part of the spectrum is rare, and absorbs essentially everything hitting it anyway, the errors in pressure scaling near the line centers are of little consequence. A bit of numerical experimentation indicates that the linear pressure scaling works quite well for pressures below 1 or 2 bars. At higher pressures (certainly by 10 bar) the lines overlap to such an extent that the whole probability distribution collapses on the median, in which case the absorption can be described as a smooth function of the wavenumber, and loses its statistical character. The other gases to be discussed have similar behavior under pressure, and so we will not repeat the description separately for each gas.

To understand a planet's radiation balance, the main thing we need to understand is where in the spectrum the atmosphere is optically thick, and where it is optically thin.

	Mod. Earth	Early Earth	Mod. Mars	Early Mars	Venus
Whole atm.	47	15 000	30	10^6	10^9
Bottom 10%	9	2900	5	$2 \cdot 10^5$	$2 \cdot 10^8$
Top 10%	0.5	160	0.3	10^4	10^7
Top 1%	0.005	1.5	0.003	100	10^5

The weighted path is based on a 100 mb reference pressure, so these paths are intended to be used with the absorption coefficients in the left hand panel of Fig. 4.17. A mean slant path $\cos \bar{\theta} = \frac{1}{2}$ is assumed. The modern Earth case is based on 300 ppmv CO_2 in a 1 bar atmosphere, while the Early Earth case assumes 10% (molar) CO_2 in a 1 bar atmosphere. Modern Mars is based on a 10 mb pure CO_2 atmosphere, while Early Mars assumes a 2 bar pure CO_2 atmosphere. The Venus case consists of a 90 bar pure CO_2 atmosphere.

Table 4.1 Pressure-weighted equivalent paths for various planetary situations, in units of kg/m^2.

Pressure exerts by far the strongest modifying influence on the standard-state absorption coefficients given in Fig. 4.17. Hence, we will first discuss the effect of CO_2 on the optical thickness of several planetary atmospheres in terms of the 260 K air-broadened coefficients using pressure-adjusted paths. Later we will make a few remarks about how self-broadening and temperature affect the results.

The strong line pressure-weighted equivalent path for a well-mixed greenhouse gas with specific concentration q is $\ell = \frac{1}{2} q \left(p_1^2 - p_2^2 \right) \big/ (p_0 g \cos \bar{\theta})$. If we take the reference pressure $p_0 = 100$ mb, the equivalent paths can be multiplied by the absorption coefficients in the left panel of Fig. 4.17 to obtain optical thickness. Some typical values of the equivalent path are given in Table 4.1. Although Mars at present has much more CO_2 per square meter in its atmosphere than modern Earth, the equivalent paths are quite similar in the two cases because the total pressure on Mars is so much lower. For Earth and Mars temperatures only the lower frequency absorption group is of interest, and within this group it is only the dominant spike in absorption near 675 cm^{-1} that contributes significantly to the absorption. Both atmospheres have optical thicknesses exceeding unity within the wavenumber band from roughly 610 to 750 cm^{-1} and are optically thin outside this band. The weak absorption shoulder occurring to the high-wavenumber side of 870 cm^{-1} has little effect for modern Earth or Mars. For any given wavenumber range, the 10% of the atmosphere nearest the ground (i.e. the lowest 100 mb on Earth) is one-fifth as optically thick as the atmosphere as a whole, meaning that a substantial part of the back-radiation of infrared to the surface attributable to CO_2 comes from the lowest part of the atmosphere. In contrast, because of low pressure aloft, the uppermost 10% of the atmosphere, which may loosely be thought of as the stratosphere, is comparatively optically thin. If CO_2 were the only factor in play, the greenhouse effect of the present Earth and Mars atmospheres would be quite similar. The two planets are rendered qualitatively different primarily by the greater role of water vapor in the warmer atmosphere of modern Earth, and by differences in vertical temperature structure between the two atmospheres (arising largely from differences in solar absorption).

On Early Earth – either during the high CO_2 phase after a long-lived Snowball, or during the Faint Young Sun period – the CO_2 may have been 10% or more of the atmosphere by molar concentration. In this case, the equivalent paths are vastly greater than at present. The Earth atmosphere in this case is optically thick from about 520 to 830 cm^{-1}, and the higher wavenumber shoulder just barely begins to be significant. It becomes rapidly more so as the CO_2 molar concentration is increased beyond 10%. This is good to keep in mind when doing radiation calculations at very high CO_2, since many approximate radiation codes designed for the modern Earth only incorporate the effect of the principal absorption feature centered on $675\,cm^{-1}$. Even with such high CO_2 concentrations, the atmosphere is optically thick in only a limited portion of the spectrum, so that the net greenhouse effect will be modest unless other greenhouse gases come into play. This remark is particularly germane to Snowball Earth, where the cold temperatures allow little water vapor in the atmosphere. Without help from water vapor, CO_2 has only limited power to warm-up a Snowball Earth to the point of deglaciation.

On a hypothetical Early Mars with a 2 bar pure CO_2 atmosphere, the equivalent path is orders of magnitude greater than the Early Earth case. This renders the atmosphere nearly opaque within the wavenumber range 500–$1100\,cm^{-1}$. However, a great deal of *OLR* can still escape through the CO_2 window regions, so continuum absorption, water vapor, and other greenhouse gases will play a key role in deciding whether the gaseous greenhouse effect can explain a warm, wet Early Mars climate.

Since the equivalent path assumes absorption increases linearly with pressure, the equivalent paths given for the bottom 10% and for the entirety of the Venus atmosphere yield overestimates of the true optical depths, given that the increase of absorption with pressure weakens substantially above 10 bars. Even assuming that the equivalent paths are overestimated by a factor of 10, the implied optical depths for the lowest 9 bars of the Venus atmosphere, and for the entire Venus atmosphere, remain so huge that these layers can be considered essentially completely opaque to infrared outside the CO_2 window regions. In fact even the top 1 bar (about 1%) of the Venus atmosphere has an optical thickness greater than unity throughout the 500–$1100\,cm^{-1}$ and 1800–$2500\,cm^{-1}$ spectral regions. Within these regions, essentially all the *OLR* comes from the relatively cold, top 1 bar of the atmosphere.

The window region, where CO_2 has no absorption lines, presents a challenge to the explanation of the surface temperature of Venus, particularly since the peak of the Planck function for the observed Venusian surface temperature lies in this region. In fact, if the 1200–$1700\,cm^{-1}$ band were really completely transparent to infrared, then emission through this region alone would reduce the Venus surface temperature to a mere $355\,K$ (assuming a globally averaged absorbed solar radiation of $163\,W/m^2$). Some energy also leaks out through the low-frequency window region, which would reduce the temperature even further. We must plug the hole in the spectrum if we are to explain the high surface temperature of Venus. The high-altitude sulfuric acid clouds of Venus play some role, by reflecting infrared back to the surface; there may also be some influence of the less common isotopologs of CO_2, since asymmetric versions of the molecule (e.g. made with one ^{16}O and one ^{18}O) can have lines in the window region. The principal factor at play is the *continuum absorption* to be discussed in Section 4.4.8. Figure 4.17 was computed taking into account only the relatively nearby contributions of each absorption line (within roughly 1000 line widths), whereas collisions can allow some absorption to occur much farther from the line centers. This far-tail absorption spills over into the window region, particularly at high pressure. It cannot be reliably described in terms of the Lorentz line shape, and therefore requires separate consideration.

The difficulties posed by Venus do not end with the window region. For planets having Earthlike temperatures, the thermal radiation beyond $2300\,cm^{-1}$ is insignificant, so we need not be too concerned about the absorption properties there. However, for a planet with a surface temperature of $700\,K$, the ground emission on the shortwave side of $2300\,cm^{-1}$ is $3853\,W/m^2$, compared with under $200\,W/m^2$ of absorbed solar radiation for Venus. In these circumstances it simply won't do to assume the atmosphere transparent at high wavenumbers. The HITRAN spectral database is a monumental accomplishment, but it is still rather Earth-centric, and lacks the weak lines needed to deal with high-temperature atmospheres; there may also be continuum absorption in the shortwave region, especially at high pressures. In order to deal with the high-wavenumber part of the Venus problem, one needs to employ specialized high-temperature CO_2 databases, which are less verified and more in a state of flux than HITRAN. Among specialized databases in use for Venus, the HITEMP database (described in the supplementary reading at the end of this Chapter) is particularly convenient, because it at least uses the same data format as HITRAN. It will be left to the reader to explore the use of this database. Suffice it to say that there appears to be sufficient absorption in the high-wavenumber region to raise the radiating level to altitudes where the temperature is low enough that one may not need to consider shortwave emission in computing *OLR*.

The preceding discussion was based on air-broadened absorption at 260 K, whereas self-broadened data would be more appropriate to the pure-CO_2 atmospheres, and in all cases one must think about whether the increase of line strength with temperature substantially alters the picture presented. A re-calculation of Fig. 4.17 for self-broadened pure CO_2 indicates that the self-broadened absorption is generally about 30% stronger than the air-broadened case, though there are a few bands in which the enhancement is as little as 13%. This is quantitatively significant, but the enhancement factor is too small to alter the general picture presented above. The temperature dependence can have a more consequential effect. In the strong line approximation, valid away from line centers, the temperature affects the absorption only in the form of the product of line strength and line width, $S(T)\gamma(p, T)$, which yields a temperature dependence of the form $\kappa_{CO2} \sim \exp(- T^*/T)$ for some coefficient T^*. The coefficient differs from line to line, but we can still attempt to fit this form to the computed temperature dependence of the median absorption within each 50 cm^{-1} band. The result is shown in Fig. 4.18. This still only gives an incomplete picture of the effect of temperature on absorption, since the other quartiles may have different scaling coefficients. Within an exponential-sum framework, however, one can do little else than pick

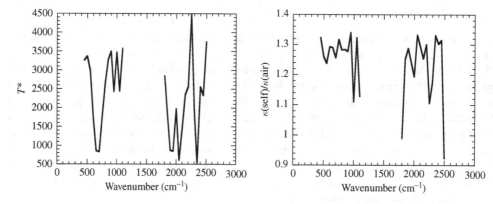

Figure 4.18 Left panel: Temperature dependence coefficient for air-broadened CO_2 at 100 mb. The coefficient is computed for the median absorption in bands of width 50 cm^{-1}, as shown in Fig. 4.17. The median absorption within each band increases with temperature in proportion to $\exp(- T^*/T)$. Right panel: The ratio of self-broadened to air-broadened median absorption coefficients.

a base temperature most appropriate to the planet under consideration, compute the probability distribution for that case, and then assume that each absorption coefficient in a band scales with the same function of temperature. This procedure in any event gives an estimate of the magnitude of the temperature effect. From the figure, it is seen that the temperature dependence varies greatly with wavenumber. It is low near the principal absorption spike at $675\,\mathrm{cm}^{-1}$, with T^* as low as $1000\,\mathrm{K}$. This value increases the absorption by a factor of 1.7 going from $260\,\mathrm{K}$ to $300\,\mathrm{K}$, and decreases absorption by a factor of 2 going from $260\,\mathrm{K}$ to $220\,\mathrm{K}$. Where $T^* = 3000\,\mathrm{K}$, the corresponding ratios are 4.7 and 0.14. Therefore CO_2 is significantly more optically thick than our previous estimates in the warmer lower reaches of the present Earth atmosphere, and significantly less optically thick in the cold tropopause regions. For the $737\,\mathrm{K}$ surface temperature of Venus, the absorption is enhanced by a factor of 12 when $T^* = 1000\,\mathrm{K}$ but by a substantial factor of 1750 when $T^* = 3000\,\mathrm{K}$. This just makes an already enormously optically thick part of the atmosphere even thicker. Outside the window regions, most of the atmosphere of Venus is in the optically thick limit where very slow radiative diffusion transfers heat; infrared radiative cooling in the deeper parts of the Venus atmosphere is determined almost exclusively by what is going on in the window regions.

Water vapor

Figure 4.19 shows the standard-state absorption spectrum for water vapor. Unlike CO_2, the H_2O molecule, which has more complex geometry, has lines throughout the spectrum, so there is no completely transparent window region. Water vapor nonetheless has two window regions where the absorption is very weak; it will turn out that continuum absorption from far tails excluded from the computation in the figure substantially increases the absorption in these window regions (Section 4.4.8). The peak absorption coefficient for water vapor has a similar magnitude to that for CO_2, but water vapor absorbs well over a far broader

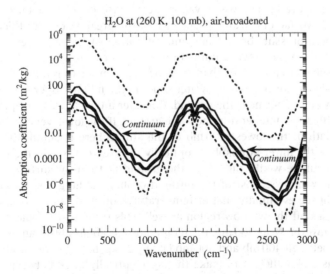

Figure 4.19 As in Fig. 4.17, but for H_2O. The continuum regions marked on the figure are regions where the measured absorption substantially exceeds that computed from the spectral lines.

portion of the spectrum than CO_2. In particular, the H_2O absorption has a peak within both the $1000\,cm^{-1}$ and longwave window regions of CO_2. This critically affects the greenhouse warming on Earth and on Early Mars, but it plays less of a role on present-day Venus, which has little water vapor in its atmosphere. It should not be concluded that water vapor overwhelms the greenhouse effect of CO_2, however. It would be more precise to say that the water vapor greenhouse effect complements that of CO_2. Carbon dioxide absorbs strongly near the peak of the Planck function for Earthlike temperatures, but the water vapor absorption is nearly two orders of magnitude weaker there. Further, for Earthlike planets, water vapor condenses and therefore disappears in colder regions of the planet; it is only the long-lived CO_2 greenhouse effect that can persist in cold parts of the atmosphere.

On a planet without a substantial condensed water reservoir, water vapor could be a well-mixed non-condensing greenhouse gas much like CO_2 on modern or early Earth. In most known cases of interest, though, atmospheric water vapor is in equilibrium with a reservoir – an ocean or glacier – which fills the atmosphere to the point that the atmospheric water vapor content is limited by the saturation vapor pressure. The prime cases of interest are water vapor feedback on Earth and on Early Mars, and the runaway greenhouse on Early Venus. The runaway greenhouse is also relevant to the ultimate far-future fate of the Earth, and the evolution of hypothetical water-rich extrasolar planets. The status of water vapor as a greenhouse gas whose concentration is limited, via condensation, by temperature does not derive from any special properties of water. One tends to focus on a role of this sort for water simply because planets that are "habitable" to the only example of life with which we are presently familiar seem to require a planet with liquid water and an operating temperature range in which the saturation vapor pressure is high enough for water vapor to be present in sufficient quantities to be active as a greenhouse gas. On present and past Mars, as well on a hypothetical early Snowball Earth, CO_2 can be limited by condensation, and on Titan today CH_4 condenses, while NH_3 and other gases have condensation layers on the gas giant planets.

First, let's think about the effect of water on OLR, supposing that the atmosphere is saturated at each altitude. The water vapor greenhouse effect is determined by a competition between two factors. Water vapor causes the greatest optical thickness near the ground, where both pressure-broadening and saturation vapor pressure are highest. However, a strong greenhouse effect requires optical thickness at higher altitudes, where the temperature is substantially colder than the surface, in order to reduce the radiating temperature of the planet. For this reason, water vapor in the mid to upper troposphere is more important than water vapor near the ground. Consider a typical Earth tropical case, on the moist adiabat with 300 K temperature near the ground. If the low-level air is saturated, then the equivalent path of the lowest 100 mb of the atmosphere is about $400\,kg/m^2$. This is sufficient to make the lower atmosphere optically thick except within the window regions, so that the atmosphere would radiate to the ground at the near-surface air temperature except within the windows. Most of the infrared cooling of the ground occurs in the windows, but we will see eventually that at temperatures of 300 K and above, the continuum substantially closes off the window region as well. This near-surface opacity doesn't much affect the OLR, however. Near 400 mb, the temperature is about 260 K, and in saturation the equivalent path between 400 mb and 500 mb is only $25\,kg/m^2$. This relatively small amount of water vapor is still sufficient to make the layer optically thick between $1350\,cm^{-1}$ and $1900\,cm^{-1}$, as well as on the low-frequency side of $450\,cm^{-1}$. This substantially reduces the OLR. At the tropopause level, in contrast, the temperature is about 200 K, the pressure is 100 mb, and the equivalent path from 100 mb to 200 mb is only $0.01\,kg/m^2$. At such low concentrations, water vapor is optically thin practically throughout the spectrum. At lower surface temperatures, the dominant greenhouse effect of water vapor comes from correspondingly lower altitudes. With a 273 K surface temperature, the 400 mb temperature on the dry air adiabat is only 210 K, and the water vapor opacity is inconsequential there. One has to go down to about 600–700 mb, where the path is about 2 kg/m^2, to get a significant water vapor greenhouse effect. On a low CO_2 Snowball Earth soon after freeze-up, where the tropical surface temperature is under 250 K, the water vapor greenhouse effect is essentially negligible. It only starts to play a role when the planet has warmed to the point that the tropical temperatures approach the melting point.

As was the case for CO_2, the absorption coefficient for water vapor decays roughly exponentially with distance in wavenumber space from each peak of absorption. This has similar consequences for *OLR* as were discussed previously in connection with CO_2. Because of the exponential envelope of absorption, doubling or halving the water vapor content of a layer of the atmosphere has approximately the same effect on the optical thickness of that layer regardless of whether the base amount being doubled or halved is very large (say 200 kg/m^2) or very small (say 2 kg/m^2). There may not be much water to work with in the Earth's mid-troposphere, but nonetheless halving or doubling the amount would have a significant effect on *OLR*. This remark is particularly significant because there are dynamical effects which in fact keep the Earth's mid-troposphere substantially undersaturated. Although there are regions with relative humidity as low as 10%, it would still significantly increase the *OLR* if the relative humidity were reduced still more to 5%, and conversely it would significantly decrease the *OLR* if the relative humidity were increased to 20%. Because of the spectral position of the absorption peaks relative to the shape of the Planck curve, the effect of water vapor concentration on *OLR* is not as precisely logarithmic as is the case for CO_2. Nonetheless, it is fair to say that the change in water vapor content *relative to the amount initially present* gives a truer idea of the radiative impact of the change than would the change in the absolute number of kilograms of water present in a layer.

With regard to water vapor, then, it is clear that subsaturation is important. However, the determination of the degree of subsaturation involves intrinsically fluid dynamical processes, and we will not have much to say about this important issue in this book. Similar considerations would apply to any radiatively active condensable substance in a planetary atmosphere.

Figure 4.20 shows the temperature dependence and the ratio of self- to air-broadened absorption for water vapor. The general range of temperature sensitivity is much the same as it was for CO_2. However, whereas self-broadened absorption for CO_2 is only a few tens of percent stronger than air-broadened absorption, the self-broadened H_2O absorption is fully five to seven times as strong as the air-broadened case. This is extremely important to the runaway greenhouse, which involves portions of the atmosphere which consist largely of water vapor. Moreover, when a species has a molar concentration in air of, say, 10% or less one wouldn't ordinarily have to worry much about self-broadening, since collisions with air are so much more common than self-collisions. However, because of the great amplification of self-broadened absorption for water vapor, the self-broadening in fact starts to become dominant even at molar concentrations of around 10%. At the Earth's surface, this concentration is achieved at a temperature of 320 K. With a dry air partial pressure of 100 mb, this concentration would be achieved at temperatures near 280 K; this situation is relevant to a hot planet on the verge of a runaway greenhouse. The strong enhancement of self-broadened absorption relative to air-broadened absorption is far in excess of what would be anticipated from simple effects associated with the different molecular weights of the

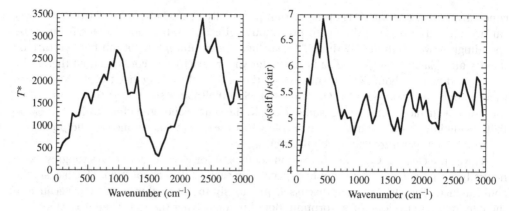

Figure 4.20 Left panel: Temperature dependence coefficient for air-broadened H_2O at 100 mb, as defined in Fig. 4.18. Right panel: Ratio of self-broadened to air-broadened median absorption for H_2O at 100 mb and 260 K. As usual, medians are computed in bands of width 50 cm^{-1}.

colliders. The sensitivity to the nature of the colliding molecule raises interesting questions about the effect of collisions with other molecules. Would CO_2 broadened coefficients be more like the air-broadened or self-broadened case? The answer to this question has some impact on the climates of Early Mars and Early Earth, which are often assumed to have had substantial amounts of CO_2 in their atmospheres. Unfortunately, laboratory measurements bearing on the subject are hard to come by.

Next we turn our attention to aspects of the absorption which govern the runaway greenhouse effect for a planet with a water-saturated atmosphere. Specifically, we revisit the question of the Kombayashi–Ingersoll limit, which is the limiting *OLR* such a planet can have in the limit of large surface temperature. As discussed for the gray gas case in Section 4.3.3, the limiting *OLR* is approximately determined by the temperature at the pressure level where the optical thickness between that level and the top of the atmosphere becomes unity. For a gray gas this characteristic pressure was independent of wavenumber, whereas for a real gas it is quite strongly wavenumber-dependent. To keep things simple, we will consider a saturated pure water vapor atmosphere. In this case, if the temperature at a given altitude is T, the corresponding pressure is $p_{sat}(T)$, determined by Clausius–Clapeyron. The equivalent path from this altitude to zero pressure is $a_{self}\frac{1}{2}p_{sat}(T)^2/(p_0 g \cos\bar{\theta})$, where p_0 is the standard reference pressure (100 mb for use with Fig. 4.19) and a_{self} is the ratio of self-broadened to air-broadened absorption (about 6). By using this path together with the absorption coefficients in Fig. 4.19, we can estimate the maximum effective radiating temperature as a function of wavenumber. Based on the median absorption in each band, the radiating temperature varies from about 245 K at 100 cm^{-1} to 278 K at 500 cm^{-1} to 350 K at the valley of the window region near 1000 cm^{-1}. The high values of radiating temperature in the window region lead to large estimates of the Kombayashi–Ingersoll limit. As a crude estimate, if we assume that the planet radiates at 350 K in the window regions between 544 and 1314 cm^{-1}, and on the high-wavenumber side of 1950 cm^{-1}, then the *OLR* would be about 520 W/m^2 even if the planet did not radiate at all in the rest of the spectrum. The Kombayashi–Ingersoll limit is strongly affected by the absorption in the window regions, and we will see in Section 4.4.8 that the absorption here is dominated by a continuum which is not captured by the nearby line contribution. The estimate of the Kombayashi–Ingersoll limit for H_2O (and CO_2) will be completed in Section 4.6, after we have discussed the continuum.

Methane

Figure 4.21 shows the standard-state absorption spectrum for methane. From the standpoint of *OLR* on modern and Early Earth, and perhaps on Early Mars, the most important absorption feature is that near $1300\,cm^{-1}$, which occurs in a part of the spectrum where water vapor and CO_2 absorption are weak, but where the Planck function still has significant amplitude at Earthlike temperatures. In contrast with water vapor, the very longwave absorption (below $1000\,cm^{-1}$) is so weak that it does not significantly affect *OLR* in any atmosphere likely to have existed on an Earthlike or Marslike planet. Titan has extremely large amounts of methane in its atmosphere, which could in principle make the very longwave group important. Even there, however, the weakening of the absorption due to the very cold temperatures makes this absorption group fairly insignificant. For atmospheres containing appreciable amounts of oxygen, methane oxidizes rather quickly to CO_2, so it is hard to build up very large concentrations. The Earth's pre-industrial climate had about 1 ppmv of methane in it, and the intensive agriculture of the past century may eventually come close to doubling this concentration. The associated equivalent paths are quite small – on the order of $0.06\,kg/m^2$. For paths this small, Earth's atmosphere has an optical thickness of only 0.14 based on the median absorption coefficient occurring at the $1300\,cm^{-1}$ peak. For methane paths typical of oxygenated atmospheres, one gets significant absorption only from the upper quartile of absorption coefficients, and a short distance from the dominant peak one only gets significant absorption very near the line centers. In this case, the ditch in *OLR* dug by methane is a very narrow feature centered on $1300\,cm^{-1}$. In an anoxic atmosphere, as Earth's is likely to have been earlier than about 2.7 billion years ago, the rate of methane destruction is much lower, and it is believed that production of methane

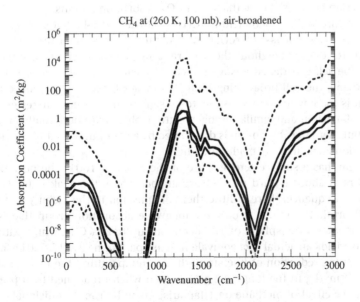

Figure 4.21 As in Fig. 4.17, but for CH_4.

by methanogenic (methane-producing) bacteria could have driven methane concentrations to quite large values. With 100 ppmv of methane in an Earthlike atmosphere, the equivalent path is about $0.6 \, kg/m^2$, and based on the median absorption coefficient the atmosphere becomes optically thick from about 1200 to $1400 \, cm^{-1}$. If the methane concentration builds up to 1% of the atmosphere, then the equivalent path is nearly $600 \, kg/m^2$, and the atmosphere becomes optically thick from 1150 to $1750 \, cm^{-1}$; for such high concentrations, the shoulder to the right of the absorption peak starts to become important. There are also some speculations that abiotic processes could have led to high methane concentrations on Early Mars, or on the prebiotic Earth.

Beyond what is shown in Fig. 4.21, methane has strong absorption bands that extend well into the solar near-IR. These are not terribly important at concentrations up to a few hundred ppmv in a 1 bar atmosphere, but when methane makes up a percent or so of the atmosphere, it can absorb most of the incident solar energy between 2500 and $9000 \, cm^{-1}$. At higher concentrations, significant absorption can extend even into the visible range.

It is often said that, molecule for molecule, CH_4 is a better greenhouse gas than CO_2. However, this is more a reflection of the relative abundances of CH_4 and CO_2 in the present Earth atmosphere than it is a statement about any intrinsic property of the gases; in fact, the absorption coefficients for the two gases are quite similar in magnitude, and CH_4 absorbs in a part of the spectrum that is less well placed to intercept outgoing terrestrial radiation than is the case for CO_2. The high effectiveness of CH_4 relative to CO_2 in the present atmosphere of Earth stems from the fact that currently there is rather a lot of CO_2 in the air (380 ppmv and rising) but rather little CH_4 (1.7 ppmv and also rising). In a situation like this, one has already depleted infrared of those frequencies that are most strongly absorbed by CO_2, so when adding CO_2 one is adding "new absorption" in spectral regions where the absorption is relatively weak. Hence, it takes a large amount of the gas to have much radiative effect. In contrast, when starting with a small amount of CH_4, when one adds more, one adds "new absorption" where the absorption coefficient is quite strong, since the strongly absorbing part of the spectrum is not yet depleted. This behavior depends crucially on the lack of significant overlap between the methane and CO_2 absorption regions.

The temperature scaling coefficient for CH_4 is shown in Fig. 4.22. It is in the same general range as for the cases discussed previously. The ratio of self- to air-broadening for CH_4 is similar to that for CO_2, throughout the spectral range of most importance for *OLR*. In the solar near-IR the self-broadened absorption can be nearly twice the air-broadened absorption. For methane, the self-broadening is mostly of academic interest, since the methane concentration is too low for self-collisions to be significant in atmospheres encountered or envisioned so far. Titan and similar cryogenic atmospheres are potentially an exception to this remark, but there the absorption is dominated by a continuum that is not clearly related to the absorption lines under consideration here.

Because Titan has a surface temperature of the order of 100 K, the peak of the Planck function occurs at about a third of the wavenumber where the peak is for Earthlike temperatures. In consequence, there is little thermal emission in the vicinity of the dominant $1300 \, cm^{-1}$ absorption group. It is only the longwave absorption group that is potentially of interest. The lower atmosphere of Titan contains up to 20% CH_4, which with Titan's low surface gravity yields an actual (*not* equivalent) mass path of nearly $15\,000 \, kg/m^2$. However, owing to the strong reduction of line strength with temperature, the median absorption is well under $10^{-6} \, m^2/kg$ in the longwave group, even when broadened by a pressure of 1.5 bar. The radiative effect of methane on Titan arises mainly from a collisional continuum of the sort described in Section 4.4.8.

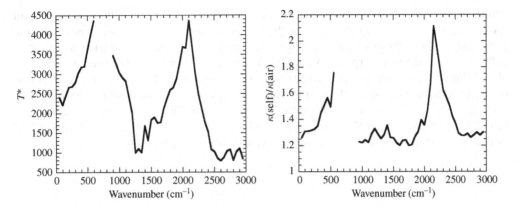

Figure 4.22 Left panel: Temperature dependence coefficient for air-broadened CH_4 at 100 mb, as defined in Fig. 4.18. Right panel: Ratio of self-broadened to air-broadened median absorption for CH_4 at 100 mb and 260 K. As usual, medians are computed in bands of width 50 cm^{-1}.

4.4.8 Collisional continuum absorption

Diatomic molecules and general considerations

A nitrogen molecule N_2 in isolation does not interact to any significant extent with infrared light; one might think that collisions do not change this picture, as N_2 has no lines to be broadened by collisions. Nonetheless, *during the time a collision is taking place* the pair of colliding molecules momentarily behaves somewhat like a more complex four-atom molecule, which has transitions that can indeed absorb and emit infrared radiation. This leads to *collision-induced absorption*, whose associated absorption coefficient is generally a smooth function of wavenumber. Because of the lack of line structure, such absorption is referred to as a *continuum*. There are many possible processes through which collisions can induce absorption. The collision can impart a temporary dipole moment to a rotation or vibration that ordinarily had none, allowing it to absorb or emit a photon. The collision can break a symmetry, allowing transitions that are otherwise "forbidden" by symmetry principles. Colliding molecules can form *dimers*, which are short-lived complexes which nevertheless persist long enough to have radiatively active transitions not present in the colliding molecules. Most of the "non-absorbing" diatomic molecules, including N_2 and H_2, exhibit significant collision-induced continuum absorption at sufficiently high densities. These are mostly associated with the induced-dipole mechanism, and therefore can to some extent be anticipated on the basis of the underlying transitions of the diatomic molecule.

Collision-induced absorption can be thought of as a trinary chemical reaction involving the two colliding molecules and a photon. The rate of "reaction" (i.e. absorption) is proportional to the product of the concentrations of the two colliding species with the photon concentration, the latter being proportional to the radiation flux. The absorption coefficient is the rate constant for the reaction. Unlike the case of collision-*broadened* line absorption, in collision-*induced* absorption there is no physical distinction between the "absorbing" molecule and the "perturbing" molecule. Both are equal partners in the process allowing absorption or emission of a photon. For this reason, it is most natural to describe collision-induced absorption in terms of a binary absorption coefficient, which expresses the proportionality between the product of the concentrations of the two colliding species and the rate of absorption of radiation. Nonetheless, in order to facilitate comparison with the previously defined line absorption coefficients, and in order to make it easier to incorporate collisional continuum absorption in radiation calculations which also take into account line absorption, it is convenient to characterize the collision-induced absorption by mass-specific absorption coefficients in which one of the colliding molecules is arbitrarily designated the "absorber," whose absorption is enhanced in proportion to the partial pressure of the "collider." For example, for an N_2–H_2 collision in a box of gas with uniform temperature T and uniform N_2 partial pressure p, the optical thickness can be expressed as

$$\tau = \frac{p_{N2}}{p_0} \kappa_{H2}(\nu, p_0, T) \ell_{H2} \tag{4.87}$$

where ℓ_{H2} is the mass path of hydrogen in the box, in kg/m^2, and p_0 is a standard pressure. The coefficient κ_{H2} has dimension m^2/kg and can be used in precisely the same way as the absorption coefficients we defined earlier for use with the line spectrum. To relate this to the binary absorption coefficient commonly defined in the spectroscopy literature, let z be the length of the path in meters, so that $\ell_{H2} = z \cdot p_{H2}/(R_{H2}T)$. Then

$$\tau = \frac{\kappa_{\nu,p_0,H2}}{p_0 R_{H2} T} p_{N2} p_{H2} z \equiv \kappa_{H2-N2}(\nu, T) p_{N2} p_{H2} z \tag{4.88}$$

which defines the binary coefficient in the case where the collider amounts are specified as partial pressures. It is more common to specify the collider amounts in terms of densities or molar densities, but the alternate forms can be readily derived from the preceding by use of the ideal gas law.

Continuum absorption may be difficult to understand from *a priori* physical principles, and difficult to measure accurately in the laboratory, but by definition the absorption coefficient for the continuum is a smoothly varying function of wavenumber. Therefore, it is relatively easy to incorporate into radiative transfer models. One only needs to determine the absorption coefficients and their pressure and temperature scaling in a set of relatively broad bands, and multiply the transmission computed from the line absorption (if any) by the corresponding exponential decay factor.

The continuum arising from diatomic molecule collisions becomes particularly important for dense, cold, massive atmospheres, of which Titan's is probably the best studied example. Figure 4.23 shows the $N_2 - N_2$ and $N_2 - H_2$ collision-induced absorption coefficients in the temperature range prevailing in Titan's atmosphere. These coefficients are based on laboratory measurements made at somewhat higher temperatures, extrapolated to colder values using a theoretical model with a few empirical coefficients fit to the data. (See the paper by Courtin listed in the Further Reading section of this chapter.) The equivalent path for Titan based on N_2 partial pressure is about $10^6 \, kg/m^2$, which yields a peak optical thickness of 40 for temperatures near those prevailing at Titan's surface. The H_2 content of Titan's atmosphere is less well constrained, but plausible estimates suggest that this gas, too, can contribute significantly to the infrared opacity of Titan's atmosphere. Note that the absorption decreases sharply with increasing temperature; this is partly due to the decrease in density with temperature, but is also affected by the shorter duration of high-velocity collisions, which apparently are less effective at inducing a dipole moment. The N_2-N_2 continuum is unimportant for Earthlike collisions because of the higher temperatures on Earth, and because Earth's atmosphere is much less massive than Titan's, per unit surface area.

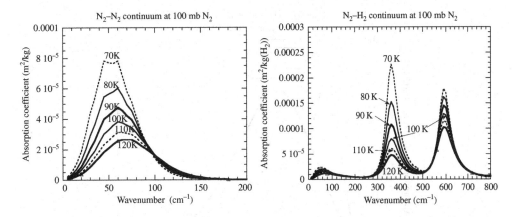

Figure 4.23 The N_2-N_2 and N_2-H_2 collision-induced continuum absorption coefficients as a function of temperature (indicated on curves). The coefficients are given at an N_2 partial pressure of 100 mb.

Hydrogen also has a significant self-broadened continuum, which provides a great deal of the infrared opacity on the gas giant planets.

Methane continuum on Titan

Methane also significantly affects the infrared opacity of Titan's atmosphere, though its effects on *OLR* are rather less than N_2 or H_2 since methane is concentrated in the warmer, lower layers of the atmosphere. Essentially all of the infrared absorption due to methane on Titan comes from a collision-induced continuum. The N_2-induced and self-induced absorption coefficients at 100 K are shown in Fig. 4.24. Like the other continuum coefficients, these too become weaker with increasing temperature. Assuming methane to be in saturation with temperature given by the methane–nitrogen moist adiabat in Titan conditions, the pressure-weighted path for self-induced absorption is in excess of 40 000 kg/m^2, while the equivalent path for N_2-induced absorption is over 170 000 kg/m^2. The self-absorption yields a peak optical thickness of about 5, while the foreign absorption gives a peak optical thickness of about 20, dropping to 10 in the higher-frequency shoulder near 200 cm^{-1}.

Carbon dioxide continuum

Given the importance of the CO_2 window regions to the high-CO_2 climates of Venus and Early Mars, it is rather surprising that the CO_2 continuum has been so little studied. The coefficients in use in most models stem from limited experiments and there is little agreement on the theoretical basis for this continuum or its temperature scaling. At the time of writing, it appears that the subject has not been re-examined in laboratory experiments since the late 1970s. The discussion below is based on absorption coefficients reported in the literature cited in the Further Reading section of this chapter.

The measured CO_2 continuum absorption, rescaled to 100 mb, is shown in Fig. 4.25. The values shown are for collisions of CO_2 in air; the self-induced continuum absorption is generally assumed to be 1.3 times that of the foreign-induced continuum. Referring to the equivalent paths in Table 4.1, we see that the continuum absorption is large enough to make the top 1 bar of the atmosphere of Venus optically thick throughout most of the

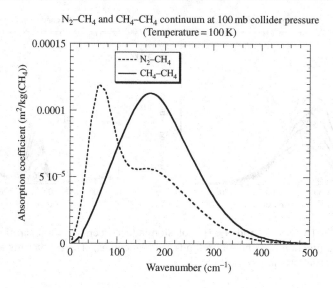

Figure 4.24 The N_2–CH_4 and CH_4–CH_4 collision-induced continuum absorption coefficients at 100 K, assuming a collider partial pressure of 100 mb. For N_2–N_2 the "collider" is N_2, while for CH_4–CH_4 the "collider" is CH_4.

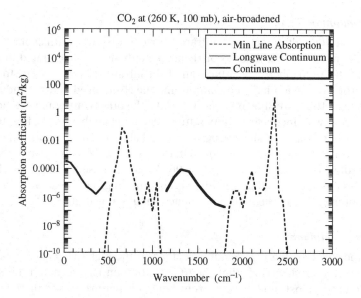

Figure 4.25 The air-induced CO_2 continuum absorption (solid lines) compared with the bandwise-minimum absorption computed from the line spectrum (dashed lines).

window region. The continuum absorption is strong enough to be important in the thick atmosphere of Early Mars, but only marginally so for the more moderate CO_2 levels present on Early Earth.

When incorporating the continuum in radiation models, it may be more convenient to work from a curve fit rather than tabulated data. The absorption coefficient for the CO_2 continuum can be fit with the function

$$\kappa_{CO2}(\nu, 300\,\text{K}, 100\,\text{mb}) = \exp(-8.853 + 0.028534\nu - 0.00043194\nu^2 + 1.4349 \cdot 10^{-6}\nu^3$$
$$- 1.5539 \cdot 10^{-9}\nu^4) \tag{4.89}$$

from 25 to 450 cm^{-1} and by

$$\kappa_{CO2}(\nu, 300\,\text{K}, 100\,\text{mb}) = \exp(-537.09 + 1.0886\nu - 0.0007566\nu^2 + 1.8863 \cdot 10^{-7}\nu^3$$
$$- 8.2635 \cdot 10^{-12}\nu^4) \tag{4.90}$$

from 1150 to 1800 cm^{-1} where ν is measured in cm^{-1}. The continuum absorption coefficients *weaken* with increasing temperature, according to the empirical power law $(300/T)^n$, with $n = 1.7$ for wavenumbers greater than 190 cm^{-1}, increasing to 1.9 at 130 cm^{-1}, 2.2 at 70 cm^{-1}, and 3.4 at 20 cm^{-1}.

The CO_2 continuum is poorly characterized experimentally, and not well understood theoretically. The continuum fits provided above are based on rather dated experiments, and there are some indications that a part of the above continuum may actually represent line spectra from asymmetric isotopologs of CO_2 such as $C^{16}O^{18}O$. This is an area where more experiments using modern techniques are badly needed.

Water vapor continuum

Since water vapor has absorption lines throughout the spectrum, it is hard to define the continuum unambiguously. Laboratory measurements clearly show, however, that in the window regions indicated in Fig. 4.19 the net absorption is far in excess of what can be accounted for by the contributions of nearby lines. The prevailing view is that this excess absorption is due to the very far tails of the stronger absorption bands flanking the window regions, rather than dimers, forbidden transitions, or collision-induced dipole moments. The theoretical and observational basis for this viewpoint is exceedingly weak, however. In the following we will confine ourselves to empirical descriptions of the laboratory measurements, without reference to underlying mechanisms. Comparisons with direct measurements of transmission in the Earth's atmosphere have confirmed that the laboratory measurements provide an adequate basis for modeling water vapor absorption in the window regions. The laboratory measurements show that the air-broadened or N_2-broadened water vapor continuum is very weak, so that the window region absorption is by far dominated by self-collisions of water vapor. The following discussion will therefore be limited to self-induced absorption; the characterization of foreign-induced absorption by CO_2 appears to be an open question at present, though it is potentially of importance to water–CO_2 atmospheres such as might have occurred on Early Mars.

There is some ambiguity in the spectroscopic literature as to how to define the water vapor continuum, given that in analyzing measurements one must be careful not to be thrown off by the strong absorption near the centers of individual lines in the window regions. Most useful definitions of the continuum amount to reading the absorption at the minima "between the lines." The results of such a measurement of the self-induced continuum are shown in Fig. 4.26. The measurements were made for water vapor in saturation at 296 K, with water vapor partial pressure of about 28 mb, but have been scaled to a standard water vapor partial pressure of 100 mb for the sake of discussion. We will focus on the lower frequency of the two window regions, since that is by far the most important for planetary climate calculations. Similar data exists for the higher-frequency window. From the figure, we see that the measured continuum absorption is several orders of magnitude stronger than the typical line contribution. To get an idea of the significance of the water vapor continuum, let's consider a layer of air of depth z, with uniform temperature T, within which

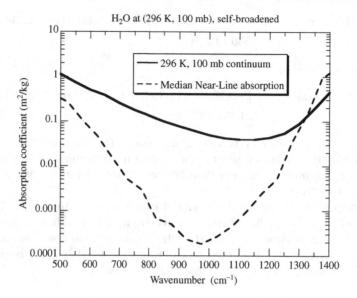

Figure 4.26 The water vapor self-induced continuum near $1000\,cm^{-1}$, compared with the median absorption coefficient computed from the Lorentz line contribution within 1000 line widths of the line centers in the HITRAN database. The continuum curve given is based on laboratory observations in saturation at 296 K, scaled up to what they would be at a water vapor partial pressure of 100 mb. See citations in the Further Reading section of this chapter for data sources.

the water vapor is at the saturation vapor pressure corresponding to T. Since the water vapor continuum is dominated by self-collisions, it matters little what the background air pressure is in this layer. The equivalent path for this layer is $(p_{sat}(T)/p_0)(p_{sat}(T)/(R_w T))z$; the first factor gives the degree of pressure-induced enhancement of absorption relative to the standard, while the second factor is the density of water vapor in the layer. Note that the equivalent path is *quadratic* in the water vapor partial pressure. For this reason, the optical thickness in the continuum region grows very rapidly with temperature. At 300 K, then, with a layer depth of 1 km the equivalent path is $9.3\,kg/m^2$. Since the minimum absorption coefficient in the window region is about $0.1\,m^2/kg$, this path gives the layer an optical thickness of unity or more in the window region. Since the absorption is even stronger outside the window region, the lowest layer of the Earth's atmosphere acts practically like an ideal blackbody at tropical temperatures. At 310 K the equivalent path increases to $27\,kg/m^2$, so the window region closes off even more. This has profound consequences for the runaway greenhouse. In fact, there would be essentially no prospect for a runaway greenhouse even in Venusian conditions were it not for the water vapor continuum.

Over the range of wavenumbers shown in the figure, the water vapor continuum absorption can be fit by the polynomial

$\kappa_{H2O}(\nu, 296\,K, 100\,mb)$

$$= \exp(12.167 - 0.050898\nu + 8.3207 \cdot 10^{-5}\nu^2 - 7.0748 \cdot 10-8\nu^3 + 2.3261 \cdot 10^{-11}\nu^4).$$

(4.91)

Like the other continua, the water vapor continuum absorption becomes weaker as temperature increases. Data on the temperature dependence is sparse, but suggests a temperature dependence of the form $(296/T)^{4.25}$. For temperatures much colder than $300\,\text{K}$, the saturation vapor pressure is so low that the details of the temperature dependence are unimportant. As temperature increases beyond $300\,\text{K}$, the exponential growth of saturation vapor pressure is far more important to the optical thickness than the rather mild decline of the continuum absorption coefficient with temperature.

Water vapor has another continuum region at shorter wavelengths, in the vicinity of $2500\,\text{cm}^{-1}$. This is not important for Earthlike temperatures, but it is a very significant factor for the hotter temperatures encountered in runaway greenhouse calculations. At temperatures much in excess of $320\,\text{K}$, there is enough emission in this region that it accounts for a significant part of the infrared cooling if the continuum is not included. The $2500\,\text{cm}^{-1}$ continuum, covering $2100\,\text{cm}^{-1} \leq \nu \leq 3000\,\text{cm}^{-1}$, can be represented by the polynomial fit

$$\kappa_{H2O}(\nu, 296\,\text{K}, 100\,\text{mb}) = \exp(-6.0055 - 0.0021363x + 6.4723 \cdot 10^{-7}x^2$$
$$- 1.493 \cdot 10^{-8}x^3 + 2.5621 \cdot 10^{-11}x^4 \qquad (4.92)$$
$$+ 7.328 \cdot 10^{-14}x^5)$$

where $x = \nu - 2500\,\text{cm}^{-1}$. The scaling in pressure and temperature can be taken to be the same as for the longer wavelength continuum, though the experimental support for the temperature dependence is somewhat weak.

The importance of the water vapor continuum to climate when temperatures approach or exceed $300\,\text{K}$ is demonstrated in Fig. 4.27. Here we present calculations of the spectrum of *OLR* and of surface back-radiation using the exponential-sum radiation code described previously, but modified to take into account the vertical variation of water vapor concentration

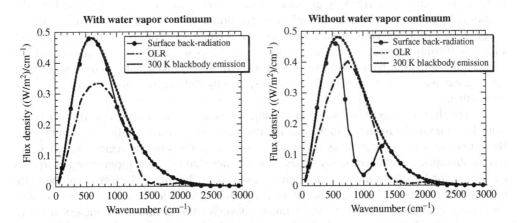

Figure 4.27 The spectra of surface back-radiation and OLR computed using an exponential-sum radiation code including the effects of the longwave water vapor continuum (left panel) and excluding the continuum (right panel). The calculation was done with the temperature profile on the water–air moist adiabat corresponding to $300\,\text{K}$ surface temperature, assuming the water vapor partial pressure to be saturated at all levels. This calculation does not take into account the temperature variation of absorption coefficients or the enhancement of self-induced absorption in the line contribution.

and the continuum. With the continuum included, the low-level atmosphere radiates to the surface practically like a blackbody; in fact, if one increases the surface air temperature slightly, to 310 K, the back-radiation becomes indistinguishable from the blackbody spectrum corresponding to the surface air temperature. In contrast, without the continuum, there is essentially no back-radiation in the window region, allowing the surface to cool strongly through the window. Likewise, without the continuum, the atmosphere can radiate to space very strongly through the window, whereas the cooling to space is very much reduced if the continuum absorption is included. These calculations were carried out for an Earthlike water–air atmosphere on a planet with $g = 9.8\,\mathrm{m/s}^2$. On a planet with weaker surface gravity, the water vapor continuum would become important at lower temperatures, because a given partial pressure corresponds to a greater mass of water. Conversely, on a planet with stronger surface gravity, the water vapor window closes at higher temperatures.

4.4.9 Condensed substances: clouds

Clouds are made of particles of a condensed substance, which may be in a liquid or solid (e.g. ice) phase. The molecules of a condensed substance are in close proximity to one another, and at typical atmospheric temperatures the collisions are so frequent that no line structure survives in the spectrum. In consequence, the absorption coefficient for a condensed substance is generally a very smoothly varying function of wavenumber. Absorption by condensed substances behaves rather like the gaseous continuum absorption we discussed in the preceding section.

Water clouds are of particular interest, since they are by far the dominant type of cloud on Earth. They would also occur on any world habitable for life as we know it, since such a world would have a repository of liquid water somewhere, and condensation of water vapor somewhere in the atmosphere would then be practically inevitable. Water clouds would also form in the course of a runaway greenhouse on a world with a water ocean, such as the primordial Venus. The absorption coefficient for liquid water is shown over the infrared range in Fig. 4.28. For comparison, we show the median absorption coefficient for water vapor at 100 mb pressure and 260 K temperature. Keep in mind that the absorption for liquid water is a true continuum, so that, unlike the median absorption curve shown for the vapor phase, the curve for liquid water displays the full wavenumber variability of the absorption.

We see that a kilogram of water in the liquid phase is a far better absorber than the same kilogram in the form of vapor. Near the peak absorption wavenumbers of water vapor, the difference can be as little as a factor of 10, but in the window regions liquid water has an absorption coefficient many thousands of times that of water vapor. The absorption coefficient for liquid water varies little enough that over the infrared range it can be quite well approximated as a gray gas (or more properly, a gray liquid). In fact, with a typical absorption coefficient of $100\,\mathrm{m}^2/\mathrm{kg}$, it takes a layer of liquid only 10^{-5} meters thick to have unit optical thickness and to begin to behave like a gray body. This is the depth of penetration of atmospheric infrared back-radiation into the surface of a lake or ocean, and it is the depth whose temperature directly determines the infrared radiation from the surface of a lake or ocean. Water ice is somewhat more transparent to infrared than liquid water, but the general features of the behavior still apply.

Because of the nonlinearity of the exponential function which determines the amount of absorption suffered by infrared as it passes through a number of particles, it matters greatly how the mass of water is distributed amongst particles of various sizes. For example, at $400\,\mathrm{cm}^{-1}$ liquid water has an absorption coefficient of $171\,\mathrm{m}^2/\mathrm{kg}$; since water has a

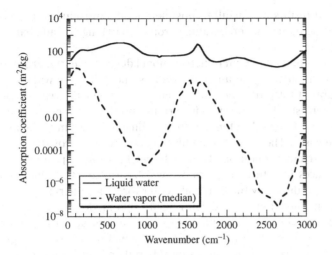

Figure 4.28 The absorption coefficient for liquid water. The median absorption coefficient for water vapor at (100 mb, 260 K) is reproduced from Fig. 4.19 for the purposes of comparison.

density of $1000 \, \text{kg/m}^3$ that means that a layer of liquid water of depth $5.8 \, \mu\text{m}$ has optical thickness of unity, and attenuates incident infrared by a factor of $1/e$. That means, roughly speaking, that a spherical droplet of radius r will remove essentially all the infrared hitting it – πr^2 times the incident flux – as long as r is $5 \, \mu\text{m}$ or more. A mere 10 grams of water is sufficient to make $1.9 \cdot 10^{10}$ particles of radius $5 \, \mu\text{m}$, which would have a total cross-section area of $1.5 \, \text{m}^2$. Distributed randomly within a column of air having base $1 \, \text{m}^2$ these particles would be sufficient to remove essentially all of the flux of infrared entering the base of the column. In other words, a mass path of liquid water as little as 10 grams per square meter is sufficient to make an optically thick cloud, provided the water takes the form of sufficiently small droplets. However, if we take the same mass of water and gather it up into a single drop of radius 1.3 cm, it would only intercept an insignificant $0.0005 \, \text{m}^2$ of the incident light.

This estimate of the absorption by cloud droplets is not quite correct, because it fails to take into account the extent to which electromagnetic radiation penetrates the droplet as opposed to being diffracted around it. The calculation will be done more precisely in Chapter 5, but the simple estimate gives the right answer to within a factor of 2 or better.

The net result is that, for the typical droplet size found in Earth's water clouds, a cloud layer containing anything more than about 10 grams per square meter of condensed water acts essentially like a blackbody in the infrared. Water is not at all typical in this regard. Other condensed cloud-forming substances, including liquid methane and CO_2 ice, are far more transparent in the infrared, and have a qualitatively different effect on planetary energy balance. The effect of such clouds will be taken up in Chapter 5.

4.5 REAL GAS *OLR* FOR ALL-TROPOSPHERE ATMOSPHERES

Calculation of *OLR* is one of the most fundamental steps in determining a planet's climate. Now that we are equipped with an ability to compute the *OLR* for real gases, we can revisit some of our old favorite problems – Snowball Earth, the Faint Young Sun, Early Mars, and so forth – but this time relate the results to the actual atmospheric composition. In this section we present results for the all-troposphere model introduced in Section 4.3.2, occasionally limiting the upper air temperature drop by patching the adiabat to an isothermal stratosphere.

The homebrew exponential-sum radiation model described in the preceding sections has the advantage of simplicity, generality, and understandability. We will use it wherever it is sufficiently accurate to capture the main phenomena under discussion. However, professionally written terrestrial radiation codes are the product of a great deal of attention to detail, particularly with regard to temperature scaling and the simultaneous effects of multiple greenhouse gases. They can give highly accurate results provided one does not stray too far from the Earthlike conditions for which they have been optimized. In the following, and at various places in future chapters, we will have recourse to one of these standard radiation models, produced by the National Center for Atmospheric Research as part of the Community Climate Model effort. We will refer to this model as the ccm radiation model. Although it uses a good many special tricks to achieve accuracy at high speed, and a lot of detailed bookkeeping to deal with the properties of a half dozen different greenhouse gases of interest on Earth, what is going on inside this rather massive piece of code is not fundamentally different from the Malkmus type band models and the exponential-sum model described previously.

A detailed discussion of surface back-radiation for real gases will be deferred to Chapter 6. Some aspects of the infrared cooling profile for real gas atmospheres will be touched on in Section 4.7. The effect of clouds on *OLR* and on shortwave albedo will be discussed in Chapter 5.

4.5.1 CO_2 and dry air

First we will compute the *OLR* for a mixture of CO_2 in dry air, with the temperature on the dry air adiabat $T(p) = T_g \cdot (p/p_s)^{2/7}$ and ground temperature equal to surface air temperature. This is a real gas version of the calculation leading to Eq. (4.33); it amounts to a canonical *OLR* computation which serves as a simple basis for intercomparison of different radiation models. Performed for other greenhouse gases, it also can provide a basis for comparing the radiative effects of the gases. The results presented here are carried out with Earth gravity and 1 bar of air partial pressure, but can easily be scaled to other conditions. The CO_2 path is inversely proportional to gravity, so 100 ppmv of CO_2 on Earth is equivalent to 1000 ppmv on a planet with 10 times the Earth's surface gravity or 10 ppmv on a planet with a tenth of the Earth's surface gravity. For fixed CO_2 concentration, surface pressure has a quadratic effect on the path, since the mass of CO_2 in the atmosphere (given fixed concentration) increases in proportion to pressure, but one gets an additional pressure factor in the equivalent path from pressure broadening. Thus, 100 ppmv of CO_2 in 1 bar of air is equivalent to 1 ppmv of CO_2 in 10 bars of air.

In Fig. 4.29 we show how *OLR* varies with CO_2 concentration for a fixed surface temperature of 273 K. This curve gives the amount of absorbed solar radiation needed to maintain the surface temperature at freezing. The CO_2 amount is expressed as molar concentration in ppmv, but for the range of concentrations considered, the difference between concentration and mixing ratio is not very significant. Results are shown for both the ccm model and the simplest form of the homebrew exponential-sum radiation code. The homebrew calculation employed 10-term sums with coefficients computed at 260 K. The path was pressure-weighted to reflect collisional broadening, but temperature weighting was neglected. Over the range of CO_2 covered, only the principal absorption region centered on 650 cm^{-1} needs to be taken into account. The homebrew calculation deviates by up to

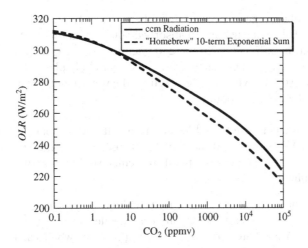

Figure 4.29 The *OLR* vs CO_2 concentration (measured in ppmv) for CO_2 in a dry air atmosphere with temperature profile given by the dry adiabat. The surface air temperature and the ground temperature are both 273 K, and the acceleration of gravity is 9.8 m/s^2. Results are shown for a simple exponential-sum radiation code without temperature weighting, and for the comprehensive ccm radiation code.

10 W/m^2 from the more comprehensive ccm calculation, but this is quite good agreement in view of the fact that the homebrew code takes up barely a page and generalizes easily to any greenhouse gas, in contrast to the rather Earth-specific ccm code, which involves several thousand lines of rather unpretty FORTRAN. Most of the mismatch arises from the neglect of temperature scaling in the homebrew code. The homebrew code slightly overestimates *OLR* for low CO_2 where the radiating level is low in the atmosphere where warmer temperatures ought to increase the infrared opacity. It underestimates *OLR* because the higher, colder atmosphere is assumed more optically thick than it should be. If one complicates the homebrew code very slightly to incorporate a band-independent temperature weighting of the form $\exp -(T^*/T - T^*/T_0)$ in the path computation, then one can reduce the mismatch to under 2 W/m^2 with $T^* = 900$ K.

As anticipated from the shape of the CO_2 absorption spectrum, the *OLR* goes down approximately in proportion to the logarithm of the CO_2 concentration. Between 10 ppmv and 1000 ppmv each doubling of CO_2 reduces *OLR* by about 4 W/m^2 based on the ccm model. At large CO_2 concentrations, the logarithmic slope becomes somewhat greater, as the weaker absorption bands begin to come into play. At 10 000 ppmv, a doubling reduces *OLR* by 6.8 W/m^2. Carbon dioxide becomes a somewhat more effective greenhouse gas at high concentrations, but never approaches the potency of a gray gas, for which each doubling would more than *halve* the *OLR* once the optically thick limit is reached. Were it not for the relatively gentle dependence of *OLR* on CO_2 caused by the highly frequency-selective nature of real gas absorption, modest fluctuations in the atmosphere's CO_2 content would lead to wild swings of temperature and almost certainly render the planet uninhabitable.

Calculations show that for fixed CO_2 concentration the *OLR* increases very nearly like the fourth power of temperature, just as in the gray gas result in Eq. (4.33). In effect, the radiating pressure remains nearly fixed as temperature varies. This makes it easy to do planetary temperature calculations. For example, let's compute what temperature the Earth would have if the CO_2 concentration were at the pre-industrial value of 280 ppmv, but there were no other greenhouse gases in the atmosphere. The *OLR* for a surface temperature of 273 K is 267 W/m^2. Balancing against an absorbed solar radiation of 240 W/m^2, the temperature is determined by $267 \cdot (T/273)^4 = 240$, yielding $T = 263$ K. Without the additional greenhouse effect of water vapor, Earth would be a very chilly place. According to Fig. 4.29, for a dry Earth CO_2 would have to be increased all the way to 24 000 ppmv just to bring the temperature up to the freezing point.

Exercise 4.9 What temperature would Venus have if it had a 1 bar air atmosphere mixed with 280 ppmv of CO_2? Assume that the planetary albedo is 30%, like that of Earth. How does this temperature compare with the temperature Venus would have without any greenhouse gases in its atmosphere?

One cannot get the Earth's temperature right without water vapor, but one can still make a decent estimate of the amount of CO_2 increase needed to offset the reduction of solar forcing in the Faint Young Sun era, or due to a global Snowball Earth glaciation. A 25% reduction of solar absorption at the Earth's orbit during the Faint Young Sun amounts to 60 W/m^2, assuming an albedo of 0.3. This is equivalent to the radiative forcing caused by increasing CO_2 from 100 ppmv to 10^5 ppmv (about 100 mb CO_2 partial pressure, or 10% of the atmosphere). From this we conclude that it's likely that it would take somewhat over 100 mb of CO_2 to keep Earth unfrozen when the Sun was dim. Next let's take a look at what it takes to deglaciate a Snowball Earth. Let's suppose that for a Neoproterozoic solar constant, the tropics freeze over when the CO_2 is reduced to 100 ppmv (this is not far off estimates based on comprehensive climate models). Icing over the Earth increases the albedo to about 0.6, leading to a reduction of almost 100 W/m^2 in absorbed solar radiation. Hence, restoring the tropics to the melting point would require well in excess of 100 mb of CO_2 in the atmosphere; calculations with the homebrew model at higher CO_2 concentrations than shown in Fig. 4.29 indicate that fully a bar of CO_2 mixed with a bar of air would be needed. Deglaciation might not require restoring the full 100 W/m^2 of lost solar forcing, but it is still clear that a great deal of CO_2 is needed to deglaciate a Snowball.

These estimates are crude, but they do get across one central idea: that because of the logarithmic dependence of *OLR* on greenhouse gas concentration, it takes a huge increase in the mass of CO_2 to make up for rather moderate changes in albedo or solar output. Aside from neglect of overlapping water vapor absorption, these estimates somewhat overstate the effect of CO_2 on *OLR* because they employ the dry adiabat rather than the less steep moist adiabat. We will revisit the estimates shortly, after we bring water vapor into the picture.

4.5.2 Pure CO_2 atmospheres: Present and Early Mars, and Venus

Figure 4.30 shows the *OLR* as a function of surface pressure for a pure CO_2 atmosphere subject to Martian gravity. The results span the range of surface pressure from those similar to the thin atmosphere of present Mars up to the thick atmospheres commonly hypothesized for Early Mars.[7] The calculations were carried out for a fixed surface temperature of 270 K, since we are primarily interested in the question of how much CO_2 there would have to be

---

[7] There is no strong reason to exclude the possibility of a substantial amount of N_2 in the Early Martian atmosphere. Addition of N_2 to the atmosphere would increase surface pressure and enhance CO_2 absorption.

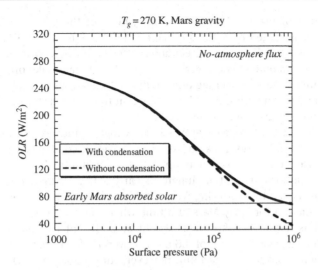

Figure 4.30 The *OLR* vs. surface pressure for a pure CO_2 atmosphere. The results were obtained with a 10-term exponential-sum code based on a 100mb reference pressure, and including the temperature scaling of both the continuum and line absorption. Calculations were performed for a ground temperature of 270 K, for Martian gravity. The dashed line shows the *OLR* for the case in which the temperature profile is on the dry ideal gas CO_2 adiabat, while the solid line incorporates the effect of condensation on the temperature profile.

in order to warm Early Mars up to near the freezing point and permit the widespread liquid water at the surface that is seemingly demanded by the surface geology of the ancient Martian terrain. Results are shown for two different variants of the all-troposphere model. In the first, the atmosphere is on the dry CO_2 ideal gas adiabat throughout its depth. This profile is inconsistent at high surface pressure, however, since it becomes supersaturated aloft. For this reason, we also include results in which the temperature profile is on the one-component condensing CO_2 adiabat, which is on the dry adiabat where unsaturated but pinned to the Clausius–Clapeyron result when it becomes saturated (as in Fig. 2.6). With condensation, the surface pressure cannot be increased beyond 35.4 bar at a surface temperature of 270 K, since the surface becomes saturated at that point and no further CO_2 can be added to the atmosphere without causing condensation. This does not pose a very significant constraint on the climate of Early Mars, however; a more important limitation is the amount of mass that could plausibly be lost from the primordial Martian atmosphere in the past four billion years. The effect of condensation aloft on *OLR*, in essence, increases the amount of CO_2 needed to warm Early Mars to the point where it is unclear that so much atmosphere could be lost.

At surface pressures comparable to that of present Mars, the CO_2 greenhouse effect reduces the *OLR* by 35 W/m² , and it would take 267 W/m² of absorbed solar radiation to maintain a surface temperature of 270 K. The required solar heating is well below the 440 W/m² solar forcing at the subsolar point. Assuming the *OLR* to scale with the fourth power of temperature for fixed surface pressure, this would support a temperature of 301 K at the subsolar point, in contrast with a temperature of 292 K in the absence of the atmospheric greenhouse effect. The atmosphere exerts only a modest warming effect on the surface temperature of present Mars. To be sure, most of the planet is much colder; the absorbed solar flux averaged over the surface of the planet is only $110\,\mathrm{W/m^2}$, which supports a temperature of $216\,\mathrm{K}$ with the greenhouse effect and $210\,\mathrm{K}$ without. Recall, however, that the planetary mean budget is not very meaningful on a planet like present Mars with no ocean and little atmosphere to average out the diurnal variations. A thin layer of the rocky surface will be quite warm within a circle centered on the subsolar point, but the nightside surface falls to temperatures well below $216\,\mathrm{K}$.

At higher CO_2, the *OLR* decreases, approximately logarithmically in surface pressure for pressures above $10^4\,\mathrm{Pa}$. At pressures above 1 bar, however, condensation becomes important, and the consequent increase in temperature aloft limits the decline of *OLR*. This limitation is quite important for the climate of Early Mars. Taking into account the relatively high albedo caused by scattering from a thick CO_2 atmosphere (see Section 5.3), the absorbed solar radiation for Early Mars at a time when the solar flux is 30% reduced compared with today is about $70\,\mathrm{W/m^2}$. This would be sufficient to sustain a $270\,\mathrm{K}$ surface temperature with a surface pressure of 3.6 bar if it weren't for the effects of condensation. When condensation is taken into account, however, fully 10 bars of CO_2 are necessary to bring the surface temperature up to $270\,\mathrm{K}$.[8] In fact, given the increase in albedo associated with scattering of solar radiation in a 10 bar atmosphere, even 10 bars is likely to prove insufficient. The effect of atmospheric scattering on albedo will be quantified in Chapter 5. In that chapter we will also discuss the potential for the scattering greenhouse effect from CO_2 ice clouds to warm Early Mars. As time progresses toward the present, and the Sun gets brighter, it becomes progressively easier to warm Mars to the point where liquid water can persist at the surface. The climate history of Mars is a race between the brightening Sun and the loss of atmosphere, and the efforts of the latter to make Mars cold and inhospitable seem to have triumphed.

The continuum absorption in the CO_2 window region is extremely important to these results. Without continuum absorption, the *OLR* for a 2 bar atmosphere would be over $50\,\mathrm{W/m^2}$ higher. The temperature scaling of the continuum affects the results by $10\,\mathrm{W/m^2}$ or more. Whether or not one can account for prevalent liquid water on Early Mars by a gaseous CO_2 greenhouse effect hinges on a matter of $10\,\mathrm{W/m^2}$ of flux or so, and therefore the importance of the continuum is disconcerting. To settle this question, one must get the CO_2 continuum right, and that is far from clear at this point, in view of the rather sparse experimental and theoretical results on the subject.

Now what about Venus? Are we finally equipped to say that we can account for the high surface temperature of Venus in terms of the CO_2 greenhouse effect? Unfortunately, not quite. The problem is that the high albedo of Venus means that the climate is maintained by a relative trickle of absorbed solar radiation, while the high surface temperature means that the infrared emission at wavenumbers higher than $2300\,\mathrm{cm^{-1}}$ would exceed $3800\,\mathrm{W/m^2}$ if the atmosphere were transparent in that spectral region. Unlike the Earthlike case, the atmospheric opacity there matters very much; however, the HITRAN database does not include the weak absorption lines needed to accurately determine the atmospheric opacity at high wavenumbers. Extensions to the database suitable for use in Venusian conditions are described in the Further Reading at the end of this chapter. We will not pursue

---

[8] The implications of CO_2 condensation for the gaseous greenhouse effect on Early Mars were first discussed in: Kasting JF 1991, *Icarus* **94**, 1-13. The reader is also referred there for a more comprehensive treatment of the radiative transfer problem, including the effects of water vapor and a stratosphere. The conclusions are broadly similar to those based on our homebrew radiation model.

detailed calculations with the extended database, since one must, after all, stop somewhere. Instead, we will make use of a highly simplified treatment of the shortwave thermal emission which at least tells us how close we are to being able to explain the temperature of Venus. Specifically, we use exponential sums based on HITRAN for the spectral region with lower wavenumbers than $2300\,cm^{-1}$, but represent the emission from higher wavenumbers by assuming that there is a radiating pressure p_{rad} such that the atmosphere radiates to space like a blackbody with temperature $T(p_{rad})$ throughout the high-wavenumber spectral region. This is essentially the same approach as we took in formulating the simplest model of the greenhouse effect in Chapter 3, except that this time we apply the radiating-level concept only to the high-wavenumber part of the emission. It is equivalent to stating that the absorptivity in the high-wavenumber region is sufficiently large to make the layer of the atmosphere between p_{rad} and the ground optically thick throughout the high-wavenumber region. Because of the shape of the Planck function, as p_{rad} is made smaller and $T(p_{rad})$ is made colder, the peak emission shifts to lower wavenumbers and in consequence the shortwave emission is sharply curtailed.

With these approximations the *OLR* can be written as $OLR_<(T_g) + OLR_>(T_g, p_{rad})$ where $OLR_<$ is computed for the low-wavenumber spectral region alone using exponential sums and

$$OLR_> = \int_{\nu_1}^{\infty} \pi B(\nu, T(p_{rad}, T_g)) d\nu \qquad (4.93)$$

where ν_1 is the frequency cutoff for the high-wavenumber region, B is the Planck function, and $T(p_{rad}, T_g)$ is computed using the dry CO_2 adiabat. The *OLR* thus computed must be balanced against the absorbed solar radiation to determine the surface temperature. At present, the solar radiation absorbed by Venus amounts to $163\,W/m^2$. Assuming a 93 bar surface pressure, this is in balance with a surface temperature of 652 K if there is no emission at all from the high-wavenumber region, i.e. if $p_{rad} = 0$. This is a limiting case giving the maximum temperature that can be obtained with CO_2 alone regardless of how optically thick the high-wavenumber region may be. The fact that this is still somewhat short of the observed 720 K surface temperature of Venus means that a modest additional source of atmospheric opacity other than CO_2 is still needed to close the remaining gap in explaining the surface temperature. If p_{rad} is increased to 10 bars, the equilibrium temperature only drops to 633 K, so it is only necessary for CO_2 to be essentially opaque in the high-wavenumber region at pressures of 10 bars or greater. On the other hand, if CO_2 were really transparent at high wavenumbers, i.e. $p_{rad} = 93$ bar, then the surface temperature would drop all the way to 461 K, which is well below the observed value. Detailed radiation modeling of the high-wavenumber region is consistent with a value $p_{rad} \approx 10$ bar, so it appears that the CO_2 greenhouse effect alone gets us almost all the way to explaining the high surface temperature. The remaining opacity needed to bring the surface temperature up to 720 K is provided by Venus' high sulfuric acid clouds, the trace of water vapor in the atmosphere, and sulfur dioxide (in order of importance). The sulfuric acid clouds exert their greenhouse effect partly through infrared scattering, as discussed in Chapter 5.

4.5.3 Water vapor feedback

For a planet like the Earth, which has a substantial reservoir of condensed water at the surface (be it ocean or glacier), if left undisturbed for a sufficiently long time water vapor would enter the atmosphere until the atmosphere reached a state where the water vapor pressure was equal to the saturation vapor pressure at all points. In a case like this, if anything happens to increase the temperature of the atmosphere, then the water vapor content will eventually increase; since water vapor is a greenhouse gas, the additional water vapor will warm the planet further. Amplification of this sort is known as *water vapor feedback*, and works to amplify cooling influences as well. In this section we'll examine some quantitative models of real gas water vapor feedback.

It turns out that atmospheric motions have a drying effect which keeps the atmosphere from reaching saturation. This is a very active subject of current research, but suffice it to say for the moment that comprehensive simulations of the Earth's atmosphere suggest that the situation can be reasonably well represented by keeping the relative humidity fixed at some subsaturated value as the climate warms or cools. That is the approach we shall adopt here, and it still yields an atmosphere whose water vapor content increases roughly exponentially with temperature. At present, there is no generally valid theory which allows one to determine the appropriate value of relative humidity *a priori*, so one must have recourse to fully dynamic general circulation models or observations. Indeed, the relative humidity is not uniform but varies considerably both in the vertical and horizontal, so there is no single value that can be said to characterize the global humidity field. Similar considerations would apply to any condensable substance, for example methane on Titan.

In these all-troposphere models, we shall also assume that the temperature profile is given by the moist adiabat. Observations of the Earth's tropics show this to be a good description of the tropospheric temperature profile even where the atmosphere is unsaturated and not undergoing convection. (See Problem 2.55.) Evidently, the regions that are undergoing active moist convection control the lapse rate throughout the tropical troposphere; there are also theoretical reasons for believing this to be the case, but they require fluid dynamical arguments that are beyond the scope of the present volume. The situation in the midlatitudes is rather less clear, but we will use the moist adiabat there as well, because it is hard to come up with something better in a model without any atmospheric circulation in it. Results are presented for Earth surface pressure and gravity, though we will make a few remarks on how the results scale to planets with greater or lesser gravity.

Since most of the big questions of Earth climate and climate of habitable Earthlike planets involve water vapor feedback in conjunction with one or more other greenhouse gases, it is in this section that we will for the first time get fairly realistic answers regarding the Faint Young Sun problem, and so forth.

The case of a saturated pure water vapor atmosphere will be treated in Section 4.6 as part of our treatment of the runaway greenhouse for real gas atmospheres. Here we will begin our discussion with water vapor mixed with Earth air, with the temperature profile on the moist adiabat for an air–water mixture. The *OLR* curve for this case is shown in Fig. 4.31. Since there is no other greenhouse gas, the *OLR* for the dry case (zero relative humidity) is just σT_g^4. For ground temperatures below 240 K there is so little water vapor in the air that the exact amount of water vapor has little effect on the *OLR*. At larger temperatures, the curves for different relative humidity begin to diverge. Even for relative humidity as low as 10%, the water vapor greenhouse effect is sufficient to nearly cancel the upward curvature of the dry case; at 320 K the *OLR* is reduced by over 130 W/m^2 compared with the dry case. At larger relative humidities, the curvature reverses, and the *OLR* as a function of temperature shows signs of flattening at high temperature, in a fashion reminiscent of that we saw in our discussion of the Kombayashi–Ingersoll limit for gray gases. This is our first acquaintance with the essential implication of water vapor feedback: the increase of water vapor with temperature reduces the slope of the *OLR* vs. temperature curve, making the climate more

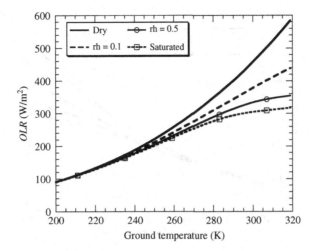

Figure 4.31 *OLR* vs. surface temperature for water vapor in air, with relative humidity (rh) held fixed. The surface air pressure is 1 bar, and Earth gravity is assumed. The temperature profile is the water/air moist adiabat. Calculations were carried out with the ccm radiation model.

sensitive to radiative forcing of all sorts – whether it be changes in the solar constant, changes in surface albedo, or changes in the concentration of CO_2. To take but one example, increasing the absorbed solar radiation from $346 \, W/m^2$ to $366 \, W/m^2$ increases the ground temperature from $280 \, K$ to $283 \, K$ in the dry case. When the relative humidity is 50%, it takes only $290 \, W/m^2$ of absorbed solar energy to maintain the same $280 \, K$ ground temperature, but now increasing the solar absorption by $20 \, W/m^2$ increases the surface temperature to $288 \, K$. Water vapor feedback has approximately doubled the climate sensitivity, which is a typical result for Earthlike conditions. The increase in climate sensitivity due to water vapor feedback plays a part in virtually any climate change phenomenon that can be contemplated on a planet with a liquid water ocean.

The influence of water vapor is strongest at tropical temperatures, but is still significant even at temperatures near freezing. It only becomes negligible at temperatures comparable to the polar winter.

One sometimes hears it remarked cavalierly that water vapor is the "most important" greenhouse gas in the Earth's atmosphere. The misleading nature of such statements can be inferred directly from Fig. 4.31. Let's suppose the Earth's climate to be in equilibrium with $256 \, W/m^2$ of absorbed solar radiation, averaged over the Earth's surface. This corresponds to an albedo of 25%, which we take to be somewhat smaller than the actual observed albedo to account crudely for the fact that part of the cloud albedo effect is canceled by cloud greenhouse effects which we do not take into account in the figure. If water vapor were the only greenhouse gas in the Earth's atmosphere, the temperature would be a chilly $268 \, K$, and that's even before taking ice-albedo feedback into account, which would most likely cause the Earth to fall into a frigid Snowball state. We saw earlier that the Earth would also be uninhabitably cold if CO_2 were the only greenhouse gas in the atmosphere. With regard to Earth's habitability, it takes two to tango. In order to maintain a habitable temperature on Earth without the benefit of CO_2, the Sun would have to be 13% brighter. It will take well over a billion more years before the Sun will become this bright.

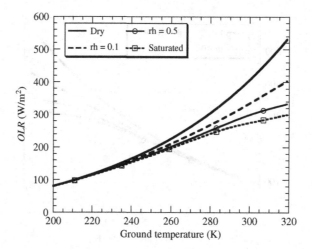

Figure 4.32 As in Fig. 4.31, but with 300 ppmv of CO_2 included. Note that the "Dry" case excludes only the radiative effects of water vapor; the moist adiabat is still employed for the temperature profile.

Now let's add some CO_2 to the atmosphere. Figure 4.32 shows how the OLR curve changes if we add in 300 ppmv of CO_2 – slightly more than the Earth's pre-industrial value. The general pattern is similar to the water-only case, but shifted downward by an amount that varies with temperature. At 50% relative humidity, the addition of CO_2 reduces the OLR at 280 K by a further 36 W/m^2 below what it was with water vapor alone. Because of the additional greenhouse effect of CO_2, the same 256 W/m^2 of absorbed solar radiation we considered previously can now support a temperature of 281 K when the relative humidity is 50%. Note that without the action of CO_2, the atmosphere would be too cold to have much water vapor in it, so one would lose much of the greenhouse effect of water vapor as well. It appears that the actual surface temperature of Earth can be satisfactorily accounted for on the basis of the CO_2 greenhouse effect supported by water vapor feedback.

Exercise 4.10 For the four moisture conditions in Fig. 4.32, determine how much absorbed solar radiation would be needed to support a surface temperature of 280 K. For each case, use the graph to estimate how much the surface temperature would increase if the absorbed solar radiation were increased by 20 W/m^2 over the original value. How does the amplification due to water vapor feedback compare with the results obtained without CO_2 in the atmosphere?

Next, let's take a look at how the OLR curve varies as a function of CO_2, with relative humidity held fixed at 50%. The results are shown in Fig. 4.33. The addition of CO_2 to the atmosphere lowers all the curves, and the more CO_2 you add, the lower the curves go. Carbon dioxide is planetary insulation: adding CO_2 to a planet reduces the rate at which it loses energy, for any given surface temperature, just as adding fiberglass insulation to a house reduces the rate at which the house loses energy for a fixed interior temperature (thus reducing the fuel that must be burned in order to maintain the desired temperature). Another thing we note is that when the CO_2 concentration becomes very large, the curve loses its negative curvature and becomes concave upward, like σT^4. This happens because

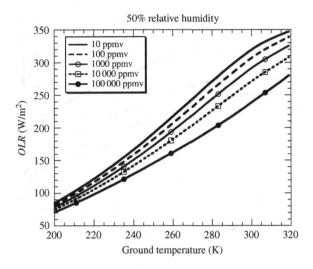

Figure 4.33 *OLR* vs. surface temperature for various CO_2 concentrations, at a fixed relative humidity of 50%. Other conditions are the same as for Fig. 4.31.

the CO_2 greenhouse effect starts to dominate the water vapor greenhouse effect, so that the flattening of the *OLR* curve due to the increase of water vapor with temperature becomes less effective over the temperature range shown. Even for very high CO_2, water vapor eventually would assert its dominance as temperatures are raised in excess of 320 K, causing the curve once more to flatten as the Kombayashi–Ingersoll limit is approached.

As an example of the use of the information in Fig. 4.33, let's start with a planet with 100 ppmv of CO_2 in its atmosphere, together with a sufficient supply of water to keep the atmosphere 50% saturated in the course of any climate change. From the graph, we see that absorbed solar radiation of $257 \, W/m^2$ would be sufficient to maintain a mean surface temperature of 280 K. From the graph we can also see that if the absorbed solar radiation is held fixed, increasing CO_2 tenfold to 1000 ppmv would increase the temperature to 285 K once equilibrium is re-established, and increasing it another tenfold to 10 000 ppmv (about 1% of the atmosphere) would increase the temperature to 293 K. A further increase to 100 000 ppmv (10% of the atmosphere) increases the temperature to 309 K. These represent substantial climate changes, but not nearly so extreme as they would have been if CO_2 were a gray gas. As another example of what the graph can tell us, let's ask how much CO_2 increase is needed to maintain the same 280 K surface temperature with a dimmer Sun. Reading vertically from the intersection with the line $T_g = 280$ K, we find that this temperature can be maintained with an absorbed solar radiation of $195 \, W/m^2$ if the CO_2 concentration is 100 000 ppmv. Thus, an increase in CO_2 by a factor of 1000 can make up for a Sun which is 25% dimmer than the base case.

Finally, we'll take a look at how the *OLR* varies with CO_2 for a fixed surface temperature (Fig. 4.34). This figure is a more Earthlike version of the results in Fig. 4.29, in that the effects of water vapor on radiation and the adiabat have been taken into account. As in the dry case, there is broad range of CO_2 concentrations – about 5 ppmv to 5000 ppmv – within which the *OLR* decreases very nearly like the logarithm of CO_2 concentration. The slope depends only weakly on the relative humidity, especially if one leaves out the completely dry case. This suggests that the effects of water vapor and CO_2 on the *OLR* are approximately additive

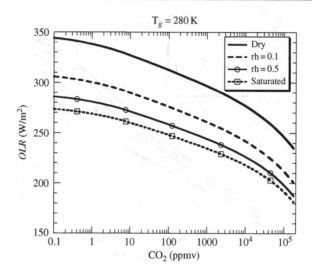

Figure 4.34 OLR vs. CO_2 for a fixed surface temperature $T_g = 280\,K$, for various values of the relative humidity. Other conditions are the same as for Fig. 4.31.

in this range. We note further that the logarithmic slope of *OLR* vs. CO_2 becomes steeper at very high CO_2, since one begins to engage more of the outlying absorption features of the CO_2 spectrum; again, CO_2 becomes an increasingly effective greenhouse gas at high concentrations. Conversely, at very low concentrations the logarithmic slope is reduced, as CO_2 absorption comes to be dominated by a relatively few narrow absorption features.

For any given CO_2 value, increasing the moisture content reduces the *OLR*, as would be expected from the fact that water vapor is a greenhouse gas. A little bit of water goes a long way. With 100 ppmv of CO_2 in the atmosphere, going from the dry case to 10% relative humidity reduces the *OLR* by 36 W/m^2. To achieve the same reduction through an increase of CO_2, one would have to increase the CO_2 concentration all the way from 100 ppmv to 10 000 ppmv. Clearly, water vapor is a very important player in the radiation budget, though we have already seen that because of the thermodynamic control of water vapor in Earthlike conditions, CO_2 nonetheless remains important. As the atmosphere is made moister, the further effects of water vapor are less dramatic. Increasing the relative humidity from 10% to 50% only brings the *OLR* down by 17 W/m^2, and going all the way to a saturated atmosphere only brings *OLR* down by a further 12 W/m^2. Further, at very high CO_2 concentrations the *OLR* becomes somewhat more insensitive to humidity, as CO_2 begins to dominate the greenhouse effect.

The information presented graphically in Figs. 4.32, 4.33, and 4.34 amount to a miniature climate model, allowing many interesting questions about climate to be addressed quantitatively without the need to perform detailed radiative and thermodynamic calculations; it's the kind of climate model that could be printed on a wallet-sized card and carried around everywhere. Calculations using this information can be simplified by presenting the data as polynomial fits, which eliminates the tedium and inaccuracy of measuring quantities off graphs. Then, what we have is a miniature climate model that can be programmed into a pocket calculator. Polynomial fits allowing the *OLR* to be calculated as a function of temperature, CO_2, and relative humidity are tabulated in Tables 4.2 and 4.3. Values of *OLR* for parameters intermediate between the tabulated ones can easily be obtained by interpolation.

CO_2	a_0	a_1	a_2	a_3
10 ppmv	258.56	2.5876	−0.0059165	−0.00013402
100 ppmv	246.13	2.5056	−0.0034095	−0.00010672
1000 ppmv	232.51	2.3815	−0.0015855	−8.3397e − 05
10000 ppmv	215.74	2.1915	0.00056634	−5.0508e − 05
100000 ppmv	189.06	1.8554	0.0044094	1.0735e − 05

Calculation carried out with 50% relative humidity.

Table 4.2 Coefficients for polynomial fit $OLR = a_0 + a_1 x + a_2 x^2 + a_3 x^3$, where $x = T_g - 275$.

h_{rel}	a_0	a_1	a_2	a_3
Dry	313.8	−6.275	−0.36107	−0.019467
0.1	277.28	−5.9416	−0.35596	−0.020237
0.5	259.52	−5.5332	−0.33915	−0.018932
Saturated	249.1	−5.183	−0.32187	−0.017367

Calculation carried out with $T_g = 280$ K for the indicated moisture conditions. h_{rel} is the relative humidity, given as a fraction.

Table 4.3 Coefficients for polynomial fit $OLR = a_0 + a_1 x + a_2 x^2 + a_3 x^3$, where $x = \ln(CO_2/100)$, with CO_2 in ppmv.

Using these polynomial fits, we can put numbers to some of our old favorite climate questions by solving $OLR(T, CO_2) = (1 - \alpha)S$ for T, given various assumptions about CO_2 and the absorbed solar radiation $(1 - \alpha)S$. To wit:

- *Global Warming:* If we assume an albedo of 22.5%, the absorbed solar radiation is 265 W/m^2. For 50% relative humidity, and a pre-industrial CO_2 concentration of 280 ppmv, the corresponding equilibrium temperature is 285 K, which is close to the pre-industrial global mean temperature. The albedo we needed to assume to get this base case is somewhat smaller than the Earth's observed albedo, because a portion of the cloud albedo is offset by the cloud greenhouse effect. Now, if we double the CO_2 to 560 ppmv, the new temperature is 287 K – a two degree warming. This is essentially the same answer as obtained by Manabe and Wetherald in their pioneering 1967 calculation, and was obtained by essentially the same kind of calculation we have employed. If we double CO_2 once more, to 1120 ppmv, then the temperature rises to 289 K, a further two kelvin of warming. The fact that each *doubling* of CO_2 gives a fixed additional increment of warming reflects the logarithmic dependence of *OLR* on CO_2; until one gets to extremely high concentrations, each doubling reduces the *OLR* by approximately 4 W/m^2.

- *Pleistocene glacial-interglacial cycles:* In the depths of an ice age, the CO_2 drops to 180 ppmv. Using the same base case as in the previous example, the temperature drops to 284 K, about a one kelvin cooling relative to the base case. This by no means accounts for the full amount of ice age cooling, but it is significant enough to imply that CO_2 is a major player. In the southern hemisphere midlatitudes, away from the direct influence of the growth of major northern hemisphere ice sheets, the CO_2 induced cooling is a half to a third of the total, indicating that either the ice sheet influence is propagated into the southern hemisphere through the atmosphere or ocean, or that the cooling we have calculated has been further enhanced by feedbacks due to clouds or sea ice.

- *The PETM warming:* How much CO_2 would you have to dump into the ocean–atmosphere system to account for the PETM warming discussed in Section 1.9.1? One answers this by first deciding how much the atmospheric CO_2 concentration needs to increase, and then making use of information about the partitioning of carbon between the atmosphere and ocean. The PETM warming has been conservatively estimated at 4 °C, and has nearly the same magnitude in the tropics as in the Arctic. The PETM event starts from an already warm hothouse climate, so for the sake of argument let's assume that the CO_2 starts at four times the pre-industrial value, yielding a starting temperature of 289 K. We need to increase the CO_2 from 1120 ppmv to 3600 ppmv (just under two doublings) to achieve a warming to 293 K. This amounts to an addition of 5340 gigatonnes of C to the atmosphere as CO_2, which is barely within the limits imposed by the ^{13}C data, but not all carbon added to the atmosphere stays in the atmosphere. In order for 5340 gigatonnes to remain in the atmosphere after a thousand years of ocean uptake, a substantially larger amount needs to be added to the system (see Chapter 8). Can such a large addition of carbon be reconciled with the carbon isotopic record? This is the essential puzzle of the PETM, and may call for some kind of strong destabilizing feedback in the climate system.

- *Deglaciation of Snowball Earth:* If we increase the Earth's albedo to 60% (in accord with the reflectivity of ice) and reduce the solar constant by 6% (in accord with the Neoproterozoic value) the absorbed solar radiation is only 128 W/m². With CO_2 of 280 ppmv, the equilibrium global mean temperature is a chilly 228 K, more or less independent of what we assume about relative humidity. To determine the deglaciation threshold, we'll assume generously that the Equator is 20 K warmer than the global mean, so that we need to warm the global mean to 254 K to melt the tropics. Assuming 50% relative humidity, increasing CO_2 all the way to 200 000 ppmv (about 20% of the atmosphere) still only brings the global mean up to 243 K, which is not enough to deglaciate. From this we conclude that without help from some other feedback in the system, CO_2 would have to be increased to values in excess of 20% of the atmosphere to deglaciate. In fact, detailed climate model calculations indicate that it is even harder to deglaciate a Snowball than this calculation suggests.

- *Temperature of the post-Snowball hothouse:* Let's assume that somehow or other the Snowball does deglaciate when CO_2 builds up to 20%. After deglaciation, the albedo will revert to 22.5%, and the absorbed solar radiation to 249 W/m². When the planet re-establishes equilibrium, the temperature will have risen to 311 K. This is hot, and the tropics will be hotter than the global mean. However, the planet does not enter a runaway greenhouse and these temperature are well within the survival range of heat-tolerant organisms, especially since the polar regions would probably be no warmer than today's tropics.

- *Faint Young Sun:* Let's consider a time when the Sun was 25% fainter than today, reducing the absorbed solar radiation by 66 W/m². How much would CO_2 have to increase relative to pre-industrial values in order to keep the global mean temperature at 280 K and prevent a freeze-out? We'll address this question by using the fit in Table 4.3. For 280 ppmv of CO_2 the *OLR* is 253 W/m² assuming 50% relative humidity. We need to bring this down by 66 W/m² to make up for the faint Sun. Using the fit, this can be done by increasing CO_2 to 240 000 ppmv (24% of the atmosphere) if we keep the relative humidity at 50%. If we on the other hand assume that for some reason the atmosphere becomes saturated with water vapor, then it is only necessary to increase the CO_2 to 15% of the atmosphere.

- *The Earth in one billion years:* According to Eq. (1.1), the solar constant will have increased to $1497 \, W/m^2$, increasing the absorbed solar radiation per unit surface area to $290 \, W/m^2$. If CO_2 is held fixed at the pre-industrial value, the Earth will warm to a global mean temperature of $296 \, K$ if relative humidity is held fixed at 50%. In order for silicate weathering to restore a temperature of $287 \, K$, the weathering would have to bring the CO_2 all the way down to $10 \, ppmv$, at which point photosynthesis as we know it would probably become impossible.

- *Temperature of Gliese 581c and 581d:* The planets Gliese 581c and 581d are in close orbits around a dim M-dwarf star. The redder spectrum of an M-dwarf would have some effect on the planetary energy balance, through changes in the proportion of solar energy absorbed directly in the atmosphere. Neglecting this effect, though, we can estimate the temperatures of these planets assuming them to have an Earthlike atmosphere consisting of water vapor, CO_2, and N_2/O_2. Gliese 581c is in an orbit where it would absorb about $583 \, W/m^2$, assuming the typical albedo of a rocky planet with an ocean. Gliese 581d would absorb only $50 \, W/m^2$. Even if CO_2 were 20% of the atmosphere, the *OLR* would be $144 \, W/m^2$ for a mean surface temperature of $254 \, K$, so Gliese 581d is likely to be an icy Snowball. On the other hand even with only $1 \, ppmv$ of CO_2 in the atmosphere the *OLR* at $330 \, K$ would be $351 \, W/m^2$, far below the absorbed solar radiation for Gliese 581c. Thus, if Gliese 581c has an ocean, it is very likely to be in a runaway state – something we will confirm when we re-examine the runaway greenhouse for real gas atmospheres. There is an additional wrinkle to the Gliese system, though, in that these planets are more massive than Earth and have higher surface gravity. The higher gravity somewhat reduces the water vapor greenhouse effect, since for a given vapor pressure the corresponding amount of mass in the atmosphere is lower, according to the hydrostatic relation. This turns out to cool Gliese 581c somewhat, but still not enough to save it from a runaway.

Exercise 4.11 About how much carbon would need to be added to the atmosphere to achieve a $4 \, K$ PETM warming if the initial CO_2 at the beginning of the event were only twice the pre-industrial value? If the initial CO_2 were eight times the pre-industrial value? (Note that in the first case we are implicitly assuming some unknown process keeps the late Paleocene warm even with relatively little help from extra CO_2.)

The above calculations include the effects of water vapor feedback, and also include the effects of changes in albedo due to ice cover, where explicitly mentioned. However, they do not incorporate any feedbacks due to changing cloud conditions. Cloud changes could either amplify or damp the climate change predicted on the basis of clear-sky physics, according to whether changes in the cloud greenhouse effect or the cloud albedo effect win out. Unfortunately, there is no simple thermodynamic prescription that does for cloud feedbacks what the assumption of fixed relative humidity does for water vapor feedbacks. We will learn more about the factors governing cloud radiative forcing in Chapter 5, but idealized conceptual models for prediction of cloud feedbacks remain elusive.

The problem of whether elevated CO_2 can account for hothouse climates such as the Cretaceous and Eocene is considerably more challenging than the other problems we have discussed above, since it requires one to answer a regional climate question: Under what conditions can we suppress the formation of polar ice? We already saw in Chapter 3 that a rather small change in radiation balance can make the difference between a planet with a small polar ice sheet and a planet which is globally ice-free. Suffice it to say at this point that an increase of CO_2 to 16 times pre-industrial values – the upper limit of what is plausibly consistent with proxy data – would yield a global mean warming of 10 K. Would this be enough to suppress formation of sea ice in the Arctic, and keep the mean Arctic temperatures around 10 °C? Would the associated tropical temperatures be too hot to be compatible with available proxy data? We will have to return to these questions in Chapter 7, where we discuss the regional and seasonal variations of climate.

4.5.4 Greenhouse effect of CO_2 vs. CH_4

There is considerable interest in the idea that on the Early Earth methane may have taken over much of the role of CO_2 in offsetting the Faint Young Sun. In part this interest is due to rather sketchy geochemical evidence that at some times in the Archean CO_2 concentrations may not have been high enough to do the trick, but regardless of whether the evidence actually *demands* a relatively low-CO_2 atmosphere, possibilities abound that in an anoxic atmosphere methane could build up to high concentrations. Even on an abiotic planet, there are possibilities for direct volcanic outgassing of methane, at a rate dependent on the state of oxygen in the planet's interior. Once biology comes on the scene, methanogens can convert volcanic CO_2 and H_2 to CH_4, or can make CH_4 by decomposing organic matter produced by anoxygenic photosynthesis.

Methane cannot build up to very high concentrations in a well-oxygenated atmosphere because it oxidizes to CO_2, but the relatively small amounts of methane in the atmosphere today (about 1.7 ppmv) nevertheless contribute significantly to global warming. There is, however, a widespread misconception that methane is in some sense an intrinsically better greenhouse gas than CO_2. A few simple calculations will serve to clarify the true state of affairs.

In order to compare the relative effects of CH_4 and CO_2 on a planet's radiation budget, we calculate the *OLR* for each case in what we have been calling the canonical atmosphere – a mixture of each gas into a dry atmosphere consisting of 1 bar of Earth air with temperature profile on the dry air adiabat, carried out with Earth's surface gravity. Results for a fixed surface temperature of 280 K, computed using the homebrew radiation model employing a constant temperature scaling coefficient $T^* = 900$ K, are shown in Fig. 4.35. This graph in essence gives the amount of greenhouse gas needed to sustain a surface temperature of 280 K, given any specified amount of solar absorption. For example, with an absorbed solar radiation of 300 W/m^2 the surface temperature can be sustained with either 464 ppmv of CO_2 or 35 600 ppmv of CH_4 (3.56% of the atmosphere by mole fraction). These results somewhat overestimate the effect of each gas as compared with an actual moist atmosphere, since a moist atmosphere would be on the less steep moist adiabat and in a moist atmosphere the water vapor absorption would compete to some extent with the CO_2 and CH_4 absorption. Still, as an estimate of the relative effect of the two gases, the story is pretty clear. Methane is, intrinsically speaking, a considerably worse greenhouse gas than CO_2. The *OLR* curve for methane is everywhere well above the curve for CO_2, so that it takes more methane than CO_2 to achieve a given reduction of *OLR*.

The common statement that methane is, molecule for molecule, a better greenhouse gas than CO_2 is true only for situations like the present where methane is present in far lower concentrations than CO_2. In this situation, the greater power of a molecule of CH_4 to reduce the *OLR* results simply from the fact that the greenhouse effects of both CH_4 and CO_2 are approximately logarithmic in concentration. Reading from Fig. 4.35, we see that for methane

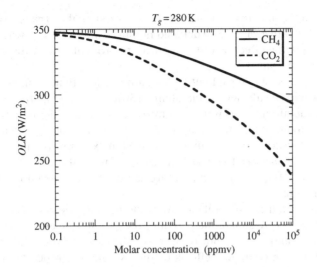

Figure 4.35 *OLR* vs. CO_2 and CH_4 concentration for each gas individually mixed with Earth air on the dry adiabat. The surface temperature is fixed at 280 K.

concentrations of around 1 ppmv, each doubling of methane reduces *OLR* by about $2 \, \text{W/m}^2$. On the other hand, for CO_2 concentrations near 300 ppmv, each doubling of CO_2 reduces the *OLR* by about $6 \, \text{W/m}^2$. Hence, to achieve the same *OLR* reduction as a doubling of CO_2 one needs three doublings of methane, but since methane starts from a concentration of only 1 ppmv, this only takes the concentration to 8 ppmv, and requires only $\frac{7}{300}$ as many molecules to bring about as were needed to achieve the same reduction using a doubling of CO_2. Equivalently, we can say that adding 1 ppmv of methane yields as much reduction of *OLR* as adding 75 ppmv of CO_2. The logarithmic slopes in this example are exaggerated compared with the appropriate values for Earth's actual atmosphere, because of the use of the dry adiabat and because of inaccuracies in the simple temperature scaling used in the homebrew radiation code. Using the ccm radiation code on the moist adiabat, with water vapor at 50% relative humidity, we find instead that each doubling of methane near 1 ppmv reduces *OLR* by $0.77 \, \text{W/m}^2$, while each doubling of CO_2 near 300 ppmv reduces *OLR* by $4.3 \, \text{W/m}^2$; in this case adding 1 ppmv of methane reduces the *OLR* by as much as adding 38 ppmv of CO_2. Nonetheless, the principle remains the same: If methane were the most abundant long-lived greenhouse gas in our atmosphere, and CO_2 were present only in very small concentrations, we would say instead that CO_2 is, molecule for molecule, the better greenhouse gas.

Joseph Kirschvink and others have proposed that the Makganyene Snowball came about through a methane catastrophe, in which oxygenation converts methane to CO_2 and reduces the greenhouse effect sufficiently to precipitate a Snowball. A methane crash, due to a reduction in methanogenic activity or some other mechanism, has sometimes been proposed as a trigger for the Neoproterozoic Snowballs as well. It is by no means easy to make these scenarios play out as they are supposed to, since methane contributes much less to the greenhouse effect than CO_2 when the two gases have similar abundances in the atmosphere. Conversion of methane to CO_2 will only reduce the greenhouse effect if methane is initially present in sufficiently small concentrations – but if there is too little methane present, the contribution of methane to the total greenhouse effect is too small to make much difference.

Let's use the data in Fig. 4.35 to illustrate how the conversion of methane to CO_2 would affect climate in a few illustrative cases. The detailed numbers would change with a more accurate radiation model, or if the effect of water vapor were brought in, but the basic conclusions would remain much the same.

For example, suppose we started out with an atmosphere that contained 36 650 ppmv of methane, which would be sufficient to maintain a 280 K surface temperature given 300 W/m^2 of absorbed solar radiation. If this were all converted to CO_2 by oxidation, then, according to Fig. 4.35, the *OLR* would plunge to 255 W/m^2. In order to re-establish radiation balance, the planet would have to warm-up to a temperature well in excess of the initial 280 K. Far from causing a Snowball, in this case the oxidation of methane would cause a hot pulse, followed by gradual recovery to the original temperature as the CO_2 is drawn down by ocean uptake and silicate weathering.

Next let's consider a more general situation, and identify the conditions necessary for a conversion of CH_4 to CO_2 to substantially reduce the greenhouse effect. Because the absorption features of CH_4 and CO_2 do not overlap significantly for the range of concentration under consideration, the combined effects of the two gases can be obtained by summing the *OLR* reduction ΔOLR for each of the gases taken in isolation. We wish to ask the question: if we have a given number of carbon atoms to use in supplying the atmosphere with greenhouse gases, how does the net greenhouse effect depend on the way we divide up those atoms between CH_4 and CO_2? Since each molecule has a single carbon, this question can be addressed by varying the molar concentration of CH_4 while keeping the sum of the molar concentrations of CH_4 and CO_2 fixed. Results of a calculation of this type are shown in Fig. 4.36. For any fixed total atmospheric carbon content, the *OLR* reduction has a broad maximum when plotted as a function of the methane ratio, and varies little except near the

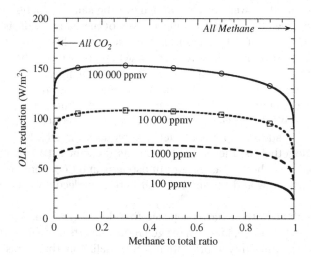

Figure 4.36 Total *OLR* reduction for the canonical atmosphere with a mixture of CH_4 and CO_2. Each curve gives the *OLR* reduction relative to a transparent atmosphere for a fixed sum of CH_4 and CO_2 molar concentrations, indicated on the curve in units of ppmv. The results are plotted as a function of the ratio of the CH_4 molar concentration to the total molar concentration for the two gases. The ratio is equal to the ratio of atmospheric carbon in the form of CH_4 to total atmospheric carbon. Larger values of the *OLR* reduction correspond to a stronger greenhouse effect.

extremes of an all-methane or all-CO_2 atmosphere. The only case in which one can get a substantial reduction in greenhouse effect by oxidizing methane into CO_2 is when the initial CO_2 concentration is very high, the initial CH_4 fraction is between about 10% and 90% of the total, and the CH_4 is almost entirely converted to CO_2. For example, with a total carbon concentration of 10 000 ppmv, reducing the CH_4 concentration from 1000 ppmv to 1 ppmv reduces the greenhouse effect from 104 W/m^2 to 80 W/m^2. Because the curve is so flat, starting from an atmosphere which is 80% methane works almost as well: in that case the greenhouse effect is reduced from 101 W/m^2 to 80 W/m^2. If we have a total of 100 000 ppmv of carbon in the atmosphere, then the maximum greenhouse effect occurs for an atmosphere which is about 25% methane, and has a value of 151 W/m^2. Reducing the methane to 1 ppmv brings down the greenhouse effect 36 W/m^2, to 115 W/m^2. In the Paleoproterozoic or Archean, when the net greenhouse effect needed to be high to offset the Faint Young Sun, it is possible that a methane crash could have reduced the greenhouse effect enough to initiate a Snowball, but it is essential that in a methane crash, the methane concentration be brought almost all the way down to zero; a reduction of methane from 50% of the atmosphere to 10% of the atmosphere would not do much to the greenhouse effect. By the time of the Neoproterozoic, when the solar luminosity is higher and less total greenhouse effect is needed to maintain open water conditions, it is far less likely that a methane catastrophe could have initiated glaciation. Some further remarks on atmospheric transitions that could initiate a Snowball will be given in Chapter 8.

4.6 ANOTHER LOOK AT THE RUNAWAY GREENHOUSE

We are now equipped to revisit the runaway greenhouse phenomenon, this time using the absorption spectrum of actual gases in place of the idealized gray gas employed in Section 4.3.3. The setup of the problem is essentially the same as in the gray gas case. We consider a condensable greenhouse gas, optionally mixed with a background gas which is transparent to infrared and non-condensing. A surface temperature T_g is specified, and the corresponding moist adiabat is computed. The temperature and the greenhouse gas concentration profiles provide the information necessary to compute the *OLR*, in the present instance using the homebrew exponential sums radiation model in place of the gray gas *OLR* integral. As before, the *OLR* is plotted as a function of T_g for the saturated atmosphere, and the Kombayashi–Ingersoll limit is given by the asymptotic value of *OLR* at large surface temperature.

We'll begin with water vapor. Figure 4.37 shows the results for a pure water vapor atmosphere, computed for various values of the surfaced gravity. The overall behavior is very similar to the gray gas result shown in Fig. 4.3: the *OLR* attains a limiting value as temperature is increased, and the limit – defining the absorbed solar radiation above which the planet goes into a runaway state – becomes higher as the surface gravity is increased, and for precisely the same reasons as invoked in the gray gas case. The result, however, is now much easier to apply to actual planets, since with the real gas calculations we have the real numbers in hand for water vapor, and not for some mythical gas characterized by a single absorption coefficient. Several specific applications will be given shortly.

As in the gray gas case the limiting *OLR* increases as surface gravity is increased. It would be useful if the real gas result could be represented in terms of an equivalent gray gas, but first one must generalize Eq. (4.39) to incorporate the increase of absorption coefficient with pressure. This is done formally in Problem 4.5, but the qualitative derivation runs as

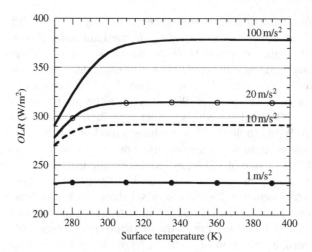

Figure 4.37 *OLR* vs. surface temperature for a saturated pure water vapor atmosphere. The numbers on the curve indicate the planet's surface gravity. The calculation was done with the homebrew exponential sums radiation code, incorporating both the 1000 cm^{-1} and 2200 cm^{-1} continua, but neglecting temperature scaling of absorption outside the continua. Twenty terms were used in the exponential sums, and wavenumbers out to 5000 cm^{-1} were included; the atmosphere was considered transparent to higher wavenumbers.

follows. We need to determine the pressure p_1 where the optical thickness to the top of the atmosphere is unity, and then evaluate the temperature at that point, along the one-component saturated adiabat. For linear pressure broadening or the continuum the optical thickness requirement implies $\frac{1}{2}\kappa_0 p_1^2/p_0 g = 1$, where p_0 is the reference pressure to which the absorption is referred (generally 100 mb for the data given in our survey of gaseous absorption properties). Substituting the resulting p_1 into the expression for $T(p)$ we infer an expression of the form

$$OLR_\infty = A'\sigma T(p_1)^4 = A'\frac{\sigma(L/R)^4}{[\ln(p^*/\sqrt{2p_0g/\kappa_0})]^4} \qquad (4.94)$$

where p^* is defined as before and A' is an order unity constant which depends on L/R. An examination of the g dependence of the calculated Kombayashi–Ingersoll limit in Fig. 4.37 shows that over the range $1\,\text{m/s}^2 \le g \le 100\,\text{m/s}^2$, the numerically computed dependence can be fit almost exactly with this formula if we take $A' = 0.7344$ and $\kappa_0 = 0.055$ (assuming $p_0 = 10^4$ Pa). Though the pressure dependence of absorption causes the limiting *OLR* to vary more slowly with g than was the case for constant κ, the limit in the real gas case otherwise behaves very much like an equivalent gray gas with $\kappa_0 = 0.055$. This is a surprising result, given the complexity of the real gas absorption spectrum. The fact that the equivalent absorption is similar to that characterizing the 2500 cm^{-1} continuum suggests that the limiting *OLR* is being controlled primarily by this continuum. Thus, the behavior of this continuum is crucial to the runaway greenhouse phenomena (see Problem 4.25). It cannot be ruled out that other continua may affect the *OLR* as temperature is increased to very high values. For example, the total blackbody radiation at wavenumbers greater than 5000 cm^{-1} is only 1.14 W/m^2 at 500 K, so it matters little what the absorption properties are in that part of the spectrum at 500 K or cooler. However, when the temperature is raised to 600 K the shortwave emission is $15 \, W/m^2$ and so the shortwave absorption begins to matter; by the time T reaches 700 K the shortwave blackbody emission is $106 \, W/m^2$ and the shortwave emission properties are potentially important. On the other hand, at such temperatures there is so much water vapor in the atmosphere that a very feeble absorption would be sufficient to eliminate the contribution to the *OLR*.

Exercise 4.12 Verify the shortwave blackbody emission numbers given in the preceding paragraph by using numerical quadrature applied to the Planck function.

Exercise 4.13 Compute the Kombayashi–Ingersoll limit for water vapor on Mars, which has $g = 3.71 \, m/s^2$. Compute the limit for Titan, which has $g = 1.35 \, m/s^2$, and Europa which has $g = 1.31 \, m/s^2$.

Now let's generalize the calculation, and introduce a non-condensing background gas which is transparent in the infrared; we use N_2 in this example, though the results are practically the same if we use any other diatomic molecule. The background gas affects the Kombayashi–Ingersoll limit in two ways. First, the pressure broadening increases absorption, which should lower the limit. Second, the background gas shifts the lapse rate toward the dry adiabat, which is much steeper than the single-component saturated adiabat. The increase in lapse rate in principle could enhance the greenhouse effect, but given the condensable nature of water vapor, it actually reduces the greenhouse effect, because the low temperatures aloft sharply reduce the amount of water vapor there. If the background gas were itself a greenhouse gas, this effect might play out rather differently. At sufficiently high temperatures, water vapor will dominate the background gas and so the limiting *OLR* at high temperature will approach the pure water vapor limit shown in Fig. 4.37. However, for intermediate temperatures, the background gas can modify the shape of the *OLR* curve.

Results for various amounts of N_2 are shown in Fig. 4.38. As expected, at large temperatures the limiting *OLR* asymptotes to the value for a pure water vapor atmosphere. This can be seen especially clearly for the case with only 100 mb of N_2 in the atmosphere; with more N_2, one has to go to higher temperatures before the water vapor completely dominates the *OLR*, but the trend is clear. A very important qualitative difference from the pure water vapor case is that the *OLR* curve for a binary mixture shows a distinct maximum at intermediate temperatures. This maximum arises because the foreign broadening of water vapor absorption features is relatively weak, while the presence of a non-condensing background gas steepens the lapse rate and reduces the amount of water vapor aloft. The hump in the curve means that the surface temperature exhibits *multiple equilibria* for a given absorbed solar radiation. For example, with 100 mb of N_2, if the absorbed solar radiation is $320 \, W/m^2$ there is a cool equilibrium with $T_g = 288 \, K$ and a hot equilibrium with $T_g = 360 \, K$. The latter is an unstable equilibrium; displacing the temperature in the cool direction will cause water vapor to condense and *OLR* to increase, cooling the climate further until the system falls into the cool equilibrium. Conversely, displacing the temperature slightly to the warm side of the hot equilibrium will cause the climate to go into a runaway state. For these atmospheric parameters, the planet is in a metastable runaway state. The climate will persist in the cooler non-runaway state unless some transient event warms the planet enough to kick it over into the runaway regime. It is only when the absorbed solar radiation is increased to the maximum *OLR* at the peak ($328 \, W/m^2$) that a runaway becomes *inevitable*. For future use, we'll note that a calculation with $g = 10 \, m/s^2$ and 1

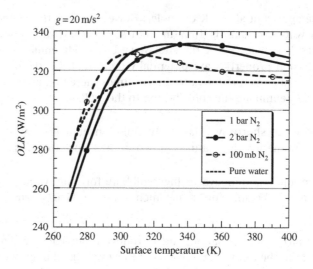

Figure 4.38 As for Fig. 4.37 but for a mixture of water vapor in N_2 on the saturated moist adiabat. Calculations were carried out with a surface gravity of $20 \, \text{m/s}^2$, for the indicated values of N_2 partial pressure at the ground.

bar of N_2 has a peak *OLR* of $310 \, \text{W/m}^2$ at $T_g = 325 \, \text{K}$, while a case with the same gravity and 3 bars of N_2 likewise has a peak at $310 \, \text{W/m}^2$ but the position of the peak is shifted to $360 \, \text{K}$. The corresponding parameters for the slightly lower surface gravity of Venus differ little from these numbers. Both cases asymptote to the *OLR* for pure water vapor when the temperature is made much larger than the temperature at which the peak *OLR* occurs.

- *Runaway greenhouse on Earth:* With present absorbed solar radiation (adjusted for net cloud effects) of $265 \, \text{W/m}^2$, the Earth at present is comfortably below the Kombayashi–Ingersoll limit for a planet of Earth's gravity. According to Eq. (1.1), as the solar luminosity continues to increase, the Earth will pass the $291 \, \text{W/m}^2$ threshold where a runaway becomes possible in about 700 million years. In 1.7 billion years, it will pass the $310 \, \text{W/m}^2$ threshold where a runaway becomes inevitable for an atmosphere with 1 bar of N_2 and no greenhouse gases other than water vapor.
- *Venus:* The present high albedo of Venus is due to sulfuric acid clouds that would almost certainly be absent in a less dry atmosphere. If we assume an Earthlike albedo of 30%, then very early in the history of the Solar System, the absorbed solar radiation of Venus would be $327 \, \text{W/m}^2$. This is just barely in excess of the mandatory runaway threshold of $310 \, \text{W/m}^2$ for a planet of Venus' surface gravity, assuming a bar or two of N_2 in the atmosphere and no greenhouse gases other than water vapor. It is thus possible that neglected effects (clouds, subsaturation, a higher albedo surface) could allow Venus to exist for a while in a hot, steamy but non-runaway state with a liquid ocean. The high water vapor content of the upper atmosphere would still allow an enhanced rate of photodissociation and escape of water to space. If Venus indeed started life with an ocean, however, it is plausible that it eventually succumbed to a runaway state, since with the present solar constant the absorbed solar radiation without sulfuric acid clouds would be $457 \, \text{W/m}^2$, well in excess of the runaway threshold.

- *Gliese 581c*: We can now improve our earlier estimates of the conditions on the extrasolar planet Gliese 581c, which has an absorbed solar radiation of $583 \, W/m^2$ assuming a rocky surface. This flux is well above the threshold of $334 \, W/m^2$ for a mandatory runaway for a planet with twice Earth's surface gravity, even allowing for 2 bar of N_2 in the atmosphere. Thus, if Gliese 581c ever had an ocean it is likely to have gone into a runaway state; if the composition of the planet included a substantial amount of carbonate in the interior, subsequent outgassing is likely to have turned it into a planet rather like Venus. It still remains, however, to assess the implications of the increased atmospheric absorption due to the greater proportion of infrared output of the M-dwarf host star.

- *Evaporation of icy moons in Earthlike orbit*: It has been suggested that icy moons like Europa or Titan could become habitable if the host planet were in an orbit implying Earthlike solar radiation. The low Kombayashi–Ingersoll limit for bodies with low surface gravity puts a severe constraint on this possibility, however. With the albedo of ice, such bodies could exist as Snowballs in an Earthlike orbit, but if the surface ever thawed, or failed to freeze in the first place, the absorbed solar radiation corresponding to an albedo of 20% would be $274 \, W/m^2$ – well above the runaway threshold of $232 \, W/m^2$ for a body with surface gravity of $1 \, m/s^2$. Small icy moons in Earthlike orbits are thus likely to evaporate away, unless they are locked in a Snowball state.

- *Lifetime of a post-impact steam atmosphere*: Suppose that in the Late Early Bombardment stage, enough asteroids and comets hit the Earth to evaporate 10 bars worth of the ocean and give the Earth a 10 bar atmosphere consisting of essentially pure water vapor (and a surface temperature in excess of 440 K, according to Clausius–Clapeyron). How long would it take for the steam atmosphere to rain out and the temperature to recover to normal? To do this problem, we assume that the atmosphere remains saturated as it cools, and loses heat at the maximum rate given by the Kombayashi–Ingersoll limit for Earth; we also need to subtract the absorbed solar radiation from the heat loss. For Early Earth conditions, the net heat loss is about $100 \, W/m^2$. On the other hand, the latent heat per square meter of the Earth's surface in a steam atmosphere with surface pressure p_s is $L p_s/g$, or $2.5 \cdot 10^{11} \, J/m^2$ for the stipulated atmosphere. To remove this amount of energy at a rate of $100 \, W/m^2$ would take $2.5 \cdot 10^9$ s, or 80 years. The rainfall rate during this time would be warm but gentle: $3.5 \, (kg/m^2)/day$, or a mere $3.5 \, mm/day$ based on the water density of $1000 \, kg/m^3$. This is the average rainfall rate constrained by the rate of radiative cooling, but it is likely that at places the local rainfall rate could be orders of magnitude greater, owing to the lifting and condensation in storms and other large scale atmospheric circulations.

- *Freeze-out time of a magma ocean*: In Chapter 1 we introduced the problem of the freeze-out time of a magma ocean on the Early Earth, and estimated the time assuming a transparent atmosphere in Problem 3.24. How long does it take for the magma ocean to freeze-out if the planet is sufficiently water-rich that the atmosphere consists of essentially pure water vapor in saturation? The time is estimated in the same way as in Problem 3.24 except that the rate of heat loss is again taken to be the difference between the Kombayashi–Ingersoll limit – giving the maximum *OLR* – and the rate of absorption of solar energy. For the Early Earth this would be about $100 \, W/m^2$, which is far smaller than the transparent atmosphere case, where the energy loss is nearly $100\,000 \, W/m^2$ based on σT^4 for the 2000 K temperature of molten magma. As a result, the freeze-out time (using the same assumptions as in Problem 3.24) increases to 3.5 million years.

The latter two estimates follow the line of thinking introduced by Norman Sleep of Stanford University, and again illustrate the principle that Big Ideas come from simple models.

Exercise 4.14 Estimate the lifetime of a post-impact pure water vapor atmosphere on Mars assuming that the planet absorbs $90\,W/m^2$ of solar radiation, per unit surface area. Estimate the precipitation rate, in mm of liquid water per day.

The above results presume that the saturated, condensable greenhouse gas is the only greenhouse gas present in the atmosphere. What happens if the atmosphere also contains a non-condensable greenhouse gas, whose total mass remains fixed as the surface temperature increases? For Earthlike or Venuslike conditions, for example, one would typically need to consider atmospheres consisting of a mixture of condensable water vapor in saturation, non-condensing CO_2, and perhaps a transparent non-condensing background gas such as N_2. Can the addition of CO_2 in this situation trigger a runaway greenhouse when water vapor alone could not support a runaway? We will not pursue detailed radiative calculations of this sort, but some simple qualitative reasoning, summarized in the sketch in Fig. 4.39, suffices to map out the general behavior. The essential insight is that, at sufficiently high temperatures, the atmosphere is completely dominated by the condensable component, whose mass increases exponentially with temperature. Hence, the Kombayashi–Ingersoll limit will be unaffected by the addition of the non-condensable greenhouse gas. However, as more and more non-condensable greenhouse gas is added to the atmosphere, one must go to ever-higher temperatures before the limiting *OLR* is approached. At lower temperatures, the addition of a large amount of non-condensing greenhouse gas brings down the *OLR* below the Kombayashi–Ingersoll limit. Whether or not this triggers a runaway depends on the details of the situation. If the *OLR* curve without the non-condensable greenhouse gas is essentially monotonic in temperature, as in the one-component cases in Fig. 4.37, then the addition of the non-condensable greenhouse gas warms the planet, but does not trigger a

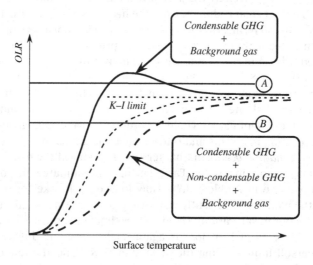

Figure 4.39 Qualitative influence of a non-condensable greenhouse gas (GHG) on the shape of the *OLR* curves. The upper curve gives the *OLR* for an atmosphere consisting of a mixture of a saturated condensable greenhouse gas with a non-condensing transparent background gas, as in the N_2/H_2O case shown in Fig. 4.38, while the lower curve illustrates how the behavior would change if a large amount of non-condensable greenhouse gas were added. The intermediate curve gives the *OLR* for a one-component saturated greenhouse gas atmosphere as in Fig. 4.37.

runaway if the absorbed solar radiation is below the Kombayashi–Ingersoll limit. However, in a case like Fig. 4.38, in which the *OLR* curve overshoots the limit and has a maximum, the addition of the non-condensable can eliminate the hump in the curve, eliminating the stable non-runaway state and forcing the system into a runaway. In the sketch, this situation is illustrated by the absorbed solar radiation line labeled "A." In that case, the addition of the non-condensing gas can indeed force the system into a runaway state. On the other hand, if the absorbed solar radiation is below the Kombayashi–Ingersoll limit, as in the line labeled "B," then the addition of the non-condensable greenhouse gas warms the equilibrium but does not trigger a runaway.

The concepts of runaway greenhouse and the Kombayashi–Ingersoll limit generalize to gases other than water vapor. For example, consider a planet with a reservoir of condensed CO_2 at the surface, which may take the form of a CO_2 glacier or a CO_2 ocean, according to the temperature of the planet. Specifically, if the surface temperature is above the triple point of 216.5 K the condensable reservoir takes the form of a CO_2 ocean; otherwise it takes the form of a dry-ice glacier. If the atmosphere is in equilibrium with the surface reservoir and has no other gases in it besides the CO_2 which evaporates from the surface, then one can use the one-component adiabat with the homebrew radiation code to compute an *OLR* curve for the saturated CO_2 atmosphere which is analogous to the water vapor result shown in Fig. 4.37. Results, for various surface gravity, are shown in Fig. 4.40. The general behavior is very similar to that we saw for water vapor, but the whole system operates at a lower temperature and the *OLR* reaches its limiting value at a much lower temperature than was the case for water vapor.

The CO_2 runaway imposes some interesting constraints on the form in which CO_2 could exist on Mars, both present and past. For Martian surface gravity, the Kombayashi–Ingersoll limit for CO_2 is a bit over 63 W/m^2. In consequence, when the absorbed solar radiation exceeds this value, a permanent reservoir of condensed CO_2 cannot exist at the surface of

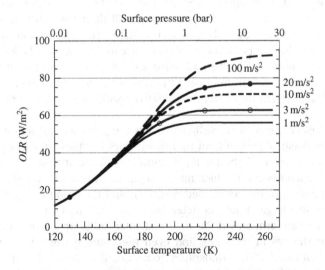

Figure 4.40 *OLR* vs. surface temperature for a saturated pure CO_2 atmosphere. Calculations were performed with the values of surface gravity indicated on each curve. The scale at the top gives the surface pressure corresponding to the temperature on the lower scale.

the planet; it will sublimate or evaporate into the atmosphere, and continue to warm the planet until all the condensed reservoir has been converted to the gas phase. At present, the globally averaged solar absorption is about $110 \, W/m^2$, so the planet is well above the runaway threshold for CO_2. From this we can conclude that Mars cannot at present have an appreciable permanent reservoir of condensed CO_2 which can exchange with the atmosphere. Note, however, that this does not preclude the temporary buildup of CO_2 snow at the surface. Such deposits can and do form near the winter poles, but sublimate back into the atmosphere as spring approaches. This situation can be thought of as arising from the fact that the *local* absorbed solar radiation near the winter pole is below the Kombayashi-Ingersoll limit for CO_2. The local reasoning applies because the thin atmosphere of present Mars cannot effectively transport heat from the summer hemisphere.

Even without the albedo due to a thick CO_2 atmosphere, Early Mars would have an absorbed solar radiation of only $77 \, W/m^2$. This is still somewhat above the Kombayashi-Ingersoll limit for a pure CO_2 atmosphere, but allowing for the scattering effects of the atmosphere and perhaps also the influence of nitrogen in the atmosphere, Early Mars could well have sustained permanent CO_2 glaciers, given a sufficient supply of CO_2. Because the planet is so near a threshold, a more detailed calculation – probably involving consideration of horizontal atmospheric heat transports – would be needed to resolve the issue.

In fact, the Kombayashi-Ingersoll limit for CO_2 is a critical factor in the determination of the outer limit of the habitable zone. It shows why you cannot make a planet in an arbitrary far orbit habitable simply by pumping enough CO_2 into the atmosphere until the greenhouse effect provides sufficient warming – if the absorbed stellar flux is below the threshold, extra CO_2 goes into the condensed reservoir instead of adding to the greenhouse warming. Early Mars is near the brink, which is why small exotic effects such as the radiative effects of CO_2 ice clouds can make the difference to habitability. Gliese 581c is in the same radiative regime. A planet in a much dimmer orbit, however, would need to get its habitability from a less condensable greenhouse gas; it is unclear whether any likely atmospheric constituent can take the place of CO_2 in bringing a distant planet to the liquid-water threshold.

One can similarly compute a Kombayashi-Ingersoll limit for methane, using the continuum absorption properties described in Section 4.4.8. This calculation would determine whether a body could have a permanent methane ocean, swamp, or glacier at the surface. Condensation of N_2 would also lead to a Kombayashi-Ingersoll limit which determines the threshold absorbed stellar radiation needed to prevent the accumulation of a surface reservoir of N_2 ice or liquid N_2. In that case, the radiative feedback would be provided by the N_2 self-continuum.

Any gas becomes condensable at sufficiently low temperatures or high pressures, and it is in fact the Kombayashi-Ingersoll limit that determines whether a volatile greenhouse gas outgassing from the interior of the planet accumulates in the atmosphere, or accumulates as a massive condensed reservoir (which may be a glacier or ocean).[9] In the latter case, additional outgassing goes into the condensed reservoir, and the amount of volatile remaining in the atmosphere in the gas phase is determined by the temperature of the planet. The condensed reservoir can form only if the absorbed solar radiation is below the Kombayashi-Ingersoll limit for the gas in question, and even then only if the total mass of volatiles available is sufficient to bring the atmosphere to a state of saturation. As an example of the latter constraint, let's suppose that Mars were in a more distant orbit, where the absorbed

---

[9] There are other places an outgassed atmosphere can go; water can go into hydration of minerals, and CO_2 can be bound up as carbonate rocks.

solar radiation were only $40 \, W/m^2$. Then, according to Fig. 4.40, the equilibrium surface temperature would be 165 K in saturation, and the corresponding surface pressure would be 5600 Pa. In order to reach this surface pressure for a planet with acceleration of gravity g it is necessary to outgas $5600/g$, or 1509 kg of CO_2 per square meter of the planet's surface. On Earth, outgassed water vapor accumulates in an ocean because the Earth is below the Kombayashi–Ingersoll limit for water vapor. With present solar luminosity Venus (without clouds) is well above the limit, so any outgassed water vapor would accumulate in the atmosphere (apart from the leakage to space). For CO_2, Earth, Mars, and Venus are all above the Kombayashi–Ingersoll limit, so outgassed CO_2 accumulates in the atmosphere (apart from whatever gets bound up in mineral form). Even if you took away the water that allows CO_2 to be bound up as carbonate, Earth would not develop a CO_2 ocean; it would become a hot Venus-like planet instead, with a dense CO_2 atmosphere.

In parting, we must mention two serious limitations of our discussion of the runaway greenhouse phenomenon. First, in computing the Kombayashi–Ingersoll limit, it was assumed that the atmosphere was saturated with the condensable greenhouse gas. However, real atmospheres can be substantially undersaturated, though the dynamics determining the degree of undersaturation is intricate and difficult to capture in simplified models. Undersaturation is likely to raise the threshold solar radiation needed to trigger a runaway state. The second limitation is that the calculations were carried out for clear sky conditions. Clouds exert a cooling influence through their shortwave albedo, and a warming influence through their effect on OLR, and the balance is again hard to determine by means of any idealized calculation. Whether clouds have an inhibitory effect on the runaway greenhouse is one of the many remaining Big Questions.

4.7 PURE RADIATIVE EQUILIBRIUM FOR REAL GAS ATMOSPHERES

Pure radiative equilibrium amounts to an all-stratosphere model of an atmosphere, and is a counterpoint to the all-troposphere models we have been discussing. Real atmospheres sit between the two extremes, sometimes quite near one of the idealizations. In this section we will focus on pure infrared radiative equilibrium. The effects of solar absorption in real gases will be taken up in Chapter 5.

From simple analytic solutions, we know essentially all there is to know about pure radiative equilibrium for gray gases. It is important to understand these things because the structure of atmospheres results from an interplay of convection and pure radiative equilibrium. A thorough understanding of pure radiative equilibrium provides the necessary underpinning for determining where the stratosphere starts, and its thermal structure. We will now examine how the key elements of the behavior of pure radiative equilibrium differ for real gases. The specific issues to be addressed are:

- For gray gases in radiative equilibrium, the minimum temperature is the skin temperature based on OLR, and is found in the optically thinnest part of the atmosphere in the absence of atmospheric solar absorption. For real gases, can the radiative equilibrium temperature be much lower than the skin temperature?
- For a gray gas atmosphere with a given vertical distribution of absorbers, the radiative equilibrium temperature profile is uniquely determined once the OLR is specified. Specifically, one can determine the temperature profile of the radiative-equilibrium stratosphere without needing to know anything about the tropospheric temperature structure. To what extent is this also true for real gases?

- For a gray gas atmosphere with a given vertical distribution of absorbers, the normalized temperature profile $T(p)/T_g$ is independent of the ground temperature. This means that the radiative equilibrium temperature profile has the same shape, regardless of the magnitude of the solar radiation with which the atmosphere is in balance. How much does this result change for real gases?
- For a gray gas, the temperature jump at the ground is greatest when the atmosphere is optically thin, and vanishes as the atmosphere becomes optically thick. For real gases, the atmosphere is optically thick in some parts of the spectrum, and optically thin in others. What determines the temperature jump in these circumstances?
- Gray gases are most unstable to convection where they are optically thick. In the optically thick limit, the slope $d\ln T/d\ln p$ equals $\frac{1}{4}$ without pressure broadening, and $\frac{1}{2}$ with pressure broadening. Hence, a dry atmosphere can go unstable near the ground only if $R/c_p < \frac{1}{4}$ without pressure broadening, or $R/c_p < \frac{1}{2}$ with pressure broadening. How do these thresholds differ for a real gas?

All of the issues except for the behavior of the static stability near the ground are well illustrated by the *semigray* model, which we referred to more poetically as "one-band Oobleck" in Section 4.4.1. In this model, we assume κ to be constant within a band of width Δ centered on frequency ν_0, and zero elsewhere. To keep the algebra simple we will make the additional assumption that Δ is small enough that B can be considered essentially frequency-independent within the band; this assumption can be easily dispensed with at the cost of a slightly more involved calculation. What makes the semigray model tractable[10] is that the infrared heating is due solely to the flux within the absorbing band, so that one can still deal with a single optical thickness without any need to sum or integrate over frequency to get the net heating.

First, let's consider the semigrey radiative equilibrium in the optically thin limit. From the results in Section 4.2.2, the infrared radiative cooling at any level is simply $2\pi B(\nu_0, T)\Delta$, whereas the heating by absorption of upwelling infrared from the ground is $\pi B(\nu_0, T_g)\Delta$. Balancing the two gives the equation determining the atmospheric temperature:

$$B(\nu_0, T) = \frac{1}{2}B(\nu_0, T_g). \tag{4.95}$$

Since the equation for T is independent of level, we conclude that in the optically thin case, the atmosphere is isothermal in infrared radiative equilibrium, just as it is for the optically thin gray case. The resulting atmospheric temperature is always lower than the ground temperature, and may be called the *semigray skin temperature*. For a gray gas, $T_{skin}/T_g = 1/2^{1/4} \approx 0.84$, but for the semigray case, the ratio depends on the frequency of the absorbing band. From the form of the Planck function, it can easily be shown that T_{skin}/T_g depends on frequency only through the ratio $h\nu_0/kT$. A simple analytic calculation shows that $T_{skin}/T_g \to \frac{1}{2}$ for small values of $h\nu_0/kT_g$ and $T_{skin}/T_g \to 1$ when $h\nu_0/kT_g$ becomes large; a simple numerical solution shows that the ratio increases monotonically between the two limiting cases (see Problem 4.23). The gray gas skin temperature ratio sits in the middle of this range, and indeed for frequencies near the peak of the Planck function, the semigray ratio does not differ greatly from the gray value. For example, at $650\,cm^{-1}$ and a ground temperature of $280\,K$, the semigray skin temperature is $233\,K$, whereas the gray skin temperature is $235\,K$.

---

[10]Approximate solutions for the semigray ("window-gray") model were presented by Sagan in 1969 (*Icarus* **10**, 290–300). The model has been rediscovered independently a number of times since.

When deriving the radiative equilibrium for gray gases, we specified the *OLR* and used that as a boundary condition for determining the thermal structure; the ground temperature is then computed from the resulting lower boundary fluxes. For real gases, it proves more convenient to specify the ground temperature T_g, and find the resulting temperature structure and *OLR*. We adopt this approach not only in our analytic solution of the semigray case, but also in our numerical solutions for actual gases. When fixing T_g and increasing optical thickness, the *OLR* goes down, and so the amount of absorbed solar radiation needed to maintain the stated T_g decreases.

The derivation of the full infrared radiative equilibrium solution for the semigray case is identical to that we used in the gray case, with the following substitutions: (1) The optical thickness τ is based on the value of κ in the absorbing band alone, (2) σT^4 is replaced by $\pi B(\nu_0, T)$, (3) the fluxes I_+ and I_- represent the flux integrated over the band Δ alone, and (4) the *OLR* appearing in the top boundary condition is no longer the net *OLR* emitted by the planet, but only the portion of *OLR* (call it OLR_Δ) emitted in the absorbing band. The assumption of infrared equilibrium still implies that the flux $I_+ - I_-$ is constant and equal to OLR_Δ, but since this flux is only the portion of *OLR* in the band, it is no longer determined *a priori* by planetary energy balance. Instead, it must be determined by making use of both boundary conditions, that $I_- = 0$ at the top of the atmosphere, and $I_+ = \pi B(\nu_0, T_g)$ at the bottom of the atmosphere (assuming the ground to have unit emissivity). The result of applying both boundary conditions is

$$I_+ - I_- = \frac{2\pi}{2 + \tau_\infty} B(\nu_0, T_g). \tag{4.96}$$

This is equal to the ground emission in the optically thin case, and approaches zero as the atmosphere is made more optically thick. Note, however, that it is only the *OLR* in the band that approaches zero; the net emission approaches a finite, and possibly quite large, lower bound, because the emission from the rest of the spectrum where the atmosphere is transparent is simply the ground emission. By substituting the expression for the net upward flux into the expression for $I_+ + I_-$, we find the following expression determining $T(\tau)$

$$B(\nu_0, T) = \frac{1 + \tau_\infty - \tau}{2 + \tau_\infty} B(\nu_0, T_g) \tag{4.97}$$

where τ is the optical thickness within the band where κ is non-zero. At the top of the atmosphere, this reduces to $B(\nu_0, T) = B(\nu_0, T_g)/(2 + \tau_\infty)$. In consequence the top of atmosphere pressure becomes progressively colder in comparison to the skin temperature as τ_∞ is made large. The behavior of the stratospheric temperature in the semigray case thus differs from the gray gas case in a fundamental way. As the atmosphere is made optically thick in the absorbing band, the stratospheric temperature approaches zero even though the net *OLR* remains finite (owing to the emission through the transparent part of the spectrum). Thus, the stratospheric temperature can be much colder than the gray gas skin temperature, resolving the quandary raised in Chapter 3. This important difference arises because, in the semigray case, the optical thickness of the lower part of the atmosphere causes the upwelling spectrum illuminating the stratosphere to be depleted in those frequencies where the stratosphere absorbs best. Nonetheless, the stratosphere continues to *emit* effectively at those frequencies, leading to very cold temperature. This depletion also has the important consequence that the stratospheric temperature is no longer independent of the existence of a troposphere, or of the tropopause height and thermal structure of the troposphere. This has the potential to affect the calculation of the tropopause height when we put radiative equilibrium together with convection.

The temperature jump at the ground is determined by $B(\nu_0, T_{sa}) = ((1 + \tau_\infty)/(2 + \tau_\infty))B(\nu_0, T_g)$, which has the same form as the corresponding expression for the gray gas case, save for the appearance of the Planck function in place of σT^4. As in the gray gas case, the jump is a maximum for optically thin atmospheres, and vanishes in the optically thick limit. The main difference here is that the jump can be made to vanish by making the atmosphere optically thick in just a limited part of the spectrum, even though the atmosphere is optically thin (in this case absolutely transparent) elsewhere. This feature will reappear in our discussion of radiative equilibrium for an atmosphere in which infrared absorption is provided by CO_2. As the atmosphere becomes optically thick in the absorption band, the unstable temperature jump at the ground diminishes, but what happens to the interior static stability of the air near the ground? To determine this, we take the derivative of Eq. (4.97) and multiply by p_s/T_{sa} to obtain the logarithmic radiative equilibrium lapse rate at the ground:

$$\frac{d \ln T}{d \ln p}\Big|_{p_s} = \frac{B(\nu_0, T_g)}{T_{sa}\frac{dB}{dT}(\nu_0, T_{sa})} \frac{\kappa(p_s)p_s/g \; \cos\bar{\theta}}{2 + \tau_\infty}. \tag{4.98}$$

In the optically thick limit, the second factor is unity without pressure broadening, or $\frac{1}{2}$ with pressure broadening. When the atmosphere is optically thick, the first factor can be evaluated with $T_{sa} = T_g$. With a little algebra, the first term can be re-written as $(\exp(u) - 1)/(u \exp(u))$, where $u = h\nu_0/(kT_g)$; this term has a maximum value of unity at $u = 0$ (high frequencies or low temperatures) and decays to zero like $1/u$ at large u (low frequencies or high temperatures). For $u = 1$ which puts the absorption band near the maximum of the Planck function, the value is about 0.63. For the semigray model then, we conclude that the degree of instability of radiative equilibrium in the optically thick limit is bounded; for example, with $u = 1$ and incorporating pressure broadening, the radiative equilibrium near the ground is statically unstable when $R/c_p > 0.315$ (vs. a threshold value of 0.5 for a gray gas). Hence, the semigray case has a somewhat enhanced instability at the ground as compared with the gray case, but the difference is not great. We will see shortly that this is one regard in which the qualitative behavior of a real gas differs significantly from the semigray case.

Finally, let's look at how the radiative equilibrium profile for the semigray atmosphere changes if we change T_g and leave everything else fixed. For a gray gas, the function T/T_g is invariant because both the surface emission and the interior atmospheric emission increase by a factor b^4 if we replace T_g by $b \cdot T_g$. For the semigray case, the emission is given by $B(\nu_0, T)$, which is no longer a simple power of T. This means that the ratio T/T_g is no longer independent of T_g. The nature of the dependence is left for the reader to explore in Problem 4.24.

Working our way up the ladder to reality, we'll now present some numerical solutions for two-band Oobleck and for mixtures of CO_2 in air. The latter are computed using our homebrew exponential-sum radiation code, incorporating the effect of pressure broadening but without taking temperature scaling into account. This is sufficient to show how the extreme range of absorption coefficients in a typical real gas affect the radiative equilibrium. The equilibria were found by a simple time-stepping method with fixed T_g. For any given initial temperature profile $T(p)$ one can calculate the infrared fluxes, and hence, by differencing in the vertical, the infrared heating rates. These are used to update

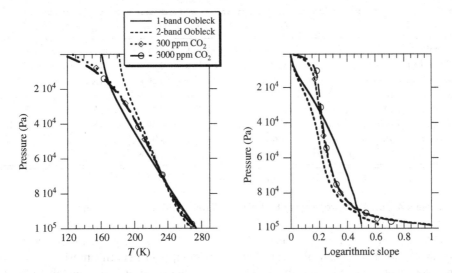

Figure 4.41 Left panel: Infrared radiative equilibrium temperature profiles for one- and two-band Oobleck, and for a CO_2–air mixture at 300 ppmv and 3000 ppmv, subject to a fixed 280 K ground temperature. For the Oobleck cases, the absorption coefficients were chosen such that the optical depth of the atmosphere as a whole is 10 in the strong absorption band (650–700 cm^{-1}) and 1 in the weak bands on the flanks (600–650 cm^{-1} and 700–750 cm^{-1}). Right panel: The corresponding logarithmic slope, $d \ln T / d \ln p$.

$T(p)$, and the whole process is repeated until equilibrium is achieved, where the infrared heating is zero and the temperature no longer changes. Since we are only interested in the equilibrium and not the time course of the approach to equilibrium, we can afford to be somewhat sloppy in our time-stepping method, so long as it is stable enough to yield an equilibrium at the end. Figure 4.41 shows the resulting equilibrium profiles for two-band Oobleck with the secondary-band absorption coefficient one-tenth the value of the primary, and for a CO_2–air mixture in an Earthlike atmosphere, at 300 ppmv and 3000 ppmv. The left panel shows the temperature profile, while the right panel gives the logarithmic slope of temperature, which determines the static stability of the atmosphere. For comparison, the analytically derived results for one-band pressure-broadened Oobleck (the semigray model) are also shown. These calculations were carried out with a ground temperature $T_g = 280$ K. Some key characteristics of the results are summarized in Table 4.4.

Comparing the curves for one-band and two-band Oobleck shows that adding in the weakly absorbing bands reduces the vertical temperature gradient except in a thin layer near the ground. In this sense, most of the atmosphere acts as if it were made more optically thin, despite the fact that we have actually made the atmosphere more opaque to infrared by adding new absorption in additional bands without taking away absorption in the original band. Indeed, the *OLR* goes down from 333 W/m^2 in the one-band case to 309 W/m^2 in the two-band case, despite the warmer temperatures aloft in the latter. The key to this behavior was pointed out back at the end of Section 4.2.2: in an atmosphere which is optically thin in some parts of the spectrum but optically thick in others, the heating rate (which in turn determines the radiative equilibrium) is dominated by the

	OLR (W/m^2)	T_{skin}	$T(0)$	$T(p/p_s = 0.5)$	$T_g - T(p_s)$
1-band Oobleck	333	232 K	161 K	206 K	6.5 K
2-band Oobleck	309	228 K	182 K	215 K	12.75 K
CO_2/air 300 ppmv	295	226 K	126 K	214 K	2.8 K
CO_2/air 3000 ppmv	275	222 K	115 K	214 K	1.75 K
Mars, CO_2, $p_s = 7$ mb	303	227 K	128 K	215 K	2.9 K
Mars, CO_2, $p_s = 7$ mb, $T_g = 250$ K	192	203 K	126 K	196 K	4.4 K
Mars, CO_2, $p_s = 2$ bar	86	166 K	102 K	208 K	1.3 K
Venus CO_2, $p_s = 90$ bar, $T_g = 700$ K	55	148 K	77 K	500 K	0.1 K

Calculations were done with $T_g = 280$ K unless otherwise noted.

Table 4.4 Summary properties of infrared radiative equilibrium solutions.

parts of the spectrum where the optical thickness is nearest unity. This is the reason the weaker absorption bands control the behavior of the temperature in the interior of the atmosphere. The associated reduction in temperature gradient warms the atmosphere aloft, at pressures below 700 mb. In essence, the weak bands allow the upper reaches of the atmosphere to capture more of the infrared upwelling from below; spreading the absorption over a somewhat broader range of the spectrum in essence makes the problem a bit more like a gray gas, and makes the upper air temperature somewhat closer to the gray gas skin temperature.

While the addition of the weak bands reduces the temperature gradient aloft and hence stabilizes the atmosphere against convection, this comes at the expense of increasing the gradient near the ground, and destabilizing the layer there. This stabilization/destabilization pattern shows up clearly (in an increase of log slope) in the plot of the logarithmic temperature derivative shown in the right panel of Fig. 4.41. Based on $R/c_p = 2/7$, the one-band Oobleck profile is unstable for pressures higher than 450 mb ($4.5 \cdot 10^4$ Pa), whereas the two-band case is only unstable for pressures higher than 760 mb ($7.6 \cdot 10^4$ Pa), though within the unstable layer the two-band case is considerably more unstable than the one-band case. In the two-band case, more of the net vertical temperature contrast of the atmosphere is concentrated in a thin layer near the ground. Overall, the atmosphere acts as if it were optically thick near the ground, but relatively optically thin aloft. The behavior near the ground results from the extremely strong absorption near the center of the absorption band. This spectral region captures the upwelling radiation from the ground, which is not yet depleted in the strongly absorbing wavenumbers. If there were much temperature discontinuity near the ground, the absorption would lead to strong radiative heating of the low-level air; hence the only way to be in radiative equilibrium is for the air temperature to approach the ground temperature. More mathematically speaking, the phenomenon arises because the low level heating is controlled largely by the boundary terms in the flux integral in Eq. (4.9), which in turn is dominated by the strongly absorbing spectral regions.

The real gas CO_2 results have many features in common with two-band Oobleck, notably the weak temperature gradient in the interior of the atmosphere and the enhancement of temperature gradient near the ground. The strong destabilization near the ground is even more pronounced for CO_2, because it has a far stronger peak absorption than the Oobleck case we considered, and because the absorption varies over a greater range of values. The associated optical thickness near the ground also keeps the unstable temperature jump between the ground and the overlying air small. An optically thin gray gas would have a large unstable temperature jump at the ground. The strong absorption bands in a real gas smooth out this discontinuity and move it into a finite width layer in the interior of the atmosphere. From the standpoint of the convection produced, there is little physical difference between the two cases.

The main differences between the CO_2 cases and the Oobleck cases shows up in the upper atmosphere, where the CO_2 cases show steep declines of temperature with height – though not so steep as to destabilize the upper atmosphere. As at the bottom boundary, the culprit is the spectral region where absorption coefficients peak. These regions lead to very strong emission in the upper layers of the atmosphere, which are poorly compensated by absorption since the upwelling infrared reaching the upper atmosphere has been depleted in most wavenumbers that absorb at all well. The strong emission causes the temperature near the top to fall far below the skin temperature. Based on the net *OLR*, the gray gas skin temperature for the two CO_2 cases is somewhat above 220 K, whereas the actual temperature at the 1 mb level is 126 K in the 300 ppmv case and 115 K in the more optically thick 3000 ppmv case. Finally, we note that increasing CO_2 by a factor of 10 has relatively little effect on the radiative equilibrium temperature profile, despite the fact that the increase lowers the *OLR* by nearly 20 W/m^2. The changes are principally seen near the ground, where the increased optical thickness in the wings has reduced the surface temperature jump. While the two-band Oobleck model reproduces the near-ground destabilization present in the real CO_2 calculation, it is unable to simultaneously represent the temperature jump at the ground. Lacking the extremely strong peak absorption of CO_2, the addition of the weak absorption wings in two-band Oobleck makes the surface budget act like an optically thinner atmosphere, increasing the surface jump.

To illustrate how the radiative equilibrium solution scales with T_g, in Fig. 4.42 we show the profiles of T/T_g for surface temperatures ranging from 240 K to 320 K. Only the case of 300 ppmv CO_2 is shown, though the other atmospheres treated in Fig. 4.41 yield similar results. For a gray gas, all the curves for a given atmospheric composition would collapse onto a single universal profile; for the reasons discussed in the semigray case, this is no longer true for real gases. However, while the temperature aloft does not scale precisely with the ground temperature, the deviations are modest enough that one can still get useful intuition about the behavior of the system by assuming that radiative equilibrium temperature scales with the ground temperature. For CO_2, the actual temperature aloft is always somewhat colder than that which one would estimate by proportionately scaling the temperature upward from a colder to warmer ground temperature. Recall that these calculations are done without temperature scaling of the absorption coefficient, so the effect shown is purely due to the shape of the Planck function.

The general features encountered in the terrestrial calculations discussed above carry over to the pure CO_2 atmospheres characteristic of present and (possibly) Early Mars. Martian radiative equilibrium solutions with a 7 mb thin atmosphere or 2 bar thick atmosphere are shown in Fig. 4.43. The most striking feature of these solutions is that increasing the mass of the atmosphere by a factor of nearly 300 causes very little increase in the vertical temperature contrast. This stands in sharp contrast to the gray gas case, for which the enormous increase in optical depth going from the 7 mb case to the 2 bar case would cause the temperature in the latter to drop to nearly zero within a short distance of the ground. As before, the reason for the relative insensitivity of the temperature profile is that the radiative heating is determined largely by the part of the spectrum where the optical depth is of order unity. For the present Mars case, this occurs in the near wings of the principal absorption peak, whereas for Early Mars it occurs within the continuum window region. The shift allows

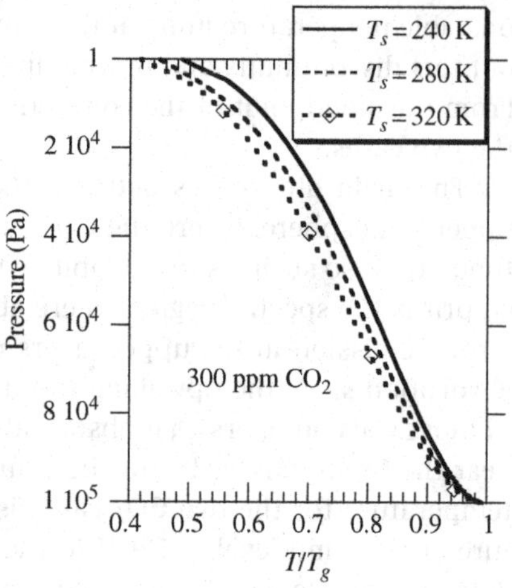

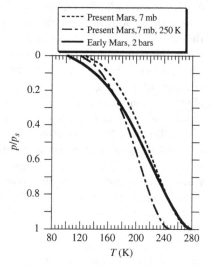

Figure 4.42 Variation of the shape of the temperature profile as a function of ground temperature. Results are shown for 300 ppmv of CO_2 in air.

Figure 4.43 Infrared radiative equilibrium for pure CO_2 atmospheres with Martian gravity. Results are shown for an Early Mars case with a 2 *bar* surface pressure and 280 K ground temperature, and for Present Mars cases with 7 mb surface pressure and 250 K and 280 K ground temperatures. To make it possible to compare the cases, the temperatures have been plotted as a function of p/p_s. The 2 bar case includes the CO_2 continuum absorption.

both cases to act roughly like a case with order unity optical depth, apart from the thin radiative boundary layers near the ground and the top. The temperature profiles are similar despite the fact that it would require $303\,W/m^2$ of absorbed solar radiation to maintain a surface temperature of 280 K in the thin present Martian atmosphere, but only $86\,W/m^2$ in a 2 bar atmosphere. For the present Mars case, we have also included a calculation with $T_g = 250\,K$, to underscore that with a realistically cold daytime surface temperature, the equilibrium temperature aloft is too cold in comparison with observations. This suggests once more an important role for solar absorption in determining the temperature structure of the present Martian atmosphere.

It is only when we go to massive atmospheres like that of Venus that the atmosphere becomes optically thick throughout the spectrum, allowing the vertical temperature contrast to increase dramatically. A simplified calculation without temperature scaling, and ignoring emission beyond $2300\,cm^{-1}$, shows that with a 700 K ground temperature, the radiative equilibrium temperature drops to 500 K at the midpoint of the atmosphere and all the way to 80 K at the 100 mb level. This yields a very strong greenhouse effect: it takes only a trickle of $55 \, W/m^2$ of absorbed solar radiation to maintain the torrid ground temperature. Note, however, that it is important that a fair amount of this trickle actually be absorbed at the ground, and not in the upper reaches of the atmosphere; otherwise the deep atmosphere becomes isothermal, and can in fact become as cold as the skin temperature in extreme cases, as discussed in Section 4.3.5. So far as the maintenance of its thermal structure is concerned, the troposphere of Venus is more like the Antarctic glacier than it is like the Earth's troposphere. The trickle of heat escaping the Earth's interior beneath the glacier – a mere $30 \, mW/m^2$ – is sufficient to raise the basal ice temperature to the melting point and create subglacial Lake Vostok precisely because the diffusivity of heat in ice is so small. So it is, too, with the atmosphere of Venus; the extremely optically thick troposphere renders the radiative diffusivity of heat very small, and allows the tiny trickle of solar radiation reaching the surface to accumulate in the lower atmosphere and raise the temperature to extreme values. Unlike the glacier, however, when the lower atmosphere becomes hot enough, it can start to convect. Convection supplants the radiative heat flux, but also establishes the adiabat, allowing the surface to be much hotter than the radiating level.

Different non-condensable greenhouse gases differ somewhat in details of their radiative-equilibrium profiles, but the general picture does not differ greatly from what we have learned by looking at CO_2. When the greenhouse gas can condense near the ground, however, the situation becomes quite different. The case of water vapor in air provides a prime example. If the ground temperature is high enough that the amount of water vapor present in saturation makes the lower atmosphere optically thick, then the temperature will decline rapidly with height above the ground, because that is what optically thick atmospheres do in radiative equilibrium. Water vapor exhibits this effect particularly strongly, since it easily makes the atmosphere optically thick everywhere outside the window regions of the spectrum, and even the windows close off above 300 K. As the temperature decreases, however, the water vapor content decreases in accordance with the limits imposed by Clausius–Clapeyron. Within a small distance above the ground, the air is so cold that there is little water vapor left, and the atmosphere further aloft becomes optically thin. As a result, most of the variation in optical thickness of the atmosphere is concentrated into a thin, radiative boundary layer near the ground, and the optical thickness (and hence the temperature) varies greatly within this layer. Because of the strong temperature gradient and high optical thickness of the boundary layer, a strong greenhouse effect is generated entirely within the boundary layer, leading to low *OLR*. If one imposes equilibrium with an Earth-like absorbed solar radiation, the ground temperature must increase to well in excess of 320 K to achieve balance, and the steep increase of saturation vapor pressure with temperature further exacerbates the high temperature gradient in the radiative boundary layer. It is a bit as if the entire optically thick atmosphere of Venus were squeezed into a boundary layer having a depth of a kilometer or less. Adding a non-condensing gas like CO_2 to the mix alters the temperature profile above the boundary layer, but does not eliminate the basic pathology of the situation. The equilibrium profile in this case is of little physical consequence because the slightest convection or other turbulent mixing would mix away the thin radiative boundary layer, warming and moistening a much deeper layer of the atmosphere. The radiative equilibrium solutions we have been studying for the non-condensing case are worthy of protracted consideration because they provide some useful insight as to the stratospheric temperature for more realistic atmospheres in which there is some low-level convection. The same cannot be said for pure radiative equilibrium in the condensing water-vapor/air system, which is an exercise in pathology having little or no bearing on the operation of real atmospheres.

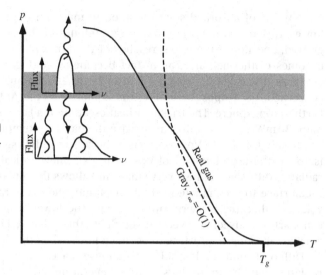

Figure 4.44 Basic features of real gas pure infrared radiative equilibrium.

Though real gas radiative equilibrium is not amenable to the kind of complete solution we enjoyed for the gray gas case, its behavior can be reasonably well captured by a few generalities, summarized in Fig. 4.44. For real gases as for gray gases heated by infrared emission from the lower boundary, the temperature decreases with distance from the boundary. This can be viewed as a kind of thermal diffusion in the sense that heat transfer is down the temperature gradient, though the process is only described by a true local diffusion in the optically thick limit. Real gases behave as if they are optically thick near the ground, exhibiting strong convectively unstable temperature gradients there and little temperature jump between the ground and overlying air. The temperature gradient weakens in the interior, but there is generally a region of strong, though stable, temperature decline near the top of the atmosphere. The upper atmosphere is considerably colder than the gray gas skin temperature, since (by Kirchhoff's law) the atmosphere radiates efficiently in the strongly absorbing parts of the spectrum, but the radiation illuminating the upper atmosphere is depleted in this portion of the spectrum. In contrast to the gray gas case, the contrast in temperature across the depth of the atmosphere is relatively insensitive to the amount of greenhouse gas in the atmosphere. As long as there are some spectral regions where the atmosphere is optically thick, and some where the atmosphere is optically thin, the radiative cooling tends to be dominated by the intermediate spectral regions. As a result, temperature tends to drop by a factor of two to three between the ground and the upper atmosphere, with only slight increases even when the greenhouse gas content is increased by many orders of magnitude. This behavior persists until there is so much greenhouse gas present that the atmosphere becomes very optically thick throughout the thermal infrared spectrum, as is the case on Venus.

The upshot of all this is that atmospheres whose temperature is maintained by absorption of upwelling infrared from a blackbody surface will never exhibit pure radiative equilibrium. There will always be a layer near the surface which is unstable to convection. If the atmosphere is optically thin, the instability is generated by a temperature jump at the surface. If the atmosphere is optically thick and subject to pressure broadening, the instability is generated by strong temperature gradients in the interior of the atmosphere near the surface. This remark even applies qualitatively to gas giants which have no surface, as the deep atmosphere is dense enough that it can begin to act like a blackbody even though there is no distinct surface. An atmosphere can be stabilized throughout its depth, however, if it is subject to atmospheric solar heating which increases with altitude in a suitable fashion.

4.8 TROPOPAUSE HEIGHT FOR REAL GAS ATMOSPHERES

The radiative equilibrium solutions discussed in the preceding section are all unstable near the ground. As convection sets in, it will mix away the unstable layer and replace it by an adiabat; the well-mixed region is the troposphere. The change in lower-level temperature profile, however, will alter the upward radiation which heats the stratosphere, and therefore cause temperature changes even above the layers reached directly by convection. When all this sorts itself out, how deep is the troposphere? This is the problem of tropopause height, which we have already touched on briefly for gray gases. Here we will offer a taste of a few of the most important aspects of real gas behavior, and lay the physical basis the reader will need for further explorations with more comprehensive models. In this section we will assume, as in the all-troposphere model, that turbulent fluxes couple the ground so tightly to the overlying air that there is no discontinuity at the ground. This assumption will be relaxed in Chapter 6.

For a gray gas, the problem of finding the tropopause height is relatively simple. Since the radiative equilibrium profile depends only on OLR – and that only via a simple formula – one starts with the radiative equilibrium profile for the desired OLR, picks a guess for the tropopause pressure, and then replaces the temperature between there and the ground with the adiabat for the gas under consideration. One then computes the actual OLR for the resulting profile, and generally will find that it is somewhat different from the OLR assumed in computing the radiative equilibrium. To make the solution consistent, one then adjusts the tropopause height until the computed OLR including the troposphere is the same as the target OLR within some desired accuracy. This is a simple problem in root-finding for a function of a single variable (the tropopause pressure), and can be solved by any number of means, Newton's method and bisection being among the most commonly employed. The behavior of the solution is explored in Problem 4.28

For a real gas, the radiative equilibrium in the upper atmosphere depends on the spectrum of the infrared upwelling from below, so we no longer have the luxury of assuming that the stratospheric temperature profile remains fixed as we vary the estimate of the tropopause height. Instead, one must simultaneously solve for both the tropopause height and the corresponding equilibrium profile aloft. This is most easily done by a modification of the time-stepping method we employed to compute the pure radiative equilibrium solutions. As in that case, it is somewhat awkward to pick an OLR and find the corresponding ground temperature T_g. Instead, we fix T_g, and compute the corresponding OLR. This can be done for a range of ground temperatures, whereafter the ground temperature in equilibrium with any specified solar absorption can be determined. We are back in the familiar business of computing the $OLR(T_g)$ curve, much as we did for the all-troposphere model, but this time taking into account the effect of a self-consistent stratosphere on the OLR.

The general problem of representing convection in climate models is a very challenging one, about which entire volumes have been written. (See the Further Reading section of Chapter 2.) For the problem at hand, there are a number of simplifying assumptions which allow us to avoid some of the more subtle aspects of the subject. First, we will be content to assume that convection instantaneously resets the profile to an adiabatic profile.

Next, given where instability occurs in the pure infrared radiative equilibrium profiles, it is safe to assume that convection occurs in a single layer extending from the ground to the tropopause height, without any possibility of multiple interleaved internal convecting and radiative equilibrium layers. Further, we will only seek an equilibrium solution, without attempting to accurately represent the approach to equilibrium. Finally, we carry out the calculation by holding T_g fixed and allowing the rest of the atmosphere to relax to the corresponding equilibrium. Under these circumstances, the elimination of unstable layers by convective mixing can be carried out through the following simple modification to the pure radiative equilibrium time-stepping algorithm: One calculates the adiabat $T_{ad}(p)$ corresponding to the ground temperature T_g and surface pressure p_s. Then at each time step, wherever $T(p) < T_{ad}(p)$, the temperature is instantaneously reset to T_{ad}. The rationale for doing this is that convection is a much faster process than radiative relaxation, and that wherever the temperature is below the adiabatic temperature, air parcels starting at the ground have enough buoyancy to reach that level, mixing air all along the way. The procedure also assumes that the turbulent coupling of the ground to the overlying air is so strong that ground and air temperature remain essentially identical at all times. The adjustment to the adiabat with surface temperature T_g in general increases the static energy (moist or dry, as appropriate) of the adjusted layer of air. This is not a source of concern if we are only using time-stepping to find the equilibrium state; if we were instead trying to represent the actual time-course of approach to equilibrium, a more sophisticated adjustment approach conserving static energy would need to be employed. Conservative adjustments would transport heat vertically by cooling the lower levels at the same time they are warming the upper levels.

Results for an Earthlike air/CO_2 atmosphere are shown in Fig. 4.45. The convection reaches to 380 mb ($3.8 \cdot 10^4$ Pa), where the temperature is about 200 K; above that, the atmosphere is in radiative equilibrium, which defines the stratosphere. Note that the temperature continues to decline even above the maximum height reached by convection, because the infrared radiative equilibrium profile also has decreasing temperature, so long as part of the spectrum is optically thick. This is the case also for Earth's real stratosphere, even though ozone heating eventually causes the upper stratospheric temperature to turn around and begin to increase. Thus, one should not take temperature decline as a signature of convection, and in a case where atmospheric heating causes upper stratospheric temperature to increase, the temperature minimum will generally be above the top of the convective layer.

The OLR for all three calculations is similar: 296.4 W/m^2 for the radiative-convective model vs. 295 W/m^2 for pure radiative equilibrium and 296.6 W/m^2 for the all-troposphere model. Thus, if the atmosphere is maintained by 295 W/m^2 of absorbed solar radiation, neither the formation of a troposphere by convection nor the formation of a stratosphere by upper-level radiative heating has much effect on the surface temperature, though the effects on the atmospheric profile are considerable.

The heating rate profile shown in the right panel of Fig. 4.45 sheds some light on the basic mechanism maintaining convection in the troposphere. The entire troposphere is subject to radiative cooling. This feature is guaranteed by the construction of the solution, since any positive heating would warm the atmosphere, causing it to exceed the adiabatic temperature and shutting off convection. Suppose convection has just occurred and reset the temperature to the adiabat. Then, in the next small interval of time, radiative cooling will cause the temperature to fall below the adiabat, triggering convection once more, which adds back the heat lost by radiative cooling and restores the adiabat. The heat is supplied by parcels of air that pick up heat from the ground, become buoyant, and carry the heat upward

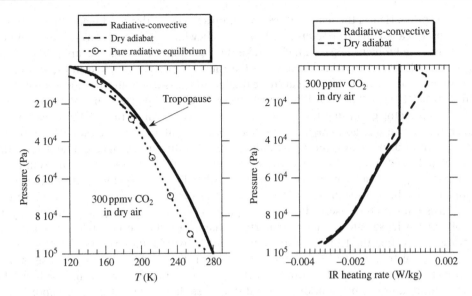

Figure 4.45 Radiative-convective equilibrium for an Earthlike dry atmosphere with 300 ppmv CO_2. The left panel shows the temperature profile in comparison with pure radiative equilibrium and the dry adiabat. The right panel shows the radiative heating rates for the radiative-convective solution and for the all-troposphere model. In all cases the ground temperature is 280 K.

to the level where it is needed. The thermal balance in the troposphere is between radiative cooling and convective heating. The addition of atmospheric solar absorption to the troposphere would have no effect on the tropospheric temperature or the tropopause height, so long as it doesn't turn the net radiative cooling at any level into a net radiative heating. Short of that happening, the sole effect of tropospheric solar absorption is to reduce the convective heating, and hence the frequency or vigor of convection. However, since infrared cooling is weakest just below the tropopause, solar absorption near the tropopause level can easily move the tropopause downward.

Using Fig. 4.45 we can compare two simple estimates of the tropopause height with the actual value. Looking at where the adiabat intersects the pure radiative equilibrium, we find an estimate of 205 mb ($2.05 \cdot 10^4$ Pa). This is somewhat too high in altitude, since the formation of the troposphere changes the upward flux and warms the stratosphere. The other way to estimate tropopause height is to look at the heating profile computed for the dry adiabat, shown in the right panel of the figure. Identifying the region of radiative heating with the stratosphere yields an estimate of 325 mb ($3.25 \cdot 10^4$ Pa), which is closer to the true value, but still too high in altitude. In the light of the discussion surrounding Fig. 4.2, we can say that the real gas atmosphere behaves rather like a pressure-broadened gray gas atmosphere with optical depth somewhat greater than unity.

How does our computed tropopause height compare to the Earth's actual tropopause? We have defined the tropopause as the height reached by convection, and in comparing this with atmospheric soundings one needs to recall that even above the convective region, the temperature continues to decrease with height, because temperature goes down with height even in pure infrared radiative equilibrium; in the Earth's atmosphere, the temperature eventually begins to increase with height because of the effects of atmospheric solar absorption. Hence, the temperature minimum seen in Earth soundings, which is sometimes loosely called the tropopause, is always somewhat above the convectively defined tropopause (see Problem 2.55). Still, if we take the position of the temperature minimum in tropical soundings as an estimate, we note that the tropopause estimated in the preceding calculation is considerably lower in altitude (higher in pressure) than the observed value of about 100 mb (10^4 Pa). What is it that makes the tropical tropopause so much higher?

The main factor governing the tropopause height is the lapse rate. If the lapse rate is weaker, then one has to go to higher altitudes in order to intersect the radiative equilibrium profile. In the warm tropics, the moist adiabat has significantly weaker gradient than the dry adiabat. In fact, a radiative-convective calculation based on the radiative effects of a dry CO_2/air atmosphere, but employing the moist adiabat in the temperature profile, yields a tropopause height of 130 mb (1.3×10^4 Pa) when the surface temperature is 300 K. This is quite consistent with the observed tropical tropopause height. This suggests that the effects of moisture on lapse rate are more important than the radiative effects of tropospheric moisture in elevating the tropical tropopause. In other words, the main reason the Earth's present tropical tropopause is higher in altitude than the midlatitude tropopause is that the tropical lapse rate is weaker, owing to the greater influence of moisture for Earthlike tropical temperatures. The additional optical thickness due to the extra water vapor in the tropics plays at most a secondary role. For colder surface temperatures, the moist adiabat deviates less from the dry adiabat, so the tropopause height approaches the lower altitude found in the dry calculation. For example, with a surface temperature of 280 K, a calculation adjusting to the moist adiabat yields a tropopause pressure of 250 mb (2.5×10^4 Pa). This reasoning suggests that the tropopause height should be lower in the midlatitudes, and indeed observations show this to be the case. Optical thickness can indeed affect the tropopause height, but it is not the main player on Earth. The calculations referred to here are carried out in Problem 4.27.

4.9 THE LESSON LEARNED

This has been a rather arduous chapter – certainly for the author, and no doubt for the reader as well, but hopefully to a lesser extent. The basic lesson, however, can be summed up in a few pithy remarks. The greenhouse effect relies on infrared optical thickness of the atmosphere and temperature decline with height. Real greenhouse gases do not make the atmosphere optically thick uniformly throughout the infrared spectrum. Rather, the optical thickness is concentrated in preferred gas-dependent spectral bands, and the main way the greenhouse effect gets stronger as the gas concentration increases is through the spread of the optically thick regions to ever greater portions of the spectrum. There are two basic ways to get the temperature decline which is necessary to translate optical thickness into *OLR* reduction: radiative equilibrium in an atmosphere that is optically thick through at least part of the spectrum, or convection in an atmosphere where the radiative equilibrium is statically unstable either at the surface or internally. The tropopause height determines the blend of the two mechanisms in force. Both mechanisms can yield a surface temperature very much in excess of the no-atmosphere blackbody temperature, but a radiatively dominated atmosphere is a very different place from a convectively dominated atmosphere, since the latter has vigorous vertical mixing that can give rise to a stew of small-scale turbulent phenomena. A mostly radiative-equilibrium atmosphere is a more quiescent place, in which mixing is dominated by the more ponderous large-scale fluid motions. A central remaining question is how the tropopause height behaves as an atmosphere is made more optically thick. When do we approach an all-troposphere atmosphere, and when do we approach an all-stratosphere atmosphere? These issues are explored in Problems 4.28 and 4.29. We will return to the problem of tropopause height in Section 5.10.

4.10 WORKBOOK

4.10.1 Basic gray gas calculations

Problem 4.1 Suppose that the temperature profile in an atmosphere is $T(p) = T_s \cdot (p/p_s)^{(1/4)}$. (This is a rather contrived temperature profile, but it is chosen to make the algebra simple.) The ground temperature is equal to T_s. Compute the *OLR* and the downward radiation into the ground, assuming the atmosphere to be a gray gas with constant absorption coefficient κ. Discuss the behavior in the limits of large and small κ.

Problem 4.2 *Gray gas OLR computations*

In this computational problem you will explore the gray gas *OLR* for the temperature profile

$$T(p) = \max\left[T_{strat}, T_{sa} \cdot \left(\frac{p}{p_s} \right)^{R/c_p} \right] \tag{4.99}$$

where T_{strat} and T_{sa} are constants. This profile follows the dry adiabat until the temperature falls to T_{strat}. At lower pressures (higher altitudes), the temperature is held constant at T_{strat}. This is a primitive model of a stratosphere, in that the temperature is prevented from falling to zero with height. You may assume that the specific absorption cross-section of the atmosphere (κ) is independent of pressure and temperature, as well as frequency. Note that this absorptivity enters the problem only through the net optical thickness of the atmosphere τ_∞. If we set $T_{strat} = 0$ the problem reduces to the all-troposphere problem discussed in the text, in which T approaches zero as p approaches zero. The ground temperature T_g may differ from the surface air temperature T_{sa}.

Write a program defining a function that computes *OLR* in the above situation by evaluating the integral form of the solution to the Schwarzschild equation using numerical quadrature. The function should return the *OLR* as a function of the arguments T_g, T_{sa}, T_{strat}, and τ_∞.

Next, use your *OLR* function to estimate the temperature of a planet in equilibrium with an absorbed solar radiation of $250 \, \text{W/m}^2$. Do this for $T_{strat} = 200 \, \text{K}$ and for optical depth $\tau_\infty = 0.1, 0.5, 1$, and 5. You may assume $T_g = T_{sa}$.

Now explore the behavior of *OLR* as the optical thickness τ_∞ is varied between small values (say, 0.1) and large values (say, 100), with $T_{strat} = 200 \, \text{K}$. Can you find a simple explanation for the limiting behavior, based on what you know about blackbodies? Now reduce T_{strat} to $100 \, \text{K}$ and try again. How does this change the behavior of *OLR* in the optically thick limit? Why? Next, set T_{strat} to zero. Compare the behavior of the *OLR* at large τ_∞ with the all-troposphere formula derived in the text. (If you stick with the value of R/c_p for air you will find it helpful to know that $\Gamma(1 + 8/7) \approx 1.07$. What if $4R/c_p$ is an integer?) Recall that the all-troposphere expression was derived assuming that τ_∞ is large. Using your numerical computation of *OLR*, discuss how the approximation breaks down as τ_∞ becomes smaller. How good is the approximate form for $\tau_\infty = 1$? What is the true behavior of *OLR* as τ_∞ is made small? Give an explanation for this behavior. Finally, set $T_{strat} = 100 \, \text{K}$ again, and compare the *OLR* with the all-troposphere formula. Where does the computed *OLR* diverge from the all-troposphere formula? Why? Under what circumstances does the choice of T_g substantially affect the *OLR*, and what is the nature of the effect?

Note that this problem is somewhat unrealistic, since we keep the stratospheric temperature fixed as optical thickness is changed. In reality, changing the optical thickness would affect the stratospheric temperature and the tropopause. Nonetheless, the problem provides some insight as to how the stratosphere affects *OLR*.

Python tips: You can do this problem using a modification of the script `GrayGasFlux.py`.

Problem 4.3 *Computation of gray gas heating rates*
In Fig. 4.2 we used a computation of net radiative flux to determine which portions of an all-troposphere atmosphere were subject to radiative cooling, and which would warm-up and form a stratosphere. Write a program to compute and plot the vertical profile of net upward flux for a gray gas with temperature profile given by the dry air adiabat throughout the depth of the atmosphere. The radiative fluxes can be most easily computed by evaluating the integral form of the Schwarzschild equation solution using numerical quadrature. Compare the results to the optically thick approximation. Check your routine by reproducing the results in Fig. 4.2, which was done assuming a pressure-independent absorption coefficient.

Now re-do the calculation allowing for pressure broadening, i.e. a linear increase of absorption with pressure. The optical thickness in this case is given by

$$\tau(p) = \frac{1}{2}\kappa(p_s)\frac{p_s}{g\cos\bar{\theta}}\left(1 - \frac{p^2}{p_s^2}\right).$$
(4.100)

Plot the results and compare with the optically thick limit (modified to allow for pressure broadening). What happens to the stratosphere as the atmosphere is made more optically thick?

Python tips: Modify `GrayGasFlux.py`.

Problem 4.4 In Eq. (4.19) for I_+ in the optically thick limit, we neglected the transmitted flux from the ground and also assumed that

$$\int_0^\tau \mathfrak{T}(\tau,\tau')d\tau' \approx \int_0^\infty \mathfrak{T}(\tau,\tau')d\tau' = 1.$$
(4.101)

Both approximations break down when τ is within a few optical depths of the ground. Find a correction to the expression for I_+ that remains valid near the ground. Show that this term yields the low-level behavior of the net flux $I_+ - I_-$ seen in the numerically computed gray gas results shown in the optically thick panels of Fig. 4.2.

Show that there is a similar correction term for I_- at the top of the atmosphere, and discuss its effect for a gray gas when $T(p)$ is on the dry adiabat all the way to $p = 0$. How would your answer change if T asymptoted to a constant at the top of the atmosphere?

Problem 4.5 *Gray gas runaway greenhouse*
Generalize formula (4.39) to the pressure-broadened case, i.e. the case in which the absorption coefficient increases linearly with pressure. Specifically, assume $\kappa = \kappa_0 \cdot (p/p_0)$, where p_0 is the reference pressure at which the absorption takes on the value κ_0. Derive an expression for the coefficient $A(L/R)$ in the pressure-broadened case, and determine its value for water vapor, CO_2, and CH_4 using a numerical quadrature.

4.10.2 Pure radiative equilibrium

Problem 4.6 A solar-transparent stratosphere in a gray gas atmosphere is in pure radiative equilibrium with upward infrared flux $I_{+,trop}$ upwelling through the tropopause. Using Eq. (4.41) and the fact that $I_+ - I_-$ is constant in the stratosphere, show that at the tropopause

$$I_+ + I_- = 2\frac{1 + \Delta\tau}{2 + \Delta\tau}I_{+,trop} \qquad (4.102)$$

where $\Delta\tau \equiv \tau_\infty - \tau_{trop}$. Use this to find the stratospheric temperature at the top of the atmosphere and at the tropopause. Show that both reduce to the skin temperature when the stratosphere is optically thin.

Problem 4.7 *Optically thick limit*
In Section 4.2.2 it was shown that radiative transfer acts like a diffusion at wavenumbers for which the atmosphere is optically thick, and the expression for the the radiative heating H_ν was given in Eq. (4.23). Suppose that the atmosphere is optically thick throughout the infrared. Show that the total infrared heating $\bar{H} \equiv \int_{ir} H_\nu \, d\nu$ still has the form of a diffusion, and write down the expression for the effective diffusivity $\bar{D}(T, p)$ as an integral. Radiative equilibrium in the absence of solar absorption is defined by $\bar{H} = 0$. Derive a differential equation describing the $T(p)$ which satisfies radiative equilibrium. The boundary condition for this equation is that the net infrared flux out of the top of the atmosphere is equal to the *OLR* (which is taken as a specified constant in this problem). Express this boundary condition as a requirement on dT/dp at $p = 0$. Note that because of the boundary condition $T = const$ is not the solution unless $OLR = 0$.

Show that this solution reduces to Eq. (4.42) for a gray gas with κ independent of p. Show that the equivalence continues to hold even if $\kappa = \kappa(p)$.

Problem 4.8 Using the results of Problem 4.7, compute the radiative diffusivity and the infrared radiative equilibrium solution for the following cases. (a) The atmosphere has constant absorption coefficient κ_0 within a narrow band of frequencies $|\nu - \nu_0| < \frac{1}{2}\Delta_0$, and is transparent elsewhere. (b) In addition to the absorption described in part (a), the atmosphere has an absorption coefficient $\kappa_1 < \kappa_0$ for $\frac{1}{2}\Delta_0 < |\nu - \nu_0| < \frac{1}{2}\Delta_1$.

In both cases you may assume that the absorption coefficient is independent of pressure and temperature, that the band is narrow enough that the Planck function can be considered independent of frequency within the band, and that the atmosphere is optically thick throughout the band where the absorption coefficient is non-zero. With a bit of cleverness, part (a) can be done analytically, but for part (b) you will probably need to find the approximate solutions to the differential equation describing $T(p)$ using numerical methods.

Fix Δ_0 at a value corresponding to a bandwidth of $25 \, cm^{-1}$, and set $\Delta_1 = 2\Delta_0$. Discuss how the solution depends on ν_0, the specified *OLR*, and κ_1/κ_0.

Problem 4.9 *Nuclear winter*
"Nuclear winter" refers to the climate following an event such as (perhaps) a global thermonuclear war, which lofts a high-altitude long-lived soot layer into the atmosphere. As an idealized form of this phenomenon, find the radiative-equilibrium temperature profile of a gray gas atmosphere below a soot cloud which is so thick that it completely absorbs the solar radiation incident upon it. You may assume that the soot cloud is isothermal, and acts like a blackbody in the infrared. Do you expect a troposphere to form in this situation? How

does the surface temperature compare to what it would be without the soot cloud? What do you think would happen to the precipitation?

Problem 4.10 Suppose that all the solar absorbers in an atmosphere are concentrated near the ground. This is typical of the case where the absorption is due to dust or water vapor. Specifically, assume that the profile of solar flux is $F_\odot = (1 - \alpha)S \cdot (1 - a\exp(-b\tau))$, where a and b are constants with $0 < a < 1$ and $b > 0$. Assuming that the atmosphere behaves like a gray gas in the infrared with pressure-independent κ, compute the radiative equilibrium $T(\tau)$ and the radiative equilibrium ground temperature T_g. Under what circumstances does the atmosphere become internally unstable near the ground? How does solar absorption affect the unstable temperature jump between the ground and the overlying air? To keep the problem simple, you may assume that the ground has zero albedo.

Problem 4.11 Consider a gray gas atmosphere with a sharply defined internal absorbing layer at pressure p_a. The profile of downward solar flux is $F_\odot = (1 - \alpha)S \cdot \frac{1}{2}(1 + \tanh((p_a - p)/\Delta p))$, where Δp is the thickness of the absorbing layer. Find an expression for the radiative equilibrium temperature profile, evaluate it numerically, and plot it. Determine analytically where the absorbing layer causes the temperature to increase with height and where it causes the temperature to decrease with height. Under what circumstances does the absorbing layer make the atmosphere unstable to convection? You may assume that the absorption coefficient is independent of pressure in this problem.

Problem 4.12 *Global warming and stratospheric cooling*
One of the signatures of anthropogenic global warming is that, in response to increases in greenhouse gas concentrations, the stratosphere cools at the same time the troposphere warms. This problem explores the phenomenon for a gray gas atmosphere.

Suppose that an atmosphere acts as an ideal gray gas in the infrared with an absorption coefficient κ independent of both pressure and temperature. Suppose that *as a function of pressure* the solar flux decays according to the law $F_\odot = (1 - \alpha)S \exp{-bp/p_s}$ for some constant b. To keep things simple, you may assume that the albedo of the ground is zero, so that all the incident solar radiation is absorbed either in the atmosphere or at the ground. Thus, you can assume that $OLR = (1 - \alpha)S$, as in the derivation of Eq. (4.56). This also means you need not worry about absorption of upwelling solar radiation reflected from the ground.

Show that in the notation of Section 4.3.5, $\tau_\infty = \kappa p_s/g$, $\tau_\infty - \tau = \kappa p/g$, and $\tau_S = \tau_\infty/b$. With these relations you can find the profile $T(p)$ using the solution given in Eq. (4.56). Since atmospheric observations are made using pressure rather than τ as the vertical coordinates, it is important to plot the temperature as a function of pressure if one wants to relate the theoretical behavior to observations.

Increasing the greenhouse gas concentration while keeping the distribution of solar absorbers fixed is equivalent to increasing τ_∞ while holding b fixed. On the same graph, plot $T(p)$ for $\tau_\infty = 0.25, 0.5, 1, 1.5, 2$, and 4, for a fixed value of b. Make graphs of this type for a range of b and explore the behavior as b ranges from very small values (e.g 0.01) where there is essentially no solar absorption to very large values (e.g 10) where all the solar energy is absorbed in the upper atmosphere. Under what circumstances does an increase of the greenhouse gas concentration warm the lower atmosphere but cool the upper atmosphere?

4.10.3 Real gas basics

Problem 4.13 Derive the result given in Eq. (4.68). To do this, substitute $y = 1/\zeta$ and then do an integration by parts. You will need the well-known result for the integral of the Gaussian, $\int_0^\infty \exp(-y^2)dy = \sqrt{\pi}/2$.

Problem 4.14 Discuss the behavior of the function $F(\ell) = \int \kappa_m^{-1} P(\kappa/\kappa_m) \exp(-\kappa\ell)d\kappa$ for: (a) $P = 2(1-x)$; (b) $P = 2x$; (c) $P = 6x(1-x)$; (d) $P = 4\max(0, 2x - 1)$. In all cases assume $P = 0$ unless $0 \leq \kappa \leq \kappa_m$. Pay particular attention to the behavior of F at large ℓ. Under what circumstances is the decay slower than exponential? What determines the decay rate at large ℓ?

Show that if $P(x) \sim x^b$ at small x, then $F(\ell)$ decays like $\ell^{-(b+1)}$ at large ℓ.

Problem 4.15 Consider a gas which is transparent outside a single frequency band, within which the transmission function is described by the Malkmus model. Compute the radiative cooling rate for an unbounded atmosphere which has a temperature discontinuity at pressure p_0, but is otherwise isothermal. Show that in the limit where the strong-line behavior dominates, there is a square-root singularity in the vicinity of the discontinuity, which tends to wipe out the temperature jump. Generalize this result by computing the radiative diffusivity for the one-band Malkmus model, in the optically thick limit.

Problem 4.16 Write a transmission function that uses the exponential-sum representation to compute the transmission function as a function of path, for a single band for which exponential sum coefficients have been tabulated. You may assume that the gas has uniform temperature along the path. Read in exponential sum data for a few bands near the $666\,\mathrm{cm}^{-1}$ principal CO_2 absorption feature, and graph the transmission as a function of path.

Exponential sum tables for various gases are found in the subdirectory ExpSumTables of the Workbook datasets directory for this chapter, available as part of the online supplement.

Python tips: It is easier to organize the calculation if you define a bandData object which stores all the necessary information for each band. Then, when you need to sum up over multiple bands, you only need to make a list of bands and loop over the bandData objects in the list. This is how the homebrew radiation code is implemented in the courseware.

Python tips: You can compute your own exponential sum tables using the function makeEsumTable in the Chapter Script PyTran.py, which makes use of the HITRAN database. A limited subset of HITRAN is included in the Workbook dataset directory for this chapter.

Problem 4.17 *Working with HITRAN*

A subset of the HITRAN database and its documentation is included in the Workbook datasets directory for this chapter, in the subdirectory hitran. The database itself, separated into files for each molecule, is in hitran/ByMolecule. The data is in plain text table format, and is not very hard to read. Read the documentation on the data format, and write a routine that reads in line strengths and line widths for CO_2. These values will be at the standard pressure and temperature for the database. To get a feel for the data, compute and plot the number of lines present in a series of bins of width $50\,\mathrm{cm}^{-1}$, covering the wavenumber range for which data is available. Compute and plot the median, maximum, and minimum line strength in each bin.

Then, assuming the Lorentz line shape, use the line strength, position, and width information to sum the contributions of each line (at standard width) and create a plot of the absorption coefficient as a function of wavenumber. When summing the contributions, you should cut off the contribution past a certain threshold distance from the line center; there are various empirical far-tail line shapes in use, but for this problem a simple truncation will do. Try truncating at a fixed number of line widths (e.g. 100). Try truncating instead at a fixed wavenumber distance (e.g. $10 \, \mathrm{cm}^{-1}$ from the line center). How much does the result depend on the cutoff prescription? These calculations will take quite a lot of computer time, even if you are quite clever at writing an efficient algorithm. You can keep down the time needed by restricting attention to just a portion of the wavenumber domain.

Next, fix the line cutoff prescription and look at how the absorption spectrum changes as you increase the pressure from 100 mb to 100 bar. You do this by rescaling line widths and peak values according to pressure, as described in the text. Note that the curves become smoother at higher pressure. Why is this? Compare results with a cutoff based on a fixed number of widths with those based on a fixed wavenumber value.

Once you have mastered these techniques, the HITRAN database can provide endless hours of enjoyable perusal of spectroscopic properties. Take a look at some infrared absorption properties for molecules not discussed in the text. Some particularly interesting ones to look at are ozone, SO_2, and the minor isotopologs of CO_2.

Python tips: The Chapter Script PyTran.py reads the line database, computes absorption coefficients, and provides functions to perform utility tasks such as generation of exponential sum tables.

4.10.4 Calculations with polynomial *OLR* fits

The following problems involve calculations with polynomial fits to the ccm model *OLR* calculation for an Earthlike atmosphere. The fitting coefficients needed are found in Tables 4.2 and 4.3 in the text.

Python tips: The Chapter Script OLRPoly.py contains the fitting coefficients given in the tables, and also contains an interpolation routine that makes it easy to compute $OLR(T_g, CO_2)$ for arbitrary values of the arguments.

Problem 4.18 For a dry air Earth atmosphere on the dry adiabat, with 1000 ppmv of CO_2 mixed in, the $OLR(T_g)$ computed from the ccm radiation model can be well fit by a polynomial $a + bx + cx^2$, where $x \equiv T_s - 275$. The coefficients $a = 273.9$, $b = 3.9436$, and $c = 0.026987$ provide a fit good to better than 1% over the temperature range 250 K to 350 K.

Compute and graph the radiating temperature T_{rad} and radiating pressure p_{rad} as a function of T_g. Also, make a graph comparing $dOLR/dT_g$ for the polynomial fit with the gray gas value $4\sigma T_{rad}^3$. Why does the gray gas estimate differ from the actual slope? (Recall that the slope is important because it provides a measure of climate sensitivity.)

Problem 4.19 Table 4.2 gives a cubic polynomial fit of *OLR* vs. T_g for a moist Earth atmosphere with fixed relative humidity, with temperature on the moist adiabat. Results are given for several different values of CO_2 concentration.

Compute and graph the radiating temperature T_{rad} and radiating pressure p_{rad} as a function of T_g. Note that since the *OLR* fit was done for an atmosphere on the moist adiabat, you will need to use the moist adiabat to compute p_{rad} from T_{rad}. Also, make a graph comparing $dOLR/dT_g$ for the polynomial fit with the gray gas value $4\sigma T_g^3$. Why does the gray gas estimate differ from the actual slope? Discuss how the temperature affects the comparison. How does $dOLR/dT_g$ compare with the dry value you obtained in Problem 4.18? (Recall that the slope is important because it provides a measure of climate sensitivity.) Carry out your analysis for several different values of CO_2 concentration.

Python tips: Use the object `MoistAdiabat` in the `phys` courseware module to help you compute p_{rad}.

Problem 4.20 Recompute the ice-albedo bifurcation diagram in Fig. 3.10 using the polynomial *OLR* fit for 50% relative humidity. Do the calculation for several different values of CO_2 concentration, and discuss the results in light of Earth's actual climate history. *Note:* This calculation neglects the effect of clouds, so for any given $L_\odot$ the temperature in the unglaciated state will tend to be too warm and the temperature in the fully glaciated state will tend to be too cold.

Problem 4.21 Recompute the ice-albedo bifurcation diagrams in Fig. 3.11 using the polynomial *OLR* fit for 50% relative humidity. In this case, use the CO_2 concentration as your control parameter instead of p_{rad}. Show results for various different values of $L_\odot$ ranging from $500\,W/m^2$ to $3000\,W/m^2$. In cases where the CO_2 concentration you need to fill out the diagram lies outside the range of validity of the polynomial fit, just show the portion of the diagram for which the fit is valid and sketch in the qualitative behavior of the missing bits of the curve. Note that the remark about clouds in Problem 4.20 applies to this problem as well.

Hint: As was the case for the bifurcation diagram using $L_\odot$ as the bifurcation parameter, it is easiest to find the control parameter which yields a given temperature T_s and then plot the control parameter (CO_2 in this case), against T_s sideways. This procedure is easier because there is a unique CO_2 corresponding to each T_s, whereas there are multiple solutions for T_s for a given CO_2. To find the CO_2 corresponding to T_s, first find the *OLR* needed to achieve that temperature, then use a Newton's method iteration to find the value of $\ln CO_2$ that corresponds to this *OLR*. You will need a function that gives you $OLR(T_s, \ln CO_2)$; this is obtained by interpolating the coefficients in Table 4.2 to the desired value of $\ln CO_2$ and then evaluating the resulting polynomial at T_s.

Problem 4.22 Consider an Earthlike planet orbiting a star with luminosity such that the solar constant is $2000\,W/m^2$ at a distance of 1 A.U. from the star. Using the information in Table 4.3, find the CO_2 needed to maintain an Earthlike habitable temperature as a function of distance from the star. Do this for various assumptions about the humidity of the atmosphere. You may assume the albedo of the planet to be 20%.

4.10.5 Real gas and semigray radiation computations

Many of the problems in this and the next section require use of the homebrew exponential sums radiation code, though the more Earthlike ones can be done using the `ccm` radiation code as an alternative. The basic building block of the homebrew code is a routine to compute the transmission function for a single band, which was done in Problem 4.16. All the necessary exponential sum tables have been precomputed for you, and are found in the workbook dataset directory for this chapter, available in the online supplement. The tables are in the subdirectory `ExpSumTables`. The rest of the procedure is described in the text. It is not too hard, but still it takes some doing to implement. Hopefully, if you are not using the ready-made Python courseware, your instructor will have done some of the work for you in advance, in whatever programming environment is being used as an alternative.

Python tips: The homebrew radiation computation is implemented in the modules miniClimt.py for a well-mixed greenhouse gas and in miniClimtFancy.py for an inhomogeneous greenhouse gas. The latter also implements a crude representation of temperature scaling of line strengths, and an allowance for the difference between self-broadened and foreign-broadened absorption. The ccm radiation model has also been made available through a user-friendly Python interface, using the climt module to interface to the fast, compiled Fortran code. Some functions useful for radiation calculations using the ccm model are defined in the Chapter Script ccmradFunctions.

Problem 4.23 The semigray skin temperature is defined by Eq. (4.95). Using Newton's method or some similar root-finder, make a plot of T_{skin}/T_g as a function of $h\nu_0/kT_g$. Analytically derive the behavior of T_{skin}/T_g in the limit of small and large $h\nu_0/kT_g$.

Problem 4.24 Numerically determine the pure-radiative equilibrium for a semigray atmosphere, in equilibrium with a specified ground temperature T_g. You find the equilibrium by numerically computing the net radiative cooling as an integral (you should already have code to do this from having done the previous problems), and then time-stepping the temperature equation until equilibrium is reached.

Let ν_0 be the frequency at the center of the narrow absorbing band of the atmosphere. Pick a ν_0 in the infrared range, and explore how the equilibrium profile $T(p)/T_g$ varies as you change T_g. How does the result depend on ν_0?

Python tips: The numerical computation can be done using a variation of the general radiative-convective equilibrium script RadConvEq.py. To do the semigray case, just edit the radmodel.bandParams line so that the one-band Oobleck data is used in the homebrew radiation model.

Problem 4.25 This problem addresses the role of the water vapor continua in determining the Kombayashi–Ingersoll limit. Consider a pure saturated water vapor atmosphere, and compute the *OLR* as a function of surface temperature, neglecting all absorption due to the two water vapor continua. In other words, recompute Fig. 4.37 without the continua. Then, re-do your calculation with only the lower wavenumber continuum region included. Finally, compute the *OLR* including the higher wavenumber continuum region, but assuming the atmosphere emits no radiation at wavenumbers outside the continuum. How well does this calculation reproduce the Kombayashi–Ingersoll limit?

Python tips: Using the homebrew radiation code as in the script RealGasRunaway.py, you can suppress the continuum by simply setting the continuum function to NoContinuum. To suppress just the higher wavenumber continuum, you need to go into the homebrew radiation module imported by RealGasRunaway.py and edit the water vapor continuum function so it only calculates the lower wavenumber continuum. To do the last part of the problem, you need to edit the radiation code so that it sums up only contributions to *OLR* from the continuum regions, and ignores contributions from all other wavenumbers.

Problem 4.26 Using the homebrew radiation model for a saturated pure water vapor atmosphere (as in Fig. 4.37), make a graph of the *OLR* as a function of surface gravity g for a fixed surface temperature $T_g = 400\,$K, and demonstrate the fit of Eq. (4.94) for the parameters given in the text.

4.10.6 Tropopause height

Problem 4.27 This problem examines the effect of lapse rate on the tropopause height for a real gas.

Consider an atmosphere whose temperature profile is governed by the moist adiabat for a mixture of water vapor in air, but whose radiative properties are determined by a mixture of 300 ppmv of CO_2 in dry air. In other words, in this problem we include the effect of water vapor on the adiabat, but neglect the radiative effects of water vapor. Using hard convective adjustment, compute the tropopause height by running a radiative-convective model to equilibrium. Carry out the calculation for surface temperatures of 280 K and 300 K. Compare the results with what you would get for these two temperatures using the dry air adiabat instead of the moist adiabat. *Note:* You can spot the tropopause height either by noting where the temperature profile departs from the adiabat, or by looking for where the radiative heating falls to essentially zero.

Python tips: This problem is done by making use of the script RadConvEq.py, uncommenting the options for enabling convective adjustment and for using the moist adiabat.

Problem 4.28 *Gray gas tropopause height vs.* τ_∞

In this problem, you will find the tropopause height for a gray gas atmosphere in radiative-convective equilibrium. In the troposphere, the temperature profile is assumed to be on the ideal gas dry adiabat. The low-level air temperature is assumed to be identical to the ground temperature. The atmosphere is transparent to solar radiation. We will solve the problem by specifying the ground temperature T_g and finding the atmosphere which is in equilibrium with this value.

You do not need to solve this problem by time-stepping. It still needs to be done with some numerical analysis, but the problem can be solved in terms of numerical quadrature and a Newton's method root finder. The solution scheme runs as follows. First, re-write the problem in terms of the optical thickness coordinate τ. τ_∞ then becomes a specified parameter of the problem. Suppose that the tropopause is located at τ_{trop}. From the dry adiabat, we then know $T(\tau)$ between the tropopause and the ground, whence the upward flux $I_{+,trop}$ can be computed as a definite integral. This is the same integral as used in the gray all-troposphere model in the text, except that it is carried out only to τ_{trop} and not all the way to $\tau = \tau_\infty$. You will need to implement a function that evaluates this integral using numerical quadrature. This is the upward flux which heats the stratosphere, since there is no solar absorption. The temperature for a stratosphere illuminated by upward flux I_+ was determined in Problem 4.6.

By evaluating this at the tropopause, you have the temperature $T_{strat}(\tau_{trop})$. In general, this will be discontinuous with the troposphere temperature just below the tropopause. If the tropopause is too low, then the temperature jump is statically unstable, so the troposphere will mix upwards by convection. The usual physical assumption governing the tropopause height is that the mixing proceeds until the unstable jump is just eliminated. Thus, the condition we impose at the tropopause is that the temperature be continuous there. This uniquely determines the tropopause height. The numerical calculation thus consists of three parts: a numerical quadrature routine to compute $I_{+,trop}$, a fairly trivial calculation of the corresponding stratospheric temperature, and a Newton's method adjustment of τ_{trop} until the temperature continuity condition is satisfied.

Write down the equations which embody the temperature continuity condition, and show that T_g drops out of the problem if all temperatures are written as T/T_g. In other words, the value of τ_{trop} which satisfies the continuity condition is independent of T_g.

Using the above procedure, make a graph of how the tropopause height depends on τ_∞ for various values of R/c_p. Can you explain the behavior in the optically thin limit analytically? Show that without pressure broadening (i.e. constant κ) the tropopause moves to the top of the atmosphere in the optically thick limit if $4R/c_p < 1$, and moves to the ground in the optically thick limit when $4R/c_p > 1$. Note that this result says that, in circumstances when the radiative equilibrium is statically stable at all levels, the atmosphere becomes all-stratosphere in the optically thick limit. How does the tropopause behave if you assume linear pressure broadening, so $\kappa = \kappa_0 \cdot (p/p_0)$?

If you are very good at evaluating definite integrals asymptotically by steepest descent, you can actually do the optically thick limit analytically.

Problem 4.29 *Real gas tropopause height vs.* τ_∞
Consider a solar-transparent dry atmosphere consisting of a mixture of Earth air with a mass concentration q of CO_2. The troposphere is assumed to be on the dry adiabat. The low-level air temperature is identical to the ground temperature. As in the text, we specify the ground temperature T_g and find the atmosphere which is in infrared radiative-convective equilibrium with the ground. Once the infrared cooling profile is computed from a radiation model, the equilibrium is found by time-stepping, applying hard convective adjustment to eliminate unstable layers, as in the examples done in the text. Any radiation model will do, but it is suggested that you use the homebrew exponential sum model. This problem requires a fairly large amount of computation time, so it is most suitable as an extended project spread over a few weeks. You can keep the computation time down by using only 20–30 points in the vertical, which should be sufficient to reproduce the basic behavior.

Take $T_g = 300\,K$ and $p_s = 10^5\,Pa$, and explore the behavior of tropopause height as q ranges from small values to unity. Note that you ought to change the value of R/c_p as q changes, but to keep things simple you may keep R/c_p constant at the value $\frac{2}{7}$ appropriate to dry Earth air. Next, explore how the behavior depends on T_g, for a few values ranging from $200\,K$ to $400\,K$. The cold case has some features in common with Snowball Earth.

Now re-do the problem for a pure CO_2 atmosphere on a CO_2 dry adiabat. In this case, you vary the optical thickness by varying the surface pressure p_s. Keeping $T_g = 300\,K$, try $p_s = 1000\,Pa$ (similar to Mars), $p_s = 2\,bar$, $p_s = 20\,bar$, and $p_s = 90\,bar$. The last of these is in the thick-atmosphere regime of Venus; recompute it at $T_g = 700\,K$. How does the tropopause height change?

Note that for the hot, high-CO_2 case you will need to use an exponential sum table that goes far enough out in wavenumber to capture the surface emission. If you need to, you may assume that the atmosphere is so optically thick in the wavenumbers past the end of the table that the fluxes there do not contribute significantly to the infrared cooling.

There are various other variants of this problem you can try. For example, you can do a problem with water vapor in air on the moist adiabat; in this case you would fix the relative humidity, and the optical thickness would increase as you went to larger temperatures, which yield higher water contents. Alternately, using the ccm radiation code you can do an atmosphere consisting of water vapor, air, and CO_2, and vary the optical thickness by varying the CO_2 with fixed T_g.

Python tips: This calculation is implemented in the chapter script RadConvEq.py, which makes use of the homebrew exponential sums radiation code.

4.11 FOR FURTHER READING

Finding spectroscopic data appropriate to a novel planetary atmosphere can be a real challenge. A wealth of specialized spectroscopic data can be found in the *Journal of Quantitative Spectroscopy and Radiative Transfer.* The reader is directed especially to their 2008 special issue on planetary atmospheres (see Rothman L 2008, *JQSRT* doi:10.1016/j.jqsrt.2008.02.002). In addition, laboratory and theoretical spectroscopy related to planetary problems can often be found in the journal *Icarus.*
The HITRAN spectroscopic database is described in

- Rothman LS *et al.* 2005: The HITRAN 2004 molecular spectroscopic database, *J. Quant. Spectrosc. Radiative Transf.*, **96**, 139–204.

In a book that one hopes will stick around for a while, there is always some risk in referring to specific means of obtaining digital data. The HITRAN database is so valuable, however, that it is sure to be available in some form more or less indefinitely. At the time of writing, the HITRAN data can be obtained over the Internet at the URL http://cfa-www.harvard.edu/hitran. The 1970s era Goody and Yung book on atmospheric radiation refers to obtaining the data on "AFGL tapes," and no doubt earlier books made reference to things like "punch cards" or "paper tapes." No doubt, our reference to the "Internet" will seem similarly quaint within a few years.

The HITRAN database does not include the very weak CO_2 absorption lines that become important for extremely massive atmospheres such as that of Venus, and moreover, the temperature dependence data of the lines that are included becomes somewhat inaccurate at Venusian temperatures. There are two databases that extend the CO_2 absorption database to cover the Venusian regime, both of which use the same data format as HITRAN. The first is the HITEMP database. At the time of writing, there is neither a convenient published document describing the database nor a generally accessible download site, but an updated version of the HITEMP database and expanded documentation are expected to be made available through the HITRAN site in the near future. In the meantime, information about the existing database can be found in

- Rothman LS *et al.* 1995: HITRAN, HAWKS, and HITEMP high-temperature molecular database, *Proc. Soc. Photo-Opt. Instrum. Eng.* **2471**, 105–111;

and the original version of the database can be downloaded by contacting the managers of the HITRAN site. A similar high-temperature, high-pressure database is described in

- Tashkun SA *et al.* 2003: CDSD-1000, the high-temperature carbon dioxide spectroscopic databank, *J. Quant. Spectrosc. Radiative Transf.* **82**, 165–196.

It is available online via ftp at ftp.iao.ru/pub/CDSD-1000.

Information on the CO_2 collision-induced continuum is very sparse. The modeling of the CO_2 continuum used throughout this book (and incorporated in the software supplement) is based on a polynomial fit to absorption coefficients described in

- Kasting JF, Pollack JB and Crisp D 1984: Effects of high CO_2 levels on the surface temperature and atmospheric oxidation state of the early Earth, *J. Atmos. Chem.* **1**, 403–428.

References to the laboratory measurements upon which the parameterization is based amount to one published paper, one NASA technical report, and one unpublished personal communication; these may be found in the above referenced article. Some theoretical developments, which have been incorporated in a few of the more recent representations of the far-infrared continuum, are described in

- Gruszka M and Borysow A 1997: Far infrared collision-induced absorption of CO_2 for the atmosphere of Venus at temperatures from 200 K to 800 K, *Icarus* **129**, 172–177,

but there seem to have been no new laboratory measurements since those discussed in the former paper.

The collision-induced continua of H_2, CH_4, and N_2 relevant to the atmosphere of Titan are given in

- Courtin R 1988: Pressure-induced absorption coefficients for radiative transfer calculations in Titan's atmosphere, *Icarus* **75**, 245–254.

Radiative transfer on gas giants is discussed in

- Guillemot T *et al.* 1994: Are the giant planets fully convective? *Icarus* **112**, 337–353.

The water vapor continuum is described in the following two papers:

- Clough SA, Kneizys FX and Davies RW 1989: Line shape and the water vapor continuum, *Atmos. Res.* **23**, 229–241.
- Grant WB, 1990: Water vapor absorption coefficients in the 8–12 μm spectral region: A critical review, *Appl. Opt.* **29**, 451–462.

The first of these is considered the standard reference at time of writing, but one must take care in reading it, as there are a certain number of typographical errors and mislabeled figures.

The full-featured ccm radiation code is described in complete and somewhat intimidating detail as part of the general description of the NCAR Community Atmospheric Model (CAM) in NCAR Technical Note TN-464+STR, available at the time of writing at

- http://www.ccsm.ucar.edu/models/atm-cam/docs/description/.

An accessible overview of an earlier version of the radiation model can be found in

- Kiehl J and Briegleb B 1992: Comparison of the observed and calculated clear sky greenhouse-effect – Implications for climate studies, *J. Geophys. Res.* **97** (D9), 10037–10049.

A version of this radiation code with a simple Python user interface is distributed as part of the software supplement to this book. The Python interface makes it easy to use the code to compute *OLR* and heating rates within a Python script, and eliminates the need for any familiarity with the FORTRAN language in which the underlying computation is written.

A comprehensive introduction to molecular absorption of infrared and the associated physics is given in

- Goody RM and Yung Y 1995: *Atmospheric Radiation*, Oxford University Press.

The correlated-k refinement of the exponential sums approximation to the transmission function is described in

- Lacis A and Oinas V 1991: A description of the correlated k-distribution method for modeling non-gray gaseous absorption, thermal emission and multiple scattering in vertically inhomogeneous atmospheres. *J. Geophys. Res.* **96**, 9027–9063.

CHAPTER FIVE

Scattering

5.1 OVERVIEW

In the atmospheres considered so far, the blackbody source term adds new radiation to the atmosphere traveling in all directions, but once present in the atmosphere radiation travels in a fixed direction; it can be absorbed as it travels, but it does not scatter into other directions. In this class of problems, the two-stream approximation consists entirely in doing the calculation for a single equivalent propagation angle $\bar{\theta}$, but does not change the essential structure of the full problem. If one wanted more information about the angular distribution, one would simply repeat the same calculation several times, with different θ, and average the results to get the net upward and downward fluxes; each calculation is independent, and has the form of a simple first order ordinary differential equation for the flux. With scattering the situation is very different, as the scattering couples the flux at one angle with the fluxes at all other angles. The full problem now takes the form of a computationally demanding integro-differential equation, with the derivative of the flux at a given angle expressed as a weighted integral over the fluxes at all other angles.

Light with wavelengths in the near-infrared or shorter is significantly scattered from molecules, though molecules are too small to appreciably scatter thermal infrared or longer wavelengths. Many atmospheres (Earth's included) contain very fine aerosol particles with diameters on the order of a micrometer or less; they are typically made of mineral dust, or of condensed substances such as sulfuric acid or other sulfur compounds. They are very powerful scatterers of solar radiation, and therefore can significantly affect a planet's albedo even when the total mass of aerosols is quite small. Cloud particles made of various condensed substances have typical diameters of 10–$100\,\mu$m. Because water clouds like Earth's absorb so strongly in the infrared, cloud scattering is often thought of primarily in terms of the solar spectrum. However, taking a broader view, cloud substances commonly found on other planets have a very important thermal infrared scattering effect. Water clouds are the exception, rather than the rule, but because of the importance of water clouds on Earth, thermal infrared scattering by clouds is a far less developed subject than is shortwave scattering.

Clouds, in their many and varied manifestations, pose one of the greatest challenges to the understanding of Earth and planetary climate. On Earth, water clouds reflect a great deal of sunlight but also have a considerable greenhouse effect. The net cloud effect is a fairly small residual of two large and uncertain terms, and the way the two effects play out against each other plays a central role in climate change problems extending from the Early Earth to Cretaceous Warmth, to ice ages, to global warming, and the distant-future fate of our climate. The high albedo of Venus is caused largely by clouds made of sulfuric acid droplets, but the very same clouds scatter and absorb infrared radiation, and help to increase the planet's greenhouse effect. On an Early Mars with a 2 bar CO_2 atmosphere, formation of clouds of CO_2 ice would play an important role in the planet's climate, both in the infrared and solar spectrum. Titan's present-day methane clouds affect the satellite's radiation budget both through infrared and visible scattering; Neptune is cold enough that methane ice clouds can form in parts of its atmosphere. The swirling psychedelic colors of Jupiter and Saturn arise from clouds of a dozen or more different types, which no doubt also affect the radiation budget. It is hard to think of a planetary atmosphere in which clouds do not play a significant role.

Regardless of whether clouds exert a greenhouse effect by scattering or by the conventional absorption/emission process, the cross-cutting issue is that clouds affect the incoming and outgoing part of the planet's radiation budget in large but opposing ways. Clouds always reflect some incoming stellar radiation, and clouds at appreciable heights above the surface almost invariably reduce the outgoing infrared radiation. It will turn out that the net radiative effects of clouds are highly sensitive to the size of the particles of which they are composed. This leads to the disconcerting conclusion that the climate of an object as large as an entire planet can be strongly affected by poorly understood processes happening on the scale of a few micrometers.

Scattering calculations play a critical role not only in determining planetary radiation balance, but also in interpreting a wide range of observations of the Earth, Solar System planets, and extrasolar planets. There is no doubt that if justice were to be served (and the reader had unlimited time) scattering should be given a treatment at least as in-depth as that which we have accorded to purely absorbing/emitting atmospheres. However, in order to maintain progress toward our primary goal of understanding the essentials of planetary climate, the treatment given in this chapter will be highly abbreviated, and focus on the minimal understanding of the subject needed to estimate planetary albedo, shortwave atmospheric heating, and the basic effect of clouds on outgoing infrared radiation and on solar absorption. In particular, we will leap directly into the two-stream approximation, without much discussion of the properties of the scattering equations in their full generality. Were it not for the position of clouds at the forefront of much research on planetary climate, we would be content to leave the discussion of scattering to a few brief remarks concerning planetary albedo.

Atmospheric absorption of the relatively shortwave light from a planet's star is often, though not invariably, significantly affected by scattering. Hence, the absorption of stellar near-infrared, visible, and ultraviolet radiation will be discussed in this chapter, together with a few implications for atmospheric structure. The effect of absorption of incident light on the temperature profile of the upper atmosphere was derived for gray gases in Chapter 4, but in the present chapter the reader will find results bearing on atmospheric heating due to absorption of incoming stellar radiation by CO_2, water vapor, and methane in a variety of planetary contexts both within and outside the Solar System. The effect of ultraviolet absorption by ozone on the stratospheric temperature structure of Earth and Earthlike planets will also be discussed here.

5.2 BASIC CONCEPTS

The atmosphere can be considered to be a mix of particles, some of which absorb, some of which scatter, and some of which do both. The particles could be molecules, or they could be macroscopic particles of a condensed substance, as in the case of cloud droplets or dust. One builds up the absorbing and scattering properties of the atmosphere as a whole from the absorbing and scattering properties of the individual particles. In keeping with the usage in the preceding chapters, we will ultimately characterize the effect of atmospheric composition on radiation in terms of scattering properties per unit mass of atmosphere, just as we did for the absorption coefficient.

Consider a parallel, monochromatic (single-frequency) beam of light with flux F in W/m^2 traveling in some specific direction. When the beam encounters a particle of finite extent, a certain amount of the flux will be absorbed, and a certain amount will be scattered into other angles. The rate at which energy is taken out of the beam by absorption and scattering can be characterized in terms of coefficients with dimensions of area, which are known as *cross-sections*. The rate of energy absorption is $F\chi_{abs}$, where χ_{abs} is the *absorption cross-section*, and the rate of energy scattering into other directions is $F\chi_{sca}$, where χ_{sca} is the *scattering cross-section*.[1] The scattering and absorption cross-sections can be quite different from the actual cross-section area of the object. The cross-sections can be thought of as the cross-section areas of hypothetical equivalent objects which absorb or scatter all light hitting the object while leaving the rest of the beam to pass by undisturbed. The ratio of scattering cross-section to the actual cross-section area of the scatterer is called the *scattering efficiency*, Q_{sca}. For a spherical particle of radius r, $Q_{sca} = \chi_{sca}/(\pi r^2)$. The *absorption efficiency* is defined similarly. For spherical particles, the cross-sections are independent of the angle at which the radiation is directed at the particle. For non-spherical particles the cross-sections for an individual particle depend on angle, but the typical physical situation involves scattering off an ensemble of particles presented with random orientations. In this case, we can average over all orientations and represent the mean scattering or absorption in terms of the cross-section for an equivalent sphere. This approach can break down if particles are not randomly oriented, as can be the case for plate-like ice crystals that become oriented through frictional drag forces as they fall. The *single-scattering albedo* for a particle is the ratio of flux of the incident beam lost via scattering to net flux lost. Using the notation ω_{p0} for the single-scattering albedo of an individual particle, we have $\omega_{p0} = \chi_{sca}/(\chi_{sca}+\chi_{abs})$. Later we will introduce the single-scattering albedo for the medium as a whole. The cross-sections for particles or molecules can be measured in the laboratory, and often can be computed from basic physical principles.

Since the radiation fields we will deal with are generally distributed over a range of frequencies and direction, instead of being monochromatic and unidirectional, we will write our equations in terms of the spectral irradiance I introduced in Chapters 3 and 4. Recall

---

[1] The more usual notation for the cross-section is σ, but in our subject matter that symbol has been reserved for the Stefan–Boltzman constant. One can think of χ as standing for $\chi\rho\omega\sigma\sigma$-section.

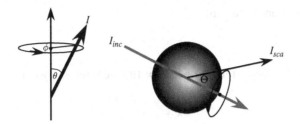

Figure 5.1 Definition of propagation angles (left) and scattering angle (right).

that if the spectral irradiance is $I((\theta, \phi), \nu)$ at a given point, then $Id\Omega d\nu$ is the flux of radiation in frequency band $d\nu$ with directions of travel within a solid angle $d\Omega$ about the direction (θ, ϕ), which passes through a plane perpendicular to the direction of travel. To apply the results of the preceding paragraph to smoothly distributed radiation, one needs only to substitute $Id\Omega d\nu$ for the incident flux F.

As in previous chapters, we'll make the plane-parallel assumption, and assume that I depends on position only through pressure. Suppose that in the vicinity of some pressure level p there are N scatterers of type i per unit mass of atmosphere, and that each scatterer has mass m_i. Suppose that the light impinging on the layer is traveling with angle θ to the vertical. Then, taking a layer of thickness dp which is small enough that multiple scattering can be neglected, the rate of energy lost by the incident beam due to absorption and due to scattering into different angles is

$$-\frac{dp}{g\cos\theta} N \cdot (\chi_{abs,i} + \chi_{sca,i}) Id\Omega d\nu = -\frac{dp}{g\cos\theta} q_i \cdot \left(\frac{1}{m_i}\chi_{abs,i} + \frac{1}{m_i}\chi_{sca,i}\right) Id\Omega d\nu \qquad (5.1)$$

where q_i is the mass concentration of the particles in question. From this we can define the absorption coefficient of the substance $\kappa_i \equiv \chi_{abs,i}/m_i$ which has units of m^2/kg. This absorption coefficient is the same quantity we defined in Chapter 4 in connection with gaseous absorption. The additional term in the above equation characterizes the energy lost from the incident beam due to scattering. We will not introduce separate notation for this term since scattering is most commonly characterized in terms of the cross-section itself.

If there is only one optically active substance i in the atmosphere, we define the optical depth in the vertical direction by the equation

$$\frac{d\tau^*}{dp} = -\frac{1}{g}\left(\kappa_i + \frac{1}{m_i}\chi_{sca,i}\right) q_i. \qquad (5.2)$$

Because the absorbing and scattering properties typically depend on wavenumber, the optical depth is generally a function of wavenumber, though we will only append a wavenumber subscript to τ^* when we wish to call attention specifically to the wavenumber dependence. If there are many types of scatterers and absorbers – which could include particles of a single substance but with different sizes – then we define the optical depth by summing over all species. Thus

$$\frac{d\tau^*}{dp} = -\frac{1}{g}\left(\kappa + \sum_i \frac{1}{m_i}\chi_{sca,i} q_i\right) \qquad (5.3)$$

where the net absorption coefficient is

$$\kappa \equiv \sum_i \kappa_i q_i. \tag{5.4}$$

We then define the single-scattering albedo for the medium as a whole as

$$\omega_0 \equiv \frac{\sum q_i \chi_{sca,i}/m_i}{\kappa + \sum q_i \chi_{sca,i}/m_i}. \tag{5.5}$$

The pair (κ, ω_0) constitutes the basic description of the absorption and scattering properties of the medium. Both are typically functions of wavelength and altitude, and may also directly be functions of pressure and temperature. If the medium consists of only a single type of particle, and the gas in which the particles are suspended neither absorbs nor scatters, then $\omega_0 = \omega_{p0}$. In general, though, the single-scattering albedo of the medium depends on the mix of absorbers and scatterers. For example, an atmosphere may consist of a mix of cloud particles which are perfect scatterers ($\omega_{p0} = 1$) with a strong greenhouse gas which is an absorber. In this case, ω_0 will go down as the greenhouse gas concentration increases, even if the cloud particle concentration is kept fixed.

Using the definition of optical depth, Eq. (5.1) for the rate of energy loss from the beam can be re-written as simply $dI = -I d\tau^*/\cos\theta$. Since the vertical component of flux is $I\cos\theta$, this expression can be recast as an expression for the rate of loss of vertical flux, namely

$$dI \cos\theta = -I d\tau^*. \tag{5.6}$$

The proportion of this lost to scattering is ω_0 while the proportion lost to absorption is $1 - \omega_0$. The fate of the energy lost to absorption is different from the fate of that lost due to scattering. The former disappears into the pool of atmospheric heat, whereas energy lost to scattering from one beam reappears as flux in a range of other directions, so we need to keep track of the two loss mechanisms separately. The beam loss in a given direction is offset by two source terms: one due to thermal emission, and one due to scattering from other directions. The thermal emission term is proportional to the Planck function, and can be treated in a fashion similar to that used in deriving the Schwarzschild equations. We'll leave the thermal emission out for now, and concentrate on scattering; the thermal emission term will be put back in in Section 5.5.

To understand better where the scattered flux goes, consider the energy budget for a box of thickness $d\tau^*$ in the vertical, shown from the side in Fig. 5.2. Since the radiation field is independent of the horizontal dimensions, the flux entering the box from the side is the same as the flux leaving it from the side, and does not affect the budget. If the base of the box has area A, an amount $A \cdot I(\tau^*)\cos\theta$ enters the box from the bottom and a somewhat lesser amount $A \cdot I(\theta, \phi, \tau^* + d\tau^*)\cos\theta$ leaves the box from the top. Taking the difference gives the loss of energy from the beam per unit time, due to scattering and absorption. Using Eq. (5.6) this can be written as simply $A \cdot I(\theta, \phi, \tau^*)d\tau^*$; it doesn't matter whether I is evaluated at $d\tau^*$ or $\tau^* + d\tau^*$ in this expression, since $d\tau^*$ is presumed small. The energy per unit time scattered and redistributed into all other directions is then $A \cdot \omega_0 I(\theta, \phi, \tau^*)d\tau^*$. Now, to write an equation for how the vertical component of flux changes between τ^* and $\tau^* + d\tau^*$, we need to find how much flux is added to the direction (θ, ϕ) by scattering from all other directions of radiation impinging on the layer. We can do this by considering the incident radiation one direction at a time, and summing up. Consider a beam of light traveling with direction (θ', ϕ'), having radiance $I(\theta', \phi', \tau^*)$. The scattering contributed to direction (θ, ϕ) comes from the scatterers in the shaded parallelogram shown in Fig. 5.2,

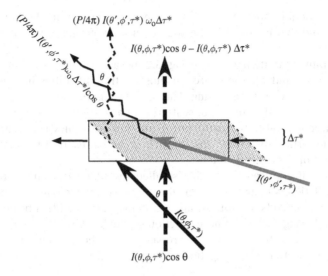

Figure 5.2 A scattering control volume, showing the flux added in the direction (θ, ϕ) due to scattering of an incident beam with direction (θ', ϕ'). Only the contribution from the slab of thickness $d\tau^*$ is considered. The incident beam illuminates the entire slab, but only the scatterers in the shaded parallelogram contribute to scattered radiation in the direction (θ, ϕ). The solid squiggly line represents scattered radiation, and the dashed squiggly line represents the vertical component of the scattered flux. The vertical straight, dashed arrows give the vertical component of the flux in the (θ, ϕ) direction, and show how it changes as the slab is traversed. Flux is lost from the (θ, ϕ) beam owing to absorption and scattering. The flux lost due to scattering shows up as scattered radiation in all other directions; these are not shown in the diagram.

which is greater than the amount of scatterer in a rectangular box by a factor of $1/\cos\theta$. Further, only a proportion of the radiation scattered from the contents of the parallelogram goes into the direction (θ, ϕ), We will write this proportion as $P/4\pi$, where P depends on both the incident and scattered directions. Thus, the radiance contributed to direction (θ, ϕ) by scattering is $A \cdot (P/4\pi)\omega_0 I(\theta', \phi', \tau^*) d\tau^*/\cos\theta$, and the vertical component of this is obtained by multiplying by $\cos\theta$, yielding $A \cdot (P/4\pi)\omega_0 I(\theta', \phi', \tau^*) d\tau^*$. This is the vertical flux contributed by scattering, and is added in to the flux leaving the top of the box. The scattering acts as a source of radiation in direction (θ, ϕ), which is added to the right hand side of Eq. (5.6). Dividing out the area of the base of the box, the flux balance for the box becomes

$$dI(\theta, \phi)\cos\theta = -I(\theta, \phi)d\tau^* + \frac{\omega_0}{4\pi}P(\theta, \phi, \theta', \phi')I(\theta', \phi')d\tau^* \qquad (5.7)$$

if one considers only the flux contributed by scattering of a single direction (θ', ϕ') . To complete the equation, one must integrate over all incident angles (θ', ϕ'). To determine the radiation field in its full generality, it is necessary to satisfy the flux balance for each direction of propagation simultaneously. Before proceeding toward that goal, we'll check Eq. (5.7) to verify that the scattered energy is conserved. Applying the control volume sketch to the *incident* beam direction, we infer that the incident beam traveling in direction (θ', ϕ') deposits energy in the control volume at a rate $I(\theta', \phi')d\tau^*$ (per unit area). A proportion

$\omega_0 P/4\pi$ of this should show up as an increase in the energy in the box propagating in direction (θ, ϕ), and that is precisely the source term appearing in Eq. (5.7). The books are indeed balanced.

It is worth thinking quite hard about Fig. 5.2, because the cosine terms that appear in such computations – and are the source of most of the difficulties in writing two-stream approximations – can be quite confusing. The cosine weights play two quite different roles. In one guise, they express the number of scatterers or absorbers encountered along a slanted path (per unit area normal to the path), but in another guise they represent the projection of the flux on the vertical direction. Most confusion can be resolved by thinking hard about the energy budget of the control volume.

The quantity P introduced in Fig. 5.2 is called the *phase function*, and describes how the scattered radiation is distributed over directions. For spherically symmetric scatterers, the phase function depends only on the angle Θ between the incident beam and a scattered beam (as depicted in Fig. 5.1). The phase function is usually expressed as a function of $\cos \Theta$. If $\hat{n}_I$ is the unit vector in the direction of propagation of the incident beam, and $\hat{n}_{sca}$ is the unit vector in the direction of propagation of some scattered radiation, then

$$\cos \Theta = \hat{n}_I \cdot \hat{n}_{sca} = \cos \theta \cos \theta' + \sin \theta \sin \theta' \cos(\phi - \phi') \tag{5.8}$$

where θ and ϕ are the direction angles of the incident beam and θ' and ϕ' are the angles of the scattered beam under consideration. The phase function for the medium as a whole can be determined from the phase functions of the individual particles doing the scattering – remember that from ω_0 and $d\tau^*$ we already know the amount of energy scattered out of a beam, so the phase function only needs to tell us how that energy is distributed amongst directions. The phase function for an individual particle is defined in such a way that the scattered flux within an element of solid angle $d\Omega'$ near direction (θ', ϕ') is $\chi_{sca} I(\theta, \phi) P(\cos \Theta(\theta, \phi, \theta', \phi')) d\Omega'/4\pi$. P is normalized such that $\int P d\Omega' = 4\pi$, so that integrating the scattered flux over all solid angles yields $\chi_{sca} I$. Note further that

$$\int P(\cos \Theta) d\Omega = 2\pi \int_{-1}^{1} P(\cos \Theta) d\cos \Theta = \int P(\cos \Theta) d\Omega' = 4\pi \tag{5.9}$$

where solid angle integrals without limits specified explicitly denote integration over the entire sphere. The final equality is a matter of definition and the other two equalities follow because one is free to rotate the coordinate system so as to define the angles with respect to any chosen axis, if one is integrating over the entire sphere. Isotropic scattering, in which the scattered radiation is distributed uniformly over all angles, is defined by $P = 1$.

If the scatterers in the atmosphere are all identical particles, then the phase function for the medium is the same as the phase function for an individual particle. If the phase functions differ from one particle to another, then the phase function for the medium is simply the average of the individual particle phase functions, weighted compatibly with Eq. (5.3). The averaging is particularly important when the particles are non-spherical. Though the phase function for any individual particle is not a function of $\cos \Theta$ alone, the particles are generally oriented in random directions, and the average phase function for an ensemble of randomly oriented particles acts like the phase function for an equivalent sphere.

If one divides Eq. (5.7) by $d\tau^*$ and integrates over all incident directions (θ', ϕ'), the equation for the vertical component of the flux due to radiation traveling in direction (θ, ϕ) is found to be

$$\frac{d}{d\tau^*} I(\cos \theta, \phi) \cos \theta = -I(\cos \theta, \phi) + \frac{\omega_0}{4\pi} \int P(\cos \Theta) I(\cos \theta', \phi') d\Omega' \tag{5.10}$$

where $\cos \Theta$ is given in terms of $(\theta, \theta', \phi - \phi')$ by Eq. (5.8). Thermal emission would add an extra source term $B(\nu, T(\tau^*))$ to the right hand side, but we shall leave that out for now. This is the full equation whose solutions give the radiation field. The integral couples together all directions of propagation; if one approximated the integral by a sum over 100 angles, for example, the equation would be the equivalent of solving a system of 100 coupled ordinary differential equations. While, with modern computers, this is not so overwhelming a task as it once might have seemed, it is still intractable in typical climate calculations, where one is doing the calculation for each of a large array of wavenumbers, at each time step of a radiative-convective model, and perhaps for each latitude and longitude grid point in a general circulation model as well. Moreover, it is always helpful to have a simplified form in hand if one's goal is understanding and not merely computing a number. Hence, our emphasis will be on reduction of the equation to an approximate set of equations for two streams of radiation, which may be thought of as the upward and downward streams. In this section we will derive some exact constraints, which will be used to obtain two-stream closures of the problem in Section 5.5.

We first need to define the upward and downward fluxes, which are

$$I_+ \equiv \int_{\Omega^+} I(\cos \theta, \phi) \cos \theta \; d\Omega = \int_{\cos \theta = 0}^{1} \int_{0}^{2\pi} I(\cos \theta, \phi) \cos \theta \; d\phi \; d\cos \theta$$

$$I_- \equiv -\int_{\Omega^-} I(\cos \theta, \phi) \cos \theta \; d\Omega = -\int_{\cos \theta = -1}^{0} \int_{0}^{2\pi} I(\cos \theta, \phi) \cos \theta \; d\phi \; d\cos \theta.$$

(5.11)

The fluxes are defined in such a way that both are positive numbers. Given that $d\Omega$ can be written as $d\cos \theta \cdot d\phi$ it is convenient to write all the fluxes as a function of $\cos \theta$, as we have done here. Henceforth we shall use Ω^+ and Ω^- as shorthand for integral over the upward or downward hemisphere, respectively. With these definitions, the net vertical flux (positive upward) is $I_+ - I_- = \int I \cos \theta \; d\Omega$, the integral being taken over the full sphere.

Solar radiation enters the top of the atmosphere in the form of a nearly parallel beam of radiation, characterized by an essentially unique angle of propagation. It is gradually converted by scattering into radiation that is continuously distributed over all angles. Because the incoming solar radiation has an angular distribution concentrated on a single direction of propagation, it is useful to divide the radiation up into a *direct beam* component propagating exactly in this direction, and a *diffuse* component, traveling over all angles. You can see the Sun as a sharply defined disk in clear sky, which shows that the direct beam solar radiation isn't completely converted into diffuse radiation by scattering, except perhaps in heavily cloudy conditions. To define the direct beam flux, let $L_\odot$ be the solar constant and ζ be the angle between the vertical and the line pointing toward the Sun; ζ is called the *zenith angle*. By convention, the zenith angle is defined as the angle of the vector pointing *toward* the Sun, rather than the direction of the rays coming *from* the Sun. Thus, if θ_{dir} is the angle of the direct beam radiation in our usual angle coordinate system, the zenith angle is $\zeta = \pi - \theta_{dir}$. The azimuth angle of the direct beam radiation ϕ_{dir} is defined in the usual coordinate system.

Now, since the direct beam flux is concentrated in a single direction, there is essentially zero probability of any scattered flux contributing back into the exact direct beam direction. That would be like exactly hitting an infinitesimal dot on a dartboard. Therefore, flux is scattered out of the direct beam but is never added into it, and the direct beam decays exponentially. Making use of the slant path, the direct beam flux is then $L_\odot \exp \left(- (\tau_\infty^* - \tau^*) / \cos \zeta \right)$. We re-write the flux as the sum of a diffuse component and the direct beam:

$$I(\cos\theta,\phi) = I_{diff}(\cos\theta,\phi) + L_{\odot}\exp\left(-(\tau_{\infty}^{*} - \tau^{*})/\cos\zeta\right)\delta(\theta - (\pi - \zeta))\delta(\phi - \phi_{dir}) \quad (5.12)$$

where I_{diff} is the diffuse flux and δ is the Dirac delta-function.[2] From now on, for economy of notation we will drop the "diff" subscript on the diffuse radiation and simply write I for the diffuse component. In typical situations, the top-of-atmosphere boundary condition states that the radiance of all downward-directed angles of the diffuse component must vanish.

Substituting into Eq. (5.10), the equation for the diffuse flux becomes

$$\frac{d}{d\tau^{*}}I(\cos\theta,\phi)\cos\theta = -I(\cos\theta,\phi) + \frac{\omega_{0}}{4\pi}\int P(\cos\Theta)I(\cos\theta',\phi')d\Omega'$$
$$+ L_{\odot}\frac{\omega_{0}}{4\pi}P(\cos\Theta(-\cos\zeta,\cos\theta,\phi - \phi_{dir}))\exp\left(-(\tau_{\infty}^{*} - \tau^{*})/\cos\zeta\right).$$
$$(5.13)$$

The scattering from the direct beam acts as a source term for the diffuse radiation. Integrating over all angles yields the following exact expression for the net vertical diffuse flux

$$\frac{d}{d\tau^{*}}(I_{+} - I_{-}) = -(1 - \omega_{0})\int I(\cos\theta',\phi')d\Omega' + \omega_{0}L_{\odot}\exp\left(-(\tau_{\infty}^{*} - \tau^{*})/\cos\zeta\right) \quad (5.14)$$

since $\int P(\cos\Theta)d\Omega = 4\pi$. In this expression, I_{+} and I_{-} now represent just the diffuse part of the flux. *Conservative scattering* – that is, scattering without absorption – is defined by $\omega_{0} = 1$. For conservative scattering the first term on the right hand side of Eq. (5.14) vanishes. Integrating the direct beam term with respect to τ^{*} just multiplies it by $\cos\zeta$, whence we are left with the result that $I_{+} - I_{-} - L_{\odot}\cos\zeta\exp\left(-(\tau_{\infty}^{*} - \tau^{*})/\cos\zeta\right)$ is a constant. Thus, for conservative scattering the sum of the direct beam vertical flux – which is negative because it is downward – with the diffuse flux is independent of height. As the direct beam is depleted, the flux lost goes completely into the diffuse component. This is as it should be, because, in conservative scattering, the flux lost has no place else to go.

Equation (5.14) provides the first of the two constraints needed to derive the two-stream approximations. The second constraint is provided by multiplying Eq. (5.13) by a function $H(\cos\theta)$ which is antisymmetric between the upward and downward hemispheres, and then performing the angle integral. The rationale for multiplying by an antisymmetric function is that we already know something about $I_{+} - I_{-}$ from the first constraint, and weighting by an antisymmetric functions gives us some information about $I_{+} + I_{-}$. Multiplying by H and carrying out the angle integral, we get

$$\frac{d}{d\tau^{*}}\int I(\cos\theta,\phi)H(\cos\theta)\cos\theta\,d\Omega = -\int IH(\cos\theta)d\Omega + \omega_{0}\int G(\cos\theta')I(\cos\theta',\phi')d\Omega'$$
$$+ \omega_{0}L_{\odot}G(-\cos\zeta)\exp\left(-(\tau_{\infty}^{*} - \tau^{*})/\cos\zeta\right)$$
$$(5.15)$$

where

$$G(\cos\theta') = \frac{1}{4\pi}\int H(\cos\theta)P(\cos\Theta(\cos\theta,\cos\theta',\cos\phi))d\Omega. \quad (5.16)$$

---

[2] The δ-function is not really a function at all, but what mathematians call a *distribution*, or *generalized function*. The object $\delta(x)$ behaves like a function which has unit area but which is zero everywhere except at $x = 0$. It can be thought of as an infinitely tall but infinitely narrow spike.

We were free to replace $\cos(\phi - \phi')$ in this expression by $\cos\phi$, since the integral is taken over all angles ϕ and so a constant shift of azimuth angle does not change the value of the integral. Since H is assumed antisymmetric, the function $G(\cos\theta')$ characterizes the up-down asymmetry of scattering of a beam coming in with angle θ'. The symmetry properties of $\cos\Theta$ imply that $G(-\cos\theta') = -G(\cos\theta')$.

Exercise 5.1 Derive the claimed antisymmetry property of G.

If the phase function satisfies $P(\cos\Theta) = P(-\cos\Theta)$ the scattering is said to be *symmetric*. For symmetric scatterers, there is no difference between scattering in the forward and backward directions. From Eq. (5.8) it follows that $\cos\Theta(\cos\theta, \cos\theta', \phi) = -\cos\Theta(-\cos\theta, \cos\theta', \phi + \pi)$. The antisymmetry of $H(\cos\theta)$ then implies that G vanishes if P is symmetric, since the contribution to the integral from $(\cos\theta, \phi)$ cancels the contribution from $(-\cos\theta, \phi+\pi)$. For symmetric scattering, Eq. (5.15) takes on a particularly simple form, since both terms proportional to ω_0 vanish. The physical content of this result is that symmetric scattering does not directly affect the asymmetric component of the diffuse radiation, since equal amounts are scattered into the upward and downward directions.

When the scattering is not symmetric, the terms involving G do not vanish, and we need a way to characterize the asymmetry of the phase function. The most common measure of asymmetry is the cosine-weighted average of the phase function

$$\tilde{g} \equiv \frac{1}{2} \int_{\cos\Theta = -1}^{1} P(\cos\Theta) \cos\Theta \, d\cos\Theta \qquad (5.17)$$

which goes simply by the name of the *asymmetry factor*. The asymmetry factor vanishes for symmetric scattering. All radiation is back-scattered in the limit $\tilde{g} = -1$, as if the scattering particles were little mirrors. When $\tilde{g} = 1$ there is no back-scatter at all, and all rays continue in the forward direction, though their direction of travel is altered by the particles, much as if they were little lenses.

The asymmetry factor $\tilde{g}$ characterizes forward–backward scattering asymmetry relative to the direction of travel of the incident beam, but some tedious manipulations with Eq. (5.8) allow one to show that the same factor characterizes cosine-weighted asymmetry in the upward–downward direction, regardless of the direction of the incident beam. Specifically, if the incident beam has direction (ϕ', θ'), then

$$\frac{1}{4\pi} \int P(\cos\Theta(\cos\theta, \cos\theta', \cos\phi)) \cos\theta \, d\Omega = \tilde{g}\cos\theta' \qquad (5.18)$$

where $d\Omega = d\phi \, d\cos\theta$ as usual. This leads to a particularly tidy result if we choose $H(\cos\theta) = \cos\theta$ in Eq. (5.15), since then $G(\cos\theta') = \tilde{g}\cos\theta'$ and the antisymmetric projection of the scattering equation becomes

$$\frac{d}{d\tau^*} \int I(\cos\theta, \phi) \cos^2\theta \, d\Omega = -(1 - \omega_0\tilde{g})(I_+ - I_-) + \omega_0 L_\odot \tilde{g}\cos\zeta \exp\left(-(\tau_\infty^* - \tau^*)/\cos\zeta\right). \qquad (5.19)$$

The integral appearing on the left hand side is sensitive only to the symmetric component of the radiance field. In order to obtain a two-stream closure, it is necessary to express the integral in terms of $I_+ + I_-$, which requires making an assumption about the angular distribution of the radiation. The same assumption applied to the right hand side of Eq. (5.14) allows one to estimate $\int I d\Omega$ in terms of $I_+ + I_-$. The different forms of two-stream approximations we shall encounter correspond to different assumptions about the angular distribution of radiance.

For other forms of H the asymmetry function $G(\cos \theta')$ has more complicated behavior that is not so simple to characterize. The other form of H we shall have occasion to deal with is

$$H(\cos \theta) = \begin{cases} 1 & \text{for } \cos \theta \geq 0, \\ -1 & \text{for } \cos \theta < 0, \end{cases} \tag{5.20}$$

which is used to derive the hemispherically isotropic form of the two-stream equations. This choice is convenient because the left hand side of Eq. (5.15) reduces to the derivative of $I_+ + I_-$, but it is inconvenient because G no longer has a simple cosinusoidal dependence on the incident angle. One could simply compute G from the phase function for the medium and use this to form the weights in the scattering equation, but given the inaccuracies we already accept in reducing the problem to two streams, it is hardly worth the effort. Instead, we will *approximate* G as having a cosinusoidal dependence as it does in the previous case. This approximation is exact if the phase function has the form $P = 1 - \frac{1}{3}b + a\cos\Theta + b\cos^2\Theta$, and one can add a third and fourth order term without very seriously compromising the representation. By carrying out the integral defining the asymmetry factor, we find that $\tilde{g} = \frac{1}{3}a$. Then, evaluating G for the assumed form of phase function we find $G = \frac{1}{2}a\cos\theta' = \frac{3}{2}\tilde{g}\cos\theta'$. With this result the antisymmetric scattering equation projection becomes

$$\frac{d}{d\tau^*}(I_+ + I_-) = -\left(1 - \omega_0\frac{3}{2}\tilde{g}\right)(I_+ - I_-) + \omega_0 L_{\odot}\frac{3}{2}\tilde{g}\cos\zeta\exp\left(-(\tau_\infty^* - \tau^*)/\cos\zeta\right). \tag{5.21}$$

The right hand side becomes precisely the same as Eq. (5.19) if we redefine the asymmetry factor to be $\frac{3}{2}\tilde{g}$. Equation 5.21 is already written in terms of the upward and downward stream, and needs no further approximation in order to be used to derive a two-stream approximation. To complete the derivation of the hemispherically isotropic two-stream equation, one need only write the integral $\int I d\Omega$ appearing in Eq. (5.14) in terms of $I_+ + I_-$ using the assumed angular distribution of radiance. If I is assumed hemispherically isotropic in forward and backward directions separately, then this integral is in fact $2(I_+ + I_-)$, which completes the closure of the problem.

There is one last basic quantity we need to define, namely the *index of refraction*, which characterizes the effect of a medium on the propagation of electromagnetic radiation. It will turn out that the index of refraction amounts to an alternate way of representing the information already present in the scattering and absorption cross-sections. For a broad class of materials – including all that are of significance in planetary climate – the propagation of electromagnetic radiation in the material is described by equations that are identical to Maxwell's electromagnetic equations, save for a change in the constant that determines the speed of propagation (the "speed of light"). In particular, the equations remain linear, so that the superposition of any two solutions to the wave equations is also a solution, allowing complicated solutions to be built up from solutions of more elementary form. The reduction in speed of light in a medium comes about because the electric field of an imposed wave induces a dipole moment in the molecules making up the material, which in turn gives rise to an electric field which modifies that of the imposed wave. The equations remain linear because the induced dipole moment for non-exotic materials is simply proportional to the imposed electric field. When the medium is non-absorbing, the ratio of the speed in a vacuum to the speed in the medium is a real number known as the *index of refraction*.

The physical import of the index of refraction is that, at a discontinuity in the index such as occurs at the surface of a cloud particle suspended in an atmosphere, the jump in the propagation speed leads to partial reflection of light hitting the interface, and deflection (*refraction*) of the transmitted light relative to the original direction of travel. The larger the jump in the index of refraction, the larger is the reflection and refraction. To a considerable extent, the refraction of light upon hitting an interface can be understood in terms of a particle viewpoint. If one represents a parallel beam of light as a set of parallel streams of particles all moving at speed c_1 in the outer medium, then if the streams hit an interface with a medium where the speed is $c_2 < c_1$, the streams that hit first will be slowed down first, meaning that the wave front will tilt and the direction of propagation of the beam will be deflected toward the normal, as shown in Fig. 5.3. The classic analogy is with a column of soldiers marching in line, who encounter the edge of a muddy field which slows the rate of march. If Θ_1 is the angle of incidence relative to the normal to the interface, and Θ_2 is the angle of the refracted beam on the other side of the interface, then the deflection due to the change in speed is described by Snell's law, which states $c_2 \sin \Theta_1 = c_1 \sin \Theta_2$, or equivalently $\sin \Theta_2 = (n_1/n_2) \sin \Theta_1$. Now, if a beam is traveling within the medium at angle Θ_2 and exits into a medium with lower index of refraction (e.g. glass to air), then the angle of the exiting beam is given by $\sin \Theta_1 = (n_2/n_1) \sin \Theta_2$; hence, the beam is deflected away from the normal, as indicated in the sketch. At such an interface, if $(n_2/n_1) \sin \Theta_2 > 1$ then there is no transmitted beam and the ray is refracted so much that it is totally reflected back into the medium – a phenomenon known as *total internal reflection*. In reality, there is always some partial reflection at an interface. Partial reflection, as well as many other phenomena we shall encounter, depends on the wave nature of light as described by Maxwell's equations, and cannot be captured by the "corpuscular" viewpoint. This was as much of a groundbreaking conceptual challenge for early optical theorists as blackbody radiation was for investigators presiding over the dawn of quantum theory.

The concept of index of refraction can be extended to absorbing media. Suppose that a plane wave propagating through the medium has spatial dependence $\exp(2\pi i k x)$, where x is the distance measured in the direction of propagation. Then, the expression for the speed of the wave in terms of its frequency and wavenumber becomes

$$\frac{\nu}{k} = \frac{1}{n} c \tag{5.22}$$

where c is the speed of light in a vacuum. Thus $k = (\nu/c)n$. Note that ν/c is the vacuum wavenumber we have been using all along to characterize radiation. For real n, k is the wavenumber in the medium, which is larger than the vacuum wavenumber by a factor of n.

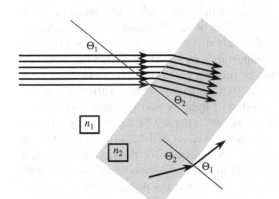

Figure 5.3 Refraction of a beam of light at an interface between a medium with index of refraction n_1 and a medium with index of refraction n_2. In the sketch, $n_2 > n_1$ so the speed of light is slower in the medium than in the surroundings, as is the case for glass or water in air.

Table 5.1 Real part of the index of refraction for selected condensed substances.

	Thermal-IR	Near-IR	Solar	UVB
Liquid water	1.40	1.31	1.33	1.43
Water ice	1.53	1.29	1.31	1.39
CO_2 ice	1.45	1.40	1.41	1.54
Liquid CH_4	1.28	1.27	1.27	1.49
H_2SO_4 38%	1.56	1.36	1.38	1.53
H_2SO_4 81%	1.41	1.51	1.44	1.58

Thermal-IR data is at $600\,cm^{-1}$, near-IR is at $6000\,cm^{-1}$, "Solar" at $17\,000\,cm^{-1}$ ($0.59\,\mu m$), and UVB at $50\,000\,cm^{-1}$ ($0.2\,\mu m$). Liquid water data was taken at a temperature of $293\,K$, water ice at $273\,K$, CO_2 ice at $100\,K$, liquid methane at $112\,K$, and H_2SO_4 at approximately $270\,K$. The percentage concentrations given for the latter are in weight percent.

If we allow n to be complex, its imaginary part characterizes the absorption properties of the medium. To see this, write

$$k_R + ik_I = \frac{\nu}{c} n_R + i\frac{\nu}{c} n_I. \tag{5.23}$$

Since the wave has spatial dependence $\exp(2\pi ikx) = \exp(2\pi ik_Rx)\exp(-2\pi k_Ix)$, the coefficient $2\pi k_I = 2\pi n_I(\nu/c)$ gives the attenuation of the light by absorption, per unit distance traveled. Note that because of the factor ν/c, the quantity $2\pi n_I$ gives the attenuation of the beam after it has traveled by a distance equal to one wavelength of the light. Hence $n_I = 1$ corresponds to an extremely strong absorption. Visible light traveling through such a medium, for example, would be almost completely absorbed by the time it had traveled $1\,\mu m$.

The absorption coefficient k_I is proportional to the absorption cross-section per unit mass which we introduced in Chapter 4, and which reappeared above in the context of absorption by particles. If the density of the medium is ρ, the corresponding mass absorption coefficient is $\kappa = 2\pi k_I/\rho = n_I \cdot (\nu/c)/\rho$. In SI units, this quantity has units of m^2/kg, and is thus an absorption cross-section per unit mass.

The real index of refraction for some common cloud-forming substances is given in Table 5.1. The index of refraction for these and similar substances lies approximately in the range 1.25 to 1.5, and is only weakly dependent on wavenumber; data also shows the index of refraction to depend only weakly on temperature. The weak dependence of index of refraction on wavelength does give rise to a number of readily observable phenomena, such as separation of colors by a prism or the droplets that give rise to rainbows, but such phenomena, beautiful as they are, are of little importance to planetary energy balance. The one exception to the typically gradual variation of the real index of refraction occurs near spectrally localized absorption features; the real index also has strong variations in the vicinity of such points. In considering the scattering of light by particles suspended in an atmospheric gas, the index of refraction of the gas can generally be set to unity without much loss of accuracy. A vacuum has $n = 1$, and gases at most densities we'll consider are not much different. Specifically, for a gas $n - 1$ is proportional to the density. At $293\,K$ and $1\,bar$, Earth air has an index of refraction of 1.0003 in the visible spectrum. Carbon dioxide in the same conditions has an index of 1.0004, and even at the $90\,bar$ surface pressure of Venus has an index of only 1.016. The resultant refraction by the atmospheric gas can be useful in determining the properties of an atmosphere through observations of refraction from the visible through radio spectrum, but it has little effect on scattering by cloud particles.

Insofar as the real index of refraction goes, it would appear that it matters very little what substance a cloud is made of. The minor differences seen in Table 5.1 are far less important than the effects of cloud particle size and the mass of condensed substance in a cloud. The absorption properties, on the other hand, vary substantially from one substance to another, and these can have profound consequences for the effect of clouds on the planetary energy budget. The behavior of the imaginary index for liquid water, water ice, and CO_2 ice is shown in Fig. 5.4. Water and water-ice clouds are nearly transparent throughout most of the solar spectrum; for these substances, n_I is less than 10^{-6} for wavenumbers between $10\,000\,cm^{-1}$ and $48\,000\,cm^{-1}$ (wavelengths between $1\,\mu m$ and $0.2\,\mu m$), though the absorption increases sharply as one moves into the far ultraviolet. In the thermal infrared spectrum, however, water and water ice are very good absorbers, having n_I in excess of 0.1 between wavenumbers of 50 and $1000\,cm^{-1}$. Such a large value of n_I implies that most thermal infrared flux would be absorbed when passing through a cloud particle having a diameter of $10\,\mu m$. For this reason, infrared scattering by water and water-ice clouds can be safely neglected, such clouds being treated as pure absorbers and emitters of infrared. This is not the case for clouds made of CO_2 ice (important on Early Mars and perhaps Snowball Earth) or liquid CH_4 (important on Titan). CO_2 ice clouds are still quite transparent in the solar spectrum, apart from strong absorption in the far ultraviolet. In contrast to water clouds, however, they are largely transparent to thermal infrared. For CO_2 ice clouds, n_I is under 10^{-4} between 1000 and $2000\,cm^{-1}$, and even between 500 and $1000\,cm^{-1}$ n_I is generally below 0.01 except for a strong, narrow absorption feature near $600\,cm^{-1}$. Likewise, liquid methane has n_I well under 0.001 between 10 and $1200\,cm^{-1}$. In both cases the infrared scattering effect of clouds can have an important effect on the *OLR*, leading to a novel form of greenhouse effect. Concentrated sulfuric acid, which makes up aerosols on Earth and the

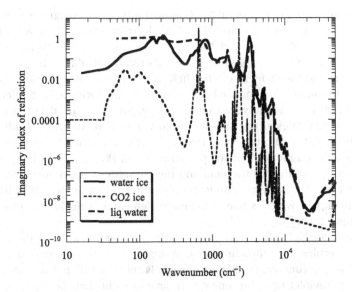

Figure 5.4 The imaginary index of refraction for liquid water, water ice, and CO_2 ice.

clouds of Venus, is quite transparent for wavenumbers larger than $4000\,cm^{-1}$ but the imaginary index of refraction increases greatly at smaller wavenumbers. In much of the thermal infrared spectrum, sulfuric acid absorbs nearly as well as water. Nonetheless, the scattering by sulfuric acid clouds has a significant effect on the *OLR* of Venus, in part because Venus is so hot that it has considerable thermal emission at wavenumbers greater than $4000\,cm^{-1}$.

The strongly reflecting character of sulfate aerosols explains the volcanic cooling of the troposphere seen in the temperature time series of Fig. 1.17, but what accounts for the accompanying stratospheric warming? Since the aerosols are largely transparent in the visible and solar near-infrared, the answer must lie in the thermal infrared effect. This seems paradoxical, since we already know that increasing the infrared opacity of the stratosphere by adding CO_2 *cools* the stratosphere. The resolution to this paradox is found in the difference in absorption spectrum between CO_2 and the aerosols. Carbon dioxide absorbs and emits very selectively and we saw in Chapter 4 that this leads to a stratospheric temperature that is considerably colder than the gray-body skin temperature. Sulfate aerosols, in contrast, act much more like a gray body. Therefore, they *raise* the stratospheric temperature towards the gray-body skin temperature. Any aerosol that absorbs broadly in the thermal infrared should behave similarly.

As a general rule of thumb, typical cloud-forming condensates tend to be very transparent in the visible and near-ultraviolet and quite transparent in the near-infrared, but vary considerably in their absorption properties in the thermal infrared. Most substances – whether gaseous or condensed – are very good absorbers in the very shortwave part of the ultraviolet spectrum, with wavelengths below $0.1\,\mu m$. For this reason, this part of the UV spectrum is often referred to as "vacuum UV," because it is essentially only present in the hard vacuum of outer space.

5.3 SCATTERING BY MOLECULES: RAYLEIGH SCATTERING

Rayleigh scattering theory is a classical (i.e. non-quantum) electromagnetic scattering theory which began life as a theory for scattering of an electromagnetic plane wave from a small sphere with real index of refraction n. "Small" in this context means small compared with the wavelength of the light being scattered. The scattering calculation is quite simple in the Rayleigh limit because the incident electric field is nearly constant over the particle, which makes it simple to compute the induced electromagnetic field within the particle. In essence, the electric field of the incident wave causes charges within the particle to migrate so that positive charge accumulates on one side and negative charge on the other, leading to a dipole moment which oscillates with the same frequency as that of the incident wave. The index of refraction is in fact a measure of the polarizability of the medium – the proportionality between the strength of the electric field and the strength of the dipole moment induced. The scattered wave in the Rayleigh limit is then simply the electromagnetic radiation emitted by an oscillating dipole, which is one of the more elementary calculations that can be done in electromagnetic theory.

Perhaps surprisingly, the Rayleigh theory works quite well as a description for scattering of light from molecules, even though molecules are not dielectric spheres. It is true that the typical size of a molecule (e.g $0.0003\,\mu m$ for N_2) is much smaller than the wavelength of visible or even ultraviolet light, but one might have thought that the quantum response of the molecule might substantially affect the scattering. Certainly, Rayleigh theory does not provide a suitable basis for computing molecular absorption of radiation, which, as we have seen in Chapter 4, is inextricably linked to the quantum nature of the molecule. We will not go further into the reasons that a classical theory works so much better for molecular scattering than for molecular absorption, but it is indeed a convenient turn of events. In practice, it works fine to use spectroscopically measured absorption coefficients to compute gaseous absorption, together with Rayleigh scattering to compute gaseous scattering.

For a spherically symmetric scatterer, Rayleigh theory yields the following formula for the scattering cross-section:

$$\chi_{sca} = \frac{8\pi}{3} \left(\frac{2\pi}{\lambda} \right)^4 \alpha_p^2 \tag{5.24}$$

where λ is the wavelength of the incident light in vacuum, and α_p is the polarizability constant of the scatterer, which expresses the proportionality between the electric field and the induced dipole moment. In practice, the polarizability constant is inferred from measurements of the scattering cross-section itself. It is only a weak function of wavelength. The very strong dependence of Rayleigh scattering cross-section on wavelength is notable; short waves (high wavenumbers) scatter much more strongly than long waves (low wavenumbers). The explanation of the blue skies of Earth is perhaps the most famous application of Rayleigh scattering: blue through violet light has shorter wavelength than the rest of the visible spectrum, and therefore dominates the diffuse radiation caused by scattering of the solar beam from air molecules. For scatterers that are not spherically symmetric – and this includes all the polyatomic molecules like N_2, H_2, and CO_2 present in most of the atmospheres we have been considering – the dipole moment is not in the same direction as the imposed electric field, and this effect slightly alters the expression for the scattering cross-section. Molecules in a gas are randomly oriented, and it can be shown that, averaged over all orientations, the modified cross-section consists of the symmetric cross-section in Eq. (5.24) multiplied by $3(2 + \delta)/(6 - 7\delta)$, where δ is the *depolarization factor*, which is a property of the molecule. The depolarization factor is zero for a spherically symmetric scatterer. For our purposes, the effect of the depolarization factor is not very consequential. It has a value of 0.054 for O_2, of 0.0305 for N_2, and 0.0805 for CO_2. These lead to only a minor increase in the scattering cross-section.

Using Maxwell's equations, it can also be inferred that the index of refraction is related to the polarizability of the molecules making up the medium via the relation

$$n = 1 + 2\pi N \alpha_p \tag{5.25}$$

where N is the number of molecules per unit volume. This is a very useful relation, as it allows one to determine Rayleigh scattering cross-sections through simple measurements of the refractive index, which can be carried out by straightforward measurement of the angle of deflection of light as it moves from a transparent solid container (e.g. glass) into the gas.

Table 5.2 gives the measured Rayleigh scattering cross-section relative to H_2 for a number of common atmospheric gases, as well as the corresponding cross-section per unit mass. The absolute value of the cross-section for H_2 is given for a number of wavelengths in the caption, allowing the actual cross-sections for the other molecules to be readily computed; the values given for H_2 in the caption deviate somewhat from the $1/\lambda^4$ wavelength scaling because of the slight dependence of index of refraction on wavelength, but generally speaking it is adequate to extrapolate to other wavelengths using the fourth-power law. Helium stands out as an exceptionally weak scatterer. Most of the rest of the molecules

	H_2	He	Air	N_2	O_2	CO_2	H_2O	NH_3	CH_4
χ_{sca}	1	0.0641	4.4459	4.6035	3.8634	10.5611	3.3690	7.3427	10.1509
χ_{sca}/m	1	0.0321	0.3066	0.3288	0.2415	0.4800	0.3743	0.8638	1.2689
τ_{ray}	0.40653	0.01305	0.12464	0.13367	0.09818	0.19513	0.15216	0.35116	0.51585

These results are based on observations of the index of refraction, and do not take into account variations in the polarization factors. The scattering cross-sections and cross-sections per unit mass for H_2 are $1.4 \cdot 10^{-38}$ m^2 ($4.215 \cdot 10^{-12}$ m^2/kg) at a wavelength of 10 μm, $8.270 \cdot 10^{-33}$ m^2 ($2.490 \cdot 10^{-6}$ m^2/kg) at a wavelength of 1 μm, and $3.704 \cdot 10^{-28}$ m^2 (0.11 m^2/kg) at a wavelength of 0.1 μm. τ_{ray} is the optical depth due to Rayleigh scattering at $\frac{1}{2}$ μm for a 1 bar atmosphere of the indicated gas under Earth gravity.

Table 5.2 Rayleigh scattering cross-sections, and cross-sections per unit mass, relative to H_2.

have scattering cross-sections per unit mass which are moderately smaller than H_2, with the exception of CH_4, which is moderately larger.

To get an idea of how important the scattering is in various contexts, we can use the cross-sections per unit mass to determine the optical depth of the entire column of an atmosphere. When the optical depth is small, the atmosphere scatters hardly at all, but when optical depth becomes large a significant amount of radiation will be scattered; in the case of the incident solar radiation, this means a lot of the incident beam will be reflected back to space. The last line of Table 5.2 gives the optical depth for an atmosphere consisting of 1 bar of the given gas under Earth gravity. One can scale this up to other planets by multiplying by the appropriate surface pressure, and dividing by the planets' gravity relative to Earth gravity. The optical depth values are given at a wavelength of $\frac{1}{2}$ μm, in the center of the visible spectrum, which is also near the peak of the solar spectrum. For Earth's present atmosphere, the optical depth is small, but not insignificant; Rayleigh scattering affects about 12% of the incident beam. For an Early Mars having 2 bar of CO_2 in its atmosphere, the Rayleigh scattering is quite strong, owing to the somewhat elevated scattering cross-section of CO_2 relative to air, to the low gravity, and to the extra surface pressure. The Rayleigh optical depth for Early Mars would be 1.03 in the visible. The associated reflection of solar radiation is a significant impediment to warming Early Mars with a gaseous CO_2 greenhouse effect. If one took away the reflective clouds of Venus, the CO_2 Rayleigh scattering would still make Venus quite reflective, since the optical depth of a 90 bar CO_2 atmosphere on Venus is nearly 20. A 1.5 bar N_2 atmosphere on Titan would have an optical depth of 1.45 in the absence of clouds. The top 10 bars of Jupiter's mostly H_2 atmosphere would have an optical depth of 1.6, and likewise scatter significantly.

The optical depths for other wavelengths can be obtained by scaling these results according to $1/\lambda^4$. Thus, at thermal infrared wavelengths which are 5 times or more greater than the one we have been considering, the optical depth is at least 625 times smaller; Rayleigh scattering is insignificant at these wavelengths, which is why it is safe to neglect gaseous scattering when doing computations of *OLR*. For similar reasons, Rayleigh scattering is a less important cooling influence for planets orbiting cool M-dwarf stars than is the case for G-class stars like the Sun. On the other hand, the Rayleigh scattering optical depths are at least 16 times greater for ultraviolet as for the visible spectrum values. We will learn how to turn these optical depth values into planetary albedos in Section 5.6.

Rayleigh scattering is not isotropic, but the phase function is symmetric between the forward and backward direction. Within the two-stream approximation, then, we do not really need any information beyond the scattering cross-section. It is worth having a look at the phase function anyway, if only to get an idea of how close to isotropic it is. Given the depolarization factor δ, the Rayleigh phase function is

$$P(\cos\Theta) = \frac{3}{2(2+\delta)}(1 + \delta + (1 - \delta)\cos^2\Theta). \qquad (5.26)$$

From this we see that Rayleigh scattering is mildly anisotropic, with stronger scattering in the forward and backward direction than in the direction perpendicular to the incident beam. For $\delta = 0$ the scattering is twice as strong in the forward and backward directions ($\Theta = 0, \pi$) as it is in the side lobes ($\Theta = \pi/2$). Increasing δ reduces the anisotropy. In fact, laboratory measurements of the intensity of side vs. forward scattering provide a convenient way to estimate the depolarization factor.

5.4 SCATTERING BY PARTICLES

Rayleigh theory tells us everything we need to know about scattering from the gas making up an atmosphere, but to deal with cloud and aerosol particles, we need to know about scattering from objects that are not small compared with a wavelength, and indeed could be considerably larger than a wavelength, as is the case for visible light scattering from water or ice clouds on Earth. The answer is provided by Mie theory,[3] which is a general solution for scattering of an electromagnetic wave from a spherical particle having uniform complex index of refraction. Mie theory reduces to Rayleigh theory in the limit of small non-absorbing particles. Ice crystals and dust particles are not spherical, and while numerical solutions are available for complex particles, for our purposes it will prove adequate to treat such cases in terms of equivalent spheres.

The Mie solution is a solution to Maxwell's electromagnetic equations which is asymptotic to a plane wave at large distances from the particle, and satisfies appropriate continuity conditions on the electromagnetic field at the particle boundary, where the index of refraction is discontinuous. Since Maxwell's equations are linear, the solution can be built up from more elementary solutions to the equations, and this is how Mie theory proceeds. It furnishes the solution in terms of an infinite sum over spherical Bessel functions, and is a real tour-de-force of early twentieth century applied mathematics. The formula is very convenient for evaluation of scattering parameters on a computer, but it is too complicated to yield any insight as to the nature of the solution. For that reason, we do not bother to reproduce the formula here; it is derived and discussed in the references section for this chapter, and a routine for evaluating the Mie solution is provided as part of the software supplement. Here we will only present some key results needed to provide a basis for cloud and aerosol scattering in the solar spectrum and infrared.

Let n_0, assumed real, be the index of refraction of the medium in which the particle is suspended. Define the Mie parameter $\tilde{r} = n_0 r/\lambda$, where r is the particle radius and λ is the wavelength of the incident light in vacuum, measured in the same units as r. The relative index of refraction is $n_{rel} = n/n_0$, where n is the index of refraction of the substance making up the particle. The main things we wish to compute from the Mie solution are the phase function, the scattering efficiency, the absorption efficiency, and the asymmetry factor. These are all non-dimensional quantities, and depend only on $\tilde{r}$ and n_{rel}.

---

[3] The theory is named for Gustav Mie (1869–1957), who published the solution in 1908, while he was a professor at Greifswald University in Germany.

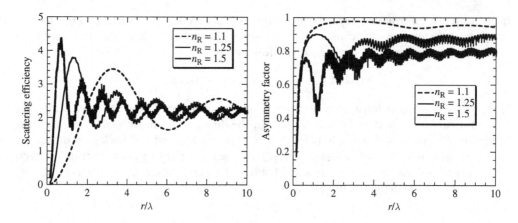

Figure 5.5 Scattering efficiency (left panel) and asymmetry factor (right panel) for Mie scattering from a non-absorbing sphere. The index of refraction of the medium is unity. r is the radius of the particle, and λ is the wavelength of the incident light measured in the same units as r.

Let's first take a look at some scattering properties in the conservative case, $n_I = 0$. In this case, $Q_{abs} = 0$. Figure 5.5 shows the scattering efficiency and asymmetry factor as a function of r/λ for several different values of the real index of refraction of the scatterer. Since $1/\lambda$ is the wavenumber, the graph can be thought of as displaying the scattering properties for increasing particle size with fixed wavenumber, or for increasing wavenumber with fixed particle size. Note that in the limit $n_R \to 1$ there should be no scattering at all, since in that case the particle is not optically distinct from the surrounding medium.

For any given n_R, the scattering efficiency becomes small when r/λ is sufficiently small; this is the Rayleigh limit. The scattering efficiency reaches its first peak at an order unity value of r/λ, and the position of the first peak gets closer to zero as $n_R - 1$ is made larger and the particle is made more refractive. In this sense, for a given size and wavelength, particles made of more refractive substances like CO_2 ice or concentrated H_2SO_4 act like smaller particles than particles made of less refractive substances like water or liquid CH_4. The first peak represents the optimal conditions for scattering. At the first peak, the scattering cross-section can be 4 times or more the actual cross-section area of the particle.

As r/λ is increased past the first peak, the scattering efficiency oscillates between values somewhat below 2 to somewhat above 2 through a number of oscillations of decreasing amplitude, asymptoting at a value of 2 when the particle is large compared with the wavelength. The limit of large r/λ is the *geometric optics* limit, familiar from schoolbook depictions of how lenses work. In the geometrics optic limit, a beam of light is represented as a bundle of independent parallel rays, each of which travels in a straight line unless deflected from its course by an encounter with the interface between the particle and the medium – once upon entering the particle, and once upon leaving it. It is surprising that, in this limit, the scattering efficiency should asymptote to 2, since one would be quite reasonable in thinking that rays that do not encounter the object would be unaffected, implying $Q_{sca} = 1$. What is missing from the geometric optics picture is *diffraction*. Light is indeed a wave, and this has consequences that cannot be captured by the ray-tracing on which geometric optics is based. The light encountering the sphere is a plane electromagnetic wave, and the scattering takes a circular chunk out of it; a wave with a "hole" in it is simply not a solution to Maxwell's equations, and as one proceeds past the obstacle the hole fills in with parts of the beam that never directly encountered the obstacle. It is the diversion of this part of the incident beam that accounts for the "extra" scattering cross-section. The nature of diffraction is far more easily understood through examination of some elementary solutions to Maxwell's equations than it is through this rather cryptic explanation, and the reader in pursuit of deeper understanding is encouraged to study the treatment in the textbooks listed in the references to this chapter.

The main thing to take away from the preceding discussion is that refractive particles made of common cloud-forming substances become very good scatterers when their radius is comparable to or exceeds the wavelength of the light being scattered. The scattering cross-section is two or more times the cross-section area, and for particles whose radius is more than a few times the wavelength, the scattering efficiency is close to 2, more or less independent of what the particle is made of, and more or less independent of wavelength. This limit applies to visible light scattering off typical cloud droplets, which have radii of 5 to 10 μm. For infrared light scattering off cloud particles (e.g. methane clouds on Titan or dry-ice clouds on Early Mars), or for visible light scattering from micrometer-sized aerosol particles, the wavelength is comparable to the particle size, and one needs to take both wavelength and index of refraction into account in order to see how much the scattering efficiency is enhanced over the geometric optics limit.

Turning to the right hand panel of Fig. 5.5, we see that the scattering becomes symmetric for particles small compared with a wavelength, but becomes extremely forward-peaked for particles of radius comparable to or larger than a wavelength. For $n_R = 1.25$, somewhat smaller than the value for water, the asymmetry factor is on the order of 0.85 for large particles. The reason for the strong forward bias is that the large particles are rather like spherical lenses, and can bend light somewhat from the oncoming path, but cannot easily reflect it into the backward direction. This feature of scattering is very important in the treatment of cloud effects on the radiation budget. The forward bias in scattering reduces the effectiveness of clouds as scatterers, and reduces their albedos well below what one would have for layers of symmetric scatterers having the same optical thickness. Without the forward scattering bias, clouds would be much more reflective of solar radiation, and planets would be much colder.

A better appreciation of the strongly forward-peaked nature of Mie scattering from large particles can be obtained by examining the phase functions shown in Fig. 5.6. Almost nothing is back-scattered, and there is a sharp spike near $\Theta = 0$ which becomes sharper as the particle is made larger. For $r/\lambda = 4$, 40% of the scattered flux is in the forward peak with $\Theta < 0.1$ radians. When the particle is comparable in size to a wavelength, the scattering is still forward-peaked, but much less so, with appreciable amounts of flux being scattered a half radian or more from the original direction. Multiple scattering in this case allows a considerable amount of light to be back-scattered relative to its original direction, though it can take several bounces before the light turns the corner. This effect allows carbon dioxide ice clouds composed of particles of size comparable to an infrared wavelength to be quite good reflectors of infrared light trying to escape to space. Visible light scattering from such clouds is much more forward-peaked, and correspondingly less efficient.

Now let's compute the optical thickness of some typical cloud and aerosol layers. Suppose that the layer is made up of non-absorbing particles with a scattering efficiency Q_{sca}. Recall that the scattering efficiency is close to 2 for particles large compared with the wavelength, and can be as large as 4 for particles with diameter comparable to a wavelength, but falls rapidly to zero as the particle size is made smaller. For light in the

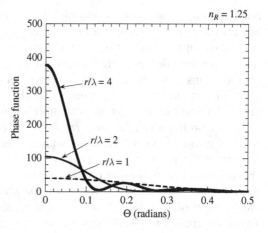

Figure 5.6 Phase function for conservative Mie scattering, under the conditions of Fig. 5.5.

solar spectrum, particles with diameters of a half micrometer or more are very efficient scatterers, with particles at the small end of this range being the most efficient. If the density of the substance making up the particle is ρ and the particle radius is r, then $\chi_{sca}/m = \pi r^2 Q_{sca}/\left(\frac{4}{3}\pi r^3 \rho\right) = \frac{3}{4}Q_{sca}/(\rho r)$. This is the factor by which one multiplies the mass path of scatterer to get the optical depth. The formula implies that, for a given mass of scatterers, small particles lead to much more scattering than large particles: 1 kg of 1 μm sulfate aerosol particles in a column of atmosphere yields as much scattering as 10 kg of 10 μm cloud droplets. The difference in index of refraction between sulfuric acid aerosols and water droplets is of far less consequence than the difference in particle size. To proceed, assume $Q_{sca} = 2$ and $\rho = 1000\,\text{kg/m}^3$. Then, for 1 μm particles, $\chi_{sca}/m = 1500$, so it takes a mere $\frac{1}{1500}$ of a kilogram – *two-thirds of a gram* – of aerosol particles added to a column of atmosphere with a one square meter base to bring the optical depth up to unity. This is the reason that tiny amounts of aerosol forming compounds can have significant effects on planetary climate. The small particle size also tends to make the albedo effect of aerosols dominate their greenhouse effect, despite the fairly strong absorption coefficient of sulfuric acid in the thermal infrared. For cloud droplets with a radius of 10 μm, it would take about 7 grams of water to achieve the same optical thickness as for the smaller aerosol particles, but this is still a tiny fraction of the water content of the atmosphere. A 1 km column of air in saturation at 280 K at Earth surface pressure contains 7.8 kg of water vapor per square meter, for example. A cloud need not weigh much in order to have a profound effect on the albedo of a planet!

Now we'll turn our attention to absorbing particles. Figure 5.7 shows the scattering and absorption efficiencies for particles with $n_R = 1$ having various non-zero values of n_I. For $n_I = 0.1$ and $n_I = 0.01$ the absorption efficiency increases monotonically with particle size, and approaches unity from below (for the latter of these cases, the absorption efficiency is 0.89 at $r/\lambda = 100$ and 0.997 at $r/\lambda = 500$, outside the range plotted in the graph). This is typical behavior for particles with n_I appreciably less than unity, for which the light takes several wavelengths to decay upon encountering the particle. Larger particles absorb more, simply because the light travels a longer distance within the particle, and has more opportunity to decay. For similar reasons, when n_I is made smaller, the particle must be made larger in order for there to be appreciable decay. In any event, when the particle is made large enough, it essentially absorbs a portion of the incident beam with area equal to the cross-section area of the particle. Perhaps more surprisingly, though, the scattering

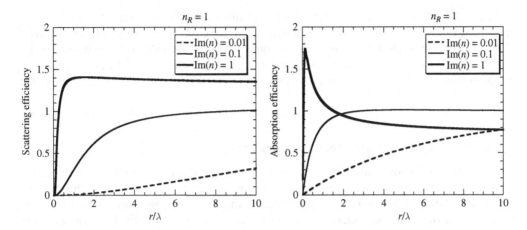

Figure 5.7 Scattering efficiency (left panel) and absorption efficiency (right panel) for Mie scattering from a partially absorbing sphere. The real part of the index of refraction is held fixed at unity while the imaginary part is varied as indicated for each curve in the figure. Other parameters are defined as in Fig. 5.5.

efficiency also rises with particle size – absorbing particles do not just absorb; they also deflect light from the incident direction. This is due to the same diffraction phenomenon we encountered previously. Taking a disk out of the incident beam inevitably causes the remaining part of the beam to be deflected from its original path. For large particles the scattering efficiency and absorption efficiency sum to 2, with half the intercepted beam being absorbed and the other half scattered.

When the particles are very strongly absorbing the behavior is somewhat different, as typified by the curve for $n_I = 1$ in Fig. 5.7. In this case the absorption efficiency actually overshoots unity for particles somewhat smaller than a wavelength. The particle is able to sweep up and absorb radiation from an area larger than its cross-section, owing to the distortion of the electromagnetic field caused by the particle itself. On the other hand, as the particle is made larger, the absorption efficiency goes down and asymptotes to a value somewhat less than unity, because the incident wave is not able to penetrate deeply into the particle, and instead skirts along near its surface. In compensation, the scattering efficiency for large particles becomes greater than unity.

The above results were for particles with $n_R = 1$, in which case there is neither scattering nor absorption when n_I approaches zero. The behavior for $n_R > 1$ is explored in Problem 5.4. When $n_R > 1$ then the scattering cross-section behavior resembles the conservative case when n_I is small, but even a small n_I damps out the ripples in $q_{sca}(r)$ for sufficiently large particles. The asymptotic value of q_{abs} is still close to unity for sufficiently large particles throughout the range of n_I, but the overshoot properties differ somewhat from the case $n_I = 1$.

As a rule of thumb, then, we can say that when particles become larger than the characteristic decay length $\lambda/2\pi n_I$, they absorb essentially everything within a disk of area equal to the cross-section area of the particle. The absorption efficiency is somewhat reduced for particles with $n_I > 0.1$, but even for $n_I = 1$, which is the largest value likely to be encountered, the reduction in efficiency is rather modest. Liquid water has $n_I > 0.1$ throughout the infrared, so any liquid water cloud droplet larger than about $10\,\mu m$ in radius will absorb nearly all the infrared it encounters. In fact, for liquid water, $n_I \approx 1$ throughout most of the infrared, so that even quite small particles are efficient absorbers. Water ice can have n_I as low as 0.05 in some parts of the infrared spectrum, so the particles need to be twice as large to be equally good absorbers in that part of the spectrum, but ice clouds observed on Earth do tend to have larger particle sizes than water clouds.

5.5 THE TWO-STREAM EQUATIONS WITH SCATTERING

The two-stream approximations to the full scattering equation are derived from Eq. (5.14) and Eq. (5.15) by constraining the angular distribution of the radiation in such a way as to allow all integrals appearing in these equations to be written in terms of either $I_+ + I_-$ or $I_+ - I_-$. In the resulting equations, flux in the upward stream is absorbed, or scattered into the downward stream, at a rate proportional to the upward stream intensity, and similarly for the downward stream. The two-stream approximations are an instance of what physicists euphemistically like to call "uncontrolled approximations," in that they are not actually exact in any useful limit but are nonetheless physically justifiable and perform reasonably well in comparison to more precise calculations. The two-stream approximations have inevitable inaccuracies because it is not, in fact, possible to precisely determine the scattering or the absorption from knowledge of the upward and downward fluxes alone. The two-stream approximations can be thought of as the first term in a sequence of N-stream approximations which become exact as N gets large. Fortunately, $N = 2$ proves sufficiently accurate for most climate problems.

The general form of a two-stream approximation for diffuse radiation is

$$
\begin{aligned}
\frac{d}{d\tau^*}I_+ &= -\gamma_1 I_+ + \gamma_2 I_- + \gamma_B \pi B\left(\nu, T\left(\tau_\nu^*\right)\right) + \gamma_+ L_\odot \exp\left(-\left(\tau_\infty^* - \tau^*\right)/\cos\zeta\right) \\
\frac{d}{d\tau^*}I_- &= \gamma_1 I_- - \gamma_2 I_+ - \gamma_B \pi B\left(\nu, T\left(\tau_\nu^*\right)\right) - \gamma_- L_\odot \exp\left(-\left(\tau_\infty^* - \tau^*\right)/\cos\zeta\right)
\end{aligned}
\tag{5.27}
$$

where τ^* is the optical depth in the vertical direction (increasing upward) including scattering loss. The coefficients γ_j depend on frequency, on the properties of the scatterers, and on the particular assumption about angular distribution of radiation that was made in order to derive an approximate two-stream form from the full angular-resolved equation. We recover the hemispherically isotropic Schwarzschild equations used in previous chapters by taking $\gamma_2 = \gamma_+ = \gamma_- = 0$ and $\gamma_1 = \gamma_B = 2$. The terms proportional to γ_B represent the source due to thermal emission of radiation, while the terms proportional to γ_+ and γ_- represent the source of diffuse radiation caused by scattering of the direct beam. The direct beam is assumed to have flux $L_\odot$ (generally the solar constant) in the direction of travel, and to travel at an angle ζ relative to the vertical. There is no upward direct beam term because it is assumed that all direct beam flux scattered from the ground scatters into diffuse radiation.

The symmetry between the coefficients multiplying I_+ and I_- is dictated by the requirement that the equations be invariant in form when one exchanges the upward and downward directions. γ_1 gives the rate at which flux is lost from the upward or downward radiation, while γ_2 gives the rate of conversion between upward and downward radiation by scattering. We can derive additional constraints on the γ_j. Subtracting the two equations gives us the equation for net vertical flux

$$\frac{d}{d\tau^*}(I_+ - I_-) = -(\gamma_1 - \gamma_2)(I_+ + I_-) + 2\gamma_B \pi B + (\gamma_+ + \gamma_-)L_\odot \exp\left(-(\tau_\infty^* - \tau^*)/\cos\zeta\right). \quad (5.28)$$

First, we demand that in the absence of a direct-beam source, the fluxes reduce to black-body radiation in the limit of an infinite isothermal medium. Since B is constant and $L_\odot = 0$ in this case, we may assume the derivative on the left hand side to vanish. Since $I_+ = I_- = \pi B$ for blackbody radiation, we find $\gamma_B = \gamma_1 - \gamma_2$. Comparing with Eq. (5.14) we also find that $\gamma_+ + \gamma_- = \omega_0$. To further exploit Eq. (5.14) we must approximate $\int I d\Omega$ as being proportional to $I_+ + I_-$. The constant of proportionality, which we shall call $2\gamma'$, depends on the angular distribution of radiation assumed. With this approximation it follows that $\gamma_1 - \gamma_2 = 2\gamma'(1 - \omega_0)$.

Exercise 5.2 As a check on the above reasoning, show that in the conservative scattering limit $\omega_0 = 1$ the sum of the diffuse vertical flux with the direct beam vertical flux is constant.

Next, we sum the equations for I_+ and I_- to obtain

$$\frac{d}{d\tau^*}(I_+ + I_-) = -(\gamma_1 + \gamma_2)(I_+ - I_-) + (\gamma_+ - \gamma_-)L_\odot \exp\left(-(\tau_\infty^* - \tau^*)/\cos\zeta\right). \quad (5.29)$$

This can be compared to the symmetric flux projection that appears on the left hand side of Eq. (5.15). Making an assumption about the angular distribution allows one to approximate $\int IH(\cos\theta)\cos\theta \, d\Omega$ as proportional to $I_+ + I_-$, and $\int IH d\Omega$ and $\int IG d\Omega$ each as being proportional to $I_+ - I_-$. In consequence, $\gamma_1 + \gamma_2 = 2\gamma \cdot (1 - \hat{g}\omega_0)$, where γ is related to the proportionality coefficient and $\hat{g}$ is a coefficient characterizing the asymmetry of the scattering. If $H = \cos\theta$, then $\hat{g}$ is in fact the asymmetry factor $\tilde{g}$ defined by Eq. (5.17), but other forms of H yield somewhat different asymmetry factors, though these tend to be reasonably close to $\tilde{g}$. For example, with the form of H given by Eq. (5.20), we showed that $\hat{g} = \frac{3}{2}\tilde{g}$ for phase functions that are truncated to their first three Fourier components. Finally, under circumstances when we can write $G = \hat{g}\cos\theta$, it follows that $\gamma_+ - \gamma_- = -2\gamma\omega_0\hat{g}$. This relation holds exactly when $H = \cos\theta$, and imposing it for other forms of H introduces errors that are no worse than other errors that are inevitable in reducing the full scattering equation down to two streams.

The general form of the set of two-stream coefficients satisfying all the above constraints is then

$$\begin{aligned}
\gamma_1 &= \gamma \cdot (1 - \hat{g}\omega_0) + \gamma' \cdot (1 - \omega_0) \\
\gamma_2 &= \gamma \cdot (1 - \hat{g}\omega_0) - \gamma' \cdot (1 - \omega_0) \\
\gamma_B &= 2\gamma' \cdot (1 - \omega_0) \\
\gamma_+ &= \frac{1}{2}\omega_0 - \gamma\omega_0\hat{g}\cos\zeta \\
\gamma_- &= \frac{1}{2}\omega_0 + \gamma\omega_0\hat{g}\cos\zeta.
\end{aligned} \qquad (5.30)$$

The coefficients γ and γ' are purely numerical factors that depend on the assumption about the angular distribution of radiation which is used to close the two-stream problem. All vertical dependence then comes in through ω_0, and possibly through $\hat{g}$ if the asymmetry properties of scattering particles vary with height. There are three common closures in use. The first is the hemispherically isotropic closure, which we used earlier in deriving the two-stream equations without scattering. In this closure, it is assumed that the flux is isotropic (i.e. I is constant) in each of the upward and downward hemispheres, but with a different

Table 5.3 Coefficients for various two-stream approximations.

	γ	γ'
Hemi-isotropic	1	1
Quadrature	$\frac{1}{2}\sqrt{3}$	$\frac{1}{2}\sqrt{3}$
Eddington	$\frac{3}{4}$	1

value in each hemisphere. The hemi-isotropic closure is derived by using the weighting function H defined by Eq. (5.20) and making use of Eq. (5.21). Given the isotropy of the blackbody source term, it is generally believed that the hemi-isotropic approximation is most appropriate for thermal infrared problems, with or without scattering. Another widely used closure is the *Eddington approximation*. The Eddington closure is obtained by taking $H = \cos\theta$ and making use of Eq. (5.19). To complete the closure, $\int I\cos^2\theta d\Omega$ is written in terms of $I_+ + I_-$ by assuming that the flux is truncated to the first two Fourier components, so $I = a + b\cos\theta$. This is probably the most widely used closure for dealing with solar radiation. It is generally believed that this closure is a good choice for dealing with both Rayleigh scattering and the highly forward-peaked scattering due to cloud particles, though the mathematical justification for this belief is not very firm. The *quadrature* approximation is similar, except that $\int I\cos^2\theta d\Omega$ is evaluated using a technique known as *Gaussian quadrature*, which yields a different proportionality constant from the Eddington closure. The defining coefficients for the three closures are given in Table 5.3.

When $\omega_0 = 0$ there is no scattering and so the upward and downward streams should become uncoupled. From Eq. (5.30) we see that this decoupling happens only if $\gamma = \gamma'$, a requirement that is satisfied for the hemi-isotropic and quadrature approximations but not for the Eddington approximation. It follows that the Eddington approximation can incur serious errors when the scattering is weak, though it can nonetheless outperform the other approximations when scattering is comparable to or dominant over absorption.

The two-stream equations form a coupled system of ordinary differential equations in two dependent variables. Therefore they require two boundary conditions. At the top of the atmosphere, there is generally no incoming diffuse radiation, so the boundary condition there is simply $I_- = 0$. At the bottom boundary we require that the upward diffuse radiation be the sum of the upward emission from the ground with the reflected direct beam and downward diffuse radiation. In general, the direct beam reflection might in part yield a reflected direct beam (as in reflection from a mirror-like smooth surface), but in the following we will assume that all reflection from the bottom boundary is diffuse. Thus, the boundary condition at $\tau^* = 0$ is

$$I_+(0) = e_g\pi B(\nu, T_g) + \alpha_g L_\odot \cos\zeta \exp\left(-\tau^*_\infty/\cos\zeta\right) + \alpha_g I_-(0) \tag{5.31}$$

where e_g is the emissivity of the ground and α_g is the albedo of the ground, both of which vary with ν; Kirchhoff's law implies that $e_g = (1 - \alpha_g)$ for any given frequency.

5.6 SOME BASIC SOLUTIONS

When the scattering and absorption properties of the atmosphere are independent of τ^*, the two-stream equations have simple exponential solutions. We will begin with an elementary solution for conservative scattering, which provides quite useful estimates of the effect of clear-sky and cloudy atmospheres on the solar-spectrum albedo of planets. We shall specify an incoming direct beam flux of solar radiation, and we seek the outgoing reflected flux at the same wavenumber. For conservative scattering, $\omega_0 = 1$. In that case $\gamma_1 = \gamma_2$. Then, Eq. (5.28) becomes

$$\frac{d}{d\tau^*}(I_+ - I_-) = L_\odot \exp\left(-\left(\tau_\infty^* - \tau^*\right)/\cos\zeta\right) \tag{5.32}$$

so

$$I_+ - I_- - L_\odot \cos\zeta \exp\left(-\left(\tau_\infty^* - \tau^*\right)/\cos\zeta\right) = C \tag{5.33}$$

where C is a constant. This equation states that, for conservative scattering, the net of the diffuse flux and the vertical component of the surviving direct beam flux is constant. As the upper boundary condition we require that the diffuse incoming radiation be zero; hence $C = I_{+,\infty} - L_\odot \cos\zeta$, and

$$I_+ - I_- = I_{+,\infty} + L_\odot \cos\zeta\left(\exp\left(-\left(\tau_\infty^* - \tau^*\right)/\cos\zeta\right) - 1\right). \tag{5.34}$$

Deep in the atmosphere, $I_+ - I_-$ becomes constant, and is equal to the difference between the top-of-atmosphere incoming minus outgoing flux, i.e. $I_{+,\infty} - L_\odot \cos\zeta$. When the atmosphere is optically thick, or when $\cos\zeta$ is small (i.e. when the sun is close to the horizon) the exponential term is significant only near the top of the atmosphere. It represents the conversion of the direct beam into diffuse radiation by scattering, which occurs within a conversion layer of depth $1/\cos\zeta$ in optical depth units.

$I_{+,\infty}$ is the reflected flux we wish to determine, and we must close the problem by applying a boundary condition at the ground. For this we need $I_+(\tau^*)$, which we obtain by using the equation for $I_+ + I_-$. Let's restrict attention to the symmetric scattering case, $g = 0$, for which $\gamma_+ = \gamma_-$ and $\gamma_1 = \gamma_2 = \gamma$. Then

$$\frac{d}{d\tau^*}(I_+ + I_-) = -2\gamma(I_+ - I_-) = -2\gamma\left(I_{+,\infty} - L_\odot \cos\zeta + L_\odot \cos\zeta \exp\left(-\left(\tau_\infty^* - \tau^*\right)/\cos\zeta\right)\right). \tag{5.35}$$

The solution which satisfies $I_- = 0$ at the top of the atmosphere is

$$I_+ + I_- = I_{+,\infty} + 2\gamma(I_{+,\infty} - L_\odot \cos\zeta)\left(\tau_\infty^* - \tau^*\right) + 2\gamma L_\odot \cos^2\zeta\left(1 - \exp\left(-\left(\tau_\infty^* - \tau^*\right)/\cos\zeta\right)\right). \tag{5.36}$$

Let's suppose that the ground is perfectly absorbing. This calculation characterizes the albedo of the atmosphere alone. Later, we'll compute how much the surface albedo enhances the planetary albedo. If the ground is perfectly absorbing, then we require $I_+ = 0$ at $\tau^* = 0$. To apply the boundary condition, we add Equations (5.34) and (5.36) and evaluate the result at the ground. Applying the boundary condition and solving for $I_{+,\infty}$ we find

$$I_+,\infty = \frac{\left(\frac{1}{2} - \gamma \cos\zeta\right)\beta_\odot + \gamma\tau_\infty^*}{1 + \gamma\tau_\infty^*}L_\odot \cos\zeta \equiv \alpha_a L_\odot \cos\zeta \tag{5.37}$$

where $\beta_\odot = 1 - \exp\left(-\tau_\infty^*/\cos\zeta\right)$; this quantity is the proportion of the direct solar beam that has been lost to scattering by the time the beam reaches the ground. The fraction multiplying $L_\odot \cos\zeta$ in Eq. (5.37) is the planetary albedo. Since any flux reaching the ground is absorbed completely, this albedo is in fact the albedo of the atmosphere alone, which we will call α_a. In the optically thin limit, $\beta_\odot \approx \tau_\infty^*/\cos\zeta$, so the albedo approaches zero like $\frac{1}{2}\tau_\infty^*/\cos\zeta$. Half of the small amount of flux scattered by the atmosphere exits the top of the atmosphere, but the other half is scattered into the ground, where it is absorbed.

As the atmosphere is made more optically thick, the albedo increases in two stages. The first stage is an exponential adjustment, as the direct beam is converted to diffuse radiation. Some simple algebra shows that the numerator of Eq. (5.37) always increases with τ^*, regardless of the value of $\cos \zeta$. However, when the incident beam is relatively near the horizon, so that $\gamma \cos \zeta < \frac{1}{2}$, the conversion term leads to an exponential increase of albedo with τ^*. The effect becomes more pronounced when the Sun is more nearly on the horizon. This effect comes from the direct scatter of the incident beam to space. When τ_∞^* becomes appreciably larger than $1/\cos \zeta$, however, the direct beam has been completely converted, $\beta_\odot \approx 1$, and the albedo no longer varies exponentially. For large τ_∞^*, the albedo approaches unity like $1 - \left(\frac{1}{2} \gamma^{-1} + \cos \zeta \right) / \tau_\infty^*$. The rather slow approach to a state of complete reflection is due to multiple scattering. In contrast to the exponential decay of the direct beam, the diffuse radiation surviving to be absorbed at the surface decays only like $1/\tau_\infty^*$ because much of the radiation scattered upward is later scattered back downward. The exponential decay of the direct beam just represents a conversion to diffuse radiation, and therefore does not materially alter the conclusion that scattering is a relatively ineffective way of preventing radiation from reaching the surface. That is why one cannot rely on Rayleigh scattering alone to shield life at the surface from harmful ultraviolet radiation, despite the fact that the Rayleigh optical thickness of an Earthlike atmosphere is quite high in the ultraviolet.

The simple albedo formula given above has many physically important ramifications. Before computing the albedo for various conservatively scattering atmospheres, though, it is necessary to bring in the effect of asymmetric scattering if we are to deal with clouds, as scattering from cloud particles is strongly forward-peaked. A non-zero asymmetry factor simply adds a direct beam source term to the equation for $d(I_+ + I_-)/d\tau^*$, since $\gamma_+ - \gamma_-$ no longer vanishes. Some straightforward algebra shows that, allowing for a non-zero asymmetry factor, the albedo formula becomes

$$\alpha_a = \frac{\left(\frac{1}{2} - \gamma \cos \zeta \right) \beta_\odot + (1 - \hat{g}) \gamma \tau_\infty^*}{1 + (1 - \hat{g}) \gamma \tau_\infty^*}. \tag{5.38}$$

This differs from the symmetric scattering form only in that the optical thickness is multiplied by $1 - \hat{g}$, which reduces the effective optical thickness when $\hat{g} > 0$. Thus, in the context of the two-stream equations, the effect of asymmetric scattering is not very surprising or subtle. Since forward scattering just adds back into the forward radiation as if scattering had not occurred at all, forward-dominated scattering simply has the effect of reducing the optical thickness of the atmosphere. The only reason one cannot get by with simply redefining the optical depth is the presence of the direct beam; the direct beam transmission factor $\beta_\odot$ is computed using the unmodified optical depth, rather than the rescaled optical depth, because even forward-scattered radiation is transferred out of the direct beam and into the diffuse component. The behavior of the albedo formula is explored in Problems 5.10 and 5.11.

In Table 5.4 we use Eq. (5.38) to compute the albedos of a number of clear-sky and cloudy planetary atmospheres, based on optical depths computed from the Rayleigh scattering or Mie scattering cross-sections in the visible spectrum. Results are shown for both the hemispherically isotropic and the Eddington approximations; the results differ little between the approximations for the symmetric Rayleigh scattering cases. In clear sky conditions, Earth's atmosphere reflects about 8% of the incoming solar energy back to space. This represents nearly a third of Earth's observed albedo, and is a significant player in the energy budget.

	Earth	Early Mars	Venus	Titan	Cloud	Aerosol
τ_∞^*	0.12	1.03	20	1.45	3	1.25
$\hat{g}$	0	0	0	0	0.87	0.76
Albedo, hemi-isotropic	0.08	0.43	0.94	0.52	0.13	0.10
Albedo, quadrature	0.08	0.43	0.94	0.51	0.17	0.13
Albedo, Eddington	0.08	0.42	0.93	0.51	0.20	0.16

Values given for Earth, Early Mars, Venus, and Titan are for hypothetical clear-sky atmospheres consisting of 1 bar of air for Earth, 2 bar of CO_2 for Early Mars, 90 bar of CO_2 for Venus, and 1.5 bar of N_2 for Titan. The water cloud case (marked "Cloud") assumes a path of 20 grams of water per square meter in droplets of radius 10 μm, having a scattering efficiency of 2. The sulfate aerosol case (marked "Aerosol") assumes a path of 1 gram per square meter in sulfuric acid droplets of radius 1 μm, having a scattering efficiency of 3. Albedos are computed at a wavelength of 0.5 μm, with a zenith angle of 45° for the direct beam.

Table 5.4 Albedos for purely scattering atmospheres.

The thick CO_2 atmosphere postulated for Early Mars has an even more significant effect on albedo, reflecting fully 43% of the solar energy. Further increases in CO_2 lead to even greater reflection, making it hard to warm Early Mars with the gaseous CO_2 greenhouse effect alone. The case of Venus is particularly interesting. We see from the table that if the thick clouds of Venus were removed, Rayleigh scattering alone would be sufficient to keep the albedo high. The value in the table is an overestimate of the albedo Venus would have in the no-cloud case, since it ignores the solar absorption by CO_2, but it suffices to show that virtually all of what escapes absorption would be scattered back to space by Rayleigh scattering. This is important to the evolution of Venus-like planets, which might not have atmospheric chemistry that supports sulfuric acid clouds like those of Venus at present; for that matter, it is not completely certain that thick clouds are a perennial feature of our own Venus. When one factors in the fact that rather little solar radiation penetrates the present thick clouds of Venus, it is evident that the strong Rayleigh scattering of the surviving flux implies that only a trickle of solar radiation reaches the surface of such a planet. It is only a trickle, but as we have seen in Chapter 4, it is a very important trickle, since the surface could not be so hot if all of the solar energy were absorbed aloft.

The cloud cases in Table 5.4 could really apply to any planet with condensable water or sulfur compounds. The cloud parameters chosen are quite typical of Earth conditions. Scattering of solar radiation from cloud particles is quite different from Rayleigh scattering because of the strong asymmetry, which makes the albedo considerably lower than one would expect on the basis of optical thickness. Nonetheless, a small mass of cloud water, or a still smaller mass of sulfate aerosol in the form of micrometer-sized droplets, leads to a very significant albedo. To put the mass path of sulfate aerosol into perspective, we note that if we assume a 10 day lifetime for aerosol in the atmosphere, the assumed mass path is equivalent to a worldwide sulfur emission of about 8 megatonnes per day, allowing for the proportion of S in H_2SO_4. Actual worldwide sulfur emissions for 1990 are estimated to have been more like $\frac{1}{3}$ megatonne per day, which is why the albedo of Earth's sulfate aerosol haze is lower than the estimate in the table, though still a significant player in the radiation budget.

Note also that when the asymmetry is as pronounced as it is for cloud particles, the albedo predicted by the Eddington approximation is significantly greater than that for the hemispherically isotropic case. Although the asymmetry factor reduces the attenuation of diffuse radiation by the cloud, the decay of the direct beam is exponential in the optical depth itself, leading to near total attenuation of the direct beam by the water cloud. This is why a thick cloud looks bright, though you cannot easily discern the disk of the Sun. It is also why it is possible to get quite thoroughly sunburned on a cloudy day.

In the preceding calculation we assumed that the ground was perfectly absorbing. If the ground is instead partially reflecting, having albedo α_g, it will reflect some of the light reaching the surface back upward. Some of this light reflected from the surface will make it through the atmosphere and escape out the top, increasing the planetary albedo. How does the albedo of the atmosphere combine with the albedo of the ground to make up the planetary albedo? Since we are assuming that there is no atmospheric absorption, if a proportion α_a of incoming light is reflected by the atmosphere, then a proportion $(1 - \alpha_a)$ reaches the ground. The proportion of this reflected back upward is $(1 - \alpha_a)\alpha_g$, and if we were to simply add this upward proportion to the part reflected directly by the atmosphere we would get a planetary albedo of $\alpha_a + (1 - \alpha_a)\alpha_g$. This simple estimate already illustrates the important result that putting a reflective (e.g. cloudy) atmosphere over an already reflective surface changes the planetary albedo much less than putting such an atmosphere over a dark surface – you can't make the planet whiter than white, as it were. The simple estimate overestimates the planetary albedo, though, since some of the upward radiation from the ground bounces back from the atmosphere whereafter some of the remainder is absorbed at the ground. The rest is reflected back upward, and ever diminishing proportions remain to multiply scatter back and forth between the atmosphere and the ground. One doesn't need to actually sum an infinite series to solve the problem; all we need to do is to correctly specify the boundary condition on the upward radiation at the ground, which requires in turn a specification of the proportion of upward radiation which is reflected back to the surface by the atmosphere. The upward radiation reflected from the ground is, by assumption, purely diffuse, so as a preliminary to this calculation we need the albedo of the atmosphere for upward-directed diffuse radiation. Since the radiation is purely diffuse, the form of this atmospheric albedo is somewhat simpler than the formula for incoming solar radiation. It is derived in Problem 5.8 , and is

$$\alpha_a' = \frac{(1 - \hat{g})\gamma\tau_\infty^*}{1 + (1 - \hat{g})\gamma\tau_\infty^*}. \tag{5.39}$$

Note that this has the same form as the expression for α_a, except that the direct beam term in the numerator, proportional to $\beta_\odot$, has been dropped. In terms of α_a', the boundary condition on upward radiation at the ground is

$$I_+(0) = \alpha_g \cdot (I_-(0) + L_\odot \cos\zeta \exp(-\tau_\infty^*/\cos\zeta)) = \alpha_g \cdot ((1 - \alpha_a)L_\odot \cos\zeta + \alpha_a' I_+(0)) \tag{5.40}$$

which can be solved for $I_+(0)$. When this boundary condition is applied to the conservative two-stream equations, the resulting expression for $I_{+,\infty}$ yields the following expression for the planetary albedo:

$$\alpha = \alpha_a + \frac{(1 - \alpha_a')(1 - \alpha_a)}{1 - \alpha_g\alpha_a'}\alpha_g$$
$$= 1 - \frac{(1 - \alpha_g)(1 - \alpha_a)}{(1 - \alpha_g)\alpha_a' + (1 - \alpha_a')}. \tag{5.41}$$

The details are carried out in Problem 5.9. When all three albedos α_a, α_a', and α_g are small, the expression reduces to the sum of the albedos $\alpha_a + \alpha_g$. Equation 5.41 has very important consequences for the net effect of clouds on the planetary radiation budget. Clouds have both a warming and a cooling effect on climate. High-altitude clouds have a warming effect, since they strongly reduce *OLR* either by absorption and emission, or by scattering, of infrared radiation. Clouds at any altitude have a cooling influence, through increasing the planetary albedo in the solar spectrum. The net effect of clouds depends on how the competition between these two factors plays out. Equation 5.41 shows that clouds increase the planetary albedo rather little, if they are put over a highly reflective surface (such as ice), or if they are put into an atmosphere which is already quite reflective (such as the dense atmosphere of Early Mars). In either case, introduction of clouds will tend to have a strong net warming effect, because the cloud greenhouse effect is relatively uncompensated by the cloud albedo effect. It is for this reason that clouds can greatly facilitate the deglaciation of a Snowball Earth, and that clouds of either water or CO_2 can very significantly warm Early Mars.

Now let's do an infrared scattering problem, one that illustrates the scattering greenhouse effect in its simplest form. Consider an atmosphere made of a gas that is completely transparent (hence also non-emitting) in the infrared. Suspended in the atmosphere is a cloud made of a substance such as CO_2 ice or liquid methane that is almost non-absorbing in the infrared; we'll idealize it as being exactly non-absorbing, and assume the scattering to be symmetric. Since neither the gas nor the cloud emits infrared, the temperature profile of the atmosphere is immaterial. This atmosphere lies above a blackbody surface with temperature T. What is the *OLR*? This problem is also a case of conservative scattering ($\omega_0 = 1$), but with different boundary conditions. Since there is no incoming infrared, the upper boundary condition is $I_- = 0$. At the ground, the upward flux boundary condition is $I_+ = \pi B(\nu, T)$. Without any direct beam or blackbody source term, $I_+ - I_-$ is a constant, which is equal to the outgoing radiation $I_{+,\infty}$ at the frequency under consideration. The equation for $d(I_+ + I_-)/d\tau^*$ then tells us that $I_+ + I_- = 2\pi B - (1 + \gamma\tau^*)I_{+,\infty}$. Finally, imposing the boundary condition that $I_- = 0$ at $\tau^* = \tau_\infty^*$, we conclude that

$$I_{+,\infty} = 2\pi B / (2 + \gamma\tau_\infty^*).\tag{5.42}$$

Hence, the infrared scattering reduces the outgoing infrared by a factor $2/(2 + \gamma\tau_\infty^*)$ relative to what it would be in the absence of an atmosphere. This increases the surface temperature of the planet in the same fashion as the *OLR* reduction from the more conventional absorption/emission greenhouse effect. However, the scattering greenhouse effect works quite differently, since it reduces the *OLR* regardless of whether the atmospheric temperature goes down with height.

The scattering greenhouse effect exerts an important warming influence on planets which form non-emissive clouds. This includes the case of CO_2 ice clouds on Early Mars and on a cold Snowball Earth. Whether the *net* effect of the clouds is to warm or cool the planet depends on how much of the scattering greenhouse effect is offset by additional reflection in the incoming stellar spectrum. The following factors tend to tilt the balance in favor of net warming:

- If the particles have a size on the order of $10\,\mu$m, then the Mie scattering efficiency is enhanced in the thermal infrared range, compared with what it is for shorter wavelengths.
- Unless the particles are very small, the asymmetry factor is much greater for the incoming shortwave radiation, which leads to inefficient scattering.
- If the planet has a high albedo to begin with, as in the case of Rayleigh scattering from a thick CO_2 atmosphere or the high albedo of a Snowball Earth, then the effect of the shortwave cloud albedo on absorption of incoming solar radiation is reduced.

- The clouds will exert a pronounced net warming effect in circumstances such as high latitude winter, in which there is little incoming solar radiation available for reflection.

Detailed calculations indicate a net warming effect on Early Mars and on Titan. The net influence of the sulfuric acid clouds of Venus presents a particularly interesting problem, because they are dynamically maintained by the sulfur cycle of the planet; it could well be that Venus has gone through periods when these clouds were absent. Would such a Venus be hotter or cooler? Venus clouds act on the *OLR* through a mix of absorption and scattering, and the estimate of their net effect depends moreover on what the solar-spectrum albedo of Venus would be if you took the clouds away. The few calculations that have been done on this problem tend to suggest that the solar albedo effect wins in this case, and Venus would become considerably hotter without clouds. This is a problem that involves many subtleties and would repay further study, particularly in view of the fact that Venus represents an archetype for the climate evolution of hot, dry planets. Further explorations of the scattering greenhouse effect and of the competition between cloud greenhouse and cloud albedo effects are pursued throughout the Workbook. See especially Problems 5.13, 5.14, 5.16, and 5.17.

Next we will extend the preceding scattering greenhouse problem to allow $\omega_0 < 1$, so the atmosphere can absorb and emit. We'll assume the atmosphere to be isothermal at the same temperature T as the ground. In this case, $I_+ = I_- = \pi B(\nu, T)$ is a particular solution satisfying the boundary condition on I_+ at the ground, though it does not satisfy the boundary condition $I_- = 0$ at the top of the atmosphere. We must add a homogeneous solution to the particular solution, which cancels I_- at the top of the atmosphere for the particular solution, but leaves the bottom boundary condition intact. The homogeneous equation is obtained by taking the derivative of Eq. (5.28) and substituting the derivative of $I_+ + I_-$ using Eq. (5.29), dropping the source terms from both. Assuming ω_0 and g to be independent of height, the homogeneous equation is then

$$\frac{d^2}{d\tau^{*2}}(I_+ - I_-) = -(\gamma_1 - \gamma_2)(\gamma_1 + \gamma_2)(I_+ - I_-)$$
$$= -4\gamma\gamma'(1 - \hat{g}\omega_0)(1 - \omega_0)(I_+ - I_-). \tag{5.43}$$

The general solution to this is $a\exp\left(-K \cdot (\tau_\infty^* - \tau^*)\right) + b\exp\left(K \cdot (\tau_\infty^* - \tau^*)\right)$, where

$$K = 2\sqrt{\gamma\gamma'(1 - \hat{g}\omega_0)(1 - \omega_0)}. \tag{5.44}$$

One term grows exponentially in optical depth, while the other decays exponentially. The solution for $I_+ + I_-$ is then obtained from the solution for $I_+ - I_-$ using Eq. (5.28), allowing us to obtain the two fluxes individually for use in applying the boundary conditions. First, the homogeneous solution we add in must not disturb I_+ at the ground, since the particular solution already satisfies the boundary condition there. To keep the algebra simple, let's assume $\tau_\infty^* \gg 1$. In that case, we approximately satisfy the boundary condition at the ground by taking the solution which decays toward the ground, i.e. $b = 0$. The boundary condition on $I_-(0)$ then determines the value of the coefficient a. Carrying out the algebra and adding the homogeneous to the particular solution, we find the outgoing radiation to be

$$I_{+,\infty} = 2\frac{\gamma'(1 - \omega_0)}{\gamma'(1 - \omega_0) + \sqrt{\gamma\gamma'(1 - \hat{g}\omega_0)(1 - \omega_0)}}\pi B. \tag{5.45}$$

In this equation, πB is the emission the planet would have in the absence of an atmosphere, and the coefficient multiplying it gives the reduction in emission due to the atmosphere.

Note that for a non-scattering atmosphere, the isothermal atmosphere assumed would have no effect whatever on the outgoing radiation. In contrast to the conservative scattering case described by Eq. (5.42), the emission in the partially absorbing case does not approach zero in the optically thick limit, but rather approaches the non-zero value given by Eq. (5.45). When there is no scattering, i.e. $\omega_0 = 0$, the atmosphere should have no effect on emission, and this limit shows the shortcomings of the Eddington approximation. For $\omega_0 = 0$, the factor reducing the emission is $2\gamma' / (\gamma' + \sqrt{\gamma\gamma'})$, which reduces to unity only when $\gamma = \gamma'$. Both the quadrature and the hemispherically isotropic assumptions satisfy this requirement, but as we have seen before, the Eddington approximation gives the wrong answer when scattering is weak, and is not suitable for such cases. For any of the approximations, as the scattering is made stronger relative to absorption, $\omega_0 \to 1$ and the emission goes to zero in proportion to $\sqrt{(\gamma/\gamma')(1 - \omega_0)/(1 - \hat{g})}$.

The basic lesson here is that sufficiently strong scattering can kill off emission from the atmosphere. When the scattering becomes strong, infrared can escape to space only from a thin layer near the top of the atmosphere; radiation from deeper layers is scattered back downwards and absorbed before it can escape.

Finally, we will use an elementary solution to show how scattering affects the vertical distribution of solar absorption. We'll suppose that thermal emission is negligible at the frequency under consideration, as is the case for solar radiation on planets at Earthlike (or even Venuslike) temperatures. The basic idea is that scattering increases the net path traveled by radiation in going from one altitude to another, because the radiation bounces back and forth many times rather than proceeding in a straight line. This allows more radiation to be absorbed within a thinner layer, as compared to the no-scattering case. At the same time, however, scattering reflects some radiation back to space before it has any opportunity to be absorbed at all. As will be shown in the forthcoming derivation, the net result is to reduce the solar absorption while at the same time concentrating it more in the upper atmosphere, as compared with a no-scattering case with the same distribution of absorbers.

The full problem with arbitrary surface albedo, optical depth, and asymmetry factor is analytically tractable so long as ω_0 is constant. The full solution is somewhat unwieldy, however, so we'll now make a few simplifying assumptions. To keep the algebra simple, in the present discussion we will assume τ_∞^* to be very large, so that the lower boundary does not affect the solution. The atmosphere is effectively semi-infinite (top but no bottom) in this solution. We'll also assume that the asymmetry factor vanishes. The solution begins with taking the derivative of Eq. (5.29) and substituting from Eq. (5.28), which gives us

$$\frac{d^2}{d\tau^{*2}}(I_+ + I_-) = K^2(I_+ + I_-) - 2\gamma\omega_0 L_\odot \exp\left(-(\tau_\infty^* - \tau^*)/\cos\zeta\right) \qquad (5.46)$$

where K is defined by Eq. (5.44) with $\hat{g}$ set to zero. A particular solution to this equation is

$$I_+ + I_- = -\frac{2\gamma\omega_0 L_\odot \cos^2\zeta}{1 - K^2\cos^2\zeta}\exp\left(-(\tau_\infty^* - \tau^*)/\cos\zeta\right) \qquad (5.47)$$

to which we have to add superpositions of the two homogeneous solutions $\exp\left(\pm K(\tau_\infty^* - \tau^*)\right)$ so as to satisfy the boundary conditions at the top and bottom of the atmosphere. So far the only assumptions we have used are that ω_0 is constant, the thermal emission is neglected, and $\hat{g} = 0$. Since the atmosphere is semi-infinite, the only admissible homogeneous solution is $a\exp(-K(\tau_\infty^* - \tau^*))$, since the other solution blows up deep in the atmosphere. It remains only to determine a, which is done by applying the condition $I_- = 0$ at the top of the atmosphere. To do this we need $I_+ - I_-$. This is obtained from

Eq. (5.29), which takes on a particularly simple form when $\hat{g} = 0$. Using the value of a thus obtained, the net vertical diffuse flux is found to be

$$
I_+ - I_- = \left[-\frac{K}{2\gamma + K} \frac{1 + 2\gamma \cos \zeta}{1 - K^2 \cos^2 \zeta} \exp\left(-K\left(\tau_\infty^* - \tau^*\right)\right) \right.
$$
$$
\left. + \frac{1}{1 - K^2 \cos^2 \zeta} \exp\left(-\left(\tau_\infty^* - \tau^*\right)/\cos \zeta\right) \right] \omega_0 L_\odot \cos \zeta.
\tag{5.48}
$$

Since $I_- = 0$ at the top of the atmosphere, the albedo is obtained by evaluating this expression at $\tau^* = \tau_\infty^*$ and taking the coefficient of the incoming flux $L_\odot \cos \zeta$. Thus,

$$
\alpha = \frac{2\gamma \omega_0}{(1 + K \cos \zeta)(2\gamma + K)}
$$
$$
= \frac{2\gamma \omega_0}{\left(1 + 2\sqrt{\gamma \gamma'(1 - \omega_0)} \cos \zeta\right)\left(2\gamma + 2\sqrt{\gamma \gamma'(1 - \omega_0)}\right)}.
\tag{5.49}
$$

In the absence of scattering, all the incident flux should be absorbed no matter how low the concentration of absorbers, since the atmosphere is assumed infinitely deep. Consistently with this reasoning the above albedo approaches zero as $\omega_0 \to 0$. For small ω_0, the albedo increases linearly with ω_0. It continues to increase monotonically as ω_0 is further increased. In the limit $\omega_0 \to 1$ where scattering becomes very strong, $\alpha \to 1$ and the atmosphere becomes perfectly reflecting; radiation is scattered back to space before it has much opportunity to be absorbed in the atmosphere.

Although strong scattering reduces the opportunity for absorption, it also reduces the depth scale over which the small amount of absorbed radiation is deposited in the atmosphere. The reason is that scattering increases absorption through multiple reflections that increase the path length. To get a better handle on what is going on, we need to examine the vertical profile of the flux as $\omega_0 \to 1$ while holding the concentration of absorbers fixed. Taking the limit this way would correspond, for example, to looking at ultraviolet absorption as we increase the amount of conservatively scattering cloud particles in an atmosphere while keeping the amount of ultraviolet-absorbing ozone fixed. This is equivalent to writing $\tau_\infty^* - \tau^* = (\kappa/(1 - \omega_0))(p/g)$, if we neglect pressure broadening, since $(1 - \omega_0)\Delta\tau$ gives the absorption in a layer of thickness $\Delta\tau$. It follows from this expression for optical thickness that all the direct beam flux is converted to diffuse flux in a very thin conversion layer if $\omega_0 \to 1$. Below the conversion layer all flux is diffuse, and the net vertical flux is then

$$
I_+ - I_- = -\frac{K}{2\gamma + K} \frac{1 + 2\gamma \cos \zeta}{1 - K^2 \cos \zeta} \exp\left(-K \frac{1}{1 - \omega_0} \frac{\kappa p}{g}\right) \omega_0 L_\odot \cos \zeta
$$
$$
\approx \sqrt{\frac{\gamma}{\gamma'}}(1 - 2\gamma \cos \zeta)\sqrt{1 - \omega_0} \exp\left(-2\sqrt{\gamma \gamma'} \frac{1}{\sqrt{1 - \omega_0}} \frac{\kappa p}{g}\right) L_\odot \cos \zeta.
\tag{5.50}
$$

Hence the flux which manages to penetrate into the atmosphere is absorbed over a layer depth which scales with $\sqrt{1 - \omega_0}$, and approaches zero as $\omega_0 \to 1$. It follows also from Eq. (5.50) that the heating rate $d(I_+ - I_-)/dp$ remains order unity in this limit, though it becomes concentrated in a thinner and thinner layer near the top of the atmosphere.

5.7 NUMERICAL SOLUTION OF THE TWO-STREAM EQUATIONS

Equation (5.27) is a coupled, linear system of ordinary differential equations requiring two boundary conditions. It takes the form of a *two-point boundary value problem*, because one specifies a boundary condition on I_+ at $\tau^* = 0$ and on I_- (generally that it vanish) at $\tau^* = \tau_\infty$. In vector form, the system can be written

$$\frac{d}{d\tau^*}\mathbf{V} = \mathbf{M}(\tau^*) \cdot \mathbf{V} + \mathbf{F}(\tau^*) \tag{5.51}$$

where

$$\mathbf{V} \equiv \begin{bmatrix} I_+ \\ I_- \end{bmatrix}, \mathbf{M} \equiv \begin{bmatrix} -\gamma_1 & \gamma_2 \\ -\gamma_2 & \gamma_1 \end{bmatrix}, \mathbf{F} \equiv \begin{bmatrix} \gamma_B \\ -\gamma_B \end{bmatrix} \pi B(\nu, T(\tau^*)) + \begin{bmatrix} \gamma_+ \\ -\gamma_- \end{bmatrix} \exp\left(-(\tau_\infty^* - \tau^*)/\cos\zeta\right). \tag{5.52}$$

The matrix $\mathbf{M}$ varies with τ^* because the single scattering albedo ω_0 and also the asymmetry parameter in general will be functions of altitude. The forcing $\mathbf{F}$ depends on τ^* on account of the variation of T with altitude and the exponential factor in the direct beam term. Because of these variations, the equations must generally be solved numerically. In most applications, the profiles of T and scatterer properties are given as functions of p, and often as tabulated values on a fixed grid rather than as functions. In order to carry out the needed integrations, the profiles must be re-expressed as functions of τ^*. The practicalities of how one goes about doing this will be discussed somewhat later. In this section we will show how to get the solution for a single frequency ν, for which the absorption is characterized by a single absorption coefficient at each pressure level. The extension to the band-averaged case where we have to deal with a distribution of absorption coefficients will be dealt with in Section 5.9.

Because the system described by Eq. (5.51) is linear, the solution satisfying the boundary conditions can, in principle, be built up from a suitable superposition of solutions obtained by numerically integrating the system starting from the lower boundary. Using a numerical differential equation integrator, one first constructs the following three solutions:

- A *particular solution* $\mathbf{V}_{\text{part}}$ which satisfies Eq. (5.51) subject to values of I_+ and I_- at $\tau^* = 0$ specified in whatever way proves convenient. We will assume $I_+ = I_- = 0$ for the particular solution at the lower boundary, but almost any other choice would do as well.
- A pair of *homogeneous solutions* $\mathbf{V}_{\text{hom},0}$ and $\mathbf{V}_{\text{hom},1}$ which satisfy Eq. (5.51) with the forcing term $\mathbf{F}$ set equal to zero. We will construct these solutions by integrating the homogeneous form of the equation subject to lower boundary conditions $I_+ = 1, I_- = 0$ for $\mathbf{V}_{\text{hom},0}$ and $I_+ = 0, I_- = 1$ for $\mathbf{V}_{\text{hom},1}$.

The general solution to the original inhomogeneous system is then

$$\mathbf{V} = \mathbf{V}_{\text{part}} + a_0\mathbf{V}_{\text{hom},0} + a_1\mathbf{V}_{\text{hom},1} \tag{5.53}$$

where a_0 and a_1 are any real numbers. At the upper boundary we require that the downward diffuse component vanish, i.e. $I_- (\tau_\infty^*) = 0$. At the lower boundary, we require that Eq. (5.31) be satisfied. The superposition of basic solutions that satisfies both boundary conditions is determined by solving the system

$$0 = I_{-,part}\left(\tau_\infty^*\right) + I_{-,hom,0}\left((\tau_\infty^*)\right) a_0$$
$$+ I_{-,hom,1}\left((\tau_\infty^*)\right) a_1$$
$$e_g \pi B(\nu, T_g) + \alpha_g L_\odot \cos\zeta \exp\left(-\tau_\infty^*/\cos\zeta\right) = (I_{+,part}(0) - \alpha_g I_{-,part}(0)) \tag{5.54}$$
$$+ (I_{+,hom,0}(0) - \alpha_g I_{-,hom,0}(0))a_0$$
$$+ (I_{+,hom,1}(0) - \alpha_g I_{-,hom,1}(0))a_1$$

for the coefficients a_0 and a_1. The conditions define a 2×2 linear system of equations which in general has a unique solution. Once the coefficients a_0 and a_1 are known, the solution to the problem is complete. The procedure can equally well be carried out by starting at the top and integrating downward instead.

This approach works well and is very efficient as long as the atmosphere is not too optically thick. However, when τ_∞^* becomes large, the method breaks down for the following reasons. The homogeneous version of Eq. (5.51) has solutions which grow or decay exponentially, with a local exponential growth or decay rate of $\pm\sqrt{y_1^2 - y_2^2}$. In the pure scattering limit $y_2 = y_1$ and the growth becomes merely algebraic, as evidenced by the analytic solutions considered previously. However, in the general case where $y_2 < y_1$, the exponentially growing solution causes numerical difficulties in the optically thick limit because the exponential growth will cause an overflow error when one attempts to integrate from the lower boundary to the upper boundary (or vice versa). A related problem is that it becomes impossible to find two linearly independent homogeneous solutions because the exponentially growing solution comes to dominate both homogeneous solutions as one integrates sufficiently far from the boundary. As a simple example of this, consider that $\exp(\tau^*) + a\exp(-\tau^*)$ eventually converges on $\exp(\tau^*)$ to within computer roundoff error for sufficiently large τ^*, regardless of the value of a. Exactly how large τ_∞^* needs to be before this problem becomes serious depends on the vertical profile of $\sqrt{y_1^2 - y_2^2}$, but, except when gaseous absorption is weak and scattering is dominant, this quantity is order unity, and so the solution method tends to break down when $\tau_\infty^* \approx 10$. Because of the exponential growth which is at the root of the problem, increasing the precision of the computer arithmetic only modestly extends this value.

Fortunately, there are a number of simple resolutions to the problem. First of all, if one only wants the *OLR*, then it is not necessary to integrate the equations all the way from the top of the atmosphere down to the ground. At a frequency where the atmosphere is optically thick by virtue of strong absorption, radiation from lower levels of the atmosphere is exponentially attenuated and does not significantly affect the *OLR*. In these circumstances, one starts the integration at a height $\tau_\infty^* - \tau_1^*$ for some suitably chosen $\tau_1^* < \tau_\infty^*$. At $\tau_\infty^* - \tau_1^*$ one can impose any order unity boundary condition on I_+ that proves convenient, since the boundary condition there will not affect the *OLR* as long as the fluxes are not made exponentially large. The lower boundary condition $I_+ = 0$ would suffice, though one could squeeze out a little more accuracy by using the optically thick limit for the boundary condition (i.e. $I_+ = \pi B(\nu, T(\tau_\infty^* - \tau_1^*))$). One carries out the rest of the procedure exactly as before, except for applying the lower boundary condition at the artificial lower boundary instead of $\tau^* = 0$. The choice of τ_1^* depends on the profile of $\sqrt{y_1^2 - y_2^2}$. If the typical value of this quantity is A, then we would take $\tau^* \approx 10/A$, based on the notion that $e^{10} + e^{-10}$ is accurately distinguishable from e^{10} with computer arithmetic having precision of at least 12 or 13 decimal places. When $\sqrt{y_1^2 - y_2^2}$ varies greatly in the vertical it can be a bit tricky determining an appropriate "typical value," and so it is generally better to start the integration from the top, in which case one can simply integrate the two homogeneous solutions downward until one or the other grows by a factor of at least e^{10}, whereupon one stops and applies the artificial lower boundary condition.

A similar idea can be applied if one wants to compute the fluxes deep in the interior of the atmosphere. The typical application of such a calculation would be to determine the radiative heating profile for use in a radiative-convective equilibrium calculation. Even in an all-troposphere model, for which the climate can be solved for knowing the *OLR* alone, one might wish to compute the interior infrared cooling rate so as to determine the magnitude of the convective heat transport required to balance radiative cooling.

Suppose one wishes to compute the fluxes in the vicinity of some level τ_0^*. The solution method exploits the fact that the fluxes near τ_0^* will be insensitive to conditions many optical thicknesses *above or below* τ_0^*, since the fluxes from distant layers will be exponentially attenuated by the time they reach τ_0^*. One then builds the solution from solutions constructed by integrating both upwards and downwards from τ_0^* as follows:

- Integrating upward, one constructs a particular solution $V_{>,part}$ and a homogeneous solution $V_{>,hom}$ which satisfy the upper boundary condition $I_- = 0$. The upper boundary condition is applied at τ_∞^* if the true upper boundary is reached before the exponentially growing mode dominates, but otherwise the integration is halted and the upper boundary condition is applied when the exponential dominance criterion is met. The particular solution satisfying the upper boundary condition is constructed from a superposition of an arbitrary particular solution and two independent homogeneous solutions, all obtained by upward integration starting from τ_0^*. Similarly, the homogeneous solution satisfying the upper boundary condition is obtained as a suitable superposition of two independent homogeneous solutions.
- Integrating downward, one constructs a particular solution $V_{<,part}$ and a homogeneous solution $V_{<,hom}$ which satisfy the appropriate lower boundary condition. If the true lower boundary $\tau^* = 0$ is reached before the exponentially growing mode dominates, then the physical boundary condition in Eq. (5.31) is applied. Otherwise, an artificial boundary condition $I_+ = 0$ is applied. As for the upward integration, the solutions are constructed through suitable superpositions of independent particular and homogeneous solutions obtained by downward integration.

The general solution satisfying the upper and lower boundary condition is then $V_{>,part} + a_0 V_{>,hom}$ for $\tau^* > \tau_0^*$ and $V_{<,part} + a_1 V_{<,hom}$ for $\tau^* < \tau_0^*$. By imposing the requirement that the fluxes I_+ and I_- be continuous across τ_0^*, one obtains a 2×2 linear system for the coefficients a_0 and a_1 much as we had before. This uniquely determines the solution. When the physical boundary isn't reached, then on the side of τ_0^* where this happens the solution will only be valid within a few optical thicknesses of τ_0^*, because the artificial boundary condition will begin to affect the solution as the artificial boundary is approached. To map out the entire flux profile, one only needs to carry out the procedure for a list of τ_0^* dense enough that the regions where the solutions are valid overlap. We'll refer to this solution method as the *piecewise ODE method* (ODE being shorthand for ordinary differential equation). Note that in the optically thick limit, the fluxes typically vary very little over a unit optical thickness, because the fluxes are mostly determined by the local values of temperature and optical constants, which vary slowly when written in optical thickness coordinates. This means that it is usually possible to compute the flux at a rather coarse grid of τ_0^* and just use interpolation to get intermediate values if they are needed.

Once the fluxes have been determined, it is easy to get the heating rate, which is proportional to $d(I_+ - I_-)/d\tau$. Rather than numerically differentiating the fluxes, one can obtain the necessary derivatives by simply evaluating the right hand side of Eq. (5.51).

The procedure may sound complicated, but it is actually rather simple to implement on the computer, since it is built from the same basic integration operation carried out several times over, supplemented by a numerical routine that solves a 2×2 linear system. An example showing the results of this procedure is given in Fig. 5.8. The solution is carried

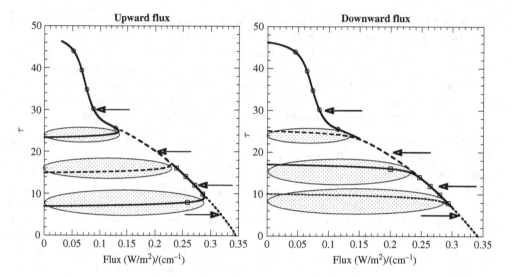

Figure 5.8 An example of the results of computing the upward and downward flux using the piecewise ODE method in the optically thick case. The arrows show the values of τ_0^* for which the upward and downward integrations were performed. The shaded ellipses show the regions where the artificial boundary condition has significantly affected the solution. This calculation was carried out for a wavenumber of $600\,cm^{-1}$ assuming the temperature profile to lie on the dry air adiabat with a ground temperature and surface air temperature of 270 K. The surface pressure is 2 bar, the absorption coefficient is $1.5 \cdot 10^{-4}\,m^2/kg$ at 100 mb and linearly pressure-broadened elsewhere. An optically thick purely scattering cloud is included, centered at an optical thickness value of 35. The calculation assumes a gravitational acceleration of $10\,m/s^2$.

out for a single infrared wavenumber ($600\,cm^{-1}$) with gaseous absorption as described in the figure caption. In addition, a purely scattering cloud has been placed by the top of the atmosphere. For this profile, carrying out the procedure for four suitably chosen τ_0^* is sufficient to map out the flux profile over the entire domain. Note that the portions of the subintegrations which are contaminated by artificial boundary conditions are different for the upward and downward fluxes; for I_+, it is the vicinity of the artificial lower boundary that is most affected, whereas for I_- it is the vicinity of the upper boundary. Note also that the subintegration carried out nearest the top of the atmosphere covers a wider range of optical thickness than the others. This is because the optical thickness near the top is dominated by the purely scattering cloud, and pure scattering does not yield solutions with exponentially growing character.

In the calculations shown in Fig. 5.8 we used $I_+ = 0$ or $I_- = 0$ for the artificial boundary conditions, so as to show more clearly the regions where the artificial conditions affect the solution. One can achieve considerably better accuracy close to the artificial boundaries by instead using the approximate optically thick solution $I_+ = I_- = \pi B(\nu, T)$ as the boundary condition, where T is evaluated at the boundary.

The piecewise ODE method is flexible, accurate, and easy to implement, and it deserves to be more widely used. A more customary approach to dealing with the numerical issues in the optically thick case is to turn the problem into a discrete set of linear equations and solve it using numerical linear algebra algorithms. To do this we set up a grid of discrete values

$\tau_j = j\Delta\tau$ for $j = 0, 1, ..., N$, and approximate the derivative in Eq. (5.51) as $(dV/d\tau^*)_j \approx (V_{j+1} - V_{j-1})/\Delta\tau$, where the integer subscripts stand for the value of the corresponding quantity at $\tau^* = j\Delta\tau$. With this approximation, Eq. (5.51) can be written as

$$V_{j+1} - V_{j-1} - \Delta\tau \mathbf{M}_j \cdot \mathbf{V}_j = \Delta\tau \mathbf{F}_j \qquad (5.55)$$

where $j = 1, 2, ..., N - 1$. The $j = 0$ and $j = N$ lines must be left out because evaluation of the approximate derivative would require data off the end of the array. These two missing lines are replaced by the statement of the boundary condition at the bottom and top of the atmosphere respectively. This results in a $2N \times 2N$ linear system if the upward and downward fluxes are merged into a single solution vector of the form $[I_{+,0}, I_{-,0}, I_{+,1}, I_{-,1}, I_{+,2}, I_{-,2}, ...]$. The corresponding matrix defining the coefficients of the system is *band-diagonal*, and has non-zero entries only within two spaces to the left or right of the diagonal. Such systems can be solved efficiently using standard methods of linear algebra available in virtually all numerical libraries.[4] We will refer to this solution technique as the *matrix method*.

Exercise 5.3 Write out the first three lines and the last three lines of the matrix problem outlined above, assuming ω_0 to be constant.

Resolving the exponential growth or decay of the homogeneous solutions requires $\Delta\tau < 1$ at least, so it might be thought that a disadvantage of the matrix method is that it requires dealing with very large matrices when the τ_∞^* is large – and with strong gaseous absorption, optical thicknesses of several thousand are not at all uncommon. However, the required resolution is not really determined by the exponential behavior, since, as noted previously, the fluxes are largely determined by local properties in the optically thick limit, but local properties such as temperature usually vary slowly when written as a function of τ^* in the optically thick case. Thus, the fluxes vary little over a unit optical depth, and the derivative of the flux can be accurately computed even if $\Delta\tau$ is quite large. Another way of seeing this is to note that when $\Delta\tau \gg 1$ the term $V_{j+1} - V_{j-1}$ in Eq. (5.55) is negligible compared with the remainder of the terms, and the solution of what is left gives the correct lowest order solution in the optically thick limit. (Recall that the optically thick local solution is determined by treating the temperature and single-scatter albedo as if they were constant.) Corrections to this approximate solution are of order $1/\Delta\tau$, and will be properly computed in the linear solution algorithm. The only caveat needed is to note that besides the correct optically thick solution the linear system also has a spurious homogeneous two-gridpoint oscillating solution of the form $I_+(j) = [1, -1, 1, -1, ...]$ and similarly for I_-. This spurious solution replaces the unresolved exponential homogeneous solution, and care must be taken to use a linear system solver that correctly suppresses contamination by the spurious mode.

There are two other small technical details that need to be taken care of when using the matrix method in the optically thick limit. The first is that the conversion layer at the top of the atmosphere is not resolved, so that the direct beam term is not correctly transformed into diffuse radiation. This is easily dealt with by eliminating the direct beam term and dumping the corresponding energy directly into $I_-(\tau_\infty^*)$ as an upper boundary condition, which explicitly captures the conversion of direct beam energy at the top of the atmosphere. Finally, if there is a significant temperature discontinuity between the ground temperature and the overlying air temperature, the exponential homogeneous solution comes into play in a kind of radiative boundary layer, and something must be done to explicitly resolve this

---

[4] See for example the routine bandec in *Numerical Recipes*.

layer, either analytically or by increasing the resolution near the boundary. When there is a temperature discontinuity, there is a strong, shallow radiative heating or cooling within a unit optical depth of the ground, and this will not be resolved by the matrix method using large $\Delta\tau$. With the ODE method, the radiative boundary layer is automatically resolved, but when using the matrix method one must generally use an analytical exponential solution to get the radiative fluxes near the ground.

Whichever method one chooses, it is often most convenient to work in optical thickness coordinates rather than transforming the equations to pressure or log-pressure space before performing the solution. Since the mapping between pressure and optical thickness is usually frequency dependent, it is necessary to express the fluxes as a function of pressure before summing up the fluxes and heating rates across frequencies. The easiest way to re-express the results is to use numerical integration to compute a suitable list of $\tau^*(p)$ from the differential equation defining τ^*, and then to use this list to define a numerical interpolation function giving $p(\tau^*)$.[5] Using this function a list of triples (I_+, I_-, τ^*) can be transformed to the list of triples (I_+, I_-, p) at the corresponding pressure levels. This list can then be used to interpolate the fluxes to a standard grid of pressure values; heating profiles are treated similarly. The interpolation function $p(\tau^*)$ fulfills an additional role when carrying out the integration for the fluxes. Namely, the temperature profile T and the scattering parameters are generally given as functions of p, or as tabulated values for a list of pressure levels. Using the interpolation function, however, one simply evaluates $T(p(\tau^*))$ and so forth to get the required expression. If $T(p)$ is specified as a table of values rather than a function, then one transforms the pressure entry to optical depth using the interpolation function, and then writes an interpolation function to get $T(\tau^*)$ from this table. When using the matrix method, it may not be necessary to write functions for T and the scattering parameters since tabulated values will generally be sufficient. When using the differential equation method, however, it is usually necessary to provide functions, since the typical high order numerical integrator needs to be able to evaluate the right hand side of Eq. (5.51) at an arbitrary value of the vertical coordinate.

Clearly, in order to carry out the above procedure, it is necessary to have at hand an efficient, accurate, and easy-to-use interpolation function. Some useful advice on interpolation methods has already been given in connection with the numerical analysis tutorial problems in the Workbook section of Chapter 1.

5.8 WATER AND ICE CLOUDS

Now we'll take a closer look at the way Earth's water and water-ice clouds affect the radiation budget, taking account of the balance between the shortwave albedo effects of clouds which act to cool the planet and the longwave cloud greenhouse effects which act to warm the planet. Much of the general behavior in evidence on Earth applies equally well to water clouds on other planets, or for that matter to any cloud-forming substance which is strongly absorbing in the infrared but fairly transparent in the solar spectrum.

To set the stage, we'll first discuss some calculations with the ccm radiation code which show how high clouds affect the albedo and *OLR* under typical tropical conditions. The

---

[5] By "interpolation function" we mean a computer-implemented function that takes a list of N values (x_j, y_j) and an arbitrary argument x, and returns an interpolated estimate of the value of y corresponding to x. By putting in a list of (y_j, x_j) instead, the same function can be used to obtain $x(y)$.

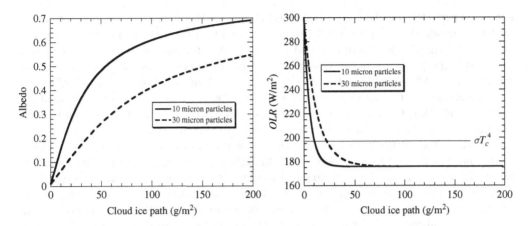

Figure 5.9 Albedo and *OLR* as a function of cloud condensed water path, for a high ice cloud with temperature $T_c = 242$ K at a pressure $p_c = 283$ mb. The temperature profile is on the moist adiabat corresponding to a surface temperature of 300 K, patched to an isothermal 180 K stratosphere. The relative humidity is 50% and the CO_2 concentration is 300 ppmv, but there is no ozone in the atmosphere. Calculations were done with the ccm radiation code, and results are shown for both 30 μm and 10 μm particles.

results are shown in Fig. 5.9. These calculations include most of the radiative effects operating in the real tropics, including solar and infrared absorption by water vapor and CO_2, though we have left ozone out of the picture. The surface albedo has been set to zero, so as to focus on the reflective effect of the cloudy atmosphere itself. In this calculation, the tropospheric temperature profile is on the moist adiabat, and we place a geometrically thin cloud with specified water content in the upper troposphere, where the pressure is 283 mb and the temperature T_c of the cloud is 243 K. At these temperatures, the cloud is primarily composed of ice, and the ccm radiation model makes use of the complex index of refraction appropriate to water-ice particles. Results are given as a function of the condensed water path of the cloud. As long as the cloud is geometrically thin enough to be essentially isothermal, the actual geometrical thickness of the cloud is irrelevant to this calculation.

For 30 μm particles, which are typical of actual tropical ice clouds, it only takes a path of 50 g/m² to make the cloud act like a blackbody. The *OLR* is about 20 W/m² below the blackbody emission σT_c^4 of the cloud itself, because there is some water vapor and CO_2 greenhouse effect in the colder air above the cloud. This effect would disappear if the cloud were placed at the minimum temperature part of the atmosphere, and would increase if the cloud were lower. In essence, a cloud that is optically thick in the infrared acts like a new "ground," radiating upward into the upper part of the atmosphere with a blackbody temperature T_c. If the particles are made smaller, it takes less cloud water in order to make the cloud optically thick, because the same mass of water yields more aggregate cross-section area of cloud particles. As a practical matter, most high clouds occurring in the vicinity of deep convection in the tropics can be considered optically thick in the infrared. The associated cloud greenhouse effect is enormous, and would lead to an uninhabitably hot planet if not compensated by shortwave albedo effects that are of similar magnitude.

The albedo effect of the cloud also increases monotonically with cloud water content, but at a much slower rate than the greenhouse effect, for the reasons discussed in Section 5.6.

For 30 μm particles, the planetary albedo has only reached 0.25 when the cloud water path is $50 \, g/m^2$. The albedo doesn't reach 0.5 until the cloud water content approaches $200 \, g/m^2$. On the other hand, the failure of cloud albedo to saturate until large cloud water paths means that the particle size can have a very important influence on albedo; reducing the particle size to 10 μm increases the albedo to 0.7 for a cloud containing $200 \, g/m^2$ of ice. As for compensation between shortwave and longwave cloud effects, taking a cloud with 30 μm particles and $100 \, g/m^2$, the albedo is about 0.4, which yields $170 \, W/m^2$ reduction in solar absorption based on typical annual average tropical insolation. This compares with a cloud greenhouse effect of $120 \, W/m^2$, so such clouds have a moderate net cooling effect. If the cloud only had a water content of $50 \, g/m^2$, though, the cloud greenhouse effect would be nearly the same but the cloud albedo effect would be reduced by nearly half, and the cloud would have a net warming effect. Similarly, if the ground were partially reflecting (owing to vegetation cover or low-lying clouds), the change in albedo due to high clouds would be reduced, shifting the balance again in favor of net cloud warming. On the other hand, reducing the particle size of the clouds makes them much brighter, making it easier for the clouds to have a net cooling effect.

For fixed particle size, the cloud altitude has relatively little effect on albedo for a given cloud condensed water path. Low altitude liquid water clouds tend to have smaller particles than ice clouds as well as larger water content (because there is more water around to condense), and are correspondingly more reflective. Because of this effect, the balance of power for mid-level and low-level clouds shifts decidedly toward a net cooling effect on the planet.

High clouds have both a warming and a cooling effect, and which one wins depends on the details of the cloud properties, including cloud temperature, particle size, and condensed water content. Colder cloud temperatures tend to favor net warming. Making particles smaller or increasing cloud water enhances the cooling effect except for very thin clouds, since it takes little cloud water to make the cloud act like a blackbody whereas the albedo continues to increase with cloud water increase or particle size decrease, for the reasons discussed in Section 5.6. The balance between albedo effect and greenhouse effect also depends on the intensity of solar radiation, since albedo matters not at all if there is no sunlight. In the polar night, clouds have an unambiguous warming effect as long as they are not right at the surface. Similarly, the cloud albedo effect depends on the albedo of the underlying surface; clouds over a reflective surface like ice (or surface clouds!) will tend to have a warming effect, as also discussed in Section 5.6. As the cloud is made lower, the cloud greenhouse effect is attenuated, because the cloud temperature is closer to the ground temperature and also because (especially in moist regions) the greenhouse effect of the clear air above the clouds masks the longwave radiative effect of the cloud itself.

Although the large and competing effects of clouds on *OLR* and albedo pose similar challenges on any planet whose atmosphere contains a condensable substance, Earth is the only case at present for which we have good observations of the net radiative effect of clouds. The first satellite mission to do this accurately was the Earth Radiation Budget Experiment (ERBE), and subsequent missions have taken a similar approach. We discussed some ERBE clear-sky results in Chapter 3, and now we will see what ERBE has to tell us about cloud effects. The ERBE mission measured the Earth's radiation budget using two sets of highly accurate broadband radiometers borne on satellites – one in the infrared spectrum and one in the shortwave (i.e. solar) spectrum. Moreover, the processing algorithm made use of the patchiness of Earth's cloud cover in order to estimate the effect of clouds on the longwave and shortwave radiation. Within each scene examined (think of a scene as a 50 km square patch of the Earth's surface) the algorithm identified those pixels which represented cloud-free clear-sky conditions, and defined "clear sky" longwave and shortwave flux as the value the flux over the scene would have if the flux of *all* pixels in the scene were replaced by the average of the clear-sky pixels. In the longwave, for example, the ERBE retrieval reports the all-sky OLR, called OLR_{all}, and the clear-sky OLR, called OLR_{clear}. The cloud longwave forcing is then defined as $OLR_{clear} - OLR_{all}$. Since clouds reduce the OLR by making the upper troposphere more optically thick, the cloud longwave forcing is positive, and represents a warming effect. Similarly, the cloud shortwave forcing is defined as $S_{abs,all} - S_{abs,clear}$, where S_{abs} is the top-of-atmosphere absorbed solar radiation – the difference between incoming and reflected solar radiation. Clouds reduce the solar absorption by increasing the albedo, and so the cloud shortwave forcing thus defined is generally negative, representing a cooling effect. The sum of the cloud longwave and cloud shortwave forcings is the net cloud forcing, with positive values representing a warming tendency and negative values representing a cooling tendency. Clear-sky and all-sky albedo can be defined similarly.

Results for clear and cloudy albedo, and for the cloud radiative forcing, are shown for the year 1988 in Fig. 5.10. Other years show a similar pattern. The ERBE dataset contains information of this sort for each month, reported on a latitude–longitude grid. Here we show only annual-mean results averaged along latitude circles. The full monthly mean dataset for all available years is provided as part of the dataset collection in the supplementary materials for this book. Turning attention first to the clear-sky albedo, we see that without clouds the albedo varies in a narrow range of 0.11 to 0.16 from 60 °S latitude to 42 °N. Poleward of 60 °S, the albedo increases sharply owing to the high albedo Antarctic ice. The values indicate that the albedo of the partially snow-covered Antarctic ice must exceed 0.7, since the atmospheric absorption makes the planetary albedo lower than the surface albedo. The clear-sky estimates near the pole are somewhat unreliable, since it is hard to distinguish between low clouds and ice. Going towards the north pole from 42 °N the albedo increases somewhat more gently, owing to the patchy distribution of sea ice and its seasonal fluctuations; the rest of the northern high-latitude albedo increase is due to winter snow cover over

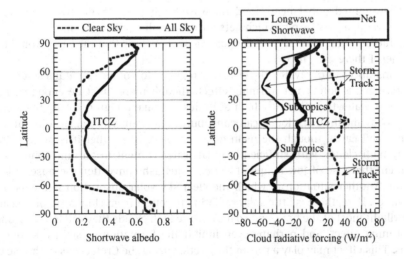

Figure 5.10 Zonally averaged annual mean clear and cloudy sky albedo (left panel) and cloud radiative forcing (right panel) measured by ERBE for the year 1988.

land. Clouds have a strong reflective effect, approximately doubling the tropical albedo and increasing the midlatitude albedo to 0.4 or more. The area-weighted mean albedo is 0.19 for clear sky and 0.33 including cloud effects. Area-weighting doesn't take into account the seasonal and latitudinal distribution of sunlight, though; a more appropriate mean albedo is based on taking the ratio of global net reflected solar radiation to the incident radiation. This estimate yields somewhat smaller values: a mean clear-sky albedo of 0.16 and a mean all-sky albedo of 0.30.

If uncompensated by the cloud greenhouse effect, the high albedo of clouds would probably be sufficient to throw the Earth into a Snowball state. In reality, the reduction of *OLR* by clouds cancels most of the cloud cooling effect, as shown in the right hand panel of Fig. 5.10. The cloud longwave forcing – i.e. the reduction in *OLR* due to clouds – is anticorrelated with the cloud shortwave forcing, and has sufficient magnitude to cancel most of the cloud shortwave forcing. The distribution of cloud forcing takes us into some consideration of aspects of the general circulation we have not introduced previously. Somewhat north of the Equator there is a region of deep convection yielding deep, thick clouds, which is manifest in the figure as a peak in both the cloud longwave and shortwave forcing (marked "ITCZ" in the figure for *Inter-Tropical Convergence Zone*, in honor of the winds which converge moisture into this region and feed the convection). The ITCZ is flanked by two subtropical regions where convection is suppressed by downward motions in the atmosphere, and is shallow or absent. Here one encounters local minima in both the cloud longwave and shortwave forcing. Throughout the tropics, the two terms sum to a net cooling effect of about $-20\,\text{W/m}^2$, which is stronger in the subtropics than near the ITCZ. The subtropical cloud cooling is in part due to near-surface clouds which are associated with the boundary layer rather than deep convection. In the midlatitudes, there is another region of deep cloud activity. This one is associated with the large-scale organized storm tracks, which loft water from the subtropical ocean and move it poleward and upward. The albedo effect of clouds more strongly dominates the greenhouse effect in this region, and even more so towards the Antarctic region, where there is strong cloud shortwave forcing associated with low-lying marine stratus clouds. As a result there is strong net cooling in the midlatitude and polar regions.

The area-weighted global mean cloud longwave forcing is $28\,\text{W/m}^2$, while the mean cloud shortwave forcing is $-47\,\text{W/m}^2$, which nets out to a cooling influence of $-19\,\text{W/m}^2$. Using a sensitivity factor of $2.2\,\text{W/m}^2\,\text{K}$ from Section 4.5, we conclude that the Earth would be about 8.6 K warmer if there were no clouds.

In circumstances under which the clear-sky regions do not absorb much solar radiation, high clouds have a potent net warming effect, though there must be enough convection around to loft water to sufficiently high altitudes to make a high optically thick cloud. As we have already mentioned, clouds can contribute significantly to deglaciation of the high-albedo Snowball Earth, though the main question there is whether it is possible to make sufficiently high clouds with sufficiently great water content in an atmosphere with low water content (because of low temperature) and sluggish convection (because of low solar absorption). Another situation in which the clear-sky solar absorption is low is the high-latitude winter. Here, there is little solar radiation to reflect from clouds simply because it is night or twilight most of the time, and so if there are high clouds they will have a pronounced winter warming effect, and perhaps even inhibit the formation of sea ice in open water conditions. This effect may play a role in the Arctic during the Cretaceous hothouse climate, since there is open water in the Arctic Ocean which can maintain a supply of relatively warm water throughout the winter to feed deep convection. It seems plausible that this mechanism would help explain the mysterious low-gradient climate of Cretaceous and similar hothouse climates, described in Section 1.9.1. General circulation models to date do not support a sufficiently strong cloud effect for clouds to be the answer to the Cretaceous puzzle, but there is much remaining to be learned about clouds, so the last word has not by any means been uttered on this topic. This potential mechanism is only viable when there is open water in the polar ocean. Over a polar continent, such as Antarctica, the ground would cool off rapidly in the winter, foreclosing any serious possibility of deep convection and the associated deep clouds.

The effect of clouds on the water vapor runaway greenhouse represents one of the most vexing and important unresolved issues in planetary climate. The observed behavior of Earth clouds can provide little guidance as to cloud effects in a much warmer atmosphere in which water vapor is the dominant component. Given the availability of water throughout the depth of the atmosphere, and how little water it takes to make a highly reflective cloud, it seems almost inevitable that the albedo will become very high. There is no simple physics, however, that can be employed to estimate the fraction of the atmosphere which will be cloudy; this is an intrinsically dynamical question. High clouds can also reduce the *OLR*, however, which has the potential to offset the albedo effect. A strong cloud greenhouse effect is likely, given that the top 100 mb of a near-runaway steam atmosphere contains far more water vapor than Earth's entire atmosphere. In order for clouds to make the radiating temperature cold enough to offset the strong cloud albedo increase, one would need optically thick clouds in regions of the atmosphere which are very cold. This is not impossible, though: a thick cloud at an upper atmospheric temperature of 200 K would reduce the *OLR* to 91 W/m^2, which is just about the same as the solar radiation Early Venus would absorb if it had an albedo of 80%. In Earth's atmosphere, optically thick tropical cirrus clouds occur at similar temperatures, so one can hardly rule them out for Venus. The question of whether the cloud greenhouse or cloud albedo effect wins out in a runaway situation simply cannot be answered by back-of-the-envelope calculations, even if we have a rather large envelope. A definitive answer must await attainment of a far better understanding of convection, cloud microphysics, and cloud fraction under near-runaway conditions.

5.9 THINGS THAT GO BUMP IN THE NIGHT: INFRARED SCATTERING WITH GASEOUS ABSORPTION

As we have already noted, infrared scattering needs to be taken into account when the planet's clouds are made of a substance that is nearly transparent to infrared in significant parts of the spectrum. This is the case for CO_2 ice clouds, liquid or solid CH_4 clouds, N_2 ice clouds, and concentrated sulfuric acid clouds. The CO_2 cloud scattering is evident even in spectra of present Mars, but it has the potential to have a major impact on the climate of an Early Mars with a thick CO_2 atmosphere. The CH_4 case is relevant to Titan, and the N_2 case would be highly important there if Titan were a bit colder. The sulfuric acid case is relevant to hot dry planets like the present Venus. All these scattering clouds, and probably more, are present to some extent in the atmospheres of gas and ice giants. This is an important cloud regime. In most of these cases, the scattering of infrared radiation by clouds acts jointly with the full complexities of gaseous absorption.

Traditionally, the problem of dealing with the joint effects of scattering and gaseous absorption has been considered to be one of the scariest problems in radiative transfer. The problem appears scary only if one intends to mount an attack on it by some kind of modification of the band-averaged transmission function approach. The problem here stems from the fact that band-averaged transmission functions do not satisfy the multiplicative property, so that the path that one feeds to the transmission function involves the entire past history of the radiation between the time it is emitted (or injected by solar radiation) and the time it leaves the atmosphere. When there is no scattering, the path is simple, but multiple reflection leads to an ensemble of very complex paths. For example, consider an isothermal layer of a non-gray gas sandwiched between a perfectly reflecting ground and a cloud that reflects half of all energy incident on it from below. Suppose the layer has a mass path ℓ. Upward radiation emitted from near the center of the layer will be attenuated according to a path length $\frac{1}{2}\ell$ by the time it reaches the cloud. Half of this will escape to space, but the other half will be reflected downward. To get out, it reaches the bottom boundary, where it is reflected upward. By the time it hits the top again, the path length is $\frac{3}{2}\ell$, and the beam has been attenuated accordingly. Half of this escapes while the other half is reflected downward, and the process continues until there is essentially no beam left. If $\mathcal{T}$ is the transmission function written as a function of the path, the escaping flux is

$$\frac{1}{2}I_0\mathcal{T}\left(\frac{1}{2}\ell\right) + \frac{1}{4}I_0\mathcal{T}\left(\frac{3}{2}\ell\right) + \frac{1}{8}I_0\mathcal{T}\left(\frac{5}{2}\ell\right) + \cdots \qquad (5.56)$$

where I_0 is the initial upward radiation emitted from the center of the layer. If the transmission function had the multiplicative property, then we would have

$$\mathcal{T}\left(\left(\frac{1}{2}+n+1\right)\ell\right) = \mathcal{T}\left(\left(\frac{1}{2}+n\right)\ell\right)\mathcal{T}(\ell) \qquad (5.57)$$

and the problem could be done as an iteration in which the fate of each reflection of flux hitting the cloud is independent of how many bounces it took before it got there. For band-averaged transmission functions, which do not retain the multiplicative property, this is not the case and one needs to perform the sum over all past histories of paths taken by radiation. For the two-reflector case this is not too bad, but when one considers a more realistic problem in which absorption and scattering occur between a continuous array of pairs of layers of the atmosphere, one does indeed begin to quake in one's boots, if just a bit.

Most of the fear can be dispelled, however, through use of the exponential sums approach and its variants. It is only the band averaging that creates the problem. For any given value of the absorption coefficient, the multiplicative property holds, which is in fact what allows us to write the two-stream equations as a differential equation. Since the exponential sum calculation is built from a number of non-averaged calculations for the individual absorption coefficients going into the sum, the essential difficulty is circumvented.

Within each band, the calculation is identical to either of the two approaches described in Section 5.7, except that one needs to do the calculation over again many times for various values of the absorption coefficient, and then sum the results with the appropriate weighting. Specifically, as in the exponential sums method without scattering, we adopt a distribution function $H(\kappa_0)$ for the absorption coefficient at a reference pressure and temperature p_0 and T_0. For each of the κ_0 in the table, we scale $\kappa(p)$ over the rest of the profile according to pressure and temperature scaling factors, perform the numerical calculation of the flux, and then sum the results weighted according to H. Note that in order to perform this sum, the fluxes for each κ_0 must be written on a common pressure grid, by interpolation or by other means. Finally, to get the net flux, the result is summed over all the bands that contribute.

To illustrate the application of the technique, we will carry out a simplified version of the Early Mars case. We assume a pure CO_2 atmosphere with a 2 bar surface pressure under the influence of Mars gravity. The temperature is taken to be on the dry CO_2 adiabat corresponding to a surface temperature T_g, until the threshold for CO_2 condensation is reached, whereafter the temperature follows the dew-point formula for the single-component saturated CO_2 adiabat. The calculations below were performed for $T_g = 275$ K, since we are principally interested in how much absorbed solar flux is needed to maintain liquid water at the surface. With this value of T_g the atmosphere hits the condensation point at a pressure level of 400 mb. We will look at how the *OLR* changes when we introduce an idealized upper-level CO_2 ice cloud in the vicinity of the condensing region. Specifically, we assume that the profile of CO_2 ice mass concentration in the cloud is

$$q(p) = q_m \exp\left(\frac{(p - p_m)^2}{\Delta p^2}\right) \tag{5.58}$$

(where p_m is the pressure in the middle of the cloud, and q_m is maximum mixing ratio). We further assume that the mass-specific cross-section of the scatterers is $\chi = 100\,\text{m}^2/\text{kg}$, which is approximately the value appropriate to $10\,\mu\text{m}$ CO_2 ice particles. This calculation ignores the wavenumber dependence due to Mie scattering properties and the variation of index of refraction with wavenumber. It also ignores the absorption of infrared by CO_2 ice. The incorporation of these additional effects presents no special technical difficulties, but they have been left out for the sake of simplicity.

Results for $p_m = 500$ mb and $q_m = 10^{-5}$ are shown in Fig. 5.11. These results show the spectrum of outgoing radiation with and without the cloud. Note that converting only a tiny proportion of the ambient atmosphere into condensed form has a profound effect on the radiation. The cloud powerfully reduces the *OLR*, particularly in the region below $500\,\text{cm}^{-1}$ where CO_2 is fairly transparent, even including the continuum. Averaging over the spectrum, introducing the cloud reduces the *OLR* from $110\,\text{W/m}^2$ to a mere $37\,\text{W/m}^2$. Recall, however, that the warming effect is partly offset by increased reflection of solar radiation, as was discussed in Section 5.6.

In dealing with multiple greenhouse gases within this framework, one no longer has the option of implementing the random overlap assumption by multiplying the transmission functions for the individual gases, since the transmission function is never explicitly

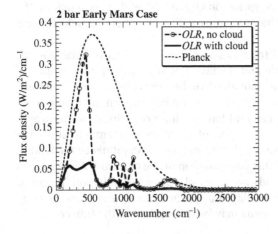

Figure 5.11 The spectrum of *OLR* for an Early Mars case with a 2 bar CO_2 atmosphere and a surface temperature of 275 K. The temperature profile is on the one-component adiabat, which is non-condensing for pressures greater than 400 mb and condensing at higher altitudes. Results are shown both for clear sky conditions and with a high-altitude CO_2 ice cloud with peak condensate mixing ratio of 10^{-5}.

computed. Instead, random overlap is implemented by forming the distribution $H(\kappa_0)$ for the mixture of gases as a suitably weighted convolution of the distributions for the individual gases. One can do even better and eliminate the random overlap assumption by computing $H(\kappa_0)$ from the sum of the actual spectrally resolved absorption coefficient data for the mixture, which is a weighted sum of the absorption coefficients for the individual species. This procedure can be slow if one does it on the fly each time the distribution for a new mixture is needed (as in the case of a highly variable substance like condensable water vapor), but the procedure can be sped up considerably by precomputing the distributions needed for a range of mixtures, and interpolating between the tabulated distributions. When many different greenhouse gases are involved, this can involve a considerable amount of storage, but computer memory is abundant and cheap, so this approach is gradually taking hold.

As with all approaches based on the exponential sums concept, the main vulnerability is the validity of the scaling assumption $\kappa(p, T) = \kappa_0 F(p, T)$. The shortcomings of scaling can be to some extent overcome using the correlated-k variant of exponential sums, but the mathematical justification for the use of correlated-k when scaling is invalid is on even more shaky grounds in the presence of scattering than it is in the non-scattering case. When dealing with novel radiative transfer problems, it is always judicious to compare selected cases with the results of line-by-line calculations.

In the above we have emphasized the joint effects of infrared scattering with gaseous absorption, but many of the same issues apply in the visible and ultraviolet range. In that case, the scattering is even important for liquid water and water ice, and and Rayleigh scattering becomes significant as well. The problems are somewhat less challenging than for the infrared case because visible and ultraviolet absorption spectra tend to have less intricate fine structure than is the case for infrared. Nonetheless, the problem presents similar challenges, which can similarly be dealt with using exponential sums or its variants just as was done for the infrared case.

5.10 EFFECTS OF ATMOSPHERIC SOLAR ABSORPTION

On the present Earth the idealized picture of climate in which all solar absorption occurs at the ground is useful, but even for the present Earth about 20% of solar radiation is absorbed within the atmosphere. For other atmospheres, the proportion absorbed in the atmosphere could be much greater. The effect of this absorption on climate depends very much on the vertical distribution of the absorption, and that is what we will explore here for selected real gases.

The two key questions we have in mind for this section are the effect of solar absorption on the stratospheric temperature profile and the effect of solar absorption on surface temperature. When is solar absorption strong enough to inhibit convection and chill the surface? Even when the primary effects are stratospheric, it should be kept in mind that indirect feedbacks through stratospheric chemistry can still have an important influence on tropospheric climate. In particular, it takes very little mass of cloud to substantially affect the planet's radiation budget, and the temperature of the stratosphere can influence the kind of clouds that form there. A prime example of this phenomenon in the current Solar System is Titan, whose stratospheric organic haze clouds are a key player in the radiation budget. It has been suggested that such clouds may have played a role in a methane-rich anoxic Early Earth atmosphere, and indeed similar phenomena may be widespread in the Universe.

For gas or ice giants, the problem of atmospheric solar absorption is even more critical, since it is the *whole* story with regard to the external energy supply of the atmosphere, there being no distinct liquid or solid surface at which to absorb solar radiation. The profile of absorption can determine whether the solar driving hinders or helps convection, and (together with heat flux from the interior) determines where the troposphere is located. These planets contain a diverse variety of condensable substances and clouds, many of which are known to absorb near-IR, visible, and UV radiation. However, little is known about the vertical distribution of absorbers, or even which ones are dominant in giant planets in our own Solar System. This is a very unsettled area about which we will have little to say; a few pointers into the literature are given in the Further Reading section. The situation with regard to stellar absorption in extrasolar gas or ice giants is of course even more unsettled, but offers wide scope for exploration of hypothetical atmospheres.

5.10.1 Near-IR and visible absorption

We'll begin with an overview of the near-IR and visible absorption characteristics of CO_2, CH_4, and water vapor, shown in Fig. 5.12 and Fig. 5.13. These are drawn from data in the HITRAN database. As was the case for thermal IR, the absorption coefficients have an intricate line structure leading to fine-scale variations with wavenumber. The figures summarize the properties by showing only the median absorption in bins of width $50\,\mathrm{cm}^{-1}$. This is sufficient to provide a general idea of where the gases absorb strongly and where they are

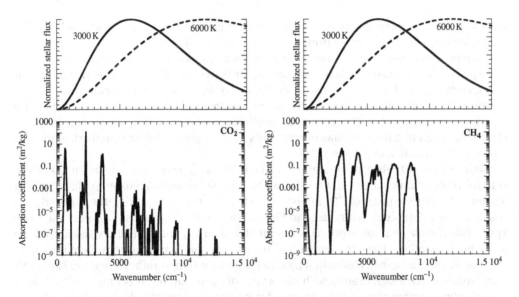

Figure 5.12 The lower panels show the median absorption coefficient in $50\,\mathrm{cm}^{-1}$ bins for CO_2 (left panel) and CH_4 (right panel). The absorption coefficients were computed with $T = 260\,\mathrm{K}$ and $p = 100\,\mathrm{mb}$. The 75th percentile coefficients are typically about an order of magnitude above the median. The upper panels show the distribution of incoming stellar radiation for a typical cool, red M-dwarf, and for a hotter yellow G-dwarf like the Sun.

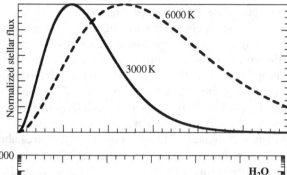

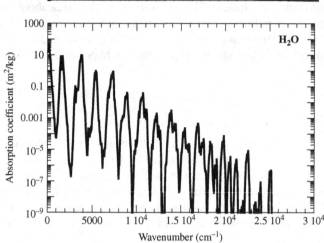

Figure 5.13 As for Fig. 5.12, but for water vapor. Note that the spectral range shown is twice that for CO_2 and CH_4, since water vapor absorbs strongly out to higher wavenumbers.

largely transparent. Results are given at a standard pressure of 100 mb, and can be scaled (approximately) linearly to other pressures as discussed in Chapter 4.

Carbon dioxide has a dense forest of absorption features in the near-IR below 8000 cm^{-1}, and sparser, weaker absorption features at higher wavenumbers. We see immediately that the spectral class of the planet's star matters very much to the importance of near-IR absorption. For a cool, red M star, the stellar spectrum overlaps very considerably with the absorption features of CO_2, whereas a much lower (though not by any means insignificant) proportion of a hotter G star output leads to absorption. The same remark applies to virtually all infrared-active gases.

Which of the CO_2 absorption features come into play depends on how much CO_2 there is in the planet's atmosphere, and for this we turn to the pressure-adjusted path, defined in Section 4.2.1. For 300 ppmv of CO_2 in 1 bar of Earth air under Earth gravity, the adjusted path based on the reference pressure used in Fig. 5.12 is about 20 kg/m^2, so it is only the three spikes that rise above an absorption coefficient of 0.01 m^2/kg that contribute significantly to atmospheric heating. (The adjusted path of the present thin Martian atmosphere is similar, at 13 kg/m^2, so the near-IR absorption picture for present Mars is rather similar to Earth.) All three spikes overlap significantly with the output of an M star, but it is only the two higher wavenumber features that contribute significantly for a G star like the Sun. If the CO_2 on Earth increases to 20% (mole fraction) of the atmosphere, then the path is over 15 000 kg/m^2, so all absorption features stronger than 10^{-5} m^2/kg come into play, which takes a fairly sizable chunk out of the incoming radiation for a G star and even more so for an M star. Carbon dioxide levels of this magnitude are typical of the Faint Young Sun period on Earth, and are also similar to the levels required to enable deglaciation of a Snowball state; in such circumstances, near-IR absorption due to CO_2 becomes a significant player in climate. If we go to a 2 bar pure CO_2 atmosphere, the adjusted path is about $2 \cdot 10^5 \, kg/m^2$ for Earth gravity, around three times as much for Mars gravity, and about half as much for a typical massive Super-Earth. At these values, the gaps where the atmosphere is transparent narrow considerably, and the atmosphere becomes optically thick in most of the spectral region below $6000 \, cm^{-1}$. A 100 bar Venuslike atmosphere would absorb nearly all the incoming stellar radiation within all the absorption regions shown in Fig. 5.12, though the net increase in absorbed flux would be modest since the atmosphere will have already absorbed most of what it can absorb in the top 2 bar.

Methane is a much more potent near-IR absorber than CO_2, and has fewer transparent window regions. This limits the potential of CH_4 as a greenhouse gas at high concentrations, since the anti-greenhouse effect arising from heating of the upper atmosphere partly offsets the surface warming due to thermal-IR opacity. At present Earth methane concentrations of around 2 ppmv, the adjusted path is only $0.06 \, kg/m^2$, so given the typical magnitude of the absorption coefficients, the near-IR absorption can be neglected. However, if the concentration rises to 1000 ppmv, as it easily could in an anoxic atmosphere, CH_4 absorbs a considerable portion of the stellar flux below $10\,000 \, cm^{-1}$. As before, the implications for climate are even more consequential for an M star than for a G star.

Water vapor has strong absorption features extending well into the visible range, though it is also fairly well supplied with relatively transparent window regions. There are three distinct archetypical planetary situations to be thought about with regard to water vapor. First, in Earthlike conditions, water vapor is a minor and condensable constituent, which is concentrated in the lower atmosphere. For example, in a saturated 100-mb-thick near-ground layer at 300 K there are $22 \, kg/m^2$ of water vapor, yielding a pressure-adjusted path of about $200 \, kg/m^2$. There are several absorption peaks below $6500 \, cm^{-1}$ that are effective for a path this large. Because water vapor in the Earthlike regime absorbs largely near the ground, it acts almost the same as reducing the ground albedo. However, if the ground would have absorbed the near-IR anyway, the net effect on climate would be minimal. Over a high albedo surface, the absorption would be more consequential. On a Snowball Earth with 250 K near-surface air temperatures, the same layer still has an adjusted path of almost $5 \, kg/m^2$, and referring to Fig. 5.13 we see that there would still be considerable near-surface absorption.

The second regime to consider is Venuslike, in which water vapor is a non-condensable well-mixed trace gas. Water vapor at 20 ppmv in the atmosphere of Venus yields a pressure-adjusted path of over $3000 \, kg/m^2$, which would yield strong absorption all the way out to $15\,000 \, cm^{-1}$, with the exception of a few narrow window regions. Water vapor thus plays a very significant role in solar absorption on Venus, just as it does in the Venus greenhouse effect. By proportion, there is not much water in the atmosphere of Venus, but because the atmosphere is so massive the amount of water adds up to a considerable value, and its absorption is further strengthened by the high-pressure environment.

The final regime to consider is a runaway greenhouse steam atmosphere. The top bar of such an atmosphere under Earth gravity has a pressure-adjusted path of $50\,000 \, kg/m^2$, and so nearly all of the star's near-IR output would be absorbed in that layer, considerably heating it and affecting the planet's energy balance. For an M star, this spectral region contains most of the star's output, and so the near-IR absorption has the potential to play a key role in the climate of a runaway greenhouse atmosphere for planets orbiting M stars.

The spectral overview we have just provided does not tell us very precisely how much incoming flux is absorbed and how the absorption is distributed in the vertical. For this, we need to compute the flux profile taking into account the full variability of the absorption coefficient. We shall do this by using the exponential sum method to compute the transmission function from the top of the atmosphere to level p in each of a set of bands covering the spectrum, and then summing up the transmission weighted by the incoming stellar flux in each band. This is in essence a simplified subset of the calculation done in the home-brew radiation code discussed in Chapter 4. In this calculation, the thermal emission from the atmosphere is negligible, so one only needs to compute the transmission of radiation entering from the top. To focus on the most essential features, we'll also assume that all flux reaching the bottom boundary is absorbed there, so we need not consider the transmission of upward-reflected stellar radiation. In most circumstances, this is a minor influence, since the parts of the spectrum where the atmosphere absorbs well are mostly depleted by the time the ground is reached. As a further simplification, we'll assume the atmosphere to be transparent outside the spectral region covered in the spectral survey, and neglect Rayleigh scattering. Rayleigh scattering is fairly weak in the near-IR, but Rayleigh scattering of visible and ultraviolet light would keep some of the incoming radiation from being absorbed at the ground. Finally, the calculations will be carried out with a fixed zenith angle having $\cos \zeta = \frac{1}{2}$, rather than averaging the zenith angle over day and season at some given latitude. The profiles to be discussed below were all computed subject to a net downward stellar radiation of $350 \, W/m^2$ coming in at the top of the atmosphere, but as the problem is linear, the flux can be readily scaled to any other value. In each case, we did the calculation for an M star with photospheric temperature of $3000 \, K$ and for a G star at $6000 \, K$. The calculations were carried out with a gravitational acceleration of $10 \, m/s^2$.

Results for a pure CO_2 atmosphere, a CO_2–air mixture, and a pure H_2O atmosphere are shown in Fig. 5.14. These were computed for an isothermal $260 \, K$ atmosphere, but the transmission is not terribly sensitive to the temperature profile. When examining fluxes plotted against a logarithmic pressure axis, it is good to keep in mind that if F is the flux, then the heating rate per unit mass is proportional to the slope dF/dp, which is $p^{-1} dF/d \ln p$. Thus, a given slope in log coordinates corresponds to a greater heating rate at low pressures

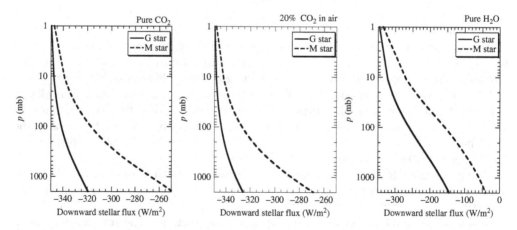

Figure 5.14 Profiles of incoming stellar flux computed using an exponential-sum transmission function for three different atmospheres: 2 bar pure CO_2 (left), 20% molar CO_2 in Earth air (middle), and 2 bar pure H_2O (right). The incoming vertical flux at the top of the atmosphere is $350 \, W/m^2$ in all cases, and results are shown for both a G star and M star spectrum.

than it does at high pressures. For well-mixed greenhouse gases, one typically finds high heating rates aloft, since the upper atmosphere gets the first chance to absorb the part of the spectrum that is absorbed very strongly.

In all cases, the absorption for the M star case is substantially greater than that for the G star case, as expected. A 2 bar CO_2 atmosphere absorbs $30\,W/m^2$ of the incident $350\,W/m^2$ for the G star, but $100\,W/m^2$ for the M star. For both kinds of stars, about a third of the total flux is deposited at pressures lower than 100 mb, which will lead to intense heating in the upper atmosphere. Still, a great deal of the flux is absorbed in the lower atmosphere. Over a highly absorbing surface like an ocean or dark land, it matters little to the tropospheric temperature whether the energy is absorbed in the troposphere or at the surface. Over a highly reflective surface, as in a Snowball state, the effect of the lower atmospheric absorption is to decrease the effective albedo, which will lead to a warming of the troposphere. In any event, the vertical distribution of absorption for CO_2 does not suggest a pronounced anti-greenhouse effect.

The situation for a mixture of 20% CO_2 in air is quite similar, and in fact leads to only slightly less total absorption, because the additional CO_2 in the pure CO_2 case is only able to absorb parts of the spectrum where CO_2 is a relatively poor absorber. This is another instance of the typical logarithmic dependence of radiative properties on greenhouse gas concentrations. The general implication is that stellar absorption by CO_2 should have only a minor effect on tropospheric climate when the surface has low albedo. Over a high albedo surface, as in a Snowball, we expect some modest tropospheric warming, which can help in deglaciation. For example, assuming a surface pressure of 1000 mb, the atmospheric absorption is $6\,W/m^2$ for a G star between the 400 mb level and the ground, or about twice as much for an M star. If the near-IR albedo of the surface is 50%, then half of the amount absorbed in the troposphere would have been absorbed at the surface anyway, so the additional radiative forcing due to tropospheric absorption is only half the stated values. Assuming a 60% surface albedo averaged over the entire solar spectrum, the surface solar absorption for a transparent atmosphere would be $140\,W/m^2$. Hence, the additional radiative forcing amounts to between 2.5% and 5% of the energy budget. This is not overwhelming, and would be partially offset by the cooling due to stratospheric absorption. Still, it is a factor that should be taken into consideration in determining the conditions for deglaciating a Snowball state.

Now let's turn to the pure water vapor case for a layer extending to a pressure of 2 bar. This can be thought of as the top 2 bar of an atmosphere undergoing a runaway greenhouse, or alternately the whole atmosphere for a world with a substantial ocean having a surface temperature of around 395 K. Since the absorption coefficients in this example are computed with a fixed temperature of 260 K, the opacity of the lower part of the atmosphere has been underestimated, but the example nonetheless suffices to demonstrate just how powerfully water vapor absorbs in the near-IR. Even on a logarithmic plot the slope at 50 mb is greater than the slope at 500 mb, indicating an extremely intense upper-level heating which should lead to pronounced warming of the upper atmosphere. Rather little stellar flux makes it to the 2 bar level – only $150\,W/m^2$ in the G star case and $40\,W/m^2$ in the M star case. Calculations with thicker atmospheres (not shown) reveal that the attenuation continues as the surface pressure is further increased. For example, in the M star case, only $15\,W/m^2$ reaches the ground for a 20 bar atmosphere and $5\,W/m^2$ for a 200 bar atmosphere. For a hot atmosphere, temperature scaling of line strengths would further reduce the penetration, though the precise effect takes us into unknown territory with regard to temperature scaling of the water vapor continua at high temperatures. Further, the pressure scaling of absorption coefficients in the exponential sum code probably underestimates the true influence of line broadening at very high pressures, so most probably the flux would be further reduced in a precise calculation. For a G star there would be less absorption, but this would be offset by greater Rayleigh scattering owing to the shorter wavelengths in the incident stellar radiation. A runaway greenhouse would be a very dark place at the surface, with hardly any radiation penetrating to the ground.

With regard to the climate implications of the limited flux penetrating the atmosphere, however, it should be kept in mind that this atmosphere is also very optically thick in the thermal infrared, so that even in the 2 bar M star case, the lower atmosphere would have to achieve a very high temperature in order to be able to lose $40 \, W/m^2$ by radiative diffusion through the highly opaque atmosphere. If the tropopause rises so high as a result of this heating that it engulfs the stellar heating region aloft, the anti-greenhouse effect will be suppressed. Even in pure radiative equilibrium one must reckon with the fact that if H_2O is a good absorber of stellar near-IR, it is also a good emitter of thermal IR, and the outcome of the competition between these two factors is not easy to resolve *a priori*. The most that can be said at this point is that near-IR stellar absorption must be considered as a serious factor in steam atmospheres, all the more so in the M star case.

Now let's take a look at the Earthlike case of saturated water vapor mixed with air, shown in Fig. 5.15. In this case, the temperature profile matters a great deal, since it determines the vertical distribution of water vapor. These calculations were done for an atmosphere on the moist adiabat. Results are shown for a tropical case with surface temperature of 300 K and a Snowball case with a surface temperature of 250 K. In contrast to the case of a pure steam atmosphere, the absorption is trapped in the lower atmosphere for the Earthlike case. This adds significantly to the heating of the troposphere, but again, over a strongly

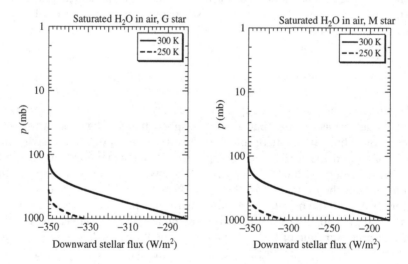

Figure 5.15 Profiles of incoming stellar flux computed using an exponential-sum transmission function for saturated water vapor in Earth air on the moist adiabat. The temperature labels indicate surface temperature. The left panel shows results for a G star spectrum, while the right shows M star results. These calculations take into account the effect of the temperature profile on water vapor, but do not incorporate the temperature scaling of absorption coefficients.

absorbing surface the absorption is only acting to radiatively deposit energy directly in the troposphere which otherwise would have been absorbed at the surface and communicated to the troposphere by convection. As we discussed for the CO_2 case, the absorption is more consequential if the surface is highly reflective. In this regard, it is important to note that the low-level absorption is substantial even when the surface temperature is only 250 K. In both the G and M star cases, the absorption is markedly greater than the corresponding low-level absorption in the CO_2 case. Since tropical temperatures during a Snowball can easily reach 250 K, solar absorption by water vapor can significantly assist deglaciation, by lowering the effective surface albedo. Any process which warms the atmosphere will further increase the atmospheric water content and thus further increase the solar absorption. This constitutes a novel sort of water vapor feedback, which operates via the effect of water vapor on the solar spectrum instead of the thermal infrared effect.

For a complete appreciation of the effect of stellar absorption on the temperature profile, one must compute the radiative-convective equilibrium in the presence of absorption. We will discuss a few such calculations now, for the case of pure CO_2 atmospheres. In Section 4.8 we carried out thermal infrared radiative-convective solutions by fixing the ground temperature T_g and finding the atmosphere that was in equilibrium with the upwelling radiation from the ground. If one then wanted to know what T_g would be supported by a given amount of absorbed stellar radiation, it was only necessary to vary T_g until the desired OLR was achieved. This procedure will not do in the presence of atmospheric stellar absorption, since the incoming flux must be known in order to compute the temperature profile. Hence, in the calculations to follow, we adopt a somewhat different procedure, specifying the incoming stellar radiation, time-stepping the temperature profile as before, but this time adjusting T_g until the top-of-atmosphere energy balance is satisfied. Where the atmosphere is statically unstable relative to the ground temperature, the temperature profile is reset to the adiabat as before.

There are various ways to approach the problem of adjusting T_g. If one were interested in reproducing the actual time evolution of the system, it would be necessary to use the surface downwelling stellar radiation and thermal infrared in order to determine the surface temperature change; then, turbulent and radiative heat fluxes would heat the low-lying air and if instability resulted, the heat would be mixed upward by convection. In our case, we are only interested in getting the equilibrium state, so any procedure that converges to the equilibrium will do. The following simple iteration works quite well in practice. The general idea is that if the net top-of-atmosphere radiation (incoming stellar minus OLR) is downward, then T_g needs to be increased in order to bring the atmosphere closer to balance; conversely, if the net is upward, T_g needs to be decreased. The stellar absorption is not very sensitive to temperature, so the change in T_g mainly affects the OLR. Thus, the main problem is to figure out the climate sensitivity, $dOLR/dT_g$. If the atmosphere is optically thick, then increasing T_g with the atmospheric temperature fixed would not change OLR, but the process we envision is that increasing T_g warms the rest of the atmosphere by convection and radiation, and this ultimately leads to an increase in OLR. The key simplification in the iteration is to estimate $dOLR/dT_g$ as if the atmosphere were a gray gas. Specifically, we compute the radiating temperature T_{rad} from $\sigma T_{rad}^4 = OLR$, since we know the OLR from the radiation calculation. Then, the climate sensitivity is estimated as $dOLR/dT_g \approx 4\sigma T_{rad}^4$. Adopting the convention that a net downward flux is negative, the iteration to be performed at each time step is

$$T_g \rightarrow T_g - \epsilon \frac{F_{toa}}{4\sigma T_{rad}^4} \tag{5.59}$$

where F_{toa} is the net top-of-atmosphere flux and ϵ is an under-relaxation factor that adjusts the ground temperature just part of the way towards the target, which improves the stability of the iteration. The following calculations were performed with $\epsilon = 0.05$, which provides a reasonable compromise between stability and rate of convergence. This iteration procedure is rather ad hoc, and no doubt there are more sophisticated schemes which rest on firmer ground. However, we have found it to serve quite well over a range of situations.

Figures 5.16 and 5.17 show radiative-convective equilibrium results for pure CO_2 atmospheres, carried out using this procedure. As in the flux profiles shown previously, the surface was assumed to be completely absorbing. The results in Fig. 5.16 are in equilibrium with $350\,W/m^2$ of incoming stellar radiation, on a planet having a surface gravity of $10\,m/s^2$. In the 2 bar G star case, the near-IR absorption moderately warms the stratosphere relative to the no-absorption case. The effect of absorption on surface temperature is hardly detectable in the figure. It amounts to a cooling of about 4 K. For the 2 bar M star case, the warming aloft is much more pronounced, and there is a significant lowering of the tropopause. In this case the surface cooling caused by absorption is 15 K, though given the high surface temperature this is hardly a very consequential effect. When the surface pressure is increased to 20 bar in the M star case, the surface temperature increases dramatically, but the stratospheric temperature changes little, with the main effect being a slight warming of the highest stratosphere, which leaves the stratosphere quite isothermal. The no-absorption case for the 20 bar atmosphere (not shown) very closely follows the adiabat and has a very high tropopause. The surface cooling caused by absorption in this case increases to 22 K, but this offsets little of the additional greenhouse warming which raises the surface temperature to 460 K.

Finally, to give an example of the situation for thin atmospheres, we show a calculation carried out in the present Mars regime in Fig. 5.17. This calculation was done

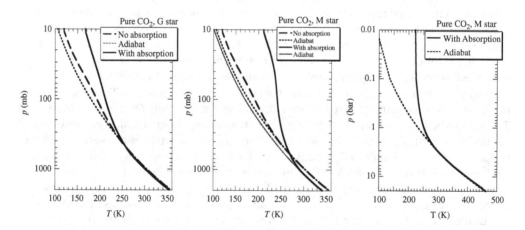

Figure 5.16 Radiative-convective equilibrium subject to an incoming stellar flux of $350\,W/m^2$ for pure CO_2 atmospheres, for a planet with $10\,m/s^2$ surface gravity. The first two panels have surface pressure of 2 bar and the rightmost panel has surface pressure of 20 bar. The surface is assumed completely absorbing. In the 2 bar cases, calculations without atmospheric stellar absorption are shown for comparison. The G star cases assume a blackbody spectrum of incoming radiation with a temperature of 6000 K, whereas the M star cases assume 3000 K.

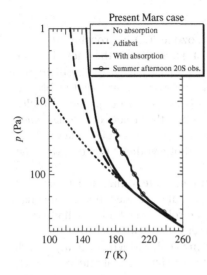

Figure 5.17 Radiative-convective equilibrium for present Mars. The incoming solar flux is $250\,W/m^2$ and has been chosen to yield a tropospheric temperature similar to the observation shown in the figure. The observation shown for comparison is from a summer afternoon tropical sounding, from the Mars Global Surveyor radio occultation dataset.

with Mars gravity, and subject to G star illumination. We see that, as in the dense atmosphere cases, the G stellar absorption causes only a moderate warming of the stratosphere. The observed stratospheric temperature is notably warmer than the calculation, which suggests that near-IR absorption alone is not able to account fully for the Martian stratospheric temperature seen in this sounding. Absorption due to dust is a likely culprit, but effects due to the global-scale stratospheric circulation may be playing a role as well.

The summary situation for pure CO_2 atmospheres is that stellar near-IR absorption causes moderate stratospheric warming for planets about G stars and more pronounced stratospheric warming for M stars, but in neither case does the stellar absorption lead to a stratospheric temperature inversion; the temperature is monotonically decreasing everywhere. The effect of absorption on the tropospheric temperature is a modest cooling. Thus, the anti-greenhouse effect is not terribly consequential for pure CO_2 atmospheres. We'll note also that all of the stratospheres in the above results are optically thin in comparison with the troposheres, which implies the happy result that reasonable estimates of surface temperature can be made on the basis of the simple and swift all-troposphere *OLR* model.

The radiative-convective behavior of thick water vapor atmospheres presents considerably more challenge because of the extreme optical thickness of water vapor both in the near-IR and thermal-IR. This is a very rich problem which entails consideration of the delicate balance between very slow radiative cooling and the small amount of stellar radiation reaching the surface, as well as the effects of condensation on the adiabat and the increase of surface pressure with temperature for an atmosphere in equilibrium with an ocean. For the hot steam atmospheres which are of greatest interest, one also comes up against the largely unknown behavior of the water vapor continua at high temperatures and pressures. We will be content to leave this deep and interesting problem as a subject for research. It is an important problem because the stellar absorption has the potential to significantly increase the threshold illumination needed to trigger a runaway greenhouse. The reader is now in possession of all the tools necessary to carry out an inquiry of this sort.

5.10.2 Ultraviolet absorption

Because of its importance to near-surface life on Earth, ozone (O_3) is probably the most familiar of all ultraviolet-absorbing gases. The interest stems from the fact that the shorter-wave and more energetic forms of UV radiation wreak havoc with key biological molecules of life as we know it, and in particular genetic information encoded in DNA. It is a highly Earth-centric view to think that an ozone shield is necessary to protect complex life in general from deadly UV radiation, but notwithstanding that issue, ozone has some very profound effects on the stratospheric temperature structure that play a key role in the prospects for detecting O_2 (and presumably oxygenic photosynthetic life) on extrasolar planets.

Ultraviolet wavelengths are customarily measured in nanometers (nm, or 10^{-9} m). The radiation begins to become harmful to Earth life at 320 nm, and wavelengths shorter than 300 nm cause extreme damage. Ozone plays a distinguished role in shielding Earth life from UVB (320–280 nm) and UVC (280–100 nm) radiation. Shorter UV wavelengths, to say nothing of solar X-rays, are even more deadly, but there are many molecules which efficiently absorb wavelengths shorter than 100 nm. For example, CO_2, which is far more abundant than O_3 in Earth's atmosphere, is every bit as absorbent as O_3 in the vicinity of 140 nm. In contrast O_3 is a potent absorber between 200 and 300 nm, whereas CO_2 and most other reasonably abundant atmospheric gases are nearly transparent there. Ozone also absorbs significantly in the 400 to 700 nm range. These wavelengths are not particularly damaging to life, but because the stellar output is abundant in this range for G-class and hotter stars, the effect on atmospheric heating is significant.

Ozone is a feature of atmospheres rich in free O_2, bombarded by UV radiation. So far, Earth's is the only known example of such an atmosphere. Ozone is a highly reactive substance with a short lifetime. Therefore, it is very inhomogeneous. In particular, in Earth's present atmosphere ozone is concentrated in a stratospheric layer near the altitude where its production rate is strongest. At earlier times, when there was less O_2 around, the ozone layer was probably found at a lower altitude. At present, maximum ozone values occur at about the 20 mb level in the tropics, and reach values on the order of 2 ppmv. The concentration drops by two orders of magnitude as the tropopause is approached. Even at such low concentrations, ozone is a very effective absorber. A concentration of 1 ppmv in the layer of atmosphere above 20 mb gives this layer an optical thickness exceeding 4.0 at 250 nm, which is sufficient to exhaust virtually all of the UV flux at that wavelength. Detailed data on the UV absorption spectrum of O_3 and other gases can be found in the resources listed in the Further Reading section.

Heating due to UV absorption by ozone has an important effect on the stratospheric temperature profile, but ozone is also a very powerful absorber in the thermal infrared range. It is the only abundant infrared absorber which is concentrated in the stratosphere, and in that sense provides a counterpoint to water vapor, which is concentrated in the troposphere.

We will now carry out some calculations with the ccm radiation model which illustrate the key effects of ozone in an Earthlike setting. We adopt an idealized ozone profile of the form

$$\eta_{o3} = \eta_m \exp\left[-(p - 20\,\text{mb})^2/(50\,\text{mb})^2 \right] \tag{5.60}$$

where η_{o3} is the molar mixing ratio of ozone and η_m is the peak value, which we take to be 2 ppmv in the following calculations. The left panel of Fig. 5.18 shows the net downward flux for an atmosphere that contains Earth air and ozone, but no other gases. The atmosphere

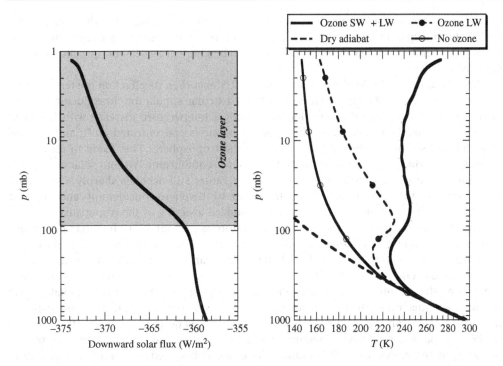

Figure 5.18 Left panel: Net downward solar flux for an atmosphere with the ozone profile described in the text. The calculation was performed for an isothermal 300 K atmosphere, but the results are essentially insensitive to temperature. The incoming solar radiation is 400 W/m², some of which is scattered back by Rayleigh scattering. Right panel: Radiative-convective equilibrium for a dry atmosphere containing 300 ppmv of CO_2, computed for three cases as follows. Thin solid curve with open circles – no ozone or solar absorption; dashed curve with filled circles – ozone thermal infrared effects incorporated; thick solid curve – ozone infrared and solar absorption incorporated. The plain dashed curve gives the adiabat for the third case. The ground temperatures differ slightly between the cases, but the difference is not visible in the figure. All calculations were performed with the ccm radiation model.

is illuminated with 400 W/m² of incoming radiation, 25 W/m² of which is reflected back by Rayleigh scattering. The ground is assumed to be perfectly absorbing. The temperature profile is immaterial, since UV absorption is nearly independent of temperature for typical planetary temperatures. Of the sunlight that is not scattered, about 15 W/m² is absorbed within the ozone layer. This includes nearly all of the harmful UVB and UVC radiation. Given the low mass of this region of the atmosphere, the absorption gives rise to a considerable heating, which should substantially warm the stratosphere. There is some weak absorption within the troposphere, which is due to O_2.

A key question is whether ozone causes the temperature to increase with height in the stratosphere, as is seen in observations of the Earth's atmosphere. The right panel of Fig. 5.18 shows radiative-convective equilibrium calculations using the ccm code, computed with a dry atmosphere containing 300 ppmv of CO_2, in which convectively unstable layers are adjusted to the dry air adiabat. The profiles shown are in equilibrium with

$400\,W/m^2$ of incident solar radiation. Results with ozone are compared with a control case for a solar-transparent CO_2–air mixture. In the control case, the radiative-convective equilibrium temperature decreases monotonically with height, just as seen in the simulations of Chapter 4.

Because of the dual role of ozone as an IR and UV absorber, its effect on the temperature profile is complex. In the circumstances of this particular simulation, introducing just the effect of ozone on infrared absorption introduces a temperature increase with height in the lower stratosphere. This arises because the ozone is concentrated aloft, and absorbs at wavelengths that escape the CO_2 effects in the troposphere. This leads to an intense heating layer, which must warm up until it comes into equilibrium. Without solar absorption by ozone, however, the upper stratospheric temperature still declines sharply with height. Introducing the solar absorption warms the upper stratosphere considerably, and causes it to increase with height. It also results in a pronounced lowering of the tropopause.

The effect of ozone on stratospheric temperature is profound, but its effect on surface temperature is modest and largely invisible in Fig. 5.18. For the control case, the surface temperature is 295.48 K. It rises to 297.8 K when ozone infrared effects are introduced, owing to the greenhouse effect of ozone. However, when the ozone solar absorption is brought into the picture, the warming of the stratosphere allows the upper atmosphere to radiate better to space, and this brings the surface temperature back down to 295.22 K. Thus, the main climatic effects of ozone are in the stratosphere, though it is quite possible that the lowering of the tropopause would have repercussions for tropospheric climate. Further, the absence of ozone in the anoxic Early Earth atmosphere would have led to a much colder stratosphere, which is important to take into account in working out the chemistry of Titanlike stratospheric haze clouds.

Water vapor, CO_2, and CH_4 all absorb ultraviolet quite strongly for wavelengths shorter than 180 nm, but the only common atmospheric constituent that competes with ozone at longer wavelengths is SO_2. This gas is abundant in volcanic outgassing, but in oxygenated atmospheres it forms sulfates which are removed by rainout if the planet supports liquid water. On dry planets or planets without oxygen, SO_2 can build up to higher concentrations, but its status as an ultraviolet shield still hinges on atmospheric chemistry. The very fact that it absorbs ultraviolet so well tends to dissociate the molecule, and the question then is whether there are chemical pathways that can restore it. There is no question, however, that SO_2 is a molecule that holds very interesting prospects as a mediator of planetary climate evolution, especially in view of the fact that it is also a potent greenhouse gas.

5.11 ALBEDO OF SNOW AND ICE

To emphasize the generality and power of the physics of scattering that has been the central theme of this chapter, we will remark in closing that the very same physical principles account for the high albedo of snow and ice. The point of commonality with scattering from cloud droplets is that any discontinuity in index of refraction will lead to scattering. In the case of snow, the discontinuity is between the crystals of condensate and the voids between them. Given how much denser most solids are than gases, it matters little whether the voids are filled with air as on Earth, or filled with near-vacuum as they would be on Europa. Similarly, as long as the snow is made of a mostly transparent solid, it matters little just what it is made of. There are variations in index of refraction amongst different ices, but all are significantly different from unity. All snow is highly reflective, whether it be N_2

snow on Triton or CO_2 snow on Mars. For ice, the scatterers are air bubbles or brine pockets, and here it matters a bit more what the composition of the freezing fluid may be. To make gas bubbles in the frozen liquid, there must be a significant amount of some gas dissolved in the fluid, and to make brine pockets there must be some suitable solute (salt in the Earth case). The rate of freezing also makes a difference to the albedo of ice, since slow freezing allows gas to be rejected before bubbles form, leading to clear, low-albedo ice. Things get even more interesting if one allows for an admixture of absorbing particles (dust or soot) with the snow or ice.

Since albedo has such an important effect on planetary radiation budgets, the physics of snow and ice albedo is a critical field of play for radiative transfer. It can be treated using essentially the same techniques that have been introduced in this chapter.

5.12 WORKBOOK

5.12.1 Scattering basics

Problem 5.1 An atmosphere consists of a mixture of scattering particles with a gray gas having pressure-dependent absorption coefficient $\kappa(p) = \kappa_0 \cdot (p/p_0)$. The mass concentration of scatterers is constant, and the scatterers have scattering efficiency $Q_{sca} = 2$. The scatterers are spherical and have the density of water. They do not absorb. Compute the profile of optical thickness $\tau(p)$ and also the profile of single-scattering albedo $\omega_0(p)$. Tabulate τ and ω_0 on a discrete grid of pressure values. Using these arrays, make use of polynomial interpolation to write a function that returns $\omega_0(\tau)$.

Python tips: The interpolation is easily done using the `interp` class.

Problem 5.2 *Partial reflection at an interface*
When a beam of radiation hits the interface of a smooth transparent body such as an ocean, lake, or clear ice, part of the beam is transmitted but part is reflected. Partial reflection of this sort is important in determining the albedo of such surfaces, most commonly for liquid bodies. The albedo for such surfaces depends rather strongly on the angle of incidence of the beam. Maxwell's equations can easily be solved to yield the reflection at a discontinuity in index of refraction. For the case where the index of refraction is purely real, the answer is given by *Fresnel's formula*. For an unpolarized beam of light hitting a smooth surface with angle Θ_i relative to the normal, the albedo is $\alpha = \frac{1}{2}(R_s + R_p)$, where

$$R_s = \left(\frac{\sin(\Theta_t - \Theta_i)}{\sin(\Theta_t + \Theta_i)} \right)^2, R_p = \left(\frac{\tan(\Theta_t - \Theta_i)}{\tan(\Theta_t + \Theta_i)} \right)^2 \tag{5.61}$$

and Θ_t is the angle of the transmitted (i.e. refracted) beam relative to the normal, given by Snell's law. For a horizontal surface such as an ocean or lake, Θ_i is the zenith angle. Stellar illumination can generally be considered unpolarized.

Using Snell's law, compute the albedo for near-normal incidence – i.e. $\Theta_i \approx \Theta_t \approx 0$ – as a function of the jump in index of refraction. Compute the albedo for visible radiation at $0.6 \, \mu m$ striking a liquid methane lake at normal incidence, and at an angle of $45°$. Do the same for a liquid water lake. How different are the results for UVB?

Problem 5.3 Compute the Rayleigh scattering optical thickness for Titan's atmosphere, idealizing it as consisting of 1.5 bar of pure N_2. Do the calculation for visible, near-infrared, and UVB wavelengths. If Titan's stratospheric haze clouds were taken away, would Rayleigh scattering contribute significantly to Titan's albedo?

Problem 5.4 This problem involves computation of the properties of Mie scattering. The computation requires the summing of a complicated infinite series whose terms require evaluation of Bessel functions. You are not expected to program this summation yourself; Mie summation routines are widely available for many programming languages, and a few are provided as part of the online software supplement.

Re-do Fig. 5.7 for a range of different real indices of refraction $n_R > 1$. (Note that $n_R = 1.3$ for liquid water.)

Python tips: The Mie series summation is provided in the chapter script `Mie.py`.

Problem 5.5 The file `co2i4a.rfi.txt` in the Workbook datasets directory for this chapter contains the real and imaginary index of refraction for CO_2 ice as a function of wavelength. Assuming spherical CO_2 ice particles with a radius of 5 μm, use the Mie scattering solution to determine the scattering efficiency, absorption efficiency, and asymmetry factor as a function of wavelength, from the thermal infrared through the visible range. Plot the results. Also, save them in a file, since you will need them in Problem 5.17 below.

Python tips: Modify `Mie.py` so that it reads in the data and loops over wavelength.

5.12.2 Use and derivation of basic two-stream solutions

Problem 5.6 Consider a planet with a 5 bar atmosphere of N_2 under a surface gravity twice that of Earth. The surface of the planet is perfectly absorbing. There are no clouds, and you may ignore absorption of stellar radiation within the atmosphere. Compute the Rayleigh scattering albedo this planet would have in orbit around stars of spectral class M, G, and A. How does your answer change if the atmosphere consists of CO_2 instead? (Note that in that case the neglect of stellar absorption is less justifiable.)

Problem 5.7 Doubling the Earth's pre-industrial CO_2 concentration would reduce the *OLR* by about $4 \, W/m^2$, requiring the atmosphere to warm up to restore balance. Suppose instead we decide to increase the albedo so as to reduce the absorbed solar radiation by $4 \, W/m^2$. How much increase of albedo does this require? Assuming stratospheric sulfate aerosols to stay aloft for just one year, how much mass of SO_2 would have to be injected into the stratosphere annually to achieve this albedo increase? Assume that the aerosols form into droplets with a radius of 1 μm, and that each SO_2 molecule injected forms one H_2SO_4 molecule, which forms a droplet which is composed of half water and half sulfuric acid.

Note: Don't try this at home! It will put your planet in a very precarious state because the extra CO_2 stays in the atmosphere around a millennium, whereas the sulfate aerosols will fall out in a few years. The planet is thus at risk of being hit with a century's worth of global warming in just a few years, if the aerosol injection should ever cease.

Problem 5.8 Derive an expression for the albedo of a planet for which the incoming radiation is diffuse and there is no incoming direct beam radiation. Assume the ground to be perfectly absorbing ($\alpha_g = 0$) and the atmosphere to contain only conservative scatterers ($\omega_0 = 1$). The derivation is similar to that used to obtain the albedo for direct beam incoming radiation, except that the direct beam source terms are dropped and I_- is no longer assumed to vanish at the top of the atmosphere, but rather takes on a specified value $I_{-,\infty}$ which defines the incoming flux. Show that the expression is the expression for α'_a given in Eq. (5.39), and state why the diffuse albedo of the atmosphere is the same regardless of whether radiation is coming in from above or bouncing off the atmosphere from below.

Use the expression for the diffuse albedo to derive Eq. (5.42) for the conservative infrared scattering greenhouse effect. *Hint:* For conservative scattering, the transmission through the atmosphere is $1 - \alpha'_a$ times the incident flux.

Problem 5.9 Solve Eq. (5.40) for $I_+(0)$ and discuss the result. Using this as the bottom boundary condition for the conservative-scattering two-stream equations with direct beam incoming radiation, derive Eq. (5.41) for the planetary albedo allowing for surface reflection. The derivation is precisely the same as the derivation used to obtain Eq. (5.38), except for the change in the boundary condition on $I_+(0)$. Note that it is not necessary to re-do the derivation from scratch. Since the problem is linear, it suffices to add the portion of I_+ transmitted through the atmosphere to the previous expression for $I_{+,\infty}$, and use the result to obtain the modification to the albedo formula.

Problem 5.10 For a conservatively scattering atmosphere over a perfectly absorbing surface, plot the albedo as a function of the zenith angle, for an optically thick case with $\tau_\infty = 10$. Show results for various values of the asymmetry factor, and compare results between the hemi-isotropic, quadrature, and Eddington approximations. When do you get spurious negative values of the albedo? Derive an analytical criterion which defines the circumstances under which the albedo goes negative.

Problem 5.11 For a conservatively scattering atmosphere over a perfectly absorbing surface, plot the albedo as a function of the optical depth for a few different values of the zenith angle. Fix the zenith angle at $45°$, and do the same for a range of values of the asymmetry parameter. Discuss the exponential and algebraic stages of increase of albedo with optical thickness. For this problem, you may restrict attention to the hemi-isotropic approximation.

Problem 5.12 Generalize the albedo formula in Eq. (5.49) to allow for finite optical depth of the atmosphere. To keep things simple, you may assume that the asymmetry factor is zero and that the ground is perfectly absorbing.

Problem 5.13 *A diabolical mirror*
Suppose you could make a mirror that is perfectly transparent to radiation with wavenumbers greater than $5000 \, cm^{-1}$ but perfectly reflective at lower wavenumbers. (Optical supply companies actually do sell mirrors that are rather like this.) You put this mirror atop a well-insulated box that does not allow any heat to leak out, so that the only energy that leaves the box is in the form of radiation that escapes through the mirror. The interior of the box has unit emissivity at all wavenumbers.

You expose this box to sunlight having a Planck spectrum corresponding to a temperature of $6000 \, K$. Once the box reaches equilibrium, what is its temperature? Does the equilibrium temperature depend on the value of the radiation *flux* incident on the box – i.e. does it matter whether you do this experiment on Mercury or Mars?

Problem 5.14 Consider a planet with a pure N_2 atmosphere and a surface that is perfectly absorbing throughout the spectrum. High N_2 ice clouds form, which have zero infrared emissivity, but are highly reflective in the thermal infrared. The albedo in the thermal infrared is α_{ir}, and the albedo in the solar spectrum which illuminates the planet is α_{sw}. What is the equilibrium temperature of the planet? You may ignore gaseous Rayleigh scattering, and assume that the atmosphere is purely transparent in the infrared.

Problem 5.15 In the text, Eq. (5.45), describing the scattering greenhouse effect in the presence of absorption and emission, was derived assuming the atmosphere to be optically thick. Generalize this result to the case where the atmosphere is not assumed to be optically thick. Also, allow the ground temperature T_g to be different from the temperature T of the atmosphere, which should still be assumed isothermal. Write your solution in terms of an atmospheric transmission coefficient A_{trans} and an atmospheric emission coefficient A_{emiss}, so that the *OLR* takes the form $A_{trans}\pi B(\nu, T_g) + A_{emiss}\pi B(\nu, T)$. Show that your result reduces to the previous one in the optically thick limit. Discuss the behavior in the limit $\omega_0 \to 1$ with fixed (but large) τ_∞^*.

Note that this solution applies as well when the absorbing/scattering layer is a high isothermal cloud deck and there is a layer of transparent non-scattering atmosphere between the ground and the cloud base.

Problem 5.16 *Net effect of Venus clouds*
Assume that the clouds of Venus are made of spherical concentrated sulfuric acid droplets with a radius of $5\,\mu m$. How much mass of sulfuric acid per square meter must be present in the cloud deck in order to give Venus an albedo of 70% in the visible spectrum at $0.5\,\mu m$? You will need to make use of conservative Mie scattering results corresponding to the index of refraction of sulfuric acid; remember to include the effect of the asymmetry factor in your albedo calculation. You may ignore all absorption in the cloud deck, and assume any radiation that passes through is absorbed in the atmosphere or at the surface. (In reality some would be bounced back by Rayleigh scattering.)

Now let's estimate the effect of the same cloud deck on *OLR*. Venus clouds are in a high cloud deck near the top of the atmosphere, and can be thought of as a partially absorbing infrared mirror acting as a leaky lid. At some infrared wavelengths, Venus clouds are strongly absorbing and emitting, and do not scatter significantly. At others, scattering is significant. As an idealization, compute the diffuse infrared albedo of the cloud deck at wavelengths of $10\,\mu m$ and $2\,\mu m$, assuming the clouds to be conservative scatterers. Compare these with the visible albedo you estimated previously. Could the scattering greenhouse effect significantly offset the high visible albedo? At the opposite extreme, assume that the clouds are so strongly absorbing that the cloud deck acts like an ideal blackbody radiating at temperature T_{cloud}. What value would this temperature have to have in order for the emission to balance the absorbed solar radiation, given the visible albedo assumed above? Compare this to the observed temperature range of the cloud deck, which is 230–260 K. Could the absorption/emission greenhouse effect also significantly offset the high visible albedo?

Finally, let's get an idea of where in the infrared spectrum the clouds are acting as absorbers and where they are acting as scatterers. At $10\,\mu m$ the index of refraction is $(n_R, n_I) = (1.9, 0.46)$, and at $2\,\mu m$ it is $(1.4, 0.0048)$. Use Mie theory to determine the scattering and absorption efficiencies, and hence the single scattering albedo ω_0 for the cloud deck. You will also need the asymmetry factors. Then plug the result into Eq. (5.45), which describes the emission from a thick, isothermal cloud deck. When does the layer radiate almost like a blackbody? When does the scattering significantly reduce the emission? *Note*: in circumstances when the scattering significantly reduces the emission, the cloud layer will also be quite reflective to infrared hitting the cloud deck *from below*. You can get a more quantitative appreciation of this by examining the solution to Problem 5.15.

Problem 5.17 *Radiative effects of CO_2 ice clouds*
This lengthy (but hopefully rewarding) problem serves as an archetype for the determination of the net climate effect of high clouds which both scatter and absorb. The problem was originally motivated by the effect of CO_2 ice clouds on Early Mars, but the occurrence of such clouds is generic to a broad class of problems involving cold, thick CO_2 atmospheres. While the specifics are worked out for CO_2 optical properties, the techniques carry over without change to other substances. The model problem described below can readily be generalized to include a non-trivial gaseous atmosphere below the cloud deck.

The model problem we consider starts with a transparent non-scattering atmosphere above a surface with temperature T_g. The surface is an ideal blackbody in the thermal infrared spectrum, but has an albedo α_g for wavelengths shorter than $1\,\mu m$. The surface temperature T_g is assumed to be low enough that the surface emission at near-IR and shorter wavelengths can be neglected. To this base state, we add a cloud deck at the top of the atmosphere. This will reflect back some of the incoming stellar radiation, and will also reduce the outgoing thermal radiation. The cloud deck is isothermal with temperature T_{cloud}. It is composed of idealized spherical CO_2 ice particles with a radius of $5\,\mu m$. The mass path of particles in the cloud deck is $0.05\,kg/m^2$. All absorption and scattering is due to the particles, and the background gas is considered transparent.

First, let's compute the albedo for incoming stellar radiation, assuming that the clouds are conservative scatterers. Use Eq. (5.41) together with your Mie scattering results for CO_2 ice from Problem 5.5 to compute the albedo as a function of wavenumber. You may assume a fixed zenith angle of $45°$. Show results for $\alpha_g = 0$, $\alpha_g = 0.3$, and $\alpha_g = 0.7$. The second of these approximates the Rayleigh scattering from a thick CO_2 atmosphere, and the last is typical of a Snowball Earth case. Finally, for each case find the mean albedo weighted according to the Planck function for a G star with $T_\odot = 6000\,K$ and for an M star with $T_\odot = 3000\,K$. Examine the absorption efficiencies over the stellar spectrum from the Mie calculation. How valid is the assumption of conservative scattering for the G star? For the M star?

Next, we'll find the effect of the cloud on the *OLR*. In this case we use the absorption as well as scattering efficiencies from the Mie calculation, and do not assume conservative scattering. Use the Mie results to compute ω_0, and then plug all the results into the formula you derived in Problem 5.15 in order to compute the emission and transmission coefficients of the cloud deck. Plot these as a function of wavenumber, and also plot the *OLR* spectrum for $T_{cloud} = 150\,K$ and $T_g = 260\,K$. Compare the results with the Planck spectra corresponding to T_{cloud} and T_g. In what parts of the spectrum is the cloud acting primarily like a scatterer? In what parts is it acting like an absorber?

Finally, put in some number corresponding to Early Mars insolation. First, assume that T_g is in equilibrium with the absorbed solar radiation given $\alpha_g = 0.3$. Then, compute how much the solar absorption goes down when you introduce the cloud, and compare to how much the net *OLR* goes down. For the latter, assume the cloud temperature is $150\,K$. Does the cloud act to warm or cool the planet? How sensitive is your result to the assumed cloud temperature?

5.12.3 Numerical two-stream solutions

Problem 5.18 One of the new numerical techniques introduced in this chapter is the solution of two-point linear boundary value problems using the ODE method, in which the solution is built up from a pair of numerically determined fundamental solutions. Use this technique to write a routine to solve $d^2Y/dx^2 - aY = F(x)$ subject to $Y(0) = Y(1) = 0$, integrating the equations all the way from one boundary to the other. Carry out the integration for $F = \sin \pi x$ and show what goes wrong when a becomes large and positive (the analog of the optically thick case described in the text). Then modify your routine to use the local-ODE method described in the text, and compare the results with the exact analytic solution.

Problem 5.19 For this problem you will need an implementation of the numerical solution to the two-stream equations, using the piecewise ODE method described in Section 5.7. This problem only deals with the solution for a single absorption coefficient corresponding to a single frequency; a problem involving exponential sums will be done later. The implementation is not intrinsically difficult but getting all the algebra right for the matching conditions takes a bit of doing, and if you are not using the Python implementation, it is to be hoped that your instructor has done some of the groundwork for you.

First, take a case with constant T and ω_0, and check the numerical results against some of the conservative and non-conservative scattering solutions given in Section 5.6. Also check the results against your answer to Problem 5.12. In that case, re-run the numerical solution with a non-zero asymmetry factor and see how the result changes. Extend the code so that it computes heating profiles, and take a look at some of those. (Recall that in this kind of calculation you do *not* need to compute heating from the fluxes by taking a finite difference. You can get them accurately by simply evaluating the right hand side of the differential equation governing the fluxes.)

Finally, let's do something a little new with the calculation. We'll do a highly simplified estimate of the emission Titan would have at $100 \, \text{cm}^{-1}$ in the absence of its high haze clouds. Assume that the atmosphere is on the dry N_2 adiabat with surface temperature $95 \, \text{K}$ and a surface pressure of $1.5 \, \text{bar}$, under Titan gravity. The gaseous absorption is dominated by the N_2 continuum at this wavenumber. The absorption coefficient is $1.3 \cdot 10^{-6}$ at $100 \, \text{mb}$ and $100 \, \text{K}$, and can be considered to scale linearly with pressure. It is also temperature dependent, but you may ignore that effect, to keep things simple. To this atmosphere, add a liquid CH_4 cloud with uniform cloud mass concentration in the lowest $500 \, \text{mb}$ of the atmosphere. The cloud is made of spherical droplets with a radius of $10 \, \mu\text{m}$. They are purely scattering, with $n_R = 1.3$. Find the *OLR* at $100 \, \text{cm}^{-1}$ as a function of the cloud mass path. For selected values, calculate the profile of the heating rate, and discuss how it is affected by the cloud.

Python tips: The piecewise ODE solution is implemented in the Chapter Script
`TwoStreamScatter.py`.

Problem 5.20 Another bit of new numerics introduced in this chapter is the idea of solving two-point boundary value problems by reduction to a matrix linear algebra problem. As a simple illustration of this technique, apply it to the solution of the boundary value problem stated in Problem 5.18. Discretized on a grid with uniform spacing Δx, the problem becomes

$$(Y_{j+1} - 2Y_j + Y_{j-1}) - a\Delta x^2 Y_j = F_j \Delta x^2 \tag{5.62}$$

using centered differences. This corresponds to a tridiagonal linear system.

Read about tridiagonal solvers in *Numerical Recipes*, and implement a solution of this system using a tridiagonal algorithm. Examine the solution for $F(x) = \sin \pi x$, and look at the convergence as the number of points in the grid is increased. Do the same for the case $F_j = 1$ for $j = m$ and $F_j = 0$ otherwise. In both cases, pay particular attention to what happens as a is made large.

Python tips: A tridiagonal solver is implemented in the Chapter Script `tridiag.py`.

Problem 5.21 Write a program which sets up the matrix describing the matrix version of the two-stream equations, given by Eq. (5.55). Read about band-diagonal solvers in *Numerical Recipes*. Implement the `bandec` algorithm, or learn to use a band-diagonal solver in the programming environment of your choice. Use this algorithm to solve the two-stream equations for the simple case in which ω_0 is constant, the atmosphere is isothermal, and there is no incident beam. Compare the numerical solution with the analytical solution to this problem, paying particular attention to what happens in the optically thick limit. Then try a case where $T(\tau)$ varies with height, and check that the fluxes have the correct behavior in the optically thick limit.

Problem 5.22 Explain the joke in the title of Section 5.9. Is it funny? If not, why not?

Problem 5.23 Evaluate the sum (5.56) for a single band using a Malkmus transmission function. Compare the *OLR* to what it would be if there were no reflecting surfaces. Attempt to derive a similar series for the case in which there is a third surface, assumed partially reflecting, at the midpoint of the atmosphere.

Problem 5.24 *Polar* CO_2 *clouds on Snowball Earth*
This problem requires an implementation of the numerical two-stream solution incorporating band-averaging via exponential sums. If you already have an implementation of the basic numerical solution for a single wavenumber and absorption coefficient, the extension to the exponential sum case is quite straightforward. The exponential sum tables you need are in the dataset subdirectory `Chapter4Data/ExpSumTables`.

Consider a Snowball Earth case with an atmosphere consisting of 100 mb of CO_2 mixed with dry air. In the polar winter regions, climate model simulations tend to show that convection shuts off and transient radiative cooling effects lead to an atmosphere that is approximately isothermal. Suppose that the atmosphere is isothermal with temperature T_a, and that the surface has temperature T_g. It is possible for the air to become so cold that CO_2 condensation occurs. To explore the effects of this, suppose that there is a CO_2 ice cloud made of spherical particles with a 5 μm radius, with uniform mass concentration q in the lowest 100 mb of the atmosphere. Using the exponential sum two-stream radiation calculation, compute the effect of this cloud on the spectrum of *OLR*, for various value of q. Discuss the effect of the cloud on the net *OLR*. You may pick any reasonable values of T_a and T_g that will help you bring out the essential features of the problem. If you are energetic, you can use the actual Mie theory scattering results to determine the wavelength-dependent scattering properties of the cloud, but to get at the basic behavior, just using a constant scattering cross-section estimated crudely from the Mie results, and ignoring cloud absorption, should be sufficient.

Python tips: The basic exponential sum calculation with scattering is implemented in the Chapter Script `TwoStreamScatterEsum.py`, which can be modified to solve this problem.

5.12.4 Absorption of stellar radiation

Problem 5.25 *Albedo of dense* CO_2 *atmospheres*
Venuslike atmospheres should be common throughout the Universe, for planets which have lost their water but still have accumulated enough carbonates to allow the creation of a thick CO_2 atmosphere. Such planets may or may not have a sulfuric acid cloud deck like that of Venus, and so it is important to understand what the albedo would be without clouds. This problem was considered in the text, but without incorporating the joint effects of gaseous absorption and Rayleigh scattering. Now you will bring together both effects.

Consider a planet with a pure CO_2 atmosphere, which has zero surface albedo throughout the spectrum. The surface pressure is p_s, and the surface gravity is $10\,m/s^2$. Carbon dioxide has significant absorption for wavenumbers below about $15\,000\,cm^{-1}$, and in this part of the spectrum you need to use exponential sums to compute the albedo. However, it is not necessary to numerically solve the two-stream equations for this problem, since you can use the analytical solution for the albedo given by the generalization of Eq. (5.49) done in Problem 5.12. You use the exponential sum coefficients to form a weighted sum of calculations for individual absorption coefficients within each band; then you sum results over bands, as usual. You should incorporate pressure broadening and the CO_2 continuum in your calculation of the absorption coefficient, but you may ignore temperature scaling, so as to keep things simple.

The tables you need are found in the Workbook datasets directory `Chapter4Data/ExpSumTables`.

For wavenumbers above $15\,000\,cm^{-1}$ you can consider CO_2 to be non-absorbing. In this part of the spectrum, you can compute the contribution to the albedo using the analytical Rayleigh scattering solution, but the albedo is wavenumber-dependent and you will need to numerically integrate it weighted by the spectrum of stellar insolation.

Using this procedure, determine the albedo of the planet for both a G star and M star case, for $p_s = 1$, 10, and 100 bar. (Note that the linear pressure scaling of the absorption coefficients may seriously underestimate the absorption in the higher pressure cases.) How much does Rayleigh scattering affect the near-IR contribution to the albedo?

Python tips: You can do this problem by modifying the script `TwoStreamScatterEsum.py`, replacing the numerical solution of the two-stream equations by the analytical albedo calculation. You will also need add in code to do the Planck-weighted sum over wavenumbers.

Problem 5.26 The dataset subdirectory `UVabsorption` contains ultraviolet absorption cross-sections for ozone, CO_2, and SO_2 as a function of wavelength. The cross-sections are given in units of cm^2 per molecule. Read these in, convert to m^2/kg and plot them. Determine and plot the ultraviolet absorption optical depth as a function of wavelength for the model ozone profile given in the text. Do the same for a 2 bar pure CO_2 atmosphere on a planet with surface gravity of $10\,m/s^2$. Do the same for a planet with a 2 bar N_2 atmosphere mixed with 10 ppmv of SO_2, and with 100 ppmv of SO_2.

Problem 5.27 *Gray radiative-convective equilibrium with solar absorption*
In Section 4.3.5 we computed the pure radiative equilibrium for a gray gas, including the effects of atmospheric solar absorption. Using techniques similar to those introduced in Problem 4.28, extend these solutions to the radiative-convective case, in which solar absorption at the surface can create a troposphere which is adjusted to a dry adiabat. Time-stepping is not required for the solution of this problem, since you can iterate on

the tropopause height until equilibrium is achieved. However, since the absorbed solar radiation (which must be balanced by *OLR*) affects the temperature profile, you can no longer do the problem by fixing the ground temperature at the desired value and computing the temperature of the rest of the atmosphere. One approach is to guess the ground temperature, compute the temperature structure and the *OLR*, and then vary the ground temperature until top-of-atmosphere balance is achieved. A better approach is to start from the radiative equilibrium solution in the stratosphere which satisfies top-of-atmosphere balance, guess a tropopause height, impose temperature continuity there, and then compute the radiative fluxes in the troposphere. You then complete the calculation by adjusting the tropopause height until the upward infrared flux is continuous across the tropopause. This is equivalent to the condition that the flux into the troposphere from above be equal to the upward infrared exiting the troposphere.

Explore the behavior of this system. When does solar absorption substantially cool the troposphere? When does it eliminate the troposphere entirely? When does increasing the infrared absorption coefficient simultaneously warm the stratosphere but cool the troposphere?

Problem 5.28 Determine the radiative-convective equilibrium for an atmosphere consisting of 1000 ppmv of CH_4 in 1 bar of N_2 subject to a surface gravity of $10 \, m/s^2$. The surface is perfectly absorbing throughout the spectrum. You may ignore Rayleigh scattering. Compare results with and without atmospheric stellar absorption, for both G and M star cases, with stellar constant $1400 \, W/m^2$ at the orbit of the planet. To what extent does stellar absorption by CH_4 inhibit its effectiveness as a greenhouse gas?

To do this problem, you will need to use an implementation of the homebrew exponential sums radiation code. The necessary exponential sum table for CH_4 is found in the Workbook dataset directory `Chapter4Data/ExpSumTables`. The problem is solved by time-stepping to equilibrium, with adjustment of ground temperature as described in the text.

Python tips: Modify `RadConvEqTrop.py`, which time-steps to radiative-convective equilibrium for a fixed incident solar flux, and has an option to incorporate atmospheric solar absorption.

5.13 FOR FURTHER READING

For general background on scattering of electromagnetic radiation, including basic concepts of refraction and diffraction, the reader is referred to

- Jackson JD 1998: *Classical Electrodynamics.* Wiley.

A complete discussion of Rayleigh scattering, including the derivation of polarizability factors, can be found in Section 7.3 of

- Goody RM and Yung Y 1995: *Atmospheric Radiation,* Oxford University Press.

The derivation of the infinite series describing Mie scattering is very intricate and involves considerable facility with manipulation of special functions and vector spherical harmonics. It is a masterful solution, but the effort required to fully understand the derivation is not commensurate with the rather limited number of additional problems one can solve using the techniques. A reasonable compromise is to read through the much simpler case of scattering from a conducting sphere, which involves many of the same concepts in a less challenging setting. This calculation is given in *Classical Electrodynamics*, referenced above. For the truly devoted, or at least those looking for a usable description of the series solution, the full Mie scattering calculation is worked out in Chapter 4 of

- Bohrens CF and Huffman DR 2004: *Absorption and Scattering of Light by Small Particles.* Wiley-VCH (available online through Wiley Interscience).

Appendix A of this book gives the full text of a computer program for evaluating the solution. Many implementations of this algorithm can be found on the web by searching for bhmie. Chapters 5 and 6 have a useful discussion of Rayleigh scattering, which can be consulted as an alternative to Goody and Yung.

For optical properties of condensed substances see

- **Liquid water, real refractive index** – Schiebener P *et al.* 1990: Refractive-index of water and steam as function of wavelength, temperature and density, *J. Phys. Chem. Ref. Data* **19**, 677–717.
- **Liquid water, absorption properties** – Hale GH and Querry MR 1973: Optical constants of water in the 200 nm to 200 μm wavelength region, *Appl. Opt.* **12**, 555–563.
- **Water ice** – Warren SG 1984: Optical constants of ice from the ultraviolet to the microwave, *Appl. Opt.* **23**, 1026–1225.
- **CO_2 ice** – Warren SG 1986: Optical constants of carbon dioxide ice, *Appl. Opt.* **25**, 2650–2674.
- **CO_2 ice (updates Warren)** – Hansen GB 2005: Ultraviolet to near-infrared absorption spectrum of carbon dioxide ice from 0.174 to 1.8 μm. *J. Geophys. Res.* **110**, doi:10.1029/2005JE002531.
- **H_2SO_4** – Myhre CE *et al.* 2003: Spectroscopic study of aqueous H_2SO_4 at different temperatures and compositions: variations in dissociation and optical properties. *J. Phys. Chem. A* **107**, 1979–1991.
  Data via http://www.kjemi.uio.no/09_spekt/Atmosfaere/OPA/OPA.html
- **H_2SO_4, esp. on Venus** – Palmer KF and Williams D 1975: Optical constants of sulfuric acid; application to the clouds of Venus? *Appl. Opt.* **14**, 208–219.
- **Liquid CH_4** – Martonchik JV and Orton GS 1994: Optical constants of liquid and solid methane. *Appl. Opt.* **33**, 8306–8317.

Many of these sources provide functional fits to the data that are convenient to use in computations. Most also give data in tabulated form, which can in some cases be found in digital form through various websites or from the authors. A selection of the data has been provided in the Workbook datasets online supplement for this chapter.

The importance of the scattering greenhouse effect for Early Mars climate was first discussed in

- Forget F and Pierrehumbert RT 1997: Warming Early Mars with carbon dioxide clouds that scatter infrared radiation, *Science* **278**, 1273–1276.

The effect of sulfuric acid cloud cycles on Venus climate evolution has been discussed in

- Bullock MA and Grinspoon DH 2001: The recent evolution of climate on Venus, *Icarus* **150**, 19–37.

The treatment of the radiative effect of removing the clouds of Venus is highly simplified in this work, and there is much room for further study.

The *MPI-Mainz-UV-VIS Spectral Atlas of Gaseous Molecules* contains a comprehensive database of ultraviolet and visible absorption cross-sections. Data and documentation can be downloaded from the site

- www.atmosphere.mpg.de/spectral-atlas-mainz

Radiative transfer on gas giants is discussed in

- Guillemot T *et al.* 1994: Are the giant planets fully convective? *Icarus* **112**, 337–353.

CHAPTER SIX

The surface energy balance

6.1 OVERVIEW

This results of this chapter are pertinent to a planet with a distinct surface, which may be defined as an interface across which the density increases substantially and discontinuously. The typical interface would be between a gaseous atmosphere and a solid or liquid surface. In the Solar System, there are only four examples of bodies having both a distinct surface and a thick enough atmosphere to significantly affect the surface temperature. These are Venus, Earth, Titan, and Mars; among these, the present Martian atmosphere is so thin that it only marginally affects the surface temperature, though this situation was probably different early in the planet's history when the atmosphere may have been thicker. Although thin atmospheres have little effect on the surface temperature, the atmosphere itself can still have interesting behavior, and the flux of energy from the surface to the atmosphere provides a crucial part of the forcing which drives the atmospheric circulation. This is the case, for example, for the thin nitrogen atmosphere of Neptune's moon Triton. Apart from the examples we know, it is worth thinking of the surface balance in general terms, because of the light it sheds on the possible nature of the climates of extrasolar planets already detected or awaiting discovery.

The exchange of energy between the surface and the overlying atmosphere determines the surface temperature relative to the air temperature. It also turns out that it determines the exchange of mass between the surface and the atmosphere (as in sublimation from a glacier or evaporation from an ocean, lake, or swamp). Because outer space is essentially a vacuum, the only energy exchange terms at the top of the atmosphere are radiative. At the surface, energy can be exchanged by means of fluid motions as well as by radiation.

The atmospheric gas in direct contact with the surface must have the same velocity as the surface; because the surface material is so much denser (and in the case of a solid so much more rigid) than the atmosphere, the atmospheric flow must typically adjust to the presence of the surface over a rather short distance. The resulting strong shears lead to random-seeming complex turbulent motions sustained by the kinetic energy of the shear flow near the boundary. We may subdivide the atmosphere into the *free atmosphere* – which is sufficiently far above the surface to be little affected by turbulence stirred up at the surface, and the *planetary boundary layer* (PBL for short) where the transfer of heat, chemical substances, and momentum is strongly affected by surface-driven turbulence. We may further identify the *surface layer*, which is the thin portion of the PBL near the ground within which all the vertical fluxes may be considered independent of height.

Given that the whole troposphere is created by convection – which is a form of buoyancy-driven turbulence – it is not at once clear why the PBL should exist as a distinct entity from the troposphere in general. The main reason one can typically distinguish the PBL is that mechanically driven turbulence is more trapped near the surface than is buoyancy-driven turbulence, and also has distinct time and space scales. On the present Earth, the effect of moisture is also important in maintaining the distinction, since moisture gives deep convection an intermittent character: most of the troposphere-forming mixing takes place in rare convective events, while most of the troposphere remains quiescent most of the time. Because dry (i.e. non-condensing) convection is typically shallower than moist convection, in planets which have both forms the dry convection can often be treated as part of the boundary layer. This is the case for Earth, and likely for other planets with a surface and an atmosphere in which latent heat release is important (Titan and perhaps Early Mars being the only other known examples so far). For planets like present Mars or Venus, where dry convection is the *only* form of convection, it is less clear that the PBL can be productively distinguished from the troposphere in general. Even in such cases, though, one can identify a constant-flux surface layer; the depth of the surface layer typically ranges from a few meters to a few tens of meters.

As in previous chapters, we let T_g be the temperature of the planet's surface. Previously, we used T_{sa} to denote the temperature of the air in immediate contact with the ground, but now we modify the definition somewhat, and allow T_{sa} to be the temperature at the top of the surface layer, assuming the air at the bottom of the surface layer (which is in contact with the ground) has the same temperature as the ground itself. A model of the PBL is necessary to connect T_{sa} to the temperature of the lowest part of the free troposphere. For many purposes, we can dispense with the PBL and patch the surface layer directly to the free troposphere. We shall adhere to this expedient in most of the following discussion.

Now let's discuss, in general terms, how the surface budget affects the climate. The state of the atmosphere and the ground must adjust so that the top-of-atmosphere and surface budgets are simultaneously satisfied. If the atmosphere is optically thick in the longwave spectrum, the top-of-atmosphere budget becomes decoupled from the surface budget, since radiation from the ground and lower portions of the atmosphere is absorbed before it escapes to space. In this case, the determination of T_g can be decomposed into two stages carried out in sequence. First one determines T_{sa} by adjusting this temperature until the top-of-atmosphere balance is satisfied, assuming that the rest of the troposphere is related to T_{sa} through the appropriate dry or moist adiabat. Then, once T_{sa} is known, one makes use of a model of the surface flux terms to determine the value of T_g which balances the surface budget with T_{sa} fixed at the previously determined value. This can be done without reference to the top-of-atmosphere budget, since the *OLR* is independent of T_g in the optically thick limit.

If the atmosphere is very optically thin in the longwave spectrum, the *OLR* is determined entirely by the ground temperature and ground emissivity. Further, since an optically thin atmosphere radiates very little, the only way the atmosphere itself loses energy is through turbulent exchange with the surface. Suppose first that the atmosphere is transparent to solar radiation. In that case, *in equilibrium* the net turbulent exchange between atmosphere and surface must vanish, since otherwise the atmospheric temperature would rise or fall, there being nothing to balance a net exchange. In consequence, the ground temperature will be just what it would have been without an atmosphere despite the presence of turbulence. In this case, one determines the ground temperature as if the planet were in a vacuum, the top-of-atmosphere budget is automatically satisfied, and then, once T_g is known, the surface budget is used to determine T_{sa}, and (via an adiabat) the rest of the atmospheric structure. It is exactly the inverse of the process used in the optically thick case. In fact, the basic picture is little altered even if the atmosphere absorbs solar radiation. In that case, the requirement that the atmosphere be in equilibrium implies that any solar radiation absorbed in the atmosphere be passed on to the surface by turbulent fluxes. The result is much the same as if the solar radiation were absorbed directly by the surface; one does the ground temperature calculation as before, but simply remembers to add the atmospheric absorption to the solar energy directly absorbed by the ground. It should be kept in mind that these considerations apply only in equilibrium. Even an optically thin atmosphere can affect the transient behavior of the surface (e.g. in the diurnal or seasonal cycle), as will be discussed in Chapter 7.

In the intermediate case, where the atmosphere is neither optically thick nor thin, one must solve for T_{sa} and T_g simultaneously, so as to find the values that satisfy both the top-of-atmosphere and surface energy budgets. We'll do this crudely in the present chapter through the introduction of atmospheric transparency factors. Generally speaking, though, when the atmosphere is not too optically thin, the surface budget will have some effect on the temperature of the ground. For Earth this temperature is of interest because the ground is where people live and where much of the biosphere resides as well; for a broad range of planets actual or hypothetical the ground temperature also affects chemical processes which determine atmospheric composition, as well as the melting of ices at the surface. We shall see, however, that it is a fairly common circumstance that the surface fluxes effectively constrain the ground temperature to be nearly equal to the overlying air temperature, so that the climate can be determined without detailed reference to how the surface balance works out.

6.2 RADIATIVE EXCHANGE

6.2.1 Shortwave radiation

The surface receives radiant energy in the form of shortwave (solar) and longwave (thermal infrared) flux. The shortwave flux incident on the surface is equal to the shortwave flux incident at the top of the atmosphere, diminished by whatever proportion is absorbed in the atmosphere or scattered back to space. We will call the shortwave flux incident on the ground S_g. The shortwave flux absorbed at the surface is then $(1 - \alpha_g)S_g$, where α_g is the albedo of the ground. The flux S_g is affected by clouds, atmospheric absorption, and atmospheric Rayleigh scattering.

6.2.2 The behavior of the longwave back-radiation

The longwave radiation striking the surface is the infrared *back-radiation* emitted by the atmosphere, which was discussed in Chapter 4. The back-radiation depends on both the greenhouse gas content of the atmosphere – which determines its emissivity – and the temperature profile. When the atmosphere is optically thick in the infrared, most of the back-radiation comes from the portions of the atmosphere near the ground, whereas in an optically thinner atmosphere it comes from higher – and generally colder – parts of the atmosphere, and is correspondingly weaker. If the atmosphere is very optically thin, the back-radiation will be weak regardless of the atmospheric temperature profile, simply because an optically thin atmosphere radiates very little. As in Chapter 4, $I_{-,s}$ will denote the back-radiation integrated over all longwave frequencies. The absorbed infrared flux is then $e_g I_{-,s}$, where e_g is the longwave emissivity of the ground. The ground loses energy by upward radiation at a rate $e_g \sigma T_g^4$. Thus, the net infrared cooling of the ground is

$$F_{g,ir} = e_g \cdot \left(\sigma T_g^4 - I_{-,s} \right). \tag{6.1}$$

According to Eq. (4.21), $I_{-,s}$ approaches σT_{sa}^4 when the atmosphere is optically thick throughout the infrared. In order to characterize the optical thickness of the atmosphere, we introduce the effective low-level atmospheric emissivity e_a, defined so that $I_{-,s} = e_a \sigma T_{sa}^4$. The value e_a depends on the temperature profile as well as the optical thickness, as illustrated by Eq. (4.21) in the optically thick limit. When $T_g = T_{sa}$ the surface cooling becomes $e_g \cdot (1 - e_a) \sigma T_g^4$, which vanishes in the optically thick limit where $e_a \to 1$. Let $e^* = (1 - e_a)$; this is the effective emissivity of the ground when the air temperature equals the ground temperature. If the air temperature is not too different from the ground temperature, we may linearize the term σT_g^4 about $T_g = T_{sa}$, which results in

$$F_{g,ir} = e_g \cdot e^* \sigma T_{sa}^4 + \left(4\sigma T_g^3 e_g \right)(T_g - T_{sa}). \tag{6.2}$$

From this equation we can define the infrared coupling coefficient, $b_{ir} = 4\sigma T_g^3 e_g$. When b_{ir} is large, a small temperature difference leads to a large radiative imbalance, and it is correspondingly hard for the ground temperature to differ much from the overlying air temperature. Later, we will derive analogous coupling coefficients for the turbulent transfers.

Figure 6.1 shows how e^* varies with temperature for an Earthlike atmosphere in which the only greenhouse gases are water vapor and CO_2, with the water vapor relative humidity held fixed as temperature is changed. In the moist case (left panel), e^* rapidly approaches zero as the temperature increases; this is because of the increasing optical thickness caused by the increase of water vapor content with temperature (owing to the fixed *relative* humidity). Increasing the CO_2 content also increases the optical thickness, correspondingly reducing e^*. At low temperatures, the CO_2 effect dominates, because there is little water in the atmosphere. However, by the time Earthlike tropical temperatures (300 K) are reached, water vapor is sufficient to make e^* essentially zero all on its own without any help from CO_2. To underscore the relative role of CO_2 and water vapor, results for a dry atmosphere are given in the right hand panel of Fig. 6.1. The value of e^* still goes down with temperature, because temperature affects the opacity of CO_2; however, the decline is much less pronounced than it is in the moist case. Even with 100 mb of CO_2 in the atmosphere, e^* falls only to about 0.4 at 320 K, and significant infrared cooling of the surface is possible. In sum, CO_2 by itself is relatively ineffective at limiting surface cooling, but the opacity of water vapor can practically eliminate surface infrared cooling at temperatures above 300 K, unless the ground temperature significantly exceeds the air temperature.

Though the results of Fig. 6.1 were computed for Earth conditions, they give a fair indication of the extent of surface radiative cooling on other planets whose atmospheres consist of an infrared-transparent background gas mixed with CO_2 and with water vapor fed through exchange with a condensed reservoir. Through the hydrostatic relation, the surface gravity g

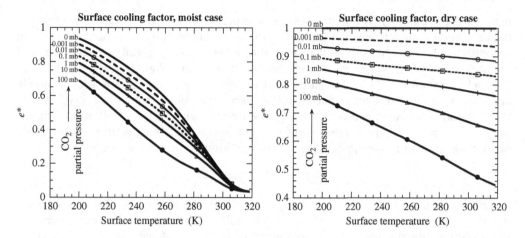

Figure 6.1 Surface cooling factor e^* for a 1 bar nitrogen–oxygen atmosphere with water vapor and CO_2. The surface gravity is that for Earth. In the left panel, the calculations were done with free tropospheric relative humidity set to 50%, and low-level relative humidity set to 80%. Results in the right panel are for a dry atmosphere (zero relative humidity, but with the temperature profile kept the same as in the moist case). In both cases, the numbers on the curves indicate the partial pressure of CO_2 in mb. The calculations were carried out with the ccm radiation model.

affects the mass of greenhouse gas represented by a given partial pressure; the lower the g the greater the mass (and hence the greater the optical thickness), and conversely. This is especially important in the case of water vapor, since in that case the partial pressure is set by temperature, through the Clausius–Clapeyron relation. Thus, for a "large Earth" with high g, it takes a higher temperature to make the lower atmosphere optically thick. For example, calculations of the sort used to make Fig. 6.1 show that with 1 mb of CO_2 in a moist atmosphere having temperature 280 K, increasing g to $100\,\mathrm{m/s^2}$ increases e^* to 0.507' (vs. 0.303 for $g = 10\,\mathrm{m/s^2}$). In the same atmospheric conditions, e^* falls to 0.102 for a "mini-Earth" with $g = 1\,\mathrm{m/s^2}$. Increasing the pressure of the transparent background gas makes the greenhouse gases more optically thick through pressure broadening. With $g = 10\,\mathrm{m/s^2}$, increasing the background air pressure to 10 bar has a very profound effect, lowering e^* to 0.094. Reducing the air pressure below 1000 mb should in principle increase e^*, but in fact it is found to reduce it very slightly, to 0.299. It appears that the reduction in opacity from less pressure broadening is offset by the changes in the moist adiabat that occur when the air pressure is reduced: the latent heat of condensation is spread over less background gas, so the temperature aloft is greater and hence the air aloft contains more water.

Without water vapor, it takes an enormous amount of CO_2 to make the lower atmosphere optically thick. This case is relevant to Venus and Venuslike planets, which may be defined as planets having a dry rocky surface and a thick, dry CO_2 atmosphere. The near-surface radiative properties can be determined using the homebrew exponential sum radiation code; for better accuracy, we did this based on exponential sum tables computed for the surface temperature and pressure conditions under consideration, so as to minimize errors due to pressure and temperature scaling of absorption coefficients. Line parameters in the HITRAN database were used to compute the absorption coefficients. For a 1 bar pure CO_2 atmosphere on a planet with the gravity of Venus, this calculation yields $e^* = 0.43$ when the surface temperature is 300 K, falling further to under 10^{-6} for pressures of 10 bar or more. The sharp decline in surface cooling between 1 bar and 10 bar arises from line broadening, which fills in the window regions in the CO_2 absorption spectrum. At the 727 K surface temperature of Venus, the surface emission shifts toward higher wavenumbers where CO_2 does not absorb as well, but the high temperature increases the line strengths while the high pressure causes the absorption to further spill over into the windows. Hence, at 92 bar and 727 K the calculation still yields a value of e^* that is under 10^{-5}, even assuming CO_2 to be completely transparent for wavenumbers higher than $10\,000\,cm^{-1}$. The estimates of e^* at high pressure should be viewed with some caution, however. At high pressures, the contribution of each line to spectral distances far removed from the line center is considerable, and there is much uncertainty about the appropriate form of line shape to be used in computing this far-field contribution. We will see shortly that it actually makes some difference to the climate of Venus whether e^* is zero or 0.05.

In the opposite extreme, atmospheres like the thin Martian atmosphere have very little effect on the surface radiative cooling. For a Martian CO_2 atmosphere on the dry adiabat with 7 mb of surface pressure, $e^* = 0.9$ at 220 K, falling only modestly to 0.86 at 280 K. Recall that, per square meter of surface, Mars actually has vastly more CO_2 in its atmosphere than the Earth has at present; allowing for the difference in gravity, a 7 mb pure CO_2 atmosphere on Mars has as much CO_2 per unit area as an Earth atmosphere with a CO_2 partial pressure of 18.5 mb at the ground. In comparison, the present Earth's atmosphere has a partial CO_2 pressure of a mere 0.38 mb (in 2006). The weak emission of the Martian atmosphere is due to the low total pressure, which yields little collisional broadening of the emission lines. If the same amount of CO_2 on Mars at present were mixed into a 1 bar atmosphere of N_2, the effective surface emissivity e^* falls to 0.75 at 230 K and 0.69 at 280 K.

Among common greenhouse gases, water vapor appears unique in its ability to make the lower atmosphere nearly opaque to infrared, even at concentrations as low as a few percent.

Clouds made of an infrared-absorbing substance such as water act just like a very effective greenhouse gas in making the lower atmosphere optically thick (making e_a close to unity). It takes very little cloud water to make the lower atmosphere act essentially like a blackbody. Infrared-scattering clouds in the surface layer, like those made of methane or CO_2, have a very different effect on the back-radiation. First, they shield the surface from back-radiation coming down from the upper atmosphere by reflecting it, rather than absorbing it; hence the shielding is accomplished without the cloud layer heating up in response to absorption. More importantly, the downwelling radiation from a reflective cloud is determined by the upwelling ground radiation incident upon it; the resulting back-radiation is then determined by the ground temperature, and is independent of the cloud temperature. As a result, the surface cannot increase its longwave cooling by warming up until it is substantially warmer than the atmosphere. This gives a scattering cloud great potency to increase the ground temperature, if it allows sufficient solar radiation to get through to the ground. Either IR-reflecting or absorbing clouds are different from a greenhouse gas, in that they also strongly increase the shortwave albedo.

6.2.3 Radiatively driven ground–air temperature difference

Now we consider the equilibrium temperature difference between the ground and the overlying air that would be attained in the absence of turbulent heat exchange. This temperature difference is important in determining the extent to which convection is driven from below, by positive buoyancy generated near the ground. We have already discussed this issue for the case in which the atmosphere itself is in pure radiative equilibrium (See Sections 3.6, 4.3.4, and 4.7). Our concern now is with what happens once convection has set in and altered the atmospheric temperature profile.

If the only heat exchange is radiative, the surface budget reads

$$(1 - \alpha_g)S_g + \sigma e_a e_g T_{sa}^4 = \sigma e_g T_g^4. \tag{6.3}$$

Since the second term on the left hand side is positive, the infrared back-radiation always drives T_g to exceed its no-atmosphere value. However, this value might be more or less than T_{sa}. To examine this difference, we linearize the surface radiation budget about T_{sa}, which results in

$$(1 - \alpha_g)S_g = \sigma e^* e_g T_{sa}^4 + b_{ir} \cdot (T_g - T_{sa}). \tag{6.4}$$

The linearized form can be immediately solved for the ground–air temperature difference. Substituting the expression for b_{ir}, we find

$$
\begin{aligned}
(T_g - T_{sa}) &= \frac{1}{4} \frac{(1 - \alpha_g)S_g}{\sigma e_g T_{sa}^4} T_{sa} - \frac{1}{4} e^* T_{sa} \\
&= \frac{1}{4} \left(\left(\frac{T_0}{T_{sa}} \right)^4 - e^* \right) T_{sa}
\end{aligned}
\tag{6.5}
$$

where T_0 is the no-atmosphere ground temperature, which satisfies $\sigma e_g T_0^4 = (1 - \alpha_g)S_g$. For planets with an optically thick lower atmosphere, the ground temperature can get extraordinarily hot relative to the air temperature if there are no turbulent fluxes to help carry away the heat. The first term on the right hand side of the upper line of Eq. (6.5) is large in tropical Earth conditions. For $(1 - \alpha_g)S_g = 300\,\mathrm{W/m^2}$ and $T_{sa} = 300\,\mathrm{K}$ with $e_g = 1$, it has the value 49 K. But in tropical Earth conditions, e^* is on the order of 0.1, so the second term subtracts little (15 K for $T_{sa} = 300\,\mathrm{K}$). Thus, the ground temperature is 34 K warmer than the overlying air temperature, or 334 K. In reality, the sea surface temperature hardly ever gets more than a few degrees warmer than the free-air temperature in the Earth's tropics.

Ironically, for planets which have such a strong greenhouse effect that the low-level air temperature is much larger than the no-atmosphere value, $T_g - T_{sa}$ can be quite small even if the lower atmosphere is optically thick enough to make $e^* \approx 0$, and even in the absence of turbulent heat fluxes. This conclusion is readily deduced from the factor multiplying T_{sa} in the second line of Eq. (6.5). For example, Venus has a small T_0 because of the highly reflective clouds which keep sunlight from reaching the surface, yet has a high T_{sa} because of its strong greenhouse effect. In consequence, this factor is only 0.0024 for Venus in the limit $e^* = 0$, whence $T_g - T_{sa} \approx 1.8\,\mathrm{K}$. If some CO_2 window region not reproduced by the procedure we used to estimate the surface radiative cooling on Venus allowed e^* to increase modestly to 0.05, then the ground temperature would actually become slightly *cooler* than the overlying air temperature, leading to a low-level temperature inversion and cutting off near-surface convection. For planets like Venus, the surface radiation budget is dominated by infrared back-radiation, and the comparatively feeble sunlight has little power to drive the ground temperature to values much greater than the overlying air temperature. It is situations like the Earth's tropics, which combine an optically thick lower atmosphere (due to water vapor in our case) with a rather modest greenhouse effect, where the radiation budget tends to drive the ground temperature to large values relative to that of the overlying air.

When the lower atmosphere is optically thin, as in the case of present Mars, the ground–air temperature difference cannot be determined without considering the top-of-atmosphere balance simultaneously with the surface balance. For an optically thin atmosphere, Eq. (6.3) tells us that T_g is just slightly greater than its no-atmosphere value, but it does not by itself tell us how T_g relates to T_{sa}. The general idea for an optically thin atmosphere is that the ground temperature is close to what it would be without an atmosphere, while the atmosphere cools down until the energy it loses by emission is equal to the energy gained by absorption of infrared upwelling from the ground (plus atmospheric solar absorption, if there is any). This generally leaves the low-level air temperature much colder than the ground, since the atmosphere loses energy by radiating out of *both* its top and its bottom. The most straightforward way to make this more precise is to consider the radiative energy budget of the atmosphere, which is the difference between top-of-atmosphere and surface energy budget.

The net infrared radiative flux into the bottom of the atmosphere is $e_g \sigma T_g^4 - e_a \sigma T_{sa}^4$, while the infrared flux out of the top of the atmosphere is the *OLR*. As discussed in Chapter 4, the *OLR* is the sum of the emission from the atmosphere itself and the portion of the upward emission from the ground which is transmitted by the atmosphere. Let a_+ be the proportion of upward radiation from the ground which is absorbed by the full depth of the atmosphere, and express the upward atmospheric emission escaping the top of the atmosphere in the form $e_{a,top} \sigma T_{sa}^4$. Then

$$OLR = e_{a,top} \sigma T_{sa}^4 + (1 - a_+) e_g \sigma T_g^4. \tag{6.6}$$

Let's assume for the moment that the atmosphere does not absorb any solar radiation. Then, in the absence of turbulent heat fluxes the atmospheric energy budget reads

$$0 = OLR - \left(e_g \sigma T_g^4 - e_a \sigma T_{sa}^4 \right) = a_+ e_g \sigma T_g^4 - (e_{a,top} + e_a) \sigma T_{sa}^4 \tag{6.7}$$

whence

$$T_{sa} = \left(\frac{a_+ e_g}{e_{a,top} + e_a} \right)^{\frac{1}{4}} T_g. \tag{6.8}$$

Note that we have not yet made use of the assumption that the atmosphere is optically thin. For an optically thick atmosphere with a very strong greenhouse effect (like Venus), $a_+ \approx e_a \approx 1$ and $e_{a,top} \approx 0$, and so we recover our previous result that $T_{sa} \approx T_g$ for such an atmosphere, provided the emissivity of the ground is close to unity. For an optically thin atmosphere, a_+, $e_{a,top}$, and e_a are all small, so one needs to know precisely how small the absorption coefficient is relative to the two emission coefficients. For an isothermal atmosphere – whether gray or not – Eq. (4.9) implies $e_{a,top} = e_a$. For a gray atmosphere, it follows in addition that $a_+ = e_{a,top} = e_a$. In this case $T_{sa} = T_g/2^{1/4}$, reproducing the result of Section 3.6. When the atmosphere is not gray, the absorption coefficient differs somewhat from the atmospheric emission coefficient, because the spectrum of the upwelling radiation from the ground is different from that of the atmospheric emission (by virtue of the difference between ground temperature and air temperature). However, the deviation from the gray gas result is typically modest for an isothermal atmosphere. For example, a 7 mb Marslike pure CO_2 atmosphere with a uniform temperature of 230 K has $a_+ = e_{a,top} = e_a \approx 0.14$

However, introduction of a vertical temperature gradient strongly affects the relative magnitude of the three coefficients. If we take the same Marslike atmosphere with the same ground temperature and pressure, but stipulate that the temperature is on the dry adiabat rather than isothermal, then a_+ and e_a are reduced slightly (to 0.116 and 0.106, respectively), but are still approximately equal. In contrast, $e_{a,top}$ is substantially reduced, to 0.043. In consequence, the temperature jump at the ground is $T_{sa} = T_g/1.28^{1/4}$ – substantially weaker than the isothermal case, but still quite unstable. Results for a dry Earth, with 300 ppmv of CO_2 in a 1 bar N_2/O_2 atmosphere having 300 K surface temperature, are similar: $a_+ \approx e_a \approx 0.14$ while $e_{a,top} \approx 0.04$. What is happening in both cases is that the atmosphere appears optically thin when averaged over all wavenumbers, but is really quite optically thick in a narrow band of wavenumbers near the principal CO_2 absorption band. The optical thickness in this range introduces a strong asymmetry in the upward and downward radiation, and also weights the absorption towards the bottom of the atmosphere (which is also where a disproportionate amount of the infrared back-radiation is coming from). A rule of thumb for such cases is that a_+ and e_a will have similar magnitudes, while $e_{a,top}$ will be smaller (but, in the optically thin case, still non-negligible); it follows that the surface temperature jump is weaker than the isothermal case, but still unstable. For an optically thin gray gas the situation is different. In that case, $e_a = e_{a,top}$ and both are less than a_+; nonetheless, the relative magnitudes are such that an unstable temperature jump can generally be sustained at the surface even if the lower atmosphere is on a dry adiabat (see Problem 6.1).

The upshot of the preceding discussion is that, in the absence of atmospheric solar absorption, the radiative balance in an optically thin atmosphere almost always drives the surface to be notably warmer than the overlying atmosphere, even if convection has established an adiabat in the atmosphere. This provides a source of buoyancy that can maintain the convection which stirs the troposphere and maintains the adiabat. A moist adiabat is more isothermal than the dry adiabat, so our conclusion is even firmer in that case. Atmospheric solar absorption, on the other hand, would warm the atmosphere relative to the surface, weakening or even eliminating the unstable surface jump.

Moving on, let's consider the temperature the ground of a planet would have in radiative equilibrium at night-time, when $S_g = 0$. In this case, there is little to be gained by linearizing the surface budget, as it reduces to simply $\sigma T_g^4 = e_a \sigma T_{sa}^4$, whence $T_g/T_{sa} = (e_a)^{1/4}$. For an optically thick lower atmosphere, the infrared back-radiation keeps the ground temperature nearly equal to the air temperature. However, when the lower atmosphere is not optically thick, the ground temperature plummets at night, or would do so if it had time to reach equilibrium. Cold climates tend to be comparatively optically thin because they cannot hold much water vapor even in saturation. For example, using the moist case in Fig. 6.1, we find that when $T_{sa} = 240$ K, $e_a \approx 0.3$ with 0.1 mb of CO_2 in the atmosphere. This implies that at night the ground temperature plunges toward the fearsomely cold value $T_g = 177$ K. Liquid surfaces like oceans cannot generally cool down rapidly enough to approach the night-time equilibrium temperature, because turbulent motions in the fluid bring heat to the surface which keeps it warm. Solid surfaces like snow, ice, sand, or rock can cool down very quickly, though, and do indeed plunge to very low temperatures at night. This situation applies to Snowball Earth and to the present-day Arctic and Antarctic. Very cold climates are of necessity dry, because of the limitations imposed by Clausius–Clapeyron. However, even relatively warm climates can be dry if the moisture source is lacking. This is why deserts can go from being unsurvivably hot in the daytime to uncomfortably cold at night. Turbulent fluxes can bring additional heat to the ground and moderate the night-time cooling somewhat, but these fluxes tend to be weak in the situation just described, because turbulent eddies must expend a great deal of energy to lift cold dense air from the ground to the outer edge of the surface layer (a matter taken up in more detail in Section 6.4).

The preceding discussion technically applies whether or not T_{sa} itself drops substantially at night, but is most meaningful in the situation where the atmosphere cools slowly enough that the atmosphere remains relatively warm as the night-time ground temperature drops. This is a fair description of the situation in the massive atmospheres of Titan, Earth, and Venus, except to some extent during the long polar night on Titan and Earth. The tenuous atmosphere of present Mars, in contrast, cools substantially throughout its depth during the night, even at midlatitudes. In this situation, the relative temperature of air and ground at night is determined by the relative rates of cooling of the two media, rather than radiative equilibrium. We will take up the issue of thermal response time in detail in Chapter 7.

6.3 BASIC MODELS OF TURBULENT EXCHANGE

Anybody who has watched dry leaves or dust blow around on a windy day has noticed that where the air comes up against the surface there arises a complex mass of turbulent eddies. In comparison, the interiors of planetary atmospheres are fairly quiescent places, except in the immediate vicinity of rapidly rising buoyant plumes and active cloud systems. The turbulent fluid motions near the planetary surface exchange energy between the surface and the atmosphere, both in the form of *sensible heat* (energy corresponding to the change of temperature in a mass) and *latent heat* (energy associated with the change of phase of a condensable substance, with fixed temperature). Representing the effects of turbulence is not like representing radiation, where we can write down some basic physical principles, then proceed through a set of systematic approximations until we arrive at a set of equations we can solve. When it comes to turbulence, the state of physics is not yet up to that challenge, and may never be. Instead, one must take a largely empirical approach from the outset, constrained by some fairly broad principles such as conservation of energy.

In this section we will derive the so-called *bulk exchange* formulae describing the flux of a quantity from the surface to the overlying atmosphere. The general idea is the same whether the quantity is a chemical tracer, sensible heat (associated with temperature fluctuation), or latent heat, so we will first present the formulae for a general tracer. The calculation will be introduced using simple physically based scaling arguments, and then will be revisited in a more precise and systematic fashion in Section 6.4.

Let c be the specific concentration of some substance, and c' be the fluctuating or "turbulent" part, usually thought of as a deviation from a time or space mean over some suitable interval. Further, let w' be the fluctuating vertical velocity at the top of the surface layer. Then, the flux of the substance, in kg/m^2, is

$$F_c = \overline{\rho w' c'} \approx \rho_s \overline{w' c'} \tag{6.9}$$

where the overbar represents a time or space average and ρ is the total density of the gas making up the atmosphere. We assume further that the surface layer is thin enough that the variation in pressure and temperature across it is small, so that the variations in density can be neglected. Thus, the density factor can be replaced by a constant typical surface density, ρ_s, and taken outside the average. The ideal gas law states that $\rho = p/RT$. If the surface layer has a thickness of a few tens of meters or less, then the hydrostatic law typically guarantees that the contribution of pressure to the density variations is small.

It is not inconceivable, however, that the temperature difference across the surface layer could reach 10% of the mean, leading to corresponding changes in the density. With a little more work, the effect of these fluctuations can be brought into the picture, but we will not pursue this refinement as the effects are probably overwhelmed by the uncertainties in the representation of turbulence itself.

Next, we must estimate the correlation $\overline{w'c'}$. We build this estimate from a typical vertical velocity δw, a typical concentration fluctuation δc, and a non-dimensional factor $0 < a < 1$ describing the degree of correlation. Thus, we write $\overline{w'c'} = a \cdot \delta w \cdot \delta c$. Next, we assume that δw is proportional to the mean horizontal wind speed U at the top of the surface layer, so $\delta w = s \cdot U$. The constant of proportionality s can be thought of as a typical slope characterizing the turbulent eddies, which is in turn roughly related to the roughness of the surface. Note that U is the wind *speed*, and is therefore positive. We then assume that the typical concentration fluctuation scales with the concentration difference between the air in contact with the ground and the edge of the surface layer, so $\delta c = f \cdot (c_g - c_{sa})$, where c_{sa} is the concentration at the edge of the surface layer, c_g is that at the ground, and f is a non-dimensional constant of proportionality. Putting it all together and lumping the proportionality constants into the *drag coefficient* $C_D \equiv a \cdot s \cdot f$, we write

$$F_c = \rho_s C_D U(c_g - c_{sa}). \tag{6.10}$$

We call C_D the *drag coefficient* because when c is taken to be the turbulent velocity itself, the flux formula gives the flux of momentum, and hence the drag force on the surface. In writing the flux in the form of Eq. (6.10), we have adopted the convention that a positive flux represents a transfer of substance from the ground to the atmosphere. The turbulent flux acts like a diffusion, transferring substance from regions of higher concentration to regions of lower concentration. It is like a bucket-brigade, with partly empty buckets being handed downstairs from the top of the surface layer to the ground, where they are filled and sent back upstairs again (or with full buckets sent downstairs to be partly dumped out on the ground). The mass of substance in a bucket being carried upstairs is proportional to $\rho_s c_g$, while the mass of substance in a bucket going downstairs is proportional to $\rho_s c_{sa}$, and $C_D U$ gives the rate at which buckets are being handed up or down the stairs.

6.3.1 Sensible heat flux

To obtain the sensible heat flux, we take $c_p T$ to be our tracer. This is essentially the dry static energy (see Eq. (2.23)), since the surface layer is thin enough that the height z can be taken to be nearly constant. With this choice of tracer, Eq. (6.10) becomes

$$F_{sens} = c_p \rho_s C_D U(T_g - T_{sa}). \tag{6.11}$$

If the ground is warmer than the air, heat is carried away from the ground at a rate proportional to the temperature difference. If the ground is cooler than the air, the sensible heat flux instead acts to warm the ground.

If C_D is independent of temperature, then F_{sens} is exactly linear in the difference between the ground temperature and air temperature. Hence the coupling coefficient b_{sens} – analogous to b_{ir} – is simply $b_{sens} = c_p \rho_s C_D U$. When the surface layer becomes stably stratified, however, C_D can be driven nearly to zero because the energy of turbulence is expended in mixing dense air upward. This effect will be quantified in Section 6.4. The consequent temperature dependence of C_D would alter the linearized coupling coefficient.

Note that the sensible heat flux becomes small when the atmosphere has low density. The "wind-chill" factor on present Mars would be exceedingly weak! Conversely, very dense atmospheres like those of Venus or Titan can very effectively exchange heat between the surface and the atmosphere. With $C_D = 0.001$, $U = 10\,\text{m/s}$, and $T_g - T_{sa} = 1\,\text{K}$, the sensible heat flux is $0.13\,\text{W/m}^2$ on present Mars, $11\,\text{W/m}^2$ on Earth, $55\,\text{W/m}^2$ on Titan, and a whopping $540\,\text{W/m}^2$ on Venus. It is for similar reasons that immersion in near-freezing water is far more life-threatening than walking about scantily clad in air of the same temperature – water is about 1000 times denser than Earth air. One must take care to distinguish thickness of an atmosphere (in terms of density) from optical thickness. An atmosphere can be thick (i.e. dense) while being optically thin, and conversely a thin (low density) atmosphere can nonetheless be optically thick if the greenhouse gas it is made of is sufficiently effective.

Now let's suppose that the sensible heat flux dominates the surface energy budget. By "dominates," we mean that the sensible heat flux due to a small departure from equilibrium (considering the sensible heat flux alone) overwhelms the other terms in the surface energy balance. This would be true if the wind speed and density were large, provided that the ground and atmosphere are dry enough that evaporation remains small. Sensible heat flux vanishes when $T_g = T_{sa}$, so this is the state that the system is driven to when sensible heat flux dominates. Taking the radiative and latent fluxes into account would cause a small deviation from this limit.

6.3.2 Latent heat flux

Whatever the condensed substance making up the surface, some of the condensed substance will transform into the vapor phase in the atmosphere contacting the surface, until it reaches the saturation vapor pressure determined by Clausius–Clapeyron. If the winds then carry away this vapor-laden air and replace it with unsaturated air, more mass will evaporate or sublimate from the surface. Since the phase change involves latent heat, a flux of mass away from the surface cools the surface by carrying away latent heat. Conversely, a flux of mass from vapor into the condensed surface will warm the atmosphere where condensation occurs. All substances will evaporate or sublimate to some extent, and whether the latent heat flux is significant is a matter of how big the saturation vapor pressure is at the typical temperature of the surface. For water ice on Titan at 95 K, the vapor pressure is under $10^{-15}\,\text{Pa}$, so the latent heat flux of water is utterly negligible. The situation is the same for basalt at 300 K on Earth, or even at 750 K on Venus. However, the vapor pressures of CO_2 on present Mars, of liquid water or water ice on Earth, and of methane on Titan are all high enough to allow substantial latent heat flux. Whatever the condensable substance in question we will use terms like "humidity" by analogy with the archetypal case of water vapor on Earth. Also, for the sake of verbal economy we will often refer simply to "evaporation" in situations where the actual process might be either evaporation or sublimation.

In dealing with latent heat flux, it is more convenient to deal with the mass mixing ratio of the condensable to dry air, rather than specific humidity. This makes it somewhat easier to treat cases where the condensable makes up a substantial part of the total mass. Thus, we use the mass mixing ratio r_w in place of the tracer concentration c in Eq. (6.10). If ρ_a is the density of dry air in the surface layer, then the mass of condensable per unit volume is $\rho_a r_w$ and this mass carries a latent heat $L\rho_w r_w$. We can write the mixing ratio r_{sa} at the edge of the surface layer as $h_{sa} r_{sat}(T_{sa})$, where h_{sa} is the relative humidity at the outer edge of the surface layer and $r_{sat}(T)$ is the saturation mass mixing ratio. In terms of saturation vapor pressure, the saturation mass mixing ratio is $(M_w/M_a)(p_{sat}(T)/p_a$, with p_a being the partial pressure of dry air in the surface layer. Now suppose that at the ground there is a reservoir of a condensed phase of the substance "w" – an ocean, lake, swamp, snow field, glacier, or the surface of an icy moon. In this case, the vapor pressure in the air in contact with the surface must be in equilibrium with the condensed phase, and must therefore follow the Clausius–Clapeyron relation evaluated at the temperature of the ground. Equivalently, we can say that $r_g = r_{sat}(T_g)$. Using the two mixing ratios, the latent heat flux becomes

$$F_L = L\rho_a C_D U(r_{sat}(T_g) - h_{sa} \cdot r_{sat}(T_a)). \tag{6.12}$$

Alternately, using the definition of the mixing ratios and assuming the partial pressure of dry air to be approximately constant within the boundary layer, Eq. (6.12) can be written

$$F_L = \frac{L}{R_w T_{sa}} C_D U(p_{sat}(T_g) - h_{sa} \cdot p_{sat}(T_{sa})). \tag{6.13}$$

The latter form of the latent heat flux demonstrates that the flux is in fact unaffected by the presence of the dry air. Assuming temperature and wind to be held constant, the evaporation from the Earth's ocean would remain unchanged even if all the N_2 were taken out of the atmosphere. This conclusion would no longer be valid if the gases in question had substantial non-ideal behavior, for then the law of partial pressures would no longer hold.

Exercise 6.1 Derive Eq. (6.13). What do you have to assume about the air temperature within the surface layer?

In situations where a major constituent of the atmosphere can condense out onto the surface or sublimate or evaporate from it, a constraint on the temperature change across the surface layer enters the problem in a significant way. The constraint arises from the fact that, since the surface layer is thin, the pressure must be nearly constant within the layer. The implications of this constraint are easiest to see when the atmosphere consists of a single condensable component; a concrete example of this situation is provided by the state of the surface layer over seasonal CO_2 frost layers on Mars. Let's suppose that the system has a layer of condensate of the atmospheric substance at the surface – an ocean or glacier. Then, since the atmosphere consists of only the one constituent, the surface pressure is fixed in terms of the ground temperature by Clausius–Clapeyron, namely $p_s = p_s(T_g)$. The pressure at the upper edge of the surface layer must be very nearly equal to this value, otherwise there would be a large unbalanced pressure gradient which would drive a strong flow that would soon transport enough mass to equalize the situation. It follows that *if the atmosphere at the upper edge of the surface layer is saturated*, we must have $T_{sa} \approx T_g$. In other words, in saturated conditions, the temperature at the ground and the temperature at the upper edge of the surface layer must adjust nearly instantaneously so as to keep the two equal. Under what circumstances can the upper edge of the surface layer be considered saturated? First note that if T_{sa} were colder than T_g, then the pressure continuity condition would require the air to be supersaturated. This situation cannot persist for long, so in a case where the free atmosphere is cooling or the ground is heating up, T_{sa} would adjust nearly instantaneously to remain equal to T_g. This adjustment does involve a transfer of latent heat, which alters the thermal response time of the system. One could treat this transfer in terms of an strong enhancement of C_D in such conditions, but there are more natural ways to deal with essentially instantaneous adjustments. The implications for the seasonal cycle of condensable atmospheres will be considered in Section 7.7.5, where such an alternate approach will be illustrated. On the other hand, a situation with $T_g < T_{sa}$ is perfectly consistent if the atmosphere aloft is subsaturated. In such situations, the transfer of latent heat flux is governed by Eq. (6.13) as usual. The transfer would act both to cool the surface, and to add mass to the atmosphere bringing it closer to saturation. However, in situations where the atmosphere remains saturated as the system cools down, the previous temperature continuity constraint applies.

From Eq. (6.12) we observe that latent heat flux carries heat away from the ground when the saturation mixing ratio at the ground is greater than the mixing ratio of the surface layer. Since typically $h_{sa} < 1$, this can happen even if the ground is colder than the overlying air. We also note that the latent heat flux becomes insignificant at sufficiently cold temperatures, since both saturation vapor pressures in the equation become small in that limit.

Sensible and radiative heat transport carry no mass away from the surface, but latent heat transport is of necessity accompanied by mass transfer. The mass flux into or out of the ground is simply F_L/L. The mass flux is needed for calculating the rate of ablation of glaciers by sublimation, the drying out of lakes or soil by evaporation, and the rate of salinity change at the surface of an ocean (since evaporation carries away the condensable but not the solute).

Now let's look at how the fluxes behave when the temperature difference between the ground and the outer edge of the surface layer is small. Carrying out a Taylor series expansion of the flux about $T_g = T_{sa}$, as we did for the infrared cooling case, we write

$$F_L = E_0 + b_L \cdot (T_g - T_{sa}). \tag{6.14}$$

Defining the characteristic flux $F_L^* \equiv C_D U p_{sat}(T_{sa})$, we find

$$E_0 = (1 - h_{sa}) \frac{L}{R_w T_{sa}} F_L^*, \quad b_L = \frac{1}{T_{sa}} \left(\frac{L}{R_w T_{sa}} \right)^2 F_L^* \tag{6.15}$$

where R_w is the gas constant for the condensable. The Clausius-Clapeyron relation has been used to substitute for dp_{sat}/dT in the expression for b_L. E_0 is the heat flux due to evaporation or sublimation that would occur with $T_g = T_{sa}$; it vanishes if the surface layer is saturated ($h_{sa} = 1$), but is positive otherwise. Both E_0 and b_L are proportional to the characteristic flux F_L^*, which vanishes vanishes as $T_{sa} \to 0$, since the saturation vapor pressure vanishes like $\exp(-L/R_w T)$ in this limit. As one might expect, latent heat flux becomes negligible at sufficiently low temperatures. How low one must go for this to be the case depends on the gas in question. As temperature increases, the characteristic flux becomes large, and hence E_0 and b_L become large as well. The increase is abetted by the fact that $L/R_w T$ is a large number at typical planetary temperatures (e.g. 18.06 for water vapor at 300 K, or 10.3 for methane at 95 K). For temperatures high enough that b_L becomes large, a modest ground–air temperature difference leads to a very large increase in latent heat flux. This tends to make it hard for the ground temperature to differ much from the free air temperature in such cases.

Table 6.1 gives some typical values of E_0 and b_L for water, carbon dioxide, and methane. In all three cases, we see that the latent heat flux rises very strongly with temperature. For water, latent heat flux is insignificant at temperatures of 230 K or lower. The feeble latent flux of a watt per square meter or so would be utterly dominated by infrared cooling of the surface, or by the sensible heat flux arising from a ground–air temperature difference of as little as 1 K. This corresponds to the situation in the Antarctic night of the present Earth,

	H_2O				CO_2		CH_4	
T_{sa} (K)	230	273	300	320	150	160	80	95
E_0 (W/m^2)	0.72	40.8	193.3	557.8	52.5	182.1	93.2	640.0
b_L (W/m^{2}K)	0.28	11.2	38.6	98.0	24.4	74.4	55.6	243

Table 6.1 Latent heat flux coefficients for various gases at selected temperatures T_{sa}. Computed with $U = 10$ m/s, $C_D = 0.001$ and boundary layer relative humidity $h_{sa} = 70\%$.

or to the daily average tropical temperatures on a Snowball Earth. However, even at the freezing point of water, the latent heat flux is quite substantial. With a 5 K ground–air temperature difference, the flux would be nearly 100 W/m^2, which is almost half of the typical midlatitude absorbed solar radiation in the ocean, and roughly equal to the typical absorbed solar radiation in ice. The latent heat flux is also comparable to the typical infrared cooling of the surface at such temperatures (inferred from Fig. 6.1). As temperature is increased further to values characteristic of the modern tropics, the flux increases dramatically; it would take about 90% of the supply of absorbed solar energy going into the ocean in order to sustain the evaporation arising from just a 2 K ground–air temperature difference. At these temperatures, the latent flux is considerably in excess of the surface infrared cooling.

For the other gases in the table, the latent heat flux becomes substantial at much lower temperatures. At temperatures comparable to the Martian polar spring, the latent heat flux due to CO_2 sublimation is comparable to the water vapor values for Earth's midlatitudes or tropics (assuming the same degree of boundary layer saturation). These fluxes are particularly consequential in light of weak supply of solar radiation on Mars, relative to Earth. Alternatively one may compare the latent flux to the infrared cooling of the surface in the thin Martian atmosphere (σT_g^4, or 37 W/m^2 at 160 K). Either way, we conclude that latent heat flux plays a key role in determining surface temperature at places on Mars where seasonal CO_2 frost is sublimating or being deposited. At Titan temperatures, latent heat flux due to methane evaporation is enormous; the solar radiation reaching Titan's surface is well under 5 W/m^2, which is two orders of magnitude less than the flux due to the methane evaporation under the conditions of Table 6.1. Somehow or other, conditions near Titan's surface must adjust until the evaporation is reduced to the point where it can be balanced by the supply of energy to the surface, but the numbers in the table tell us that methane latent heat flux is the dominant constraint on the adjusted state. Ironically, Titan at 95 K is like an extreme form of the Earth's tropics, in that evaporation dominates the surface energy budget to an even greater extent than it does in Earth's tropics. If the temperature of the Earth's tropics were raised to 320 K, as might happen in the high CO_2 world following deglaciation of a Snowball Earth, then E_0 on Earth, too would greatly exceed the available solar energy, though not to such an extent as it does on Titan. The way the surface conditions adjust to accommodate this state of affairs will be taken up in the Section 6.5.

When the surface is sufficiently cold relative to the air, vapor from the air can be deposited on the surface in the form of dew or frost. In this case the latent heat flux is negative, and carries energy from the atmosphere to the ground. If the boundary layer is saturated ($h_{sa} = 1$) then frost or dew deposition occurs whenever $T_g < T_{sa}$. If the boundary layer is unsaturated, deposition will not occur until the ground temperature is made sufficiently cold that the saturation vapor pressure there falls below the partial pressure of the condensable in the overlying atmosphere (a temperature known as the "dew-point" or "frost point"). When latent heat is being carried to the surface – as it is during the seasonal polar CO_2 frost formation on Mars – the rate of condensation is limited by the rate at which the surface can get rid of the deposited latent heat. Since the surface is colder than the atmosphere during deposition, sensible heat flux carries heat the wrong way to balance the budget, so it is only infrared cooling of the surface that can sustain frost or dew. Otherwise, the surface will simply warm in response to the deposited latent heat until it is no longer cold enough for frost or dew to form.

Over land, there are two further complications that must be considered. The first is that land, unlike a deep ocean or lake or a thick glacier, can dry out. If the land surface is a mix of condensable and (essentially) non-condensable substance, the latent heat flux can exhaust the supply of condensable, whereafter the boundary condition $r_g = r_{sat}(T_g)$ is no longer appropriate. In the absence of further supply of condensable at the ground, the latent heat flux must fall to zero. In such a case, one must keep track of the mass of the condensable reservoir at the ground, and zero out the latent heat flux when the reservoir is exhausted. This would be the case for thin snow cover, scattered puddles, or soil moisture on Earth, for CO_2 frost layers on Mars, and for liquid methane swamps on Titan. For soil moisture, a common simple model is the *bucket model*, in which each square meter of soil surface is treated as a bucket whose capacity is determined by its porosity and depth. The bucket is filled by rainfall, and emptied by evaporation. Once the bucket is full, any additional rainfall is assumed to run off into rivers (which may or may not be tracked, according to the level of sophistication of the model). As long as the bucket has some water in it evaporation is sustained, but when the bucket is empty, latent heat flux is zeroed out and only radiative and sensible heat transfers at the ground are allowed. The bucket model may serve also as a model of conditions at Titan's surface, which may consist not only of liquid methane puddles but also bogs consisting of beds of granular water-ice sand or pebbles whose pores are saturated with liquid methane.

The second complication over land concerns the effect of land plants. At present, Earth's climate provides the only example where this must be taken into account. Plants actively pump water from deep storage, at rates determined by their own physiological requirements. This is known as *transpiration*, and given that moisture flux over vegetated land is always some mix of transpiration and evaporation, the joint process is called *evapotranspiration*. In this case, the moisture boundary condition at the ground may be more appropriately represented as a flux condition determined by plant physiology rather than setting the moisture mixing ratio at the ground. The moisture flux may be limited by the rate at which trees pump moisture, and not by the rate at which turbulence carries it away. The mixing ratio at the ground still cannot exceed saturation, so when the transpiration becomes strong enough to saturate the air in contact with the ground, one can revert to the previous model of conventional evaporation. Yet a further complication in vegetated terrain is the very notion of ground and ground temperature. Is "the ground" the forest floor or the elevated leaf canopy? Is the ground temperature that of the leaf surface or the soil? How do we take into account the mix of illuminated hot leaves and relatively cool leaves in shade? A proper treatment of these factors requires a detailed model of the microclimate in the vegetation layer, which is beyond what we aspire to in this book. One need not abandon all hope of estimating conditions over vegetated terrain, however. As a rule of thumb, dense forests that get enough rainfall to survive in the long term tend to act more or less like the ocean, save for an elevated C_D caused by greater surface roughness. Grasslands, shrub, tundra, and prairie can be crudely modeled using the bucket model.

When evaporation dominates the surface budget, equilibrium requires $F_L = 0$, or equivalently $p_{sat}(T_g) = h_{sa}p_{sat}(T_{sa})$. Since p_{sat} is monotonically increasing in temperature, this relation requires $T_g < T_{sa}$ if the boundary layer air is unsaturated ($h_{sa} < 1$). Thus, evaporation or sublimation drives the ground temperature to be *colder* than the overlying air temperature. However, the ground and surface could also achieve equilibrium by transferring enough moisture to the surface layer that it becomes saturated ($h_{sa} = 1$), in which case $T_g = T_{sa}$ in equilibrium, as for the case of sensible heat flux. The extent to which equilibrium is attained by adjusting temperature vs. humidity depends on the competition between the rate at which moisture is supplied to the boundary layer and the rate at which dry air from aloft is entrained into the boundary layer. Observed boundary layers on Earth and Titan are significantly undersaturated, leading to the conclusion that the ground temperature would be considerably less than the air temperature, if other fluxes did not intervene. Using the linearized form of the latent heat, the equilibrium ground–air temperature difference is $T_g - T_{sa} \approx -E_0/b_L$. For the conditions of Table 6.1, this is $-2.6\,K$ for Titan at $95\,K$. For a hot Earth at $320\,K$, the difference is about $-5.7\,K$. There are currently no observations of the state of saturation over the sublimating Martian CO_2 frost cap, but given the saturation assumed in the table the equilibrium occurs with $T_g - T_{sa} \approx -2.4\,K$ when the air temperature is $260\,K$. Thus, even when evaporation dominates, the equilibrium ground temperature does not differ greatly from the overlying air temperature. This was also found to be the case when the surface budget is dominated by sensible heat flux. It is only the radiative terms that can drive the ground temperature to be substantially different from the overlying air temperature.

6.4 SIMILARITY THEORY FOR THE SURFACE LAYER

The surface layer theory based on dimensional analysis tells us most of what we need to know, but it does not tell us how the drag coefficient depends on the height at which the top-of-layer conditions are applied, nor does it say precisely how the coefficient depends on stratification or the surface roughness. We will now re-do the surface layer theory using a more precise form of the similarity assumption. The most important thing we will get out of this is a quantification of the suppression of turbulent mixing in stable conditions. This exerts a very important control on the fluxes at night-time and over ice or snow, where surface layer conditions are often stable. In particular, when ice or snow is melting the temperature is pinned at the freezing point, so if the atmospheric temperature is significantly above freezing, the surface layer is very stable, and this limits the delivery of heat available for melting.

Let c be any quantity whose flux we wish to determine in the surface layer. It might be temperature, water vapor, methane, or some other chemical tracer. We will also consider the flux of horizontal momentum (proportional to horizontal wind u) using the similarity theory. Though we are not attempting to do much dynamics in this book, we will nevertheless need to talk a bit about momentum flux since this is what will tell us how the mean wind varies with height within the surface layer.

Within the surface layer, the fluxes of tracer and momentum are constant, by definition. This allows us to define the following velocity and tracer scales:

$$u_*^2 \equiv \overline{w'u'}, \quad c_* \equiv \overline{w'c'}/u_*. \tag{6.16}$$

As a consequence of the second definition the tracer flux is just $u_* c_*$. The velocity scale u_* is called the *friction velocity*, and is taken to be positive by convention. When the flux of tracer is upward, then the tracer fluctuation scale c_* is positive.

Next we derive equations for the vertical gradient of mean tracer and mean wind. We'll first consider the neutrally stratified case, in which buoyancy forces are negligible. Strictly speaking, this case only applies when the density within the surface layer is constant. Any situation with heat transport would involve some temperature fluctuations, and hence some density fluctuations. In practice, though, when the temperature difference across the surface layer is sufficiently weak, the neutrally stratified calculation yields accurate results. The definition of "sufficiently weak" will be made precise later, when we come to incorporate buoyancy forces.

When buoyancy is insignificant in the surface layer, the only length scale appearing in the problem is the height z above the ground. Since the only tracer scale is c_* and the only velocity scale is u_*, dimensional analysis then tells us that the equations for the vertical gradients must be

$$K_{vk}\frac{d\overline{c}}{dz} = -\frac{1}{z}c_*$$
$$K_{vk}\frac{d\overline{u}}{dz} = \frac{1}{z}u_*$$
(6.17)

where K_{vk} is a non-dimensional constant called the *von Karman constant*. In principle, the non-dimensional constant appearing in the tracer equation could be different from that in the momentum equation, but laboratory experiments indicate that in fact the same constant applies to both. The von Karman constant has been measured in a wide range of turbulent laboratory experiments, which indicate that $K_{vk} \approx 0.4$. The sign choice in the tracer equation is dictated by the physical requirement that the tracer flux be upward when the concentration is greater at the surface than it is aloft.

The similarity equations allow us to relate the flux of tracer to the difference in tracer concentration between the ground and the upper edge of the surface layer, and similarly for momentum. Let z_1 be the upper edge of the surface layer. Integrating from a smaller height z_* to z_1 and assuming that the winds vanish at the height z_*, we find

$$K_{vk}(\overline{c}(z_1) - \overline{c}(z_*)) = -c_* \ln\left(\frac{z_1}{z_*}\right)$$
$$K_{vk}\overline{u}(z_1) = u_* \ln\left(\frac{z_1}{z_*}\right).$$
(6.18)

The height z_* at which we set the lower limit of integration cannot be set to zero because of the logarithmic divergence in that limit. In fact, it has a physical meaning, and is called the *roughness height*. It corresponds to the height at which the airflow is so perturbed by the irregularities in the boundary that the mean flow is essentially zero. The roughness height corresponds loosely to the typical height of the bumps on the surface, but is generally smaller than one would intuit from the physical height of the bumps. In practice, it is determined by fitting the observed mean wind profile with the logarithmic form. Over open water, the roughness length is on the order of 0.0002 m, though at strong wind speeds the wind-driven waves increase the roughness significantly. Over ice or smooth land, the roughness length is more like 0.005 m, increasing to 0.03 m if there is grass or low vegetation, 0.5 m for low forest and 2 m for large forests or urban areas.

The logarithmic profile of wind and concentration is called *the law of the wall*, and has been verified in a great variety of turbulent flows, ranging from wind and water tunnel experiments to atmospheric measurements to velocity profiles in tidal surges in the Bay of Fundy. Next, the tracer and velocity equations can be combined to give the tracer flux

$$\overline{w'c'} = u_* c_* = \frac{K_{vk}^2}{\left(\ln\left(\frac{z_1}{z_*}\right)\right)^2} \overline{u}(z_1)(\overline{c}(z_*) - \overline{c}(z_1)) \tag{6.19}$$

from which we identify the drag coefficient

$$C_D = \frac{K_{vk}^2}{\left(\ln\left(\frac{z_1}{z_*}\right)\right)^2}. \tag{6.20}$$

This formula allows us to compute the drag coefficient explicitly given the roughness height and the height at which one chooses to apply the boundary condition at the upper edge of the surface layer; one is free to choose z_1 as a matter of convenience, so long as it is low enough that the fluxes are constant within the surface layer. Though different roughness lengths are sometimes applied for moisture and momentum, it is generally adequate to use the same drag coefficient for all mixed quantities. Using the previous values for roughness length and assuming $z_1 = 10\,\text{m}$, we get $C_D = 0.0014, 0.0028, 0.0047, 0.018, 0.062$ for open water, ice or smooth land, grassland, low forest and large forest, respectively.

Now let's introduce the effects of buoyancy. To do this, we must first define buoyancy quantitatively. Let ρ_g be the mean density at the ground. Then the net force (per unit volume) on an air parcel with density ρ will be $g \cdot (\rho_g - \rho)$ while the parcel is near the ground. The acceleration of the air parcel is obtained by dividing by the force by the mass, and is thus $g \cdot (\rho_g/\rho - 1)$. The buoyancy acceleration is a form of *reduced gravity*, reflecting the fact that buoyancy forces cancel part of the gravitational forces, leading to a reduction in acceleration. We will refer to the buoyancy acceleration as simply "buoyancy" for short, and denote it by the symbol β. The buoyancy is affected both by the temperature and the composition of the atmosphere. For uniform composition, warm air will be positively buoyant when surrounded by colder air. However, air that is rich in a low molecular weight substance will be buoyant when surrounded by air that has lower concentration of that substance, even if the temperature is uniform. For example, since water vapor has lower molecular weight than dry Earth air, moistening an air parcel adds to its upward buoyancy and drying it tends to make it sink. The same can be said for adding methane to N_2 in Titan's atmosphere. Similarly, the Martian atmosphere contains a few percent of N_2 on average, which has a lower molecular weight than CO_2. Thus, when CO_2 condenses onto the Martian CO_2 seasonal frost cap, the remaining air will be positively buoyant in the background mixture of CO_2 and N_2. One can imagine a variety of situations in which an atmospheric constituent is released from or absorbed into the surface, but the most common situation involves a condensable substance which condenses onto or sublimates/evaporates from a reservoir at the surface. That could be N_2 ice on Triton, liquid CH_4 on Titan, CO_2 ice on Mars, or solid or liquid H_2O on Earth. Using the ideal gas law, the density is

$$\rho = \frac{p}{R_a T}(1 - \eta_c) + \frac{p}{R_c T}\eta_c = \frac{p}{R_a T}\left(1 + \left(\frac{M_c}{M_a} - 1\right)\eta_c\right) \tag{6.21}$$

where M_a and M_c are the molecular weights of the non-condensable background gas and the condensable component, respectively. Since the surface layer is thin, p can be assumed nearly constant within the layer. The buoyancy is then

$$\beta = g \cdot \left(\frac{T}{T_g} \frac{1 + (\epsilon - 1)\eta_{c,g}}{1 + (\epsilon - 1)\eta_c} - 1\right) \tag{6.22}$$

where $\epsilon \equiv M_c/M_a$ and $\eta_{c,g}$ is the molar concentration of the condensable substance at the ground. When $\eta_c = \eta_{c,g} = 0$, or when $\epsilon = 1$, buoyancy is simply $g \cdot (T_g - T)/T_g$. In the general case, if $\epsilon < 1$ then increasing η_c makes the parcel more positively buoyant, while if $\epsilon > 1$, increasing η_c makes the parcel more negatively buoyant. In general, the buoyancy is a nonlinear function of temperature and concentration, but in the special case where both $(T - T_g)/T_g$ and η_c are small, the buoyancy takes on the simple form

$$\beta \approx g\frac{T - T_g}{T_g} + g \cdot (\epsilon - 1)(\eta_{c,g} - \eta_c) \tag{6.23}$$

in which the buoyancy is the sum of a temperature contribution and a composition contribution. When using either form of the buoyancy, it is assumed that the buoyancy-generating tracer is saturated at the ground, in the typical case where its flux is maintained by a condensable reservoir there. Thus, $\eta_{c,g} = p_{sat}(T_g)/p$, where p_{sat} is given by Clausius-Clapeyron.

We can treat the buoyancy flux and mean buoyancy profile much as we did the tracer flux and tracer profile in the neutral case, defining $\overline{w'\beta'} = u_*\beta_*$. However, the buoyancy scale β_* is now a dynamically significant quantity with dimensions of acceleration, which can affect the profile. This has the important consequence that z is no longer the only length scale that enters into the problem – in addition we can define the *Monin-Obukhov length*

$$\ell \equiv \frac{1}{K_{vk}}\frac{u_*^2}{|\beta_*|}. \tag{6.24}$$

The inclusion of the von Karman constant in the definition of the Monin–Obukhov length is purely a matter of convention, and has no particular physical significance. The Monin-Obukhov length is on the order of the height to which a negatively buoyant plume with velocity u_* would rise before exhausting its kinetic energy, or the height to which a positively buoyant plume initially at rest would rise before attaining velocity u_*. For distances much closer to the boundary than ℓ, the turbulence is dominated by the kinetic energy of the wind shear, as in the neutral case. For heights much greater than ℓ buoyancy suppresses or enhances the turbulence. Using the Monin-Obukhov length, we define the non-dimensional depth $\zeta \equiv z/\ell$. In contrast to the neutral case, the equations for the gradient of wind, tracer, or buoyancy can each depend on some function of ζ; in principle, because the function has a dimensionless argument, one can take data in the field or laboratory, and determine the function once and for all, as if it were a sine or cosine. The test of the validity of this bold assumption is to evaluate the functions from a wide variety of different field and laboratory datasets, and see if one gets essentially the same result from each. This is the assumption upon which the similarity theory rests, and it seems to work out well in practice. In terms of the similarity functions, the equations for buoyancy and wind gradient become

$$K_{vk}\frac{d\overline{\beta}}{dz} = -\frac{1}{z}\beta_* F_\beta(\zeta)$$
$$K_{vk}\frac{d\overline{u}}{dz} = \frac{1}{z}u_* F_u(\zeta). \tag{6.25}$$

In general, one should allow for different scaling functions for the buoyancy and momentum equations, and indeed this is sometimes necessary to provide a good fit to data. For the stably stratified (i.e. negatively buoyant) case, it has been found that using the same scaling function for both equations is adequate. Some remarks on the positively buoyant case will be given later. In any event, let's assume $F_u = F_\beta$. In non-dimensional form, the equations become

$$K_{vk}\frac{d\overline{\beta}}{d\zeta} = -\frac{1}{\zeta}\beta_* F(\zeta)$$
$$K_{vk}\frac{d\overline{u}}{d\zeta} = \frac{1}{\zeta}u_* F(\zeta) \tag{6.26}$$

which integrate out to yield the relations

$$K_{vk}(\overline{\beta}(\zeta_1) - \overline{\beta}(\zeta_*)) = -\beta_* G(\zeta_1)$$
$$K_{vk}\overline{u} = u_* G(\zeta_1) \tag{6.27}$$

where $G(\zeta_1) \equiv \int_{\zeta_*}^{\zeta_1} (F(\zeta)/\zeta)d\zeta$. To eliminate the buoyancy and velocity scales, we divide the first equation by the square of the second and multiply by $z_1 - z_*$, which results in

$$-(z_1 - z_*)\frac{(\overline{\beta}(z_1) - \overline{\beta}(z_*))}{\overline{u}^2} = \frac{\zeta_1 - \zeta_*}{G(\zeta_1)} \tag{6.28}$$

in which we have re-written the mean buoyancies as a function of the dimensional height. The left hand side is a non-dimensional number called the *bulk Richardson number*, denoted henceforth by Ri. The Richardson number can be computed in terms of known quantities at the upper and lower edges of the surface layer, and gives the relative importance of potential and kinetic energy; when $|Ri| \ll 1$, buoyancy forces are negligible, and the surface layer can be treated as if it were neutral.

Given Ri, Eq. (6.28) can be solved either explicitly or iteratively for ζ_1, which allows β_* and u_* to be determined from Eq. (6.27). The buoyancy flux is then

$$\overline{w'\beta'} = u_*\beta_* = \frac{K_{vk}^2}{G(\zeta_1)^2}\overline{u}(\overline{\beta}(\zeta_*) - \overline{\beta}(\zeta_1)) \tag{6.29}$$

from which we identify the drag coefficient

$$C_D \equiv \frac{K_{vk}^2}{G(\zeta_1)^2}. \tag{6.30}$$

This gives the drag coefficient for the buoyancy flux, but what we really want is the drag coefficient for sensible and latent heat flux. When the atmosphere has uniform composition, the buoyancy flux is proportional to the sensible heat flux, and so the above-derived drag coefficient can be unambiguously used for sensible heat flux. When the atmosphere has non-uniform composition contributing to buoyancy, however, the drag coefficients for sensible and latent flux could in principle be different from that for buoyancy. This is sometimes handled by introducing separate empirically determined scaling functions for moisture and heat flux. Refinements of the theory are quite straightforward, and can be found in the references given in the Further Reading section for this chapter. There is a fair amount of data pertinent to similarity functions for moisture flux on Earth, where the concentration of moisture reaches a few percent of the total atmosphere. These can almost certainly be applied to other buoyancy-generating substances at similar concentrations. However, behavior of the similarity functions when the buoyancy-generating component accounts for a substantial fraction of the atmosphere, as is the case for methane on Titan, is essentially unexplored. In our calculations, we will be content to use the same drag coefficient for all fluxes. In the stably stratified case the resulting errors are probably not too consequential, since we will see shortly that the main effect of the stratification is to choke off essentially all turbulent fluxes when the Richardson number exceeds a critical value; the refinements to the theory only modify the fluxes in the rather narrow window between neutral conditions and nearly complete suppression.

To proceed further, we need to specify an explicit similarity function $F(\zeta)$. In the stably stratified case ($Ri > 0$) field and laboratory experiments can be adequately fit by functions of the form $F(\zeta) = 1 + \zeta / Ri_c$, with $Ri_c \approx 0.2$. With this definition,

$$G(\zeta_1) = \frac{1}{Ri_c}(\zeta_1 - \zeta_*) + \ln \frac{\zeta_1}{\zeta_*} = \frac{1}{Ri_c}(\zeta_1 - \zeta_*) + \ln \frac{z_1}{z_*}. \tag{6.31}$$

With this simple form of the similarity function, Eq. (6.28) can be analytically solved for ζ_1 in terms of Ri, though for more complicated functions commonly in use a numerical iteration is generally required. With this form of G it is a straightforward matter to solve Eq. (6.28) for $\zeta_1 - \zeta_*$ in terms of R_i, evaluate $G(\zeta_1)$, and then compute C_D from Eq. (6.30). Note that with the assumed form of G the right hand side of Eq. (6.28) has a maximum value of Ri_c when the argument approaches infinity. Thus, there is no consistent solution when $Ri > Ri_c$. It is assumed that the turbulence is completely suppressed, and that the turbulent fluxes vanish, for more stable values of Ri. Complete suppression of turbulence is somewhat unrealistic, and some formulations use alternate forms of G so as to allow a bit of flux to persist into the very stable case. However, the applicability of Monin–Obukhov theory to very stable conditions is a matter of considerable dispute.

Carrying out the above procedure, we find that the drag coefficient is

$$C_D = \begin{cases} \frac{\kappa_{vk}^2}{\left(\ln \frac{z_1}{z_*}\right)^2}\left(1 - \frac{Ri}{Ri_c}\right)^2, \text{ for } 0 \le Ri \le Ri_c \\ 0, \text{ for } Ri > Ri_c \end{cases}. \tag{6.32}$$

Note that this reduces to the previously derived neutrally stratified result when $Ri = 0$. As the surface layer is made more stable, the drag coefficient goes down monotonically, and approaches zero as $Ri \to Ri_c$.

Now let's do a few examples illustrating the effects of stable surface-layer physics on turbulent heat flux. First consider a melting slab of sea ice or glacier ice, in an environment where the air temperature is 280 K. Since the ice is melting, the ground temperature is pinned at the freezing point, namely 273.15 K. Since the air is warmer than the ice, there will be a flux of sensible heat from the air to the ice, which will help sustain melting. At these temperatures, it is safe to neglect the contribution of water vapor to buoyancy. Suppose that the wind is 5 m/s, the air temperature and wind have been specified at 10 m above the ice surface, and the roughness height is 0.005 m. Then, for a neutrally stratified boundary layer, $C_D = 0.0028$ and the sensible heat flux would be 125 W/m^2. With the specified parameters, the Richardson number is 0.1, and incorporation of buoyancy effects on the turbulence bring C_D down to 0.0007, and the sensible heat flux falls to 32 W/m^2. If we increase the air temperature to 285 K, the Richardson number increases to 0.17, and the increasing stability reduces the sensible heat flux to a mere 4.8 W/m^2, whereas neutral surface layer theory would have led us to expect a substantial increase in flux.

Exercise 6.2 How long would the sensible heat fluxes computed above take to melt through a 5 meter thick layer of water ice, if all the energy is used to melt ice?

Next, let's consider the effect on the night-time temperature inversions appearing in cold, dry climates such as Antarctica or the tropics on Snowball Earth. The radiative balance for such cases was discussed towards the end of Section 6.2.3. For example, in Earthlike conditions with $T_{sa} = 240$ K, the equilibrium surface temperature is 177 K if the only coupling of ground to atmosphere is via infrared. Using the same assumptions as in the previous example, except for the new temperatures, neutral surface layer theory would predict a sensible heat flux of over 1500 W/m^2, which of course would mean that the turbulent transfers would keep the inversion from getting nearly as strong as it would be in the purely radiative case. However, with such a large temperature difference, the Richardson number is 1.4, which leads to a complete suppression of turbulence and allows the extremely strong inversion to be realized, if there is enough time for the surface to cool down to equilibrium. An interesting aspect of this problem, however, is that the sensible heat flux is not a monotonic function of $T_{sa} - T_g$. As T_g is decreased from T_{sa}, the flux first increases, reaches a maximum, then decreases to zero as the critical Richardson number is approached. This means that there is the possibility of multiple equilibrium states – one with turbulence and heat flux, and another with turbulence suppressed. Whether or not this happens depends on the slope of the radiative flux, but even when there are not multiple equilibria, there tend to be abrupt transitions between turbulent and non-turbulent states as a control parameter such as solar absorption is continuously varied. This behavior is explored in Problem 6.6.

The inclusion of a light, condensable substance like methane or water vapor has an important effect because it allows the surface layer to remain neutrally stratified even when the ground temperature is significantly lower than the air temperature at the upper edge of the surface layer – provided that the air there is appreciably undersaturated. To get a feel for the numbers, let's do an example involving water vapor in air. Suppose that the air temperature is 300 K and the relative humidity is 70%. Then, using the formulae for the Richardson number and for buoyancy, we find that the surface layer is neutrally buoyant ($Ri = 0$) when $T_g = 299$ K. Without the effect of water vapor on buoyancy, the Richardson number would be 0.013 assuming $\bar{u} = 5$ m/s, so in this case the suppression of turbulence caused by neglect of the moisture contribution to buoyancy is small. The effect increases sharply at higher temperature, though. For $T_{sa} = 340$ K, the surface layer remains neutral down to 335.5 K, and the Richardson number without the moisture contribution to buoyancy would be 0.05. The maintenance of buoyancy by light vapor will be important in our estimates of precipitation rates on hot planets, in Section 6.8.

Finally, let's take a quick look at the unstable case, where the surface layer is positively buoyant. In this case, the buoyancy-driven turbulence adds to the mechanically driven turbulence due to wind shear across the surface layer. Buoyancy-driven turbulence is particularly important when the mean winds at the top of the surface layer are weak. When $\bar{u} = 0$ the neutral theory would predict that there are no turbulent fluxes, but if the surface layer has upward buoyancy, then convection should in fact be able to sustain turbulence. The case $\bar{u} = 0$ with upward buoyancy is the *free convection limit*. In this limit, there is no longer an intrinsic velocity scale separate from that defined by the buoyancy scale, and there is no longer any intrinsic length scale such as enters the Monin–Obukhov theory. Instead, we can define a velocity scale $\sqrt{\beta_* z}$, which is the order of magnitude of the upward velocity attained by a buoyant plume when it reaches height z. Since there is no longer a characteristic length scale, the buoyancy profile $\bar{\beta}(z)$ scales with z_*, just as for the neutral case, and this allows us to relate the buoyancy gradient to the difference in buoyancy between the upper and lower edge of the surface layer. In this case, however, the z-dependence of the velocity scale requires β_* to vary in z if the flux is to stay constant, and this leads to a power law rather than the logarithmic profile. The corresponding buoyancy flux in the free convection limit can be written

$$\overline{w'\beta'} = a \cdot (z_* \Delta\overline{\beta})^{1/2} \Delta\overline{\beta} \equiv C_{D,neut} U_{free} \Delta\overline{\beta} \tag{6.33}$$

where $C_{D,neut}$ is the usual neutral drag coefficient, U_{free} is a characteristic buoyancy velocity, and a is a non-dimensional constant whose value has been empirically determined to be about 0.24. To apply this result to the flux of other quantities, such as latent or sensible heat, we use the same drag coefficient and U_{free}, but replace $\Delta\overline{\beta}$ with the difference in the quantity whose flux we wish to obtain. Note that the buoyancy flux scales like the $\frac{3}{2}$ power of buoyancy. However, by casting the flux formula in the above form we can see that it is just like the neutral case, but with a buoyancy velocity replacing the mean wind. As an example, consider a dry case with ground temperature of 305 K, air temperature 300 K, Earth gravity, a surface layer thickness of 10 m and a roughness length of 0.001 m. With these parameters $U_{free} = 1.5$ m/s, so buoyancy-driven turbulence becomes a significant player when the mean wind is 1.5 m/s or weaker. Based on this estimate, it is clear that convection will lead to large sensible and latent heat fluxes whenever the ground temperature tries to get much bigger than the overlying air temperature. The upshot is that it is quite easy for the ground to get a lot colder than the overlying air, because of the inhibition of turbulence in stable conditions, but it is harder for the ground to get much hotter than the overlying air.

The general unstable case with non-zero mean wind can be treated similarly to the way we treated the stable case, though it is necessary to adopt different scaling functions for wind and momentum and the form of the functions is sufficiently complicated that a Newton's method iteration is generally needed in order to solve for ζ_1. In addition, the scaling functions most commonly in use do not, in fact, reduce to the correct free-convection limit when the mean wind becomes weak. This point, together with a resolution of the problem, is discussed in the paper by Delage and Girard given in the Further Reading section for this chapter. A simple expedient for dealing with the general unstable case, however, would be to compute the turbulent fluxes for both the free-convection and neutral limits, and then take whichever of the two is greater. This procedure by definition gives the correct free-convection limit, and also eliminates the chief shortcoming of the neutral theory, namely the spurious vanishing of fluxes when mean winds become very weak. One can easily implement this formulation by computing fluxes using the usual neutral C_D, but replacing $\overline{u}$ with U_{free} when $\overline{u} < U_{free}$.

6.5 JOINT EFFECT OF THE FLUXES ON SURFACE CONDITIONS

Including turbulent heat fluxes, the surface energy budget can be written

$$0 = F_{rad} - F_{sens} - F_L \tag{6.34}$$

where F_{rad} is the net radiative flux into the surface, given by

$$F_{rad} = (1 - \alpha_g)S_g + \sigma e_a e_g T_{sa}^4 - \sigma e_g T_g^4. \tag{6.35}$$

Without turbulent fluxes, the surface budget would be $F_{rad} = 0$. In isolation, F_{rad} can drive the ground temperature to be either larger or smaller (and perhaps *much* larger or smaller) than the air temperature, according to the circumstances discussed in Section 6.2. Sensible heat flux always drives the ground temperature and air temperature to become identical, whereas latent heat flux drives the ground temperature to be colder than the air temperature, by an amount that depends on the relative humidity of the boundary layer. When all three fluxes act in concert, the resulting behavior depends on the relative importance of the fluxes.

We will begin our tour of the range of possible behaviors by discussing how the surface balance is accomplished for typical conditions in the Earth's tropical oceans. Take $T_{sa} = 300\,\mathrm{K}$, $C_D = 0.0015$, $U = 5\,\mathrm{m/s}$, and $h_{sa} = 80\%$. We'll assume the absorbed solar radiation $(1 - \alpha_g)S_g$ is $320\,\mathrm{W/m^2}$, which is typical of clear-sky conditions over the tropical ocean. To determine the back-radiation, we need e_a. At tropical temperatures in the moist case, this coefficient is not very sensitive to CO_2, and has a value of about 0.9. The terms making up the surface balance are shown in the left panel of Fig. 6.2. As noted previously, the equilibrium ground temperature would be exceedingly large without turbulent heat flux. In the figure, the no-turbulence equilibrium occurs where F_{rad} crosses zero, at around $336\,\mathrm{K}$. Adding sensible heat flux to the budget makes the slope of the flux curve more negative, and brings the equilibrium ground temperature down to $316\,\mathrm{K}$. Adding in evaporation steepens the curve yet more, and brings the ground temperature down to $303\,\mathrm{K}$, which is only slightly warmer than the $300\,\mathrm{K}$ temperature of the overlying air. At the equilibrium point, the dominant balance is between the evaporation ($206\,\mathrm{W/m^2}$) and the absorbed solar radiation ($320\,\mathrm{W/m^2}$), leaving only $114\,\mathrm{W/m^2}$ to be balanced by the other terms. The sensible heat flux is weak because the ground temperature and air temperature are nearly identical, which also makes the net infrared cooling of the surface weak given that $e_a \approx 1$.

Next we'll discuss a typical set of Earth polar or midlatitude winter conditions. We set the absorbed solar flux $(1 - \alpha_g)S_g$ to $100\,\mathrm{W/m^2}$, taking a low value on account of the high albedo of snow or ice and the reduced solar flux received at high latitudes. We'll set $T_{sa} = 265\,\mathrm{K}$, in which case $e_a \approx 0.6$ with $300\,\mathrm{ppmv}$ of CO_2 in the atmosphere. The remaining parameters are held at the same values used in the tropical case. The main differences from the tropical case are that in the cold case the latent heat flux and the infrared back-radiation are weaker – the latter doubly so because of the lower air temperature and the lower e_a. The right panel

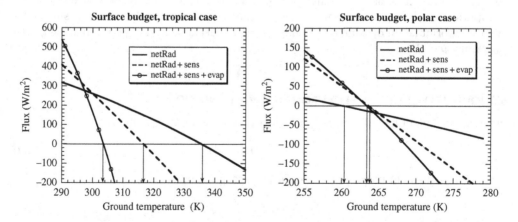

Figure 6.2 Terms in the surface balance for conditions representing the tropical oceans on Earth (left panel) and cold polar conditions on Earth (right panel). See text for the parameters of the calculation.

of Fig. 6.2 shows that because of the weak solar radiation and the weak back-radiation, the radiative equilibrium ground temperature is nearly 5 K colder than T_{sa}, in contrast to the tropical case. The situation here is a less extreme version of the night-time radiative equilibrium temperature considered in Section 6.2.3. Since e_a is fairly small the temperature plummets at night when $(1 - \alpha_g)S_g = 0$. In the present case, S_g doesn't vanish, but its weak value is insufficient to warm-up the ground temperature to the point where it exceeds the air temperature. This is the typical daytime condition in high latitude winter over ice and snow. Warm air imported from low latitudes helps to keep T_{sa} from getting too cold in the polar and midlatitude winter, but the weak sunlight and weak back-radiation leave the ground colder.

Since the radiative equilibrium ground temperature in the cold case is colder than the air temperature, adding in sensible heat flux conveys heat from the atmosphere to the ground, warming the ground up to just over 263 K. The sublimation is weak at such cold temperatures, and causes little additional change in the surface temperature. While the dominant balance in the tropical case was between solar heating and evaporative cooling, the dominant balance in the cold case is radiative, with slight modifications due to sensible heat flux. For any given air temperature, the amount by which the ground temperature departs from the air temperature depends on the absorbed solar radiation, but the sensible heat flux always pulls the ground temperature back towards equality with air temperature. For example, at higher latitudes or deeper in the winter or near sundown, we might take $(1 - \alpha_g)S_g = 50\,\text{W/m}^2$. In this case the radiation-only ground temperature is 246.6 K, which is substantially below the air temperature; however, addition of sensible heat flux brings the ground temperature back up to 260 K. Nearer to noon, or as summer approaches, we might have $(1 - \alpha_g)S_g = 150\,\text{W/m}^2$. In this case, the radiation-only ground temperature is 271.8 K; again addition of sensible heat flux brings the ground temperature closer to air temperature, in this case by cooling the ground to 267.1 K, rather than warming it. Note that in these calculations of the effects of sensible heat transfer, the drag coefficient C_D was held constant. Incorporating the inhibition of turbulence in stably stratified layers has the potential to reduce the warming effect of sensible heat fluxes substantially, particularly when the absorbed solar radiation is weak, since the inversion is strongest in those cases. This is explored in Problem 6.6.

Next let's estimate the maximum daytime temperature over a subtropical desert on Earth. Solid surfaces like sand or rock take little time to reach equilibrium, and so the maximum temperature can be estimated by computing the equilibrium temperature at local solar noon. Using the present Earth solar constant and a relatively high albedo of 0.35 (typical of Sahara desert sand), the absorbed flux is about $890\,\text{W/m}^2$. Over the interior of a dry desert, there should be little moisture in the boundary layer, so set $e_a = 0.72$ corresponding to a boundary layer relative humidity of 20%. Finally, we take $T_{sa} = 300\,\text{K}$. In these circumstances the radiative equilibrium ground temperature is a torrid 383 K – hot enough to boil water. When sensible heat flux is added into the budget, heat is transferred from the ground to the air, moderating the surface temperature. Taking a relatively high drag coefficient $C_D = 0.003$ on account of the roughness of land surfaces, the equilibrium ground temperature is brought down to 330 K if the surface layer wind speed is 5 m/s. The temperature approaches the radiative temperature as the wind is made weaker; for example when the wind is reduced to 2.5 m/s the temperature increases to 349 K. Consistent with these estimates, the hottest satellite-observed ground temperatures do indeed occur in subtropical deserts, and are near 340 K. With a wind of 5 m/s, making the ground moist and turning on evaporation brings the equilibrium temperature down from 330 K to 306 K. The general lesson is that dry surfaces

Table 6.2 Some typical surface flux coupling coefficients.

	T	b_{ir}	b_{sens}	b_L
Water+air	250	3.54	21.00	2.76
Water+air	280	4.98	18.89	19.72
Water+air	300	6.12	17.99	57.95
Water+air	320	7.43	17.84	147.0
$CH_4 + N_2$	85	0.14	95.07	161.36
$CH_4 + N_2$	90	0.17	92.55	287.29
$CH_4 + N_2$	95	0.19	91.88	365.75

The "Water+air" cases are done under Earthlike conditions, with a 1 bar non-condensable background of Earth air. The $CH_4 + N_2$ cases are done under Titanlike conditions, with a 1.5 bar non-condensable N_2 background. Both cases were done with 70% relative humidity at the top of the surface layer, $U = 10\,m/s$ and C_D held fixed at 0.0015. Units for all the coupling coefficients are W/m^2K.

heat up greatly during the daytime. Their maximum temperature can greatly exceed the overlying air temperature, especially when the wind is light. This can contribute to the *urban heat island* effect, since constructed environments often replace moisture-holding surfaces with low albedo impermeable surfaces like asphalt, which hold little water and dry out quickly. The surface heating also leads to amplified climate change over land, in circumstances where a formerly moist soil becomes dry, or *vice versa*.

We conclude this section with a discussion of the linearized form of the surface balance, which enables simple, explicit solutions for the temperature jump across the surface layer. Using the linearized flux coefficients defined previously, the temperature jump is simply

$$\Delta T \equiv T_g - T_{sa} = \frac{(1 - \alpha_g)S_g - e_g \cdot e^* \sigma T_{sa}^4 - E_0}{b_{ir} + b_{sens} + b_L}. \tag{6.36}$$

The numerator in this expression is the energy imbalance the surface would have if the ground temperature were equal to the overlying air temperature. It can be either positive or negative, and its sign determines the sign of ΔT, since all three terms in the denominator are positive. The denominator is a stiffness coefficient. For any given magnitude of the numerator, the denominator determines how much the ground and air temperature differ. In other words, when the denominator is large, the ground and air temperature are very tightly coupled, but when the denominator is small, they can vary independently. The coupling constants will prove useful in making simple models of the seasonal and diurnal cycle of temperature, as we shall do in Chapter 7.

Table 6.2 shows some typical coupling coefficients for Earth and Titan. Since these are derived by linearizing around $T_g = T_{sa}$, the effects of buoyancy on C_D do not affect b_{sens}. Moreover, since both methane and water vapor are positively buoyant in the background gas, the surface layer is in the unstably stratified regime, so that suppression of turbulence does not enter into the picture. The unstable buoyancy effects do cause C_D to increase slightly with T_g, and this would slightly increase b_L. Note that b_{sens} is nearly independent of temperature; the slight variation is due to the effect of the composition on mean specific heat and on surface layer density. The radiative coupling coefficient b_{ir} increases gently with temperature, but the latent heat coefficient b_L increases sharply, owing to the exponential behavior of Clausius–Clapeyron. For Earth, the sensible heat transfer dominates the coupling in cold conditions, which apply near the poles in climates like the present and globally for Snowball conditions. As temperature increases, latent heat fluxes increasingly come to dominate the coupling. In tropical conditions for the present Earth climate, evaporation accounts for 71% of the total coupling coefficient of $82.1\,\mathrm{W/m^2K}$. To get an idea of how tightly coupled the ground temperature is to the overlying air temperature in this case, we note that an increase of $40\,\mathrm{W/m^2}$ in absorbed solar radiation at the ground (arising perhaps from a drastic decrease in cloudiness) could be accommodated by an increase of ground temperature by under a half a kelvin. In these circumstances, the most effective way to increase the surface temperature is not to alter the surface energy budget, but rather to increase the temperature of the atmosphere. This is the main way in which increases in greenhouse gases increase the ground temperature. As temperature increases beyond modern tropical values, evaporation becomes even more dominant, and coupling becomes even tighter.

In the Titan case, the infrared coupling is almost completely insignificant. It is interesting, however, that the sensible heat coupling coefficient is quite significant, owing to the high density of Titan's atmosphere. We have already noticed that on Titan evaporation easily dominates the weak absorbed solar radiation. Evaporation makes the numerator of Eq. (6.36) strongly negative if the relative humidity is appreciably less than 100%, which drives a temperature inversion at the surface. The dominant balance in this case is between evaporative cooling of the ground and transfer of sensible heat from the warmer atmosphere to the colder ground. In this situation, the suppression of turbulence by stable boundary layer effects can play an important role in determining the strength of the inversion (see Problem 6.5). The Earth's tropics is not hot enough to be in this regime, but on a much hotter Earth the increase of water vapor would lead to very similar effects.

6.6 GLOBAL WARMING AND THE SURFACE BUDGET FALLACY

A common fallacy in thinking about the effect of doubled CO_2 on climate is to assume that the additional greenhouse gas warms the surface by leaving the atmospheric temperature unchanged, but increasing the downward radiation into the surface by making the atmosphere a better infrared emitter. A corollary of this fallacy would be that increasing CO_2 would not increase temperature if the lower atmosphere is already essentially opaque in the infrared, as is nearly the case in the tropics today, owing to the high water vapor content of the boundary layer. This reasoning is faulty because increasing the CO_2 concentration while holding the atmospheric temperature fixed reduces the *OLR*. This throws the top-of-atmosphere budget out of balance, and the atmosphere must warm-up in order to restore balance. The increased temperature of the whole troposphere increases all the energy fluxes into the surface, not just the radiative fluxes. Further, if one is in a regime where the surface fluxes tightly couple the surface temperature to the overlying air temperature, there is no need to explicitly consider the surface balance in determining how much the surface warms. Surface and overlying atmosphere simply warm in concert, and the top-of-atmosphere balance rules the roost.

Arrhenius properly took both the top-of-atmosphere and surface balances into account in his estimate of the effect of doubling CO_2, though he did so using a crude one-layer model of the atmosphere. Guy Stewart Callendar (1938) and Gilbert Plass (1959) employed more sophisticated multilevel models, but when it came to translating their radiation results into surface temperature change both got mired in the surface budget fallacy. The prime importance of the top-of-atmosphere balance was emphasized

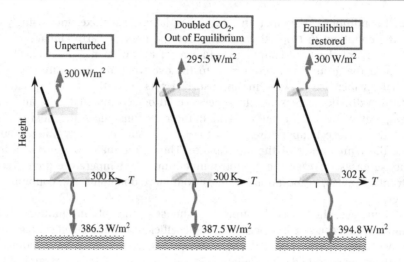

Figure 6.3 Changes in top-of-atmosphere and surface radiative fluxes upon doubling CO_2. Calculations were carried out with the ccm radiation model assuming the atmosphere to be on a moist adiabat patched to an isothermal 180 K stratosphere. The low-level relative humidity is fixed at 80%, while the relative humidity in the free troposphere is 50%.

with crystal clarity in Syukuro Manabe's work of the early 1960s, but one still encounters the surface budget fallacy in discussions of global warming from time to time even today.

Figure 6.3 shows how the budgets change when CO_2 is doubled from 300 ppmv. The case shown is typical of the present Earth's tropics, for which water vapor makes the boundary layer optically thick. The system starts off in balance, at a surface temperature of 300 K. If CO_2 is immediately doubled, the downward radiation into the surface increases by a mere 1.2 W/m². However, the *OLR* goes down by over 4 W/m². The atmosphere–ocean system is receiving more solar energy than it is losing, and so it warms up. The top-of-atmosphere balance is restored when the surface air temperature has warmed to 302 K. This increases the radiation into the ground by an additional 7.3 W/m². Part of this increase comes from the fact that the warmer boundary layer contains more water vapor, and therefore is closer to an ideal blackbody. Most of the increase, however, comes about simply because the low-level air temperature T_{sa} increases, and hence σT_{sa}^4 increases along with it. This increase occurs even if the boundary layer is an ideal blackbody – i.e. completely opaque to infrared. In addition, the increase of T_{sa} would increase the latent and sensible heat fluxes into the surface if the surface temperature were to stay fixed, and this increase also contributes to the warming of the surface.

There are a few situations in which the detailed surface balance could have a significant effect on surface warming. This can only happen in the weakly coupled regime. In that regime ΔT can be fairly large, and changes in ΔT can add to whatever warming is directly caused by the atmospheric warming that comes from satisfying the top-of-atmosphere budget. For example, if land dries out, the loss of evaporation will cause ΔT to increase. Conversely, if a formerly dry area becomes moist, ΔT would decrease, moderating the surface warming. The weakening of the low-level inversion in Antarctica can play a crucial role in Antarctic surface climate change. Finally, it should be noted that when the atmosphere is optically thick, ΔT does not affect the *OLR*. However, when the atmosphere is somewhat transparent to infrared from the surface, an increase in ΔT increases the *OLR* a bit, so that the atmosphere does not have to warm-up quite as much as one thought in order to bring the top-of-atmosphere budget into balance.

The relative roles of the surface budget and the top-of-atmosphere budget in determining surface temperature change upon doubling of CO_2 are further explored in Problems 6.2 and 6.18.

6.7 MASS BALANCE AND MELTING

When the surface consists of a solid ice which can undergo melting, the surface balance works in a somewhat different way once the ground has warmed to the melting temperature. We can no longer solve Eq. (6.34) for T_g, since T_g cannot rise above the melting temperature so long as there is any ice left to melt. Instead, we compute the surface residual at the melting temperature T_f, i.e.

$$F_{net}(T_{sa}, T_f) = F_{rad}(T_{sa}, T_f) - F_{sens}(T_{sa}, T_f) - F_L(T_{sa}, T_f). \qquad (6.37)$$

If $F_{net}(T_{sa}, T_f) > 0$, then the energy flux F_{net} is available for melting. In that case, the mass melted per unit time per unit area is given by F_{net}/L_f, where L_f is the latent heat of fusion. Often this is converted to liquid equivalent depth per unit time by dividing by the density of the liquid phase. Expressed that way, the rate corresponds to the rate of growth of liquid layer that would be caused by the melting, if the liquid did not run off to some other place. Melting is a very powerful means of ablation of ice, be it mountain glacier, ice sheet, or sea ice. The latent heat of fusion is much smaller than the latent heat of sublimation, so a given amount of energy can turn ice into rapidly removable form much more rapidly by melting than by sublimation. For example, the ratio of latent heats for water is 0.118, so a given amount of energy can get rid of 8.5 times as much ice by melting as it can by sublimation. Sublimation always carries the vapor away from the ice surface, but for melt to become realized ablation, the meltwater needs to go away somewhere. It may flow to the base of a glacier through an abyss called a *moulin*, or it may run away to a melt pond and form a temporary lake. It may also percolate into the snow and re-freeze, releasing latent heat in the process. In that case the melting actually constitutes an energy transport mechanism rather than a true ablation.

Figure 6.4 shows the melting rate as a function of air temperature for three different values of surface absorbed solar radiation (see the caption for the rest of the conditions). To illustrate the importance of stable surface layer physics, each calculation is done in two ways: using Monin–Obukhov theory and using neutral surface layer theory. Note that melting can begin even when the air temperature itself is below freezing; this is simply because the energy balance allows the ground to be warmer than the air, if there is a sufficient supply of solar absorption. Note also that even for a fixed air temperature, the melting increases with solar absorption. This shows that the ablation of a glacier can be affected by the solar radiation, even if air temperature does not change. Various processes can change the absorbed solar radiation, among them clouds, changes in the Earth's orbital parameters (see Section 7.6), and the fact that fresh snow is more reflective than old ice. For any given amount of absorbed solar radiation, the melting rate increases dramatically as air temperature is increased, because this increases the delivery of heat to the surface by sensible heat flux and infrared heat flux. The development of a stable surface layer when the air temperature gets large sharply limits the increase of melting. In this regime, increasing the wind

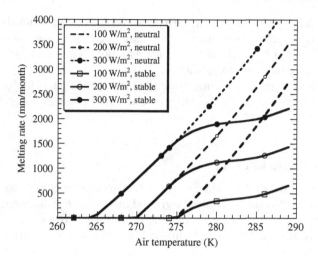

Figure 6.4 Melting rate in liquid water equivalent as a function of air temperature. Results are given for three different values of absorbed solar radiation. The calculations were performed with $U = 5\,\text{m/s}$, $z_* = 0.00033\,\text{m}$, and $h_{sa} = 0.8$.

speed very strongly increases melting, since higher winds favor lower Richardson numbers and higher drag coefficients.

If some geological indicator of past glacial behavior tells us that a mountain glacier was more extensive at some particular time in the past, should we conclude that it was colder then or that it simply snowed more at that time? Similarly, if we see mountain glaciers retreating worldwide at present, should we take that as an indication of a warming climate, or of a reduction in snowfall? The sensitivity of melting rate to temperature has a great bearing on this question. Consider the case with $200\,\text{W/m}^2$ of absorbed solar radiation in Fig. 6.4. Increasing air temperature from 270 K to 274 K increases the melting rate from zero to 650 mm/month. To offset such an increase in melting, the snow accumulation rate would thus have to increase by 650 mm/month. This is a very substantial increase. To put it in perspective, the required *increase* in precipitation is over three times the *maximum* monthly precipitation rate observed in the past few years in central Iceland – which is a very snowy place. Recall, too, that the melting rate we have stated is in liquid water equivalent depth. Snow has low density, and the actual thickness of the corresponding snow accumulation would be 6.5 meters per month or more. It is not impossible for glacier extent to be affected by precipitation, but in situations where melting occurs, it takes a truly enormous change in precipitation to have the same impact as a rather modest change in temperature.

Melting is a very nonlinear process, which acts as a kind of rectifier capable of turning a fluctuating temperature signal into a secular growth or decay of a mountain glacier or continental ice sheet. Melting turns on when the air temperature approaches freezing, and greatly accelerates as temperature becomes warmer. On the other hand, the melting turns off sharply when the air temperature falls much below freezing. In consequence, ablation of ice cares little about just how cold it gets during the depth of winter, but a great deal about the length and warmth of the summertime melt season. This makes the growth and decay of midlatitude and polar mountain glaciers or continental ice sheets very sensitive to what is going on during the melt season.

6.8 PRECIPITATION–TEMPERATURE RELATIONS

There is more to climate than temperature, and for atmospheres that contain a condensable substance – e.g. water on Earth or methane on Titan – the precipitation rate is of as much interest as the temperature. Aside from the role of rainfall in making land habitable on Earth, there are many reasons for being interested in precipitation. For example, it is the precipitation of snow that feeds the growth and flow of glaciers. Precipitation of water on Earthlike planets exerts a controlling influence on the chemical weathering processes that ultimately control atmospheric CO_2, as will be discussed in Chapter 8. In the long term, the rate at which a condensable substance precipitates from an atmosphere must be balanced by the rate at which that substance evaporates or sublimates from a reservoir at the planet's surface – an ocean or glacier. Since latent heat flux can be turned into mass flux upon dividing by the appropriate latent heat, much can be learned about the evaporation or sublimation rate by a careful examination of the surface energy budget. To streamline the prose, we'll generally use the term "evaporation" to refer to evaporation or sublimation in the following, with the understanding that when the surface in question is ice the phase change is actually sublimation.

We begin by writing the surface balance in the form

$$F_L(T_{sa}, T_g) = \left[(1 - \alpha_g) S_g - \sigma e^* T_{sa}^4 \right] + c_p \rho_s C_D U (T_{sa} - T_g) + \sigma \left(T_{sa}^4 - T_g^4 \right). \tag{6.38}$$

In this equation we have assumed $e_g = 1$ for simplicity. The functional form of F_L is given by Eq. (6.13). As usual, this equation must be solved for T_g given T_{sa} and the other parameters affecting the surface budget. We can distinguish two regimes: the weak evaporation regime and the strong evaporation regime. The weak evaporation limit is defined by the condition $F_L(T_{sa}, T_{sa}) \ll (1 - \alpha_g) S_g$. This condition guarantees that F_L will be negligible compared with the surface solar flux as long as the solution does not require that T_g be enormously greater than T_{sa}. Given the form of the Clausius–Clapeyron relation, the weak evaporation regime applies at sufficiently low temperatures, since in that case the atmosphere can carry little vapor even when it is saturated. The notion of "low" vs. "high" temperature must be understood with reference to the volatility of the substance undergoing the phase change. For methane, 95 K is a "high" temperature in this sense, but for water vapor even 250 K is a "low" temperature.

In the weak evaporation limit, we can set the left hand side of Eq. (6.38) to zero when solving for T_g, and then use the resulting ground temperature to evaluate the evaporation by plugging it into the formula F_L. Since the latent heat flux is small, leaving it out of the surface balance causes only small errors in T_g. In this limit, the evaporation is not significantly constrained by the energy supply. In the regime where $T_g \geq T_{sa}$ – i.e. when the term in square brackets in Eq. (6.38) is positive – the exponential increase of saturation vapor pressure with temperature yielded by Clausius–Clapeyron leads to a roughly exponential increase of the latent heat flux with temperature, as T_{sa} is increased. Even when the surface absorbed solar radiation is so weak that an inversion forms in the surface layer, the control exerted by Clausius–Clapeyron is so strong that one still tends to get an exponential increase with temperature, unless the inversion gets very strong. The behavior is explored quantitatively in Problems 6.9 and 6.10.

At high temperatures, when the energy carried away by latent heat flux has a strong effect on the ground temperature, the behavior is very different. Intuitively, one expects that the evaporation cannot increase beyond the point where the entirety of the absorbed solar radiation goes into evaporating material from the surface. It's not quite as simple as that, owing to the effect of sensible and radiative heat fluxes, but a constraint very similar to this does come into play. Let's assume that the condensable substance is like water vapor and makes $e^* \approx 0$ in warm conditions. Then the surface balance becomes

$$F_L(T_{sa}, T_g) = (1 - \alpha_g)S_g + c_p \rho_s C_D U(T_{sa} - T_g) + \sigma \left(T_{sa}^4 - T_g^4 \right). \qquad (6.39)$$

If the second two terms on the right hand side were not there, we would have the result that the latent heat flux is equal to the surface absorbed solar radiation. The two additional terms are positive when $T_g < T_{sa}$, and can thus allow the latent heat flux to somewhat exceed the available solar forcing under circumstances when an inversion can form at the surface. The strength of this inversion determines the amount of "excess evaporation" that can be sustained.

We can show that an inversion must always form in the strong evaporation limit, and also put a strict bound on the temperature difference across the inversion. First, note that when $T_g = T_{sa}$ the last two terms in Eq. (6.39) vanish, while by definition of the strong evaporation limit F_L is much greater than the remaining solar term. Thus, when $T_g = T_{sa}$, the left hand side exceeds the right hand side. Next, recall that latent heat flux vanishes when the saturation vapor pressure corresponding to the ground temperature equals the vapor pressure at the top of the surface layer. Thus, if the relative humidity is below 100%, $F_L(T_{sa}, T)$ will vanish at some temperature $T < T_{sa}$ which we will call T_0. Actually, it is possible that stable boundary layer physics will extinguish the turbulence at some temperature that is somewhat warmer than the temperature at which the gradient of vapor pressure vanishes. In any event, there is a $T_0 < T_{sa}$ where $F_L(T_{sa}, T_0) = 0$. This temperature will be a function of T_{sa} and the other surface layer parameters, as well as the thermodynamic properties of the atmosphere. Because the left hand side vanishes at $T_g = T_0$, while the right hand side is positive, it follows that the left hand side is less than the right hand side. Together with the previous result, we now know that there is a solution to the surface balance equation with $T_0 < T_g < T_{sa}$. The maximum strength of the inversion is $T_{sa} - T_0$, and this determines the maximum excess evaporation, through determining the maximum possible size of the second two terms on the right hand side of Eq. (6.39). It can be shown that $T_{sa} - T_0$ increases very slowly with T_{sa} (Problem 6.11), whence we conclude that the excess evaporation must increase only slowly with temperature.

In Fig. 6.5 we show the latent heat flux as a function of T_{sa}, obtained by solving Eq. (6.38) for T_g using a simple Newton's method iteration. The calculation was carried out for a water/air atmosphere on Earth with $(1 - \alpha_g)S_g = 200\,\mathrm{W/m^2}$, 80% relative humidity in the free atmosphere $U = 5\,\mathrm{m/s}$, and a constant $C_D = 0.0015$. As expected from our analysis of the weak and strong evaporation limits, the flux grows approximately exponentially at low temperatures, but the growth levels off and becomes much weaker once the latent heat flux exceeds the absorbed solar radiation. Notably, the latent heat flux grows by more than two orders of magnitude as the air temperature is increased from 220 K to 300 K, but hardly doubles as it is increased an equal amount to 380 K. An examination of the dry Richardson number (proportional to $T_{sa} - T_g$) shows that an inversion develops in this circumstance, and becomes stronger as the temperature increases. If water vapor were not positively buoyant, the inversion would become stable, limiting the turbulence and reducing the evaporation even more. However, the buoyancy of water vapor allows the surface layer to remain unstable despite the inversion, especially at high temperatures. In other circumstances permitting a stronger inversion, the surface layer can become stable despite water vapor buoyancy (see Problem 6.7).

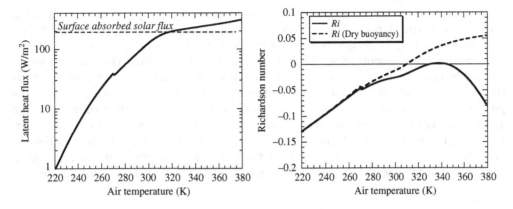

Figure 6.5 Left panel: Latent heat flux computed from the surface energy balance assuming surface absorbed solar flux = 200 W/m², relative humidity of 80%, C_D = 0.0015, and U = 5 m/s. Right panel: Corresponding Richardson number computed with and without the contribution of water vapor to buoyancy.

The preceding results shed some light on precipitation rates of water in both cold and warm climates. To turn the latent heat fluxes into precipitation rates, note that $1\,W/m^2$ of latent heat flux is equivalent to 1.21 cm/yr of liquid water equivalent precipitation if the flux is due to sublimation, or 1.26 cm/yr if the flux is due to evaporation. It is sometimes erroneously supposed that in the cold conditions of a Snowball Earth, the hydrological cycle shuts down. Let's estimate the precipitation rate for the Snowball tropics. Taking the mean surface absorbed solar flux to be $130\,W/m^2$ as is reasonable for ice subject to tropical insolation, and $T_{sa} = 240\,K$, in equilibrium we find that the ice surface temperature is 241.28 K. With these temperatures, the latent heat flux is $1.81\,W/m^2$, which translates into a precipitation rate of 2.2 cm/yr liquid water equivalent. This may not sound like much, but the Snowball can last a very long time. Given 100 000 years to accumulate, this trickle of snowfall can build a glacier 2.2 km high, which is high enough to flow significantly. Thus, there is no essential incompatibility with cold Snowball conditions and geological evidence for active glacier flow. The instantaneous noontime precipitation can be much higher, because it is driven by greater solar flux. Only the mean is relevant for building glaciers, but the relatively heavy noontime sublimation, followed by snowfall as night approaches, can be important in modifying the surface conditions and covering dusty, dark ice with fresh, reflective snow during part of the day.

Turning attention next to the warmer conditions of the Earth's present tropical oceans, we take the absorbed solar radiation to be $200\,W/m^2$ and assume $T_{sa} = 300\,K$. Under these circumstances $T_g = 301\,K$ and the latent heat flux is $125\,W/m^2$, which translates into a precipitation rate of 156 cm/yr. This is reasonably close to the observed tropical precipitation rate. Now suppose that we introduce a high cloud which reflects a lot of sunlight back to space, but which has such a strong greenhouse effect that the change in *OLR* compensates, leaving the top-of-atmosphere radiation budget unchanged; as we saw in Chapter 5, Earth's actual high tropical clouds do something approximating this idealization. Since the air temperature is determined primarily by the top-of-atmosphere balance in the optically thick limit, we can keep T_{sa} fixed at 300 K as in the previous case. However, the surface absorbed solar will be reduced, say to $100\,W/m^2$, while the downwelling infrared is essentially unaffected by the cloud because the tropical atmosphere is optically thick. Under these circumstances, with reduced surface solar flux the surface temperature falls only modestly, to $T_g = 299$ K. However, the latent heat flux falls dramatically, to 57 W/m^2, by about the same proportion as the reduction in surface solar flux. This example shows that in the optically thick warm regime where the surface is tightly coupled to the air by latent heat exchange, the surface energy budget has little influence on temperature. For the purposes of estimating temperature, we could do pretty well by simply assuming that the ground temperature equals that of the overlying air. However, changing terms in the surface budget – as we did here by reducing the surface solar flux – has a profound effect on precipitation. In brief, in warm tropical conditions, the surface energy budget tells us about precipitation, while the top-of-atmosphere energy budget determines the temperature.

6.9 SIMPLE MODELS OF SEA ICE IN EQUILIBRIUM

There are many circumstances in which one would like to know the thickness attained by a layer of ice floating on an ocean once it comes into a state of thermal equilibrium. This problem is relevant to the state of the sea ice cover which forms in the polar regions on Earth today and in other icy climates. It is also relevant to the thick-ice regimes prevailing on a globally glaciated Snowball Earth. Another application is the determination of the ice-crust thickness for icy moons such as Europa, which consist of a crust of water ice floating on a deep brine ocean. In this section, we'll lay out some elementary models which shed light on the determination of ice thickness. While the physics set forth here would apply equally well to the freezing of any liquid whose solid form has lower density than the liquid form, in practice the condition that the ice floats for the most part restricts the applicability to water. An important exception to this is the determination of the thickness of the crust of rocky planets, which can be viewed a a form of "rock-ice" supported by a more fluid interior.

In order to proceed we need to know a bit about the flux of heat through an immobile solid. Heat is conducted through such a substance through collisions of molecules with one another, which propagate information about changes in temperature in one part of the body to the remainder of the body. Both experiment and theory show that in most circumstances, the flux of heat is proportional to the temperature gradient, and the flux is in the direction opposite to the gradient. In other words, the heat flows in such a way as to try to wipe out temperature gradients and make the body isothermal. The constant of proportionality is called the *thermal conductivity*, and we shall call it κ_T. The thermal conductivity is determined by molecular properties of the solid, the specific heat, and the density. Low density substances generally have lower thermal conductivity, since there is less mass available to transmit heat. The thermal conductivity of various substances will be discussed at greater length in Chapter 7. For now, it will suffice to know that for water ice $\kappa_T \approx 2.24$ W/m $\cdot$ K, and that of new fluffy snow is $\kappa_T \approx 0.08$ W/m $\cdot$ K.

If there are no internal sources or sinks of heat within the ice layer, then the heat flux must be constant once the system reaches equilibrium. The value of the flux is set by the heat flux delivered to the bottom of the ice. In some cases, for example the heat flux through a quiescent ocean for a globally glaciated Snowball Earth, the flux would be just the geothermal heat flux escaping the interior of the planet. In other cases, for example in the case of sea ice or shelf ice abutting an open ocean, the flux would be the much larger value delivered by dynamic ocean heat transport – the delivery of heat in the form of relatively warm water by ocean currents. Regardless of the source, we will call this flux F_i. In this section, we'll denote the temperature profile within the ice as $T(z)$, with $T(0)$ being the temperature of the ice or snow surface (also called T_g) and $T(-h)$ being the temperature at the base of the ice, where h is the ice thickness. Then, the constant flux requirement yields the following differential equation for the temperature profile:

$$-\kappa_T \frac{dT}{dz} = F_i. \tag{6.40}$$

Note that this equation remains valid even if the thermal conductivity varies with depth. For example, snow has much lower conductivity than ice, owing to its low density and the immobilization of air in the pore space. Hence, if a layer of ice were blanketed with snow, the temperature gradient within the snow layer would be much steeper than the temperature gradient within the rest of the ice. Suppose that the ice layer has not frozen all the way to the rock, so that the ice is floating on a layer of the same liquid (generally water) which freezes to make the ice. Where the ice is in contact with the liquid, the temperature of the ice must equal the freezing point of the liquid, which we will call T_f. Thus, the bottom boundary condition on temperature is $T(-h) = T_f$. Note that the freezing point of sea water or any other brine is lower than the freezing point of pure water.

The energy balance at the ice surface imposes another condition on the temperature profile. This energy balance is identical to the surface energy budget given in Eq. (6.34), except that one must add in the contribution from the heat flux through the ice. Thus, at $z = 0$ we require

$$0 = F_i + F_{rad} - F_{sens} - F_L. \tag{6.41}$$

The internal flux is usually small, and makes a negligible contribution to the surface balance. In most cases, we can drop the term and compute the ice surface temperature as if it were not there. This equation determines T_g (which is now the ice surface temperature, called $T(0)$) as before, and the inclusion of F_i would only make the ice surface temperature ever so slightly warmer than it otherwise would have been without heat diffusion through the ice.

Sublimation takes away mass as well as heat, and in general this mass loss must be taken into account when formulating the conditions for equilibrium thickness. Let's assume first that there is no net mass loss or gain at the surface. This could be because the temperature is so low that sublimation is negligible, or it could be because all the sublimated mass precipitates back out onto the surface locally. In this case, since there is no mass loss at the surface, there is no freezing at the base of the ice once equilibrium is attained, and hence no latent heat release there, and the only heat flux that needs to escape through the ice is F_i. If we divide Eq. (6.40) by κ_T and integrate over the ice layer, we find

$$T_f - T_g = F_i \int_{-h}^{0} \frac{dz}{\kappa_T} = \frac{F_i h}{\overline{\kappa_T}} \tag{6.42}$$

where $\overline{\kappa_T}$ is the harmonic mean of the thermal conductivity, that is,

$$\overline{\kappa_T} \equiv \left(\frac{1}{h} \int_{-h}^{0} \frac{dz}{\kappa_T} \right)^{-1}. \tag{6.43}$$

Solving for the ice thickness, we find

$$h = \frac{\overline{\kappa_T}(T_f - T_g)}{F_i}. \tag{6.44}$$

This determines the ice thickness once T_g is known. The physical content of this statement is that the ice thickness grows until it is just thick enough to let through the amount of heat delivered to the bottom of the ice. Ice can exist in equilibrium if $T_g < T_f$, and the ice thickness approaches zero as T_g warms to the freezing point. Increasing the heat flux or decreasing the thermal conductivity would also thin the ice. The fact that it is the *harmonic* mean of the thermal conductivity that appears in this equation has important consequences. The harmonic mean gives greatest weight to regions of small conductivity, with the consequence that even a thin layer of very small conductivity can drive the harmonic mean to small values, and hence require thin ice. In particular, a relatively thin blanket of snow can hold in the heat diffusing through the ice, and cause the ice to thin dramatically – all other things being equal. The following exercise gives some feel for the numbers.

Exercise 6.3 (a) The mean geothermal heat flux on Earth is about $0.03\,\text{W/m}^2$, and it was not much greater back in the Neoproterozoic. Snowball Earth simulations indicate tropical ice surface temperatures on the order of 230 K for a globally glaciated planet. How thick do you expect the tropical ice to be if the entire layer has the thermal conductivity of ice? How would the thickness change if you added a 1 meter layer of snow at the top?

(b) In a situation like the present Earth with a great deal of open water, ocean currents can deliver heat to the bottom of an ice shelf or sea ice layer at a rate much greater than the geothermal heat flux. Suppose that ocean currents transport a mean flux of $2\,\text{W/m}^2$ to the bottom of a polar ice layer which has a surface temperature of 250 K. How thick is the ice in equilibrium?

(c) The atmosphere of Europa is so tenuous that it has essentially no radiative effect, so the temperature is determined by a balance between absorbed solar radiation and blackbody emission. Suppose that near the equator the annual mean absorbed solar radiation is $5\,\text{W/m}^2$. The internal heat flux for Europa is not well constrained. Compute the equilibrium ice thickness assuming a heat flux of 0.01, 0.1, and $1\,\text{W/m}^2$.

The second example in the exercise shows that a rather small amount of heat delivered to ice by oceanic heat fluxes can be very effective in melting back sea ice. This is true because essentially all the heat so delivered can be used in melting. In contrast, if one delivered the equivalent of $1\,\text{W/m}^2$ to high latitude regions by heat transport in the atmosphere, a great deal of the heat would be lost by radiation to space, and only a small portion would actually be usable for melting ice.

Note that although the insulating properties of a snow layer thin the ice if the ice surface temperature is held constant, this effect is offset by the fact that snow has a considerably higher albedo than ice, which reduces the surface temperature. Moreover, the low thermal conductivity and low density of snow allow it to cool very rapidly at night, particularly when a stable inversion forms. The daytime warming tends to be not so extreme, owing to stronger turbulent heat fluxes in a neutral or unstable surface layer. This process tends to reduce the daily-mean snow surface temperature, which again has a thickening effect on the ice.

Now let's bring in the effects of mass loss at the surface. In equilibrium, mass loss from the top must be balanced by freezing at the base. The latent heat of fusion adds to the flux delivered to the base of the ice. Hence, if *all* the mass sublimated from the surface is carried away and precipitated elsewhere, F_i is replaced by $F_i + (L_f/L_{sub})F_L$, where L_f is the latent heat of fusion and L_{sub} is the latent heat of sublimation. This makes the ice thinner than it was in our previous estimate. In conditions cold enough to form ice we are generally in the weak evaporation limit, so F_L is small; moreover, for water $L_f/L_{sub} = 0.118$, which brings down the additional flux even more. The net flux delivered to the base is on the order of $1\,W/m^2$ or less in typical conditions, and does not significantly increase the ice surface temperature. Thus, we can get the ice thickness including sublimation by simply replacing F_i in Eq. (6.44) with the modified basal heat flux. The latent heat flux F_L can be estimated using the results of Section 6.8 in the weak evaporation limit. For $F_L = 1\,W/m^2$, which is typical of the cold Snowball Earth tropics, the basal flux increases from $0.03\,W/m^2$ to $0.148\,W/m^2$, which thins the ice by a factor of 5. Clearly the effect of sublimation on ice thickness is very significant, and can allow the tropical ice on Snowball Earth to be much thinner than it otherwise might have been, though even with sublimation taken into account the ice is over $300\,m$ thick when the mean surface temperature is $250\,K$. With stronger sublimation, as would happen as CO_2 increased and the ice surface warmed towards the freezing point, the effect is even more pronounced.

If some regions experience net ablation of ice through sublimation, others must be experiencing net accumulation, since the vapor that sublimates must ultimately precipitate out somewhere else. What happens to ice thickness in regions of net accumulation? To a point, accumulation at the surface can be balanced by melting at the base. However, the only supply of heat available to melt ice at the base is the geothermal heat flux, and a flux of $0.03\,W/m^2$ can sustain a melt of only $3\,mm$ of ice per year. If the accumulation exceeds this tiny rate, then if no other process intervenes the ice thickness increases until it freezes to the bottom. In reality, the generation of regions of thick ice would drive a flow of ice into regions where it can ablate by sublimation, thickening tropical ice and thinning polar ice. Things that flow are beyond the scope of this book.

We'll now consider one last variation on the theme of ice thickness. In the preceding calculations, it was assumed that all solar radiation that was not reflected was absorbed at the surface of the ice. In reality, some radiation will penetrate into the ice and be absorbed in the interior. If the penetration is significant, this can have a powerful effect on thinning the ice, because the low thermal conductivity of ice means that heat buried in the ice has a hard time getting out, and therefore accumulates until a considerable degree of warming has been achieved. To model this process, we modify the steady state thermal diffusion equation to allow for internal heating, which we represent as the vertical gradient of the downward solar flux $F_\odot$. The equation becomes

$$\frac{d}{dz}\kappa_T\frac{dT}{dz} = -\frac{d}{dz}F_\odot \qquad (6.45)$$

or, upon integrating once,

$$\kappa_T\frac{dT}{dz} + F_\odot = const. \qquad (6.46)$$

This says simply that the sum of the diffusive and radiative heat flux must be constant. The constant of integration is determined by the requirement that the net flux out of the surface of the ice must equal the total solar absorption in the ice layer plus the heat flux delivered to the base of the ice. We'll simplify the problem by assuming that there is no applied flux, and that the ice is thick enough that all the solar radiation is absorbed with none penetrating through into the ocean. In this case, the heat flux out of the top of the ice, $-\kappa_T dT/dz$, must equal the solar flux $F_\odot(0)$, whence the constant of integration is zero. In this case, the temperature profile within the ice is

$$T(z) = T_g + \int_z^0 \frac{F_\odot}{\kappa_T}\,dz. \qquad (6.47)$$

The temperature increases with depth in the ice, and becomes uniform at depths where $F_\odot \approx 0$, i.e. below the layer within which most of the solar radiation is absorbed. The deep ice temperature becomes larger as the solar flux penetrates deeper into the ice. To make this more clear, suppose that κ_T is constant and $F_\odot = (1 - \alpha_g)S_g \exp(z/H_\odot)$. Then, the deep ice temperature is

$$T_\infty = T_g + (1 - \alpha_g)S_g H_\odot / \kappa_T. \tag{6.48}$$

This temperature increases without bound as the penetration depth $H_\odot$ increases. For sufficiently large $H_\odot$, the deep ice temperature increases to the melting point, which in that case essentially limits the ice thickness to the solar penetration depth. A precise calculation of the ice thickness is done in Problem 6.12. Note that much of what we have said about the effect of internal absorption of solar radiation applies equally well to other internal heat sources, notably tidal heating arising from flexing of the ice crust. This heat source may be particularly important in determining the ice crust thickness on Europa, or on Snowballs in close orbits about dim M-dwarf stars.

Exercise 6.4 Assume that $T_g = 240\,\mathrm{K}$ and $(1 - \alpha_g)S_g = 100\,\mathrm{W/m^2}$. If the decay of solar flux is exponential, how great does the penetration depth have to be in order to bring the deep ice temperature to the freezing point?

The interest in mechanisms for thinning tropical ice in Snowball conditions arises from two challenges facing the Snowball hypothesis. The first is the obvious need to find a way to exit from the globally glaciated state. Accumulation of CO_2 can warm the planet, but it is not clear that this process is actually sufficient to deglaciate thick ice, or that the necessary levels of CO_2 can be achieved. Thin ice can help make deglaciation easier, especially if solar radiation can penetrate the ice and warm the underlying ocean. The second challenge is that photosynthetic eukaryotes, which are rather fragile creatures in comparison with cyanobacteria, seem to have made it through the Neoproterozoic Snowball without any evidence of a major crisis (such creatures were not yet around at the time of the putative Paleoproterozoic Snowball, and so pose less of a problem then). Thin ice allows more fractures and leads, which can provide open water refugia. If the ice is thin and clear enough, enough solar radiation may even be able to penetrate the ice to support photosynthetic life beneath the ice layer.

Though we have mainly had floating water ice in mind in the preceding derivation, the heating due to solar penetration is generically applicable to planets with an icy crust, whatever the ice may be made of. For example, burial of solar heating is thought to lead to cryovolcanism in the nitrogen-ice crust of Neptune's moon Triton. The interior heating of ice is a form of solid greenhouse effect, in that the solar energy penetrates well, but the heat can only escape slowly. In this case, the heat transfer through the ice is by molecular diffusion rather than infrared radiation, but the general principle is the same.

The problem of solar absorption within ice is a radiative transfer problem nearly as challenging as that confronted for clouds. It depends on scattering off air bubbles and brine pockets, and therefore requires some understanding of the distribution of these. The absorption and scattering is wavenumber-dependent, so spectrally resolved radiative transfer should ideally be used, and as is the case for the atmosphere, the spectral absorption features generally lead to non-exponential attenuation of the solar beam. These issues are all at the frontier of climate research.

6.10 WORKBOOK

6.10.1 Basic surface balance calculations

Problem 6.1 Using Eq. (4.30), show that for an optically thin gray gas the emission and absorption coefficients e_a, $e_{a,top}$, and a_+ defined in Section 6.2 are given by $a_+ = \tau_\infty$ and $e_a = e_{a,top} = \tau_\infty (\bar{T}/T_{sa})^4$, where $\bar{T}$ is the mean atmospheric temperature given by Eq. (4.31). Compute $(\bar{T}/T_{sa})^4$ and $a_+/(e_a + e_{a,top})$ for the following three cases:

- An all-troposphere ideal gas atmosphere on the dry adiabat, without pressure broadening;
- The same with linear pressure broadening;
- The same with linear pressure broadening, but with the temperature patched to an isothermal stratosphere at high altitudes.

Under what circumstances is the radiatively driven temperature jump at the ground unstable to convection?

Problem 6.2 *Surface global warming fallacy, Part 1*
Suppose we hold the atmospheric temperature fixed on the moist adiabat having a low-level air temperature of 280 K as we increase CO_2 from 300 ppmv to 600 ppmv. Assuming a moist atmosphere, interpolation from the data in Fig. 6.1 yields $e^* = 0.314$ for the original CO_2 and $e^* = 0.307$ for doubled CO_2. Using these values in a surface energy balance model, determine how much the surface would warm if we ignore turbulent fluxes and assume that the surface balance is purely radiative. Then, assuming the surface to be dry so that there is no evaporation, re-do the calculation including sensible heat fluxes. Now assume the surface is wet and include evaporation as well. For the turbulent fluxes, assume that the surface wind speed is 5 m/s and that the boundary layer relative humidity is 80%. Keep C_D fixed at 0.0015, and assume that the surface absorbed solar radiation is 200 W/m^2.

Next, for the case including all turbulent fluxes, compute how much the surface warms if you increase the air temperature by 2 K at the same time CO_2 is doubled. How much does the answer change if you leave the CO_2 fixed at 300 ppmv?

Finally, re-do all the above calculations assuming the low-level air temperature to be 300 K, as is typical of the tropics. For this case, $e^* = 0.1286$ at 300 ppmv and $e^* = 0.1270$ at 600 ppmv.

Summarize all your results in a table. What do you conclude about the importance of increase in atmospheric temperature vs. the direct effect of CO_2 on downwelling radiation into the surface? Discuss the role of evaporation in determining how much the surface temperature changes.

Problem 6.3 *Making ice in the desert*
You are out in the desert at night, and you would like to make some ice. There is plenty of dry air around. You fill a trough with water. The trough is open at the top so evaporation can occur, but it is well insulated so that no heat is exchanged through the walls. Suppose that the relative humidity of the air is h and the temperature of the air is T_a. Find the conditions on h and T_a under which ice can be made, i.e. the conditions under which the surface temperature of the water can be brought down to freezing if the wind blowing over the water is strong enough. If you are free to make the wind as strong as you like, does your answer depend on the surface radiative cooling factor e^*?

Next, fix T_a at 290 K and pick a value of h that allows the temperature to go below freezing in the strong-wind limit. Compute the value of ρU needed to make ice for

$e* = 0.3, 0.5,$ and $0.7.$ (In reality, $e*$ would be determined by h with drier values yielding higher $e*$.) First do the calculation with fixed $C_D = 0.0015$, and then do it incorporating Monin–Obukhov stable surface layer physics.

Problem 6.4 The surface of Neptune's moon Triton and the former planet Pluto are believed to be composed primarily of N_2 ice. The atmospheres of these bodies are so thin that their effect on infrared radiation can be neglected. Therefore, the temperature of the surface can be determined from the surface energy balance. At the subsolar point, the surface will heat up until the absorbed solar radiation equals the energy carried away by blackbody radiation (simply σT_g^4 in this case) plus the latent heat of sublimation of the N_2 ice. The saturation vapor pressure of N_2 over the relatively warm subsolar point is much greater than the pressure over the colder parts of the body, and this drives a mass flow away from the subsolar point, carrying latent heat with it. You may assume that the albedo is 60% .

What is the surface pressure of N_2 at the subsolar point if the latent heat flux is negligible? (*Hint:* The gas must be in equilibrium with the solid surface with which it is in contact.) Now suppose that you are given the flux of mass (per unit area) out of the subsolar point. Compute the temperature and pressure as a function of this mass flux. Is there a maximum possible mass flux? Speculate on what processes might determine the mass flux. Can you use the latent heat flux formula derived in the text for this class of problems?

Some relevant data: The solar constant at Triton's orbit is $1.5\,\mathrm{W/m^2}$; at Pluto's it is $0.89\,\mathrm{W/m^2}$ on average. The latent heat of sublimation of N_2 ice is $2.44 \cdot 10^5\,\mathrm{J/kg}$.

Problem 6.5 Consider a Titanlike situation in which the dominant balance at the surface is between latent and sensible heat fluxes, and all other terms are negligible. Allow for stable boundary layer effects, so that C_D is a function of the Richardson number. Show that if C_D is non-zero, the solution becomes independent of $C_D U$, and derive an expression for the strength of the surface inversion. Show that there is another solution, with $Ri = Ri_c$ and $C_D = 0$, and derive an expression for the strength of the inversion in this case.

Put in numbers corresponding to a CH_4/N_2 atmosphere on Titan, assuming $h_{sa} = 0.7$, $U = 5\,\mathrm{m/s}$, $z_1 = 10\,\mathrm{m}$, and $z_* = 0.001\,\mathrm{m}$, for temperatures in the vicinity of 95 K. Do the same for an Earthlike water/air atmosphere with temperatures in the vicinity of 350 K. How valid is the neglect of infrared radiation in the Earthlike case?

Problem 6.6 Consider the surface energy budget in a cold climate, in which evaporation can be neglected and the budget consists only of solar absorption, infrared, and sensible heat flux terms. Include the Monin–Obukhov theory for stable surface layers, and set the parameters to Earthlike values.

Plot the surface energy budget as a function of T_g for fixed T_{sa} so as to determine the equilibrium surface temperature. Try this for various values of the absorbed solar radiation $(1 - \alpha_g)S_g$, and discuss the strength of the inversion and the effect of turbulence suppression. Are there multiple equilibria? Are there sharp transitions between strong-inversion and weak-inversion cases as the solar absorption is changed? Repeat this calculation for some much colder and much warmer values of T_{sa}, so as to alter the importance of the infrared radiation term.

To keep things simple, you may assume that e_a is held constant at a value of 0.2 throughout this problem, and that the ground has unit emissivity. *Hint:* You should first get a feel for the problem by making a set of energy balance graphs analogous to Fig. 6.2, but at some point you may wish to write a simple Newton's method routine to solve the surface budget equations iteratively. You can use the iteration to implement a function Tg(Tsa) using your favorite programming language.

6.10.2 Evaporation and sublimation

Problem 6.7 Re-do the graph in Fig. 6.5 for parameters which make the surface layer more stable: smaller U, smaller relative humidity, and/or smaller solar absorption.

Problem 6.8 Sublimation steals energy that could be more effectively used for melting. Study this by computing the latent heat flux for the cases in Fig. 6.4 and determining how much additional melting could be sustained if this energy went into melting instead. Study the issue further by re-computing the figure with lower and higher relative humidity. Increasing the wind speed increases both sensible and latent heat fluxes; does the net effect enhance or retard melting?

Problem 6.9 Use the linearized form of the surface energy budget to compute $T_g - T_{sa}$ in the weak evaporation limit, and use the resulting expression to discuss how the latent heat flux increases with temperature. Assume the boundary layer to be governed by neutrally stratified theory, so that C_D can be regarded as independent of temperature. For simplicity, you may also assume $e^* = 0.8$ throughout this problem, and neglect variations with temperature.

Problem 6.10 This problem explores the behavior of the weak evaporation limit when the surface layer becomes stable enough to significantly suppress turbulence. You may assume that the atmosphere is also cold enough that the buoyancy is determined by temperature alone, and is unaffected by the concentration of the condensable vapor. First determine the ground temperature $T_{g,0}$ at which turbulence is completely suppressed. You do this by setting $C_D = 0$ in the surface balance equation and finding an expression for the resulting radiatively determined T_g. Then use this value to determine the circumstances under which $Ri = Ri_c$ for this temperature. This determines when the assumption $C_D = 0$ is consistent. Finally, assuming this condition to be satisfied for some given air temperature $T_{sa,0}$, discuss how T_g varies as T_{sa} is increased beyond $T_{sa,0}$ (holding other parameters fixed), retaining terms out to second order in $T_{sa} - T_{sa,0}$. Use the result to say how the evaporation behaves as a function of temperature in the weak evaporation limit. Illustrate your result by putting in some numbers characteristic of a cold Earthlike climate with weak absorbed solar radiation, with an atmosphere consisting of air and water vapor. For simplicity, you may assume $e^* = 0.8$ throughout this problem, and neglect variations with temperature.

Problem 6.11 As discussed in the text, the limiting strength of the inversion in the strong-evaporation limit is given by the temperature T_0 where $F_L(T_{sa}, T_0) = 0$, i.e. where the latent heat flux vanishes. Using the expression for F_L, show that this occurs when $p_{sat}(T_0) = h_{sa}p_{sat}(T_{sa})$, where h_{sa} is the relative humidity of the air at the upper edge of the surface layer. Use the simplified exponential form of Clausius–Clapeyron to derive an expression for $T_{sa} - T_0$. How rapidly does this increase with T_{sa}? Use the results to put an upper bound on the growth of evaporation with temperature in the strong evaporation limit. Put in some numbers corresponding to Titan, and to a very hot Earth.

6.10.3 Ice

Problem 6.12 Determine equilibrium ice thickness for a slab of ice which allows solar penetration into the ice as discussed in text, but with the following extensions: (1) Assume that a geothermal heat flux F_i is applied at the base of the ice; and (2) allow for the possibility that the ice is so thin or transparent that some solar radiation penetrates the ice and escapes into the ocean below.

First do the calculation assuming that solar radiation that leaves the bottom of the ice is carried away in the ocean and does not need to re-escape through the ice. Then say what happens if the energy is absorbed locally and heats the water below the ice, leading to the heat being delivered to the base.

For the purposes of this problem, you may neglect the effects of sublimation and accumulation at the top of the ice.

Note that aside from its possible application to tropical ice on Snowball Earth, this "thin, clear ice" calculation applies to ice-covered Antarctic Dry Valley lakes, which some see as a modern analog for Snowball Earth tropical conditions.

Problem 6.13 Consider sea ice of thickness d on a Snowball Earth, in a region where the surface temperature is in equilibrium with an absorbed solar flux of $50\,W/m^2$. The ice has unit infrared emissivity. What is the surface temperature ignoring the heat flux through the ice? Then, assuming that the base of the ice has a temperature of $270\,K$, compute the surface temperature as a function of d, taking into account the heat flux through the ice. You may assume that the temperature profile within the ice is in equilibrium. Ignore the effects of solar flux penetration into the ice.

Problem 6.14 In the text we determined the effect of solar flux penetration into ice assuming the decay rate to be exponential. In reality, the decay rate is weaker than exponential, because of wavelength dependence of the absorption coefficient. Re-do the calculation assuming that the solar flux decays according to a power law of the form $1/(1 + a \cdot z^b)$, where z is distance from the surface and $b > 0$ is a decay exponent. You may assume the ice to be infinitely thick. What is the deep ice temperature?

Problem 6.15 The current permanent glaciers on Mars are made of water ice, but suppose that at some past time Mars had CO_2 ice glaciers. Let the thickness of the glacier be h at some point. In order for melting at the base to be possible, the basal pressure has to be above the triple point pressure. What is the critical height? Suppose that the surface temperature of the glacier is $200\,K$, which is approximately in equilibrium with a 1.5 bar CO_2 atmosphere. Assuming the height to be above the critical height, compute, as a function of height, the magnitude of geothermal heat flux needed to melt the base. At $200\,K$ the thermal conductivity of CO_2 is is about $0.5\,W/m \cdot K$.

Basal melting is important, because it lubricates the glacier and allows it to surge into regions that are warm enough to ablate it rapidly.

6.10.4 Radiative-convective equilibrium coupled to a surface budget

Problem 6.16 In Problem 5.27 you computed the radiative-convective equilibrium for a gray gas atmosphere assuming that the surface coupling was so strong that the ground temperature T_g was always equal to the overlying air temperature T_{sa}. Using the same basic solution approach, generalize the problem to the case in which T_g differs from T_{sa}, and the relation between the two temperatures is determined by the surface energy balance. To keep things simple, you may assume that the surface balance involves only radiative and sensible heat fluxes. Further, you may assume that the atmosphere is transparent to solar radiation. Note that even though the atmosphere absorbs no solar radiation, you cannot do this problem by fixing the surface temperature and computing the *OLR* afterwards, since the surface budget depends explicitly on the surface absorbed solar radiation $(1 - \alpha_g)S_g$. You will need to use the solution technique outlined in Problem 5.27.

Note that when doing convective adjustment in a problem like this, the convective adjustment should be based on the low-level air temperature T_{sa}, not the ground temperature. The ground cannot convect upward! Heat must first be transferred to the air by turbulent and radiative fluxes. As long as the troposphere does not disappear completely, one does not need to do anything explicitly to take this heat transfer into account. It is implicitly mixed throughout the troposphere by convection, and the transfer of solar heating of the ground into the troposphere is implicitly taken care of when one imposes the condition of net (solar plus IR) radiation balance at the tropopause.

Assume $R/c_p = \frac{2}{7}$, and that the gray absorption coefficient increases linearly with pressure. Fix the absorbed solar radiation at $250\,W/m^2$ and explore the behavior of the system as a function of $\rho C_D U$ and τ_∞. At large $\rho C_D U$ the solution should reduce to the previous one, with $T_g \approx T_{sa}$. How does the surface budget affect the tropopause height? How much does it affect T_{sa}? How much does it affect the stratospheric temperature? Make sure to cover both optically thick and optically thin cases. Discuss the magnitude of the surface temperature jump in the optically thin (Marslike) case. Finally, look at the warming in T_g and T_{sa} you get when you increase τ_∞ from 1 to 1.1. How important is the surface budget in determining the amount of low-level warming?

Problem 6.17 Extend the inquiry in Problem 6.16 to the real gas case. Specifically, use the homebrew exponential sums radiation code to compute radiative-convective equilibrium for CO_2 mixed with dry air on the dry air adiabat. You may ignore temperature scaling of the absorption coefficients and the effects of atmospheric solar absorption. Include only radiative and sensible heat terms in the surface budget, and assume a constant C_D. You solve the problem by time-stepping to equilibrium in a fashion almost identical to that described in Section 5.10. The only exception is that instead of adjusting T_g at each time step so as to move towards a state of top-of-atmosphere balance, you adjust the air temperature T_{sa} and then determine the corresponding T_g by solving the surface energy balance equation using the downwelling infrared from the atmosphere as one of the inputs.

Carry out a calculation of this type, and explore how the surface temperature, air temperature, stratospheric temperature, and tropopause height depend on the surface wind speed. Try some very low CO_2 values where the atmosphere is very optically thin, and try partial pressures up to 100 mb, where the atmosphere is more optically thick. You can keep the surface air pressure fixed at 1000 mb for these calculations, and assume Earth gravity. Remember that convective adjustment should be done based on T_{sa}, not on T_g.

If you can run the ccm radiation code, you can alternatively do this problem by time-stepping a ccm-based radiative-convective model to equilibrium. In that case, you can include the joint effects of CO_2 and water vapor, determining the latter assuming a fixed relative humidity. In that case, it also makes sense to include evaporation in the surface budget.

Python tips: The radiative-convective calculation for the homebrew model is done in the Chapter Script `RadConvEqTrop.py`, and that for `ccm` is done in `ccmRadConv.py`. Both are included in the Chapter Scripts directory for Chapter 5. These scripts do not include a surface balance, but need only slight modifications in order to allow T_g to be computed by solving the surface balance at each time step.

Problem 6.18 *Surface global warming fallacy, Part 2*

In this problem you will extend the inquiry of Problem 6.2 by simultaneously determining the change in ground temperature and air temperature when CO_2 is increased. This is done by simultaneously satisfying the top-of-atmosphere and surface balance conditions. We'll do this based on real gas radiation calculations from the `ccm` radiation model incorporating both CO_2 and water vapor, but to keep things simple we'll employ an all-troposphere model so that the calculations can be done using polynomial fits to the radiation computations. With this approach, you do not need to be able to run the `ccm` model yourself, but more importantly the simplification makes it much easier to understand what is going on.

In Table 4.2 we gave polynomial fits to $OLR(T_g, CO_2)$ for fixed relative humidity, assuming that $T_g = T_{sa}$. With the surface budget incorporated, we must allow the two temperatures to differ, and because the atmosphere is not completely optically thick, this affects the OLR. We can write the OLR using a variant of Eq. (6.6) as follows

$$OLR(T_{sa}, T_g) = OLR(T_{sa}, T_{sa}) + (1 - a_+(T_{sa})) \left(\sigma T_g^4 - \sigma T_{sa}^4 \right) \tag{6.49}$$

where a_+ is the atmospheric absorption factor discussed in the text. The first contribution to OLR, i.e. $OLR(T_{sa}, T_{sa})$, is obtained using the polynomial fit in Table 4.2. This function, as well as a_+, depends on CO_2 in addition to temperature. To keep things simple, assume the atmosphere to be transparent and non-scattering, so that the OLR is in equilibrium with an absorbed solar flux $(1 - \alpha_g)S$. This balance condition gives you one relation between T_g and T_{sa}.

The surface balance condition imposes an additional relation between T_g and T_{sa}, which closes the problem. This is the same surface balance condition as discussed in the text, and which you used in Problem 6.2. It requires that you know the surface cooling factor e^*, and the way it depends on temperature. In the surface budget, you may assume that C_D is fixed at 0.0015. Include both sensible and latent heat fluxes in your surface budget.

To complete the calculation, you need to be able to determine how $(1 - a_+)$ and e^* vary as a function of T_{sa} and CO_2. Tabulations giving the variation are provided in the datasets `ccmTransWet.txt` and `SurfEstarWet.txt` in the Workbook datasets directory for this chapter. These were computed using the `ccm` radiation model with 50% relative humidity aloft. Values for arbitrary temperature and CO_2 within the range of the table can be found using interpolation.

The solution procedure is simple. You guess T_{sa}, and then solve for T_g using the surface energy balance. Solving for T_g requires a small Newton's method iteration if you use the full form of the surface budget, but if you are in a regime where the surface budget can be linearized then you can actually do this step analytically, as discussed in the text. Either way, once you have T_g, you can compute the OLR. You then iterate on T_{sa} until the top-of-atmosphere budget is satisfied.

Note that when $(1 - a_+(T_{sa}))$ is small or when the surface coupling is so strong that $T_g \approx T_{sa}$, then the OLR expression allows T_{sa} to be determined nearly independently from the surface budget. Except for optically thin atmospheres with weak coupling, the surface budget usually imposes only a small correction on the T_{sa} you get by ignoring the second term in the *OLR* expression above.

Write a program which implements this procedure. First fix CO_2 at 300 ppmv, and find values of $(1 - \alpha_g)S$ that give you values of T_{sa} near 260 K, 280 K, and 300 K with evaporation included and a wind $U = 5$ m/s. Then double CO_2 and re-compute the temperatures. How much of the increase in T_g is due to the increase in the air temperature? How much of the increase is due to the direct effect of CO_2 on downwelling radiation for fixed air temperature? Compare the results with the temperature increase you get by fixing $T_g = T_{sa}$.

Then, re-do the series of calculations for a case with significantly weaker winds and a case with significantly stronger winds. Finally, with $U = 5$ m/s, how do your results change if you eliminate the latent heat fluxes, as would be the case over a dry surface?

If you like you can also do this problem by time-stepping the full ccm calculation to radiative-convective equilibrium. See Problem 6.17 for hints on how to do that.

Python tips: Remember that the interp class in ClimateUtilities makes polynomial interpolation easy. To interpolate to an arbitrary temperature and CO_2 from the table, first do an interpolation to the temperature you want for all the CO_2 columns, and then do an interpolation in $\log(CO_2)$. The transparency and surface cooling coefficients are computed in the Chapter Script ccmSurfrad.py, but if you re-compute the coefficients with different gravity, surface pressure, or humidity you will need to make sure to use an *OLR* polynomial which is computed consistently.

6.11 FOR FURTHER READING

For a comprehensive reference to the planetary boundary layer and atmospheric turbulence, see

- Garrett JR 1994: *The Atmospheric Boundary Layer.* Cambridge.

Measurements of the hottest ground temperature on Earth are discussed in:

- Mildrexler DJ, Zhao M and Running SW 2006: Where are the hottest spots on Earth? *EOS* **87**, doi: 10.1029/2006EO430002.

A discussion of Monin–Obukhov scaling functions for the unstable case, with particular attention to the free-convection limit, can be found in

- Delage Y and Girard C 1992: Stability functions correct at the free convection limit and consistent for both the surface and Ekman layers, *Boundary-Layer Meteorology* **58**, 19–31.

This journal is the primary source for results about turbulent fluxes in the surface layer and the planetary boundary layer.

CHAPTER SEVEN

Variation of temperature with season and latitude

7.1 OVERVIEW

Why is the Earth generally hotter near the Equator than at the poles? Why is it generally hotter in summer than in winter, especially outside the tropics? Would this be true on other planets as well? How would the pattern change over time, as features of the planet's orbit vary? Would a very slowly rotating planet lose its atmosphere to condensation on the nightside? Would a planet whose rotation axis was steeply inclined relative to the normal to the plane of the orbit, or a planet in a highly elliptical orbit, have such an extreme seasonal cycle that it would be uninhabitable? The answers to these questions are to be found in the way the geographic and temporal pattern of illumination of the planet plays off against the thermal response time of the atmosphere, ocean, and solid surface of the planet. Generally speaking, in this section we seek to understand the features of a planet that determine the magnitude and pattern of geographic and seasonal variations in temperature.

Most of the discussion of temporal variability will focus on seasonal rather than diurnal variations, but much of the same considerations apply to both cycles, and so some remarks will be offered on the diurnal cycle as well. It should be kept in mind that the distinction between diurnal and seasonal cycle is meaningful only for bodies such as the Earth, Mars, or Titan whose rotation period is short compared with the period of orbit about the Sun. For a planet whose length of day is a significant fraction of its year, one should think instead of a hybrid seasonal/diurnal cycle. The formulation developed below is sufficiently general to handle that case.

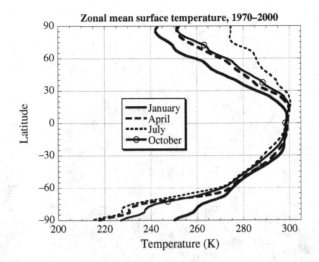

Figure 7.1 Observed zonal mean surface air temperatures for January, April, July, and October. Computed from NCEP data for 1970–2000.

7.2 A FEW OBSERVATIONS OF THE EARTH

First, let's take a look at how the Earth's surface temperature varies with the seasons. Figure 7.1 shows the zonal-mean air temperature near the surface for representative months in each of the four seasons. The first thing we note is that the temperature is fairly uniform in the tropics (30° S to 30° N), but declines sharply as the poles are approached. The temperature difference between the Equator and 60° N is 39 K in the winter but only 12 K in the summer. The southern hemisphere has a much weaker seasonal cycle, except over the Antarctic continent: The temperature difference between the Equator and 60° S is 26 K in the winter and 22 K in the summer. However, over Antarctica, poleward of 60° S the seasonal cycle is extreme. Noting that the northern hemisphere has more land than the southern hemisphere, the data imply that the oceans have a strong moderating effect on the seasonal cycle. The temperature patterns in Figure 7.1 are what we seek to explain in terms of the response of climate to the geographically and seasonally varying solar forcing.

An even better appreciation of the effect of land masses on the seasonal cycle can be obtained by examining the map of July–January temperature differences, shown in Fig. 7.2. This map shows that the strongest seasonal temperature contrast occurs in the interior of large continents, and that the ocean temperature varies by at most a few kelvin over the year – and even less in the tropics. The strong seasonal cycle of the northern hemisphere continents extends very little beyond the coastlines, and the seasonal cycle of the northern oceans has similar magnitude to that of the more extensive southern oceans.

7.3 DISTRIBUTION OF INCIDENT SOLAR RADIATION

The geographical variations of temperature are driven by variations in the amount of sunlight falling on each square meter of surface, and also by variations in albedo. Seasonal variations are driven by changes in the geographical distribution of absorbed sunlight as the planet proceeds through its orbit. Therefore, the starting point for any treatment of

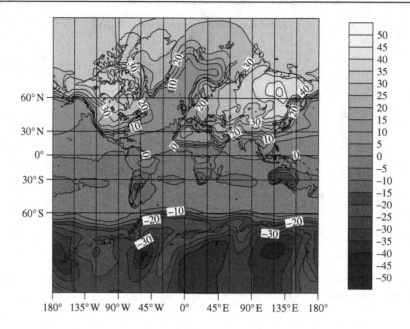

Figure 7.2 Map of July–January surface air temperature difference, in kelvins.

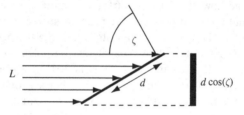

Figure 7.3 Sketch illustrating calculation of the flux per unit surface area for a beam incident on a slanted surface.

seasonal and geographical variation must be the study of how the light of a planet's sun is distributed over the spherical surface of the planet. This section deals only with the distribution of incident sunlight, or *insolation*. The geographical distribution of the amount of sunlight absorbed is affected also by the distribution of the albedo. The albedo variations can also affect the seasonal distribution of solar forcing through seasonal variations in ice, snow, cloud, and vegetation cover.

It will help to first consider an airless planet, so that we don't at once have to deal with the possible effects of scattering of the solar beam by the atmosphere. If our planet is far from its sun, as compared with the radius of the sun, the sunlight encountering the planet comes in as a beam of parallel rays with flux $L_\odot$. Even if the surface of the planet is perfectly absorbing, the sunlight the planet intercepts is not spread uniformly over its surface; per unit area, parts of the planet where the sun is directly overhead receive a great deal of energy, whereas parts where the sunlight grazes the surface at a shallow angle receive little, because the small amount of sunlight intercepted is spread over a comparatively large area, as shown in Fig. 7.3. The nightside of the planet, of course, receives no solar energy at all.

   To obtain a general expression for the distribution of incident solar radiation per unit of surface area, we may divide up the surface of the planet into a great many small triangles, and consider each one individually. The solar energy intercepted by a triangle is determined by the area of the shadow that would be cast by the triangle on a screen oriented perpendicular to the solar beam. To compute this area, suppose that one of the vertices of the triangle is located at the origin, and that the two sides coming from this vertex are given by the vectors $\vec{r_1}$ and $\vec{r_2}$. By the definition of the cross product, the area of the triangle is given by $2A\hat{n} = (\vec{r_1} \times \vec{r_2})$ where $\hat{n}$ is the unit normal to the plane containing the triangle. To obtain the area of the shadow cast by the triangle, we apply the cross product to the projection of the vectors $\vec{r_1}$ and $\vec{r_2}$ onto the plane. These projections are given by $\vec{r_1} - \hat{z}\vec{r_1} \cdot \hat{z}$ and $\vec{r_2} - \hat{z}\vec{r_2} \cdot \hat{z}$, where $\hat{z}$ is the unit vector pointing in the direction of the sun. The cross product of these two vectors is

$$(\vec{r_1} - \hat{z}\vec{r_1} \cdot \hat{z}) \times (\vec{r_2} - \hat{z}\vec{r_2} \cdot \hat{z}) = \vec{r_1} \times \vec{r_2} - (\hat{z} \times \vec{r_2})(\vec{r_1} \cdot \hat{z}) - (\vec{r_1} \times \hat{z})(\vec{r_2} \cdot \hat{z}). \tag{7.1}$$

Now, the cross product of two vectors in the xy plane must point in the direction of the z axis. Hence, we can obtain the magnitude of the above vector by taking its dot product with $\hat{z}$. This is very convenient, since the dot product of $\hat{z}$ with the second two terms vanishes, leaving us with

$$2A_{\perp} = \hat{z} \cdot (\vec{r_1} \times \vec{r_2}) = 2A\hat{z} \cdot \hat{n} = 2A\cos\zeta \tag{7.2}$$

where $A_{\perp}$ is the area of the shadow and ζ is the angle between the normal to the patch of surface and the direction of the sun. This is known as the *zenith angle*. When the zenith angle is zero, the sun is directly overhead, and when it is 90° the sunlight comes in parallel to the surface and leaves no energy behind. Zenith angles greater than 90° are unphysical, since they represent light that would have to pass through the solid body of the planet in order to illuminate the underside of the surface; these are on the nightside of the planet. If one draws a line from the center of the planet to the center of the sun, the zenith angle will be zero where the line intersects the surface of the planet; this is the *subsolar point*. At any given instant, the curves of constant zenith angle make a set of concentric circles centered on the subsolar point, with a zenith angle of 90° along the great circle which at the given instant separates the dayside of the planet from the nightside. If the surface were in equilibrium with the instantaneous incident solar flux, the subsolar point would be the hottest spot on the planet, with temperature falling to zero with distance away from the hot spot. As the planet rotates through its day/night cycle, a given point of the surface is swept through a range of distances from the hot spot, leading to a diurnal temperature variation. As the planet proceeds through its orbit in the course of the year, the diurnal cycle will change as the orientation of the planet's rotation axis changes relative to the sun. Insofar as the surface actually takes a finite amount of time to heat up or cool down, the diurnal cycle will be attenuated to one extent or another.

   As the next step toward realism, let's now consider a rapidly rotating planet whose axis of rotation is perpendicular to the line connecting the center of the planet to the center of its sun. If the axis of rotation is in fact perpendicular to the plane of the orbit, this situation prevails all year round; otherwise, the condition is met only at the equinoxes, and indeed the condition defines the equinoxes. We assume that the planet is rotating rapidly enough that the day-night difference in solar radiation is averaged out and the corresponding temperature fluctuations are small. In other words, the length of the day is assumed to be short compared with the characteristic thermal response time of the planet's surface, a concept which will be explored quantitatively in Section 7.4. Consider a small strip of the planet's surface near a latitude ϕ, of angular width $d\phi$. If a is the planet's radius, then the area of this strip is $2\pi a^2 \cos \phi d\phi$, if angles are measured in radians. The cross-section area of the strip seen edge-on looking from the sun determines the amount of solar flux intercepted by the strip. This area is $2a^2 \cos^2 \phi d\phi$ when $d\phi$ is small. In consequence, the incident solar radiation per unit area at latitude ϕ is $L_\odot(\cos \phi)/\pi$. At the planet's equator, the solar radiation per unit area is $L_\odot/\pi$, which is somewhat greater than the value $L_\odot/4$ which we obtained in Chapter 3 by averaging solar radiation over the *entire* surface of the planet. If the planet has no atmosphere to transport heat or create a greenhouse effect, the equilibrium temperature as a function of latitude is

$$T = \left(\frac{L_\odot \cos \phi}{\pi \sigma} \right)^{\frac{1}{4}}. \tag{7.3}$$

The temperature has its maximum at the equator, and falls to zero at the poles.

Exercise 7.1 For the geometric situation described above, derive an expression for the cosine of the zenith angle as a function of latitude and longitude. Re-derive the expression for the daily-average distribution of solar absorption by averaging the cosine of the zenith angle along latitude circles.

Now we turn to the general case, in which the axis of rotation of the planet is not perpendicular to the plane containing the orbit. The angle between the perpendicular to the orbital plane and the planet's axis of rotation, is known as the *obliquity*, and we shall call it γ. It can be regarded as constant over the course of a planet's year, though there are longer-term variations which will be of interest to us later. The near-constancy of γ arises from the conservation of angular momentum in the absence of torques; small torques on the planet give rise to the long-term variations.

For a moon, it is the obliquity of the rotation axis relative to the plane of the orbit of the host planet that counts. The typical behavior of the major Solar System moons is to be orbiting very nearly in the plane of the host planet's equator, and to be tide-locked to the planet. This is the case for Titan, for example. For such moons, the obliquity of the moon (for the purposes of determining the seasonal cycle) is essentially the same as the obliquity of the host planet. The obliquity of Neptune's moon Triton is the principal known exception to this simple behavior. Its orbit is highly inclined to Neptune's equator, and this leads to a very complex seasonal cycle. In any event, a moon's "year" for the purposes of computing illumination is essentially the same as the host planet's year, since the radius of a moon's orbit is inevitably small compared with the distance of its planet to the host star.

The task now is to determine the solar zenith angle as a function of latitude, position along the latitude circle, and time of year. Let the point P be the center of the planet, and S be the center of the sun. If we draw a line from P to S, it will intersect the surface of the planet at a latitude δ, which is called the *latitude of the sun*, or sometimes the *subsolar latitude*. It is a function of the orientation of the planet's axis alone, and serves as a characterization of where we are in the march of the seasons. If the obliquity of the planet is γ, then δ ranges from γ at the northern hemisphere summer solstice to $-\gamma$ at the southern hemisphere summer solstice. Let Q be a point on the planet's surface, characterized by its latitude ϕ and its "hour angle" h, which is the longitude relative to the longitude at which local noon (the highest sun position) is occurring. For radiative purposes, we just need to compute the zenith angle ζ, defined previously. To get the zenith angle, we only need to take the vector dot product of the vector $\vec{QS}$ and the vector $\vec{PQ}$. To do this, it is convenient to introduce a local Cartesian coordinate system centered at P, with the z axis coincident with the axis of rotation, the x axis lying in the plane containing the rotation axis and $\vec{PS}$, and the y axis orthogonal to the other two, chosen to complete a right-handed coordinate system.

First, note that by the definition of the dot product,

$$\cos \zeta = \frac{\vec{PQ} \cdot \vec{QS}}{|PQ||QS|}. \tag{7.4}$$

Further, $\vec{PQ} + \vec{QS} = \vec{PS}$, so

$$\cos \zeta = \frac{\vec{PQ} \cdot \vec{PQ}}{|PQ||QS|} + \frac{\vec{PQ} \cdot \vec{PS}}{|PQ||QS|} \approx \frac{\vec{PQ} \cdot \vec{PS}}{|PQ||QS|} \tag{7.5}$$

where we drop the first term based on the assumption that the radius of the planet is a small fraction of its distance from the sun. For the same reason, $|QS|$ in the denominator can with good approximation be replaced by $|PS|$, leaving the expression in the form of a dot product between two unit vectors. Letting $\hat{n}_1 = \vec{PQ}/|PQ|$ and $\hat{n}_2 = \vec{PS}/|PS|$, the unit vectors have the following components in the local Cartesian coordinate system:

$$\hat{n}_1 = (\cos \phi \cos h, \cos \phi \sin h, \sin \phi), \hat{n}_2 = (\cos \delta, 0, \sin \delta). \tag{7.6}$$

whence

$$\cos \zeta = \cos \phi \cos \delta \cos h + \sin \phi \sin \delta. \tag{7.7}$$

When $\cos \zeta < 0$ the sun is below the horizon. Note that this formula just amounts to a transformation from a latitude–longitude coordinate system with the pole placed at the subsolar point (in which case the zenith angle formula becomes very simple) to a coordinate system with the pole at some other point. For a rotating planet it is convenient to use a coordinate system with the poles lined up along the rotation axis, but the formula would be equally valid if one for some reason wanted to adopt a different coordinate system. The longitude in Eq. (7.7) is not geographically fixed, since the hour angle is defined relative to the longitude of local solar noon, but one can introduce a *longitude of the sun*, which we'll call $\lambda_\odot$, which is the longitude of the subsolar point measured in a geographically fixed coordinate system such as the conventional latitude–longitude system in use at present on Earth. If λ is the longitude measured in this coordinate system, then the hour angle is given by $h = \lambda - \lambda_\odot$. Both δ and $\lambda_\odot$ are functions of time, but $\lambda_\odot$ depends on the rotation of the planet as well as its position in its orbit.[1]

The cosine of the zenith angle attains a maximum value $\cos (\phi - \delta)$ when $h = 0$, and a minimum value $-\cos (\phi + \delta)$ when $h = \pm\pi$. Both values are above the horizon when $|\phi| > |\pi/2 - \delta|$, corresponding to the perpetual polar summer day. Both values are below the horizon when $|\phi| > |\pi/2 + \delta|$, corresponding to the perpetual polar winter night. At the solstices, δ takes on its extreme values of $\pm\gamma$. Therefore, perpetual day or night is experienced at some time of year for latitudes poleward of $\pi/2 - \gamma$. These circles are known as the *Arctic* and *Antarctic* circles on Earth. Apart from the case of perpetual day or night, there is a *terminator* which separates the illuminated from the dark side of the planet. The position of the terminator is given by

$$\cos h_t = -\tan \phi \tan \delta. \tag{7.8}$$

---

[1] The reader should be cautioned that the "longitude of the subsolar point" defined here differs from the "ecliptic longitude of the Sun" in common use in astronomy. The latter uses a celestial reference frame for longitude.

We shall adopt the convention $h_t = 0$ in the case of perpetual night, and $h_t = \pm\pi$ for perpetual day.

If the planet's day is short compared with the time required to orbit the star, the position and axis orientation of the planet can be considered nearly fixed as the planet rotates about its axis. In this case, h_t translates directly into the length of the daylight period. Let Ω be the angular velocity of rotation of the planet relative to the fixed stars, so that the duration of the planet's *stellar day* is $T_{day} = 2\pi/\Omega$. Then the duration of daylight in the rapidly rotating case is close to $2h_t/\Omega = (h_t/\pi)T_{day}$.[2]

Exercise 7.2 For a given latitude ϕ, what δ yields the fewest hours of daylight? What δ yields the most hours of daylight? The Earth's present obliquity is 23.5 degrees, and its length of day is 23.94 hours. Sketch a plot of the maximum and minimum hours of daylight vs. latitude for the Earth. In all these calculations, you should make use of the rapidly rotating approximation.

When the planet's stellar day is an appreciable fraction of its year, then the distinction between the stellar day and solar day becomes important. In such cases, the time variations of δ and of $\lambda_\odot$ over the course of the day must be taken into account when computing the duration of daylight and the diurnal variation of insolation. Some calculations of this sort are carried out in Problem 7.9.

The diurnal variations of the zenith angle lead to hot days and cold nights. Where the thermal response time is long enough to average out an appreciable portion of the diurnal temperature variation, the daily mean incident solar flux is an informative statistic. Since the incident solar flux per square meter of surface is $L_\odot \cos\zeta$, where $L_\odot$ is the solar constant in W/m^2, one can obtain the daily mean flux by averaging $\cos\zeta$ over a rotation period of the planet. This results in a non-dimensional flux factor f, by which one multiplies the solar constant in order to obtain the daily mean solar radiation incident on each square meter of the planet's spherical surface. The daily average can be performed analytically, resulting in

$$f(\phi,\delta) = \frac{1}{2\pi} \int_{-h_t}^{h_t} \cos\zeta\, dh$$

$$= \frac{1}{\pi}[\cos\phi\cos\delta\sin h_t + \sin\phi\sin(\delta)h_t] \tag{7.9}$$

where h_t is determined by Eq. (7.8). This derivation of the daily average assumes that the length of the day is much less than the length of the year, so that δ and $\lambda_\odot$ may be regarded as constant over the course of the day. If the length of the day is a significant fraction of the length of the year, as is the case for slowly rotating planets like Mercury or Venus, the expression still gives the correct instantaneous average along the latitude circle, but this average is no longer identical to the time average over a day. In that case, however, the distinction between the diurnal and seasonal cycle is no longer meaningful, and the time variations of insolation from Eq. (7.7) should be used directly, without diurnal averaging.

[2] The stellar day is slightly different from what is misleadingly called the *sidereal day*, since the latter takes into account the effect of the precession of a planet's rotation axis. Precession will be discussed in Section 7.6. It is usually a very slow process, so for the purposes of computing a planet's climate, the distinction between sidereal and stellar day is almost always immaterial. In the context of extrasolar planets the approved term "stellar day" becomes very confusing, since it refers to the distant stars and not the planet's host star; pending some agreement on better terminology, we'll use the term "solar day" to refer to the diurnal variation of incident radiation regardless of what star the planet is orbiting.

During the equinoxes, $\delta = 0$ and $f = \cos{(\phi)}/\pi$, independent of the obliquity. This agrees with the result we obtained earlier by direct geometrical reasoning. At other times of year, the daily mean flux is governed by two competing factors: the varying length of day, which tends to produce higher fluxes near the summer pole, and the average zenith angle, which tends to produce high fluxes near the subsolar latitude (which remains near the equator if the obliquity is not too large). The latitude where the maximum daily mean insolation occurs is always between the subsolar latitude δ and the summer pole. For $\delta = 0$ the maximum occurs at the equator, and a little numerical experimentation shows that the latitude of the maximum increases to about 43.4° when $\delta = 23.4°$ (and similarly, with reversal of signs, in the southern hemisphere). For larger δ, the length-of-day effect wins out over the slant angle effect at the pole, and the maximum occurs at the summer pole itself. This state of affairs just barely happens at the solstice for the present obliquity of Earth and Mars; as a result, the summer hemisphere solstice inso-lation is fairly uniform in these two cases. It is also useful to note that the daily mean insolation at the summer pole exceeds the daily mean insolation at the equator when $|\delta| > 17.86°$.

To obtain a general appreciation of the seasonal cycle, recall that δ varies from $-\gamma$ during the southern hemisphere summer solstice to γ at the northern hemisphere summer solstice, taking on a value of zero at the equinox which lies between the two solstices. Consider a planet with uniform albedo, so that the absorbed solar radiation is determined by the distribution of incident solar radiation. Suppose further that the thermal response time is long enough to average out the diurnal cycle, but short compared with the length of the year. If the obliquity is below 23.4°, the "hot spot" starts some distance poleward of the equator in the southern hemisphere, moves to the equator as the equinox is approached, and then migrates a similar distance into the northern hemisphere as the northern summer solstice is approached. If the obliquity is greater than 23.4°, the hot spot starts at the south pole, discontinuously jumps to $-43.4°$ at the point in the season where the subsolar latitude crosses $-23.4°$, smoothly migrates through the equator and on to 43.4° when the subsolar latitude approaches 23.4°, and then discontinuously jumps to the north pole. Note that in either case, the hot spot crosses the equator twice per year, at the equinoxes; the two solstices are the coldest times at the equator. The climate at the equator has a periodicity that is *half* the planet's year.

It only remains to express δ as a function of the position of the planet in its orbit. The planet is spinning like a top, and if there are no torques acting on the planet (an assumption we will relax later) its angular momentum is conserved. Hence the rotation axis keeps a fixed orientation relative to the distant stars throughout the year. This is why Polaris is the northern hemisphere pole star all year around. Let κ be the angle describing the position of the planet, as shown in Fig. 7.4. We shall adopt the convention that $\kappa = 0$ occurs at the northern hemisphere summer solstice. We shall refer to κ as the *season angle*, though it is closely related to what astronomers call the *ecliptic longitude of the Sun*. In our case, we have defined the season angle relative to the northern hemisphere summer solstice, but other choices are also common, for example defining it relative to the northern winter solstice or the spring equinox. When discussing the progression through the seasonal cycle on planets other than Earth, a season angle is almost universally used to describe where the planet is in its annual cycle, since this description obviates the need to make up names for months for each planet. If we project the rotation axis onto the plane of the ecliptic (i.e. the plane containing the planet's orbit), then the angle made by this vector with $\vec{PS}$ is equal to κ. The rotation axis projected onto the plane of the ecliptic acts like the hand of a clock, which

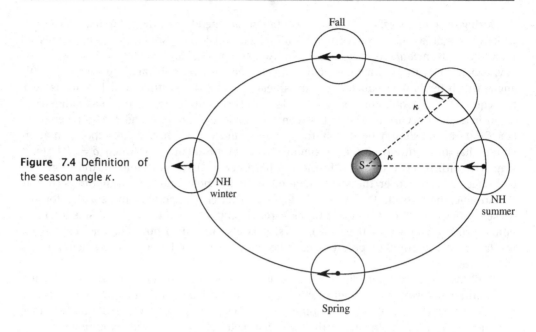

Figure 7.4 Definition of the season angle κ.

rotates around the clock face once per year, though at a non-uniform rate if the orbit is not perfectly circular.

Let $\hat{n}$ be the unit normal vector to the plane of the ecliptic, and $\hat{n}_a$ be the unit vector in the direction of the rotation axis. Introduce a new Cartesian coordinate system with x pointing along $\vec{PS}$, z pointing along $\hat{n}$, and y perpendicular to the two in a right-handed way. Then $\hat{n}_a = (\cos\kappa\sin\gamma, \sin\kappa\sin\gamma, \cos\gamma)$ and the latitude of the sun is the complement of the angle between $\hat{n}_a$ and the x axis, whence

$$\sin\delta = \cos\left(\frac{\pi}{2} - \delta\right) = \cos\kappa\sin\gamma. \tag{7.10}$$

In the limit of small obliquity, this equation reduces to $\delta = \gamma\cos\kappa(t)$. For a circular orbit, $\kappa(t) = \Omega t$, where Ω is the orbital angular velocity (2π divided by the orbital period). In this special case, the subsolar latitude varies cosinusoidally over the year, with amplitude given by the obliquity. This is actually not a bad approximation even for the roughly $23°$ current obliquity of Earth and Mars, agreeing with the true value to two decimal places. At the opposite extreme, when $\gamma = 90°$, the subsolar latitude is given by $\delta = \pi/2 - \kappa$, which is not at all sinusoidal.

Exercise 7.3 Compute the length of day as a function of the time of year for the latitude at which you are currently located. Compare with data for the current day, either observed yourself or presented in the newspaper weather report. Compute the length of a shadow that would be cast by a tall, thin skyscraper of height $100\,\text{m}$, as a function of the time of day and time of year at your latitude.

Contour plots of the diurnally averaged flux factor for various obliquities are shown in Fig. 7.5. These plots assume the orbit to be perfectly circular, so that there is no variation in distance from the sun in the course of the year. Over the course of the year, the hot spot moves from south of the equator to north of the equator, and back again, passing over the

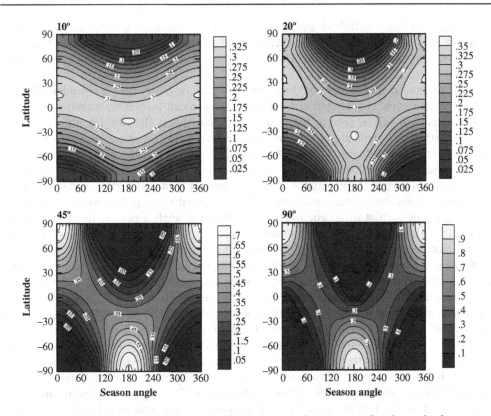

Figure 7.5 The seasonal and latitudinal distribution of daily-mean flux factor for four different values of the obliquity. In these plots, a circular orbit has been assumed. To obtain the daily mean energy flux incident on each square meter of the planet's surface, one multiplies the flux factor by the solar constant. For example, if the solar constant is 1000 W/m², the incident solar flux at the pole during the summer solstice is about 700 W/m² if the obliquity is 45°.

equator at the equinoxes. The amplitude of the excursion increases with obliquity, and goes all the way from pole to pole for sufficiently large obliquity. Earth, Mars, Saturn, Titan, and Neptune, with present-day obliquities of 23.5°, 24°, 26.7°, 26.7°, and 29.6° respectively, are qualitatively like the 20° case. The pattern of variation of incident solar radiation which forces the seasonal cycle is similar in all these cases. However, the nature of the seasonal cycle will differ amongst these planets because the differing nature of the atmospheres and planetary surfaces will lead to different thermal response times. In the case of gas giant planets, another variable is the proportion of energy received from solar energy vs. that received by transport from the interior of the planet. Insofar as the latter becomes dominant, the role of solar heating, and hence the prominence of the seasonal cycle, becomes less. Jupiter has a low obliquity (3.1°), which, compounded by a fairly high proportion of internal heating (11%) should lead to a minimal seasonal cycle. At the opposite extreme is Uranus, which has an obliquity of nearly 90°, and apparently insignificant internal heating. Venus is so slowly rotating that its obliquity is of little interest. Obliquity is not constant in time; it varies gradually over many thousands of years. We will see in Section 7.6.1 that relatively slight variations in the Earth's obliquity are believed to contribute to the coming and going of the ice ages. The obliquity of Mars varies more dramatically, and perhaps with greater consequence; at various times in the past it could have reached values as high as 50° and as low as 15°.

If the thermal response time of the planet is a year or more, then a considerable part of the seasonal cycle is averaged out and the annual mean insolation becomes an informative statistic. It will be seen in the next section that this is the case for watery planets like the Earth. The annual mean flux factor is shown in Fig. 7.6. When obliquity is small, the poles receive hardly any radiation. As obliquity is increased, the polar regions receive more insolation, at the expense of the equatorial regions. For Earthlike obliquity, the maximum insolation occurs at the equator, which is why this region of Earth's surface tends to be warmest. When the obliquity exceeds 53.9°, the annual mean polar insolation becomes greater than the annual mean equatorial insolation. For such a planet, the poles will be warmer than the tropics, provided that the thermal response time is long enough to average out most of the seasonal cycle. Consider a planet with 20° obliquity, zero albedo, and a very long thermal response time. If the planet were put in Earth's orbit about the Sun, the solar constant would be $1370\,W/m^2$, yielding equatorial insolation of $422\,W/m^2$ and polar insolation of $149\,W/m^2$, based on the flux factors given in Fig. 7.6. In the absence of any greenhouse effect or lateral energy transport by atmospheres or oceans, the equatorial temperature would be 294 K and the polar temperature would be 226 K. If one takes into account the clear-sky greenhouse effect of an Earthlike atmosphere with 300 ppmv CO_2 and 50% relative humidity using the *OLR* results given in Chapter 4, the polar temperature rises to 237 K, but the equatorial temperature becomes problematic: The annual mean equatorial solar flux is near or above the runaway greenhouse threshold discussed in Chapter 4, leading to extremely high or even unbounded equatorial temperatures. Lowering the relative humidity to 20% to reflect the fact that much of the tropical troposphere is very dry still leaves the tropics with temperatures in excess of 350 K. Part of the problem lies in the neglect of albedo. Simply using the observed planetary albedo in the tropics gives the wrong answer, because almost all of the cloud albedo is offset by the cloud greenhouse effect in the present climate (Section 5.8). Using an albedo of 0.15, based on the observed tropical clear-sky albedo, reduces the equatorial solar absorption to $360\,W/m^2$, which is in balance with a tropical temperature of 318 K assuming a relative humidity of 20%. This is still well in excess of the observed tropical temperature. In the real atmosphere, heat transport due

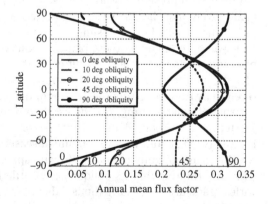

Figure 7.6 The annual mean flux factor for various obliquities, assuming a circular orbit.

to large-scale atmospheric and oceanic motions removes some of the heat from the tropics and deposits it at high latitudes, reducing the tropical temperatures and increasing the polar temperatures. Since incorporation of ice-albedo effects would reduce the polar temperatures below the estimates given above, such transports are also needed to bring the polar temperatures up into the observed range. Some elementary models of heat transport will be discussed in Chapter 9, though a proper treatment of the subject requires a full understanding of the fluid dynamics governing atmospheric transport.

If we put the same planet at the orbit of Mars the temperatures become 238 K and 183 K without any greenhouse effect. Since there is little water vapor feedback at such low temperatures, the greenhouse effect is less dramatic in this case. Addition of an Earth-like atmosphere with 300 ppmv CO_2 would increase the equatorial and polar temperatures to 256 K and 192 K, respectively, based on a linear OLR fit in the range 200 K to 250 K (specifically, $OLR = 80.6 + 1.83(T - 200)$). Thus, a waterworld placed at the orbit of Mars would require a much stronger greenhouse effect than the Earth's to avoid succumbing to a Snowball state.

The effects of obliquity on the seasonal and latitudinal pattern of insolation may be summed up as follows. Increasing obliquity increases the intensity of the seasonal cycle at mid to high latitudes. The summer insolation gets steadily higher relative to the global mean, and a greater area of the winter hemisphere is exposed to cold perpetual night or low insolation. Increasing obliquity also increases the annual mean polar insolation, though the way this affects polar climate depends on the thermal response time of the atmosphere-surface system. The increase might show up as very hot summers and bitterly cold winters, or as year-round warming, accordingly as the response time is fast or slow.

The preceding results on incident solar radiation have been derived in the absence of an atmosphere, but can still be used if there is an intervening atmosphere which may absorb or scatter solar radiation before it reaches the surface. The general geometry is illustrated in Fig. 7.7. In this case, one suspends an imaginary sphere at an altitude above which the atmosphere is too thin to have a significant effect on the solar radiation. The preceding results then give the solar flux entering each square meter of the surface of this sphere, and the angle at which the light enters the atmosphere. This is all that is needed as input to one-dimensional scattering models of the sort discussed in Chapter 5. One simply divides up the atmosphere into a series of patches near each latitude and longitude point, within which the properties are considered uniform, and applies a one-dimensional column model to each of these patches. As illustrated in Fig. 7.7, if the horizontal size of the patches is large compared with the depth of the atmosphere, the energy loss by horizontal scattering from one patch to another can be neglected, and each patch can be considered energetically closed, so far as radiation is concerned. The atmosphere has three effects, which can be

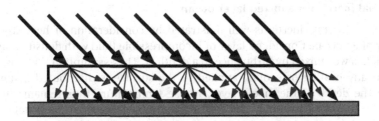

Figure 7.7 Schematic effect of atmospheric scattering on insolation.

inferred from the column radiation model: (1) some of the incident solar radiation reaches the surface in the form of diffuse radiation at a continuous distribution of angles, rather than at the zenith angle; (2) some of the solar radiation is absorbed in the atmosphere, rather than at the surface; and (3) some of the incident solar radiation is reflected back to space instead of entering the climate system. Of the three effects, it is the last – the effect of the atmosphere, including its clouds, on planetary albedo – that is most important for determining the climate. Diffuse radiation and atmospheric absorption do not change the amount of energy entering a column, but only the place and angle with which it enters. Often, this is of little consequence, so one can get a good estimate of the planet's temperature if one can obtain an estimate of the planetary albedo from one means or another.

The above reasoning can even to some extent apply to gas giant planets which have no surface. One can still define the imaginary sphere through which radiation enters the system, as before, but the problem comes in defining a characteristic depth scale. For the purposes of solar radiation, it suffices to consider the depth of atmosphere over which most of the solar radiation is absorbed, in effect a "photic" zone. This is typically shallow compared with the prodigious size of the gas giants. The full problem, including internal heat sources and dynamical motions, might require consideration of a deeper layer. Whatever the depth of this "active layer," the preceding reasoning applies provided one can sensibly model the large-scale aspects of the climate on the basis of averaging over patches of horizontal extent that is large compared with the characteristic depth scale. For giant planets, as for the Earth or any other planet, the essential difficulty is that clouds, temperature, water vapor, and other climate variables are manifestly not uniform over length scales comparable to or longer than the depth of the atmosphere. One makes progress by boldly assuming that one can represent the effects of these fluctuating quantities by their large-scale averages. It is an assumption that is difficult or impossible to justify mathematically, and in some cases may not even be true. With the present state-of-the-art, one can only make progress by proceeding on the basis of the averaging assumption, and seeing how things work out.

7.4 THERMAL INERTIA

At several points in the preceding discussion, we have needed to make reference to the thermal response time of the system. In the present section, we shall make this notion precise. The heat storage in the planet's solid or liquid surface, and in its atmosphere, means that it takes time for the system to heat up or cool down. The strength of this effect, known as *thermal inertia*, determines the extent to which the seasonal and diurnal fluctuations are averaged out in the climate response.

7.4.1 Thermal inertia for a mixed layer ocean

The concept of thermal inertia is well illustrated by consideration of heat storage in the mixed layer of an ocean. Consider a layer of incompressible fluid with density ρ and specific heat c_p, which is well mixed by turbulence to a depth H. The assumption that it is well mixed implies that any heating or cooling applied to the surface is distributed instantaneously throughout the depth of the mixed layer, whose temperature thus remains uniform. Let $(1 - \alpha)S(t)$ be the absorbed solar flux heating the mixed layer. We have deliberately left off subscripts denoting whether this is a top-of-atmosphere or surface flux, since we will eventually see that there are circumstances in which the energy budget of the surface can be determined directly in terms of a top-of-atmosphere budget, without the necessity of considering in detail just how the flux is communicated to the surface. Assume further that the cooling of the mixed layer by infrared radiation or other means can be written as a function of temperature, which we shall call $F(T)$. For example, if the atmosphere above the ocean has no greenhouse effect and carries no heat away from the surface by turbulent transport, the cooling is the radiative cooling $F(T) = \sigma T^4$. The energy balance equation for the mixed layer is then

$$\frac{d}{dt}\rho c_p HT = (1 - \alpha)S(t) - F(T). \tag{7.11}$$

Exercise 7.4 Consider a planet with a 50 m deep mixed layer water ocean ($c_p = 4218\,\text{J/Kg\,K}$, $\rho = 1000\,\text{kg/m}^3$). Suppose that the atmosphere for some reason has no effect whatsoever on the surface energy budget. (Why would this situation be hard to arrange, even for a pure N_2 atmosphere?) Hence $F(T) = \sigma T^4$. Suppose that the temperature of the polar ocean is 300 K when the sun sets and the long polar night begins. Find a solution to Eq. (7.11) for this situation, and use it to determine how long it takes for the ocean to fall to the freezing point (about 271 K for salt water).

We may define a thermal inertia coefficient $\mu = \rho c_p H$ for the mixed layer ocean. If an amount of energy ΔE is added to or removed from a column of the ocean having a cross-section of one square meter, the corresponding temperature change is $\Delta E/\mu$. For a 50 m mixed-layer water ocean, $\mu = 2.1 \cdot 10^8\,\text{J/(m}^2\text{K)}$, so that an energy flux of 100 W/m^2 out of the surface would lead to a cooling rate of $100/\mu = 4.74 \cdot 10^{-7}\,\text{K/s} = 0.04\,\text{K/day}$. Clearly, a rather shallow layer of well-mixed water can buffer a considerable surface flux imbalance. The Earth's ocean is several kilometers deep, but it is only the upper few tens of meters that are well mixed on short time scales; 50 m is in fact a reasonable approximation to the overall mixed layer depth of Earth's ocean, though there are geographical variations. Most other liquids would do about as well as water at storing heat. It is primarily the mixing depth that determines the thermal buffering effect of a planet's ocean.

Atmospheres also have thermal inertia, which can be considered in a fashion analogous to a mixed layer. The entire mass of the troposphere is well mixed, and this often makes up most of the mass of an atmosphere. For a well-mixed atmosphere the temperature profile can be tied to the temperature at any convenient fixed level (usually the ground), and we need to determine how much energy it takes to change this index temperature by one kelvin. This is where the notion of moist or dry static energy, introduced in Chapter 2, comes into its own. For simplicity, let's consider a non-condensing atmosphere, for which case the dry static energy (per unit mass) $c_p T + gz$ is independent of height within the troposphere. Hence, by evaluating its value at $z = 0$, the *dry static energy per unit mass* can be written $c_p T_{sa}$, where T_{sa} is the near-surface air temperature. The mass per unit area of the atmosphere is p_s/g, so the energy required to change the surface air temperature by 1 K while keeping the dry static energy well mixed in the vertical is simply $\mu \equiv c_p p_s/g$. When the atmosphere has a condensable component, one needs to take into account latent heat storage as well. Exploration of that aspect of the problem will be relegated to the Workbook section of this chapter. The following discussion is most valid for non-condensing atmospheres, but will nonetheless apply approximately to condensable atmospheres undergoing fluctuations in which the latent heat storage does not change too much. Determining the thermal inertia of atmospheres always involves some assumption about the vertical structure of temperature and condensable substances, and in the most general case one needs to model the full vertical profile to determine the extent of the atmosphere in which heat storage is changing significantly.

It is convenient to express the atmospheric thermal inertia coefficient μ in terms of the depth of a water mixed layer ocean, H_{eq}, which would have the same thermal inertia coefficient. For the Earth atmosphere, $H_{eq} = 2.4$ m, which is insignificant in comparison to the mixed-layer depth of the ocean. Hence, one expects the Earth's atmosphere to come into equilibrium much more quickly than the ocean. The current 6 mb CO_2 atmosphere of Mars has $H_{eq} = 0.03$ m, while the massive atmosphere of Venus has H_{eq} in excess of 155 m. Neither Mars nor Venus has an ocean to buffer the seasonal cycle, but the Venus atmosphere alone can be expected to have a considerable moderating effect, whereas present Mars should behave more or less as if each point of the globe is in instantaneous equilibrium. Early Mars (circa 4 billion years ago) may have had a 2 bar CO_2 atmosphere, which would translate into a 10 m equivalent mixed layer. This is considerably greater than that of Earth's atmosphere, but still not enough to have much moderating effect, in view of the fact that Mars' year is about twice as long as Earth's. Titan has a mostly N_2 atmosphere with a surface pressure only slightly in excess of Earth's, but its weak gravitational acceleration of 1.35 m/s² means that this pressure translates into a much greater mass of atmosphere per square meter of planetary surface. Thus, the Titan atmosphere has an equivalent mixed layer depth of about 24 m. Given the low temperature of Titan, and consequent low rate of energy loss by infrared emission, this value is expected to yield a very considerable buffering effect on Titan's seasonal cycle, regardless of whether there is a liquid ocean at the surface. For example, based on a typical surface temperature of 90 K, blackbody emission would cool the planet at a rate of only about 1 K per 300 Earth days, if insolation were completely shut off.

Exercise 7.5 The specific heat of liquid methane is 3450 J/kg K. How deep would a well-mixed methane ocean on Titan have to be for it to have thermal inertia comparable to Titan's atmosphere?

We shall now consider some simple solutions to the mixed layer model, keeping in mind that this model applies to atmospheres as well as oceans, with a suitable choice of the equivalent mixed layer depth. At this point we assume that $\rho c_p H$ is constant, though models with a time-varying mixed layer depth are possible. Without any loss of generality we may write the insolation and temperature in the form

$$S = S_0 + S'(t), \quad T = T_0 + T'(t) \tag{7.12}$$

where S_0 and T_0 are the time means of S and T and the deviations have zero time mean. Now, suppose that $T' \ll T_0$ for whatever reason; this need not require that $S' \ll S_0$, since the temperature fluctuations might be small by virtue of a slow response time of the system. Because the temperature fluctuations are small, the surface cooling can be expanded about T_0 and approximated by a linear function:

$$F(T) = F(T_0) + bT'(t), b = \frac{dF}{dT}(T_0). \tag{7.13}$$

Now we choose T_0 to be the equilibrium temperature corresponding to the mean insolation S_0, i.e. $F(T_0) = (1 - \alpha)S_0$. With these assumptions the equation for the temperature fluctuation becomes

$$\frac{dT'}{dt} = \frac{1}{\rho c_p H}((1 - \alpha)S'(t) - bT') \tag{7.14}$$

or equivalently, if we define the relaxation time $\tau = (\rho c_p H)/b$,

$$\frac{dT'}{dt} + \frac{T'}{\tau} = \frac{1}{\rho c_p H}(1 - \alpha)S'(t). \qquad (7.15)$$

We can distinguish two limiting cases for Eq. (7.15). When the time scale over which S' varies is slow compared with τ, then the first term on the left hand side is negligible compared with the second, whence the solution becomes $T' = \tau \cdot (1 - \alpha)S'(t)/(\rho c_p H)$, which is $(1 - \alpha)S'(t)/b$. In other words, the system acts as if it is in equilibrium with the instantaneous solar radiation at each time. In the opposite limit, the time scale of the solar fluctuation is rapid compared with τ, in which case it is the second term on the left hand side that may be neglected. Thus,

$$T'(t) = \frac{1}{\rho c_p H} \int_0^t (1 - \alpha)S'(t')dt'. \qquad (7.16)$$

In this case, the temperature is out of phase with the heating, and represents a time average of the fluctuating heating. The peak temperature occurs later than the peak solar heating, since it takes time for the mixed layer to respond to the accumulating heating. Further, in this case, the seasonal temperature fluctuation becomes small as the mixed layer depth is made large, since the mixed layer becomes more and more efficient at averaging out the seasonal fluctuations of solar flux.

The variations in solar radiation over the course of a year are not sinusoidal, but we can nonetheless gain some further insight into the seasonal cycle by writing $S' = S_1 \cos(\omega t)$. For this form of forcing, Eq. (7.15) can be solved most easily by using complex exponentials. Since $S' = S_1 Real(\exp(-i\omega t))$, the solution may be written $T' = Real(A\exp(-i\omega t))$. Substituting this form of solution into Eq. (7.15) we find

$$A = \frac{(1 - \alpha)S_1}{\rho c_p H} \frac{1/\tau + i\omega}{1/\tau^2 + \omega^2} = |A|e^{i\Delta} \qquad (7.17)$$

where the phase and amplitude are

$$\Delta = \arctan(\omega\tau), |A| = \frac{(1 - \alpha)S_1}{\rho c_p H} \frac{1}{\sqrt{(1/\tau^2 + \omega^2)}}. \qquad (7.18)$$

With these definitions, the solution can be written

$$T'(t) = |A| \cos(\omega t - \Delta). \qquad (7.19)$$

The character of the response depends on the period of the forcing relative to the characteristic response time of the system. This determines both the amplitude of the fluctuation and the phase shift relative to the forcing. For $\omega\tau << 1$ we have $\Delta = 0$ and $|A| = (1 - \alpha)S_1/b$. For $\omega\tau >> 1$ we have $\Delta = \pi/2$ and $|A| = (1 - \alpha)S_1/(\rho c_p H\omega)$. Note that in this case the temperature fluctuation becomes weak in inverse proportion to the frequency of the solar forcing fluctuation. These are special cases of the limits discussed previously, but we now have the further advantage of an explicit formula showing how the phase and amplitude of the seasonal cycle vary between the two extreme cases.

So far, we have not specified the flux which is to be used for the solar heating term $((1 - \alpha)S)$ and the heat loss term $F(T)$ in Eq. (7.11). One possibility is to use the top-of-atmosphere fluxes, which are purely radiative. The other is to use surface fluxes, which include turbulent as well as radiative heat exchange between the planetary surface and the atmosphere, as discussed in Chapter 6. There are two thermal reservoirs at play – the atmosphere and the ocean – and a full treatment would require writing separate energy budgets for each, and accounting for the fluxes between them. There are several important cases, however, in which the thermal inertia of one or the other reservoir dominates, making it possible to make do with a one-layer model. The appropriate flux to use in such a one-layer model depends on which reservoir dominates. There are four cases to consider.

There are two circumstances in which the top-of-atmosphere fluxes are the appropriate ones to use. First, if the time scale under consideration is long enough that the surface budget stays near equilibrium, then the net solar flux transmitted by the atmosphere and absorbed at the surface is equal to the net turbulent and infrared flux passing from the surface into the atmosphere. In this case, the bottom boundary is energetically closed, and the energy budget of the atmosphere–ocean column can be determined from the top-of-atmosphere fluxes. In this case, the thermal inertia is provided by the atmosphere, and one uses the atmosphere's equivalent mixed layer depth in the mixed layer model equations. In this limit, the atmosphere is considered to be the mixed-layer "ocean," and the underlying surface simply acts to hand back to the atmosphere any energy it receives.

Alternately, if the response time of the atmosphere is short enough compared with the time scale under consideration, the energy budget of the atmosphere comes into equilibrium. In this case, by the definition of equilibrium, the net flux (solar plus infrared) entering the top of the atmosphere must equal the net flux (solar plus infrared plus turbulent) leaving the bottom of the atmosphere. In this case, one can use the top-of-atmosphere absorbed solar radiation for the heat gain term and the OLR for the heat loss term in the surface energy budget, obviating the need to know the detailed physics behind the surface-to-atmosphere energy transfer. In this case, the thermal inertia is provided by the heat capacity of the mixed layer ocean.

In either case, one can compute $OLR(T)$ using a radiation model and some assumption linking the temperature and humidity profile to surface temperature, or one can use one of the linear or polynomial fits to the OLR curve discussed in Chapter 4. For example, with a linear fit to the OLR curve for a terrestrial atmosphere with 300 ppmv CO_2 and 50% relative humidity, b is about $2(W/m^2)/K$ in the range 250 K to 310 K. The corresponding relaxation time τ is 1200 days for a 50 m mixed layer, or 60 days for the 2.4 m mixed layer which is equivalent to the thermal inertia of the Earth's atmosphere. In consequence, the seasonal cycle is expected to be strongly attenuated on the ocean-covered parts of the Earth (apart from coastal effects). The atmosphere alone does not have enough thermal inertia to damp out the seasonal cycle, but it does have enough thermal inertia to keep the atmospheric temperature roughly constant in the course of the diurnal cycle. Colder temperatures tend to make the relaxation time longer. For example, in an Earthlike atmosphere with 300 ppmv CO_2, the relaxation time roughly doubles at 160 K. As noted earlier, Titan has a very long relaxation time owing to its thick atmosphere and low temperature; now we can make the statement more precise. Ignoring the greenhouse effect and setting $b = 4\sigma T^3$, $T = 90$ K, we find a relaxation time of 20 Earth years, based on the equivalent 24 m mixed layer depth of Titan's atmosphere. Since Titan's year (which is the same as Saturn's year) is about 30 Earth years, the seasonal cycle on Titan is expected to be considerably damped, though not so much so as the seasonal cycle over the Earth's oceans. The weak greenhouse effect from methane in Titan's atmosphere would somewhat enhance the damping. In contrast, a similar calculation for the thin atmosphere of present Mars gives a relaxation time of only 0.8 Earth days, based on $T = 200$ K. Since a Mars day is approximately the same as an Earth day, the thermal inertia of the Martian atmosphere at present has relatively little damping effect on the diurnal cycle.

The thermal relaxation process is different if the time scale under consideration is short compared with the response time of the atmosphere, but long compared with the response time of the surface. In this case, which constitutes the third basic type of behavior, the atmospheric temperature remains approximately constant while the surface temperature fluctuates. This is the way the diurnal cycle works on Earth over ice or land. The relaxation time of surface temperature is then determined using the turbulent and radiative surface-atmosphere flux formulae discussed in Chapter 6, rather than the *OLR*. Because of the great thermal inertia of Titan's atmosphere and the low thermal inertia of the mostly solid surface, the seasonal cycle of Titan is somewhat in this regime as well. The situation of present Mars is not in this regime, since the atmosphere has little thermal inertia. There, the diurnal cycle affects the entire depth of the atmosphere, and the diurnal response is approximately governed by the *OLR* and the thermal inertia of the surface, much as for the Earth's seasonal cycle.

The fourth class of behavior in which a one-layer model suffices arises when the ocean mixed layer is so deep that the surface temperature can be regarded as nearly constant throughout the year. If there were no solar absorption in the atmosphere, the atmosphere would not have any seasonal temperature variation either, but solar absorption will drive a seasonal cycle in atmospheric temperature, for which the thermal inertia is provided by the atmosphere. Even if the solar absorption is small, the resulting temperature variations can be considerable if the atmosphere has fairly low thermal inertia, as is the case for Earth. This leads to a paradoxical situation in which the atmospheric temperature cycle is in phase with the insolation, even though the deep ocean mixed layer might lead one to think that the seasonal cycle should be a quarter year out of phase. In a seasonal cycle of this type, neither the atmosphere nor ocean is near equilibrium. One fixes the ocean surface temperature and then computes the turbulent and radiative fluxes out of the bottom of the atmosphere using a surface budget model. One also needs to compute the net radiative fluxes into the top of the atmosphere. The difference, which is a function of the atmospheric temperature, gives the heating that drives the atmospheric seasonal cycle. This case is explored quantitatively in Problem 7.16. It is a case of real interest, since some aspects of the southern hemisphere seasonal cycle of atmospheric temperature appear to fall into this category.

We are now equipped to compute the seasonal cycle for an idealized Earthlike water-world subject to realistic seasonal insolation. The object of this calculation is to provide a more quantitative feel for the extent to which oceans damp the seasonal cycle. The time series of insolation $S(t)$ is far from sinusoidal, particularly in the polar regions of planets with non-zero obliquity. The solution for this insolation cannot be obtained analytically, but it is an easy matter to numerically integrate the mixed layer equation subject to solar forcing obtained from the time dependence of the zenith angle given by Eq. (7.7) or its diurnally averaged version in Eq. (7.9). As we discussed in the context of the annual mean case, the temperatures obtained by doing a purely local balance are far too hot at the equator and far too cold at the poles, compared with Earth observations. This is because the real atmosphere takes some heat away from the tropical regions and deposits it in the polar regions. In order to represent this transport and get the model in a roughly realistic temperature range, one needs to put in the effect of the heat transport by *fiat*, in the absence of a fully dynamical model of heat transport. Similar considerations apply to any planet with an atmosphere that is dense enough to transport a significant amount of heat. In the present calculation, we will represent the effects of the heat transport by simply adding a latitude-dependent dynamic flux F_d to the column budget at each latitude. We will assume this flux to be independent of time. The flux F_d is negative in the tropics, where dynamics takes heat out of

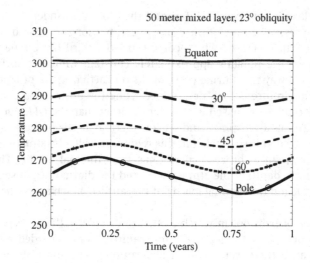

Figure 7.8 Numerically computed seasonal cycle for a 50 m mixed layer ocean subject to a realistic seasonal cycle of insolation. The planet is in a circular orbit having an obliquity of 23°. Infrared cooling was calculated from top-of-atmosphere balance with a CO_2 concentration of 300 ppmv and fixed relative humidity of 50%. The time axis represents one Earth year. The northern hemisphere summer solstice occurs at time zero, and the northern winter solstice is at time 0.5 years. See text regarding treatment of dynamical heat fluxes.

the column, and it is positive in the polar regions. For the present Earth, the magnitude of F_d can be determined by examining satellite-derived top-of-atmosphere energy budgets, or more simply it can be set at a value that gives approximately the right annual-mean temperature at each latitude. For the calculation to follow, we set F_d to $-35\,W/m^2$ at the Equator, $-20\,W/m^2$ at $\pm30°$ latitude, zero at $\pm45°$ latitude, $40\,W/m^2$ at $\pm60°$ latitude, and $80\,W/m^2$ at the poles. These values are somewhat weaker than the observed dynamical fluxes on Earth, but given the other simplifications we have made in this calculation, they suffice to keep the temperatures in a reasonably Earthlike range.

The calculations were carried out for a 50 m mixed layer ocean, which is roughly consistent with observed mixing in Earth's ocean, though the real mixed layer depth does vary somewhat both geographically and seasonally. For this depth of mixed layer, the ocean dominates the thermal inertia at seasonal time scales, and the atmosphere can be considered to be in equilibrium. Hence, it is appropriate to use top-of-atmosphere fluxes to drive the model. We'll neglect the albedo variations that would be caused by the cloud distribution and the formation of sea ice, and adopt a uniform top-of-atmosphere value of 0.2. The infrared cooling was computed based on the clear-sky polynomial OLR fit in Table 4.2, for a CO_2 concentration of 300 ppmv. Results are shown for Earthlike obliquity, using a circular orbit of period one Earth year, in Fig. 7.8. These results use the approximate diurnally averaged form of the flux function.

The seasonal cycle at the equator is so weak as to be barely detectable in the figure, but the amplitude grows steadily as one moves towards the poles. At 45° the peak-to-trough amplitude is 6 K, which agrees quite well with the observed midlatitude amplitude within extensive ocean basins. The weak tropical seasonal cycle is also consistent with observations. Note that from 30° through 60° latitude, the peak and trough is shifted a quarter year relative to the solstices, as is expected in the case of strong thermal inertia. Also, at these latitudes the seasonal cycle in temperature looks quite sinusoidal; this is in part because the insolation is fairly sinusoidal at these latitudes, but also because the mixed layer acts as a low-pass filter, damping down the response to higher harmonics of the insolation time series.

Because ice albedo feedback is not included, the high-latitude results cannot be expected to yield a realistic seasonal temperature pattern once ice forms. The calculations do give an indication of how strong the high-latitude seasonal cycle would be if the climate were warmed up enough to suppress the formation of ice, and also provide a reasonable estimate of temperature variations during the ice-free part of the year. It is significant that despite the long polar night, the ocean is only able to cool by about 10 K from its summer peak. Therefore, with a moderate-depth mixed layer, formation of midwinter polar ice is far from inevitable.

The formation of polar ice is very sensitive to the values chosen for the high-latitude dynamical heat flux and the albedo of the ice-free ocean; the latter of these is strongly influenced by low cloud cover. For the conditions of the present case, the ocean gets cold enough to form ice after the winter solstice at 60° latitude, while at the pole the ocean is below freezing the entire year. Once ice forms, not only does the albedo increase, but also the effective thermal inertia is drastically reduced, since the solid ice layer insulates the atmosphere from the heat storage of the mixed layer. A discussion of these effects is deferred to Section 7.7, where we take up the very challenging question of how to account for the absence of high-latitude ice in a hothouse climate like the Cretaceous.

Although this calculation has been tuned to Earthlike conditions, it gives a pretty good picture of the general state of affairs for moderate-obliquity waterworld planets in nearly circular orbits. The picture also applies to planets with atmospheres so thick they have equivalent mixed-layer depths comparable to 50 m. The general picture to take away is that such planets have relatively weak seasonal cycles in circumstances where polar ice fails to form; when polar sea ice forms, the high-latitude seasonal cycle can become much stronger. All other things being equal, planets with a deeper equivalent mixed layer will have a seasonal cycle that is even more moderate than the one shown here, while planets with a shallower equivalent mixed layer will exhibit more extreme seasonal variations. The results can be scaled to planets with different atmospheric radiative damping constants and different year lengths using the fact that the mixed layer equations with linearized radiative flux depend on the mixed layer depth and length of year only through the combination τ_y/τ, where τ_y is the duration of the year and τ is the response time derived previously. Since τ is proportional to mixed layer depth, a planet with twice the length of Earth's year but a mixed layer of depth 100 m would have a seasonal cycle of similar amplitude to Earth's (other parameters being equal), and so forth. Halving the radiative damping coefficient $dOLR/dT$ would have the same effect as doubling the mixed layer depth. A detailed exploration of the behavior of the mixed layer equations for various planetary configurations is carried out in Problem 7.31. Determining the appropriate dynamical heat flux for an arbitrary planet, however, requires either doing a full dynamical calculation, or invoking observations to constrain the annual mean temperature or radiation balance.

7.4.2 Thermal inertia of a solid surface

Heat diffuses slowly through a non-metallic solid, so when the underlying surface is solid it is typically necessary to consider the continuous distribution of temperature as a function of depth within the solid. To a good approximation, heat flux within a solid is proportional to the temperature gradient; the proportionality constant is called the *thermal conductivity*, which we shall call κ_T. Balancing the rate of change of heat content against the convergence of heat flux yields the *diffusion equation*

$$\partial_t \rho c_p T = \partial_z (\kappa_T \partial_z T). \tag{7.20}$$

In this equation it is assumed that there are no internal heat sources. The surface heat budget enters the problem through the boundary condition at the surface ($z = 0$), which states that the diffusive heat flux into the surface equals the net heating of the surface by insolation and radiative and turbulent heat transfers. Using the same notation as we employed for the mixed layer case, this boundary condition reads

$$\kappa_T \partial_z T|_{z=0} = (1 - \alpha)S(t) - F(T). \tag{7.21}$$

When S is a constant S_0, the problem is solved with a constant temperature T_0 satisfying $(1 - \alpha)S_0 = F(T_0)$, just as for the mixed layer case. Linearizing the boundary condition about T_0 and substituting the complex exponential form for S' yields

$$\kappa_T \partial_z T'|_{z=0} = (1 - \alpha)S_1 e^{-i\omega t} - bT'. \tag{7.22}$$

If ρc_p is constant, this boundary condition can be satisfied by a solution of the diffusion equation of the form

$$T' = A e^{i(kz - \omega t)}, k = -\sqrt{\frac{\omega}{D}} \frac{1 + i}{\sqrt{2}} \tag{7.23}$$

where A is a constant and D is the diffusivity $\kappa_T/(\rho c_p)$. The complex vertical wavenumber k has been determined by substitution of the exponential form of T' into the diffusion equation and requiring that the solution decay as $z \to -\infty$. The constant A will be determined by substitution of the solution into the boundary condition, but before doing so it is worth pausing to make some remarks on the solution Eq. (7.23). This solution was first obtained by Fourier, in his study of diurnal and seasonal variations of temperatures in the interior of the Earth. Equation (7.23) shows that the characteristic depth to which temperature fluctuations penetrate is $\sqrt{(D/\omega)}$. Low-frequency fluctuations penetrate to a greater depth than high-frequency fluctuations, because heat has a longer time to diffuse before the surface temperature reverses. Note also that the phase lag of the time of maximum temperature with depth also reflects the time required for the surface conditions to penetrate to the interior. For the diffusivity of water ice (Table 7.1) the characteristic depth is 12 cm for the diurnal period, 2.4 m for the annual period, 24 m for a century and 76 m for a millennium. Solid rock yields similar numbers. Hence, the temperature profile within ice or rock still contains information about temperatures centuries or even millennia in the past, albeit in a rather smoothed and degraded form. This fact has been exploited in reconstructions of past temperatures.

Exercise 7.6 You are designing a lunar colony to be placed at a lunar latitude where the Sun is directly overhead at noon. The Moon has an albedo close to zero, and the response time of the surface is rapid, so that the noontime surface temperature is close to the instantaneous equilibrium temperature of 394 K (re-derive this temperature yourself). At night, the equilibrium temperature would be absolute zero, but there is not enough time to reach equilibrium; still the night-time temperature plummets to 100 K. Since the Moon is tide-locked to the Earth, the lunar day is 28 Earth days. The diffusivity of the lunar regolith ("soil") is about 10^{-8} m^2/s.

	ρc_p(J/m^3 K)	Conductivity (W/(m^{2}K))	Diffusivity (m^2/s)
Water ice	$1.93 \cdot 10^6$	2.24	$1.16 \cdot 10^{-6}$
Fresh snow	$0.21 \cdot 10^6$	0.08	$0.38 \cdot 10^{-6}$
Old snow	$1.0 \cdot 10^6$	0.42	$0.42 \cdot 10^{-6}$
Sandy soil	$1.28 \cdot 10^6$	0.3	$0.24 \cdot 10^{-6}$
Clay soil	$1.42 \cdot 10^6$	0.25	$0.18 \cdot 10^{-6}$
Peat soil	$0.575 \cdot 10^6$	0.06	$0.1 \cdot 10^{-6}$
Rock	$2.02 \cdot 10^6$	2.9	$1.43 \cdot 10^{-6}$
Lunar regolith	$1 \cdot 10^6$	0.01	$0.01 \cdot 10^{-6}$

Table 7.1 Thermal properties of some common surface materials.

Approximate the day–night temperature variation by a sinusoidal curve. What would be the constant temperature far below the surface (neglecting internal heat sources)? How deeply would the colony habitat have to be buried in order for the ambient diurnal temperature fluctuations to be less than 1 K?

Hint: Given the low diffusivity of the regolith, your main difficulty is likely to be getting rid of the heat generated by energy use (biological and otherwise) within the colony.

Now we substitute Eq. (7.23) into the boundary condition (7.22). The result is

$$
\begin{aligned}
A &= \frac{(1-\alpha)S_1}{b + \rho c_p \sqrt{\omega D}\frac{1-i}{\sqrt{2}}} \\
&= \frac{(1-\alpha)S_1}{b} \frac{1}{1 + \sqrt{\omega \tau_D}\frac{1-i}{\sqrt{2}}} \\
&= \frac{(1-\alpha)S_1}{\rho c_p \sqrt{D/\omega}} \frac{1}{\frac{1}{\tau_1} + \frac{1-i}{\sqrt{2}}\omega}
\end{aligned}
\tag{7.24}
$$

where $\tau_D = (\rho c_p)^2 D/b^2$ and $\tau_1 = \rho c_p \sqrt{D/\omega}/b$. Upon comparison of the third line of this equation with the solution for the mixed layer model, it is seen that the solid case acts somewhat like a mixed layer model with frequency-dependent layer depth $\sqrt{D/\omega}$. For low-frequency forcing, $\omega \tau_D \ll 1$, the surface temperature follows the instantaneous equilibrium, $A = (1-\alpha)S_1/b$, just as for the mixed layer case. For high-frequency forcing, the amplitude of the surface temperature fluctuation decays like $1/\sqrt{(\omega)}$. This is slower than was the case for the fixed-depth mixed layer, since the layer determining the thermal inertia now gets thinner as frequency is increased. Note also that the phase lag of surface temperature relative to insolation differs from the mixed layer case. For the diffusion equation, the surface temperature lags the insolation by $\pi/4$ radians in the high-frequency limit rather than $\pi/2$.

Apart from some exceptional circumstances, the thermal inertia of a solid surface has little effect on the seasonal cycle, though it can substantially moderate the diurnal cycle. This can be seen easily through the evaluation of τ_D in a few typical cases. First we consider the case of Antarctic or Arctic ice-covered regions. The flux coefficient based on a linear *OLR* fit in the temperature range 240 K to 270 K is $b = 2.16\,\text{W}/(\text{m}^2\,\text{K})$. Using the heat capacity and thermal diffusivity for water ice, given in Table 7.1, we find $\tau_D = 11$ d. At latitudes somewhat away from the poles, the diurnal cycle of insolation becomes significant, particularly during the equinoxes. Since the time scale for the surface is shorter than that for the atmosphere, it would be more appropriate to use surface flux coefficients than OLR in analyzing the terrestrial diurnal cycle. As noted in Chapter 6, the turbulent heat transfer is strongly inhibited at night-time, when the boundary layer is statically stable. In this case, the flux coefficient is dominated by the radiative term $4\sigma T^3$ based on surface temperature. For temperatures around 255 K this yields an even shorter response time, $\tau_D = 4$ d. In the midlatitudes and tropics, the estimate differs only in the use of the slightly larger values of b appropriate to the warmer temperatures, and the somewhat different thermal properties of rock or soil, but the result remains that τ_D is on the order of a few days or less. For Mars, one may use $b = 4\sigma T^3$ based on $T = 200$ K, given the thin atmosphere. This yields $\tau_D = 15$ d, which is still not sufficient to appreciably affect the seasonal cycle. It is only at the extremely cold temperatures of Titan that the response time of a solid ice surface becomes significantly longer (roughly 1300 Earth days), but even there the effect is of little interest, owing to the much longer response time of Titan's atmosphere. In sum, a solid surface can generally be considered to be in equilibrium for the purpose of computing temperature fluctuations on the seasonal time scale.

It should not be concluded from the above estimates that the thermal inertia of solid surfaces is sufficient to eliminate the diurnal cycle. The variation of insolation between noon and night-time is huge; on Earth, at a latitude where the Sun is overhead at noon, the amplitude of the variation is $1370\,\text{W}/\text{m}^2$, which leads to an undamped temperature fluctuation of 685 K based on a flux coefficient $b = 2\,\text{W}/(\text{m}^2\text{K})$. Even damped by a factor of 20, this amounts to a very considerable diurnal fluctuation. Similar considerations apply to the Martian diurnal cycle.

It should further be noted that while solids do not provide much thermal inertia for the surface temperature, they do provide substantial thermal inertia for the subsurface temperatures. Consider a nearly airless body, like the Moon or Mars or Triton. One sees a diurnal cycle corresponding nearly to instantaneous equilibrium at the surface, but at a sufficient depth the diurnal cycle is filtered out and one sees only the seasonal cycle. At still greater depths the seasonal cycle is filtered out and one sees only the annual mean. At yet greater depths, what remains are the effects of long-term climate variations. A calculation illustrating this behavior is carried out in Problem 7.21.

As a complement to the periodically forced solution, Figs. 7.9 and 7.10 show the solutions for the diffusion equation in water ice which is initialized at a uniform temperature of 300 K and allowed to cool without solar heating subject to a flux upper boundary condition. The heat loss from the surface was computed using an Earthlike OLR fit $OLR(T_s) = 48.461 + 1.5866(T_s - 180) + 0.0029663(T_s - 180)^2$. A quadratic fit was used so that the fit would remain accurate over a large temperature range. Except for the high initial temperature, which turns out to be inconsequential, this problem can be thought of as representing the cooling of the Antarctic ice cap after permanent winter night closes in. Figure 7.9 illustrates the progressive penetration of the surface cooling into the depth of the ice; at time t, the cooling has penetrated to a depth on the order of $\sqrt{Dt}$, where D is the thermal diffusivity of the ice. Figure 7.10 shows that there is an extremely rapid initial cooling, owing to the thin layer of ice affected at short times. After a half day, the temperature has already fallen below freezing. Thereafter, the temperature drop becomes slower, as the depth of the ice layer involved becomes greater. The reduction in OLR as temperature drops also contributes

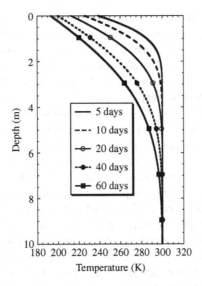

Figure 7.9 Temperature vs. depth at various times, for an ice layer subject to temperature-dependent heat loss at the surface. See text for specification of the heat loss rate.

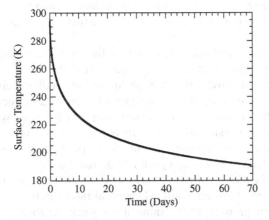

Figure 7.10 Time evolution of surface temperature for the solution shown in Fig. 7.9.

to the reduction in cooling rate. Nonetheless, after two months, the temperature has fallen to 190 K, which is well below the 235 K minimum temperature observed at the south pole. Incorporation of the atmosphere's thermal inertia reduces the cooling rate somewhat, but does not much increase the extremely cold temperature encountered at the end of the winter. Clearly, the Antarctic interior relies on heat transport from warmer latitudes to limit its winter temperature drop.

We conclude this section with a few remarks on the special effects of snow and ice (whether from water, CO_2, or some other substance) on the seasonal and diurnal cycle. Snow has a profound effect on the diurnal cycle, because of its very low thermal conductivity, which is nearly an order of magnitude lower than that of ice (see Table 7.1 for the case of water snow). The low thermal conductivity arises from the high proportion of the snow's volume which consists of air trapped in pores which are too small to allow the air to flow; since air itself has extremely low thermal conductivity, heat must primarily make its way through the contorted pathways of snow crystals in contact with each other. Other gases, trapped in snows made of other substances, have a similar effect. The low conductivity greatly reduces the characteristic response time of the surface, even for a snow layer of modest thickness. In the Antarctic case discussed above, τ_D drops to a mere 60 minutes for old snow, and 20 minutes for fresh snow. At night, the temperature of the snow surface plunges almost instantaneously to its equilibrium value. In the case of the Earth, the atmosphere has sufficient thermal inertia that it does not cool much at night, above the boundary layer. Given the suppression of turbulent flux in the stable nocturnal boundary layer, the night-time equilibrium temperature is maintained mainly by the downwelling infrared flux from the atmosphere, as discussed in Chapter 6. When the low-level air temperature is $255\,K$, the downwelling infrared flux is about $120\,W/m^2$, maintaining a snow surface temperature of $214\,K$. On present Mars, the atmosphere cools down markedly at night, and in any event is too thin to provide much downwelling flux, so it is less obvious what limits the night-time temperature drop over the CO_2 snow fields that form in the winter hemisphere. One relevant consideration is that the flux coefficient b drops dramatically at very cold temperatures, leading to an increase of the relaxation time; when the surface temperature falls to $150\,K$, τ_D increases to 23 hours even over snow. However, at such low temperatures the saturation vapor pressure of CO_2 is only $1.26\,mb$, well below the ambient surface pressure. Hence, the night-time temperature minimum is likely to be governed by the latent heat release due to CO_2 condensation, which sets in at surface temperatures near $160\,K$.

Snow cover on any planet can change rapidly in the course of the seasons, and on Earth, sea ice cover similarly expands and retreats. Since snow and ice have higher albedo than the surfaces they generally cover, this has an important feedback effect on the seasonal cycle. It enhances the wintertime cooling once ice or snow begin to accumulate, delays the springtime warming, but then accelerates the warming once ice or snow begin to retreat. The albedo feedback of snow is especially pronounced, since snow has a much higher albedo than ice. For water snow, for example, the albedo of fresh snow averaged over the solar spectrum can exceed 0.85, whereas a typical albedo for sea ice is on the order of 0.6. The high albedo of snow, like its low diffusivity, arises from its highly porous nature which offers many opportunities for light to encounter discontinuities in index of refraction, leading to scattering. It is a generic property of the snow of any weakly absorbing substance. Note that the concept of "sea ice" is peculiar to planets with water oceans. On a planet with a liquid methane or CO_2 ocean, "sea ice" would sink, and not have any chance to affect the surface albedo until the ocean were frozen to the bottom.

The presence of a solid phase on the surface of the planet also introduces a new form of thermal inertia, associated with the latent heat of phase change from the solid to liquid form. Where there is ice, whether it be in the form of sea ice or land glaciers, the surface temperature cannot rise above the triple point (the "melting point") until all the ice has been melted. The phenomenon is familiar from an experiment commonly performed in elementary school science classes, in which one tries to boil a pot of water containing ice cubes, and finds that the temperature does not start to rise above freezing until all the ice is gone. Thermal inertia effects due to growth and decay of ice are discussed further in Section 7.7 and Problem 7.17.

7.4.3 Summary of thermal inertia effects

The preceding discussion has revealed two limiting forms of behavior that a planet can exhibit in the course of its seasonal cycle. A "waterworld," having high thermal inertia in the ocean–atmosphere system, responds primarily to the annual average insolation. Such a world will be coldest at the poles and warmest at the equator, unless the obliquity exceeds about 54°, in which case the warmest climates will be found near the two poles. A "desert-world," having little thermal inertia in either the surface or the atmosphere, responds to the instantaneous insolation at each time of the year. The location of the highest temper-ature moves from some latitude north of the equator to the same latitude south of the equator, and back again, in the course of the year. The hot spot crosses the equator twice per year, leading to a basic rhythm for tropical climate which is twice as fast as higher-latitude regions. For small obliquity, the poles are frigid throughout the year, and the hot spot executes modest excursions about the equator. For obliquities greater than about 18°, the excursion goes all the way from pole to pole, assuming a uniform albedo. Geographi-cal and temporal albedo variations alter this picture. Formation of permanent ice or snow cover near the poles will tend to keep the polar regions cold throughout the year; this effect is assisted by the thermal inertia implied by the latent heat required to melt or sublimate ice, which limits the summertime temperature increase. The desertworld case is explored in Problem 7.7.

Thermal inertia sufficient to moderate the seasonal cycle can be provided either by a thick atmosphere or a well-mixed liquid layer at the surface. A liquid layer need only have a depth of some tens of meters to have a significant moderating effect, though the precise amount of thermal inertia needed to moderate the seasonal cycle must always be consid-ered relative to the length of the planet's year. Heat storage provided by non-melting solid surfaces is almost never sufficient to have a significant affect on the seasonal cycle, though it can substantially moderate the diurnal cycle for planets with rotation periods on the order of a few Earth days or less.

The Earth shows some characteristics of both limiting cases, with extreme continental cli-mates and equable maritime climates. In the deep tropics, the twice-per-year rhythm of land temperature contrasts with the nearly constant ocean temperature, leading to monsoonal circulations driven by the strong land–sea temperature gradient.

7.5 SOME ELEMENTARY ORBITAL MECHANICS

Sir Isaac Newton showed that the orbit of a single planet revolving about its star takes the form of an ellipse, with a focus of the ellipse at the center of mass of the system. Since stars are typically much more massive than their planets, the center of mass for most purposes is identical to the center of the star. The elliptical nature of orbits has an important effect on the seasonal cycle, since the planet is farther from its sun at some parts of the year than it is at others. This makes the solar "constant" $L_\odot$ a function of time of year. On Earth, we don't notice this effect too much because our orbit is nearly circular. Nonetheless, the effect has an important influence on the long-term evolution of climate. On other planets, it can be even more important.

The distance of closest approach of a planet to its star is called the *periastron*, which we shall call r_p. The greatest distance is called the *apastron*, which we shall call r_{ap}. The *semi-major axis* is then $a = (r_p + r_{ap})/2$. Let κ_1 be the angle made by the line between the star and

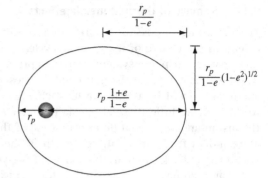

Figure 7.11 Geometry of an elliptical orbit with eccentricity $e = 0.66$.

the planet, defined so that $\kappa_1 = 0$ at the periastron. Then, in polar coordinates, the equation of the elliptical orbit is

$$r = a\frac{1 - e^2}{1 + e\cos(\kappa_1)} \tag{7.25}$$

where e is the *eccentricity* of the orbit, which lies in the interval $[0, 1]$. Eccentricity $e = 0$ yields a circular orbit, while the ellipse becomes progressively more elongated as $e \to 1$. Specifically, the periastron is $(1 - e)a$, the apastron is $(1 + e)a$ and the ratio of the distance at apastron to the distance at periastron is $(1 + e)/(1 - e)$. To get the semiminor axis, we maximize $r(\kappa_1)\sin(\kappa_1)$, yielding $a\sqrt{1 - e^2}$. Hence, the ratio of the minor to major axis is $\sqrt{1 - e^2}$. The geometry of the orbit is summarized in Fig. 7.11.

Exercise 7.7 What eccentricity would yield an ellipse with a 3:1 axis ratio? Sketch such an ellipse, indicating the correct location of the Sun relative to the orbit.

The variation of the solar constant is then given by

$$L_\odot = \frac{1}{4\pi}\frac{\mathcal{L}_\odot}{r^2} \tag{7.26}$$

where $\mathcal{L}_\odot$ is the power output of the star (about $3.8 \cdot 10^{26}$ watts for the Sun at present). The annual variation in distance from the Sun leads to "distance seasons," which are synchronous between the hemispheres. This contrasts with the "obliquity seasons" (dominant for Earth and Mars) which are out of phase between the hemispheres, one hemisphere enjoying winter while the other suffers under the torrid heat of summer. In the limit of small eccentricity, the ratio of solar constant at apastron to that at periastron is $1 + 4e$. This represents a very considerable variation, even for modest eccentricity. For the present eccentricity of the Earth (0.017), it amounts to 6.8%, or $93\,\mathrm{W/m^2}$ difference in the solar constant between periastron and apastron. To turn this flux into a crude temperature estimate, we divide by 4 to account for the averaging over the Earth's surface, and apply a typical terrestrial $OLR(T)$ slope of $2\,\mathrm{W/(m^2K)}$, yielding a temperature difference of more than $11\,\mathrm{K}$ between periastron and apastron. This represents the amplitude of the distance seasons. For Mars, with its present eccentricity of 0.093, the effect is even greater. The periastron to apastron flux variation is 37%, or $219\,\mathrm{W/m^2}$. For Martian conditions, where the atmosphere has a weak greenhouse effect, this translates into an amplitude of $30\,\mathrm{K}$.

To determine the time dependence of r, we must know $\kappa_1(t)$. Because the orbit is no longer circular, the angular velocity is no longer constant; the planet moves faster when it is close to the sun than when it is farther away. There is no analytic expression for the time variation of the orbital position. However, it can be easily computed by numerically solving a first order differential equation, which can be derived either from Kepler's equal-area law, or directly from angular momentum conservation. We shall take the latter route. Let $v_\perp$ be the component of velocity perpendicular to the line joining the planet to its star. Then, by conservation of angular momentum, $rv_\perp = J$ is independent of time. However, the angular velocity of the orbit is simply $v_\perp/r$, so the angle satisfies the equation

$$\frac{d\kappa_1}{dt} = \frac{J}{r^2} = \frac{J}{a^2} \frac{(1 + e\cos\kappa_1)^2}{(1 - e^2)^2}.$$ (7.27)

This equation shows that the angular velocity of the planet speeds up as it approaches periastron, and slows down as it approaches apastron. In consequence, the planet spends less time near the sun than it does at greater distances, and "distance summer" is shorter than "distance winter."

The average of the solar constant over the course of the year can be written

$$< L_\odot > = \frac{1}{4\pi} \frac{\mathcal{L}_\odot}{a^2} \left\langle \frac{a^2}{r^2} \right\rangle = L_a \left\langle \frac{a^2}{r^2} \right\rangle$$ (7.28)

where angle brackets denote the average over the planet's year and L_a is the solar constant evaluated at a distance equal to the semi-major axis of the orbit. We can take advantage of the fact that the same $1/r^2$ factor appears in Eq. (7.27) to relate the mean solar constant to the non-dimensionalized duration of the planet's year. Specifically, integrating Eq. (7.27) over one year and dividing by the length of the year yields

$$< L_\odot > = \frac{1}{\tau_y^*} L_a$$ (7.29)

where $\tau_y^* = \tau_y/(2\pi a^2/J)$, τ_y being the length of the year in dimensional terms. The quantity $2\pi a^2/J$ is the length of year for a circular orbit with radius a. Numerical integration of Eq. (7.27) shows that the non-dimensional year defined in this way decreases as the orbit becomes more eccentric. For $e = 0.1$, τ_y^* is .995, for $e = 0.25$, τ_y^* is 0.968, and $e = 0.5$, τ_y^* is 0.866.

Most Solar System planets have nearly circular orbits; leaving out Mercury and Pluto, the planets have current eccentricities ranging from 0.007 to 0.093. Even Pluto only has a value of 0.244, though other large Kuiper Belt Objects have higher eccentricity. The most nearly habitable of the close-orbiting Super Earths in the Gliese 581 system have orbits that are quite eccentric by Solar System standards: 0.16 for Gliese 581c and 0.38 for Gliese 581d, based on the best estimates available in the year 2009. Planetary systems with more highly eccentric orbits are common, and appear at present to represent the rule rather than the exception (see Section 1.3). This makes it important to understand the seasonal cycle of planets with highly eccentric orbits, though most of the detailed exploration of that regime will be left to the Workbook problems.

Note that the difference between periastron and apastron distance is of order e ($O(e)$), whereas the ratio of major to minor axes deviates from unity by only $O(e^2)$. Hence, for small e, the orbit still looks like a circle, but with the Sun displaced from the circle's center by $O(e)$. For small e, Eq. (7.27) can be solved approximately by a straightforward expansion in e. Substituting

$$\kappa_1(t) = \frac{J}{a^2}[t + eF(t) + e^2 G(t)]$$ (7.30)

into the equation and matching like terms in e yields the solution

$$\kappa_1 = 2\pi t^* + 2e\cos 2\pi t^* + e^2 \left[\pi t^* + \frac{5\pi}{2} \sin 4\pi t^* \right] + O(e^3) \qquad (7.31)$$

where $t^* = tJ/(2\pi a^2)$. The first order term causes an $O(e)$ variation in the orbital angular velocity over the course of the year, but this term by itself does not alter the length of the year. Taking into account the second order term, it may be inferred that the non-dimensional length of the year is approximately $\tau_y^* = 1 - e^2/2$. In consequence, the annual mean insolation varies very little from what it would be for a circular orbit with radius equal to the semimajor axis. For $e = 0.1$, close to the present value for Mars, the eccentricity increases mean insolation by only 0.5%. For $e = 0.02$, similar to Earth at present, the increase is a meager 0.02%, or $0.274\,\text{W/m}^2$. Except in very unusual cases, orbital eccentricity affects the climate through the intermediary of the seasonal cycle, and not through any effect on the annual mean radiation budget.

The consequences of orbital eccentricity for a planet's climate derive from the way the distance seasons interact with the tilt seasons. Each of these types of seasons has a period of one planetary year, so the nature of the interaction is governed by the position in the orbit at which the northern hemisphere summer solstice occurs, measured relative to the position of the periastron. This can be measured by an angle, called the *precession angle* or *precession phase*. We will define the phase such that when it is zero, the northern hemisphere solstice occurs at the periastron. It is also common to define the phase as the angle between the periastron and the northern hemisphere spring ("vernal") equinox. When the precession angle is zero, the distance seasons make the northern hemisphere seasonal cycle stronger, since "northern tilt summer" happens when the planet is closest to the Sun and "northern tilt winter" happens when the planet is farther from the Sun. Conversely, the southern hemisphere seasonal cycle is attenuated when the precession angle is zero. When the precession angle is 180°, the situation is reversed between the hemispheres, with the southern hemisphere getting very hot summers and very cold winters, and the northern hemisphere experiencing more moderate seasons. When the precession angle is 90° or 270°, the solstices conditions are no longer modulated by the distance seasons, but instead the vernal equinox becomes warmer than the autumnal equinox, or vice versa.

Figure 7.12 illustrates the effect of eccentricity and precession on the seasonal cycle of insolation. These results were computed by numerically solving Eq. (7.27), and substituting $\kappa_1(t)$ into the flux distribution function given by Eq. (7.10) and Eq. (7.9), after shifting its phase to account for the precession angle. Given $\kappa_1(t)$, we also know $r(t)$. Using this, we multiply Eq. (7.9) by $(a/r(t))^2$ to account for the variations in orbital distance. This is the flux factor plotted, at selected latitudes, in Fig. 7.12. One multiplies this flux factor by the solar constant at a distance equal to the semi-major axis, in order to obtain the actual insolation in W/m^2. Using the symmetries of Eq. (7.9), the results for precession angles of 180° and 270° can be obtained from those shown in Fig. 7.12 by simply shifting the curves shown by a half year, and interchanging the two hemispheres, so these cases do not require separate discussion.

For both eccentricities, we see that the northern hemisphere extratropical seasonal cycle is made more extreme when the precession angle is 0°, while that in the southern hemisphere is moderated. At the equator, the two equinoxes have identical insolation, but the time of maximum equatorial insolation is shifted towards the northern summer solstice, which is also the time of periastron in this case. For the larger, Marslike, eccentricity ($e = 0.1$), the maximum equatorial insolation in fact occurs at the solstice. For the case

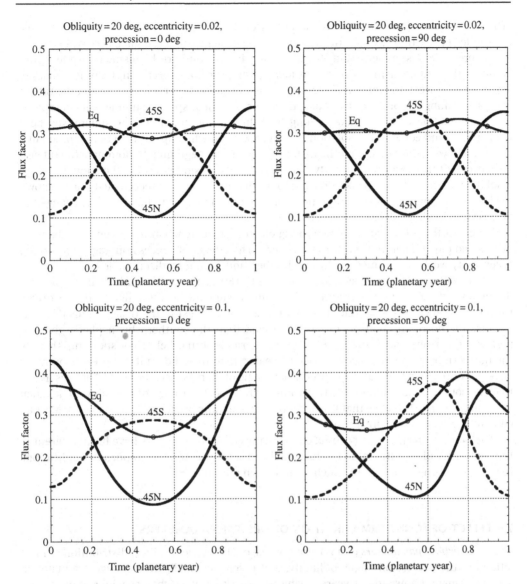

Figure 7.12 The seasonal cycle of solar flux factor for a planet with 20° obliquity, at the equator, 45° N and 45° S. To obtain the insolation at any given time of year, this flux factor is multiplied by the solar constant at the semimajor axis of the orbit. Results are shown for an Earthlike eccentricity of 0.02 (top row), and a Marslike eccentricity of 0.1 (bottom row). The left column gives results for a precessional phase of zero degrees, while the right gives results for 90 degrees, both measured relative to the northern hemisphere summer solstice.

of 90° obliquity, the extratropical seasonal cycle has identical strength in both hemispheres, but the equinox conditions now differ from each other, the autumnal equinox receiving less insolation than the vernal (spring) equinox. Also, the time of maximum and minimum extratropical insolation is also significantly displaced from what it would be for a circular orbit. The effect of orbital velocity variations on the seasonal cycle is just barely visible for the lower, Earthlike, eccentricity, but it is prominent for the higher eccentricity case. For 0° precession, summer is longer than winter in the southern hemisphere, while winter is longer than summer in the northern hemisphere; for 90° precession, there is a marked asymmetry between the rate of increase of insolation going into each season, and the rate of decrease coming out of it. For example, in the northern hemisphere, summer sets in rapidly, but the transition to winter takes a long time. In fact the northern hemisphere, southern hemisphere, and equatorial insolation maxima are all bunched up within a period of about a quarter of a year, indicating that the distance seasons are beginning to dominate the tilt seasons even at this modest eccentricity. The effect of precession phase on the annual average insolation at each latitude is insignificant; for both the high and low eccentricity cases shown in Fig. 7.12, changing the precession phase leaves the annual mean flux factor unchanged to at least four decimal places.

Note that the precession angle has a big effect on climate when the eccentricity is large, but has no effect when the eccentricity is zero. The effects of precession angle and orbital eccentricity work in conjunction with each other, and cannot be disentangled.

At present, Earth's precession angle is close to 180°, so that the southern hemisphere is driven towards hotter summers and colder winters, while the northern hemisphere is driven towards a weaker seasonal cycle. This pattern is not manifest in the observations (Fig. 7.1) because the northern hemisphere has more land than the southern hemisphere, giving it a stronger seasonal cycle, owing to its lower thermal inertia. Relatively speaking, though, the northern hemisphere seasonal cycle is weaker than it would be if the precession angle were 90° or 0°. Coincidentally, the precession angle of Mars is also about 180° at present, so that the southern hemisphere Martian winters are expected to be considerably colder than those in the north. Evidence that this indeed occurs, and its broader implications for Martian climate, will be taken up in Section 7.7.

The precession angles and orbital eccentricities of Earth and Mars have been different in the past, and will be different in the future. This has some extremely important implications for the evolution of climate, to which we now turn our attention.

7.6 EFFECT OF LONG-TERM VARIATION OF ORBITAL PARAMETERS

The three *orbital parameters* that govern the seasonal and geographical distribution of insolation are the precession angle, obliquity, and eccentricity. All three change gradually on a scale of many thousands of years, owing to basic laws of mechanics which apply to any planet in any solar system.

The evolution of the precession angle derives from a fairly elementary property of the mechanics of rigid-body rotation. The rotation axis of a rotating body subject to a net torque executes a rotation at constant rate about a second axis whose orientation is determined by the torque. The *precession rate* is determined by the magnitude of the torque and the angular momentum of the rotating body. The phenomenon of precession can be easily observed on a tabletop, by setting down a toy gyroscope with its axis inclined from the vertical. The top will precess, because there is a torque caused by the Earth's gravity and the force of the tabletop pushing up on the point of the top. For planets, the torque instead is provided by the slight deviations of the mass distribution from spherical symmetry. The equatorial bulge caused by rotation is a major player, but other asymmetries, including those due to the distribution of ice, and of major geographic features, are also of consequence.

Obliquity variations also stem from the basic properties of rigid-body rotation, but these variations arise from fluctuations in the torque on the planet, rather than the mean torque. The obliquity cycle is inextricably linked with the precessional cycle, which modulates the orientation of the aspherical planet with respect to the non-uniform gravitational field caused by the sun, the planet's moon(s) (if sufficiently massive), and all the other planets.

Eccentricity evolves because the periodic elliptical orbit is a solution only of the two-body problem, consisting of a planet and its star in isolation. Although the gravity of the Sun greatly dominates that of the other planets in our Solar System (and most likely in other planetary systems as well) the relatively small tugs of the planets on each other cause eccentricity to change gradually. Early in the history of this subject, it was shown by Laplace and Lagrange that the semi-major axis remains very nearly constant in the course of such eccentricity changes. The results of the preceding section therefore imply that eccentricity cycles have only a weak effect on annual mean insolation, since the mean insolation changes little if the semimajor axis is held fixed, except for extremely non-circular orbits.

Tiny deviations of the stellar gravity field from the ideal $1/r^2$ law add up to significant effects on obliquity and eccentricity over sufficiently long periods of time. The fact that the Sun is not perfectly spherical enters the problem, and even general relativistic deviations from Newtonian gravity have major effects.

Eccentricity modulates the distance seasons, and precession determines whether they constructively or destructively interfere with the tilt seasons. Meanwhile, obliquity variations modulate the strength of the tilt seasons. The net result is a rich variety of rhythms and patterns in insolation, which may lead to dramatic cycles in the state of a planet's climate.

In the following we discuss Milankovic variations for Earth and Mars, but similar Milankovic cycles should be a generic feature of planetary systems.

7.6.1 Milankovic cycles on Earth

Earth's precessional cycle is shown in Fig. 7.13. The precession angle increases at a nearly constant rate, completing a cycle every 22 000 years. Though the variation in rate is not evident over any one cycle, the rate is not exactly constant, and therefore the phase drifts over the course of hundreds of thousands of years.

The precessional cycle is very rapid, and the precession angle has changed markedly even over historical times. Eight thousand years ago, when the first Sumerians poured into the valleys of the Tigris and Euphrates, the star we now call Polaris (the "Pole Star," in the tail of the Little Bear) was about 40° of arc away from the star that the north polar axis then pointed to, and about which the constellations rotated at the time. The consequences of precession for change in seasonality are potentially highly consequential. In Fig. 7.13, the July insolation at 65° N is shown as a general indication of the magnitude of the seasonality effect; high northern July insolation in the precessional cycle goes with low January northern insolation, weak southern January (summer) insolation, and relatively strong southern July (winter) insolation. Ten thousand years ago, the northern hemisphere summer insolation was fully 40 W/m^2 greater than at present, and so the northern summers should have been considerably warmer than today, while the northern winters should have been considerably colder. The effect should show up especially over land, which is dark enough to absorb most of the solar radiation and has low enough thermal inertia to respond nearly instantaneously to seasonal changes. The climate system in its full glory is nonlinear and complex, so the response of climate to this change in seasonality could show up in any number of

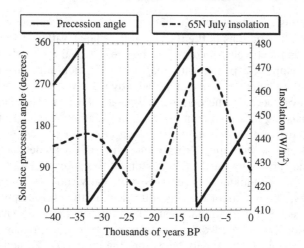

Figure 7.13 Evolution of precession angle relative to the northern hemisphere summer solstice, and the associated July insolation at 65° N. Data taken from Berger and Loutre (1991). BP, before present.

unexpected ways, and not simply as an enhancement of the northern hemisphere seasonal cycle over land.

The event which is most likely to be a recent manifestation of the precessional cycle is the "Climatic Optimum," covering the period of about 5000 to 7000 years ago (see Chapter 1). The term is most often used to refer to a period of generally warmer Eurasian temperatures. The "optimum" is sometimes said to be about 1–2 K warmer than present, but it is difficult to get reliable estimates of global mean temperatures, or even annual means. What is certain is that some regions during some seasons were warmer than they were at recent pre-industrial times. At about the same time, the Sahara, which is now a torrid desert, experienced a period of greening, with currently dry riverbeds ("wadis") filled with water, and a teeming variety of animal life and flora not found there at present. The greening of the Sahara is thought to be associated with atmospheric circulation systems known as "monsoons," forced to a greater extent by the enhanced heating of northern hemisphere subtropical land. A central question, though, is why the greening of the Sahara and the Climatic Optimum occurred several thousand years after the precessional peak in northern hemisphere insolation. There are some indications that the warming may have *begun* as much as 10 000 years ago, but the question of the physics accounting for the time delay in response remains unsettled. Candidates for the necessary inertia in climate response include vegetation adaptation, land ice, and deep ocean heat storage.

Looking further back in time, the obliquity and eccentricity variations become significant, though of course the precession cycle also continues to have a large effect. The Earth's obliquity and eccentricity cycle is shown in Fig. 7.14. The amplitude of the obliquity cycle varies considerably over time, but its dominant period is on the order of 40 000 years. The Earth's obliquity varies narrowly in a range from about 22° to 24.5°. At present, the Earth is in the middle of its obliquity range. Eccentricity varies on a longer time scale of approximately 100 000 years. However, in Fig. 7.14 there are also hints of a 400 000 year cycle of eccentricity, whose fingerprint consists of two high eccentricity cycles followed by two low eccentricity cycles. This visual impression is borne out by spectral analysis. Currently, the

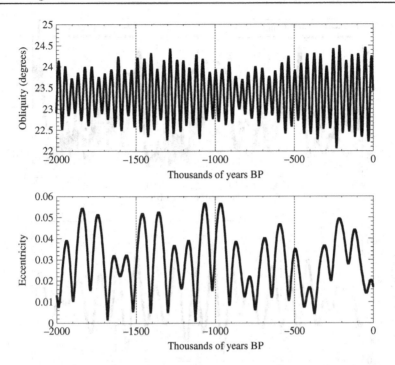

Figure 7.14 Evolution of the Earth's obliquity and eccentricity. Data taken from Berger and Loutre (1991).

Earth is near the low end of its eccentricity range, though it has been quite close to zero during the past two million years. At the other extreme, Earth's eccentricity has reached as high as 0.055, or more than half that of Mars.

The idea that ice ages are due to changes in Earth's orbital parameters is nearly as old as the discovery of ice ages themselves. The idea has gained currency, but it is nearly as hard to justify today on basic physical principles as it was when first proposed. The main reason for its acceptance is circumstantial, in that increasingly detailed data on the observed rhythm of the ice ages shows the unmistakable imprint of the calculated rhythm of the orbital forcing. James Croll first proposed in the 1870s that changes in the Earth's eccentricity led to ice ages, and his idea was refined a half century later by Milutin Milankovic, whose name is now generally attached to the theory. The centerpiece of Milankovic's idea is that ice ages require the accumulation of snow on land, and that this in turn is favored by mild summers (limiting melting of old snow and ice) and warmer, but still sub-freezing, winters (favoring snow accumulation, since warmer air contains more water). The gaping hole in Milankovic's theory is that it predicts that ice ages should follow the precessional cycle. In particular, the northern hemisphere and southern hemisphere should have ice ages in alternation every 10 000 years, with the severity of the ice ages modulated by the eccentricity cycle. This is not at all what is observed. Figure 7.15 shows the Antarctic temperature record for the past 400 000 years, together with eccentricity and the July insolation at 65° N. Numerous other temperature proxies worldwide show that the northern hemisphere temperature, and global glacier ice volume, are nearly in phase with the Antarctic temperature record, so that the Antarctic temperature can be taken as an index of when the world is in an ice age. The

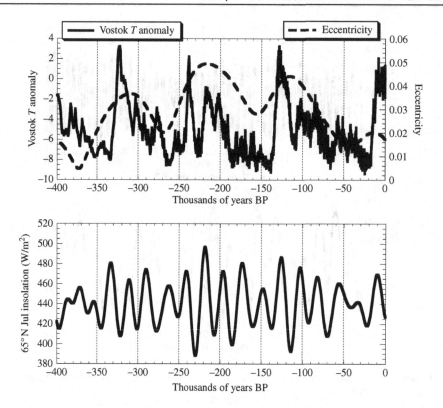

Figure 7.15 Comparison of the Earth's eccentricity cycle with the Antarctic temperature reconstructed from deuterium measurements made on the Vostok ice core. The bottom panel shows the corresponding July insolation at 65° N. Temperature is given as deviation from the mean modern value. Vostok temperature data was taken from Peteet *et al.* (1999) (see Chapter 1).

dominant signal in the climate response is an approximately 100 000 year spacing in the major interglacial warm periods, and a similar spacing in the coldest glacial periods. Crudely speaking, each interglacial corresponds to a peak in eccentricity, and a time within which (during parts of the precessional cycle) the northern hemisphere seasonality is unusually strong. This is somewhat reminiscent of the Milankovic mechanism, but what filters out the high-frequency precessional cycle? Why does the entire Earth fall into an ice age at the same time, rather than alternating between hemispheres? A closer examination of the 65° N July insolation strongly suggests that major deglaciations occur when the northern hemisphere seasonality is weak, suggesting that the Earth listens to the northern hemisphere forcing more than the southern, in deciding when to have an ice age. This probably has something to do with the fact that the northern hemisphere has more land, and hence more seasonality, than the southern, but the precise way this asymmetry influences global glaciation remains largely obscure.

The problem is not that the amplitude of radiative forcing associated with Milankovic cycles is small: it amounts to an enormous $100 \, W/m^2$, with the amplitude determined by the eccentricity cycle. The problem is that the forcing occurs on the fast precessional time scale, whereas the climate response is predominately on a much slower 100 000 year time scale. One does not so much need an amplifier of Milankovic forcing, as a "rectifier," which is sensitive to the *amplitude* of the precessional variation, rather than to its mean. Recall that atmospheric CO_2 is observed to vary on the glacial–interglacial time scale. Certainly, this is a major piece of the puzzle, since the drop in CO_2 during glacial times is sufficient to account for a major portion of the cooling of the climate, particularly in the southern hemisphere (see Chapter 4). Carbon dioxide is a globalizing effect, and (insofar as it is linked to the glacial–interglacial physical climate changes) an amplifying feedback. The circumstantial role of CO_2 in ice ages is also a reprise of an old idea. The nineteenth-century physicist Tyndall, whose work on infrared spectroscopy is at the foundations of our current understanding of the greenhouse effect, was primarily interested in explaining the ice ages, and the association reappeared later in the work of Chamberlain. The mechanism of the CO_2 cycle is not known, but almost certainly involves CO_2 storage in the deep ocean. The lack of a theory for the glacial–interglacial CO_2 cycle is the central impediment to a theory of the ice ages. The presence of ice does seem to be a prerequisite for a strong climate response to orbital forcing. Before the onset of permanent polar ice at the beginning of the Pleistocene, response to orbital forcing was weak (see Chapter 1). Besides CO_2, ocean circulations can potentially play a major role in globalizing and rectifying the northern hemisphere signal, through direct heat transport as well as indirect effects on CO_2. The answer to the mystery of the ice ages lies somewhere in the space: ice, ocean, CO_2; but how the system works its miracles to yield a 100 000 year cycle is still unknown.

In recent years, an alternate picture of the way the joint precesssional/eccentricity cycles affect glaciation and deglaciation has emerged. This picture proceeds from the observation that when the tilt seasons line up with the distance seasons, Kepler's law implies that the intense summers in the in-phase hemisphere are also short, while the moderate summers in the opposite hemisphere are long. Then, if one assumes that deglaciation is primarily responsive to some measure of integrated summer melt energy – to be thought of as the net energy input for that part of the season where the glaciers are brought to the melting point – the variations of duration of the seasons for an Earthlike obliquity and eccentricity range largely cancel out the variations of the peak intensity. This makes the precessional and eccentricity cycles mostly drop out of the picture, leaving obliquity variations as the main player. In this view, the obliquity cycle is the natural rhythm for glacial–interglacial cycles, and the early Pleistocene, in being dominated by the obliquity cycle, is responding in the simplest and most readily understood way. The longer cycles that emerge in the later Pleistocene are seen not as a consequence of the eccentricity variations, but rather as a matter of skipping of obliquity cycles because of some as-yet unknown nonlinearity in glacial dynamics. What we see as a 100 000 year cycle is thought of instead as an irregular blend of 80 000 year cycles coming from a single skip, and 120 000 year cycles coming from a double skip. The glacial–interglacial CO_2 variations are still an important amplifying feedback for this mechanism, and indeed could play a role in the mechanism accounting for skipping of obliquity cycles.

7.6.2 Milankovic cycles on Mars

As expected from general mechanical considerations, Mars has Milankovic cycles analogous to those of Earth. Mars' cycles differ in some key respects, because of the lack of a massive moon, and because of the proximity of Jupiter.

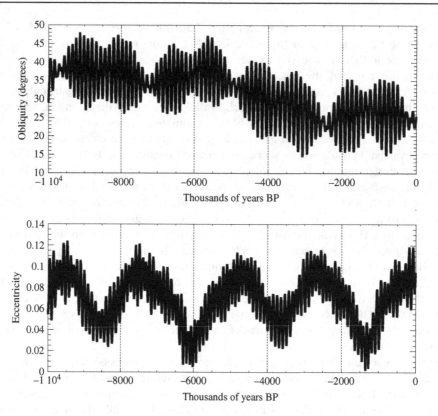

Figure 7.16 Evolution of Mars' obliquity and eccentricity. Data taken from Laskar *et al.* (2004).

As for Earth, the precession angle of Mars increases at a nearly constant rate. However, because Mars does not have a moon as massive as Earth's, the precession is dominated by solar gravity, and is slower. The Mars precessional cycle has a period of approximately 50 000 Earth years. The current precession phase is 145°, and will reach 180° in about 5000 years.

The obliquity and eccentricity variations are shown in Fig. 7.16. Obliquity has short term variations with amplitude on the order of 20°. The period is not visible in the figure, but a finer-scale examination of the data shows that the period is about 125 000 Earth years in recent times. The amplitude is markedly larger than that of Earth's obliquity cycle, but what is even more remarkable is that the obliquity drifts to values as large as 47° over 10 million years. The extreme obliquity variations are directly linked to the absence of Earth's massive moon, which can be shown to provide a considerable damping effect on obliquity. This raises the intriguing possibility that a massive moon may be a necessary condition for a planet to avoid extreme climate fluctuations that could compromise its habitability. Calculations of the Earth's obliquity have also been carried out for tens of millions of years, and do not yield any greater variations than have been encountered in the past million years.

Mars is close to its maximum eccentricity at present, though it can get somewhat larger. The eccentricity of Mars undergoes quasiperiodic large amplitude cycles with a period on the order of 3 million years. In addition, there are short period, lower amplitude eccentricity variations with a period on the order of 100 000 years, rather similar to Earth's. In contrast, the very long period variations are not found in Earth's eccentricity.

Mars has no ocean, little thermal inertia, and a thin atmosphere that has a relatively modest effect on the planet's surface temperature. These features should lead to a different, and perhaps simpler, response to orbital forcing on Mars as compared with Earth. The predicted climate changes have been simulated in detail using comprehensive climate models, but we will confine ourselves here to some general remarks. The main effect of Martian Milankovic cycles is likely to be the redistribution of water deposits, in the form of either glaciers or permafrost. There are two aspects to this redistribution. On the short precessional time scale, the asymmetry between the northern and southern polar ice caps should reverse. For example, about 25 000 years ago, the southern hemisphere should have had milder summers and winters, while the northern had cold winters and hot summers; the default reasoning would imply that at such times, the southern ice cap should be large and be composed mainly of water ice, whereas the northern ice cap should become smaller and experience massive seasonal CO_2 snow deposition. On the time scale of millions of years, the obliquity of Mars becomes much greater, leading potentially to a situation where water may migrate from poles that are seasonally very hot, and redeposit in the tropics. At times of much lower obliquity, permafrost ice may migrate to both poles. The migration of water deposits and changes in patterns of deposition of CO_2 snow probably leaves some imprint on the surface geology of Mars, and the growth and decay of glaciers certainly does. These offer some prospects for reconstructing the consequences of Milankovic cycles on Mars. Even better information would be obtained through analyzing cores of the polar ice caps, much as is done in Antarctica and Greenland. It is very exciting that the technology for doing this robotically on Mars is already under development. With respect to Mars, we are more or less at the stage of Croll or Milankovic, who thought they had found the key to Earth's ice ages. Data showed they were on the right track, but that the climate system is much more intricate than they imagined. Given that we do not yet have a satisfactory theory leading from orbital variations to climate response on Earth, one can look forward to many surprises once data on the Martian climate response becomes available.

7.7 A PALETTE OF PLANETARY SEASONAL CYCLES

This section provides a sampler of the many ways in which thermal inertia and orbital characteristics can combined to yield seasonal cycles with interesting consequences. These are presented as a series of vignettes, which are primarily meant to serve as impetus to further inquiry.

7.7.1 Formation and inhibition of polar sea ice

For continental configurations similar to Earth's present state, the processes governing high-latitude climate variations differ greatly between the north and south polar regions. At present there is an ocean at the north pole, so ice can form there through freezing of ocean water, but in the absence of ice the seasonal cycle would be moderated by the thermal inertia of the ocean. The south pole is surrounded by the Antarctic continent, so that ice can accumulate only through snowfall; moreover, the seasonal cycle there will be strong even if there is no glacier. The conditions for formation of ice in a polar ocean are germane to Earth's climate variations at times going back at least to the Cretaceous, and there are numerous other times in Earth's history when one or the other pole was surrounded by ocean. Since most planets with an ocean will undergo some periods when there is no continent near one of the poles the problem is of general interest for exoplanet climate as well.

At the time of writing, the high northern latitudes are covered by sea ice for most or all of the year, though that appears to be changing rapidly as a result of anthropogenic global warming. In the hothouse climates of the Eocene and Cretaceous, these regions were nearly or completely ice-free, and deep ocean temperatures as well as other proxies indicate annual mean temperatures of at least 10° C, and perhaps at times as high as 20° C. What conditions could lead to such warm polar seas, given that we know the Earth can also support icy polar conditions?

There are two points of reference to keep in mind when thinking about the formation of polar ice. Both depend on the dynamic heat flux and ice-free ocean albedo, as well as the insolation parameters and atmospheric greenhouse effect. The first point is whether the temperature that would be in equilibrium with the annual mean absorbed solar radiation for an ice-free ocean is above the freezing point of sea water. This temperature is the temperature that would be attained for an infinitely deep mixed layer. If the mean equilibrium temperature is below freezing, then it is inevitable that sea ice will form at least seasonally, since the seasonal cycle can only make the winter temperature fall below the mean. If the mean equilibrium temperature is above freezing, then it is possible that sea ice can be entirely suppressed. That would happen, for example, if the seasonal cycle were weak enough that the winter temperature never hit freezing. However, even if the winter temperature fell to freezing, sea ice might still be suppressed if the ocean stratification were such that the dense, cold water at the surface sank deep into the ocean. This case is equivalent to having a very deep mixed layer, and since there would not be time in a single winter to freeze the deep ocean, ice would never form. Salt stratification plays a big role in the formation of sea ice on Earth, since relatively fresh Arctic surface water (coming from ice melt, river runoff, and net rainwater input) can hit the freezing point without triggering oceanic deep convection.

The second point of reference is whether the midsummer instantaneous equilibrium temperature attained by an ice surface is above the melting point. If it is, there is the possibility (but not the certainty) that the ice can be melted away during the summer. If it is not, then it is inevitable that ice will survive from one winter to the next, leading to perpetual ice cover.

The ice evolution occurring in modern times, as well as the transition to ice-free states in the Eocene, suggests that the Earth is not deeply in the perpetual-ice regime, but rather hovering at the boundary between where perpetual ice is conditionally possible, and the regime where the annual mean conditions rule it out. With simple models – in fact even with very complex models – it is only possible to make the case that the radiative conditions can be made compatible with such a state. Let's take conditions at 80° N as an example. The satellite-observed annual mean dynamical heat flux at that latitude is about $110 \, \text{W/m}^2$ at present. Let's assume that the albedo of an ice-free ocean would be 0.33, which allows for partial cover by low clouds. In these circumstances, for Earth's present obliquity the annual mean temperature for a deep mixed layer would be 270 K at a CO_2 concentration of 300 ppmv – just below freezing. At 1200 ppmv, it rises to 273.5 K – just above freezing. On the other hand, even at 300 ppmv the equilibrium ice temperature at the summer solstice for an ice albedo of 0.6 is over 300 K including the effects of dynamic heat flux, far in excess of what is needed to sustain summer melt. In fact, it would only take about $20 \, \text{W/m}^2$ of dynamic heat flux to bring the summer ice surface to the melting point. This is consistent with the

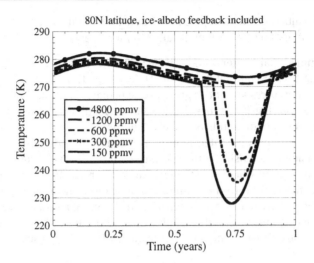

Figure 7.17 Numerically computed seasonal cycle for a mixed layer ocean subject to a realistic seasonal cycle of insolation at 80° N latitude. The albedo for ice-free conditions is 0.33, which allows for partial low-cloud cover. The dynamic heat flux is 110 W/m^2, which is close to the observed annual mean value at this latitude. This calculation includes an idealized representation of the feedback of sea ice on albedo and thermal inertia (see text for details). Results are shown for a range of values of the CO_2 concentration.

observation that high-latitude northern hemisphere temperatures hover around freezing in the summer, but it also says that apart from extreme changes in obliquity, summer melt is inevitable. The real question is whether ice starts to form at all, and whether it becomes thick enough to last through the summer.

To give these ideas a more quantitative expression, let's extend the basic mixed layer model to include some idealized sea-ice feedbacks. We will assume that whenever the ocean temperature drops below the freezing point of sea water (say, 271 K), ice forms and the albedo is increased to the albedo of sea ice (say, 0.6). Further, in order to crudely represent the low thermal inertia of ice, we'll decrease the effective mixed layer depth to 1 m when ice is present. Finally, we'll ignore the time it takes to melt ice, and instantaneously remove the ice when its surface temperature reaches the melting point. Results for this model, adopting the dynamic heat flux and ocean albedo values stated previously, are shown in Fig. 7.17. The results show that there is a quite extensive and very cold winter ice season when the CO_2 is at 300 ppmv (somewhat above pre-industrial levels) or at 150 ppmv (somewhat below ice-age values). In neither case, though, does the ice cover become perpetual. This failing arises from the unrealistic assumption that ice melts instantaneously. In reality, the longer and colder the winter icy period, the thicker the ice will become, and the longer it will take to melt (see Problem 7.17). The following exercise shows that the time required for melting could easily extend the icy period to the rest of the year.

Exercise 7.8 Suppose that partway through winter, when the sea ice is at its thickest, the ice has a thickness of 5 m. Once the ice surface reaches the melting point, the temperature will stay fixed until all the ice has melted, and this delays the disappearance of the sea ice.

Assuming that the albedo of melting ice is 0.5 and that the incident solar radiation at the surface is $200\,W/m^2$, estimate how long it takes to melt the ice.

With the assumed parameters, the ice season shortens when CO_2 is increased to 600 ppmv and disappears completely when CO_2 exceeds 1200 ppmv. Even at 4800 ppmv, the mean polar temperatures are not as great as the proxies suggest, though, indicating that some additional amplifying factor besides ice-albedo feedback must come into play. The winter warming effect of high clouds is a likely candidate.

The calculation we have presented is only a plausibility argument. It only shows that, without going outside a defensible range of parameters, the basic model of the seasonal cycle can be put in a state where a CO_2 increase comparable to what is plausible for the Eocene or Cretaceous can lead to the elimination of sea ice. The actual proof that the Earth system is near this transition depends on myriad details governing ocean stratification, clouds, realistic sea ice physics, and dynamical heat flux. Indeed, comprehensive general circulation models vary considerably in the threshold CO_2 at which sea ice is suppressed.

7.7.2 Continental climates on hothouse Earth

One of the toughest problems regarding the nature of hothouse climates like the Cretaceous is accounting for the absence of cold winters in the interiors of continents. There is abundant evidence for mild high-latitude winters, in the form of fossils of creatures (notably alligators) and plants (notably palms) that cannot tolerate freezing conditions. The problem is that there is little local solar flux in winter, while the thermal inertia of land is too weak to keep temperatures from plummeting. For example, with $23°$ obliquity, the flux factor at $50°\,N$ latitude at the winter solstice is only 0.063. Even ignoring snow cover, the absorbed solar radiation for Earth is only $61\,W/m^2$, based on an albedo of 0.3. With 300 ppmv of CO_2, the local equilibrium temperature would be well below $220\,K$. In fact, even increasing the CO_2 concentration all the way to 10^5 ppmv is insufficient to bring the local equilibrium temperature up to $220\,K$, let alone the freezing point. Including the albedo of typical winter snow cover at this latitude, the absorbed solar flux drops further to $34\,W/m^2$, leading to still more frigid conditions. A more complete exploration of continental temperature is carried out in Problem 7.7.

Actual continental winters never get nearly as cold as the equilibrium temperature. At $50°\,N$ latitude in the coldest part of Eurasia and North America, monthly mean winter temperatures hover around $250\,K$. Continental interior temperatures are warmed by lateral heat transports from the warm upstream ocean and from the warm tropics. Under clear sky conditions, the *OLR* with a surface temperature of $250\,K$, for the conditions of Table 4.2, is $178\,W/m^2$ with 300 ppmv of CO_2; it falls modestly to $169\,W/m^2$ at 2400 ppmv. The difference between either of these numbers and the absorbed solar radiation gives the rate at which heat needs to be imported. Given the slight absorbed solar flux in winter for a snow-covered surface, it is clear that dynamic heat transports are far and away the *dominant* factor in determining winter high-latitude temperatures in the continental interiors. A similar calculation tells us that the total flux needs to be brought up to $222\,W/m^2$ in order to bring the surface temperature up to $273\,K$ when the CO_2 is increased to 2400 ppmv, requiring an increased flux of $53\,W/m^2$ over and above the help provided by increasing CO_2. You could get about $27\,W/m^2$ of this from the albedo reduction that occurs when the snow disappears, but where does the other $26\,W/m^2$ come from? The problem is inherently dynamical, and requires detailed consideration of fluid dynamical heat transports.

Current general circulation models do not generate enough heat transport to resolve the paradox, even if CO_2 is increased to the point where the upstream oceans have temperatures similar to paleoclimatic reconstructions. It is hard to see that there could be any major effect missing from the large-scale dynamics, but low-level inversions form in the winter, as discussed in Chapter 6, and it is possible that some flaw in the representation of the boundary layer spuriously weakens the heat transfer from the atmosphere to the underlying surface.

In the wintertime there is little solar flux to reflect and a fairly high surface albedo. In these circumstances mid- and high-level clouds could exert a substantial warming effect. Deep convection does not occur in the winter extratropics, and cloud formation in this regime is inextricably linked to large-scale dynamics. Current general circulation models do not form enough high clouds to solve the problem of continental hothouse winter, but given uncertainties in the representation of convection and clouds, there is plenty of room for creative thinking in this area.

7.7.3 Snowball Earth

The high albedo of a Snowball state leads to very low temperatures in comparison with an unglaciated state for the same orbital position, but from the standpoint of the seasonal cycle, the most important effect is that replacing the liquid ocean surface with a solid ice/snow surface practically eliminates surface thermal inertia from the system. For a planet with a very thick atmosphere, the atmospheric thermal inertia could still moderate the seasonal cycle, but for an Earthlike atmosphere, the surface temperature would be nearly in equilibrium with the instantaneous diurnally averaged insolation in the course of the seasonal cycle, in the absence of lateral atmospheric heat transport. The equilibrium surface temperatures for various obliquities are computed in Problem 7.7.

For Earthlike obliquity, the insolation gradients are weak in the summer hemisphere, so the temperature will be fairly uniform there and one does not expect a big role for dynamical heat transport. In the winter hemisphere, however, dynamical heat transports will moderate the temperature contrast implied by the extreme gradients in insolation found there. These expectations are borne out by simulations of the Snowball Earth state using full general circulation models (see the Further Reading). For much higher obliquities the summer pole begins to become substantially hotter than the summer tropics, and for some orbits and surface albedos could even seasonally reach the melting point.

Overall, the Snowball Earth climate has more in common with the climate of Mars than it does with the present Earth. The high albedo reduces the solar forcing to values similar to those of rocky Mars, while the frozen surface yields a low thermal inertia similar to that for the dry, rocky surface of Mars. The cold conditions also reduce the role of water vapor and water clouds in climate, while making CO_2 condensation possible at the winter pole. The principal differences between the climate states of Snowball Earth and Mars arise from the surface pressure of the atmosphere and the size of the planet; the latter of these comes in exclusively through dynamical effects.

Note that while the surface temperature undergoes extreme seasonal variations, the results of Section 7.4.2 show that the deep ice temperature will be equal to the annual mean surface temperature. This temperature will be similar to, though not exactly equal to, the surface temperature that would be in equilibrium with the annual mean flux; for Earthlike obliquity, this temperature has its maximum at the equator. For thick ice, deglaciation requires that the deep ice temperature reach the melting point, so that it is the annual mean conditions that primarily determine when deglaciation can occur. The formation of seasonal summer melt ponds at the surface can indirectly affect the deep ice temperature, however, through reducing the surface albedo and delaying the onset of autumn freezing conditions.

7.7.4 Venus

The deep atmosphere of Venus has an enormous thermal inertia, both because of its great mass and because it is so optically thick that heat escapes only slowly via sluggish convection and slow radiative diffusion. The slow rotation of Venus means that the nightside atmosphere has a long time to cool down, but it also makes it easier for the atmosphere to transport heat effectively from the dayside to the nightside. Observations confirm that there are essentially no geographic or seasonal variations of surface temperature apart from those that can be attributed to surface height variations.

This does not mean that the seasonal cycle of Venus is without interest, for the top bar of its atmosphere has as much mass as Earth's entire atmosphere, and this portion of Venus exhibits a lively mixed seasonal/diurnal cycle. The cycle here is driven by *in situ* atmospheric solar absorption, rather than by heating from below – a situation rather similar to that which arises owing to ozone heating in the Earth's stratosphere. The dayside is considerably hotter than the nightside at these altitudes, and this drives a circulation carrying heat and constituents from one side to the other. Moreover, for rather subtle dynamical reasons, the upper atmosphere takes on a rotation of its own even though the surface of the planet is hardly rotating. The upper atmosphere rotation is not a rigid-body rotation, but it carries atmosphere along latitude circles with a time scale on the order of 5 Earth days. Because the sense of this circulation is in the same sense as the equatorial surface motion, but faster, it is called a *super-rotation*. The super-rotation of the upper atmosphere of Venus gives the seasonal cycle there many of the characteristics familiar from more rapidly rotating planets.

Venus can be considered to be the archetype for a probably common class of planets with thick CO_2 atmospheres containing effective solar absorbers. Understanding the seasonal cycle in this regime has important ramifications for the interpretation of exoplanet observations, since most observations of Venus class planets would only be able to see fluctuations in the infrared and reflected solar radiation from the upper portions of the atmosphere. Determination of the seasonal cycle of cloud albedo is a key part of this story, and while that certainly involves both dynamics and atmospheric chemistry, the seasonal temperature variations provide the context against which both effects take place.

7.7.5 Mars, present and past

The seasonal cycle of present Mars typifies a generic class of seasonal cycle involving condensation of a major constituent of the atmosphere. This kind of seasonal cycle is likely to be common throughout the Universe, and over time for a broad class of planets. Figure 7.18 shows the seasonal cycle of surface pressure for the Viking 2 lander, which sat in the northern hemisphere midlatitude of Mars. The surface pressure varies from 7.3 mb to 11 mb over the course of a year, amounting to a 20% variation around the mean. Earth's surface pressure typically varies by no more than 1% on large scales, or perhaps as much as 5% if one includes the minimum pressure at the center of strong hurricanes. The difference between Earth and Mars is that only a small portion of the Earth's atmosphere is

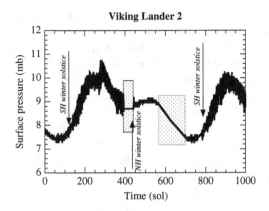

Figure 7.18 Surface pressure time series for the Viking 2 lander, which was located at 47.97° N latitude. The time scale is in Mars days (*sol*) since landing. Solstices are marked. The shaded areas indicate missing data.

condensable, whereas the primary constituent of the atmosphere of Mars can condense at the winter poles. If the major pressure variations on Mars are similar over the globe, then the pressure fluctuation implies that up to 16% of the maximum mass of the atmosphere can be sequestered in condensed form at one of the winter poles. A significant feature of the Martian seasonal pressure cycle is that the minimum pressure occurs somewhat before the southern hemisphere winter solstice, and that the secondary minimum at the northern hemisphere winter solstice is less pronounced. This is due to the high eccentricity of the present Martian orbit, and the current phase of its precessional cycle, which conspire to create southern winter conditions that are far colder than northern winter conditions.

The simplest model of the seasonal cycle of pressure would be to posit that the global pressure is equal to the saturation vapor pressure corresponding to the temperature at the winter pole. But how does one determine that temperature? Thermal inertia, and perhaps also dynamical heat fluxes, must still be involved since the polar night equilibrium temperature would be very nearly absolute zero. We have seen that at temperatures of 150 K or less, even a solid surface can provide significant thermal inertia, since the radiative relaxation times become so long at low temperatures. Another factor is that the latent heat of condensation must be radiated away to space in the polar regions, and this provides a form of thermal inertia limiting the condensation rate. For example, at a surface temperature of 150 K, one radiates only 28 W/m^2 to space, and using the latent heat of sublimation of CO_2 one could only condense at a rate of $4.8 \cdot 10^{-5}$ kg/m^{2}s. Carried out over a tenth of the Martian surface, this would condense out about a quarter of the Martian atmosphere in the course of one winter. If the polar temperature were to get much colder, however, polar condensation would practically halt, even though the surface saturation vapor pressure might be very low compared with the global mean pressure. If one tried to make the condensation rate much higher, however, the latent heat release would raise the surface temperature to the point where the surface pressure became subsaturated, and condensation would have to halt. The thermal inertia due to latent heat release is therefore a very significant player in Marslike seasonal condensation cycles. Similar considerations would come into play for the rate of condensation of atmosphere on the nightside of a tide-locked planet. In both cases, the condensation rate could also be limited by the dynamics governing mass transport into the condensing region.

The greenhouse effect of the thin atmosphere of present Mars is slight, but in cases with a more substantial atmosphere, massive seasonal pressure fluctuations due to polar condensation have even more interesting consequences. In that case, the polar condensation amounts to an amplifying feedback on the seasonal cycle. During each solstice, sequestration of CO_2 in polar regions reduces the greenhouse effect *globally*, leading to very cold conditions in the winter hemisphere and moderated temperatures in the summer hemisphere. The situation is somewhat akin to water vapor feedback, in that Clausius–Clapeyron controls the mass of a greenhouse gas present in the atmosphere. It is less local than water vapor feedback, however, since the reservoir of condensate is localized in polar glaciers, rather than spread out over a nearly ubiquitous ocean.

With a more massive atmosphere, the polar condensation engages interesting glaciological questions as well. A massive CO_2 glacier would flow, particularly in view of the fact that it would be rather easy to liquefy CO_2 at the base. The glacier flow would bring condensed CO_2 to warmer parts of the planet, where it could be recycled to the atmosphere.

Returning to present Mars, let's take a look at the seasonal and diurnal cycle of temperature. Some typical midlatitude summer and winter diurnal cycles are shown in Fig. 7.19. As expected from the low thermal inertia of the system, the seasonal and diurnal cycles are large. The daytime peak summer temperature is 63 K warmer in summer than in winter, while the seasonal cycle in night-time temperature is a more moderate 32 K. Because of the low thermal inertia, night-time temperatures are almost as low in summer as they are in winter. The diurnal cycle in summer, at about 50 K, is nearly as large as the seasonal cycle. Note also that the diurnal cycle in winter is weaker than it is in summer. This is because the low temperatures lead to longer radiative relaxation time, and also because the daytime peak insolation is weaker. A more quantitative analysis of the seasonal cycle in this dataset is carried out in Problem 7.38.

The reader has no doubt noticed that the winter diurnal cycle is much less regular than the summer cycle. This is a general feature, and is not peculiar to the snippets of data shown in Fig. 7.19. The irregularity of the winter diurnal cycle is due to the presence of wavy instabilities of the Martian winter jet stream – an analog of Earthlike midlatitude storms. These are absent in the summer, for reasons that are inherently dynamical.

For Early Mars, an additional set of questions present themselves. Let's suppose that the planet has a thick CO_2 atmosphere of perhaps 2 bar, but that the radiative effect of this atmosphere is not sufficient to bring the annual mean temperature above freezing at any point. Suppose further that Early Mars does not have extensive deep oceans, so that the high albedo of a Snowball state does not come into play. Under these circumstances, the surface would have low albedo and little thermal inertia. Could the seasonal cycle lead to seasonal summer glacier melt sufficiently strong to account at least in part for the ancient river beds seen on Mars? The extreme obliquity cycle of Mars accentuates this picture, since the summer hemisphere would become particularly hot during high obliquity phases, especially when they line up with periastron during a high eccentricity phase.

Earlier, we estimated that the heat capacity of a 2 bar Early Mars atmosphere was equivalent to a 10 m water mixed layer. This is far less than Earth's mixed layer depth, but to complete the calculation of thermal response time we also need to know the radiative damping coefficient for the atmosphere. Using the homebrew radiation model for the 2 bar pure CO_2 atmosphere, for the conditions of Fig. 4.30, we find that $dOLR/dT$ is about $1 \, W/m^2 K$ for

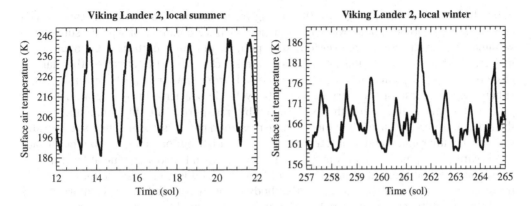

Figure 7.19 A segment of the near-surface air temperature time series for the Viking 2 lander showing the diurnal cycle in summer (left panel) and winter (right panel).

surface temperatures near 273 K. With the effective mixed layer depth, this yields a thermal response time of 484 Earth days, which is notably less than the Martian year of 687 Earth days. Therefore, the thermal inertia of the Martian atmosphere would have some moderating effect on the seasonal cycle, but the seasonal cycle would be far less strongly attenuated than Earth's. Therefore, one can expect quite hot summers, especially during high obliquity stages, perhaps even leading to extensive seasonal melt. The relatively weak thermal inertia is a two-edged sword, however. During periods of high obliquity, the long dark winter will be exceptionally cold, and if enough CO_2 condenses at the winter pole, the drawdown of atmospheric pressure could limit the greenhouse effect. Precisely how much CO_2 condenses in these circumstances depends on the moderating effect of CO_2 ice clouds discussed in Section 5.9, and on the rate of heat and mass transport from the summer to the winter hemisphere. These, together with the volume of summer melt, constitute outstanding Big Questions.

7.7.6 Nearly airless bodies

At first glance, it might be thought that the seasonal cycle of nearly airless solid bodies like the Moon or Europa or Triton would be fairly dull. With little thermal inertia at the surface, and no significant atmospheric effect on surface temperature via heat transport or greenhouse effect, the seasonal cycle would seem to be a simple matter of the hot spot moving along with the subsolar point. On reflection, though, it is easily seen that the problem offers a rich variety of novel features worthy of the attention of the most discriminating planetary climatologist.

First of all, the thermal inertia of even a solid surface can become significant when temperatures become cold enough, since the radiative cooling decreases so rapidly with temperature. Thus, the temperature of the unilluminated part of the Moon is determined mainly by heat diffusing up from the subsurface and by the amount of time that has passed since the most recent illumination. For very cold bodies such as Europa or Triton, the thermal inertia becomes significant even under the subsolar point, so that rather than a simple hot spot, one should see a warm tail trailing the hot spot along the past path of the subsolar point. There can be interesting patterns arising from variations in surface thermal inertia and surface albedo, and moreover, the transport of trace amounts of volatile substances sublimated from the crust can, over long time periods, modulate both the albedo and thermal inertia. Mars is practically in the nearly airless class, and the effect of the Martian glaciers provides a prime example of the feedback from slow transports in a thin atmosphere. In the course of the Martian obliquity cycle, the glaciers slowly migrate from the poles to the low latitudes, and this is mediated by transport in the thin atmosphere. Even the Moon and Mercury have interesting volatile migration patterns; there appears to be water ice near the lunar poles, and an interesting seasonal redistribution of sodium on the surface of Mercury. Triton has a complex and young surface, which is probably due at least in part to sublimation and redistribution of N_2 and CH_4 ices, and it is even possible that the redistribution of water ice on Europa affects that body's albedo. Europa has a 3.5 d diurnal cycle but little obliquity, so the main redistribution of water should take the form of sublimation and redeposition in the equatorial belt, plus some leakage of snow into the midlatitude and polar regions. This would be a very slow process, but over the course of hundreds of millions of years could result in important evolution in the surface.

Moreover, as one goes deeper below the surface, the temperature is determined by the surface temperature averaged over longer and longer time scales. Thus, at a moderate depth, the temperature will be given by the diurnal average, while still deeper layers will have a temperature determined by the annual mean. To some extent, subsurface temperature can be probed via radio frequency emissions, and the variations can be used to recover information about composition and structure.

For icy bodies composed of substances that have sufficiently elevated vapor pressures when exposed to temperatures encountered over the course of the seasonal cycle, things get even more interesting. This is the case for N_2 and CH_4 ice on Triton and Pluto. In the former case, subsurface heating even leads to spectacular cryovolcanism. The "solid greenhouse effect" arising from penetration of solar radiation into ice, as discussed in Section 6.9, can lead to even more elevated subsurface temperatures and correspondingly more dramatic phenomena.

Finally, even if the atmosphere is so thin that it feeds back little on surface temperature, that does not mean that its dynamics is uninteresting or unimportant. It only means that the problem of atmospheric dynamics becomes simpler to think about. On Earth one needs to determine the surface temperature simultaneously with atmospheric temperature, both of which are then strongly affected by dynamical heat transports. On Mars, to a good approximation one can determine the surface temperature evolution as if the atmosphere were not there at all, and then use the resulting time-varying temperature pattern as the lower boundary condition to drive the atmospheric circulation. The feedback of the Martian atmosphere is not completely negligible, but as the atmosphere gets thinner, this approach becomes more and more accurate. The transient N_2 atmosphere of Triton is a case in point, and is all the more novel in that its dynamics involves supersonic expansion of sublimated gas into a near-vacuum.

7.7.7 Titan

Titan is one of the few bodies in the Solar System having both a thick atmosphere and a solid surface, and in this regard is more Earthlike than present Mars, whose thin atmosphere makes it a candidate for inclusion in the "nearly airless" category. Like most moons, Titan is tide-locked to its primary, orbits in the plane of the primary's equator, and therefore shares the obliquity and the year of the primary. Titan's day is its orbital period about Saturn, or 16 d.

As already mentioned in Section 7.4, the distinguishing feature of Titan's seasonal cycle is provided by the great thermal inertia of its cold, thick N_2 atmosphere. One expects very little seasonal fluctuation in the interior of the atmosphere, but there can be a considerable seasonal cycle in surface temperature, owing to the low thermal inertia of the mostly solid surface. As for Venus, one expects a considerable seasonal cycle in the thin outer portions of the atmosphere. Even though the interior atmosphere temperature may not change much in the course of the season, modeling studies show that the variations in surface temperature modulate convection, which affects the circulation, methane precipitation, and methane cloud patterns of Titan's troposphere. Since Titan's year, like that of Saturn, is 29 Earth years, one will have to wait a decade or so to see whether the various predictions that have been made regarding the evolution of these features are borne out.

7.7.8 Gas and ice giants

Jupiter has very low obliquity and a nearly circular orbit, so there is not much to drive a seasonal cycle there. However, Saturn and Neptune have obliquities in excess of 26° and Uranus has the highest obliquity of all, at 97°. This demonstrates that high obliquity can be imparted to a gas or ice giant either during the process of formation or at some point thereafter. The gas and ice giants in the Solar System all have quite circular orbits, but high eccentricities are very common among extrasolar gas/ice giant planets. In addition, some of the extrasolar giants are in sufficiently close orbits that the mean rotation may be slow or tide-locked, leading to interesting hybrid seasonal/diurnal cycles. The main issue concerning the expression of the seasonal cycle on giant planets stems from the thermal inertia of the atmosphere and the depth of penetration of solar radiation. The very deep atmosphere receives little illumination, and is massive enough and opaque enough in the infrared that little seasonal variation should be expected; this situation is much the same as that for the deep layers of the atmosphere of Venus. However, there will always be some layer near the top of the atmosphere which will have low enough thermal inertia to respond seasonally. The depth of this layer will depend on the composition of the atmosphere, and the observational expression will depend on the nature of cloud-forming substances, if any. For a fairly clear atmosphere, the seasonal variations would be most apparent through variations in infrared emissions tied to temperature. With clouds, modulations in temperature could also be expressed as variations in cloud albedo and cloud pattern. For the case of extrasolar planets, understanding such seasonal variations is particularly important, since the observable quantities are tied to emission and reflection from the outer portions of the atmosphere, and one would like to know how deep a layer of the atmosphere these observables are probing. Extrasolar giant planets abound in orbits where they receive solar flux at rates similar to or even greatly exceeding Earth's solar constant. Higher fluxes and higher temperatures generally lead to larger seasonal fluctuations. Opportunities abound for exploring the novel seasonal cycles of giant planets operating in a thermal range far hotter than any seen for giants in the Solar System.

Gas giants are primarily made of H_2, which does not become a good infrared absorber until quite high pressures are reached. The thermal emission contribution to thermal inertia is therefore likely to be dominated by minor constituents, both in the form of gases and cloud particles. Ice giants have a more complex composition, and a correspondingly greater range of possible optically active constituents. A detailed consideration of the depth of the seasonally active layer of known and hypothetical planets is beyond the scope of what we wish to attempt here, particularly in view of considerable uncertainties regarding the vertical distribution of solar-absorbing constituents.

The study of seasonal cycles of giant planets is in its infancy, perhaps in part because one has to be very patient indeed to observe the seasonal changes. Saturn, with its orbit of 29 Earth years, presents the best near-term observational possibility in our own Solar System. Some hints of a seasonal cycle on Saturn have already been observed. Between 1980 and 2002 the equatorial winds in the visible cloud layer of Saturn have decreased by about 40%, and there have also been substantial changes in the cloud patterns. It has been suggested that in the case of Saturn, the shading provided by the ring system could be playing a significant role in the seasonal insolation pattern. At the time of writing, the southern hemisphere of Saturn is just coming into midsummer. It will be interesting in the coming years to track the evolution of Saturn's atmosphere through its full seasonal cycle. Along with Titan, Saturn provides a rare opportunity to test our understanding of seasonal cycles by attempting to predict the evolution in the coming decade, with little guidance from past observations.

The case of Uranus should be even more dramatic, but given an orbital period of 84 Earth years, one will have to wait two decades or so in order to see a hint of what is going on. Uranus is a fairly featureless planet, so it is difficult to observe changes in the atmospheric circulation. The observing systems in place for the past two decades do not provide a suitable basis for probing the seasonal changes of Uranus so far, but improvements in orbital telescopes should make it possible to do better in the coming decades.

7.7.9 Habitability of planets with extreme orbital configurations

Outside the Solar System, planets with highly eccentric orbits are common. A massive moon can stabilize a planet's obliquity cycle. If the Earth did not have the benefit of its large Moon, its obliquity would probably undergo the same extreme fluctuations as that of Mars. Either of these situations could potentially lead to such extreme seasonal cycles as to pose a threat to habitability, even if the planet had an Earthlike composition and annual mean insolation. The high-eccentricity cases would seem particularly inhospitable, as there would be short periods at periastron with intense solar radiation which could lead to perilously hot conditions, followed by long periods far from the star when the oceans could well freeze over, and perhaps even the atmosphere could snow out.

If the planet had a mostly solid surface, these would indeed be severe habitability barriers, but even a moderately deep mixed layer ocean can average out a quite extreme seasonal variation of insolation. The essential issue, as always for seasonal cycle problems, is the length of the planet's year relative to the thermal response time of the ocean. A 50 m mixed layer can average out most of the seasonal cycle for a planet with high obliquity or a fairly high eccentricity even if the planet's year is as long as 5 Earth years. The critical point to consider is whether there is enough time at apastron for the ocean to freeze over, since if that happens one loses the benefit of the ocean's thermal inertia, and extremely cold conditions can ensue. This problem has much in common with the problem of formation of polar sea ice on Earth. In particular, ocean stratification comes into play, since the mixed layer will deepen and prevent freezing even for long winters, in cases where the ocean stratification permits deep oceanic convection in near-freezing conditions. That situation could preserve habitability even for planets with quite long years and very high eccentricity, provided the ocean is reasonably deep.

The question of ocean stratification and sea ice suppression is inherently dynamical, but one can at least get an idea of the depth of mixed layer needed to preserve habitable conditions for any given orbital configuration. Another thing to keep in mind in this connection is that planets with a close-in periastron tend to adopt a variety of slowly rotating spin states, the simplest of which is quasi-tide-locked in the sense that the planet always presents the same face to the host star at periastron. Such planets will be in the regime where there is a hybrid seasonal/diurnal cycle, rather than two distinct cycles.

Some simple calculations pertaining to habitability of extreme orbital configurations are developed in Problem 7.34.

7.8 WORKBOOK

Python tips: The Chapter Scripts `solar.py`, `plot_solar.py` and `FluxExplorer.py` will be useful in doing the problem in this Workbook section. The script `solar.py` includes routines to carry out the integration needed to determine the season angle as a function of time, and can be imported into other scripts that need insolation calculations. The function `getOrbit(...)` defined in `FluxExplorer.py` retrieves the orbital parameters for Earth from a data file given the time expressed in thousands of years relative to the present.

Some of the problems call for use of a graphics tool which has not been previously introduced, namely contour plotting. In the `Python` courseware, the contour plots can be done using the function `contour(...)` in the `ClimateUtilities` module.

7.8.1 Distribution of insolation

Problem 7.1 Using Eq. (7.8), estimate the duration of daylight at the latitude where you are living, at the time of year at which you are solving this problem. Use the current obliquity of Earth, which is 23.45°. Compare your results with observations, preferably taken by yourself. Using Eq. (7.9), estimate how much solar power could be generated by a one square meter, 10% efficient solar panel pointed straight up at the sky at this time, assuming that 80% of the solar energy incident on the atmosphere makes it through to the ground. Thirty thousand years ago, the obliquity was 22.25°. How would your results have been different at that time?

You may make use of the approximation that the length of the day is short compared with the length of the year, and approximate the orbit as being circular.

Problem 7.2 Determine the latitudinal pattern of diurnally averaged insolation for a planet with small but non-zero obliquity, assuming that the planet's length of day is short compared with its year. Assuming a circular orbit, average this over the course of a year to obtain the annual mean insolation pattern. Compare the annual mean results to the more precise numerically determined flux factor for obliquities of 10°, 20°, and 30°. Make sure to retain enough terms in your approximation to see the lowest order deviation from the zero obliquity case.

Problem 7.3 Estimate what the summer temperature of Antarctica would be if you took away the ice sheet, leaving bare rock. Use the polynomial *OLR* fit in Table 4.2 to determine the infrared cooling, and neglect horizontal heat redistribution by dynamical heat transports.

Problem 7.4 The Earth's obliquity fluctuates between 22° and 24.5° on a time scale of roughly 40 000 years. Assuming that the albedo is 20%, compute the latitudinal profile of the *difference* in absorbed solar radiation between these two states. Convert these fluxes into temperature change using a climate sensitivity of $2 \, W/m^2 K$. This gives you an estimate of the temperature change over extensive oceans, including the effect of water vapor feedback but neglecting the additional sensitivity due to possible melting of sea ice or land glaciers.

Problem 7.5 *High obliquity waterworlds*
This problem examines the climate of high obliquity waterworlds. These are of interest because a planet without a large moon such as Earth's would generally pass through high obliquity episodes at various times. Would such planets be habitable?

Consider a waterworld planet in a circular orbit, whose thermal response time is sufficiently long that the seasonal cycle is weak and the surface temperature responds mainly to the annual average insolation. The albedo is 0.2, and the local solar constant is $L_\odot$. The planet has an Earthlike atmosphere, whose *OLR* can be approximated using the polynomial fit in Table 4.2. For a planet in an Earthlike orbit, with $L_\odot = 1370 \, W/m^2$, plot the surface temperature as a function of latitude for obliquities of 45°, 50°, 55°, and 60°, assuming a CO_2 value of 1000 ppmv. Then, for a fixed 60° obliquity, examine the range of values of CO_2 and $L_\odot$ which yield a globally ice-free climate. Determine the threshold $L_\odot$ below which the planet would definitely be in a Snowball state (as indicated by below-freezing temperatures everywhere), with CO_2 at 10^5 ppmv.

For the purposes of this problem, you may neglect horizontal heat transports and assume the planet to be in equilibrium with the absorbed solar radiation at each latitude.

Problem 7.6 The obliquity of Mars undergoes more extreme variations than that of Earth. Consider a time when Mars has an obliquity of 55°. Assume that the planet has no oceans, and that the polar glaciers have dissipated so that the albedo is 20% at the poles. What is the temperature of the summer pole, assuming that there is no greenhouse effect and that the polar temperature is in equilibrium with the local solar absorption (i.e. that there is no horizontal heat transport)? If you put a water ice cap with 60% albedo there, would it melt or stay below freezing? How cold would it be? (In reality, even an ice cap that is too cold to melt may sublimate and be redeposited elsewhere on the planet.) For the glaciated case, use the *OLR* data in Fig. 4.30 to determine the surface pressure of a pure CO_2 atmosphere that would be needed in order to bring the glacier to the melting point.

For this problem, you may assume Mars to have a circular orbit, though in reality the eccentric orbit of Mars will episodically render one of the summer poles much warmer than the estimates obtained here.

Problem 7.7 Consider a planet with a solid surface, having an atmosphere which is thick enough to average out the diurnal cycle but which has a response time short compared with the length of the planet's year. In this case, the temperature at any given latitude is in equilibrium with the absorbed solar flux computed from Eq. (7.9), provided that the lateral heat transport in the atmosphere is neglected. This calculation also yields a reasonable approximation to what the temperature would be in the interior of large continents, in the absence of lateral heat transports.

Using the *OLR* polynomial fit for fixed relative humidity in Table 4.2, determine the temperature profile of this planet at the solstice and equinox, assuming Earth's present solar constant and an obliquity of 23°. Do the calculation for a rocky planet with albedo 0.2, and a Snowball state with albedo 0.6, and a snow-covered Snowball state with albedo 0.8. Examine how the results change as you increase the CO_2 from 100 ppmv to 10^5 ppmv (100 mb). Then, fix the CO_2 at 1000 ppmv and look at how the temperature pattern changes as you vary the obliquity from 0° to 90°.

You should cut off the calculation where the temperature falls below 200 K, or goes above 350 K, since that is beyond the range of validity of the polynomial fit. In reality, the hot cases would go over into a runaway state, if the relative humidity continued to be constant and if the unrealistic neglect of lateral heat transport really were valid.

Problem 7.8 *Triton*
Neptune's moon Triton has probably the most unusual seasonal cycle in the Solar System. Like most moons, it is tide-locked to its planet, in Triton's case with a period of 5.67 d (in the retrograde direction, though that does not matter for the purposes of this problem). Uniquely among moons, however, its orbit is in a plane which is quite steeply inclined to Neptune's equator. The orbital plane precesses about the normal to Neptune's orbital plane with a period estimated to be about 637 Earth years, at the same time that Neptune orbits the Sun with a period of 165 Earth years. Putting the two effects together, the latitude of the subsolar point follows the empirical formula

$$\sin \delta = 0.4636 \cos(a_0 - a_1 t) + 0.3495 \sin(a_0 - b_0 + (b_1 - a_1)t) + 0.0251 \sin(a_0 + b_0 - (a_1 + b_1)t)$$

(7.32)

where $a_0 = 1.332$, $a_1 = 0.006650$, $b_0 = 0.3229$, $b_1 = 0.0001721$. In this formula, angles are given in radians and t is the time given in Earth years AD. Thus, the formula can be used to determine the latitude of the Sun on Triton occurring at an actual date. Because Triton is an icy body rich in N_2 and CH_4 ices that can warm-up sufficiently to give off significant amounts of gas when illuminated, the complex motion of the subsolar point leads to a rich variety of atmospheric and surface evolution features on Triton.

The solar constant at Neptune's orbit is $1.51 \, W/m^2$. Using Eq. (7.9), plot the time series of diurnally averaged flux on Triton at a both poles, at the equator, and at latitudes of ±45°. Plot the time series of surface temperatures that would be in equilibrium with these fluxes, and use the Clausius–Clapeyron relation to plot the saturation vapor pressure of N_2 and CH_4 that would be in equilibrium with these fluxes.

Because heat takes time to diffuse through a solid, to a first approximation the subsurface temperature can be determined by averaging the surface temperatures obtained above over longer periods; the longer the averaging period, the deeper the layer the temperature corresponds to, with averaging times of an Earth year yielding the ice temperature at around a meter of depth, increasing to around 30 m for an averaging period of 100 Earth years and around 100 m for 1000 Earth years. Plot time series of the averaged temperature at selected latitudes, using averaging periods of 1, 10, 100, and 1000 Earth years. Also show the N_2 and CH_4 saturation vapor pressures at these periods. The pressures are of interest because when sufficiently high pressures build up it can fracture ice and lead to cryovolcanism. Note that the subsurface temperatures you have estimated are underestimates of the actual subsurface temperature, since the penetration of solar radiation into ice can lead to considerably elevated temperatures, as discussed in Section 6.9.

Problem 7.9 *Path of the subsolar point*

Show that for a non-rotating planet with a latitude–longitude coordinate system based on an axis with obliquity γ, the longitude λ_0 of the subsolar point is given by

$$\tan \kappa = (\cos \gamma)(\tan \lambda_0) \tag{7.33}$$

where κ is the season angle. When solving this formula for λ_0, one must take care to choose the branch of $\tan^{-1}$ which yields an angle which increases smoothly from 0 to 2π radians in the course of the orbit. *Hint:* Project the equatorial circle, with all its longitude labels, onto the plane of the orbit. The result will be an ellipse, labeled by longitude. This ellipse rotates along with the season angle, and you now only need to find the longitude of the point where the ellipse intersects a line segment extending from its center to the center of the sun.

Discuss the limit of zero obliquity, as a check that the equation is behaving sensibly. What should the behavior be when the obliquity is close to 90°? Compare your expectation with the behavior of the above equation for obliquity close to 90°.

If the planet is rotating about the axis with constant angular speed Ω, then the longitude of the subsolar point becomes $\lambda_\odot = \lambda_0(t) - \Omega t$.

The *solar day* is defined as the time elapsed between successive local noons – i.e. the time between successive crossings of the local longitude by the subsolar point. Note that this is independent of latitude. Find a formula for $\lambda_\odot(t)$ for the case $\gamma = 0$. Show that if the year contains n stellar days, then it contains $n - 1$ solar days.

Find an approximate expression for the duration of the solar day assuming the duration of the stellar day to be short compared to the orbital period. Use this formula to determine how Earth's solar day would vary over the year if its orbit were circular with a period of 365.25 d, given a present stellar day of 86 164 s. Then, numerically determine the sequence of solar day lengths for a planet in a circular orbit with period 400 d and a stellar day of 100 d. Show results for obliquities of 0°, 20°, and 45°.

Using the formula for $\lambda_\odot$ together with the formula for the latitude of the subsolar point ($\delta(t)$) given in the text, plot the trajectory of the subsolar point in the longitude–latitude plane for the following cases, assuming the planet to be in a circular orbit: (a) A planet with an orbital period of 400 d and a stellar day of 40 d, and (b) a planet with an orbital period of 400 d and a stellar day of 100 d. Show results for obliquities of 20°, 45°, and 80°.

Problem 7.10 Consider a planet in a circular orbit, with an orbital period of 400 d and a stellar day of 100 d, with 45° obliquity. Use the results of Problem 7.9 together with the expression for the zenith angle in Eq. (7.7) to make a contour plot of the diurnally averaged flux factor in the longitude–latitude plane, for each of the three solar days in the planet's year. Then show the annual-average flux factor. Should the annual mean depend on longitude? How do these results compare with what you would get by ignoring within-day variations in the latitude and longitude of the subsolar point, and simply averaging over the hour angle?

Hint: You evaluate the time-dependence of $\cos \zeta$ by computing $\delta(t)$ and $\lambda_\odot(t)$ and plugging the values into Eq. (7.7), writing the hour angle as $h = \lambda - \lambda_\odot$, where λ is the longitude. The time average at any given latitude and longitude can no longer be done analytically, and must be done instead by numerical quadrature. Remember to zero out the flux for times when the sun is below the horizon.

Problem 7.11 *A walk in the sun*

As preparation for doing this problem, first read Geoff Landis's story "A Walk in the Sun".

Astronaut Katie Mulligan has crashed on the Moon, right at the subsolar point. She can be rescued, but it will take 28 days (a lunar day) for the rescue mission to get there. At the time of the crash, the latitude of the subsolar point is ϕ_0. Her space suit can maintain life support over that time period *but* it is solar powered and has essentially no battery storage capability, so she will need to stay in sunlight for the whole time. How fast does she have to walk if she chooses to walk along her initial latitude circle? Suppose instead she walks along a longitude line directly toward the summer pole and waits there. Under what circumstances is this a better strategy?

Recall that the Moon is tide-locked to the Earth and orbits very nearly in the Earth's equatorial plane. Therefore, it shares Earth's obliquity. Note also that the latitude of the subsolar point will change somewhat in the course of a lunar day, so you should take this into account, though the effect is slight.

7.8.2 Thermal inertia

Problem 7.12 *Climate sensitivity and equilibration time*

In circumstances where the thermal inertia is determined by the top-of-atmosphere energy budget, the radiative damping coefficient $b = dOLR/dT$ is the same coefficient that determines climate sensitivity. This observation links the ultimate amount of warming due to CO_2 increase with the time it takes to approach this equilibrium. All other things being equal, a more sensitive climate will also have a greater time lag between forcing and response. This link was first pointed out by James Hansen in the 1980s.

Considering only annual-mean conditions, suppose that the Earth is in equilibrium at a temperature T_0, for some given CO_2 concentration. Then, CO_2 is instantaneously doubled, leading to a reduction in OLR of $4\,W/m^2$ for any given temperature; assume that this leaves the slope b unchanged. Find and plot the subsequent time evolution of temperature for a standard case with $b = 2\,W/m^2K$ and for a case with stronger amplifying climate feedbacks having $b = 1\,W/m^2K$. In both cases, assume a mixed layer depth of $50\,m$.

Problem 7.13 Solve the problem

$$\rho_0 c_{p0} H \frac{dT}{dt} = (1 - \alpha)S - \sigma T^4 \tag{7.34}$$

analytically, for arbitrary initial temperature T_0. Analyze the behavior of the system, and compare the analytical results to the approximate results obtained by either linearizing about the equilibrium temperature or holding the infrared cooling rate fixed at σT_0^4. *Hint:* This is essentially an intricate exercise in the use of partial fractions. You need to factor something of the form $1/(A + T^4)$ into a sum of four simpler fractions. You can plot the results analytically if you plot $t(T)$ instead of $T(t)$.

Note that this problem can be solved numerically by ODE integration with only a trivial amount of programmer time, and microseconds or less of computer time. It probably takes less computer time to solve the problem numerically than it does to evaluate the expression yielded by the analytic solution. However, it is good to exercise one's skills at algebraic manipulation from time to time.

Problem 7.14 *Thermal inertia for a shallow troposphere*

When the tropopause is high, then convective mixing allows heat to be stored in essentially the entire mass of the atmosphere. When the convection extends only over a shallow layer, however, less mass is available to store heat and the thermal inertia is correspondingly lower. As a simple example of this consider a dry atmosphere with an isothermal stratosphere. Specifically assume the troposphere is on the dry adiabat from the ground up to the point where the temperature equals a fixed stratospheric temperature T_{strat}, and is isothermal at higher altitudes. Compute the thermal inertia coefficient $\mu = dE(T)/dT$ for this atmosphere, and discuss how it varies with T.

Problem 7.15 *Latent heat and thermal inertia*

In the text we determined the thermal inertia of a dry atmospheric column using the dry static energy. If the atmosphere instead is in equilibrium with a condensable reservoir such as a water ocean, then it takes more energy to increase the surface temperature by 1 K because one has to take into account the increase in latent heat storage of the atmosphere. This problem considers this effect in two simple limits.

In the first limit, we consider the condensable substance to be dilute, as is the case for water vapor on the present Earth. In this case we can use the dilute moist static energy formula given in Section 2.7.3. Suppose that the stratosphere has negligible heat storage, and that the moist static energy can be considered constant for the purpose of computing thermal inertia. Suppose that the condensable substance is saturated at all heights. By evaluating the moist static energy at the ground, find an expression for the energy storage as a function of surface temperature. The thermal inertia coefficient μ is the derivative of this with respect to surface temperature. Evaluate this at 260 K, 280 K, 300 K, and 320 K for saturated water vapor on Earth. Evaluate it for saturated methane in an N_2 atmosphere on Titan, for a suitable range of temperatures. Express your answer in terms of an equivalent mixed-layer depth of liquid water. How do the values compare with what you would get in the same situations using dry static energy? Note that when the atmosphere is unsaturated, the moist static energy per unit mass isn't constant, and the determination of the thermal inertia coefficient is correspondingly more complicated.

In the opposite extreme, we take an atmosphere that is on the single-component condensing adiabat, with temperature profile given by the dew-point formula. This is an appropriate model to use in conjunction with the formation or sublimation of a CO_2 ice cap in the polar winter and spring of present Mars or Early Mars, and it can also apply to a steam atmosphere undergoing a runaway greenhouse. For this kind of atmosphere, the energy storage is just the latent heat times the mass per unit area of the atmosphere, which by Clausius–Clapeyron can be written as a function of surface temperature. Find the thermal inertia coefficient by taking the derivative of heat storage with respect to surface temperature. How does the gravity affect the thermal inertia? Evaluate the thermal inertia coefficient as a function of temperature, and express it in terms of an equivalent water mixed layer depth for the case of a CO_2 atmosphere under Mars gravity. Do the same for a pure water vapor atmosphere under Venus gravity, at temperatures up to 400 K.

Problem 7.16 *Atmospheric thermal response time over a deep mixed layer*

For an atmosphere over a very deep mixed layer the surface temperature can be considered essentially fixed at a specified temperature T_0 as the atmospheric temperature varies in the course of the seasonal or diurnal cycle. Find an expression for the thermal relaxation time appropriate to this case. Aside from using a thermal inertia coefficient appropriate to the atmospheric mass, the only difference from the calculations of relaxation time carried out in the text is in the form of the flux coefficient b, which determines how the atmospheric heating or cooling depends on the atmospheric temperature. In this case the atmospheric heat budget involves not only $OLR(T)$, but also the surface flux terms. The latter are treated using the linearized flux coefficients discussed in Chapter 6 (see esp. Table 6.2 and the associated discussion). The linearization should be done about an atmospheric temperature chosen such that the atmosphere is in equilibrium. In determining this temperature, allow for the effect of a time-independent absorbed solar flux in the atmosphere.

Discuss the behavior of the relaxation time as T_0 and surface wind is varied. Note particularly the implications of the increasing importance of latent heat flux as T_0 increases. How does the atmospheric absorbed solar radiation affect the result?

To keep this problem simple you may assume that the thermal inertia coefficient μ has its dry adiabatic deep-troposphere value $c_p p_s/g$. At warmer temperatures where latent heat flux is significant, for consistency one should actually consider the effects of latent heat storage in the atmosphere. When the atmosphere is saturated, this can be done using the results of Problem 7.15, but for an unsaturated atmosphere finding a proper formulation is quite challenging, since the moist static energy is no longer uniform and the effect of moisture depends a lot on whether or not it is condensing.

Use your estimate of the relaxation time together with the analytic solution for a mixed-layer seasonal cycle driven by sinusoidally varying forcing to estimate the amplitude of the seasonal cycle of atmospheric temperature at latitude 45° over a deep mixed layer. Assume that the atmosphere absorbs 20% of the incident solar radiation. Determine the phase lag between the insolation and the temperature.

Problem 7.17 *Sea ice growth and decay*
The growth and decay of a sea ice layer provides a form of thermal inertia, since the latent heat of fusion must be taken into account in determining the heat storage in the atmosphere–ocean column. Assume that all freezing takes place at the sea ice base, and that the diffusion equation is nearly in equilibrium within the ice layer so that the temperature profile is linear. Suppose the base has temperature T_f and the ice surface has temperature T_s, both of which are constant. As an idealization, assume that all the heat released by freezing at the base must be taken away by diffusing through the ice layer, and that this is the limiting factor in ice growth. (In reality much of the heat in the early stages escapes through open-water leads in the ice layer, and it is also possible that ocean currents could take away some of the heat.) Show that the growth of the ice thickness h is governed by the equation

$$\rho_i L_f \frac{dh}{dt} + \Phi_0 = \kappa_T \frac{T_f - T_s}{h} \tag{7.35}$$

where κ_T is the thermal conductivity, L_f is the latent heat of fusion, ρ_i is the density of the ice, and Φ_0 is the heat flux delivered to the ice base by geothermal heating or other means. Generally, T_f can be taken to be the freezing point of sea water.

Find the solution to this equation. With $\Phi_0 = 0$, how long does it take for ice to grow to a thickness of 5 m for $T_f - T_s = 5\,$K? For $T_f - T_s = 20\,$K? Next set Φ_0 to the present Earth geothermal heat flux, and determine how long it takes to reach the equilibrium ice thickness for a typical annual mean Snowball Earth polar temperature of 200 K, and a typical Snowball tropical temperature of 240 K.

The preceding considerations determine how much ice is built up during the growth season. The resulting thermal inertia comes into its own during the melt season, which has different physics since ice can melt from the top. Using the surface energy balance models developed for melting ice in Section 6.7, determine the time required to melt a layer of sea ice of thickness d, as a function of the surface solar flux and the air temperature. You may assume that the albedo of melting ice is 0.5, and that the wind speed is 5 m/s. Discuss the difference between melting rates with Monin–Obukhov vs. neutral surface layer physics. Put in some numbers corresponding to polar summer conditions, and determine how long it takes to melt a 5 m layer.

7.8.3 The diffusion equation and its uses

Problem 7.18 You are spending the winter inside a hemispherical igloo with radius of 3 m. The walls are 0.25 m thick. The floor is insulated with caribou skins, so that the heat loss through the floor can be neglected. The leakage of air through the entrance is also negligible. The outside temperature is 220 K. You and your companion put out a total of 300 W on average (because of all the food you eat), and you burn a seal oil lamp that puts out another 100 W. What is the temperature inside the igloo if it is made of snow? If it is made out of blocks of ice?

Problem 7.19 *Numerical methods for the diffusion equation*
In this chapter, we made use of some new numerical techniques which have not been introduced earlier, namely the methods for numerical solution of the time-dependent diffusion equation in one dimension. Read about methods for the diffusion equation in *Numerical Recipes*. For the calculations we need, the simplest algorithm (second-order centered finite differences, with leapfrog or midpoint-method time-stepping) will suffice. Implement the method in the computer language of your choice. First, test your implementation on a basic configuration consisting of an ice layer of thickness 100 m and uniform diffusivity, resting on a water layer which keeps the lower boundary of the ice at a fixed temperature of 271 K. Impose an insulating boundary condition ($\partial_z T = 0$) at the upper boundary, and examine the approach to an isothermal state assuming the ice temperature is initially 200 K independent of depth, except for the lower boundary point.

Next, try a few extensions of the basic problem. Re-do the calculation assuming that there is a 5 m deep layer of low-diffusivity snow on top of the ice, with the insulating boundary condition applied at the snow surface. Describe how the time evolution changes, concentrating on the time required for the ice temperature to reach equilibrium. Finally, replace the insulating upper boundary condition with the assumption that the surface loses heat by radiating at a rate σT_s^4, where T_s is the surface temperature. What equilibrium surface temperature do you expect with and without the snow layer? How long does it take to reach equilibrium? Plot a time series of the heat transfer from the ocean into the ice, and use this to estimate the rate at which the ice thickness would grow if you hadn't artificially kept the thickness fixed. *Python tips:* The basic solution of the initial value problem for the diffusion equation is implemented in the Chapter Script `diffusion.py`.

Problem 7.20 Suppose that the temperature at the subsolar point on the Moon is initially in equilibrium at 370 K. Suddenly, a lunar eclipse occurs, which eliminates the input of solar radiation. Use the diffusion equation to describe the temperature drop over the course of the next several hours, assuming (unrealistically) that the initial state is isothermal at 370 K. As a crude estimate of the night-time temperature of the Moon, compute how much the point would cool down if it remained unilluminated for 14 d. As a somewhat more realistic model of lunar cooling, assume instead that the subsurface is isothermal at 370 K only down to a depth h, and that below that it is isothermal with a temperature of 250 K. Describe how the time series of temperature over 14 d depends on h. You can estimate an appropriate value of h for the actual Moon by estimating how far heat diffuses into the lunar regolith in the course of a lunar daylight period (about 14 d). What is this depth?

Problem 7.21 *Diurnal cycle on rocky airless planets*
Solve the diffusion equation for a thick layer of the crust of a nearly airless rocky planet, subject to radiation from the surface at a rate σT_S^4 and insolation by the actual diurnally and seasonally varying time series of insolation at a latitude of 45°. You may assume an insulating boundary condition at the lower boundary of the rock layer. If you make the layer thick enough, your result will not depend on the layer thickness. Based on the basic behavior of heat diffusion, how thick do you think the layer needs to be? Test your estimate against the actual behavior of your numerical calculation. For the purposes of this problem, you should put in numbers corresponding to the solar constant, obliquity, albedo, length of day, and length of year corresponding to present Mars, but you may assume the orbit to be circular. You may assume a constant diffusivity corresponding to the dry rock of your choice.

How does the amplitude of the surface diurnal cycle compare with the surface seasonal cycle? At what depth do you see a significant diurnal cycle? At what depth do you see a significant seasonal cycle? At what depth is the temperature essentially constant over the course of the year? How does the deep subsurface temperature compare with the surface temperature that would be in equilibrium with the annual mean absorbed solar radiation? Why do the two numbers differ?

Problem 7.22 Jupiter's moon Europa has essentially zero obliquity. It shares Jupiter's year and solar constant, and since it is tide-locked to Jupiter its day is equal to its orbital period about Jupiter, or 3.55 d. Jupiter's orbit is essentially circular. Europa's crust is made largely of water ice, and there is essentially no atmosphere. Its albedo is about 0.67. Numerically solve the diffusion equation for a point on the equator of Europa, and describe the time series of surface and subsurface temperature. What is the deep-ice temperature? You may use an insulating lower boundary condition.

7.8.4 Orbital mechanics and eccentricity

Problem 7.23 The planet Gliese 581d has an orbital period of 66.8 d ($1 \, \text{d} \equiv 86\,400 \, \text{s}$). Its eccentricity is 0.38, and the parent star has a mass of 0.31 times that of the Sun. What is its semi-major axis? How does this compare with the result you would get assuming the planet to be in a circular orbit? The luminosity of Gliese 581 is 0.013 times that of the Sun. What is the instantaneous incident stellar flux (the "solar constant") seen by the planet at periastron? At apastron? What is the annual mean?

Do the same for the planet HD222582b, which is a 5 Jupiter-mass gas giant in orbit about a class G star having a mass and luminosity approximately equal to that of the Sun. The eccentricity of this planet is 0.76 and the orbital period is 572 d.

Problem 7.24 Discuss the effect of the precessional cycle on the annual mean solar flux distribution for a planet in an eccentric orbit. Specifically, show the solstice and equinox flux factors for a planet having 25° obliquity, at several different stages of the precessional cycle. You may hold the eccentricity fixed at 0.1. For Earth's present obliquity and eccentricity (23.45° and 0.017), how much does the precessional cycle affect the annual mean incident solar flux at 65° N latitude?

Problem 7.25 *Summer melt energy*
The *summer melt energy* is defined as the incident solar flux at a given point, integrated over the period of time for which the flux exceeds some threshold value F_0. This is meant to serve as an index of the amount of energy available to melt ice, and so F_0 is chosen to approximate the flux needed to bring the ice surface to the melting point. The appropriate choice of F_0 depends on the ice albedo, meridional heat transport, surface flux conditions and various other things. Given a suitable choice of F_0 (perhaps guided by observations, or by examination of simple or comprehensive climate models), the melt energy statistic is simple to compute and provides a valuable guide as to the trade-off between the Kepler's law effect on duration of seasons and the precessional effect on intensity of seasons. There are actually two variants of the melt energy floating around. In the original one, the entire flux is integrated over the period where F_0 is exceeded. In a more recent alternate form, one only integrates the portion of the flux that is in excess of F_0, rather in the spirit of "heating degree days" used to estimate demand for home heating fuel. In the following, you should compute both variants and compare. Note that the melt energy concept can be useful for studying ablation of mountain glaciers as well as for large continental ice sheets.

Compute the summer melt energy at latitudes 65° and 40° as a function of the phase of the precessional cycle for planets in orbits having eccentricity 0.02, 0.05, 0.1, and 0.5. Try a variety of different values for F_0 ranging from zero to the seasonal maximum. You may hold the obliquity fixed at 25° for this problem, but if you are energetic you may wish to explore the behavior for both higher and lower obliquities; the high obliquity case is relevant to Martian Milankovic cycles.

Hint: You do not need to specify the stellar constant $L_\odot$, since the melt energy statistic can be formulated in terms of thresholds on the non-dimensional flux factor $f(t)$. To obtain values for any given planet, the resulting statistic is simply multiplied by the value of $L_\odot$ appropriate to the orbit's semi-major axis.

Problem 7.26 Mercury has an orbital period of 87.967 d, and a stellar day of 58.646 d. These are almost exactly in the ratio 3:2, and the state represents one of the stable spin states for a planet in a fairly eccentric orbit. (Up until 1965, it was thought that Mercury was approximately tide-locked to the Sun.) The current eccentricity of Mercury's orbit is 0.206. Mercury's obliquity is essentially zero.

Make a graph of the time variation of the longitude of the subsolar point over the course of several Mercury years, taking into account the modulations of orbital speed caused by eccentricity. Are there any times when the Sun appears to "go backwards?"

At any given latitude and longitude, the mean insolation over a stated period at that point is determined by the time average of $L_\odot(t) \cos \zeta$, where $L_\odot(t)$ is the time-varying incident solar flux given by the orbital position; in taking the average, you must remember to set the flux to zero when the Sun is below the horizon. Make contour maps of annual-mean insolation patterns for several successive Mercury years. Does the pattern repeat? Show the patterns for an averaging period of 3 Mercury years.

The surface temperature for illuminated regions is largely determined by the instantaneous position of the subsolar point. However, the subsurface temperature depends on longer-term averages, and the temperature of dark regions depends on how recently and how strongly they were illuminated, since that determines how much time they have had to cool down.

Problem 7.27 *Quasi-tide-locked spin states*
A planet in an eccentric orbit cannot be truly tide-locked, since the angular rotation rate of the planet's spin about its axis is nearly constant while the angular velocity of the planet's orbit about the star increases as the planet gets closer to the star. However, such a planet can still enter a quasi-synchronous spin state, in which the planet's rotation period is equal to its year. In such a state, the planet always presents the same face to its star at periastron. It has been conjectured that close-orbit eccentric extrasolar planets like Gliese 581c are in such a state. In a truly tide-locked configuration, the subsolar point is anchored to a single point on the planet's surface, whereas in a quasi-synchronous spin state, the subsolar point wobbles east and west over the course of the orbit.

Discuss the pattern of insolation on a planet in a quasi-synchronous orbit, for various values of the eccentricity. By how many degrees of longitude does the subsolar point wobble? What portion of the equator never receives any illumination? What portion is perpetually illuminated? (Without loss of generality, you may put the planet's equator in the plane of the planet's orbit; obliquity is zero in this class of problems.) Plot the behavior of the flux factor along the equator at the periastron, apastron, and two intermediate points in the orbit. Plot the annual mean flux factor. If you know how to make contour plots, make a series of maps showing the seasonal cycle of the flux factor over the course of the year.

Problem 7.28 Mercury is in a quite eccentric orbit, but its spin state is not quasi-synchronous. Instead, it is locked in a *3:2 spin–orbit resonance*, which is just a fancy way of saying that it rotates about its spin axis three times for each two orbits about the Sun. The obliquity is essentially zero. Determine the time series of Mercury's insolation for a geographically fixed point on the equator. Mercury's eccentricity is 0.2. It is possible that eccentric extrasolar planets may also be in 3:2 spin states rather than being quasi-synchronous.

Problem 7.29 The directory `MilankovicMars` in the Workbook datasets directory for this chapter contains orbital parameters for Mars over the past 21 million Earth years, in increments of a thousand years. The data itself is in the file `INSOLN.LA2004.MARSe.txt`. Note that the precession angle in this file is given relative to the spring equinox, and needs to be shifted so it is measured relative to the northern hemisphere solstice as used in the text. Both the obliquity and precession angle in this dataset are given in radians.

Using the orbital parameter data, compute a time series of the diurnally averaged Martian insolation at 65° N, 30° N, and the equator over the past 21 million years; plot the annual maximum, annual minimum, and annual mean values. Then compute a time series of temperature at these latitudes, assuming that the albedo is 20%, that the surface is in instantaneous equilibrium with the absorbed solar radiation and that the atmosphere transports no heat away from the surface. You can also ignore the greenhouse effect of the atmosphere. (These assumptions are reasonable so long as the atmosphere remains roughly as thin as it is at present.) Plot and discuss the annual maximum, annual minimum, and annual mean temperatures, with particular attention to regions that may become warmer than the freezing point of water and to the relative warmth of the low latitudes vs. high latitudes. Although the seasonal cycle has large amplitude, the annual mean surface temperature is still a relevant statistic, since it determines the deep subsurface temperature. When the annual mean tropical region becomes colder than the poles, ice can be expected to migrate from the poles to the tropics.

Recompute the annual mean temperature by finding the temperature which is in equilibrium with the annual mean flux. How much does this differ from your previous result? Why does it differ? If Mars had a thicker atmosphere or a deep mixed ocean would you expect the two different ways of computing the annual mean to differ so much?

Note that with a thin atmosphere the Martian diurnal cycle will be extreme, just as is observed today. The diurnally averaged temperature still is relevant, because it is the temperature of a layer somewhat below the surface. However, computing the diurnally averaged temperature by balancing radiation against the diurnally averaged insolation results in some error in the mean temperature, because σT^4 is nonlinear. Estimate the magnitude of the error in terms of the diurnal temperature range ΔT, which you can take as a given. A more precise way to solve this problem would be to compute that actual diurnal cycle of temperature using the heat diffusion equation, as was done in Problem 7.21.

Problem 7.30 *Earth orbital parameter variations*
The subdirectory Milankovic in the Workbook datasets directory for this chapter contains data on the variations of Earth's eccentricity, obliquity, and precession angle over the past five million years, together with associated documentation. The orbital parameters are in the text file orbit91. Angles in this file are in degrees; the precession angle is defined relative to the spring equinox, and needs to be shifted to yield the angle relative to the northern hemisphere summer solstice as used in the text.

Using the data in this file plot a time series of the annual mean insolation at the Equator, $45°$ N, $45°$ S, $65°$ N, and $65°$ S. Since oceans respond primarily to the annual mean, these give an indication of the response of the oceans to Milankovic forcing. Discuss the relative role of eccentricity, obliquity, and precession in the variations of the fluxes.

Over land, Milankovic variations have a strong effect on the seasonal cycle of temperature. Compare the seasonal cycle of diurnal mean insolation at $65°$ N and $15°$ N between the present, 6000 years ago (the Holocene Thermal Maximum), and 11 000 years ago. The differences are quite subtle, so you can make the variations more apparent by plotting your results as deviations from the modern cycle.

Python tip: The routine getOrbit(...) in FluxExplorer.py retrieves the orbital parameters for you, and takes care of shifting the precession angle so that it is measured relative to the solstice.

7.8.5 Simulation of planetary seasonal cycles

Python tip: The required calculations in this section can be done by modifying the Chapter Script SeasonalCycle.py.

Problem 7.31 Reproduce the calculation shown in Fig. 7.8 and compare the results with the analytic solution based on sinusoidal time-dependence of the insolation. Then re-do the numerical calculation with the assumed dynamical heat fluxes zeroed out. Put the dynamical heat fluxes back in, and repeat the calculation with a mixed layer depth of 5 m. Finally, for a 50 m mixed layer depth see what happens if you double and halve the length of year.

You may assume the orbit to be circular throughout this problem.

Problem 7.32 *Southern hemisphere atmospheric seasonal cycle*
The Earth's southern hemisphere is mostly water, and because of the great thermal iner-tia of the ocean, to a first approximation the sea surface temperature can be regarded as nearly fixed in the course of the seasonal cycle. However, the atmosphere has lower thermal inertia, so that absorption of solar radiation by the atmosphere can drive a significant atmo-spheric seasonal cycle even if the sea surface temperature is fixed. This situation serves as a prototype for seasonal cycles on all waterworlds.

Explore the seasonal cycle in this regime by formulating and numerically solving the differential equation for the seasonal cycle of air temperature making the following assump-tions: (a) The surface temperature remains fixed at a specified value T_s; (b) The atmosphere absorbs 20% of the incident solar energy, and the rest is either reflected back to space or passed on to the ocean surface; (c) The latitude is 45° S and at this latitude the horizontal dynamical heat export from the atmospheric column vanishes; (d) The atmospheric temper-ature structure remains on the moist adiabat; and (e) The *OLR* is given by the polynomial fit in Table 4.2 for Earth conditions with a CO_2 concentration of 300 ppmv. Try various values of T_s ranging from near freezing to 310 K, to explore the effect of increasing evaporation. You may assume the orbit to be circular if you like. Ideally, the thermal inertia of the atmo-sphere should be computed taking the latent heat storage into account (see Problem 7.15), but to keep things simple you may compute the atmospheric thermal inertia as if it were on the dry adiabat, i.e. $\mu = c_p p_s / g$.

A peculiarity of the formulation of this problem is that T_s has been specified indepen-dently of CO_2. In general, the top-of-atmosphere fluxes will be out of balance even in the annual mean. Compute the annual mean top-of-atmosphere imbalance for each of your cases. What effect does the imbalance have on the mean atmospheric temperature? Try to rectify this problem by adjusting the CO_2 for each T_s so as to bring the annual mean top-of-atmosphere budget into approximate balance.

Hint: The atmosphere can be treated like a mixed layer ocean with an equivalent thermal inertia. The mixed layer equations only need to be modified to allow for the fact that only a portion of the solar flux is absorbed in the atmosphere, and to allow for the fact that the atmosphere exchanges energy with the surface as well as cooling to space. The surface fluxes can be computed using the linearized or fully nonlinear surface flux formulae given in Chapter 6. (See also Problem 7.16.) In computing the surface fluxes, assume a fixed surface wind speed of 5 m/s and a roughness length appropriate to the ocean. You may use neutral boundary layer theory and ignore Monin–Obukhov stable boundary layer effects.

Problem 7.33 *Temperature response to Earth Milankovic cycles*
This problem explores the response of the Earth's temperature over atmosphere and ocean to Milankovic variations in insolation. The problem is treated through driving a mixed layer model with the actual seasonal cycle of insolation taking into account the variations in obliq-uity, eccentricity, and precession angle. All of the calculations should be done using the polynomial *OLR* fit for pre-industrial CO_2 values (280 ppmv) so as to focus on Milankovic effects; in reality, the glacial–interglacial CO_2 variations also contribute to temperature vari-ations. For oceanic cases, use a mixed layer depth of 50 m. As an approximation to thermal inertia over land on the seasonal time scale, take a mixed layer depth of 1 m. Assume the oceanic albedo to be 0.2 and the land albedo to be 0.3. You may assume that the atmospheric heat budget is in equilibrium, so that the net radiative heat loss from the column is given simply by the *OLR*. In the oceanic cases you may ignore the effects of sea ice formation, but you should note cases in which sea ice can form.

To attack this problem first write a function that takes obliquity, eccentricity, and precession angle as inputs, and returns the annual mean, annual maximum, and annual minimum temperature by solving the differential equation describing the mixed layer model.

Using this function, plot the annual maximum, minimum, and mean temperatures for the past two million years at increments of one thousand years, using the time series of orbital parameters for the past. Make plots showing the behavior at latitudes of 60°, 45°, and 30°, for land and oceanic conditions. In order to keep the mean temperature in a realistic range, you will need to impose a dynamical heat export from the atmospheric column. Assume that this has a fixed value of $-20\,W/m^2$ at $\pm 30°$ latitude, zero at $\pm 45°$ latitude, and $40\,W/m^2$ at $\pm 60°$ latitude. Discuss your results. To what extent do Milankovic cycles affect the annual mean temperature? To what extent do they affect the amplitude of the seasonal cycle?

Note that for each set of orbital parameters you only need to run the simulation for a few years until the seasonal cycle settles down. You do not need to run a simulation with time-varying orbital parameters over the full two million year period!

See Problem 7.30 for information on the dataset of Earth's Milankovic orbital parameter variations.

Problem 7.34 *Extreme orbits*
Is high obliquity or high eccentricity a threat to habitability? This problem shows that a mixed layer ocean is in fact quite effective at averaging out extreme seasonal variations of insolation, thus keeping the temperature variations from getting too big. For each of the configurations stated below, numerically solve the equations for the seasonal cycle with a mixed layer ocean and show the time series of temperature at the equator, $\pm 45°$, and the poles. Assume a mixed layer of depth $50\,m$ and a uniform albedo of 0.2. For radiative cooling use the *OLR* polynomial fit in Table 4.2 for 1000 ppmv of CO_2. You may neglect the effects of sea ice formation, but you should note and discuss conditions under which ice would form. Carry out your calculations with $L_\odot = 1400\,W/m^2$; for elliptical orbits, this should be interpreted as the flux at a distance from the star equal to the semi-major axis of the orbit. Use the diurnally averaged form of the flux factor valid for planets that are rapidly rotating relative to the length of the year.

To carry out the simulation you will also need to make an assumption about the dynamical heat flux ΔF exported from the column. As a very simple representation of this, assume $\Delta F = a_F \cdot (1 - \alpha)L_\odot \cdot \left(\frac{1}{4} - \bar{f}(\phi) \right)$, where $\bar{f}(\phi)$ is the annually averaged flux factor at the latitude in question and a_F is a constant between zero and unity measuring the effectiveness of the horizontal mixing of heat. When $a_F = 0$ each latitude is locally in equilibrium, and when $a_F = 1$ the annual mean temperature is nearly uniform over the globe. Carry out your calculations with $a_F = 0.5$, but feel free to explore the effects of varying this parameter.

Carry out your calculations for two different cases: (a) a planet with a circular orbit having an obliquity of 70°, and (b) a planet with 25° obliquity in an orbit with eccentricity 0.4.

If you are energetic, you can also explore the effect of variations in $L_\odot$. How large does it have to be in order to suppress ice formation? At what point do you approach a runaway greenhouse threshold? How low does it have to be in order to form ice at 45° latitude (a situation that is likely to provoke a Snowball once ice-albedo feedback is factored in)? Note that when there is no ice the only nonlinearity in the problem comes from $OLR(T)$, so in circumstances where this function can be linearized without loss of accuracy, all temperatures simply scale in proportion to $L_\odot$. Discuss the effects of the deviation from linearity.

7.8.6 A little data analysis

Problem 7.35 *Earth seasonal time series*
The subdirectory NCEP of the Workbook datasets directory for this chapter contains the file
SeasonalTseries.txt, which provides a 40 year record of monthly mean temperature at
selected points on the Earth's surface.

Plot up and take a look at the two oceanic time series in the dataset. Model them using
a mixed layer model driven by the actual insolation computed for these location and using
the polynomial *OLR* fit for 300 ppmv CO_2 concentration. Optimize your fit between model
and data by varying the mixed layer depth, albedo, and the amount of heat exported from
the column by dynamic horizontal heat transports.

Since the dataset covers several decades, it gives you the chance to see how much the
seasonal cycle varies from one year to the next. Carry out your analysis for a few individual
years of your choice, seeing how much your best-fit model parameters need to be varied.
Also, you can form composite seasonal cycles for various 10 year periods by averaging all
the Januaries, all the Februaries, etc. within the period.

Problem 7.36 The data file described in Problem 7.35 also contains a time series from the
Australian interior desert and one from a high-latitude Eurasian continental interior. Take a
look at these, and see if you can make sense of the behavior in light of what you know about
seasonal cycles over land. In the Eurasian case, can you see any imprint of thermal inertia
due to the energy needed to melt snow in the springtime?

Problem 7.37 An interesting way of showing the effect of thermal inertia is to do a scatter
plot of the surface temperature against the time series of insolation at the same point. If
there were no thermal inertia, the temperature would just be a function of the insolation
and the data would lie along a curve enclosing essentially no area. The "opening" of the loop
shows the amount of memory in the system (due to both local and remote effects).

Carry out this analysis for the monthly mean temperature time series described in Problem 7.35. To connect the observations with theoretical expectations, plot some output from
a mixed layer model with various layer depths in the same way; use the same insolation time
series as you used in the data analysis.

Problem 7.38 *Viking Mars temperature analysis*
Data from the Viking landers VL1 and VL2 can be found in the MarsViking subdirectory of
the Workbook datasets directory for this chapter. VL1 was located at 22.48° N and VL2 at
47.97° N.

Pick a short winter period and a short summer period (of a few days each) for each
lander, and see how well you can model the diurnal cycle and mean temperature using an
equivalent mixed layer model to represent the thermal inertia of the solid surface. Drive
the model using a time series of actual Martian insolation at the lander sites, computed
from the present orbital parameters of Mars and formulae in the text. Optimize your fit
by varying $c_p H$ and the surface albedo. You may represent the surface cooling as being
purely due to infrared emission at a rate σT_g^4. There will be some error in the results due to
the neglect of the Martian greenhouse effect. There will be some additional mismatch since
Viking measured the air temperature at lander height, which can be somewhat different
from the surface temperature itself.

Then, if you are feeling ambitious, re-do the calculation using the diffusion equation in the solid surface in place of an equivalent mixed layer model. In this case, you vary the thermal conductivity to optimize the fit rather than $c_p H$.

7.9 FOR FURTHER READING

The Milankovic cycle of Earth's orbital parameters is discussed in

- Berger A and Loutre MF 1991: Insolation values for the climate of the last 10 million years, *Quat. Sci. Rev.*, **10**, 297–317.

The cancellation of precessional forcing by Kepler's Law effects, and the concept of summer melt energy are discussed in the following three papers, among others (see especially Table 1, p. 87 of Berger and Pestiaux (1984)). The essence of the idea was already noted by Milankovic, whose calculation was updated and extended in Berger (1975). Good ideas often need to be rediscovered and restated in alternate ways many times before they gain permanent currency, the latest step in this process being Huybers (2006).

- Berger A 1975: Terrestrial insolation and elliptic integrals. Détermination de l'irradiation solaire par les intégrales elliptiques. *Annales Société Scientifique de Bruxelles, series 1, Sciences Mathématiques, Astronomiques et physiques*, Tome **89(1)**, 69–91.
- Berger A, Pestiaux P 1984: Accuracy and stability of the Quaternary terrestrial insolation. In: *Milankovitch and Climate*, A Berger, J Imbrie, J Hays, G Kukla, B Saltzman (Eds), Reidel Publ. Company, Dordrecht, Holland, pp. 83–112.
- Huybers, P 2006: Early Pleistocene glacial cycles and the integrated summer insolation forcing, *Science*, **313**, 508–511.

The Martian Milankovic cycle, and its climatic expression, are discussed in

- Laskar J, Correia ACM, Gastineau M *et al.* 2004: Long term evolution and chaotic diffusion of the insolation quantities of Mars, *Icarus* **170**, 343–364.
- Levrard B, Forget F, Montmessin F and Laskar J 2007: Recent formation and evolution of northern Martian polar layered deposits as inferred from a global climate model. *J. Geophys. Res. Planets* **112**, E06012.

The seasonal cycle on a globally glaciated Snowball Earth, as well as many other aspects of the hard Snowball climate, is discussed in

- Pierrehumbert RT 2004: High levels of atmospheric carbon dioxide necessary for the termination of global glaciation, *Nature* **429**, 646–649.

Triton's exotic seasonal cycle is discussed in

- Trafton L 1984: Large seasonal variations in Triton's atmosphere, *Icarus* **58**, 312–324.

CHAPTER EIGHT

Evolution of the atmosphere

8.1 OVERVIEW

As we emphasized back in Chapter 1, atmospheres are not static. The mass and composition of an atmosphere evolves over time, as a result of a great variety of chemical, physical, and biological processes. Now it is time to survey those processes in greater detail, and to put numbers on them to the extent possible in the limited space available in this chapter.

Throughout the following we will need to refer to some constituents of a planet as *volatiles*. These are "not rocks" – things that can become gases to a significant extent. The concept of a volatile is relative to the temperature of a planet. On Earth, water is a volatile but on Titan it is basically a rock, as is CO_2, though N_2 and CH_4 remain as volatiles even at the low temperatures of Titan. On Earth, sand (SiO_2) is a rock, but on a roaster – a hot extrasolar Jupiter in a close orbit – it could be a volatile.

For planets in which some atmospheric volatiles exchange with a condensed reservoir, as in the case of Earth's ocean and glaciers, the whole atmosphere–ocean–cryosphere system is best treated as a unit for many purposes, and we will refer to this as the *volatile envelope*. In other cases, the portion of the volatile envelope which resides in the atmosphere plays a distinguished role. Only the atmospheric portion provides a greenhouse effect, and volatiles must first enter the atmosphere before they can escape to space as gases.

The main factors that govern atmospheric evolution of rocky terrestrial type planets or icy bodies with a thick solid crust are as follows. First, there is the matter of what, if anything, outgasses from the planetary interior, and at what rate. The composition of the outgassing depends on the chemistry and physics of the planet's interior; for example, the early segregation of the Earth's core took iron out of the rest of the planet which allows oxygen to react with other elements. This favors the outgassing of oxidized gases such as CO_2, SO_2, and water vapor, though limited amounts of H_2 and CH_4 do exit the interior. On Titan it is speculated that ice-volcanism can release NH_3 and possibly CH_4 to help maintain the atmosphere, and on Venus it is speculated that the sulfuric acid clouds are maintained by outgassing of SO_2 (though no active volcanism has yet been observed there).

Next, whatever enters the volatile envelope is subject to a number of further alterations. Atmospheric constituents are lost to space, either by gradual mechanisms or in catastrophic events such as giant impacts. Different constituents are generally lost at different rates, leading to evolution of the composition. Then, too, chemical reactions between the volatile envelope and the solid crust can bind up constituents in mineral form, selectively removing material from the volatile envelope. If the planet has no way to engulf bits of the crust and close the cycle by cooking out the volatiles, then the volatile envelope will eventually come into equilibrium with the static upper part of the crust, and this means of evolution will mostly cease, though changes in climate could still lead to changes in partitioning between the volatile envelope and the crust. This is the situation on Mars at present. If the planet is tectonically active, as is the case for Earth, then crustal material is mixed down into the interior, and there is instead a dynamic geochemical equilibrium involving a much greater proportion of the mass of the planet. The question of whether crustal recycling occurs on Venus and Titan is one of the current Big Questions of planetary science. The discovery of extrasolar planets of the Super-Earth class raises the Big Question of whether plate tectonics or some other form of resurfacing becomes more or less likely on rocky planets larger than the Earth or Venus.

The main energy sources that sustain plate tectonics or other forms of resurfacing are fossil heat left over from the formation of the planet and heat release by radioactive decay in the interior. These heat sources are small, but they can have a big effect on interior temperature because the diffusivity of heat through solids is so low. The importance of radioactive decay introduces another dependence of climate evolution on planetary composition, since planets can be formed with a greater or lesser endowment of radioactive materials than the Earth. Further, for planets in close orbits about their stars, or satellites in close orbits about giant planets, interior heating by tidal stresses could be important. Such processes drive spectacular volcanism on Jupiter's satellite Io, and could well play a role on Super-Earths in the habitable zone of M-dwarfs, which is quite near to such stars.

Finally, chemical reactions within the atmosphere determine how the elemental composition is arranged into molecules. These are usually fast processes on geological time scales, requiring seconds to a few million years to operate. They nevertheless affect the long-term evolution by affecting what can escape to space, what can react with the crust, and whether the elements are arranged as greenhouse gases (e.g. oxygen in the form of CO_2) or not (oxygen in the form of O_2). Some examples include the breakup of water vapor by ultraviolet light, the oxidation of methane or hydrogen which limits their atmospheric concentration, and the breakup of methane on Titan followed by resynthesis into ethane.

Life profoundly alters virtually every aspect of atmospheric evolution. Through the use of complex enzymes, life can break stable bonds and synthesize compounds in low-temperature, low-energy environments where inorganic processes do very little. Nitrogen fixation and oxygenic photosynthesis are two prime examples. We will see in Section 8.7 that the oxygen generated in the latter process actually raises the temperature of Earth's outermost atmosphere from about 250 K to over 1000 K, affecting the escape of gases to space. Life can synthesize methane at rates far greater than the methane flux produced by volcanic outgassing. Life also alters the chemical environment at a planet's surface, altering the rate of reaction of atmospheric components with the crust.

The processes involved in atmospheric evolution are summarized in Fig. 8.1. These are the processes and concepts that will be made quantitative in the rest of this chapter.

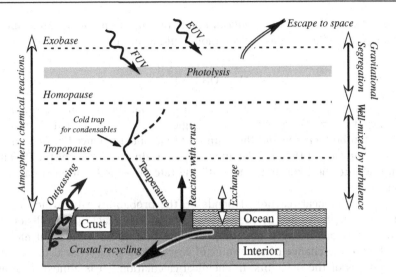

Figure 8.1 Summary of processes involved in atmospheric evolution. Outgassing from the planet's interior adds mass to the atmosphere, while atmospheric constituents are removed via both reaction with the crust and escape to space. Constituents bound up in the crust may be recycled to the atmosphere if there are processes which engulf bits of crust into the hot interior and cook out gases which ultimately escape to the atmosphere. Atmospheric constituents also exchange with the liquid ocean, if there is one. Escape from the thin outermost portions of the atmosphere is energized by absorption of extreme ultraviolet (*EUV*) radiation. Below the *homopause*, the atmosphere is well-mixed by turbulence; non-condensing, non-reactive substances will have uniform mixing ratio there. Above the homopause, transport is by diffusion, and atmospheric constituents segregate gravitationally, with the lighter constituents rising to the top. At high altitudes, atmospheric molecules are broken up by absorption of far ultraviolet (*FUV*) radiation, in a process called *photolysis*. Chemical reactions between photolysis products can occur throughout the atmosphere, and some of these re-form the photolyzed molecules. Temperature goes down with height within the troposphere and lower stratosphere. This leads to a *cold trap* which can limit the flux of condensable substances to high altitudes where they would photolyze and possibly escape to space.

8.2 ABOUT CHEMICAL REACTIONS

The next few topics will require some basic knowledge about how chemical reactions work, so we will pause here to provide some elementary background. Consider, for example, two hypothetical substances A and B which can react to form a product C. The reaction is written as

$$A + B \rightleftharpoons C. \tag{8.1}$$

The double arrow symbol, $\rightleftharpoons$, is there for an important reason: it reminds us that chemical reactions proceed in both directions. When a molecule of A encounters one of B, with a certain probability it will react to form C. However, from time to time, a molecule C will also break up into its components A and B. The net reaction depends on the competition between the forward reaction and the back reaction, and when the two rates are equal the system is in *chemical equilibrium* and concentrations of the substances do not evolve. As an example, in any glass of common liquid water, the following reaction is taking place:

$$H^+ + OH^- \rightleftharpoons H_2O. \tag{8.2}$$

The superscripts indicate that in this particular reaction, the reactants are *ions*, having a positive or negative charge (in this case, a single charge equal to that of an electron or proton). Note that both charge and the count of atoms must balance between the left hand and the right hand side of the reaction. In water, a certain proportion of the H_2O molecules will decompose into the indicated ions, until the rate of recombination equals the rate of production.

The rate of a chemical reaction depends on the probability with which molecules of the reactants collide multiplied by the probability that they react upon collision. In simple cases, such as reactions between molecules in a gas or a dilute solution of molecules dissolved in a liquid, the chance of encounter of reactants is proportional to the product of the concentrations of the reactants. In chemical calculations, it is almost invariably most convenient to measure concentrations as Molar concentrations, rather than mass concentrations. For liquid phase reactions between solutes, Molar concentration and Molar density (e.g. Moles of solute per liter of solvent) are practically the same thing, since the density of the liquid varies little. Recall that we use the capitalized term "Moles" to represent kilogram-moles, as discussed in the Preface. For gases, it is often more convenient to represent the quantity of a substance in terms of its partial pressure instead of its mass or Molar density.

It is commonly the case that the availability of a molecule to participate in reactions is not simply proportional to its concentration. This might happen, for example, because other molecules in a solution cluster around a solute molecule, partially shielding it from reactions. To deal with this, chemists have introduced the notion of *activity*, as a generalization of concentration. The representation of activity, when it is something other than a simple concentration, is particular to the class of reactions under consideration. To distinguish the activity of a substance, which is a number, from the abstract symbol denoting the substance itself, the activity is written in square brackets: [A] is the activity of substance A. However the activities are defined, the reaction rate is expressed as the product of all the activities multiplied by a rate coefficient. By convention, we'll call the rate coefficient k_+ for the forward reaction and k_- for the back reaction. Thus, the reaction rate for the forward reaction in Eq. (8.1) is $R_+ = k_+[A][B]$. The product of activities represents the probability of encounter of the two reactants, while the rate coefficient represents the probability of reaction given a collision. For the unary back reaction involving decomposition of C, the rate at which the decomposition proceeds is proportional to the amount of substance C present, so we write $R_- = k_-[C]$.

By way of example, let's consider the net of forward and back reactions in Eq. (8.1), in the simple case where the activities are just concentrations, in which case the reaction rates are the time derivatives of the concentrations. The evolution of the concentrations is then given by

$$\frac{d[A]}{dt} = \frac{d[B]}{dt} = -k_+ \cdot [A][B], \quad \frac{d[C]}{dt} = -k_- \cdot [C]. \tag{8.3}$$

The decay rates of [A] and [B] are equal because each reaction consumes one particle of A and one of B, producing a particle of C. The rate depends on the product of the activities of the two reactants, since the product gives the probability of particles of the reactants encountering each other. Now, the first equation given only determines the decay of A owing to the reaction with B. This will be the net reaction when there is no C present, but after the reaction proceeds a while in a closed vessel, some C will accumulate and the decomposition of C back to A and B needs to be taken into account. Thus, when the activity is simply a concentration or density, the full system is governed by

$$\frac{d[A]}{dt} = -k_+ \cdot [A][B] + k_- \cdot [C] \tag{8.4}$$

which will come into equilibrium when the left hand side is zero, namely when

$$\frac{[C]}{[A][B]} = \frac{k_+}{k_-} \equiv k_{eq}. \tag{8.5}$$

The quantity on the right hand side is the *equilibrium constant* for the reaction. This equation constrains the relative proportion of the three substances once equilibrium has been achieved, but how one uses this information depends on what is specified in the setup of the problem. For example, if for some reason we know the activity [A], then we immediately know the ratio [B]/[C], though we don't know the absolute amounts unless something in addition is specified. As a slightly more complicated example, suppose we put 2 Mole/m^3 of A and 1 Mole/m^3 of B into a closed vessel which initially contains no C. Then, after equilibrium has been reached x Mole/m^3 of C will have been produced, which depletes each of [A] and [B] by x Mole/m^3, since it takes one of each to produce a particle of C. The equilibrium equation then tells us that

$$\frac{x}{(1-x)(2-x)} = k_{eq}. \tag{8.6}$$

Given knowledge of the equilibrium constant, this allows us to solve for x using the quadratic equation.

Exercise 8.1 Carry out the algebra to determine x in the above example, and describe how it behaves as k_{eq} is varied from very small to very large values. Do you ever completely use up the reactants?

When the activity is something other than simply concentration or density, the reaction rate is no longer the time derivative of activity, and the forward and back reaction rates should be written abstractly as R_+ and R_-. These are measured in various ways, depending on the nature of the reaction. For example, if the reaction is between a gas or dissolved substance and the surface of a solid, it would be typical to characterized the reaction rate in terms of Moles per unit time, per unit surface area of the reacting solid. The characterization of reaction rate affects how one calculates the approach to equilibrium, but it does not affect the equilibrium itself, since the equilibrium is defined by $R_+ + R_- = 0$, which yields the same equilibrium conditions on activity as before.

As a concrete example of the use of equilibrium coefficients, let's consider the dissociation reaction that occurs in liquid water.

$$H_2O \rightleftharpoons H^+ + OH^-. \tag{8.7}$$

This reaction is of ubiquitous importance in aqueous chemistry. Only a tiny fraction of the water molecules dissociate, so it is customary to set the activity of H_2O to unity in writing the equilibrium relation

$$[H^+][OH^-] = k_w. \tag{8.8}$$

The activities of the ions are not identical with concentrations, but it is customary to express the activities as a non-dimensional coefficient times the actual concentration, so that activities and concentrations have the same units. In the units of moles per liter (that's *gram-moles*) that aquatic chemists like, the equilibrium constant k_w is very close to 10^{-14} at room temperatures. If we take the $\log_{10}$ of the equilibrium equation written in these units, then

$$(- \log_{10} [H^+]) + (- \log_{10} OH^-) = 14. \tag{8.9}$$

This leads to the definition pH $\equiv - \log_{10} [H^+]$, assuming the activity to be expressed as moles per liter. If water dissociates in the absence of any solute that changes the H^+ or OH^- ion concentrations, then equal amounts of the two ions are produced, and the equilibrium relation in the form Eq. (8.9) implies that pH = 7. This case, without an excess of protons, is referred to as *neutral*. If a substance like H_2SO_4 dissociates and adds H^+ ions without adding OH^- ions, then Eq. (8.8) implies that the concentration of OH^- must go down, whence pH falls below 7 resulting in an *acid*, which has an excess of protons. The opposite situation, with a deficit of protons, is a *base* and has pH > 7. If one knows the pH, then the activity of H^+ expressed in moles per liter is 10^{-pH}, and if the activity coefficients are near unity this is nearly the actual concentration. To convert to units of mole fraction (molecules of substance per total molecules) that we favor, one must divide this concentration by 55.56, since that is the number of gram-moles in a liter of water. Note also that, like all equilibrium constants, k_w depends on temperature, so that the H^+ concentration corresponding to a neutral solution is somewhat temperature-dependent; the definition of pH, however, is always tied to the H^+ concentration itself.

Rate and equilibrium equations generalize to reactions involving more reactants and more products in the obvious way. For example, a reaction of the form A + 2B + 3C has a rate proportional to $[A][B]^2[C]^3$. The equations for equilibria are modified correspondingly. Sometimes one of the "reactants" is a molecule that participates in the reaction only to the extent of providing some extra energy by collision. A reaction like that is written, for example, as A + B + M $\rightleftharpoons$ C + M, where M is a generic colliding molecule; in kinetics, its activity [M] would generally be the number density of molecules of any sort available for collision with the other reactants. For example, an atmospheric reaction between substances A and B might proceed more rapidly in a background of N_2 even though N_2 does not itself react with either A or B. It is also common in atmospheric chemistry for one of the "reactants" to be a photon, usually one within a designated range of frequencies. A photon is designated in reactions by the symbol $h\nu$, as in the dissociation reaction AB + $h\nu$ $\rightleftharpoons$ A + B. The symbol used for photons is meant to be reminiscent of the energy carried by a photon of frequency ν. The activity of the photons is usually the number flux of photons having the correct energy range to react. This is obtained by dividing the energy flux at frequency ν by the energy of an individual photon $h\nu$, then summing up the result over all frequencies involved in the reaction.

When one or more of the reactants is in the form of a pure, solid body – for example a lump of solid substance A reacting with a gaseous substance B within a given volume – the quantity of the solid within the volume is not the limiting factor in determining the availability of molecules of the solid substance to react. Rather, in this case it is the area of solid exposed to other reactants that counts. In such a case, as a matter of convention the activity of the pure solid substance is set to unity, and the effect of available surface area for reaction is taken into account by expressing the reaction rate as Moles per unit area of contact per unit time. If the pure phase is reacting with a gas, the reaction rate expressed this way will depend on the partial pressure of the gas, but not on the amount of solid present (so long as there is some present and it is in contact with the gas). Equilibrium at any given temperature yields a unique value of partial pressure, just as Clausius-Clapeyron yields a unique vapor pressure in equilibrium with the solid or liquid form of a substance, regardless of how much of the solid or liquid form is present. The same behaviour applies when a pure solid phase reacts with a substance B dissolved in a fluid in contact with the solid – the reaction rate per unit area will depend on the concentration of B, but not on the amount of solid present, and equilibrium at any given temperature will yield a unique value of [B], regardless of how much solid phase is present, as long as there is some. We will encounter this situation soon, in our study of weathering reactions.

For all reactions considered in this chapter, the activity will be either a concentration (or its equivalent, such as a partial pressure), or it will be unity for pure condensed phase reactants. The reader will not need to be concerned with more exotic expressions for activity, though it is good to be aware that they exist.

The equilibrium and rate constants depend strongly on temperature. Most of the temperature dependence of the rate constants k_+ and k_- is captured by the Arrhenius law, which states that $k(T) = A \cdot \exp(-E/R^*T)$ where A is a constant and E is a quantity known as the *activation energy*, measured in J/Mole when the Arrhenius law is written in this form. The temperature dependence can be fit a bit more accurately if the constant A is replaced by a power law in T, but for most of our purposes the unmodified Arrhenius law suffices. Since the equilibrium constant is k_+/k_-, it follows that the equilibrium constant has temperature dependence $A_{eq}\exp(-\Delta E/R^*T)$, where ΔE is the difference in the activation energy between the reactants and the products, and A_{eq} is the ratio of the prefactors of the forward and back reactions. While activation energy must be positive, and hence all reactions speed up with temperature, ΔE may be either positive or negative; therefore the equilibrium constant may either increase or decrease with temperature, accordingly as whether ΔE is positive or negative.

The activation energy in the Arrhenius law is a generalization of the concept of latent heat, which we have discussed in connection with phase transitions, and which appears in the Clausius-Clapeyron equation in a manner similar to the way activation energy appears in the Arrhenius law. Indeed, a phase transition between, say, the gas and liquid forms of a substance can be considered as a binary reaction in which two molecules of "gas" substance collide and react to form one molecule of "liquid" reactant. To relate the latent heat L to the activation energy in the Arrhenius law, we need only observe that Clausius-Clapeyron was written using the gas constant specific to the gas in question, rather than the universal gas constant. Re-writing in terms of the universal gas constant, the temperature dependence of saturation vapor pressure becomes proportional to $\exp(-(L \cdot M)/R^*T)$, where M is the molecular weight of the gas. Thus, the activation energy for condensation of water vapor into liquid water is $4.49 \cdot 10^7$ J/Mole.

The constants also depend on the total pressure at which the reaction takes place, but the pressure dependence is usually weak over the range of pressures of interest in atmospheric problems.

8.3 SILICATE WEATHERING AND ATMOSPHERIC CO_2

Carbon dioxide is one among many greenhouse gases that can be present in a planetary atmosphere, but it plays a distinguished role in the evolution of atmospheres of rocky planets like Earth, Mars, and Venus because of its participation in chemical reactions that allow it to be exchanged between the atmosphere and minerals in the crust and planetary interior. What's more, acting over sufficiently long time scales, the dependence of the interchange on temperature can potentially act as a thermostat and help to keep the planet in the habitable range despite considerable changes in solar luminosity and other factors. Understanding the precision with which this feedback mechanism controls planetary temperature, the limits within which it can operate, and the circumstances under which it can break down, is one of the most central of the Big Questions.

It is likely that there are other gases important to climate which undergo similar exchanges, but CO_2 exchanges have been far more extensively studied than any other case, and so this problem serves as a template for thinking about more exotic possibilities, some of which will be mentioned at the end of this section.

Carbon dioxide undergoes exchange with the solid crust and interior of a rocky planet through reaction with silicate minerals to form carbonates. The class of reactions involved in this exchange is typified by the idealized *Ebelmen–Urey reactions*

$$MgSiO_3(s) + CO_2(g) \rightleftharpoons MgCO_3(s) + SiO_2(s)$$
$$CaSiO_3(s) + CO_2(g) \rightleftharpoons CaCO_3(s) + SiO_2(s) \qquad (8.10)$$
$$FeSiO_3(s) + CO_2(g) \rightleftharpoons FeCO_3(s) + SiO_2(s).$$

In all of these reactions, the carbon in gaseous CO_2 exchanges with the silicon in solid silicate minerals, to form a solid carbonate mineral plus solid silica (SiO_2, of which quartz is one form, and which is also common in beach sand). The reader should keep in mind that the actual silicate minerals involved in the formation of carbonates can be considerably more complex than the simple chemical compounds referred to in the above reactions, and may have different equilibrium and kinetic properties. The feldspar family of minerals is one of the most important players in silicate weathering in Earth's present crust. These minerals are aluminum silicates involving varying amounts of sodium, potassium, and calcium; their weathering products include a broad variety of clay minerals, rather than just simple silica. Nonetheless, the Ebelmen–Urey reactions are often taken as indicative of what is going on in more realistic cases.

First, let's take a look at the equilibria that would be reached in the Ebelmen–Urey reactions after a sufficiently long time has passed. Imagine, for example, putting a pile of powdered $MgSiO_3$ in a closed chamber filled with a large quantity of CO_2 gas, and holding the entire apparatus at constant temperature T. The gas will react with the silicate to form carbonate and silica, drawing down the pressure until the products have built up to the point that the rate of recombination of carbonate plus silica is equal to the rate of reaction of CO_2 with silicate, at which point equilibrium has been reached. As long as the silicate has not been used up, the pressure will equilibrate at a value that depends only on T. Since the activities of the solid phases in Eq. (8.10) are unity, the only activity that can vary is that of the gaseous CO_2, and this activity can be characterized by the partial pressure of the gas. As long as none of the solid reactants has been exhausted, chemical equilibrium for any one of the reactions in Eq. (8.10) is described entirely in terms of the way the partial pressure of CO_2 gas (pCO_2 in chemical parlance) depends on temperature, when in the presence of the three solid reactants in the equation. In this situation, when there is only one activity which can vary, the equilibrium constant can be taken to be pCO_2 itself, and its temperature dependence follows the Arrhenius law

$$pCO_2 = p_1 \exp\left(-\frac{\Delta H}{R^* T}\right) \qquad (8.11)$$

where p_1 is some constant, R^* is the universal gas constant, and ΔH is a characteristic energy, specifically the difference in *enthalpy of formation* between the products and reactants. For the Ebelmen–Urey reactions, the reaction releases energy (i.e. is *exothermic*), so ΔH is positive. If the gas constant R^* is given in J/Mole $\cdot$ K then ΔH has units of J/Mole. The equation for partial pressure looks just like the Clausius-Clapeyron relation, and this is no accident since the thermodynamic formalism for the temperature dependence of equilibrium pressure is identical in the two cases: formation of a condensed phase from a gas is just a form of chemical reaction involving a single substance. The latent heat of fusion for water is 158 kJ/Mole, which compares with ΔH values of 79 496, 88 700, and 64 852 kJ/Mole for the Mg, Ca, and Fe reactions measured in the laboratory at 298 K.

The equilibrium pressures for the three reactions are shown as a function of temperature in Fig. 8.2. The measured temperature dependence of ΔH has been taken into account in calculating these pressures. First, we note that at Earthlike temperatures the equilibrium partial pressures for Mg and Ca minerals are very low. For the Mg case at 300 K the equilibrium has $pCO_2 = 1.9 \cdot 10^{-5}$ bar, or equivalently 19 ppmv in a 1 bar background atmosphere. For the Ca case, it is even lower, amounting to a mere 0.1 ppmv. In equilibrium, weathering of Mg and Ca silicates would draw down atmospheric CO_2 to such low values that it would have little greenhouse effect. At present (and probably for most of Earth history), the pCO_2 is well in excess of these equilibrium values, so silicate weathering is always trying to reduce atmospheric CO_2 to nearly zero, though it never gets especially close to equilibrium because the system is kept out of equilibrium by outgassing of CO_2 from the interior. Weathering of iron silicates leads to a somewhat different story line. At 300 K the equilibrium for the Fe case is 0.006 bar, or 6000 ppmv. This is well in excess of the present pCO_2, and probably in excess of any value attained in the past half billion years. The implication is that even if iron silicates were prevalent at the surface of the Earth, the weathering of Mg and Ca silicates keeps the atmospheric CO_2 too low for iron carbonates to form. On the other hand, we know that during the era of the Faint Young Sun the CO_2 must have been well in excess of 0.006 bar if the CO_2 greenhouse effect was to be strong enough to keep the Earth unfrozen. Under these circumstances, iron carbonates should have formed. Therefore, the presence of iron carbonates in ancient deposits serves as a proxy for high CO_2. The lack of iron carbonates in certain Archean formations is often taken as evidence that CO_2 alone could not have been the answer to the Faint Young Sun Paradox, but this interpretation should be treated

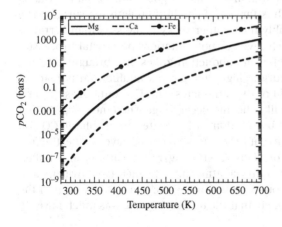

Figure 8.2 Equilibrium CO_2 partial pressure as a function of temperature for equilibrium with magnesium (Mg), iron (Fe), and calcium (Ca) silicates.

with caution, since so little Archean surface rock is preserved and the temporal record is very sporadic.

The equilibrium pressures rise sharply with temperature. At 700 K the equilibria are 1200 bars, 45 bars, and 15 000 bars for the Mg, Ca, and Fe cases, respectively. Because the Earth's interior temperature exceeds 700 K not too far below the surface, this immediately implies that as carbonates and silica are engulfed by plate tectonics and subducted into the interior, CO_2 will be cooked out of the rocks and outgas through volcanoes and fissures, allowing the carbon to be returned to the atmosphere. Things are likely to work similarly for any planet with plate tectonics and a rocky crust, and there may be other (as-yet unknown) means of episodically engulfing crustal material and bringing it to a high enough temperature to release CO_2. The high-temperature equilibria also have implications for the state of the atmosphere and crust of Venus. The atmospheric surface pressure of Venus is about 90 bars of nearly pure CO_2; this is well below the equilibrium pressures for Mg and Fe carbonates, so these minerals would be unstable at the surface of Venus, given sufficient SiO_2. In contrast, the equilibrium pressure for Ca carbonates reaches 90 bars at 737 K, which is quite close to the actual surface temperature of Venus. It thus appears possible that the atmosphere of Venus is in equilibrium with a crustal reservoir of calcium carbonate.

There are no atmosphere-bearing planets in the Solar System at present with surface temperatures intermediate between those of Earth and Venus. Venus may have experienced such temperatures in the past in the course of a near-runaway greenhouse state, and extrasolar planets could well have an orbit and composition to put them in this range. At 400 K the Ca equilibrium is still only 650 ppmv, though that for the Mg case amounts to 5% of a 1 bar atmosphere. By the time one gets to 500 K, however, large amounts of CO_2 are inevitably left in the atmosphere – 6 bars for the Mg case and 0.12 bar for Ca case. Reasoning by analogy from the state of the present Earth atmosphere, when the system is out of equilibrium by virtue of outgassing from the interior, the actual atmospheric pCO_2 will be considerably in excess of the equilibrium value, though it is reasonable to conjecture that at high temperatures equilibrium might be approached more rapidly, which would allow the system to stay closer to equilibrium even in the presence of substantial outgassing.

The chemical equilibrium behavior of the silicate/carbonate/CO_2 system is straightforward, but it is probably the only thing that is straightforward about silicate weathering and its role in climate evolution. The picture of the CO_2 in an atmosphere as being in equilibrium with rocks near the surface may perhaps be valid for some planets with a high temperature surface, but in general it is a poor representation of what is going on. Certainly, this is the case for Earth, the atmosphere of which is far out of equilibrium with the surface. The inorganic carbon cycle on Earth and probably many other rocky planets with an Earthlike temperature is a dynamic equilibrium which involves both interior and crustal processes. The carbonates near the surface are engulfed in subduction zones and brought into the interior of the planet, where the temperature is high enough that equilibrium is reached quickly and CO_2 is driven almost entirely out of the carbonates. This CO_2 outgasses through volcanoes and through submarine features like the mid-ocean ridge where new ocean crust is born. If no new carbonates were formed by reaction with silicates, atmospheric CO_2 on Earth would build up to very high levels. Current CO_2 outgassing rates have been estimated at 0.1 gigatonnes (Gt) of carbon per year, or about $2 \cdot 10^{-4}$ kg/m^2 of carbon. In 4 billion years this would pump 784 215 kg/m^2 of C into the atmosphere, which is equivalent to a CO_2 surface pressure of over 280 bars – about a million times the amount observed in the atmosphere today. Since it is extremely unlikely that the outgassing rate was much lower in the past ages – indeed it was probably greater when the Earth was younger and had lost less interior heat – there clearly must be an effective removal mechanism which takes CO_2 out of the atmosphere. Formation of carbonates by reaction with silicates is by far the most likely candidate.

Rather than building up until the supply of interior carbon is exhausted, the atmospheric CO_2 only builds up to the point where the rate of formation of carbonates equals the rate of outgassing. For any given outgassing rate, then, determination of the amount of CO_2 in the atmosphere requires that we quantify the way the carbonate formation rate depends on CO_2 levels, temperature, and other aspects of the climate. This is unfortunately not a matter of simple chemistry. For low temperature planets like Earth, the carbon dioxide pressure at the surface is generally well in excess of the equilibrium value corresponding to surface temperature and composition, but the rate at which carbonates form and bring the system back toward equilibrium is not determined primarily by the kinetics of the chemical reaction. The quantity we need to understand is the weathering rate W which is the number of Moles of CO_2 per unit time which is converted to carbonate by reactions with silicates over the entire surface of the planet. We need to determine how W depends on temperature, precipitation, and other aspects of climate, as well as the nature of the surface of the planet we seek to understand.

As with so many facets of planetary climate, silicate weathering involves a conspiracy of CO_2 and water, and this is of central importance in the determination of weathering rates. Although the reactions as written in Eq. (8.10) do not involve water, at Earthlike temperatures the reactions occur in solution when water comes into contact with rock, and should be thought of as a kind of dissolution of silicate minerals in the presence of the weak carbonic acid formed when CO_2 dissolves in liquid water. At low temperatures – certainly below 300 K and perhaps below 400 K – the dry reaction proceeds too slowly to be of significance over a time scale of a few billion years.[1] Given that the reaction is aqueous, it would be natural for the reader to conclude that silicate weathering on Earth would be dominated by undersea processes; after all, there is plenty of water there as well as plenty of silicate in the ocean floor. Contrary to expectations, however, the best estimates indicate that at present sea-floor weathering amounts to only a tenth of the global total. The factors limiting sea-floor weathering at present include the degree of acidity of the ocean, the low temperatures of the deep ocean, the composition of the ocean crust, and the sluggish delivery of ocean water to new reactable surfaces (occurring primarily in hydrothermal systems today). One should not over-generalize from this state of affairs, since it is highly dependent on the current state of the climate, and in a radically different climate with much higher CO_2 and higher temperatures things could be quite different. Further, on Snowball Earth or on a waterworld with no crust exposed to the atmosphere, sea-floor weathering would be dominant because it is the only form of weathering there is. Moreover, it is quite difficult to estimate the rate of sea-floor weathering even in present conditions, and there are credible estimates suggesting that it accounts for a considerably higher proportion of the total than the standard picture would indicate. If sea-floor weathering were to prove to be a considerable fraction of the total, then most of what we shall have to say subsequently

---

[1] The kinetics of the dry phase reaction do not appear to have been quantified to any great degree. There is some laboratory evidence that CO_2 can be formed in dry reactions between carbonate and silica at temperatures above 500 K, and indeed the dry phase reaction must be taking place in Earth's interior to sustain outgassing. It is generally believed, though, that even at high temperatures carbonates cannot form at a significant rate without water.

about silicate weathering and climate regulation would be called into question. Climate evolution under the dominant control of sea-floor weathering represents largely unexplored territory.

We shall adopt the conventional picture that silicate weathering for planets in a regime something like that of the Earth occurs primarily over land, as a result of rain washing over silicate-bearing rocks. As a result, the rate of carbonate formation is expected to increase with the rate at which rain falls over weatherable silicate rocks; the rain both accelerates the reaction and carries away the soluble carbonates, exposing fresh silicate for further reactions. The functional form of this relation cannot be measured in the laboratory, and depends on the mineral, its physical structure, and the presence of vegetation and other biological activity. There have been numerous attempts to estimate the precipitation dependence from various kinds of field measurements of weathering, but the process is still poorly constrained.

Laboratory measurements of aqueous phase silicate weathering show clearly that the rate of carbonate formation increases with temperature according to the Arrhenius law, $\exp(-E/R^*T)$, where E is an empirically determined activation energy. When the range of temperatures about some base temperature T_0 is not too large, the Arrhenius law can be simplified to the exponential form $W(T_0)\exp((T - T_0)/T_U)$, where $T_U = T_0^2 R^*/E$. The coefficient T_U varies amongst minerals and also depends on other conditions such as the acidity of the environment in which weathering occurs. Typical values in widespread use in weathering models lie in the vicinity of $10\,K$, for temperatures within a few tens of kelvin of Earth's current surface temperature. It is generally assumed that the temperature-dependent reaction rates measured in the laboratory lead to the same temperature dependence of net weathering rate in the field, though it is far from clear that this should be the case. If rain remains in contact with weatherable rock for only a short time, then it would be expected that increasing the reaction rate would indeed increase the amount of carbonate formed; this is the prevailing view of what is going on in Nature. However, in circumstances where water remains in contact long enough for the reaction to come to equilibrium, the kinetics becomes irrelevant and the weathering rate should not be directly dependent on temperature, though it will still depend indirectly on temperature through the effect of temperature on the precipitation rate. Because of the steep exponential dependence of the reaction rate, the chemical kinetic effect is likely to become far less important as temperatures become much hotter than that of the present Earth. For example, with $T_U = 10\,K$, a reaction that takes one day to reach equilibrium at $300\,K$ would equilibrate in only 4 seconds at $400\,K$ and 178 microseconds at $500\,K$. Given the slow kinetics at Earthlike temperatures and below, it does seem a reasonable assumption that something like the Arrhenius law applies for such temperatures. For very hot conditions, as in a near runaway, the direct temperature dependence of W would need to be reconsidered.

A more problematic issue is whether W also depends directly on the partial pressure of CO_2 in the atmosphere. Note that this is a separate question from the *indirect* effect of CO_2 concentration on weathering, mediated by its effect on temperature and by the effect of temperature on precipitation. Normally, for gas–solid reactions, it would be expected that the reaction would proceed more rapidly if there were more molecules of the reactive gas around. This is precisely what happens in laboratory measurements of the weathering of specific silicate minerals (mostly feldspar) at temperatures of $400\,K$ or more. The directly measured high-temperature dependence is often extrapolated down to lower temperatures using the Arrhenius law. However, laboratory measurements at Earthlike temperatures, conducted in the presence of organic acids thought to act similarly to those produced by land plants, very clearly show that weathering rate is essentially independent of the amount of CO_2; such experiments have been conducted for CO_2 partial pressures ranging from about 0.3 mb all the way up to 1 bar. The prevailing view in this subject seems to be that without land plants (or perhaps, without bacterial life modifying silicate surfaces) there is a power law dependence of weathering rate on CO_2 partial pressure, but that this dependence disappears once land plants have appeared on the scene. It is quite unclear whether lichens or bacterial life are sufficient to cause the transition in behavior, and it is even more unclear whether the supposed abiotic pCO_2 dependence really exists at low temperatures. One can also wonder whether abiotically produced acids could also eliminate the direct pCO_2 dependence.

The weathering rate is affected by a number of other processes going on at the surface. Notably, once land plants are on the scene, changes in precipitation and temperature distributions will affect the distribution of land plant cover, and this will feed back on the weathering rate; this is a very difficult feedback to model. Besides that, the weathering rate depends on the availability of weatherable surface area. This is affected by physical erosion rates, and can be greatly enhanced by mountain-building such as the rapid uplift of the Himalayas; erosion rates are also affected by glacier flow and by freeze–thaw cycles which can fracture rock. Volcanism is also important in providing fresh weatherable surfaces. On Venus today, weathering is likely to be slow even for reactions that can take place without liquid water, because most of the erosional processes that produce fresh weatherable surface are absent or weak.

Note also that the conventional picture of silicate weathering on relatively cool planets like the present Earth presumes that the equilibrium atmospheric CO_2 is far below the prevailing atmospheric value, so that the weathering can be thought of essentially driving the atmospheric CO_2 towards zero. If, in contrast, the equilibrium has a significant amount of CO_2 left in the atmosphere, then one needs to take into account the actual equilibrium toward which the weathering reactions are driving the system. This becomes more and more of a significant factor as the temperature increases, and the importance of the effect also varies with the surface mineralogy, since the equilibria are dependent on what kinds of reactions are taking place. The following development adopts the conventional cool-planet formulation in which weathering reactions are slow enough relative to outgassing rates that the atmospheric CO_2 is always far above the equilibrium value.

Suppose that we have somehow managed to write the weathering rate as a function $W(P, T, pCO_2)$, where P is the rate at which precipitation falls over land. Then, if Γ is the outgassing rate, measured in the same units as W, equilibrium is determined by $W(P, T, pCO_2) = \Gamma$. The rate Γ may change gradually over geological time, leading to long-term changes in climate. The problem is then closed if CO_2 is the dominant greenhouse gas controlling climate, since then T and P can be determined as functions of CO_2, given other pertinent data such as the solar (or stellar) constant at the planet's orbit and the configuration of the continents. Continental configuration can strongly affect the weathering rate, because the size and placement of continents affects how much of the global precipitation falls on land and sustains continental weathering. We are led to a condition $W(P(pCO_2), T(pCO_2), pCO_2) = \Gamma$, which can be solved for pCO_2. Once that is known, everything else is known, and one can then explore the dependence of the resulting climate on slowly varying parameters such as the stellar luminosity, the continental configuration, and the outgassing rate. This is the basic idea behind all CO_2 weathering feedback models, and in such models we seek to determine the extent to which the weathering feedback can control temperature and keep it in a habitable range.

Without even writing down a specific form of W, we can obtain a very important general result in a special case. Namely, if the weathering rate is not *explicitly* dependent on pCO_2, and if the precipitation depends only on temperature and not directly on the supply of absorbed stellar energy, then the equilibrium condition is simply $W(P(T), T)) = \Gamma$. This means that for any given outgassing rate and stellar luminosity, *the temperature must adjust to a fixed value.* In particular, this temperature does not change as the star gets brighter over time. In this case the weathering feedback acts to control climate perfectly, and the planet's temperature changes only as a result of either the outgassing rate or the continental configuration changing. What happens in this regime is that the requirement of weathering balance fixes T, and then pCO_2 must take on whatever value is necessary to achieve this T. For example, when the Sun is dimmer, pCO_2 must take on a higher value so as to make up for the reduced solar energy. Conversely, when the Sun is brighter, pCO_2 must take on a lower value. The thermostat can break down at the cold end if the required amount of CO_2 exceeds the supply. The temperature regulation may also break down at the cold end if the planet falls into a Snowball state. Conventional wisdom has it that silicate weathering ceases in this regime, allowing CO_2 to build up to where the planet thaws. However, sea-floor weathering will continue, and if it is effective enough it could prevent the planet from ever warming up sufficiently to escape the Snowball state. On the hot end, the thermostat can break down if the amount of CO_2 required to achieve temperature T falls all the way to zero. At this point, further increases in luminosity will cause the temperature to increase. The planet does not immediately go into a runaway greenhouse at this point, but if the planet has an ocean (as it must, if we are to have silicate weathering at all) then the water vapor feedback will ultimately lead to a runaway once the luminosity exceeds the runaway threshold.

Now let's return to the general case, including the direct effect of pCO_2 on weathering. Putting all the effects together, the weathering rate can be represented by the empirical expression

$$\frac{W}{W_0} = \left(\frac{P}{P_0}\right)^a \left(\frac{p}{p_0}\right)^b \exp \frac{T - T_0}{T_U} \tag{8.12}$$

where W is the weathering rate, P is the rate at which rain falls over rocks (the runoff), p is the partial pressure of CO_2 in the atmosphere, and T is the temperature of the surface at which weathering is taking place. W_0 is the weathering rate for the reference state with runoff P_0, carbon dioxide partial pressure p_0 and temperature T_0, and a, b and T_U are empirically determined constants. The last of these represents the direct temperature sensitivity of the Ebelmen–Urey reactions. Based on values that have appeared in the literature, we adopt $a = 0.65$, $b = 0.5$, and $T_U = 10\,\mathrm{K}$.

To complete the determination of the weathering rate, we need to determine T as a function of the absorbed stellar radiation and pCO_2. In the calculations to follow, we do this using the polynomial OLR fits described in Chapter 4. The OLR curve chosen is based on Earth gravity, with an assumed relative humidity of 50%. In determining the absorbed stellar radiation, we will assume an albedo of 0.2, which approximates Earth's, adjusted for the greenhouse effect of clouds. We also need to know how P depends on temperature. One reasonable choice would be to make it increase in proportion to Clausius–Clapeyron. However, calculations with comprehensive dynamic climate models tend to indicate that precipitation increases less rapidly than Clausius–Clapeyron, even before the limitation due to surface shortwave absorption is encountered. Thus, we'll take $P/P_0 = 1 + a_P \cdot (T - T_0)$, with $a_P = 0.03$. With these specifications we can solve $W/W_0 = \Gamma/\Gamma_0$ and determine how the pCO_2 and temperature change as the luminosity is changed. We will see how effectively the weathering thermostat can offset the Faint Young Sun, now that we have included the explicit pCO_2 dependence. According to the conventional wisdom, we are now studying the case of a planet for which the continents are abiotic.

The temperature and CO_2 as a function of the stellar constant are shown in Fig. 8.3. The reference temperature T_0 is taken to be near the Earth's present-day temperature, and the outgassing rate is held constant at the value that balances weathering at this temperature. These calculations do not include the effects of ice-albedo feedback, which properly should enter in the cold conditions encountered when the star is faint. For comparison, the figure also shows what the equilibrium temperature would be if there were no silicate weathering thermostat and the CO_2 remained fixed at its baseline value. The results show that the CO_2 rises to very high values at early times when the star is faint – nearly 100 mb, or about 10% of an Earthlike atmosphere. However, because the weathering increases so much at high pCO_2, the balance can be achieved with quite low temperatures. The temperature falls to 270 K, which is only slightly warmer than the 250 K value it would have in the absence of a weathering thermostat. It thus appears that, without some modification of the picture by land plants or some other way to get rid of the explicit pCO_2 dependence of weathering, the weathering thermostat does not go very far to resolve the Faint Young Sun Paradox. *With the abiotic pCO_2 dependence, the silicate weathering thermostat leaves the Early Earth in a very cold state, unless the outgassing rate is considerably higher at that time.*

One can also see the moderating effect of the weathering thermostat on the warm side, as the star gets brighter. When the star is 30% brighter than its baseline value, the pCO_2 has fallen all the way to 0.002 mb, or 2 ppmv, and the temperature has risen to 311 K. This compares to the temperature of 323 K that would occur in the absence of the weathering thermostat. At this point, there is hardly any CO_2 left in the atmosphere, and there is limited scope for the weathering thermostat to defend the planet against further increases in the luminosity of its star.

If it really is true that land biota affect the explicit pCO_2 dependence as much as conventional wisdom suggests, then the implication is that land biota play a crucial role in planetary climate regulation. With land biota suppressing the explicit pCO_2 dependence

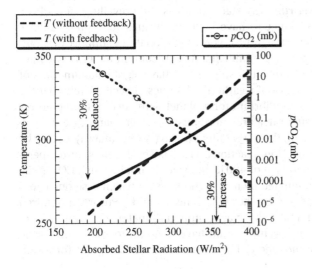

Figure 8.3 Effect of the abiotic form of the silicate weathering feedback on the variation of surface temperature with stellar luminosity. The luminosity in this graph is translated into absorbed stellar radiation per unit surface area of the planet, allowing for a 20% albedo. The pCO_2 curve gives the value of pCO_2 in equilibrium with a fixed outgassing rate for the case including the weathering thermostat.

of weathering, the climate regulation is nearly perfect. Without land biota, the silicate weathering thermostat moderates the influence of stellar luminosity on temperature change, but the regulation is not particularly tight, and on Earth one would have to invoke an increase in outgassing or some other effect in order to account for why the planet was not frozen during the Faint Young Sun – though, to be fair, the thermostat does bring the planet much closer to the temperature where a Snowball could be avoided. For Earth, the question of when the thermostat became very efficient is tied in with the question of whether the suppression of pCO_2 dependence requires land plants (which only entered the picture about 300 million years ago), or whether bacterial colonization of land would be sufficient. The latter would extend the portion of Earth history for which the thermostat is efficient. The overall scenario including the evolution of land biota would go something like this: Early in the planet's history, there are no land biota and the thermostat is only moderately effective. The Faint Sun period would be very cold, though perhaps not completely frozen. As the luminosity increases, the planet would get considerably warmer, perhaps to conditions much warmer than the present. Then land biota evolve, greatly increasing the efficiency of weathering and eliminating the explicit pCO_2 dependence. At this point the planetary temperature drops sharply, but the reduction of weathering in cold, dry conditions prevents a Snowball state. Thereafter, the weathering thermostat becomes very efficient, and further changes in temperature are suppressed, until the required pCO_2 drops so low that temperature regulation ceases.

Another potentially important issue is the temperature–precipitation relation assumed in the calculations displayed in Fig. 8.3. In Section 6.8 it was shown that for sufficiently high temperatures, precipitation becomes limited by the absorbed solar radiation. Basically, we found that it is hard for the latent heat flux to exceed the surface absorbed solar radiation, which puts a cap on the high-temperature precipitation. This limitation affects weathering, and also can make the weathering rate directly sensitive to things such as clouds or atmospheric shortwave absorption. The implications of this for climate regulation are largely unexplored. Some preliminary investigations are suggested in Problem 8.5.

In Chapter 4 we discussed the classic dry runaway greenhouse, in which the entire ocean evaporates into the atmosphere. However, if a planet fails to meet the criterion for a total dry runaway (perhaps owing to cloud or subsaturation effects), it may nevertheless get hot enough that it can lose a great deal of water vapor to space, despite retaining a liquid (but still hot) ocean. This is known as a *wet runaway* state, and it is the prevailing view of the path taken by Venus. Supposing this planet to have enough silicates to support silicate weathering, and supposing that there is enough reservoir of carbonate to support CO_2 outgassing from the interior, what do we expect the atmospheric CO_2 content to be? Do we have a steam-dominated atmosphere, or would the atmosphere also have significant amounts of CO_2 in it? Given the exponential dependence of weathering kinetics, it seems likely that the weathering reaction would be driven to equilibrium for a planet with surface temperatures much above 400 K, unless the outgassing rates are enormously greater than those prevailing on Earth today. Thus, one can estimate the CO_2 content during a wet runaway simply by looking at the equilibria for the Ebelmen–Urey reactions. The answer depends on temperature and surface mineralogy. At a temperature of 400 K there would not be much CO_2 in the atmosphere unless the surface was dominated by iron carbonates, in which case one could have around 10 bar of CO_2. By the time one reaches 600 K there could be nearly a hundred bar of CO_2 even if the surface is dominated by magnesium carbonates. Calcium carbonates are very stable, however, so the temperature would have to go up to 700 K before one had some tens of bars of CO_2 in the atmosphere. This is above the critical point for water, at which point the distinction between liquid water and water vapor disappears. This renders the distinction between a dry and wet runaway moot, and raises the interesting (and evidently unresolved) question of how the Ebelmen–Urey reactions behave in the presence of supercritical water. Is it like the dry reaction, or like the aqueous reaction, or is it something completely different?

On a planet with oxygenic photosynthesis, the atmospheric CO_2 can also be affected by the burial of organic carbon and the release of organic carbon by oxidation of the organic carbon pool – an example of oxidative weathering. Burial of organic carbon produced by oxygenic photosynthesis converts CO_2 (which is a greenhouse gas) into O_2 (which is not). This would cool the planet if it happened rapidly enough that the silicate weathering thermostat could not keep up. On a well-oxygenated planet like the current Earth, organic carbon burial is relatively inefficient, since bacteria have had a few billion years to get very good at extracting energy by oxidizing any organic carbon that may be around. During the Great Oxygenation Event near the dawn of the Proterozoic, however, it is conceivable that oxygenic photosynthesis took over the planet so quickly that huge amounts of organic carbon were buried, drawing down atmospheric CO_2 and precipitating the Makganyene Snowball event. It is hard to imagine circumstances where something like this could happen under heavily oxygenated conditions. Massive release of CO_2 by oxidative weathering, on the other hand, can occur under modern oxygenated conditions. It is likely that the CO_2 release during the PETM event 55 million years ago came from oxidative weathering of the land carbon pool, and what is the present era of anthropogenic CO_2 release other than a form of oxidative weathering due to a particularly exotic form of biology? One ought to worry about whether the resulting warming could trigger an additional PETM-like land carbon release, which would add to the direct anthropogenic CO_2 release and compound our climate woes.

On a planet such as the Early Earth with little or no O_2 in its atmosphere, the greenhouse effect of CO_2 could be considerably augmented or even largely supplanted by that of biologically produced CH_4. The silicate weathering thermostat can to a certain extent accomodate this situation by reducing CO_2 so that the temperature does not become too high. The consequences are explored in the Workbook, in Problem 8.6. For a methane-rich atmosphere, however, one also needs to keep in mind the possibility of cooling due to the formation of high-altitude organic haze clouds.

The CO_2 weathering feedback is but one of many possible climate feedbacks involving atmospheric reactions with crustal minerals, though to date it is the only one that has been worked out in detail. Other cycles that have been proposed include release of CH_4 from organic carbon by methanogenic bacteria on the young Earth, the regulation of SO_2 on Venus from dry reactions with surface carbonates, and the regulation of SO_2 on Early Mars and Early Earth by aqueous reactions producing sulfite minerals at the surface. On a planet without a substantial oceanic reservoir of water, the exchange of water with hydrated minerals could exert an important influence on atmospheric water vapor content; insofar as water affects the fluidity and melting point of minerals, this can even feed back on the plate tectonics that affects the recycling of crustal material. Nitrogen does not easily form minerals on rocky planets, and so is unlikely to participate in major climate cycles, though it has been suggested that the biological formation of the ammonium ion NH_4^+ allows some drawdown of N into the Earth's mantle; this would somewhat affect the climate via Rayleigh scattering, pressure-broadening, and lapse rate effects. On an icy body like Titan, however, N_2 readily forms "minerals" – they just happen to be called "ices" instead. Indeed, cryovolcanism based on various mixed NH_3–H_2O ices may well play a role in determining the amount of N_2 in Titan's atmosphere. The methane cycle on Titan is also likely to involve crustal and interior exchange processes. Photochemistry converts Titan's atmospheric methane to liquid ethane and various tarry sludges on the surface. At the time of writing the search is on for some way of closing the cycle by converting these substances back into methane.

8.4 PARTITIONING OF CONSTITUENTS BETWEEN ATMOSPHERE AND OCEAN

How does the presence of a liquid ocean affect the composition of a planet's atmosphere? In this section, we will consider the exchange of a substance such as the carbon in atmospheric CO_2 with the liquid ocean, neglecting any geochemical processes such as sea-floor weathering which could remove dissolved components from the liquid and put them in long-term storage as solids in the oceanic crust. The primary example we have in mind is exchange of CO_2 between atmospheres and a water ocean, but some of the lessons apply more generally, especially to other reactive or non-reactive gases dissolving in water. The same general principles would apply to solvents other than water as well, though the general nature of ocean chemistry for oceans made of liquids other than water is at present essentially unexplored. The importance of oceans as a reservoir resides in the fact that they have vastly more mass than the typical atmosphere (hence a lot of room to hold substances), but mix rapidly and can therefore exchange atmospheric constituents on relatively short time scales of a millennium or so. This contrasts with the much more sluggish mixing associated with the solid crust and mantle, which prevents the still-greater mass of the rest of the planet from exchanging substances except on much longer time scales.

Within the general scenario outlined in the preceding section, the oceanic storage does not affect the atmosphere when the system is in equilibrium. In equilibrium, the amount of atmospheric gas entering the ocean equals the amount leaving, since – by definition – the oceanic reservoir is not changing when the system is in equilibrium. For atmospheric CO_2 concentration in equilibrium, for example, the ocean is just a pass-through reservoir, and in the traditional picture the atmospheric CO_2 concentration can be determined by silicate weathering on land without reference to how much of the planet's net carbon pool resides in the ocean water. When the system is out of equilibrium, however, uptake and release of CO_2 by the ocean can have important consequences. For example, at the time of writing industrial humanity is releasing about 9 Gt of carbon per year into the atmosphere in the form of CO_2. How much of this stays in the atmosphere and adds to the greenhouse effect, and how much disappears into the ocean? To the extent that it disappears, how fast is the process and what is the rate-limiting step? Similarly, during the PETM event (see Section 1.9.1) an estimated 5000 Gt of carbon was released as CO_2 through oxidation of terrestrial organic carbon; if all of this CO_2 remained in the atmosphere, the resulting warming would be much greater than if a substantial portion disappeared into the ocean. The partitioning of carbon between atmosphere and ocean has similar consequences for deglaciation of Snowball Earth. In 10 million years, about 300 000 Gt of carbon in the form of CO_2 can be outgassed from the Earth's interior. As in the PETM case, the amount of warming resulting depends on the fraction which remains in the atmosphere. The Pleistocene glacial–interglacial CO_2 cycles present a more complex form of transient exchange of CO_2 with the ocean, involving both uptake and release, probably assisted by the export of photosynthetically produced organic carbon from the upper ocean to the deep ocean.

As a warm-up to the full problem, let's first consider the simple case of a gas which does not undergo any chemical reactions after it dissolves in the ocean. The case of N_2 dissolving in a water ocean conforms well to this idealization. The solubility of a gas in a liquid is governed by *Henry's law*. For a gas A, with partial pressure p_A, Henry's law states that in equilibrium

$$p_A = K_H(T)c \qquad (8.13)$$

where c is the concentration of substance A in the liquid, T is the temperature of the liquid, and K_H is the Henry's law constant, which has been measured and tabulated for a wide variety of substances. A small value of the Henry's law constant implies a higher solubility of the gas in liquid, since it takes less partial pressure to force a given concentration into solution. The units of K_H depend on the way the atmospheric content of A is represented and the way the concentration is represented; we will measure the gaseous partial pressure in Pa, and the concentration as mole fraction (number of molecules of A divided by number of molecules of solvent). At 298 K, for example, the Henry's law constant for N_2 dissolving in water is $9 \cdot 10^9$ Pa using these units, since mole fraction is dimensionless. That means that our present atmospheric N_2 pressure of $8 \cdot 10^4$ Pa would lead to a molar concentration of N_2 of $8.9 \cdot 10^{-6}$, or 8.9 ppmv, within any body of 298 K water that has been in contact with the atmosphere long enough to come into equilibrium.

The temperature dependence of the Henry's law constant follows an Arrhenius law, similar to chemical equilibrium constants or Clausius–Clapeyron. Thus,

$$K_H(T) = K_H(T_0)\exp\left(-C_H \cdot \left(\frac{1}{T} - \frac{1}{T_0}\right)\right). \qquad (8.14)$$

The coefficient C_H is invariably positive. For N_2 in water, $C_H = 1300$ K. Note that since K_H *increases* with temperature, the concentration in solution goes *down* as temperature increases, with partial pressure held fixed. In other words, the solubility of a gas goes down with temperature. This is somewhat non-intuitive, and is just the opposite behavior one has come to expect from the more familiar experiment of things like sugar dissolving in hot tea vs. cold lemonade. For a given atmospheric partial pressure, cold water holds more gas than hot water. This is why trout like cold streams – they are more oxygen-rich than warm waters.

Henry's law is another one of those magically universal thermodynamic relations, which applies across a vast range of circumstances. It applies as well for dilute solutions of N_2 in a liquid CH_4 ocean, or CO_2 dissolving in molten silicate, as it does for atmospheric gases dissolving in a conventional water ocean.

As an example, let's consider the partitioning of N_2 between Earth's present atmosphere and its ocean. The ocean contains $7.8 \cdot 10^{19}$ Mole of water, so at 298 K the ocean would contain $6.9 \cdot 10^{14}$ Mole of N_2 using the equilibrium concentration calculated above. Now, using the surface pressure, the hydrostatic law, the surface area of the Earth and a mean molecular weight of 29 for air, we find that the atmosphere contains $1.8 \cdot 10^{17}$ Mole, of which about 80%, or $1.4 \cdot 10^{17}$ Mole are N_2. Thus, the oceans at 298 K would contain only 0.005 of the amount of N_2 in the atmosphere – a trivial proportion. This is an underestimate, since the oceans are on average closer to 275 K than 298 K, but even allowing for the temperature effect, the presence of the oceans has little effect on the amount of N_2 in the atmosphere, as shown in the following exercise.

Exercise 8.2 Using the Henry's law temperature scaling coefficient for N_2, compute the proportion of N_2 in the ocean assuming a mean ocean temperature of 275 K.

Henry's law data for a selection of gases is given in Table 8.1. The main distinction is between polar molecules like NH_3, SO_2, and H_2S, whose undisturbed state has a dipole moment, and non-polar molecules like the rest in the table. Non-polar gases are not very

	N_2	O_2	CO_2	CH_4	H_2	Ar	NH_3	SO_2	H_2S
K_H (Pa)	$9.1 \cdot 10^9$	$4.3 \cdot 10^9$	$0.16 \cdot 10^9$	$4.0 \cdot 10^9$	$7.1 \cdot 10^9$	$4.0 \cdot 10^9$	$9.1 \cdot 10^4$	$4.0 \cdot 10^6$	$6.39 \cdot 10^7$
C_H (K)	1300	1500	2400	1600	500	1500	4200	2900	2100

Values of K_H are given at 289 K.

Table 8.1 Henry's law constants for selected gases dissolving in water.

soluble, and have solubilities similar to N_2. Carbon dioxide is significantly more soluble than the rest, but not orders of magnitude so; we shall see shortly that there are other factors that have a far more profound impact on CO_2 uptake by oceans. The polar molecules are incredibly soluble, on the other hand, so on a planet with an ocean these substances would exist dominantly in aqueous solution, in raindrops, lakes, and the ocean.

Carbon dioxide behaves very differently from N_2, because once it dissolves in water it reacts to form carbonic acid (H_2CO_3) which dissociates to form bicarbonate ion (HCO_3^-) and a proton. The bicarbonate further dissociates into carbonate ion (CO_3^{--}) and another proton. The reactions in question are

$$CO_2(aq) + H_2O \rightleftharpoons HCO_3^- + H^+$$
$$HCO_3^- \rightleftharpoons CO_3^{--} + H^+ \tag{8.15}$$

whose equilibria are described by the equations

$$\frac{[HCO_3^-][H^+]}{[CO_2(aq)]} = K_1(T)$$
$$\frac{[CO_3^{--}][H^+]}{HCO_3^-} = K_2(T). \tag{8.16}$$

Water is essentially inexhaustible as a reactant, so the activity of water is absorbed into the definition of K_1. The activities indicated in square brackets differ somewhat from concentrations, but we shall ignore that refinement and take the activities to be simply the concentrations of the respective substances. Most conventionally, the concentrations are written in units of moles per liter, which for water is almost the same as moles per kilogram; the equilibrium constants would be stated in corresponding units. Here, we will instead measure concentrations as mole fraction, which is the ratio of the number of molecules of solute to the number of molecules of water in a given sample.

Like all equilibrium constants, K_1 and K_2 are temperature-dependent, and follow an Arrhenius law; both constants increase with temperature. Values of the equilibrium constants plus their temperature dependence coefficients are given in Table 8.2. The constants also increase somewhat with increasing pressure. Sea water on Earth contains a great many ions from dissolved salts, measured in the aggregate as "salinity." These ions have a very strong effect on the activity of carbonate and bicarbonate ions (particularly the former). In the table and in subsequent calculations, we avoid dealing explicitly with activity coefficients by absorbing them into effective equilibrium constants, which depend on salinity. With these effective equilibrium constants, calculations can be done as if the activities in the equilibrium relations were simply concentrations. It should be recognized, however, that the equilibrium constants are quite strongly affected by the salinity. For example, the effective value of K_2 for fresh water is an order of magnitude smaller than the value for typical sea

	K_1	c_1	K_2	c_2	K_{sp}
35‰, 2 m	$1.83 \cdot 10^{-8}$	508 K	$1.38 \cdot 10^{-11}$	1910 K	$1.54 \cdot 10^{-10}$
35‰, 3 km	$2.28 \cdot 10^{-8}$	1218 K	$1.61 \cdot 10^{-11}$	2407 K	$2.56 \cdot 10^{-10}$

The equilibrium constants are given at 298 K, assuming that concentrations are stated as mole fraction. The temperature dependence constants c_j are in K; the temperature dependence is of the form $\exp -c_j(1/T - 1/298)$. Temperature constants are not given for the solubility product constant K_{sp} because this quantity is not significantly dependent on temperature in the range 0–30 °C. Rows of the table give results at the stated depths in an ocean subject to Earth gravity. Values are given for sea water with a salinity of 35‰.

Table 8.2 Equilibrium constants and their temperature scaling parameters for carbonate–bicarbonate equilibrium reactions and calcium carbonate dissolution.

water, which renders carbonate deposition in freshwater environments very different from what goes on in our ocean. The equilibrium constants might conceivably be affected even by the composition of sea water, which is something worth keeping in mind when thinking about oceans of the distant past, such as the very sulfate-rich anoxic oceans that could have prevailed at times during the Proterozoic and Archean.

Henry's law gives the CO_2 concentration in terms of of the atmospheric partial pressure pCO_2, and pH gives the concentration $[H^+]$. Equation (8.16) then states that for any given pH, the bicarbonate concentration is proportional to the dissolved CO_2 concentration, and in turn the carbonate concentration is proportional to the bicarbonate concentration; by Henry's law, the dissolved CO_2 concentration is itself proportional to pCO_2, so in the end the concentrations of all carbon-bearing species are proportional to pCO_2, given a fixed pH. The partitioning amongst species is profoundly affected by the pH, though. The constant K_1 is a very small number so the concentration of bicarbonate will be negligible compared with dissolved CO_2 unless the concentration of $[H^+]$ is sufficiently small. For example, under fairly acidic conditions, with pH= 5, the proportionality constant $K_1/[H^+]$ is only 0.06 at 273 K. Since K_2 is even smaller than K_1, under these circumstances the concentration of carbonate is still smaller than that of bicarbonate. Thus, under sufficiently acidic conditions, essentially all of the dissolved inorganic carbon in the ocean is in the form of dissolved gas. As the pH rises, however, the concentration of bicarbonate rises. At pH= 6.2 the bicarbonate concentration equals that of the dissolved gas, though the carbonate concentration is still a factor of 0.0006 smaller than that of bicarbonate. At pH= 8.2, which is the value prevailing in the present ocean, bicarbonate concentration is 100 times that of the dissolved gas, and carbonate concentration rises to 5.6% that of bicarbonate; in this case, most of the ocean inorganic carbon storage is in the form of bicarbonate. At still greater pH, carbonate comes to dominate the ocean carbon storage. A small change in the ocean pH makes a big change in the carbon storage in the ocean because the hydrogen ion concentration is exponential in pH. The situation is summarized in the left panel of Fig. 8.4.

Because CO_2 as a gas is not very soluble in water, one cannot store much carbon in the ocean when the pH is neutral or acidic. As pH increases, more of the dissolved CO_2 gets converted to bicarbonate and carbonate, and so the proportion of carbon in the ocean rather than the atmosphere rises dramatically. If there are N_0 Moles of water in the ocean, then the number of Moles of carbon in the ocean is

$$pCO_2 \frac{N_0}{K_H}(1 + K_1/[H^+] + K_1 K_2/[H^+]^2) \qquad (8.17)$$

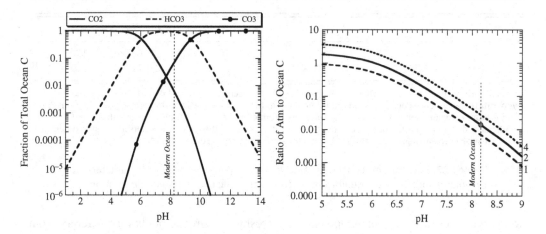

Figure 8.4 Left panel: The partitioning of total ocean dissolved inorganic carbon into dissolved CO_2, bicarbonate, and carbonate, as a function of pH. Right panel: Ratio of atmospheric carbon to ocean inorganic carbon as a function of pH. The numbers to the right of the curves give the values of f_0, the atmosphere vs. ocean size parameter defined in the text. Smaller values correspond to a larger ocean, all other things being equal. For Earth's ocean $f_0 \approx 1.87$. All calculations were done with an ocean temperature of 275 K, and for near-surface pressures.

Now, if A is the surface area of the planet and g is its surface gravity, then the hydrostatic relation tells us that the number of Moles of carbon in the atmosphere is $pCO_2 A / \overline{M} g$, where $\overline{M}$ is the mean molecular weight of the atmosphere. Hence the atmospheric fraction, or ratio of atmosphere to ocean carbon, is

$$f_{atm} = \frac{K_H A}{\overline{M} g N_0} \frac{1}{1 + K_1/[H^+] + K_1 K_2/[H^+]^2}. \tag{8.18}$$

The non-dimensional coefficient at the front of this expression, which we shall call f_0, is a measure of the "size" of the atmosphere relative to the size of the ocean. It depends on pCO_2 only through the mean molecular weight of the atmosphere. When pCO_2 is small compared with the pressure of the rest of the atmosphere, $\overline{M}$ is fixed at the non-CO_2 mean value (29 in the case of present Earth air). As pCO_2 becomes large, $\overline{M}$ approaches 44, the value of a pure CO_2 atmosphere. The coefficient f_0 is the limiting atmospheric fraction in the case when bicarbonate and carbonate concentrations are negligible. For the Earth, $f_0 = 2.03$ when CO_2 concentration is low, decreasing to 1.34 in the limit of a pure CO_2 atmosphere. An acidified ocean degasses CO_2. It fizzes, just like when you pour vinegar on baking soda (sodium bicarbonate), and for much the same reasons. As pH increases, the atmospheric fraction decreases sharply, as shown in the right panel of Fig. 8.4. In modern conditions on Earth, the atmosphere contains only about 2% of the total atmosphere/ocean carbon inventory. Adding carbon to the inventory generally changes the pH, so we are not yet fully equipped to say how much of the *added* carbon stays in the atmosphere.

To complete the solution to the problem we need to determine the pH of the ocean. This is the tricky part, since the answer ultimately depends on the supply of carbonate ion to the ocean from dissolution of carbonate on land and the sea floor. In order to determine the pH, we need to satisfy *charge balance* – the constraint that the net charge of positive and negative ions sums to zero. We will consider the simple case in which the only additional supply of carbonate comes from dissolution of $CaCO_3$ (limestone), which supplies both positive Ca^{++} ions and negative carbonate ions. In this case, the charge balance can be written

$$2[Ca^{++}] = [HCO_3^-] + 2\,[CO_3^{--}] + [OH^-] - [H^+] \qquad (8.19)$$

assuming the activities written as square brackets are concentrations of the respective species. If pH is specified, then $[OH^-]$ can be computed from the water dissociation equilibrium relation in Eq. (8.8) (taking care to convert to the mole fraction units we shall use here). Further, if pCO_2 is given, the pH and carbonate/bicarbonate equilibrium relations determine $[HCO_3^-]$ and $[CO_3^{--}]$, whence the right hand side is known as a function of pH and pCO_2. This must balance the net charge of calcium ions. For any given $[Ca^{++}]$ and pCO_2, one can then iterate on pH using Newton's method until the charge balance is satisfied. Note that if there were no external source of carbonate from dissolution of limestone, then $[Ca^{++}] = 0$ and sea water in equilibrium with 300 ppmv of atmospheric CO_2 would be mildly acidic, with a pH of 5.3 (See Problem 8.12); inorganic carbon in the ocean would be overwhelmingly in the form of dissolved CO_2, and hence the ocean would store relatively little carbon. The actual pH of the present ocean is 8.17, and this alkaline state is maintained by the supply of dissolved carbonate and accompanying calcium ions, which wash in from the land and dissolve from the sea floor. This supply of carbonate is thus crucial to the ocean's ability to store carbon; without it, the ocean would be acidic and the bulk of dissolved inorganic carbon in the ocean would take the form of CO_2, whence a much greater fraction of the total atmosphere/ocean inventory would reside in the atmosphere.

Suppose next that we start with a situation in which charge balance is satisfied and then change pCO_2. This changes the value of the right hand side of Eq. (8.19), and disrupts the charge balance. In order to re-establish balance, the pH must change; how much it must change depends on what happens to $[Ca^{++}]$. Let's first consider the case in which the atmospheric CO_2 increases so rapidly that there is no time for new carbonate to be added to the ocean by dissolution on land or on the sea floor. This approximates the present situation, in which CO_2 has increased very rapidly owing to unrestrained burning of fossil fuels. In that case, $[Ca^{++}]$ remains fixed. To put some numbers to it, suppose that burning of fossil fuels were to add one trillion tonnes of C in the form of CO_2 to the atmosphere–ocean system. (This is a rather optimistic assessment of where we might hold the line with aggressive carbon emissions controls.) That amounts to just under 2 kg, or 0.17 Moles, of carbon per square meter of Earth's surface. It would increase the CO_2 concentration by 454 ppmv if it were to all stay in the atmosphere, increasing the partial pressure pCO_2 by 45.4 Pa. Suppose, though, that the added CO_2 re-equilibrates with the ocean, without benefit of input of additional limestone from dissolution. To solve this problem, we begin by computing the $[Ca^{++}]$ needed to maintain charge balance in an unperturbed ocean in equilibrium with 280 ppmv (28 Pa) atmospheric pCO_2, using the observed ocean pH as a constraint. Holding $[Ca^{++}]$ fixed, we then compute the total mass of carbon in the ocean–atmosphere system as a function of atmospheric pCO_2, imposing all the carbonate/bicarbonate equilibrium constraints in the ocean.

Carrying out the calculation, we find that the unperturbed pre-industrial ocean-atmosphere system contains 46.64 trillion tonnes of carbon, almost entirely in the ocean. This increases by 1 trillion tonnes when the atmospheric pCO_2 is increased to 38.4 Pa, or 384 ppmv. Thus, when the equilibrium is achieved, only 104 ppmv of the added carbon, or 23%, remains in the atmosphere. At this point, the pH of the ocean has decreased more than a full unit, to 7.08. The resulting acidification of the ocean can have a severe impact on the vast array of microscopic and macroscopic ocean life that uses carbonate to make shells. Details of the calculation, and further explorations, are carried out in Problem 8.14.

Naively, one might have thought that since the ocean initially contains 98% of the carbon in the ocean–atmosphere system, the ocean would take up all but 2% of the carbon added to the atmosphere. This doesn't happen because the additional CO_2 depletes the supply of carbonate ion, allowing the pH of the ocean to go down. Carbonate ion is thus the bottleneck; it could almost be said to be "anti-CO_2." If one were to add a further trillion tonnes of carbon to the atmosphere, the ocean would acidify further, and a greater proportion of the second trillion would stay in the atmosphere, as compared with the first trillion (see Problem 8.14).

Although nearly 80% of the first trillion tonnes of carbon we add to the atmosphere will eventually work its way into the ocean, transforming the environmental problem from one of global warming to one of ocean acidification, it will take 500 years or more for this equilibrium to be achieved. The time scale is set by the time required to mix dissolved carbon species into the deep ocean, since one needs the full mass of the ocean in order to deal with such a large amount of carbon. The deep ocean time scale becomes important because the exhaustion of carbonate ion keeps the upper ocean alone from being able to take up much additional carbon (see Problem 8.13). The important role of the carbonate ion in limiting ocean carbon uptake, and its consequences for the magnitude of the global warming problem, was first recognized by Bolin and Ericsson. This breakthrough is commonly misattributed to Revelle and Suess (see Chapter 1), who in fact completely misinterpreted the effect and thought the ocean could handily take care of any likely amount of carbon humankind could throw at it. It will take about another 10 000 years for carbonate ion to be resupplied by dissolution, allowing the pH to rise gradually and the ocean to take up additional carbon over that span of time.

Finally, we'll consider the case in which the concentration of dissolved limestone is held in equilibrium with a reservoir of solid $CaCO_3$, rather like water vapor in equilibrium with ice at the saturated vapor pressure. This is approximately the state of affairs in the ocean on sufficiently long time scales. Dissolved $CaCO_3$ is held in a state of saturation by precipitation of solid or dissolution of solid, the latter of which might actually occur during rainfall over land, with the dissolved material washed into the oceans by rivers. Dissolution of calcium carbonate is described by the reaction and corresponding equilibrium constant

$$CaCO_3(s) \rightleftharpoons Ca^{++} + CO_3^{--}$$
$$K_{sp} = [Ca^{++}][CO_3^{--}]$$

(8.20)

where K_{sp} is known as the *solubility product constant*. The activity of $CaCO_3$ does not appear in the equilibrium expression since, by convention, the activity of a pure phase is set to unity. The value of K_{sp} increases with pressure, but it is only very weakly dependent on temperature. Some representative values are given in Table 8.2. If there were no source of carbonate or calcium ion other than dissolution of limestone, then $[Ca^{++}] = [CO_3^{--}] \equiv x$, whence the equilibrium expression implies $x = \sqrt{K_{sp}}$. This is the concentration of dissolved limestone, once the water has come into equilibrium with the solid phase. Higher K_{sp} implies greater solubility.

In the real ocean, CO_2 also is a source of carbonate ion. As CO_2 goes up, the ocean acidifies and the concentration $[CO_3^{--}]$ goes down. However, this reduces the product $[Ca^{++}][CO_3^{--}]$, which means that if the unperturbed system was saturated, the new system is unsaturated, and so more $CaCO_3$ will dissolve to bring the system back into saturation.

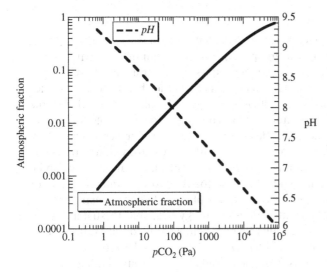

Figure 8.5 Atmospheric fraction and pH for a system buffered by dissolution of $CaCO_3$, which is presumed to be in a state of saturation. Calculations were carried out for Earthlike conditions, assuming an atmosphere whose non-CO_2 component has a molecular weight of 29. The number of Moles of non-CO_2 air in the system is held fixed as CO_2 is changed, at a value corresponding to 1 bar surface pressure in the absence of CO_2 (note that this pressure differs from the partial pressure of air in the mixture when pCO_2 is large).

This will reduce the acidity of the ocean, *buffering* the pH. Limestone is nature's antacid, which is why it is the main component of antacid tablets used to combat overactive stomach acid. The buffering of pH allows the ocean to hold more carbon than it otherwise would, once sufficient time has passed for enough limestone to dissolve that a state of saturation is restored. Figure 8.5 shows how the atmospheric fraction and ocean pH vary as a function of pCO_2, taking into account the buffering effect of limestone. We see that, despite the buffering effect, the atmospheric fraction rises with increasing pCO_2. For example, under pre-industrial equilibrium conditions with $pCO_2 \approx 28\,Pa$, the atmosphere contains only on the order of a percent of the total carbon in the atmosphere-ocean system. However, if pCO_2 rises to $10^4\,Pa$, which is on the order of the level needed to deglaciate a Snowball Earth, then the atmosphere contains about a third of the amount of carbon in the ocean, or a quarter of the total carbon. Were it not for this rise in atmospheric fraction, the amount of CO_2 that would need to be pumped into the system to attain an atmospheric pCO_2 of $10^4\,Pa$ would be so huge as to definitively prevent deglaciation of the Snowball.

Other gases that form acids upon dissolving in water can have effects analogous to those we have been discussing in connection with CO_2, though none of these have been studied to nearly the same extent as the CO_2–bicarbonate–carbonate case. Sulfur dioxide is a case of notable interest, because it is a greenhouse gas and is also a significant component of volcanic outgassing. Upon dissolving in water, SO_2 establishes an equilibrium with HSO_3^- (bisulfite) and SO_3^{2-} (sulfite), and forms a moderately strong acid. Compared with CO_2, SO_2 gas is extremely soluble in water, so oceans can take up a vast quantity of this substance even without conversion into bisulfite; the conversion has the potential to increase the uptake even further. Dissolution of sulfite minerals would play a role similar to that played by limestone in the carbonate system. If carbonate is also present in the system, it interacts in an interesting way with the sulfite system, since the dissolved SO_2 acidifies the ocean, prevents carbonate from precipitating and allows a greater proportion of the CO_2 to stay in the atmosphere. It has been suggested that the sulfite system, together with accompanying analogs of silicate weathering, could have played a role in the climate of Early Mars, and perhaps even of Early Earth. The study of climate-relevant geochemical cycles beyond the CO_2 system is in its infancy, and offers ample opportunities for the creative mind.

8.5 ABOUT ULTRAVIOLET

Extreme ultraviolet (*EUV*) consists of electromagnetic radiation with wavelengths between 0.12 μm and 0.01 μm. EUV makes up only a tiny part of the spectrum of stars with photospheric temperatures under 10 000 K – Main Sequence stars of spectral class B, A, F, G, K, and M. (Recall that our Sun is a class G star.) It nonetheless fuels the chemistry and physics of the outer atmosphere. EUV photons have sufficient energy to break up otherwise stable atmospheric compounds, allowing their components to combine into less stable forms that would not otherwise exist in appreciable quantities in the atmosphere. Further, because EUV photons are energetic enough to penetrate and interact with the electron clouds of atoms and molecules, the absorption cross-section is so high that significant heating rates can be sustained despite the low EUV flux. This is not the case for the more abundant visible or near-ultraviolet photons, to which the tenuous outer atmosphere is largely transparent. For this reason, EUV absorption is an important source of energy available to sustain atmospheric escape. Because of the nearly ubiquitous role of EUV in what is to follow, it is useful to pause at this point and provide some general background on EUV radiation.

Extreme ultraviolet is not produced by blackbody radiation in a star's photosphere. Instead, it is produced high in the star's thin outer atmosphere – its *corona* – where temperatures are brought to extremely high values by a variety of arcane and poorly understood heating mechanisms. The production of EUV in the hot corona has three important consequences. First, it allows the EUV energy flux to be far in excess of the blackbody value corresponding to the photospheric temperature. For example, at the Earth's orbit the photospheric blackbody flux from the Sun in the wavelength range from 0.055 μm to 0.060 μm would be a mere $7 \cdot 10^{-11}$ W/m^2, whereas the observed solar EUV flux in this range is $9 \cdot 10^{-5}$ W/m^2. The second important consequence is that the EUV fluctuates considerably in response to the solar activity cycle – loosely speaking, the sunspot cycle – because activity is intimately connected with the cycle of a star's magnetic field. The magnetic environment, in turn has a great effect on the corona. Thus, while the total solar luminosity varies very little over the course of the 11 year solar cycle, the net EUV flux varies by a factor of two or more – typically between 0.003 W/m^2 and 0.007 W/m^2 at Earth's orbit. In some sub-bands of EUV the variation is even more pronounced. The third consequence is that cool stars such as M-dwarfs can nonetheless have high sporadic EUV output, because the processes determining the coronal temperature are quite distinct from those controlling the photospheric temperature. Some M-dwarfs in particular have strong activity cycles, which give rise to the kind of events that effectively produce EUV. The strong and sporadic EUV output of such stars should give rise to novel aspects of escape and chemistry of the atmospheres of planets orbiting these stars. Not all M-dwarfs are active stars in this sense, but it is believed that all M-dwarfs undergo an active stage when they are young.

One should also bear in mind that the Faint Young Sun could have had EUV output out of proportion to the dimness of the rest of the spectrum. In fact, while total stellar energy output increases with time, it is generally agreed that EUV output is higher for young Main Sequence stars. Solar EUV output is often assumed to be 3–6 times greater than at present in the first two billion years of the Sun's life as a star, though neither stellar models nor observations of other stars offer the possibility of precise estimates at present. This period of enhanced EUV output is distinct from the much shorter *T-Tauri* stage which stars undergo before fusion ignites and they enter the Main Sequence. For solar mass stars, the T-Tauri stage only lasts on the order of 100 million years; it would be shorter for more massive, brighter stars and longer for less massive, dimmer stars. Observations of T-Tauri stage stars indicate that the Sun's EUV output during that stage could have been as much as 10 000 times greater than its present output, which would powerfully erode the atmospheres of any planets that had already formed by that time, provided that the remnants of the protoplanetary nebula were not so thick as to provide shielding.

EUV is so strongly absorbed by the outer atmosphere that it can only be observed from outer space. The EUV output of stars other than our Sun is of extreme importance, but it was long thought that EUV astronomy was a dead end, on account of absorption by interstellar hydrogen. It turns out, however, that there are sufficient fluctuations in the interstellar hydrogen density to permit a great deal of useful information to be obtained about stellar EUV output. Results from this emerging science will have a great bearing on the evolution of atmospheres of extrasolar planets. Even where EUV from other stars cannot be directly observed, correlations between EUV output and X-ray or radio spectrum output can be used to draw useful inferences.

Middle UV (0.3 μm to 0.2 μm) and far UV (0.2 μm to 0.12 μm) are both important players in the photochemistry of the lower parts of the Earth's atmosphere. The output of the Sun in this part of the spectrum also deviates from the photospheric blackbody spectrum, though less markedly than is the case for EUV. In this case, the output is actually somewhat *below* the blackbody value, by a factor of two or so.

8.6 A FEW WORDS ABOUT ATMOSPHERIC CHEMISTRY

Atmospheric chemistry influences climate evolution through determining the extent to which certain greenhouse gases can accumulate in the atmosphere. If there is a source of a gas such as CH_4 either from biological production or volcanic outgassing, the atmospheric concentration must be determined by a balance between source rate and chemical destruction rate. In addition, atmospheric chemistry plays a crucial role in the runaway greenhouse scenario, since permanent water loss requires water to be broken up into its constituent parts, and the hydrogen to escape before it can recombine into water or other substances too heavy to escape.

It would, of course, be preposterous to pretend that anything like an adequate treatment of atmospheric chemistry could be provided within the compass of a single section of a single chapter. This is a subject that, like atmosphere–ocean dynamics, requires a volume of its own. Indeed, many tomes have been published on chemistry of planetary atmospheres, and many more pertinent specifically to present and past atmospheres of Earth, without coming close to exhausting this rich and important subject. Still, as a problem in chemistry, atmospheric chemistry has certain peculiarities of its own, and the few words we provide here may help the orient the reader toward further study of the subject. This brief section is also intended to highlight where chemical reactions internal to the atmosphere fit into the bigger planetary climate picture portrayed in this book.

8.6.1 Photodissociation stirs the pot

All atmospheric chemistry is photochemistry. Except perhaps for hot, exotic atmospheres such as found in close-orbit "roasters," the thermal energy in planetary atmospheres is not sufficient to break the bonds of the relatively small molecules which typically make up most of an atmosphere. This can only be done by energetic photons, primarily in the UV and EUV range. This hailstorm of energetic photons shatters simple molecules into reactive components which can recombine into compounds that would not otherwise be present, though these compounds eventually degrade and are resupplied by the reactions amongst newly dissociated molecules. The *photodissociation* of atmospheric molecules due to absorption of energetic photons maintains chemical disequilibrium within the atmosphere. The role of photodissociation is worth keeping in mind when thinking about atmospheric chemistry on planets orbiting stars different from the Sun, since the differences in flux of energetic photons will have a profound effect on the character of the disequilibrium. M-dwarfs present a particularly interesting case; being cooler, they should generally have low UV and EUV output, but they also tend to be very active stars, at least episodically and especially when young, in which case they can actually have greater UV and EUV output in proportion to their brightness than the Sun.

Photodissociation is not the only source of disequilibrium, but it is the only one that happens within the atmosphere itself, and other sources of disequilibrium almost invariably act more sluggishly. Life is a major source of disequilibrium, having a remarkable ability to use enzymes to break bonds and synthesize new compounds at temperatures where this would ordinarily be impossible. Oxygenic photosynthesis is one example, as is the much older ability of nitrogen-fixing bacteria to synthesize nitrate from atmospheric N_2, a molecule so stable that it is hardly even subject to photolysis on Earth. There is a lesser and more sluggish source of disequilibrium due to outgassing of compounds released in the hot interior of the Earth – for example H_2, which methanogens can use as a source of energy by combining it with CO_2.

Photodissociation occurs when a sufficiently energetic photon is absorbed by a molecule, leading to vibrations strong enough to break one of the bonds holding the molecule together. It is typically a one-way reaction, occurring in circumstances in which the chance that the two pieces recombine and emit a photon can be neglected; hence the reaction is usually written with a single rather than a double arrow. For example, the photodissociation of NH_3 is written as

$$NH_3 + h\nu \rightarrow NH_2 + H. \tag{8.21}$$

Writing the photon in the form $h\nu$ is simply a matter of convention, which serves to remind us that it is the energy added to the molecule by the photon that causes the dissociation. A certain minimum amount of energy, known as the *dissociation energy*, is needed to break a bond, and a photon has to have at least this much energy in order to have a chance to cause dissociation. The required energy is related to the frequency (and hence the wavelength) of the photon by the relation $E = h\nu$. Thus, one can define a threshold wavelength above which absorption cannot cause dissociation. Threshold wavelengths for a selection of molecular dissociations are given in Table 8.3. With the exception of ozone (O_3), dissociation can only occur within the far UV and EUV spectrum. At very high energies, absorption of a photon causes *ionization* (liberation of an electron) instead of dissociation. Such photons are extremely rare, but they are responsible for the formation of a planet's *ionosphere*.

The photodissociation threshold does not tell the whole story regarding dissociation rates, because the probability of absorption can be low. This probability is described by the absorption cross-section χ, which has units of area, and can be thought of as the area of the target a molecule provides when exposed to a beam of photons. Following common practice in photochemistry, we will measure cross-sections in per-molecule units. The cross-section is a function of the frequency of the photon; often it is a rather smooth function, but in some cases it can exhibit well-defined lines such as we have extensively discussed in the

N_2	O_2	H_2	CO	CO_2	O_3	NH_3	H_2O	SO_2	CH_4
0.1265	0.2400	0.2743	0.1113	0.2247	1.1222	0.2637	0.2398	0.2168	0.2722

Table 8.3 Photodissociation threshold energies, expressed as the wavelength a photon must have in order for its energy to be equal to the energy of the bond being broken. Wavelengths are in μm. Photodissociation breaks off an O for CO_2, O_3, and SO_2, and an H in the remainder of the non-diatomic cases.

case of infrared absorption. Suppose that $J_\nu \Delta \nu$ is the flux of photons with frequency near ν, within a small band of width $\Delta \nu$. The photon flux is determined by dividing the energy flux in the band by the photon energy, $h\nu$. The dissociation rate due to photons in this band is then $\chi_\nu J_\nu \Delta \nu$. This expression can in fact be taken as the definition of the cross-section. In SI units, the expression has units of $1/s$. The number of molecules in a thin layer of gas exposed to photon flux in the band would decay exponentially at this rate. The layer must be assumed thin enough that consumption of photons by absorption does not appreciably deplete the flux J_ν as the beam traverses the layer.

The wavelength dependence of the cross-section typically provides a more severe limit on the net rate of dissociation than does the dissociation threshold. For example, H_2 does not significantly dissociate until the wavelength to which it is exposed falls below $0.1 \, \mu m$, despite the fact that much longer wavelengths have sufficient energy. This is because H_2 does not have internal quantum states that permit absorption of the longer wavelength photons. Methane dissociation is also severely limited by a paucity of suitable quantum states, and requires wavelengths of $0.15 \, \mu m$ or less in order to dissociate. In contrast, NH_3 begins to have a significant cross-section already at $0.23 \, \mu m$, H_2O at $0.2 \, \mu m$ and CO_2 at $0.2 \, \mu m$, all of which are much closer to the energy threshold. At altitudes where the supply of suitable photons has not been too depleted, dissociation rates (also called *photolysis* rates) are typically very fast, corresponding to decay times on the order of minutes to days. For example, the cross-section of CH_4 is about $10^{-21} \, m^2$ for wavelengths below $0.12 \, \mu m$, and at the Earth's orbit the photon flux in that wavelength range is about $5 \cdot 10^{14}/m^2 s$ averaged over a spherical surface. This yields a photolysis rate of $5 \cdot 10^{-7}/s$, or a decay time of about 23 days. Molecules which have significant cross-sections at longer wavelengths have much shorter photochemical lifetimes, because the supply of photons increases steeply at longer wavelengths.

The rapid decay rates give a misleading impression of how rapidly photolysis can transform an atmosphere. Except for very thin atmospheres, the net rate of photolysis in an atmosphere is not limited by the cross-section, but by the supply of photons. All dissociating molecules must compete for the same photons, so the supply of photons at any given altitude depends on what has been used up by all dissociating molecules of all types at all higher altitudes; since the dependence of cross-section on wavelength varies greatly between species, this accounting must be done separately for many different wavelength bands. Molecules high in the atmosphere get the first chance to absorb photons. An abundant species, or a species with a very high cross-section, can deplete the photon supply and protect other species from dissociating. This is known as *shielding*. Shielding is central to the way photodissociation plays out in planetary atmospheres. For the most part, the net flux of energetic photons determines the number of molecules that dissociate per second throughout the atmosphere, but it is shielding that determines which species dissociate, and at which altitude.

The cross-section determines the depth of atmosphere over which photolysis occurs, while the photon flux determines the rate at which molecules in an atmospheric column are dissociated. The depth of atmosphere over which photons of a given frequency are depleted can be estimated in terms of the optical thickness, τ_ν. Let n_A be the number of molecules of gas A in an atmospheric column, per unit surface area of the column. If A is well mixed and has molar concentration η_A, then $n_A = 6.02 \cdot 10^{26} \eta_A (p_s/gM)$, where p_s is the surface pressure and M is the mean molecular weight of the atmosphere. The optical thickness is then given by $\tau_\nu = n_A \chi_\nu$. When $\tau_\nu > 1$, almost all of the photons at frequency ν are absorbed before reaching the ground, and the dissociation reaction is photon-limited at that frequency. If the gas is well mixed, then the photons are absorbed in a layer at the top of the atmosphere of of thickness $(\Delta p)_\nu$ given by

$$(\Delta p)_\nu \approx p_s/\tau_\nu = 1.67 \cdot 10^{-27} \frac{gM}{\chi_\nu \eta_A}. \qquad (8.22)$$

For typical cross-sections and concentrations, photolysis occurs only in a very thin region in the uppermost atmosphere. If there is appreciable shielding of a minor species by more abundant species or species with higher cross-sections, then the rate of conversion of the entire atmosphere would be limited by the rate of transport of the minor species into the photolysis layer. This contrasts with the (admittedly unrealistic) situation in which there is a single photolyzing species, in which case the substance would first be destroyed in the uppermost layer, and the wave of destruction would gradually work its way downward into the deeper atmosphere.

As an example of how photolysis works, let's consider the photochemical lifetime of NH_3 in an anoxic Early Earth atmosphere. Early attempts at resolving the Faint Young Sun problem for the early abiotic Earth relied on the greenhouse effect of NH_3 and CH_4, but these proposals quickly fell by the wayside once it was realized that the photochemical destruction rate of these compounds could not keep up with any likely sources. In reality the photolysis rate of NH_3 depends on what other UV-absorbing gases there are in the atmosphere, but here we will consider an idealized case in which only NH_3 is dissociating, to get a general feel for the time scales and depth scales involved. The cross-section for NH_3 reaches $10^{-26}\,m^2$ at $0.23\,\mu m$, and rises steeply to a peak of about $10^{-21}\,m^2$ at $0.2\,\mu m$, staying generally above $10^{-22}\,m^2$ between $0.21\,\mu m$ and $0.15\,\mu m$. In the latter wavelength range, the top-of-atmosphere photon flux is about $5 \cdot 10^{16}/m^2 s$, leading to a local photochemical lifetime of between 5 hours and 50 hours. This does not mean that all the NH_3 in the atmosphere is destroyed in a few days, however. If the molar concentration of NH_3 is $100\,ppmv$ then a 1 bar Earth atmosphere contains $2 \cdot 10^{25}$ molecules of NH_3 per square meter of the planet's surface. With this mass path, Eq. (8.22) then implies that the entire supply of photons is consumed in the top $50\,Pa$ of the atmosphere, when the cross-section is $10^{-22}\,m^2$. Photons do not reach the ground in appreciable numbers until the cross-section falls to $5 \cdot 10^{-26}\,m^2$, which occurs at wavelengths slightly shorter than $0.23\,\mu m$. The time scale for photolyzing all the NH_3 molecules is then the number of molecules in the column divided by the photon flux in the relevant wavelength range. This is roughly $(2 \cdot 10^{25}/m^2)/(1.5 \cdot 10^{17}/m^2 s)$, which works out to somewhat under 5 years.

Once NH_3 dissociates, the resulting NH_2 can further break up and form N_2 and H, or it can react with itself to form N_2H_4 (hydrazine – a rocket fuel). Hydrazine is very soluble in water and would be quickly removed by rainfall.[2] The reactions that re-form NH_3 from the photolysis products are inefficient, so to a good approximation it can be assumed that when a molecule of NH_3 photodissociates, it is gone for good. In this case, the net photolysis rate indeed gives the rate of loss of NH_3 from the atmosphere, and determines the rate at which it would need to be resupplied. To maintain 100 ppmv of NH_3 in the atmosphere would require the entire atmospheric inventory to be replaced every 5 years or so, demanding an extremely potent source. Shielding by other dissociating gases would increase the photochemical lifetime of NH_3, but detailed calculations confirm the difficulty of maintaining climatologically significant NH_3 concentrations against photolysis.

Photodissociation breaks up molecules, but the pieces react to form other compounds. This is the hardest part of atmospheric chemistry, since the possibilities for reactions are myriad, and data on reaction rates can be hard to come by. There can be several hundred reactions of importance in a planetary atmosphere, involving different combinations of reactants at different altitudes. The behavior at all levels is coupled together by vertical transport of reactants and competition for photons. In the face of all this essential complexity, it is hard to come up with idealized models that have any bearing on reality.

The importance of the back reactions which can re-form the photodissociating substance is well illustrated by photolysis of CO_2. This is a more stable molecule than NH_3, but it still has a cross-section which exceeds 10^{-25} m^2 for wavelengths shorter than $0.17\,\mu$m. This is more than enough to make an Earth atmosphere with 100 ppmv of CO_2 optically thick, or to make a 600 Pa pure CO_2 Mars atmosphere optically thick. Based on the mass path in the two cases and a photon flux of $6 \cdot 10^{15}$/m^{2}s for Earth and about half this rate for Mars, the Earth's CO_2 should dissociate into CO and O in only a century, while the entire Martian atmosphere should decompose in about 20 000 years. In reality, rather little CO is found on either planet. For Earth one could perhaps invoke shielding, but there is little to shield CO_2 from photodestruction on Mars. The answer lies in the back reactions that re-form CO_2. A brief study of this process also serves to illustrate the typical time scales involved in atmospheric binary reactions. In the following example, we will limit attention to the Mars case.

The simplest reaction for re-forming CO_2 is

$$CO + O + M \rightarrow CO_2 + M \tag{8.23}$$

where the M stands for collision with any third molecule, which provides necessary kinetic energy. The rate of formation of CO_2 by this process is $k[CO][O][M]$. If the activities are measured as molecules per cubic meter, as is typically done for gaseous reactions, then $k = 3 \cdot 10^{-49}$ m^6/s at 200 K. From the laws of chemical kinetics discussed earlier, the rate of change of CO concentration is

$$\frac{d}{dt}[CO] = -(k[O][M])[CO]. \tag{8.24}$$

This is also the rate at which CO_2 is produced by the reaction, since each molecule of CO that reacts produces one CO_2. The factor in parentheses on the right hand side is the exponential decay rate of [CO], which also gives the regeneration rate of CO_2. Suppose that the whole Martian atmosphere had decomposed into a well-mixed atmosphere consisting of CO and O.

---

[2] Actually, for a planet with an appreciable ocean, the extremely high solubility of NH_3 is probably by itself sufficient to preclude much accumulation in the atmosphere, independently of the photolysis issue.

At a surface pressure of 600 Pa and surface temperature of 200 K, the density of O is $2.2 \cdot 10^{23}$ molecules per cubic meter, while the density of all molecules (used for [M]) is twice that. That yields a CO decay time of a mere 35 seconds, which is how long it would take for the lower atmosphere to mostly recombine into CO_2. This time is vastly shorter than the photochemical destruction time of the atmosphere, and so the reaction would proceed until the concentrations of CO and O had fallen to the point that the net production rate of CO_2 equals its net photolysis rate. Like all multimolecular reactions, the regeneration of CO_2 proceeds at a greater rate at low altitudes, where the densities of the reactants are greater. This leads to a typical photochemical situation, in which photolysis occurs at high altitudes while critical parts of the resynthesis occur at lower layers, with the two regions connected by vertical mixing; the rate of vertical mixing and rate of reaction together determine the depth scale over which the photolysis products recombine.

The preceding discussion of the back reaction is oversimplified. To do the calculation properly, one also needs to take into account the reactions $O + O \rightarrow O_2$ and $O_2 + h\nu \rightarrow O + O$. In the Martian regime, the more complete reaction system succeeds in halting atmospheric breakdown and keeping the CO concentration down to a few percent, but this is still much more CO than is observed in the Martian atmosphere. In fact, there is a far more effective set of reactions involving the small quantities of OH produced by photolysis of water. Such reactions are of ubiquitous importance in planetary atmospheres, and will be taken up in Section 8.6.2.

Next, let's consider photolysis of methane in an idealized anoxic Early Earth atmosphere, and on Titan. An oxygenated atmosphere produces a lot of OH, and we will see in Section 8.6.2 that, as in the case of oxidizing CO to CO_2, the OH can catalyze very efficient destruction of CH_4. In an anoxic atmosphere (especially a dry one like Titan's where there isn't even a source of O and OH from photolysis of water), photolysis of CH_4 is a very significant sink. Methane is a relatively stable molecule, which dissociates at appreciable rates only for relatively short wavelengths. At a wavelength of 0.15 μm the cross-section is about 10^{-25} m^2, which is sufficient to give an optical thickness of 2.1 for a well-mixed Earthlike atmosphere containing 100 ppmv of CH_4. At Earth's orbit, the photon flux at wavelengths shorter than 0.15 μm is $2.2 \cdot 10^{15}$/m^{2}s, which would take 305 years to destroy all the methane in such an atmosphere, neglecting shielding by other molecules. A very potent methane source would be needed to compete with this photolysis rate; outgassing from the Earth's interior cannot do it, and it is unclear whether a bacterial source could do the trick either unless the photolysis rate is brought down considerably by shielding. Titan has vast amounts of methane in its atmosphere, and is subject to a much lower photon flux. Scaling the photon flux to Titan's orbit, we can estimate that it would take 300 million Earth years to photolyze all the methane in Titan's atmosphere. This is a long time, but it is still short compared with the age of the Solar System, so it means that either Titan's present methane content is a transient stage of the atmospheric evolution or that there is some means of resupply of methane, presumably by outgassing from the interior.

In an anoxic atmosphere, the main fate of the photolysis products of CH_4 is to turn into C_2H_6 (ethane). There are various pathways by which this happens, but for every two CH_4 molecules that are photolyzed, we get one C_2H_6 plus two hydrogens, which can be either in molecular or atomic form. The subsequent hydrocarbon chemistry depends on how much hydrogen sticks around. If the hydrogen largely stays in place, as it does on Jupiter, it "cracks" ethane and more complex hydrocarbons. This limits the formation of higher hydrocarbons, and can also provide a significant pathway for re-formation of CH_4. This is what happens on Jupiter and Saturn. On Titan, however, much of the hydrogen escapes, owing to the low gravity. In that case, formation of complex hydrocarbons, including those in the high-altitude haze clouds, becomes essentially irreversible. These hydrocarbons (including ethane) condense and gradually settle out onto the surface, and there is no known way to convert them back into methane. Over the lifetime of the Solar System all the methane in Titan's atmosphere should have been converted to a surface reservoir of ethane and higher hydrocarbons. Why this has not happened constitutes one of the Big Questions regarding Titan's atmosphere. The amount of hydrogen that might have been present in the Early Earth atmosphere is a matter of considerable dispute at present, and the answer has an important bearing on the fate of methane photolysis products on an anoxic Early Earth. If there is much hydrogen, then methane can re-form and in fact it can be hard to keep methane from accumulating to high concentrations. If there is little hydrogen, methane photolysis should produce ethane. Earthlike temperatures are too high to permit much ethane condensation and rainout, so ethane would accumulate in the atmosphere. It is likely that this would be accompanied by synthesis of a variety of higher hydrocarbons and organic hazes as well.

High-altitude organic haze clouds, such as those on Titan, intercept sunlight and cool the surface. They constitute a potentially important feedback on climate. Such hazes could also occur in other methane-rich atmospheres, including that of the anoxic Early Earth, but the presence of sufficient quantities of CO_2 can lead to additional reactions which prevent haze formation. The understanding of the circumstances in which haze clouds form is evolving rapidly, particularly with regard to the ratio of CO_2 to CH_4 above which haze formation is inhibited.

There are many, many other chemical cycles in planetary atmospheres that have important repercussions for climate. Notable among these is the sulfur cycle on Venus, which begins with SO_2 outgassing, and enters the realm of atmospheric chemistry upon SO_2 photolysis in the upper atmosphere, leading ultimately to a sink of sulfur in the form of various compounds that settle to the surface and may or may not be recycled. Because the sulfuric acid clouds that give Venus its high albedo result from conversion of SO_2 to sulfate and thence to sulfuric acid clouds, the sulfur cycle plays directly into the radiation budget of the planet. This is likely to be the case on any hot, dry planet which has sustained outgassing of SO_2 or other sulfur-bearing gases. Things would be different on a wet planet, since both SO_2 and sulfate are highly soluble and can easily be washed out.

Finally, let's consider the dissociation of H_2O in the steam atmosphere of a planet undergoing a runaway greenhouse. To lose water permanently from a planet with gravity at least as great as Venus or Earth, the light hydrogen must be broken off and lost to space. The question arises as to whether the loss rate is seriously limited by the rate at which H_2O dissociates. The dissociation cross-section for H_2O begins to rise sharply at 0.21 μm, and at Venus' orbit, the top-of-atmosphere photon flux shorter than this wavelength is about $6 \cdot 10^{16}/m^2s$. The photon supply can dissociate all the water in a 200 bar atmosphere in only 35 million years, so photon supply is not in itself a major limiting factor in escape of a massive ocean from Venus. Given a similar inventory of water, photon flux is unlikely to be the limiting factor for any planet in a close enough orbit that a runaway greenhouse is possible. The main issue is the subsequent fate of the dissociation products, H and OH. This will be taken up in Section 8.6.2.

Note that even for a cross-section as small as $10^{-25} m^2$, typical of the wavelengths where photolysis first becomes significant, the photolysis layer is only 3 Pa deep; over most of the far UV spectrum, the layer is orders of magnitude thinner. Since other photolyzing species also exhaust the photon supply over a thin layer, H_2O can only be photolyzed at an appreciable rate if the atmospheric transport processes and atmospheric structure permit a high water vapor concentration in the uppermost atmosphere. At temperatures where water vapor is a minor constituent, as on the present Earth, condensation at the cold tropopause severely limits the supply of water vapor to the upper atmosphere.

For the most part, the atmospheric chemistry we have been discussing takes place in the top few Pa of an atmosphere, since that is where the photon supply is consumed, and the slow vertical mixing in the upper atmosphere tends to assure that binary back reactions occur within a layer not terribly far below the photolysis layer. The upper atmosphere is not as well mixed as the stratosphere or troposphere, and the composition in this region can differ greatly from that of the bulk of the atmosphere. One should think of atmospheres as having a thin photochemically active layer near the top, fed by a supply of reactants from a comparatively well-mixed reservoir in the stratosphere and troposphere. Many reactions can in fact be limited by the rate of transport of fresh unreacted species upward. The vertical transport process is thus important, but unfortunately also very difficult to determine from first principles.

8.6.2 OH: a radical's life

Methane burns in an oxygen atmosphere. It's what happens every time you light your kitchen stove. Hydrogen and oxygen are an explosive mixture – remember the Hindenburg! In the face of this common wisdom, it would seem that it would be a simple matter to get rid of oxidizable substances like methane in an oxygen-rich atmosphere. It's not that simple. The reaction between oxidizable substances and oxygen proceeds extraordinarily slowly at room temperature, despite the fact that the reaction releases energy; it takes a very high ignition temperature to make the reaction self-sustaining. At typical planetary atmosphere temperatures, oxidation is almost entirely mediated by reactions involving OH. The species OH is a *radical* – a highly reactive neutral molecule that does not live for very long before it reacts with something. It must be distinguished from the OH^- *ion* which results from dissociation of water in the liquid phase, as in Eq. (8.7). Biology accomplishes a similar miracle of low temperature combustion of organic carbon ("respiration") through the employment of the eight enzymes that mediate the *Krebs cycle*.

As an example of the role of OH in oxidation, let's consider the oxidation of CH_4, which is the dominant sink in atmospheres containing appreciable quantities of oxygen. The initial steps of the process consist of the reactions

$$CH_4 + OH \rightarrow H_2O + CH_3$$
$$CH_3 + O_2 + M \rightarrow CH_3O_2 + M. \tag{8.25}$$

The peroxymethyl radical (CH_3O_2) participates in a wide variety of subsequent reactions, leading ultimately to CO_2 and a variety of other products and consuming one additional OH along the way. The rate-limiting step is the first reaction, which thus determines the lifetime of CH_4. If k is the rate constant for the reaction and [OH] is the density of OH in molecules per cubic meter, then the exponential decay rate of CH_4 is $k[OH]$. At $250\,K$, $k = 1.77 \cdot 10^{-21}\,m^3/s$, so an [OH] density of $1.78 \cdot 10^{12}$ molecules per cubic meter is sufficient to bring the lifetime of CH_4 down to $10\,years$. Based on the air density at $50\,mb$, this amounts to just over one part per *trillion*. This is a reasonable approximation to the state of affairs in the present Earth atmosphere, but it must be kept in mind that the lifetime depends on the full chain of atmospheric reactions that determine the OH concentration, and could easily change as circumstances change.

The OH radical is present in atmospheres in quantities so small as to be essentially unmeasurable, but it is so reactive it takes very little to have a profound impact on atmospheric chemistry, as the methane oxidation example illustrates. Because of its extreme reactivity, the OH radical has a short lifetime in the atmosphere. It is not made "to go" but is consumed on the spot without being significantly transported. Where does OH come from? Modest amounts can be made by photolysis of water vapor, but in an oxygen-rich atmosphere, the radical can be efficiently made by pathways involving ozone. Ozone can be produced by a variety of means in an oxygenated atmosphere, and, once present, photolysis by UV in the 0.3 μm to 0.36 μm range dissociates O_3 into two electronically excited states of oxygen, called $O_2(^1\Delta)$ and $O(^1D)$. The latter of these is a very important radical in oxygenated atmospheres, and is pronounced "O singlet D." Most of the $O(^1D)$ collide with air molecules and are knocked into the ground state, losing their internal energy to kinetic energy of the molecules. Some will collide with H_2O and undergo the reaction

$$O(^1D) + H_2O \rightarrow OH + OH. \tag{8.26}$$

Because ozone can dissociate at relatively long UV wavelengths which penetrate deeper into the atmosphere, the presence of free O_2 and hence ozone in an atmosphere allows OH mediated oxidation to occur at much lower altitudes than would otherwise be possible.

The reaction that destroys CH_4 is not *catalytic*. It consumes two OH radicals for each CH_4 oxidized to CO_2. The OH thus consumed must be regenerated by other fast reactions, and if it is not, the OH will be depleted and the lifetime of CH_4 will rise dramatically, allowing it to accumulate to greater concentrations for any given source rate. In contrast, the following fast pathway to regenerate CO_2 from CO and O on Mars is catalytic:

$$
\begin{aligned}
CO + OH &\rightarrow CO_2 + H \\
H + O_2 + M &\rightarrow HO_2 + M \\
O + HO_2 &\rightarrow O_2 + OH.
\end{aligned}
\tag{8.27}
$$

The net result of this transforms one CO and one O back to CO_2, leaving the OH supply unchanged. The rate-limiting step in the sequence is the first reaction. As in all OH reactions, the rate constant is very large, so a small amount of OH leads to a high reaction rate. Since so little OH is required, photolysis of water vapor can produce enough to make the reaction go swiftly despite the rather small amount of water vapor present in the Martian atmosphere. The HO_2 appearing in this reaction chain is another radical of widespread importance in atmospheric chemistry. Like all radicals, its high reactivity allows it to have a profound effect on atmospheric chemistry even at very low concentrations.

Finally, we are prepared to return to the subject of water vapor photolysis and hydrogen escape. This, too, intimately involves OH and other radicals, which are involved both in the breakup of water into hydrogen and oxygen, and in the back reactions which can oxidize hydrogen back to water. The reaction of H with OH to re-form water is so slow in comparison with other relevant reactions that standard compilations of atmospheric reaction rates do not even bother to list a rate coefficient. Instead, the following sequence in the net breaks up one water molecule into its component H and O:

$$
\begin{aligned}
2(H_2O + h\nu &\rightarrow H + OH) \\
OH + OH &\rightarrow H_2O + O.
\end{aligned}
\tag{8.28}
$$

This takes two dissociated H_2O molecules, re-forms one H_2O and leaves 2H and one O. The resulting H atoms will mostly combine to make H_2, and the O atoms will combine to make O_2. Left to its own devices, this sequence of reactions would turn the water vapor atmosphere into an H_2/O_2 atmosphere, with some traces of H, O, and OH mixed in. However, the accumulation of oxygen makes it easy to oxidize hydrogen back to water. One of the many reaction sequences that does this is

$$2(OH + H_2 \rightarrow H + H_2O)$$

$$H + O_2 + M \rightarrow HO_2 + M \tag{8.29}$$

$$HO_2 + H \rightarrow H_2O + O.$$

In this reaction chain we again see the HO_2 radical, which we encountered earlier in connection with the oxidation of CO back to CO_2. The reaction between HO_2 and H actually preferentially produces 2OH, but these quickly react to form H_2O and O, so the end result is much the same. If much oxygen accumulates, then additional reactions involving O_3 and its photolysis products can also assist in the oxidation of H and H_2 back to water. Thus, a water vapor atmosphere exposed to far UV will not decompose into its constituents, but instead will reach a steady state consisting of some mix of H_2O, H_2, and O_2, with much smaller admixtures of the highly reactive species O, H, OH, and HO_2, plus a variety of other radicals we have not discussed. The amount of H_2 left in this mixture determines the reservoir of hydrogen which is available to escape to space. The main chemical limitation on hydrogen escape is the resulting accumulation of oxygen in the remaining atmosphere. If there is no sink of oxygen, the concentration of O_2 builds up, which increases the rate of production of HO_2 from H, which in turn increases the rate of conversion of H to H_2O. Escape to space and re-formation of H_2O are both sinks of H, and both must compete for the supply of H produced by photolysis. If the re-formation rate becomes fast enough, then escape to space cannot compete effectively, and hydrogen escape will be severely limited.

The solution to the full photochemical problem bearing on irreversible water loss is complex and can involve species other than those produced by photolysis of water; moreover, it must be solved simultaneously with the vertical transport and far UV radiative transfer problems. Nonetheless, it is instructive to flesh out the operation of the reactions in Eq. (8.29) with a few numbers. Let k_1 be the rate constant for the second reaction in Eq. 8.29, and k_2 be the rate constant for the third. For simplicity, we'll assume that there is enough OH around that the first reaction is not rate-limiting. If there is no other source or sink of HO_2, then the production rate by the second reaction must equal the consumption rate by the third, so $k_1[H][O_2][M] = k_2[H][HO_2]$, which allows us to determine the HO_2 density as $[HO_2] = (k_1/k_2)[O_2][M]$. Substituting this into the rate equation for the third reaction we find

$$\frac{d[H]}{dt} = -(k_1[O_2][M])[H] \tag{8.30}$$

from which we identify the decay rate of H as $k_1[O_2][M]$. This increases linearly with oxygen density and also increases with pressure because [M] increases with pressure. The rate of consumption of H is actually twice that indicated in Eq. (8.30) because each H consumed in the third reaction also requires one H to be consumed to make the required HO_2 in the second reaction. At 300 K, $k_1 \approx 1.2 \cdot 10^{-43}\,\mathrm{m^6/s}$. If we assume a modest rate of vertical mixing such that most of the re-formation reaction occurs at a pressure of 10 Pa and put in the appropriate densities assuming a temperature of 300 K there, then a 1% mixing ratio of O_2 in an H_2O-dominated atmosphere yields an H decay time of a mere 72 s. This rate takes into account the consumption of an additional H in the formation of HO_2.

Now, on Venus the flux of photons energetic enough to dissociate water is $6 \cdot 10^{16}/m^2$ s, which is approximately the rate of production of H_2, allowing for dissociation from the OH reaction and recombination of 2H into H_2. If there were no escape to space this rate of production of hydrogen would be balanced against the rate of H_2 consumption by re-formation of water. Using the rate at 10 Pa and assuming that the first reaction in Eq. (8.29) is not rate-limiting, one gets an H_2 molecular density of $8.6 \cdot 10^{18}/m^3$ (basically the destruction time scale times the production rate by photolysis). Based on the total molecular density of a water-dominated atmosphere at 10 Pa and 300 K, this amounts to only 355 ppmv of H_2. Hydrogen escape would drive this value lower if it happened rapidly enough, but recall that photolysis is not the limiting factor in losing an ocean. At the orbit of Venus, a 200 bar ocean can be lost in under half a billion years even if only 10% of the hydrogen from photolyzed water escapes rather than reacting with oxygen to re-form water. With such a small escape sink, the H_2 concentration would not be driven to be much below the photochemical steady-state value. The main way that chemistry limits hydrogen escape is by driving the hydrogen density to low values. In Section 8.7 we will take up the calculation of escape fluxes of H and other species, and see whether the required escape flux can be maintained with such a low hydrogen density. Generally speaking it is quite difficult to do so. The oxygen accumulating from losing only 1% of the hydrogen in the water inventory is already a significant impediment to escape, though it is not clearly precluded that an ocean could be lost if the oxygen concentration were held to around 1%. To lose half of the water inventory is much harder. Greater vertical mixing allows the re-formation reaction to occur at higher pressures, greatly increasing the reaction rates and further inhibiting the escape of hydrogen.

In order to keep oxygen accumulation from choking off the massive hydrogen loss required to lose an ocean in a runaway greenhouse, there must be a strong sink of oxygen. For planets as massive as Venus, or perhaps even Mars, oxygen is too heavy to escape to space at the required rate, except perhaps under conditions of extraordinarily elevated exposure to EUV radiation. The only known way to get rid of the required amounts of oxygen is to react it with suitable minerals in the crust, assuming such minerals exist. This may well be the weakest link in the runaway greenhouse theory of irreversible water loss. The oxygen must be transported at a sufficient rate from the upper atmosphere where it is created to the surface where it can react, but more importantly there must be mechanisms that continually expose fresh unreacted minerals to the atmosphere at a sufficient rate. It is not obviously impossible for this to happen, but the processes involved must be regarded as dimly understood at best.

8.6.3 A planet's first CO_2

There are two important exceptions to the general idea that only ultraviolet photons provide enough energy to break stable bonds and sustain chemical disequilibrium. The high temperatures occurring in the interior of planets (especially gas giants) and the high temperatures encountered during formation of planets by planetesimal collisions can provide enough thermal energy to provoke resynthesis. The hot early stages of a planet's lifetime are particularly important to the problem of how a planet acquires its supply of CO_2. Given the myriad roles of CO_2 in climate, this is a matter of the highest importance.

Carbonates are not an abundant component of the debris from which planets are made. Most carbon starts out as elemental or organic carbon. To form CO_2, the carbon must be brought together with oxygen in H_2O or other compounds. This may happen as the planet accretes, or during the formation of rocky or icy planetesimals. Once the CO_2 has formed, it becomes possible to form carbonate in the presence of water and silicate minerals, and this provides a way to build up a storehouse of CO_2 in a planet's interior, which can be outgassed later. The following exercise shows that it would require exceedingly high temperatures to directly dissociate stable chemical constituents.

Exercise 8.3 Using the relation $\Delta E = h\nu$ convert the photodissociation wavelengths in Table 8.3 into energies. Find the temperatures for which the thermal energy kT is equal to the dissociation energy threshold for each of the species.

Hence, it is not likely that dissociation of H_2O in a post-impact cloud or planetary atmosphere with a temperature of a few thousand kelvin would provide enough oxygen to oxidize carbon into CO_2. However, at temperatures of a few thousand kelvin, the reaction

$$C + H_2O \rightarrow CO + H_2 \tag{8.31}$$

can proceed swiftly. The CO can then be gradually oxidized into CO_2 using photolytically produced OH. The *water gas shift reaction*

$$CO + H_2O \rightleftharpoons CO_2 + H_2 \tag{8.32}$$

can accomplish the same thing thermally. These reactions are in fact the same ones used industrially in coal gasification. In order to produce a large inventory of CO_2 the necessary reactants must be brought together and held at a high temperature for a sufficiently long time. The range of possible outcomes of this process during atmospheric formation has not been much explored, and is a fruitful area for further inquiry.

Another familiar industrial process, however, could intervene and prevent the formation of CO_2. In the *Fischer–Tropsch process*, metallic catalysts including iron, cobalt, nickel, or ruthenium catalyze the production of hydrocarbons from the CO–H_2 mixture. If this happens, it would form organic carbon or methane at the expense of CO_2, with attendant consequences for the subsequent fate of the planet's evolving atmosphere.

8.7 ESCAPE OF AN ATMOSPHERE TO SPACE

Atmospheric escape calculations play a role in determining how a planet got to be the way we see it today, and what kind of atmosphere it might have had in its past; escape calculations can also indicate how long a planet can maintain a habitable climate, and can inform investigations into the mechanisms needed to maintain an atmosphere in a given state. For example, does there need to be a source of N_2 outgassing to maintain Titan's largely N_2 atmosphere? The answer to that hinges primarily on how rapidly a small, cold body like Titan can lose a relatively heavy gas like N_2. Similar questions apply to maintenance of CH_4 on Titan, since the processes which turn CH_4 into ethane and other compounds which accumulate on the surface liberate H_2, which must escape to space or accumulate in the atmosphere. Likewise, if Venus ever had an ocean comparable to Earth's, and went into a runaway state, both the hydrogen and oxygen in the water must have been lost somehow, since there is very little water in Venus' atmosphere today. Could the hydrogen escape to space at a sufficient rate? The oxygen? The escape of water to space is what makes a runaway greenhouse essentially irreversible, so it enters into the habitability question for Venus and other planets subject to a runaway greenhouse. One picture of the climate of Early Mars posits a warm, wet climate supported in part by the greenhouse effect associated with a CO_2

atmosphere with a surface pressure of around 2 bars. How much of such an atmosphere could be lost to space in the available time? How much would have to be lost into chemical reaction with surface rocks instead? Is it possible that a body as small as Mars could remain habitable for billions of years, or does escape to space inevitably doom such a planet to the long chill? The contrast between Mars and Titan is stark. We expect atmospheres to be in some sense bound by gravity, but why is it then that Titan, with far weaker gravity than Mars, retains a 1.5 bar atmosphere while Mars retains practically none? One feels it must have something to do with the lower temperature of Titan, but that notion begs to be quantified. Even for Earth, escape mechanisms are important: before the rise of atmospheric oxygen, there is the possibility that large amounts of hydrogen could accumulate in the atmosphere, if it does not escape rapidly to space. This is important because hydrogen and carbon dioxide can serve as feedstocks for synthesis of many prebiotic organic molecules.

Atmospheric escape may also play a critical role in determining the habitable zones around M-dwarf stars. These stars have low luminosity, and therefore a planet needs to be in a close orbit in order to be warm enough to be in the habitable zone as conventionally defined with respect to liquid surface water. However, M-dwarfs episodically have much higher EUV output relative to net energy output than G stars, so such a planet would be blasted by so much high energy radiation that it would risk eroding even higher-molecular-weight components of the atmosphere, such as N_2 or CO_2.

8.7.1 Basic concepts

The problem of permanently removing a molecule from a planet's atmosphere is much the same as the problem of sending a rocket from Earth to Mars: one must impart enough velocity to the object, and in the right direction, to allow the object to overcome the potential energy at the bottom of the gravitational well, and still have enough kinetic energy left over to allow the object to continue moving away. This leads to the central concept of *escape velocity*, which is the minimum velocity an object needs in order to escape to infinity, provided no drag forces intervene. The escape velocity is obtained by equating initial kinetic energy to the gravitational potential energy. Let m be the mass of the object, v its speed, and r its initial distance from the center of the planet. Let M_P be the mass of the planet and G be the universal gravitational constant. Then, equating kinetic to potential energy we find that $\frac{1}{2}mv^2 = GM_Pm/r$, so $v = \sqrt{2GM_P/r} = \sqrt{2gr}$ where g is the acceleration of gravity at distance r from the planet's center. In many situations of interest, the altitude from which the molecules escape is sufficiently close to the ground that g is only slightly less than the surface gravity. Important exceptions include small bodies with a massive atmosphere, such as Titan, or such as the Moon would be if it were given a massive atmosphere. In this section, we will use g to represent the actual radially varying acceleration $g(r)$, and will use the symbol g_s when we need to refer specifically to the surface gravity. The acceleration at distance r is then $g(r) = g_s(r_s/r)^2$, where r_s is the radius of the planet's surface.

In order to allow a molecule to escape, enough energy must be delivered to the molecule to accelerate it to the escape velocity. The study of atmospheric escape amounts to the study of the various ways in which the necessary energy can be imparted. Since the kinetic energy of a molecule with mass m is $\frac{1}{2}mv^2$, light molecules like H_2 will escape more easily than heavier molecules like N_2, given an equal delivery of energy. Dissociation of a molecule like CO_2 or H_2 into lighter individual components also aids escape. Using the formula for escape velocity, we can define the *escape energy* of a molecule with mass m as mgr. For escape of N_2

from altitudes not too far from the Earth's surface, this energy is $2.9 \cdot 10^{-18}$ J, or 2.9 attoJ.[3] For H_2 the escape energy is only 0.2 attoJ.

In the end, like so many things in planetary climate, atmospheric escape is all about energy. There are four principal sources of energy that could potentially feed atmospheric escape:

- The general thermal energy of the atmospheric gas, which ultimately comes either from absorbed solar radiation or from heat leaking out of the interior of the planet.
- Direct absorption of solar energy in the outer portion of the planet's atmosphere, which may energize particles to escape velocity either directly or through indirect pathways, or which may manifest itself in a hydrodynamic escaping current of gas. The solar radiation responsible for these mechanisms is typically in the extreme ultraviolet (EUV) portion of the solar spectrum because there is so little mass in the regions involved that absorption is weak, whence there is a premium on absorbing individual photons which have a great deal of energy.
- Collisions with the energetic particles (usually protons) of the stellar wind which streams outward from the atmosphere of the planet's star.
- Kinetic energy imparted to the atmosphere by the impact of large objects.

For a portion of the atmosphere where collisions are frequent enough to maintain thermodynamic equilibrium, each degree of freedom gets an energy of $\frac{1}{2}kT$ on average. Some molecules will have more energy than this, and some will have less, but if the mean energy is considerably less than the escape energy, only a very small proportion of molecules will have sufficient energy to escape. For $T = 300$ K, the typical energy is only 0.002 attoJ. At this temperature only a small fraction of H_2 molecules would have sufficient energy to escape from Earth, and the escape rate becomes only moderately higher if the temperature of the escaping gas is raised to 1000 K. Heavier molecules like N_2 could hardly escape from a planet with Earth's or Venus' gravity at all, if the only source of energy were thermal motions. Even on a light body like Titan, the escape energy for N_2 is still 0.16 attoJ based on surface gravity, so escape will not be easy. Escape due to the energy associated with the thermal motions of particles in thermodynamic equilibrium is called *thermal escape*, or sometimes *Rayleigh–Jeans* escape. The ratio of escape energy to kT, namely $\lambda_c \equiv mgr/kT$, is an important parameter in the theory of thermal escape. In this formula, r is the radius of the atmospheric shell from which particles escape, g is the acceleration of gravity at that radius, and T is the typical temperature there.

For a particle to escape, it is not enough to have reached escape velocity. It must also have a reasonable chance of escaping the gravitational well of the planet without suffering many collisions, since each collision will divert the particle from its outward path and rob it of some of its velocity. If the particle undergoes many collisions, it will undergo a random walk, leading to slow diffusion of the substance rather than rapid outward streaming. The portion of the atmosphere where particle collisions are so infrequent that a particle with sufficient energy has a good chance of escaping without collision is called the *exosphere*. In order to define the exosphere quantitatively, we first introduce the notion of *mean free path*, which is the mean distance traveled by a molecule between collisions. The mean free path depends on the total number density of particles, n, and the effective cross-section area of the molecules, χ. For simplicity, we'll assume for the moment that all the molecules

---

[3] One frequently sees the unit *electron volt* used for measuring small quantities of energy in a context like this. 1 eV = 0.16 attoJ.

in the gas are identical. Two molecules are considered to collide if their centers approach within a distance of $2\sqrt{\chi/\pi}$. To estimate the mean free path, construct an imaginary cylinder with axis aligned with the direction of travel of the particle we are tracking, and having a radius $\sqrt{\chi/\pi}$. A collision is inevitable if this cylinder contains a particle from the rest of the gas, which becomes likely when $nV = 1$, where V is the volume of the cylinder. Writing $V = \ell\pi a^2 = \ell\chi$ we find that the mean free path is $\ell \approx 1/n\chi$. This estimate would be precise if the particles being collided with were stationary. In reality, the distance moved by the test particle before collision is affected by the fact that some of the other particles are moving toward the test particle, while others are moving away from it; a more precise calculation making use of the actual distribution of particle velocities in thermodynamic equilibrium yields the slightly modified result

$$\ell = \frac{1}{n\chi\sqrt{2}} \tag{8.33}$$

when all the particles have the same mass. When the background particles are very massive compared with the particle we are tracking, then they can be regarded as essentially stationary and the $\sqrt{2}$ factor in the denominator can be dropped; for most of the uses to which we will put the mean free path, this effect is of little importance.

The effective particle collision cross-section depends on the pair of molecules which are colliding, and is also a weak function of the energy of the collision, since molecules are not hard spheres but rather can be penetrated when the collision energy is large enough. Still, the effective collision radius does not differ greatly between one molecule and another. The collision radius can be inferred from measured diffusion rates, and a few typical values are given in Table 8.4. For the most part, the effective collision radius is between 100 and 200 picometers (pm). An important exception is atomic hydrogen (H), which has an effective radius on the order of a mere 10 picometers. Curiously, the effective binary collision radius remains this small even when H is colliding with a much larger molecule. Evidently, the H atoms are like little bullets, which punch right through the outer electron clouds of the bigger molecules. This effect gives atomic hydrogen an anomalously large mean free path for a given density, which has a number of important consequences. Data for atomic oxygen (O) is hard to come by, but neon is believed to be an analog and should have a similar collision radius. Air at the Earth's surface with a temperature of 300 K has a particle density of $2.4 \cdot 10^{25}/m^3$. Based on a collision radius of 125 pm the mean free path is about 0.6 μm. Adopting a scale height of 8 km, the particle density at an altitude of 100 km falls to $9.0 \cdot 10^{19}/m^3$ and the mean free path increases to 16 cm – about the width of a hand.

The mean free path increases exponentially with altitude, because the particle density in a gravitationally bound atmosphere decreases exponentially with altitude, at a rate given by the density scale height (RT/g for the isothermal case). The exosphere is said to begin where the mean free path becomes sufficiently large that further collisions before escape are unlikely; this critical altitude is called the *exobase*. Commonly, it is defined as the altitude where the mean free path becomes equal to the scale height, since the exponential increase of mean free path with height means that if a particle doesn't collide with another within the first scale height, it is basically home free and unlikely to find another to collide with. When the exobase is not too far above the ground in comparison to the planet's radius, g can be approximated by the surface gravity and one can immediately estimate the exobase altitude in terms of the temperature of the exosphere and the scale height for the dominant constituent of the exosphere. In applying this procedure, one must keep in mind that the temperature of the exosphere could be quite different from the temperature of the lower

Table 8.4 Effective collision radius for various binary collisions, computed from diffusion data.

	100 K	300 K	1000 K
H–H$_2$	11.08	6.02	3.09
H–air	12.26	6.34	3.08
H–CO$_2$	13.63	7.46	3.85
H$_2$–air	144.50	125.96	108.36
H$_2$–CO$_2$	157.24	137.06	117.91
H$_2$–N$_2$	146.65	123.01	101.46
H$_2$O–air	194.71	142.21	100.78
CH$_4$–N$_2$	177.57	154.87	133.31
Ne–N$_2$	139.88	122.40	105.74
Ar–air	166.62	145.32	125.09

The radius is given for three different temperatures, in units of picometers. (1 pm = 10^{-12} m). The collision cross-section for collision radius a is $\chi = \pi a^2$.

atmosphere, and the composition could differ greatly from the bulk composition of the atmosphere. We will have more to say about both these aspects of the exosphere a bit later. For the most part, one is interested in the exobase particle density, since that is what will determine the flux of particles to space. When one at least knows that the exobase is low enough that $g \approx g_s$, the exobase density is immediately given by the requirement that $\ell = H$, which implies that the exobase particle density is

$$n_{ex} = \frac{g_s}{\chi R T_{ex} \sqrt{2}} = \frac{m g_s}{\chi k T_{ex} \sqrt{2}} \tag{8.34}$$

where T_{ex} is the exobase temperature, m is the actual mass (in kilograms) of the molecule which makes up most of the exosphere and k is the Boltzmann thermodynamic constant.

For the more general case, one must extend the hydrostatic relation to account for the decay of gravity with altitude. This case is important for light constituents like H or H$_2$, which have a large scale height because low molecular weight implies large gas constant R. The exosphere can also be very extended even for heavier molecules, for small bodies with low surface gravity, such as Titan. Allowing for the inverse-square reduction of gravity with distance r from the center of the body, the equation of hydrostatic balance becomes

$$\partial_r p = -\rho g_s \frac{r_s^2}{r^2}. \tag{8.35}$$

As in the treatment of the hydrostatic relation in Chapter 2, the equation is closed by using the ideal gas law, $p = \rho R T$, where R is the gas constant for the mixture making up the uppermost part of the atmosphere. In escape problems it is often more convenient to deal with particle number density rather than mass density. The particle density $n(r)$ is obtained by dividing ρ by the mass of a molecule, m. If the upper atmosphere is isothermal with temperature T_0, then the solution expressed as number density is:

$$n(r) = n(r_0) \exp\left(\frac{r_s}{H_s} \left(\frac{r_s}{r} - \frac{r_s}{r_0} \right) \right), \; H_s \equiv \frac{R T_0}{g_s} \tag{8.36}$$

where r_0 is some reference distance, generally presumed to be in the upper atmosphere. From this equation, it follows that the local scale height at radius r is $H(r) = (r/r_s)^2 H_s$.

Because of the attenuation of gravity, the density no longer asymptotes to zero at large distances from the planet. The limiting value is $n_\infty = n(r_0) \exp -(r_s/H_s)(r_s/r_0)$. Since this formula neglects all gravity but that of the planet, it is not surprising that one can fill infinite space with a finite density and have it stay put, since there is essentially zero gravity out there. To do so would require infinite mass, though, which means that if the planet is initially endowed with a finite-mass atmosphere, it will all leak away to space, given sufficient time. The estimate of that time scale is the objective of this section. In reality, the limiting density is not achieved, because the atmosphere is truncated by particle loss from the exobase. Still, it is important that there is a non-zero limiting density, since this implies that there is a non-zero limiting mean free path. Combined with the fact that the scale height increases with r, this can remove the exobase to infinity, meaning that there is no altitude at which the atmosphere can be considered collisionless. This situation will not arise for heavy constituents on reasonably massive bodies, since n_∞ is exceedingly small. Using a collision radius of 125 pm, the limiting mean free path is over 10^{167} m for O on Earth, based on a temperature of 300 K. This is well in excess of the size of the Universe. The limiting value for heavier constituents is even greater, and considerations for Venus turn out similarly. Even for Titan, the limiting mean free path for N_2 is about 10^{44} m based on a temperature of 100 K. The situation for the light species H and H_2 is more ambiguous. For example, if Titan had a pure H_2 atmosphere with a temperature of 100 K, the limiting mean free path would be only 850 m even if the surface pressure were a mere 0.1 Pa. This would certainly preclude a collisionless exosphere. Since H and H_2 typically appear in the atmospheres of terrestrial-type bodies as minor constituents mixed in with a heavier gas, further discussion is best deferred until after we have considered inhomogeneous atmospheres.

Exercise 8.4 Show that Eq. (8.36) reduces to the conventional hydrostatic relation derived in Chapter 2 when $r = r_s + z$ with $z/r_s \ll 1$, r_s being the radius of the planet.

To determine the exobase height, we need a model of the atmospheric structure which gives us the total number density $n(r)$ as a function of position. This can be challenging to do precisely, since $n(r)$ depends on the temperature and composition profile of the atmosphere; often, exobase heights for present-day Solar System planets are calculated from measured rather than theoretical density profiles. The other factor involved is the scale height at the exobase, which depends on the exosphere composition, position (through gravity) and temperature. Higher temperature increases the scale height and hence tends to move the exobase further out. The temperature of the exosphere is determined by a balance between heating by absorption of solar radiation (mainly ultraviolet), heating by absorption of outgoing thermal infrared from deeper in the atmosphere, and cooling by emission of infrared radiation. In some cases, there can be energy gain from collision with solar wind particles, and there can also be energy loss by outward streaming of mass in *hydrodynamic escape* (see Section 8.7.4). The radiative transfer in the exosphere is simplified by the fact that the atmosphere is optically thin, but is complicated by the fact that it is so tenuous that local thermodynamic equilibrium (and hence Kirchhoff's law) is not accurate. Still, when the exosphere is made of a good infrared emitter, the temperature tends to be on the order of a skin temperature, augmented a bit by solar absorption. Thus, the CO_2-dominated exobase of Venus has temperature between 200 K and 300 K, and that of Mars is somewhat higher. The Earth's exosphere is unusually hot, since it is dominated by atomic oxygen which comes from photodissociation of the large O_2 concentration of the lower atmosphere. Atomic oxygen is a good ultraviolet absorber, but

Planet	p_s (bar)	T_0	T_{ex}	z_{ex} (km)	t_{loss} (Gyr)	$w_{J,H}$ (m/s)
Earth, N_2	1.0	300	300	221	$7 \cdot 10^{286}$	$4.97 \cdot 10^{-7}$
Earth, O	0.2	300	1000	401	$4 \cdot 10^{42}$	7.94
Venus, CO_2	90.0	500	300	304	$> 10^{300}$	$1.7 \cdot 10^{-5}$
Venus, H_2O	1.0	400	300	542	$7 \cdot 10^{147}$	$3.458 \cdot 10^{-5}$
Mars, CO_2	2.0	250	300	352	$3 \cdot 10^{81}$	36.41
Mars, O	1.0	250	1000	1312	4000	808.13
Titan, N_2	1.5	100	200	774	$3 \cdot 10^{15}$	268.83
Titan, N	1.5	100	200	2454	5000	365.18
Moon, N_2	1.0	260	300	6145	4	613.54

The planet and the atmospheric composition are given in the leftmost column. In the other columns, p_s is the surface pressure in bars, T_0 is the effective mean temperature of the atmosphere below the exobase, T_{ex} is the exobase temperature, z_{ex} is the altitude of the exobase above the surface (in kilometers), t_{loss} is the time needed to lose the atmosphere by thermal escape (in billions of years), and $w_{J,H}$ is the Jeans escape coefficient for atomic hydrogen (in m/s). The hydrogen escape coefficient assumes that hydrogen is a minor constituent at the exobase. The mean free path was computed using a fixed molecular collision radius of 125 pm in all cases.

Table 8.5 Characteristics of the exosphere and loss rates for various hypothetical single-component atmospheres.

radiates infrared poorly, leading to high temperatures. Exosphere temperatures for N_2 or N-dominated exospheres are a delicate matter, since N_2 neither absorbs nor emits well, and slight contamination by infrared emitters or good solar absorbers can make a big difference. Titan's N_2-dominated exosphere has an observed temperature of about 200 K.

Once the parameters of the outer atmosphere are settled, the exobase position is determined by solving $\ell(r)/H(r) = 1$ by iteration, where $H(r)$ is the scale height at position r and the mean free path $\ell(r)$ is inversely proportional to $n(r)$. To get some numbers on the table for discussion, let's adopt a simple model atmosphere in which $n(r)$ is computed based on Eq. (8.36) with a uniform equivalent lower atmosphere temperature T_0 all the way down to the planet's surface, where the surface pressure p_s is specified. The exobase temperature is specified separately. Exobase altitudes for some hypothetical single-component atmospheres are given in Table 8.5. The atomic oxygen case for Earth is meant to serve as a crude representation of the oxygen-dominated exosphere of Earth. In this approximation, the atmospheric structure is computed as if all the Earth's oxygen were in the form of O, and ignores the fact that the O_2 is only converted to O at altitudes above 100 km as well as ignoring the effect of other gases on the vertical structure. We'll be able to do better later when we take up mixed atmospheres, but it is interesting to note that the exobase height of 400 km in this approximation is not too far from the true exobase height (500 km) computed on the basis of the observed O density in Earth's upper atmosphere. The N_2 case may be thought of as approximating an Early Earth situation in which there is little O_2 available to feed an oxygen-dominated exosphere. The two Venus cases represent approximations to the present state of Venus, and a hypothetical past near-runaway state with a pure steam atmosphere. The first Mars case assumes a dense CO_2 atmosphere such as might have prevailed on Early Mars, while the second is an approximate to the situation where the exosphere is dominated by atomic oxygen arising from photodissociation of CO_2. The first Titan case approximates the present, while the second gives some indication of what would happen if the N_2 were to dissociate into atomic nitrogen (a somewhat implausible situation, but one which is included to allow us later to put a generous bound on nitrogen loss from Titan).

Finally, the N_2 lunar atmosphere gives us an indication of what an atmosphere on Earth's Moon might have looked like if it had retained or gained an atmosphere after formation. The lower atmosphere temperature approximates the temperature the Moon would have with little or no greenhouse gas in its atmosphere.

Relative to the planetary radius, the estimated exobases are all fairly close to the ground with the exception of the atomic oxygen case on Mars, the N_2 case on Titan, the atomic nitrogen case on Titan, and the N_2 case on the Moon. In the first two cases, the altitude of the exobase is on the order of a third of the planetary radius, but in the latter two cases the exobase extends far out into space. The effect is particularly pronounced in the lunar N_2 case. Note that the exobase extends much farther out than in the Titan N_2 case, even though the Moon has somewhat higher surface gravity than Titan. This happens because we have assumed a greater atmospheric temperature for the lunar case, consistent with its closer proximity to the Sun. This remark underscores the importance of lower atmospheric temperature in determining the characteristics of atmospheric escape: perfectly apart from the exospheric temperature, a hotter lower atmosphere has a larger scale height, and therefore can extend further out to where the gravity is lower and the atmosphere can escape more easily. This is not much of an issue for bodies as massive as Earth or Venus but for smaller bodies it can be quite a significant effect.

Let's take stock of what we know so far. To determine the rate of escape of a constituent, we need to know the height of the exobase, the number density of that constituent at the exobase, and the proportion of particles whose energy exceeds the escape energy computed at the exobase. The definition of the exobase involves the temperature at the exobase, through the definition of scale height, so we must know a temperature for the exobase as well. This temperature might or might not also serve to characterize the distribution of particle velocity, depending on circumstances. Note that the concept of "exobase" is itself a severe idealization. The picture this calls to mind is of a distinct surface separating lower altitudes where collisions are frequent enough to maintain thermal equilibrium and higher altitudes where particles undergo ballistic trajectories without collision. This would be nearly the case for evaporation of a liquid into a vacuum, since there is a near-discontinuity in density in that situation. For gases, the transition is gradual, and it would be better to talk in terms of an "exobase region" involving a continuous profile of collision frequency and some escape to space from each layer – more toward the top, less toward the bottom. Modern calculations of atmospheric escape do indeed employ this level of sophistication, but the refinement alters estimates based on the ideal picture only by a factor of two or so. We'll see soon that this is not a serious threat to our main conclusions.

To proceed further, we need a probability distribution for molecular energy. The simplest case is one in which the molecules near the exobase can be regarded as being in thermodynamic equilibrium. This leads to what is called *Rayleigh-Jeans* or *thermal* escape. It is by far the simplest theory of atmospheric escape, but it is also the most useless; its main utility is to show that thermal escape is not a significant means of removing atmospheric constituents with the possible exception of light species such as He or molecular hydrogen (and even those only to a limited extent and in limited circumstances). The calculation proceeds as follows. For a gas in thermodynamic equilibrium at temperature T, the probability of a molecule with mass m having speed v is proportional to $\exp\left(-\frac{1}{2}mv^2/kT\right)$. This is the *Maxwell-Boltzmann distribution*. Note that if the gas is a mixture of molecules with various m, this formula still applies for the velocity distribution of each species separately, with the corresponding m used in the formula. The Maxwell-Boltzmann distribution has the important property that the proportion of molecules with energy much greater than kT becomes exponentially small. To determine the escape flux, one must integrate the Maxwell–Boltzmann distribution to determine the proportion of particles that have enough energy to escape, taking into account also the fact that particles are moving isotropically in all directions and that it is only the part of energy associated with radially outward motion that contributes to escape. If $n_{ex,m}$ is the number density at the exobase of a species whose molecules have mass m, then the flux of particles to space is $w_{J,m} n_{ex,m}$, where

$$w_{J,m} = \frac{1}{2\sqrt{\pi}}(1 + \lambda_c(m))e^{-\lambda_c(m)}\sqrt{\left(\frac{2kT_{ex}}{m}\right)},\tag{8.37}$$

in which $\lambda_c(m) = mg(r_{ex})r_{ex}/kT$ is the escape parameter defined previously. The Jeans flux coefficient $w_{J,m}$ has the dimensions of a velocity, and consists of the typical thermal velocity at the exobase reduced by an exponential factor that accounts for the proportion of molecules whose energy exceeds the escape energy. The total escape flux from the planet, expressed as molecules per second, is $4\pi r_{ex}^2 w_{J,m} n_{ex,m}$. For calculations of the lifetime of an atmospheric constituent, it is convenient to introduce the escape flux per unit surface area of the planet, for which we will use the notation Φ. Thus, $\Phi = w_{J,m} n_{ex,m}(r_{ex}/r_s)^2$.

The penultimate column of Table 8.5 gives the characteristic loss time of the dominant constituent of the hypothetical atmosphere by Jeans escape. This loss time is obtained by dividing the Jeans loss rate for the dominant constituent into the total number of particles in the atmosphere. With the exception of the lunar N_2 case, all the loss times are far too long to allow significant loss over the lifetime of the Solar System. For that matter, most of the loss times are well in excess of the lifetime of the Universe, and not even the extreme assumption of total decomposition of the atmosphere into lighter atomic constituents changes this conclusion. For the most part, the main constituents of terrestrial-type atmospheres cannot escape to any significant degree by thermal means. In particular, it is impossible to lose a primordial Venusian ocean by Jeans escape of H_2O. Even if we split off the O, the Earth atomic oxygen case says that it would be impossible to lose any significant quantity of the oxygen in the water by Jeans escape. The one case in which thermal escape is of interest for a heavy constituent is the warm lunar case. This case may seem somewhat fanciful but it is of considerable relevance to the question of habitable moons, such as might belong to extrasolar gas giants which orbit their primaries at Earthlike distances. Jeans escape is a real threat to the atmospheres of small moons if they are warm enough to support liquid water.

To get a feeling for the magnitude of the thermal escape of atomic hydrogen in the regime where H is a minor constituent of the exosphere, let's suppose that the molar concentration of H is 10% at the exobase. Later we'll learn how to relate the exobase concentration to the composition of the lower atmosphere. Let's take first the Earth case with an O-dominated exosphere. Using the assumptions of Table 8.5 the total number density at the exobase, obtained by plugging the exobase position and gas constants into Eq. (8.36), is $2.4 \cdot 10^{14}/m^3$, whence the H number density is $2.4 \cdot 10^{13}/m^3$. Multiplying this by the Jeans escape coefficient from the table and normalizing to surface area, we find an H escape flux $\Phi = 2.15 \cdot 10^{14}/m^2 s$. If the H ultimately came from decomposition of sea water, then each two atoms of H that escape account for the loss of one water molecule and the generation of one atom of oxygen. Converting this to mass, we find that in four billion years you could lose about $365\,000\,kg$ of water from each square meter of the Earth's surface, equal to a depth of about $365\,m$. You could not come close to losing an ocean on Earth or Venus this way, even with a hot exosphere; with a colder exosphere such as on Venus or the Early Earth, the loss rate would dwindle to practically nothing. If we want to get rid of a primordial Venusian ocean, we'll have to look at hydrogen-dominated exospheres, and find means other than thermal escape to pump the hydrogen into space.

Hydrogen, in the form of H_2, is one of the substances commonly outgassed from volcanoes on Earth and probably on other geologically active planets. In Earth's present highly oxygenated atmosphere, this hydrogen rapidly oxidizes into water, so there is little opportunity for free hydrogen to accumulate. On the anoxic Early Earth, however, the accumulation of hydrogen would be limited by the rate of escape to space. For a cold N_2-dominated exosphere, the exobase density of atomic hydrogen is $6.7 \cdot 10^{15}/m^3$ if the molar concentration is 10%. Using the Jeans escape coefficient from Table 8.5, the hydrogen escape flux would be a mere $\Phi = 3.32 \cdot 10^9/m^2s$. It has been estimated that the volcanic outgassing rate of H_2 on the Early Earth could have been on the order of 10^{15} molecules per second per square meter of Earth's surface, which is many orders of magnitude in excess of the Jeans escape flux. Thus, if Jeans escape were the only escape mechanism for hydrogen, hydrogen would accumulate to very high concentrations on the Early Earth. In reality, it would only accumulate to the point where the exosphere became hydrogen-dominated, whereupon other, more efficient escape mechanisms would take over. Still, we have learned from this exercise that hydrogen has the potential to build up to high values on an anoxic planet, that a cold exobase plays an important role in allowing this to happen, and that the exosphere is likely to have been hydrogen-dominated.

Another case of interest is hydrogen loss from Titan. In this case, the hydrogen is supplied by decomposition of CH_4 in the atmosphere, and the loss is important because the atmospheric chemistry would change quite a bit if hydrogen stuck around in the atmosphere. Let's suppose a 10% atomic hydrogen concentration at an N_2-dominated exobase on Titan. From Table 8.5, we see that the Jeans escape coefficient for atomic hydrogen is very large in the Titan case, owing to the low gravity. The exobase particle density is about $1.3 \cdot 10^{14}/m^3$ based on the table, whence the assumed hydrogen density is $1.3 \cdot 10^{13}/m^3$ and the escape flux is $\Phi = 6 \cdot 10^{16}/m^2s$. Assuming a CH_4 molar concentration of 30% over a layer 15 km deep in Titan's lower atmosphere, there are $1.3 \cdot 10^{30}$ hydrogen atoms per square meter of Titan's surface, stored in the form of CH_4. The calculated Jeans escape rate would be sufficient to remove this entire inventory in under a million years. The precise rate of hydrogen loss depends on the rate of decomposition of methane and the rate of delivery of hydrogen to the exobase, but it seems quite likely that Jeans escape can get rid of the hydrogen resulting from methane decomposition.

Before taking on more complicated and effective means of escape, there is one more basic concept we need to take on: diffusion and gravitational segregation of atmospheric species. Let's track the position of a molecule of species A moving with typical speed v, and colliding with background molecules from time to time. The typical distance the molecule moves between collisions is the mean free path ℓ computed earlier. If we idealize the collisions as causing a randomization of the particle's direction, then the particle will undergo a random walk. For particles undergoing random motions of this sort, the flux is proportional to the gradient of particle concentration; the process is called *diffusion*, and the proportionality constant is the *diffusion coefficient*, which we shall call D. It is closely related to the heat diffusion coefficient we have introduced in previous chapters. In addition, molecules or atoms in a gravitational field will accelerate downward under the action of gravity until the drag force due to collisions with the rest of the gas equals the gravitational force. This equilibration happens quickly, so that the particles attain a *terminal fall velocity* w_f. The terminal velocity is proportional to the local acceleration of gravity, and leads to a downward particle flux which is the product of the fall speed with the particle density.

The diffusion coefficient has units of length squared over time, and by dimensional analysis must be proportional to the product of mean free path ℓ with the typical thermal velocity $\sqrt{kT/m}$ where m is the particle mass of the species we are tracking. Since ℓ is inversely proportional to the *total* particle density n, the diffusion coefficient increases in inverse proportion to n. For this reason, it is often expressed in terms of a *binary diffusion parameter* b, via the expression $b \equiv Dn$. For any given pair of species, b is a function of temperature alone. For ideal hard-sphere collisions between particles with masses m_1 and m_2 and radii r_1 and r_2, the binary collision parameter is given by the expression

$$b = \frac{3\sqrt{2\pi}}{64}\frac{v}{\chi} \tag{8.38}$$

where χ is the collision cross-section area based on radius $(r_1 + r_2)/2$ and $v \equiv \sqrt{kT/\bar{m}}$ is the thermal velocity based on the harmonic mean of the masses, $\bar{m} = m_1 m_2/(m_1 + m_2)$. For an ideal hard-sphere gas the binary parameter, and hence the diffusion coefficient, increases with the square root of temperature. For actual gases, however, the effective collision diameter goes down somewhat with temperature, leading to other empirical temperature scaling laws, generally with temperature exponents in the range of 0.7 to 1. As with collision radius, atomic hydrogen is an exception, having a temperature scaling exponent between 1.6 and 1.7 for collisions with most species. The magnitude of the binary parameter for atomic hydrogen is also greater than one would expect from hard-sphere theory, since the effective collision radius is that of atomic hydrogen even when it is colliding with a substantially larger particle. Thus, atomic hydrogen has anomalously large diffusion, which increases anomalously strongly with temperature. The diffusion coefficient of atomic hydrogen is even larger than one would expect solely on the basis of its low mass and hence large thermal velocity.

Next, we remark that the fall speed scales with the velocity acquired by the particle through gravitational acceleration in the time between collisions. Thus w_f scales with $g\ell/\sqrt{kT/m}$. The ratio D/w_f thus is proportional to $(kT/m)/g = R_A T/g$, where R_A is the gas constant for the species A. This is just the isothermal scale height H_A that the species would have in isolation. In fact a more detailed calculation shows that for an isothermal ideal gas, the ratio D/w_f is not just proportional to H_A, but is actually exactly equal to it. (As a short cut to this result, we may argue that it is implied by the requirement that the scale height be equal to the usual hydrostatic result when the species A dominates the atmospheric composition.) In the following, we will keep things simple by restricting attention to the isothermal case, which will be sufficient for our purposes.

Let n_A be the particle density of species A, which may be one of many constituents of the gas. Putting together the flux due to diffusion and the gravitational settling, the net flux of the species (upward positive) is

$$F_A = -w_f n_A - D\frac{dn_A}{dr}. \tag{8.39}$$

Equilibrium is defined by a state of zero flux. In that case, the particle density is governed by the equation

$$\frac{dn_A}{dr} = -\frac{1}{H_A}n_A, \qquad H_A \equiv \frac{w_f}{D} = \frac{R_A T}{g(r)} = \frac{kT}{m_A g(r)} \tag{8.40}$$

where m_A is the mass of a molecule of species A. Thus, the particle density decays exponentially with scale height H_A. Note that this is identical in form to the particle density given by hydrostatic balance, except that the equation we have just derived applies to the particle density of each species separately, and not just to the particle density of all species together.

In fact, in equilibrium each species acts as if it were in hydrostatic equilibrium separately, and has the same hydrostatic scale height as if the other gases were not there at all. Since we are assuming thermodynamic equilibrium, all species are characterized by the same temperature, and of course all species are subject to the same gravitational acceleration. Thus, the scale height varies inversely with the molecular weight of the species. Since the density of light species decays less sharply with altitude than the density of heavier species, the atmosphere will tend to sort itself out in the vertical according to molecular weight. Light constituents will congregate near the top of the atmosphere like escaped helium balloons at the top of a circus tent.

This can only happen, however, if the mixing is dominated by molecular diffusion. In the lower atmosphere, mixing is overwhelmingly due to turbulent fluid motions, which treat all species equally and keep the mixing ratios uniform, in the absence of strong sinks or sources by chemistry or phase change. Since the molecular diffusivity is inversely proportional to total particle density, it will increase roughly exponentially with altitude, and will therefore come to dominate turbulent mixing at sufficiently high altitudes. The altitude where diffusive segregation begins to set in is called the *homopause* (sometimes *turbopause*) and the lower part of the atmosphere where mixing ratios of non-reactive substances are uniform is called the *homosphere*. It is very difficult to get an *a priori* estimate of turbulent mixing rates – indeed at one time it was expected that the Earth's stratosphere would be diffusively segregated. More often than not, observations of atmospheric composition provide the most reliable estimate of the degree of turbulent mixing. For the present Earth, the homopause is near 100 km, at which point the observed particle density is $n_h(Earth) = 1.2 \cdot 10^{19}/m^3$.

Above the homopause, then, atmospheric constituents segregate in the vertical according to molecular weight. The region above the homopause is also typically (though not necessarily) where atmospheric molecules begin to be exposed to ultraviolet photons sufficiently energetic to break up even the more stable components into lighter constituents, which also will stratify according to molecular weight. For Earth the scale height for N_2 is 9.1 km, for CO_2 5.8 km, for O_2 8.0 km, for atomic oxygen 15.9 km, for H_2 127.3 km, and for atomic hydrogen 254.5 km, all based on a temperature of 300 K. Numbers for Venus are similar. The first implication of these numbers is that the scale height for hydrogen is so large that even a small concentration of hydrogen at the homopause can cause the atmosphere to become hydrogen-dominated a small distance above the homopause. For example, if an N_2/H_2 atmosphere contains 1% molar concentration of H_2 at the homopause, then 50 km up the concentration has risen to 63% and 70 km up it is 93%. If the H_2 is converted to atomic hydrogen above the homopause, the segregation is even more effective. Similarly, if we take an Earthlike atmosphere that is 80% N_2 and 20% O_2 at the homopause, and then convert the O_2 into atomic oxygen, we wind up with 33% atomic oxygen near the homopause. By 20 km up, the concentration reaches 56%, and at 40 km it is 76% and thoroughly dominates the atmospheric composition. Finally, if we take an Early Earth CO_2/N_2 atmosphere consisting of 10% CO_2 at the homopause, then the CO_2 concentration falls to 0.5% at 50 km up, whereby we expect the outer atmosphere to be N_2-dominated. It is possible that the dissociation of CO_2 into CO and O could lead to an exobase dominated by atomic oxygen, but the fact that this does not happen on Venus today suggests strongly that the dissociation is too weak for this to happen, or the recombination of the two species is too efficient.[4]

---

[4] Venus at present does have a region above the exobase which is dominated by atomic oxygen, but this layer is too tenuous to affect the exobase height. The question of the circumstances in which oxygen can build up to a hot Earth-type exobase is a delicate and difficult one, which hinges on

Given an estimate of the turbulent diffusivity D_{turb}, the homopause density can be estimated directly from the scaling of the diffusion coefficient. Specifically, since $D \approx \ell(kT/m)^{\frac{1}{2}}$ then setting $D = D_{turb}$ and using the expression for the mean free path ℓ implies

$$n_h \approx \frac{1}{\chi D_{turb}} \left(\frac{kT}{2m}\right)^{\frac{1}{2}}. \tag{8.41}$$

In cases where no observations bearing on D_{turb} are available, assuming D_{turb} to be the same as for Earth is probably as good an assumption as any. In that case, the homopause density is related to Earth's value by the formula

$$n_h \approx \frac{\chi(Earth)}{\chi} \left(\frac{T}{T(Earth)} \frac{m(Earth)}{m}\right)^{\frac{1}{2}} n_h(Earth). \tag{8.42}$$

Given the weak variation of the factors multiplying $n_h(Earth)$ in typical cases, a good rule of thumb for use in making crude estimates of escape fluxes is simply to assume the homopause density to be the same as Earth's, though of course where observations are available it is better to use the observed value. For most escape calculations, it is not necessary to know the homopause altitude, though it can be estimated from the lower-atmosphere scale height if it is desired. The homopause altitude only becomes important when it is high enough that gravity is significantly attenuated relative to surface gravity.

The homopause density provides the essential point of reference for most atmospheric escape problems involving multicomponent atmospheres. Since most of the atmospheric mass is in the troposphere, in order to understand how long it takes to lose some part of an atmosphere, we need to relate the escape flux to the tropospheric composition, which in turn requires us to relate the composition of the exobase to the tropospheric composition. This proceeds via the intermediary of determining the homopause composition. To take the simplest case first, consider gases that do not undergo condensation or significant chemical sinks or sources within the lower atmosphere – for example O_2 and N_2 on Earth. These will be well mixed throughout the homosphere, so if there is about 20% molar O_2 in the troposphere, there will be about 20% molar O_2 at the homopause. To get the O_2 particle density, we multiply this ratio by the homopause density, which we know how to determine. This then gives the supply of O_2 molecules, which dissociate above the homopause into atomic oxygen, allowing us to determine the atomic oxygen concentration at the exobase using the scale height for atomic oxygen alone to extrapolate from the homopause to the exobase. (A more precise calculation would require modeling of dissociation and reaction rates above the homopause.) Things would work similarly for other gases that are non-reactive/non-condensing in the lower atmosphere.

For condensing gases, such as water vapor mixed with air on Earth, CH_4 mixed with N_2 on Titan, or water vapor mixed with CO_2 on Early Venus, there is an additional step on the way to determining the composition of the homopause. Condensable gases do not have uniform concentrations in the homosphere, because of the limitations imposed by Clausius–Clapeyron. Let's take water vapor on Earth as an example. Water vapor makes

details of atmospheric chemistry and atmospheric composition. An early water-rich Venus atmosphere would have another source of oxygen through dissociation of water vapor, which could conceivably lead to an oxygen-dominated exobase. Calculations performed to date do not seem to bear out this possibility, but the situation has not been thoroughly explored and there is plenty of room for surprises.

up a few percent of the lower atmosphere at present, but most water vapor entering the stratosphere must make it through the cold tropical tropopause. The temperature there is around 200 K, and the corresponding water vapor mixing ratio, by Clausius–Clapeyron, is $1.6 \cdot 10^{-5}$, given a tropopause pressure of 100 mb. Though there are slight additional water vapor sources in the stratosphere from oxidation of methane, the tropopause concentration is a good estimate of the water vapor concentration that will be found at the homopause. The tropopause acts as a *cold trap*, dehumidifying the upper atmosphere and strongly limiting the opportunity for water vapor to escape or for hydrogen to build up in Earth's upper atmosphere through decomposition of water vapor.

The cold trap temperature is defined as the lowest temperature encountered below the homopause, and to determine it precisely, one must carry out a full radiative–convective calculation of the atmospheric structure. On an adiabat, temperature would go down indefinitely with height until absolute zero were reached; it is the interruption of temperature decay by the takeover of radiative equilibrium in the stratosphere that usually determines the cold trap temperature. In the absence of a full radiative–convective equilibrium calculation, the skin temperature of the planet often provides an adequate crude estimate of the cold trap temperature. Once the cold trap temperature is known, the maximum possible partial pressure of the condensable at the cold trap is given by Clausius–Clapeyron. However, it is the molar concentration of the condensable we need, since this is the quantity that is preserved as air is mixed up toward the homopause without further condensation. To determine the molar concentration we need the partial pressure of the non-condensable gas at the cold trap. This is obtained using the tools provided in Chapter 2. One computes the adiabat starting from a specified surface pressure, surface temperature, and surface condensable concentration – following the dry adiabat with constant condensable condensation until the atmosphere becomes saturated, and following the moist adiabat thereafter until the cold trap temperature is reached. The usual procedure for computing the moist adiabat then yields the necessary molar concentration. All other things being equal, as more non-condensable is added to the atmosphere, the cold trap concentration goes down owing to greater dilution of the condensable substance. The precise functional form of the dilution depends on the thermodynamic constants of the condensable and non-condensable substances under consideration.

As an example, let's look at the cold trap water vapor concentration that would be encountered during a dry runaway greenhouse in a CO_2–H_2O system. Recall that in a dry runaway the surface gets so hot that the entire ocean is evaporated into the atmosphere, and there is no liquid water at the surface; in this case, the shutoff of silicate weathering should allow any outgassed CO_2 to accumulate in the atmosphere, resulting in atmospheres consisting of CO_2 and water vapor in proportions determined by the abundance of these substances in the planetary composition (less whatever water may have already escaped). Results for various sizes of oceans and various CO_2 abundances are given in Table 8.6, based on a cold trap temperature of 200 K. The saturation vapor pressure and the adiabat were computed using the ideal gas equation of state and the idealized exponential form of Clausius–Clapeyron; these are not quantitatively accurate for the pressures and temperature under consideration, but they suffice to delineate the general behavior of the cold trap concentration. We see that, for any given inventory of water, the cold trap concentration approaches unity (pure steam) when there is little CO_2 present, but that the cold trap concentration falls to very small values as the CO_2 inventory approaches values similar to that of Venus. or the CO_2 equivalent of Earth's crustal carbonates. Also, for any fixed partial pressure of CO_2 at the surface, the cold trap concentration increases as the water inventory

	Partial pressure CO_2					
	0 bar	1 bar	10 bar	30 bar	60 bar	90 bar
25 bar	1	0.83	0.23	0.016	$3.3 \cdot 10^{-4}$	$5.7 \cdot 10^{-5}$
50 bar	1	0.90	0.42	0.11	0.0090	$9.3 \cdot 10^{-4}$
100 bar	1	0.95	0.61	0.28	0.092	0.024

The column headers give the partial pressure of CO_2 at the surface. For each row, the mass of the water vapor inventory (which has come from evaporation of an ocean) is held fixed at the indicated amount. The water vapor inventory is expressed as the pressure that would be exerted by the ocean if the water were condensed out into a liquid layer. For a planet with $g = 10 \, m/s^2$, a 100 bar ocean corresponds to a mass of $10^6 \, kg/m^2$, or a depth of about 1 km. The 25 bar and 50 bar cases were computed with a surface temperature of 540 K, while the 100 bar case was computed at 570 K so as to allow for a more massive water content without bringing the surface too close to saturation. Note that, as discussed in Chapter 2, the equivalent pressure of ocean differs somewhat from the partial pressure of water at the surface, since the mixing ratio of water is not uniform above the altitude where condensation first occurs.

Table 8.6 Table of water vapor molar concentrations at a 200 K cold trap, for a CO_2–water atmosphere.

increases. Still, for a 90 bar ocean (about half the mass of Earth's), and with a 90 bar inventory of CO_2, the cold trap concentration is only 2.4%. Thus, unless the CO_2 inventory on a planet is very low or the water inventory is very high, the cold trap is likely to impose a significant barrier to water loss during a dry runaway scenario. Even if the water inventory is initially high, as water is lost the cold trap becomes a progressively more severe impediment, making it hard to lose the last 90 bars worth of ocean, and even harder to lose the last 50 bars.

For a single-component condensing atmosphere such as a water-vapor-dominated runaway atmosphere on Venus or a condensing CO_2 atmosphere on Mars, one no longer has to consider the cold trap issue, however. If there is only a single atmospheric component, then perforce knowing the total homopause density tells us the particle density of the atmospheric substance, regardless of how much condensation it has undergone in the troposphere.

Besides condensation traps, there can be chemical reactions which affect the homopause concentration. Notably, H_2 has little chance to escape in the modern oxygenated Earth, because it oxidizes to the heavier, condensable H_2O before it has a chance to reach the homopause.

Now let's revisit the problem of hydrogen loss from Early Earth, a runaway-state Venus, and Titan. We will assume a mixture of hydrogen with some other gas in a known proportion at the homopause, and then use the scale heights of the two gases to compute the changing composition as the exobase is approached. This allows us to say when hydrogen dominates the exobase, and what the resulting exobase height is. An important complication is the anomalously small collision cross-section of atomic hydrogen, and we must remember to take this into account when computing the mean free path for hydrogen-dominated exospheres.

For the anoxic Earth, we wish to determine how high the homopause concentration has to be in order for the escape flux to equal the volcanic outgassing. We simplify the problem by assuming H_2 to be well mixed below the homopause, but to dissociate into atomic hydrogen just above the homopause. Thus, if we know the homopause concentration of atomic hydrogen, the well-mixed tropospheric H_2 density is half this value. Start by assuming the atomic hydrogen density at the homopause to be 20%, and the balance of the atmosphere to be N_2. Using the scale heights for the two gases, when we compute the exobase position taking into account the varying composition with height, we find that the exobase is completely hydrogen-dominated, and that the exobase has moved out to an altitude of 1853 km (based on an exobase temperature of 300 K). The escape flux from this extended pure hydrogen exobase is $\Phi = 10^{12}/m^2s$, which is still three orders of magnitude below the estimated volcanic outgassing rate of H_2. Unless some more effective escape mechanism intervenes, hydrogen should build up to extremely high concentrations in the lower atmosphere.

For Venus, we assume an all water-vapor lower atmosphere. We take the homopause density to be $1.2 \cdot 10^{19}/m^3$ and assume that one-half of the water vapor there breaks up into atomic hydrogen and oxygen. To avoid dealing with a three-component atmosphere, we'll somewhat arbitrarily ignore the resulting oxygen (perhaps it recombines into O_2 which has such a small scale height that not much of it reaches the exobase) and compute the exobase from a homopause composition consisting of one-third water vapor and two-thirds atomic hydrogen; in addition, we'll assume a 300 K exobase temperature. The exobase is again found to be hydrogen-dominated, and at the relatively high altitude of 3050 km above the surface. The escape flux is $\Phi = 1.1 \cdot 10^{14}/m^2s$, which would remove the hydrogen in one bar of water vapor in 200 million years. This is significant, but in 2 billion years one could only remove 10 bars of ocean. By this means one could get rid of an ocean only about a tenth the mass of Earth's, though one could get rid of more if one could justify using a higher exobase temperature. Assuming that the water vapor at the homopause dissociates completely into atomic oxygen and atomic hydrogen changes these numbers very little, since the exobase is still hydrogen-dominated.

It should be remarked that it is hard enough to get rid of an ocean's worth of hydrogen on a runaway Venus, but getting rid of an ocean's worth of oxygen by escape to space is completely out of bounds and none of the other escape mechanisms we will consider come close to closing the gap. The only hope of getting rid of the oxygen resulting from runaway followed by hydrogen escape is to react the oxygen with crustal rocks. Even this is problematic, since a great volume of crustal rock must be made available in order to take up the oxygen from an appreciable ocean. Whether this is indeed possible is one of the outstanding Big Questions. There is no data that absolutely forces us to assume that Venus indeed started with an ocean, so it remains possible that Venus was quite dry from the very beginning.

In our earlier calculation of hydrogen loss from Titan we found that a 10% hydrogen concentration at the exobase was sufficient to sustain a large thermal escape rate. How low does the homopause concentration have to be in order to keep the exobase N_2-dominated? To answer this, we again make use of the scale heights of the two gases to compute the exobase composition simultaneously with the exobase height. In this case, we find that with a 300 K exobase, the homopause mixing ratio of hydrogen must be 10^{-6} or less in order to keep Titan's exobase N_2-dominated. With that homopause concentration, the escape flux is $\Phi = 8.8 \cdot 10^{15}/m^2s$, which is somewhat less than our previous estimate (mainly because of the different means of estimating the exobase density). The main conclusion to be drawn from this exercise is that it only takes a tiny hydrogen concentration at the homopause to sustain the large escape rates we computed earlier. If the hydrogen concentration is increased to the point that the exosphere begins to become hydrogen dominated, then the exobase in fact moves out to infinity, because of the large scale height and low gravity. In that regime, hydrogen is likely to escape hydrodynamically (Section 8.7.4) rather than thermally.

8.7.2 Diffusion-limited escape

The efficiency of escape of material that reaches the exobase is not necessarily the controlling factor determining atmospheric mass loss. For mass to escape from the exobase, it must first be delivered to the exobase, and in many circumstances the rate of transport of mass to the exobase is the limiting factor. When a minor constituent of an atmosphere is escaping, it must first diffuse through the dominant component on its way to the exobase, and even if the escape from the exobase is very effective, mass cannot escape faster than the rate with which it can diffuse up to the exobase. In such cases we can put an upper bound on the rate of escape without knowing much about the precise means of escape from the exobase. This upper bound is the rate of *diffusion-limited escape*. It has the virtue that it can be computed in a very simple and straightforward fashion.

We consider the diffusion in a gravitational field of a substance A with number density $n_A(r)$ through a background gas with density $n(r)$ satisfying $dn/dz = -n/H$, where H is the scale height of the background gas. The equilibrium distribution of A was determined earlier by setting the flux to zero, but now we will determine its distribution assuming a constant non-zero flux. If we let b be the binary diffusion parameter for substance A in the background gas, then Eq. (8.39) for the flux can be re-written

$$F_A = -\frac{1}{H_A}\frac{b}{n}\cdot n_A - \frac{b}{n}\frac{dn_A}{dr} = -b\frac{n_{A,e}}{n}\frac{d}{dr}\frac{n_A}{n_{A,e}} \tag{8.43}$$

where $n_{A,e}$ is the equilibrium distribution of substance A, which satisfies $dn_{A,e}/dr = -n_{A,e}/H_A$. As expected, the flux vanishes when $n_A = n_{A,e}$.

For the density distribution to be time-independent, the net flux through a spherical shell, $4\pi r^2 F_A(r)$, must be independent of r. We'll normalize this constant flux to the surface area, writing $\Phi = (r/r_s)^2 F_A$ as we did for the Jeans flux. For a given constant Φ, Eq. (8.43) defines a first order differential equation for n_A. The upper boundary condition for this equation is applied at the exobase, and states that the flux delivered to the exobase must equal the escape flux from the exobase. The escape flux can be written $w_* n_A(r_{ex})$, where w_* is the escape flux coefficient associated with Jeans escape or some other mechanism. Thus, the upper boundary condition can be written $(r_{ex}/r_s)^2 w_* n_A(r_{ex}) = \Phi$. This determines $n_A(r_{ex})$ in terms of Φ, and we must then solve the equation to get $n_A(r)$ and adjust Φ so that the lower boundary condition on n_A at the homopause is satisfied. Now, when w_* becomes very large, molecules are removed essentially instantaneously when they reach the exobase. In this case $n_A(r_{ex}) \to 0$ and we can take a shortcut to determine the limiting flux.

For simplicity we'll assume that the layer is isothermal, so that b is constant. Multiply Eq. (8.43) by $n/n_{A,e}$ and integrate from the homopause to the exobase to yield

$$\Phi \cdot \int_{r_h}^{r_{ex}} \frac{r_s^2}{r^2}\frac{n}{n_{A,e}}\,dr = b \cdot \frac{n_A(r_h)}{n_{A,e}(r_h)}, \tag{8.44}$$

which makes use of the assumption $n_A(r_{ex}) \approx 0$. The integral involves only known quantities, so this equation defines the limiting flux in terms of the homopause density of the escaping substance. Let's suppose that the exobase has low altitude in comparison with the radius of the planet. In this case, gravity is nearly constant and $n/n_{A,e}$ varies like $\exp(-(1/H - 1/H_A)z)$ where z is the altitude. The ratio decays exponentially if the scale height of the background gas is less than the scale height of minor constituent, i.e. if the minor constituent is lighter than the background gas. If, moreover, the layer between homopause and exobase is thick enough that the ratio decays to zero at the exobase, then the integral is simply $n/n_{A,e}$ at the homopause divided by $(1/H - 1/H_A)$. Therefore, under these assumptions, which are quite widely applicable, the diffusion-limited escape flux takes on the simple form

$$\Phi = b \frac{n_A(r_h)}{n(r_h)} \cdot \left(\frac{1}{H} - \frac{1}{H_A} \right) = D n_A(r_h) \cdot \left(\frac{1}{H} - \frac{1}{H_A} \right) \qquad (8.45)$$

where D is the diffusivity at the homopause. The escape of the minor constituent cannot exceed this flux no matter how effective the escape mechanism may be.

Exercise 8.5 Derive an expression for the diffusion limited flux in the case when the diffusing constituent has greater molecular weight than the background gas. How does the limiting flux depend on the layer depth in this case?

As a first simple example of diffusion-limited escape, let's take a look at the escape of hydrogen from an anoxic Early Earth. Suppose that H_2 diffuses through pure N_2 above the homopause. At $300\,K$, the binary parameter for this pair of species is $2 \cdot 10^{21}/ms$. Then, noting that n_{H_2}/n is the molar concentration η_{H_2} of hydrogen at the homopause and that the scale height for H_2 is much greater than the scale height for N_2, Eq. (8.45) implies that $\Phi \approx (2 \cdot 10^{21}/H_{N_2})\eta_{H_2} = 2.2 \cdot 10^{17}\eta_{H_2}/m^2s$. We'll assume that H_2 has nearly uniform concentration below the homopause, as is reasonable in the absence of oxygen. With this assumption, we can directly determine how high the concentration has to go in order to lose the volcanic hydrogen source, assuming the loss to be diffusion limited. Thus, $\eta_{H_2} = 10^{15}/2.2 \cdot 10^{17} = 0.0045$. Therefore, while Jeans escape would allow H_2 to build up to very high concentrations in the troposphere, diffusion-limited escape of H_2 could hold the concentration to well under a percent. Of course, for the diffusion limit to be reached, we would need a much more efficient means than Jeans escape of removing H_2 once it reaches the outer atmosphere. If the H_2 were to dissociate into atomic hydrogen near the homopause, the diffusion-limited escape flux would be much greater and the equilibrium diffusion-limited concentration would be much lower, since the binary coefficient for atomic hydrogen diffusing through N_2 is $10^{24}/ms$ at $300\,K$.

Exercise 8.6 Suppose that an N_2/H_2 atmosphere has some initial hydrogen concentration which is escaping at the diffusion-limited rate, but which is not being replenished by volcanic outgassing or any other source. Compute the exponential decay time of the hydrogen in the atmosphere assuming (a) H_2 diffuses through N_2 above the homopause, or (b) H_2 dissociates and diffuses as H through N_2 above the homopause.

As a second example, let's consider diffusion-limited water escape for a dry runaway state on a planet like Venus. For the dry runaway, we assume that the entire ocean is evaporated into the atmosphere as water vapor, and that in the absence of liquid water a dense CO_2 atmosphere accumulates because of weak or absent silicate weathering. In this case, water may escape in the form of H_2O diffusing through CO_2. The binary parameter for this pair is $8.4 \cdot 10^{20}/ms$ at $300\,K$ based on the hard sphere approximation with $140\,pm$ radii. Plugging in the appropriate scale heights for Venus, Eq. (8.45) implies that $\Phi \approx 7.8 \cdot 10^{16}\eta_{H_2O}/m^2s$, where η_{H_2O} is the water vapor concentration at the homopause. For the present application, we are interested in how long it takes to lose the hydrogen in an ocean, assuming it diffuses upwards as water vapor which then dissociates at high altitudes. We have seen earlier that η_{H_2O} is determined by the cold trap concentration in a dry runaway, which can range from minuscule values of a few parts per million to values approaching unity as wet runaway conditions are approached. As a point of reference, let's assume that the cold trap concentration is 10%, which is typical of hot, moist conditions. Then, in one billion years, $2.4 \cdot 10^{32}$ molecules of water could be lost per square meter of the planet's surface. This amounts to $7.2 \cdot 10^6 \, \text{kg/m}^2$, or just over a 7 km depth of ocean. It would thus appear that diffusion limitation is not a major impediment to loss of an ocean if the cold trap concentration is 10% or more. When the cold trap concentration falls below 1%, though, the diffusion limitation becomes a serious bottleneck. On the other hand, if water dissociates near the homopause and hydrogen escapes by diffusing as H or H_2, the diffusion-limited escape rate becomes much greater, and lower cold trap concentrations can be tolerated without making it difficult to lose an ocean, as is illustrated in the following exercise

Exercise 8.7 For the conditions given in the preceding paragraph, compute the diffusion limited escape flux assuming water dissociates into H_2 at the homopause, so that $\eta_{H_2} = \eta_{H_2O}$, the latter being the cold trap concentration. Compute the lowest value of η_{H_2O} that permits the hydrogen in a 7 km deep ocean to escape. Do the same for the case in which water dissociates into H (in which case $\eta_H = 2\eta_{H_2O}$). In both cases you may use the hard-sphere formula in computing the binary parameter b, assuming a collision radius of 140 pm for the H_2 case and a collision radius of 7 pm for *both* colliding species for the atomic hydrogen case.

Further examples of diffusion-limited escape are developed in the Workbook for this chapter.

The concept of diffusion-limited escape applies in a straightforward fashion only to the escape of a minor constituent, which makes up a small proportion of the diffusing layer. The removal of substantial quantities of a major constituent from the top of the diffusing layer has the potential to create large, unbalanced pressure gradients, which would drive an upward mass flux far in excess of the diffusive flux. There is no magic concentration threshold defining a "major" constituent, but certainly one should begin to worry about the induced flow when the concentration exceeds 10% or so. This regime is largely unexplored territory, but is somewhat related to the hydrodynamic escape mechanism which we shall discuss a bit later in this section.

8.7.3 Non-thermal escape

Using Planck's constant, the energy of an EUV photon with wavelength 0.05 μm is $4 \cdot 10^{-18}$ J. This is sufficient to dissociate the components of many molecules, and to knock off electrons from just about anything – a process called *ionization* which produces charged particles in the outer atmosphere. In fact, ionization is the principal means by which absorbed EUV heats the outer atmosphere, since ejection of an electron increases the kinetic energy of the ion left behind, as well as imparting energy to particles with which the electron subsequently collides. When easily ionized species are used up, the heating is correspondingly reduced.

The energy of such a photon is somewhat in excess of the escape energy for atomic oxygen on Earth, and so if this energy can be converted into kinetic energy of an atom, it can lead to escape at rates far higher than one would get from Jeans escape. This idea underlies the subject of non-thermal escape.

The term *non-thermal escape* represents a whole zoo of mechanisms united only by the common theme that they are not thermal – that they rely on the strong deviations from the Maxwell–Boltzmann distribution that are possible when collisions are infrequent. The study of non-thermal escape is rather like, to paraphrase a remark of Stanislaw Ulam, the study of the physiology of non-elephants. We will discuss a few examples to give the general flavor of the issues involved, but the reader should be aware that we are barely scratching the surface of this difficult and interesting subject.

The Maxwell–Boltzmann distribution corresponding to a given temperature is maintained through frequent collisions, which redistribute energy amongst the molecules making up the gas. If an individual particle in the gas acquires some new kinetic energy as a result of absorption of an EUV photon or a chemical reaction which releases energy, the energy may initially be much greater than the typical energy kT characteristic of the temperature of the gas. It is only after several collisions that the extra energy is equitably redistributed amongst the other particles – a process known as *thermalization*. If collisions are infrequent, the thermalization time can be very long, leading to a large population of particles that have anomalously large energy. Indeed, in such a case, the energy distribution deviates greatly from the Maxwell–Boltzmann distribution, and the gas is no longer characterized by a temperature in the usual sense of equilibrium thermodynamics. The anomalously energetic atoms or molecules are often referred to as the "hot" population, as in "hot oxygen" or "hot hydrogen." The hot atoms may have enough energy to escape before they thermalize, or they may remain gravitationally bound but later impart their energy to lighter species which can escape. This is the general idea of non-thermal escape, and the reader can probably understand already why there are many ways in which it can happen.

When the non-thermal escape is an indirect result of the accumulation of gravitationally bound hot atoms, the calculation of escape rate is very complicated, since one needs first to model the number of hot atoms, and the distribution of their energy. Ultimately, the energy is supplied by EUV absorption (or perhaps solar wind interactions, to be discussed separately), so that the flux of EUV provides an upper limit to the rate at which any constituent can escape. However, the actual rate depends on how the delivered energy is spread amongst the escaping constituents. If the energy is concentrated in relatively few particles, than an escape flux can be sustained, whereas if the energy is spread too thinly or is wasted on heavy particles, there may be little escape. Another important consideration is that only EUV delivered above the exobase can lead to non-thermal escape; energy deposited at lower altitudes will instead thermalize through collisions. Determining the proportion of EUV which is deposited above the exobase requires consideration of the number of particles above the exobase and the EUV absorption cross-section of the species making up the exosphere.

Photons have little momentum, so the absorption of an EUV photon cannot directly increase the kinetic energy of the molecule or atom absorbing it to any great degree. Rather, the increase of kinetic energy takes place through *ionization* or *dissociation*. In the first case, a fast electron is ejected in one direction while the heavier positive ion moves more slowly in the opposite direction, with such a speed as to satisfy the momentum balance. It takes energy (the *ionization energy*, which differs according to species) to pry loose an electron, so the total kinetic energy supplied to the electron and ion is the energy of the photon minus the ionization energy. Dissociation is similar, except that the molecule breaks apart into neutral or charged heavy constituents instead. Dissociation also requires energy, so that the energy available to increase the kinetic energy of the fragments is the energy of the photon diminished by the dissociation energy.

As a concrete and relatively simple example, let's consider photodissociation of N_2 on Mars, which has been suggested as a possible means of losing the nitrogen in a hypothetical nitrogen-rich primordial Martian atmosphere. This problem is of interest because Mars at present has little N_2 in its atmosphere, and we would like to know whether this means that Mars must have formed with very little nitrogen (unlike Earth or Venus), or whether the smaller size of the planet allowed its initial nitrogen endowment to escape.

Assuming the Martian exobase to be reasonably close to the ground, an N atom requires $2.92 \cdot 10^{-19}$ J of energy to escape. This is much greater than the typical thermal energy $kT = 4. \cdot 10^{-21}$ J. Now, the dissociation energy of N_2 is about $1.94 \cdot 10^{-18}$ J, so assuming the excess absorbed energy to be equally distributed between the two dissociated atoms, a photon with energy in excess of $1.94 \cdot 10^{-18}$ J $+ 2 \cdot 2.92 \cdot 10^{-19}$ J, i.e. $2.52 \cdot 10^{-18}$ J, can cause escape. Using Planck's constant, this corresponds to EUV photons with wavelengths shorter than 0.078 μm. Next, we must determine the rate at which such photons collide with N_2 molecules and cause dissociation. For simplicity, we'll assume a pure N_2 atmosphere, so we need not worry about collisions with other molecules. If the exosphere is optically thin in the EUV, then the proportion of incident photons which are absorbed is the product of the absorption cross-section with the number of particles per square meter in the exobase. The latter is approximately the product of the exobase density with the scale height, i.e. $1/\chi\sqrt{2}$, where χ is the molecular collisional cross-section area. We are left with the rather tidy result that *the proportion of photons that are absorbed in the exosphere is the ratio of the EUV cross-section to the molecular collision cross-section*. In the relevant part of the EUV the absorption cross-section for N_2 is about $3 \cdot 10^{-21}$ m^2, so the proportion of incident photons which are absorbed in the N_2 exosphere is 4.6%. Note that this fraction is independent of the total mass of the atmosphere, so that we cannot increase the rate of escape by increasing the total amount of N_2 in the atmosphere. In general, this is one of the chief factors limiting the effectiveness of non-thermal escape due to photodissociation.

Based on satellite observations, the flux of sufficiently energetic EUV photons at Earth's orbit is currently about $8 \cdot 10^{14}$/m^{2}s, which scales to about $4 \cdot 10^{14}$/m^{2}s at the orbit of Mars. Allowing for the proportion which are absorbed in the exosphere, this leads to an escape of $4.6 \cdot 10^{12}$ N atoms per square meter of the planet's surface per second or $1.45 \cdot 10^{29}$ atoms per square meter of surface in a billion years. One bar of N_2 on Mars contains $1.16 \cdot 10^{30}$ atoms per square meter, so we conclude that EUV-induced non-thermal escape has the potential to remove about a half a bar of N_2 from Mars over the lifetime of the Solar System, or more if the EUV flux were higher in the early Solar System.

The above calculation is sufficient to show that escape from EUV-induced dissociation is a potentially important factor worthy of more sophisticated study, but it is highly over-simplified and leaves out many considerations that could substantially reduce the escape. First, the dissociated N atoms are not always in the lowest energy electron configuration (the ground state). The energy that goes into excited electron states is not available to feed kinetic energy sustaining escape. To crudely take this into account, in the above calculation we used the energy for dissociation into the most favored excited electron configuration; the dissociation energy into ground-state N atoms would be only $1.57 \cdot 10^{-18}$ J, though such dissociations are believed to be rare. Second, photons with energy higher than $2.5 \cdot 10^{-18}$ J can cause ionization of the N_2 molecule rather than dissociation. Only a small proportion of the molecules will directly dissociate. Some of the N_2^+ ions will later dissociate and lead to escape when they recombine with the free electrons released by ionization – a process called *dissociative recombination* – but calculation of the proportion that do so is quite involved. Third, in an atmosphere that is not pure N_2 there is a stew of other dissociation products, both neutral and ionized, that need to be considered, as well as the reactions between them.

On the other hand, dissociation due to EUV photons with energy less than $2.52 \cdot 10^{-18}$ J leads to a population of gravitationally bound N atoms which have typical energies much larger than the typical thermal energy. As mentioned earlier, such populations of energetic particles are referred to as "hot," as in "hot nitrogen" or "hot oxygen," even though they are not in thermodynamic equilibrium and their temperature is not strictly defined. Sometimes a "temperature" is defined for such populations just as a means of characterizing the energy. For example, if a photon with energy $2.3 \cdot 10^{-18}$ J causes dissociation, it leaves behind two N atoms with an energy of $0.17 \cdot 10^{-18}$ J each. Setting kT equal to this energy yields a "temperature" of over $12\,000$ K. This is merely a way of summarizing how the energy of the hot population compares with the much lower background thermal energy (characterized by a temperature of around 300 K for the Martian exosphere). The hot population does not itself escape, but it represents a reservoir of energy that can be transferred to other species (especially lighter ones), and which can lead to the escape of those. In this mode, the outer atmosphere acts like a storage-beam particle accelerator, accumulating energetic particles for later use.

Other dissociations can also lead to escape or accumulation of hot atoms. Nitrogen (N_2) has an unusually large dissociation energy. The reactions $O_2 + h\nu \rightarrow O + O$, $CO_2 + h\nu \rightarrow CO + O$, $H_2O + h\nu \rightarrow H + OH$, and $OH + h\nu \rightarrow O + H$ all have dissociation energies of around $0.8 \cdot 10^{-18}$ J. Dissociation of these by EUV leaves more energy left over to feed escape.

Dissociative recombination, particularly involving oxygen, is an important indirect means by which EUV absorption can lead to escape or accumulation of hot atoms. For example, when the recombination of the O_2^+ ion with an electron leads to the dissociation of the result into two O atoms in their ground states, then $1.12 \cdot 10^{-18}$ J are released, or $0.56 \cdot 10^{-18}$ per atom. At present, such reactions take place on Venus and Mars as a consequence of the photodissociation of CO_2, but in earlier epochs on Venus water vapor could also have been involved. The escape energy for O on Venus is $1.43 \cdot 10^{-18}$ J, so dissociative recombination does not directly lead to escape of O on Venus, but rather to the accumulation of hot oxygen that can later allow a smaller part of the oxygen (or a greater part of a lighter species) to escape. Significant amounts of oxygen are indeed observed escaping at present from Venus, but it is not thought that the non-thermal mechanisms could have gotten rid of any significant part of the oxygen in a primordial Venusian ocean. On Mars, however, dissociative recombination directly leads to the escape of an oxygen atom, since the escape energy is only $0.33 \cdot 10^{-18}$ J.

A further set of complications arises from the fact that charged particles interact with the planet's magnetic field, if it has one. Charged particles tend to spiral tightly along magnetic field lines, and so they can escape easily only when they are on the relatively limited proportion of field lines that are *open* – that have one end leading out into outer space. When the magnetic field is important, there is thus a great premium in generation not just of energetic particles in general, but energetic *neutral* particles. For this reason, a great deal of attention in work on non-thermal escape has been lavished on processes that allow energetic ions to deliver their energy to neutral particles.

8.7.4 Hydrodynamic escape

Hydrodynamic escape is basically a more efficient means of deploying the energy available to the atmosphere in order to assist escape. The energy involved still comes from EUV absorption or the general thermal energy of the atmosphere, but instead of this accumulating in a more or less random set of motions, in some circumstances the energy can sustain a mean outward escaping flow which carries fluid to space without wasting energy on motions directed toward the planet or on a population of molecules with velocities too small to escape. In these circumstances, there is no longer an exobase from which particles escape directly to space. Instead, there is an outflow that acts like a collisional fluid out to distances so great that the atmosphere is no longer gravitationally bound. Hydrodynamic escape plays a central role in hydrogen-rich outer atmospheres (including those that arise from dissociation of water vapor), and so we will accord it a great deal of attention. The phenomenon can also play a role in escape of heavier species in the case of small bodies or very hot planets. Hydrodynamic escape is a fascinating and important subject, and one which is very ripe for further research. It is also a subject where the student can attain a nearly complete level of understanding on the basis of some simple principles of thermodynamics and mechanics. Therefore, it is a subject which we will delve into at some considerable length.

Equations for 1D compressible transonic flow

The starting point for our discussion is Newton's second law – force equals mass times acceleration – written for the radial direction measured outward from the planet's center. Let r be the radial position, and suppose that the only non-vanishing velocity is the radial velocity $w(r)$. We suppose further that the system is in a steady state, so that winds, temperature, and so forth are time-independent when measured at any fixed position. This does not mean that acceleration vanishes, however, since the acceleration must be measured following the path of an outward-moving fluid particle, whose position can be written $r(t)$. Since $w \equiv dr/dt$, the acceleration following a fluid parcel is $dw/dt = (dw/dr)(dr/dt) = w(dw/dr)$. Let r_s be the radius of the planet's surface,[5] and $g_s = g(r_s)$ be the surface gravity. Then Newton's law (expressed per unit volume) becomes

$$\rho w \frac{dw}{dr} = -\frac{dp}{dr} - \rho g_s \frac{r_s^2}{r^2}.$$ (8.46)

When $w = 0$ this reduces to the hydrostatic balance given in Eq. (8.35). Atmospheres are never exactly at rest, and so the hydrostatic *approximation* we have been using throughout this book amounts to an assumption that the accelerations on the left hand side of the equation are negligible compared with the individual terms on the right hand side. What we are up to now is the business of figuring out what happens when the radial acceleration becomes large enough to disrupt the hydrostatic balance. Basically, we only need to solve the radial momentum equation subject to suitable boundary conditions; the resulting solution determines the outward mass flux. However, there are a number of subtleties concerning the circumstances in which a steady solution can exist, and the nature of the boundary conditions that can be applied. Therefore, we proceed to the solution through a number of intermediate steps.

To obtain a solution, the momentum equation must be supplemented by mass conservation and thermodynamic relations. For steady flow, conservation of mass requires that the mass flux must be independent of r. Defining the area of the shell at radius r as $A(r) = 4\pi r^2$, the mass flux $\rho(r)w(r)A(r)$ is constant. It is convenient to define the mass flux per unit surface are of the planet,

$$\Phi \equiv (\rho w A(r))/A(r_s) = \rho w (r/r_s)^2,$$ (8.47)

which is of course also constant. The thermodynamic relations needed consist of the equation of state ($p = \rho RT$ in the present discussion) and the corresponding equation

---

[5] Any other convenient reference radius can be used in place of r_s.

for potential temperature (θ) or entropy ($c_p \ln \theta$). In the adiabatic case, the entropy is independent of r and is fixed by the boundary conditions. In the general case including heating, we need an equation for the radial variation of entropy, which we will bring in later.

The transition from flow speeds slower than that of sound (subsonic flow) to flow speeds greater than that of sound (supersonic flow) plays an important role throughout the following, so we will need to know the speed of sound. The sound speed will be denoted by the symbol c in this section, since there is little risk of confusion with the speed of light here. For an ideal gas with gas constant R and temperature T, the speed of sound is given by $c^2 = \gamma RT$, where $\gamma = c_p/c_v$ (see Problem 8.24). With a little manipulation, Eq. (8.46) can then be re-written to yield a powerful constraint on the circumstances in which a one-dimensional flow can smoothly make the transition from subsonic to supersonic. We start by dividing the equation by ρ and working with the resulting pressure gradient term $(1/\rho)dp/dr$. Using the definition of potential temperature, this term can be re-written

$$\frac{1}{\rho}\frac{dp}{dr} = \frac{p}{\rho}\frac{d\ln(p/p_0)}{dr} = c^2\frac{d\ln(\theta)}{dr} - c^2\frac{d\ln(\rho)}{dr}. \tag{8.48}$$

Next, we need to use mass conservation to eliminate ρ. Specifically, we take the log of $\Phi = \rho w A(r)/A(r_s)$, then take the derivative with respect to r, and solve for $d\ln(\rho)/dr$. Upon substitution of the result into the momentum equation and rearranging terms we find

$$\left(1 - \frac{c^2}{w^2}\right)w\frac{dw}{dr} = c^2\frac{d\ln(A/\theta)}{dr} - g_s\frac{r_s^2}{r^2}. \tag{8.49}$$

The ratio w/c is the *Mach number*, for which we will use the symbol M. Equation (8.49), called the *transonic rule*, implies that the right hand side must vanish at the point where $M = 1$, if we require that dw/dr be finite there. This relation is valid even in the presence of diabatic heating due to radiation, thermal diffusion, or any other means; diabatic heating causes θ to vary with r, and the entropy equation needs to be used to obtain this gradient in terms of the heating rate. The point where $M = 1$ is called the *sonic point*, or sometimes the *critical point*.

If gravity is set to zero, Eq. (8.49) also constrains the flow of fluid in a tube with cross-section area $A(r)$, with r being the distance along the axis of the tube (see Problem 8.25). In that context, it implies that if one wants to create an adiabatic supersonic jet by feeding subsonic flow into one end of a tube, then things must be arranged so that the sonic point occurs at a constriction of the tube where the area has a local minimum. In particular, one cannot make a supersonic nozzle in the intuitive shape of a cone with the point snipped off. This realization was the basis of the design of the de Laval nozzle.[6]

If the heating vanishes in the vicinity of the transonic point, θ is constant there and the transonic condition becomes

$$c^2 = \frac{1}{2}g_s\frac{r_s^2}{r} = \frac{1}{4}w_{esc}^2(r) \tag{8.50}$$

---

[6] Gustaf Patrick de Laval (1845–1913) was a Swedish inventor who developed the de Laval nozzle as a way of making a more powerful steam turbine. Subsequent developments led to rotary separation of oil and water, which found even greater commercial applications in the dairy industry, to the problem of cream separation. Centrifugal cream separators were the mainstay of his company, Alfa Laval, which exists to this day. De Laval also invented the first commercially viable milking machine. The de Laval nozzle was first used for rocketry by Robert Goddard, and makes a cameo appearance in Homer Hickam's book, *Rocket Boys* (which became *October Sky* when it reached the silver screen).

where w_{esc} is the escape velocity at radius r. Thus, the transonic rule states that at the point of transition between subsonic and supersonic flow, the speed of sound must be half the escape velocity. Using the expression for c^2, this condition determines the temperature at the sonic point once the position is given. Except for small bodies with low surface gravity, the sonic point must be very far out from the planet if one is to avoid temperatures far higher than are likely to be sustainable given the supply of energy. For example, with H_2 on Earth the sonic point temperature would be over 4800 K if it were placed at 1.1 Earth radii from the planet's center. At 30 radii out, where the gravity is weaker, the sonic point temperature falls to 177 K. For heavier molecules, the temperatures are much higher. For N_2 the sonic point temperature would be 2500 K even at 30 Earth radii. Numbers for Venus would be similar. Because it is difficult to sustain such high temperatures, hydrodynamic escape of gases much heavier than H_2 from Earth- or Venus-sized bodies is not likely, except perhaps for planets in orbits where they receive much more radiation from the primary star than does Earth or Venus. For smaller bodies, escape of heavier gases starts to come more within the realm of possibility; for N_2 on Mars, the sonic point temperature is 503 K at 30 Mars radii, and on Titan it is 139 K at 30 Titan radii. Before long, we will learn how to compute how much absorbed solar radiation is necessary to sustain such temperatures.

The next step is to derive an energy equation, which we do by re-writing the pressure gradient term in the momentum equation in a different form from that used in the preceding discussion. The First Law of Thermodynamics states that $c_p dT - 1/\rho \, dp = \delta Q$, where δQ is the heat added per unit mass, as defined in Chapter 2. If we divide by dr then the First Law can be used to re-write the pressure gradient term $(1/\rho) dp/dr$ in the momentum equation. Assuming c_p to be constant, the result can be put into the form

$$\frac{d}{dr}\left[\frac{1}{2}w^2 + c_p T - \frac{1}{2}2g_s r_s \frac{r_s}{r}\right] = \frac{\delta Q}{dr}. \tag{8.51}$$

Note that $2g_s r_s$ is the square of the escape velocity from the surface, and when multiplied by r_s/r it becomes the square of the escape velocity from radius r. Let us defer for the moment the business of making sense of the term $\delta Q/dr$ – which is a bit of a mathematical monstrosity – and explore the adiabatic case, for which $\delta Q = 0$. In this case, the expression

$$E = \frac{1}{2}w^2 + c_p T - \frac{1}{2}2g_s r_s \frac{r_s}{r} \tag{8.52}$$

must be independent of r. Here, E is the energy per unit mass of the fluid, the three terms representing kinetic energy, internal thermal energy, and gravitational potential energy. Note that when the kinetic energy is negligible and the gravitational term is expanded about r_s by writing $r = r_s + Z$ with $Z/r_s \ll 1$, E reduces to the dry static energy $c_p T + gZ$ derived in Chapter 2. The neglect of w is in fact why this quantity is called *static* energy.

Equation (8.52) is much like the energy balance we used to determine the escape velocity, except this time the thermal energy $c_p T$ is also in play. The energy must be positive at infinity for escape to be possible, so the threshold for escape is defined by $E = 0$; evaluating this at $r \approx r_s$ and assuming w to be small there, we find that escape is possible when $c_p T > g_s r_s$ near r_s. For atmospheres having temperature low enough that Jeans escape is small, the thermal term is negligible compared with the gravitational term, though, since it can be easily shown that $c_p T/g_s r_s$ is the same order of magnitude as the escape parameter λ_c defined earlier.

Exercise 8.8 Write down the relation between $c_p T/g_s r_s$ and λ_c.

The large-r behavior is important, since it provides our boundary condition where the atmosphere meets the near-vacuum of outer space. The combination of mass conservation and energy conservation yields two possible kinds of large-r behavior in the adiabatic case. Since $\rho w r^2$ is constant, then if w remains finite at large r, $\rho \to 0$ there. If the far-field flow is adiabatic, then ρ is proportional to $(p/p_0)^{1/\gamma}$, and so vanishing density implies vanishing pressure, which in turn (by the formula for potential temperature) implies that $T \to 0$ at large r. Since w is finite but T gets small, the speed of sound approaches zero and the flow is supersonic at infinity. Thus, the branch that is supersonic at infinity has vanishing pressure, density, and temperature at infinity, and patches smoothly to outer space. Moreover, since the flow is supersonic, information cannot travel upstream back toward the surface of the planet, so conditions at infinity do not affect conditions lower down in the escaping atmosphere. So far we have not used the energy conservation equation directly, but only the adiabatic assumption; the supersonic conditions and vanishing of temperature and pressure at infinity can survive the addition of a moderate amount of heating in the exterior region. If there is no heating there, then using Eq. (8.52) energy conservation tells us that if w is non-zero at infinity, it in fact becomes *constant* there. At large r, both the temperature and gravitational potential energy vanish, meaning that all the energy of the lower atmosphere is converted to kinetic energy of the escaping flow.

On the other hand, w may vanish at large r. In this case, Eq. (8.52) immediately implies that T asymptotes to a constant at large r, whence density and pressure also asymptote to constants. Finite temperature with vanishing velocity implies that this solution is subsonic at infinity. This is not a viable steady state, because it requires that the interplanetary medium exert a back-pressure on the atmosphere to hold it in. The solution also has the unphysical property of filling interplanetary space with gas having a finite density. What happens if we try to set up a subsonically escaping atmosphere? Note that for subsonic flow, information can propagate upstream towards the planet's surface. Therefore, if there is no back-pressure to hold back the flow, it is likely that the escaping outer region of the atmosphere will accelerate to supersonic conditions, while sending a signal upstream that modifies the upstream boundary condition in such a fashion that the transonic rule is satisfied.

Adiabatic escape

For an escaping atmosphere with non-zero outward mass flux, it is therefore generally believed that the solution must be supersonic at large r in order to be physically realizable. If the atmosphere starts with small w at the base of the escaping flow (as is generally required by the large density there), then the relatively high temperature at the base yields a large sound speed, implying that the lower atmosphere is subsonic. *Therefore, an escaping atmosphere will have a sonic point, where the transonic rule will need to apply.* This allows us to compute the temperature at the base of an adiabatic escaping atmosphere. There is no exobase for atmospheres undergoing hydrodynamic escape, but we will take the base to be at a position where a single gas, light enough to escape hydrodynamically, dominates the atmospheric composition. This position, which we will denote by r_b, could be at the planet's surface in some cases, but more typically would be somewhat above the homopause. In any event, it is generally not terribly far above the planet's surface, as compared with the planet's radius, so r_b/r_s is typically of order unity. We equate the energy at r_b (assuming kinetic energy $\frac{1}{2}w^2$ to be negligible there) to the energy at the critical point r_c as follows:

$$c_p T_b - \frac{1}{2} 2 g_s r_s \frac{r_s}{r_b} = \frac{1}{2} c(T_c)^2 + c_p T_c - \frac{1}{2} 2 g_s r_s \frac{r_s}{r_c}$$
$$= \frac{5 - 3\gamma}{4(\gamma - 1)} \frac{1}{2} 2 g_s r_s \frac{r_s}{r_c}. \tag{8.53}$$

The second equality makes use of the transonic rule in order to eliminate the temperature and sound speed at the critical (i.e. sonic) point. This determines T_b in terms of the critical point position, the size and gravity of the planet, and the characteristics of the gas. It is a remarkable fact that for a gas made of spherical atoms with no internal degrees of freedom, the base temperature T_b becomes independent of the critical point position, since $\gamma = \frac{5}{3}$ for such gases. This is particularly important given that hydrodynamic escape of atomic hydrogen is of primary interest.

For more complex gases, T_b decreases gently as the sonic point is moved outward, approaching a limiting value at large distances. Assuming $r_s/r_c \ll 1$, and $r_s/r_b \approx 1$, then $T_b \approx g_s r_s/c_p$. Note that this limiting basal temperature is precisely the same threshold temperature for escape which we computed earlier on the basis of simple energetic grounds. Eq. 8.53 becomes invalid if we move the critical point all the way to r_b, since we assumed the kinetic energy to be negligible at low levels when deriving the equation. In the limit where the critical point approaches the base, we get the base temperature directly from the transonic rule, which tells us $c^2 = \gamma R T_b = \frac{1}{2} g_s r_s (r_s/r_b)$, or $T_b = \frac{1}{2}(\gamma - 1)^{-1} g_s r_s/c_p$ if $r_s/r_b \approx 1$. Hence, the maximum base temperature exceeds the minimum by a factor of $\frac{1}{2}(\gamma - 1)^{-1}$, which is 1.3 for H_2, 1.25 for N_2, and 1.7 for CO_2. However, the high end of the temperature range where the critical point approaches the base is of little physical relevance, since it corresponds to the case where the lower atmosphere has somehow directly been given enough kinetic energy to escape.

Except for small bodies, the threshold temperature for escape is very high: over 3000 K for atomic hydrogen on Earth, or 4386 K for H_2 – rising to an outrageous 60 000 K for N_2. The required temperatures are high simply because the potential temperature is constant, so that the atmosphere is expanding and cooling along the dry adiabat; in order still to be hot enough to match the temperature at the sonic point, it must thus start with an exceedingly high temperature. On Titan, where gravity is weaker, adiabatic escape of atomic hydrogen requires a basal temperature of only 167 K, while H_2 requires 244 K and N_2 requires 3352 K. For the Moon, escape of H_2 would require a minimum temperature of 198 K; this is somewhat cooler than the Titan case because the greater surface gravity of the Moon is more than made up for by its smaller radius. Adiabatic hydrodynamic escape of hydrogen from Titan or the Moon seems within the realm of possibility, but for larger, denser bodies it would be possible only in exotic high energy conditions. In fact, high-temperature adiabatic escape is much more relevant as a theory for the solar wind than it is for escape of most planetary atmospheres. Indeed, the fluid dynamics we have outlined above was first developed to explain the solar wind, and only later adapted for planetary purposes. For Earth or Mars sized bodies, it is conceivable that the required temperatures could be attained for H or maybe H_2 in the aftermath of a giant impact, or perhaps while a magma ocean is still extant. In general, though, hydrodynamic escape of planetary atmospheres must be sustained by absorption of solar (or stellar) radiation at lower altitudes, which offsets the adiabatic cooling of the expanding atmosphere and allows the sonic condition to be met without unreasonably high temperatures at the base. The absorption acts like the boost phase of a rocket, launching the atmosphere toward escape velocity.

At this point we encounter a troubling difficulty regarding the lower boundary condition: for adiabatic hydrodynamic escape, we have little or no ability to control the temperature at the base of the escaping atmosphere, since the basal temperature is nearly (or completely) insensitive to the position of the sonic point. The density (or equivalently, the pressure) at the sonic point is a free parameter, but its value affects only the density at the base, not the temperature. We have already shown that when the atmosphere is too cold, it runs out of energy if it tries to flow outward, so that the system must instead settle into the hydrostatically balanced solution with no outflow. But what happens if the atmosphere is too *hot*? What goes wrong when the lower atmosphere is too hot is that the entire atmosphere is too hot to have a sonic point, so it is either supersonic everywhere or subsonic everywhere. The first solution violates the condition that the velocity in the lower atmosphere starts off small, while the second solution fails to meet the customary boundary condition at infinity. Yet, it is implausible that an atmosphere with a somewhat lower temperature that is happily escaping will suddenly lose its ability to escape if more energy is added to the system by increasing the temperature. Recognizing that in subsonic flow, the conditions at infinity can send a signal upstream modifying conditions near the planet's surface, it seems most likely that in the too-hot case, a supersonic flow will develop at infinity, which will reduce the air temperature near the planet's surface in such a way as to allow the transonic rule to be satisfied. This situation does not seem to have been explored by numerical simulation at the time of writing, however.

To illustrate how the calculation of escape flux works, let's consider adiabatic blowoff of a pure H_2 atmosphere from the Moon, assuming the surface pressure to be 1 bar. In this case, we assume the atmosphere is heated from below by solar absorption at the lunar surface, idealizing the atmosphere as being transparent to solar radiation. The surface temperature is just the equilibrium no-atmosphere blackbody temperature, since H_2 is nearly transparent to infrared at these pressures. Under these conditions, we can take the base of the escaping atmosphere to be right at the surface. Now, Eq. (8.53) tightly constrains the allowable range of surface temperatures for an escaping atmosphere. Since w was assumed small at the base in the derivation of Eq. (8.53), the formula becomes invalid if we move the sonic point all the way to the surface. However, if we put it fairly close, at $r/r_s = 1.5$, then $T_b = 270.5$ K, while the transonic rule tells us that the sonic point temperature is $T_c = 163$ K. As T_b approaches the minimum temperature of 197.7 K, the sonic point moves to infinity and T_c falls to zero. We'll assume the Moon to be in an orbit for which the surface temperature lies in this range; interestingly, the position of the actual Moon would satisfy this condition if the Moon had enough atmosphere to even out temperature fluctuations. Now, to determine the escape rate, we need the density at the sonic point. Since potential temperature is constant, the temperature ratio T_c/T_b determines the pressure at the sonic point in terms of the surface pressure, and from this and the temperature we get the required density, which is 0.026 kg/m^3 when $T_b = 270.5$ K and the sonic point is at $r_c/r_s = 1.5$, or 0.00064 kg/m^3 when $T_b = 208.7$ K and the sonic point is at $r_c/r_s = 10$. The velocity at the sonic point is just the sound speed there (known because we know the temperature), and the mass flux per unit planetary surface area is $\Phi = \rho_c(\gamma RT_c)^{\frac{1}{2}}(r_c/r_s)^2$. This is 56.8 kg/m^{2}s for $T_b = 270.5$ K and 24.0 kg/m^{2}s for $T_b = 208.6$ K. Note that for any given r_c, the critical point density ρ_c is proportional to the surface pressure, so changing the surface pressure just changes the escape flux proportionately. These are very large escape fluxes. The mass of the hypothetical lunar atmosphere is about 62 000 kg/m^2, and would be lost in just over a half hour in the hotter case and just over an hour in the colder case. We have thus learned that for a Moon-sized body in an orbit where the solar radiation is similar to that received at Earth's orbit today, H_2 would be lost by adiabatic blowoff almost at once. As the temperature decreases toward the threshold temperature, then the sonic point moves out to infinity, and the escape flux very gradually reduces; in fact, it can be easily shown that the escape flux approaches zero as the surface temperature approaches the threshold temperature. Hydrodynamic escape is a process with dramatic thresholds: for a surface temperature of 208 K the atmosphere is lost in a matter of hours, whereas the escape flow shuts off completely when the temperature drops below 198 K (though other escape mechanisms may still cause loss of the atmosphere).

Exercise 8.9 Following the reasoning in the preceding paragraph, for an arbitrary body derive a formula for the escape flux Φ in terms of the surface temperature and surface density. Show that the escape flux is proportional to surface density, and that the escape flux approaches zero as the surface temperature approaches the minimum surface temperature for adiabatic hydrodynamic escape.

Re-interpretation of the transonic rule in terms of energetics

We can gain a better appreciation of the meaning of the transonic rule, and of what happens when it is violated, by expressing the energy conservation relation in terms of the Mach number. This requires expressing c^2 in terms of the Mach number, or equivalently writing T in terms of the Mach number, which can be done by making use of mass conservation, the ideal gas equation of state, and the expression for potential temperature. In particular, if we let (p_0, T_0, ρ_0, M_0) be the state of the atmosphere at some point r_0, then the four relations

$$T = \theta \cdot \left(\frac{p}{p_0}\right)^{R/c_p}, \rho = \rho_0 \frac{T_0}{\theta}\left(\frac{p}{p_0}\right)^{1/\gamma}, \Phi = \rho c M \frac{r^2}{r_s^2} = \rho_0 c_0 M_0 \frac{r_0^2}{r_s^2}, \frac{c}{c_0} = \left(\frac{T}{T_0}\right)^{\frac{1}{2}} \tag{8.54}$$

can be solved for temperature, yielding

$$T = \theta \cdot \left[\left(\frac{\theta}{T_0}\right)^{\frac{1}{2}} \frac{r_0^2}{r^2} \frac{M_0}{M}\right]^{\beta} \tag{8.55}$$

where $\beta \equiv 2(\gamma - 1)/(\gamma + 1)$. This result is valid whether or not the flow is adiabatic. In the adiabatic case, θ is independent of r and $\theta = T_0$ if we define the potential temperature with regard to the pressure p_0 at the reference point. With the relation 8.55 in hand, Eq. (8.52) can be written

$$E = c^2 \cdot \left(\frac{1}{2}M^2 + \frac{1}{\gamma - 1}\right) - \frac{1}{2}2g_s r_s \frac{r_s}{r}$$

$$= \gamma R\theta \cdot \left[\left(\frac{\theta}{T_0}\right)^{\frac{1}{2}} \frac{r_0^2}{r^2} \frac{M_0}{M}\right]^{\beta} \cdot \left(\frac{1}{2}M^2 + \frac{1}{\gamma - 1}\right) - \frac{1}{2}2g_s r_s \frac{r_s}{r}. \tag{8.56}$$

The right hand side can be considered to be a function $E(M, r/r_s)$ with $2g_s r_s$ and conditions at r_0 as parameters. This expression is valid even in the presence of heating, in which case θ is a function of r. For the sake of generality, we have written the expression in terms of an arbitrary reference point r_0, but most commonly we will take r_0 to be a sonic point, at which $M_0 = 1$ by definition and at which moreover the transonic rule is satisfied. The transonic rule gives T_0 in terms of r_0 and the gravitational acceleration, so in this case the only free parameters governing the shape of the curve are r_0 (which we'll call r_c in this case) and θ. When the flow is moreover adiabatic throughout the escaping atmosphere, $\theta = T_0$ and the curve is governed by a single parameter, for any given planet.

Recall from Chapter 2 that y is determined by the number of excited degrees of freedom of the atoms or molecules making up the gas. For spherical particles with no internal degrees of freedom, $y = 5/3$, while $y \rightarrow 1$ for complex molecules with very many excited degrees of freedom, though for most atmospheric gases $y > 1.29$. Hence, $0 < \beta \leq \frac{1}{2}$, with the lower limit not being very closely approached in practice. Because $\beta < 2$, it follows that $E(M, r/r_s) \rightarrow \infty$ as $M \rightarrow \infty$. In addition, since β is positive, it follows that $E(M, r/r_s) \rightarrow \infty$ as $M \rightarrow 0$. For fixed r/r_s, the function has a single minimum at $M = 1$, as can be verified by differentiation of the expression with respect to M. It is the shape of the energy curve, and the occurrence of the minimum at $M = 1$, that underlies the transonic rule.

Exercise 8.10 Carry out the suggested differentiation, verify that the minimum of E occurs at $M = 1$, and write down the expression for $E_{min}(r/r_s)$.

Some representative energy curves are sketched in Fig. 8.6. For any given energy E_0, there are three possible situations regarding the flows that satisfy $E(M, r/r_s) = E_0$:

- There can be a pair of solutions, one of which is subsonic and the other supersonic.
- There can be a single solution, with $M = 1$.
- There can be no solutions at all.

Now let's consider what happens if we start with a subsonic solution and track its evolution as we increase r, which changes the energy curve. The gravitational potential term shifts the curve upward without changing its shape, while the r-dependent factor in the first term of Eq. (8.56) flattens the curve and moves it downward as r is increased. The situation is depicted in Fig. 8.6, where we start with $T = 300\,\mathrm{K}$ and $M = 0.01$ at $r/r_s = 1.1$. We already know from our previous results that this temperature is too low to allow the transonic rule to be satisfied for Earthlike gravity, so what happens as r is increased? Examining the intersection points between the initial energy and the energy curve, we see that the Mach number increases until it reaches the sonic point at $r/r_s = 1.167$. When r is further increased, the energy curve continues to move upward, however, so there are no solutions compatible with the initial energy. This is the generic behavior when we start from too low a temperature and the transonic rule is not satisfied. The atmosphere can still approach $M = 1$ from either the subsonic or supersonic side, but the upward movement of the energy curve through the

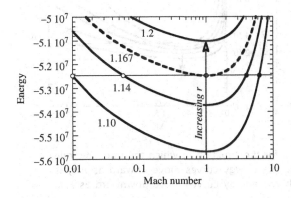

Figure 8.6 A family of curves of of energy vs. Mach number. Numbers on the curves indicate the distance from the center of the Earth in units of Earth radii. These curves depict the situation at low base temperatures, when the transonic condition is not satisfied. Open circles indicate subsonic solutions while filled circles indicate supersonic solutions. The curve for the critical radius, where the transonic point is reached, is dashed. The calculation was carried out for H_2 under Earth gravity, assuming $T = 300\,\mathrm{K}$ at $r/r_s = 1.1$; $M = 0.01$ at that point.

sonic point means that the gradient of $M(r)$ and $w(r)$ have square-root singularities there – a consequence of the violation of the transonic rule. From a physical standpoint, what is important is not so much the singularity as the fact that solutions cease to exist altogether once r is moved past the sonic point.

An examination of Eq. (8.55) shows that the temperature decreases as the sonic point is approached; this increases the Mach number by decreasing the speed of sound. In fact, where the Mach number is small enough that kinetic energy is negligible, the energy equation reduces to conservation of dry static energy $c_p T + gz$, and tells us that the atmosphere ascends along the dry adiabat. This means it cannot ascend far, because the temperature falls to zero at a finite height for the dry adiabat. Incorporation of the kinetic energy term causes the atmosphere to run out of energy before the temperature falls to zero. This is the situation we are fighting if we try to make an atmosphere hydrodynamically escape while starting from a realistic base temperature: without some additional supply of energy, the atmosphere runs out of energy before it gets very far. In this case, energetics requires the atmosphere to settle into a state of rest without a mean outflow. When starting from a low temperature on a planet with Earthlike gravity, energy must be deposited in the upper atmosphere in order to allow it to escape. Generally, this energy is supplied in the form of extreme ultraviolet.

Next we'll examine the geometry of the energy curves in a case where the transonic rule is satisfied and adiabatic hydrodynamic escape is possible. The situation is depicted in Fig. 8.7. To pass from a subsonic to a supersonic state, the energy curve first comes up to where the subsonic and supersonic solutions coalesce at the sonic point, but then moves down again as r is further increased, allowing the solution to continue on the supersonic branch. The transonic rule is equivalent to the requirement that when expressed as a function of r, the minimum of the energy curve $E(M = 1, r/r_0)$ has a maximum at $r = r_c$, so that the curve first goes up to the sonic point then turns around and heads back downwards again.

Exercise 8.11 Verify this property by using the expression for $E(M = 1, r/r_0)$ and locating its minimum by taking the derivative with respect to r. To keep things simple, you can restrict attention to the adiabatic case $\theta = $ const.

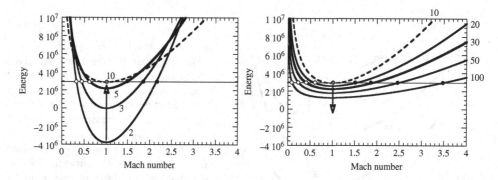

Figure 8.7 Sequence of energy curves for the situation in which the transonic rule is satisfied for $r/r_s = 10$. Left panel: The energy curves move upward as r/r_s is increased from 2 to 10. Right panel: The energy curves move downward as r/r_s is increased further from 10 to 100.

Escape driven by EUV heating

It is now time to bring heating into the picture. Heating alters the preceding construction in two ways: First, instead of the expression E being independent of r, it will vary on account of the deposition of energy from solar radiation or other sources. This moves the energy curve up or down with r, without changing its shape. Second, heating will cause θ to vary with r, which changes the shape of the energy curve. It remains true, however, that the energy curve for any r has a unique minimum where the Mach number is unity, and that in order to effect the transition from subsonic to supersonic flow, the line defining the available amount of energy at any given r must start above the minimum, be brought to tangency at the minimum (either by moving the curve upward or moving the line downward or some combination of the two), and then be moved to some distance above the minimum as r is increased further.

In order to incorporate heating into the energy equation, we re-write the diabatic term on the right hand side of Eq. (8.51) as follows:

$$\frac{\delta Q}{dr} = \frac{\rho \delta Q/dt}{\rho \, dr/dt} = \frac{\dot{q}}{\rho w} = \frac{(r/r_s)^2 \dot{q}}{\Phi} \tag{8.57}$$

where $\dot{q}$ is the heating rate *per unit volume*. Now let $F(r)$ be the flux of energy due to means other than the bulk fluid flow, with the convention that inward fluxes are taken as positive. We will mostly deal with radiative flux, but F could equally well represent flux due to molecular diffusion of heat. In terms of the flux, the heating rate is given by $\dot{q} = r^{-2}d(r^2 F)/dr$, so given that Φ is constant the heating term in the energy equation becomes

$$\frac{\delta Q}{dr} = \frac{d}{dr}\left[\frac{r^2}{r_s^2}\frac{F}{\Phi}\right]. \tag{8.58}$$

Since this is the gradient of a flux, Eq. (8.51) tells us that it can be combined with the previous expression for E to yield the revised conservation law

$$\frac{1}{2}w^2 + c_p T - \frac{1}{2}2g_s r_s \frac{r_s}{r} - \frac{r^2}{r_s^2}\frac{F}{\Phi} = E_0. \tag{8.59}$$

where E_0 is a constant. Note that F/Φ is an energy flux divided by a mass flux, and therefore has dimensions of velocity squared. For a given variation in energy flux, reducing the mass flux Φ leads to a greater radial variation in the energy density E (the first pair of terms in the equation), because fluid has more time to accumulate energy when it is moving slowly.

The treatment of the radiative part of the heating is a bit tricky. As in our earlier calculations of planetary temperature, we are doing a globally averaged energy budget assuming uniform atmospheric conditions over the globe. That requires figuring the amount of EUV flux intercepted by the *planet's atmosphere*, and distributing it uniformly over the sphere. The assumption of effective redistribution of solar heating over the outer atmosphere is highly questionable, but it is quite customary in hydrodynamic escape calculations and is in any event the only approximation which allows us to make headway without very extensive and complex numerical simulations. That notwithstanding, there is an additional issue that did not arise in our earlier treatments of globally averaged energy budgets. As usual, the planet intercepts a disk of light of some radius, and the intercepted flux is spread uniformly over the surface of sphere of the corresponding radius. The difference in the present case is that the portion of the atmosphere that is dense enough to absorb significant amounts of

EUV radiation can extend many planetary radii out from the surface. Therefore, the radius of the disk of intercepted radiation depends on the optical thickness (hence density) of the atmosphere. Let's suppose that the atmosphere is transparent to EUV for radii farther out than some radius r_{abs}, but that the closer-in atmosphere absorbs strongly. Then, if $F_\odot$ is the incoming flux of EUV from the sun (analogous to the solar constant), the intercepted power is $\pi r_{abs}^2 F_\odot$. If we want this to be the total power entering the system, then the corresponding uniform *radial* flux $4\pi r^2 F(r)$ through a shell of radius r must be held constant at the value $\pi r_{abs}^2 F_\odot$ for $r > r_{abs}$. Therefore, we model radiative absorption by stipulating that $r^2 F(r) = \frac{1}{4} r_{abs}^2 F_\odot$ for $r > r_{abs}$ while allowing the flux to decay to zero as radiation is absorbed deeper in the atmosphere.

Equation (8.59) imposes a powerful constraint on the mass flux that can be sustained by a given level of heating. We will suppose as usual that w is small at the base of the escaping atmosphere, but this time we will also assume that the base is cool, so that $c_p T$ is negligible compared with the gravitational term at the base. This is the typical situation, in which the atmosphere is strongly gravitationally bound and the escape parameter λ_c is small. Unlike the adiabatic escape case, we do not endow the base of the atmosphere with enough thermal energy to sustain the outflow, but rather deposit it gradually through absorption of stellar flux, generally in the extreme ultraviolet spectrum. Equating energy at the base (where radiative flux falls to zero) and at infinity yields the relation

$$\frac{1}{\Phi}\left(\frac{1}{4}\frac{r_{abs}^2}{r_s^2}F_\odot\right) = \frac{1}{2}w_\infty^2 + \frac{1}{2}2g_s r_s \frac{r_s}{r_b}. \tag{8.60}$$

Since $w_\infty^2 > 0$, this imposes an upper bound on the mass flux Φ for any given amount of radiative absorption. Note that this energetic constraint survives the addition of heat diffusion to the flux term, since diffusion only redistributes energy in the vertical and does not add new energy to the system. If $(r_{abs}/r_s)^2 \approx 1$ and $r_s/r_b \approx 1$ as is typically the case, than the constraint is simply $\Phi \leq \frac{1}{4}F_\odot/g_s r_s$. It is important to recognize that this bound on the escape flux applies only in the low temperature limit, in which $c_p T$ at the base is negligible compared with the gravitational potential. For any finite temperature, the escape flux can exceed the limiting flux, by an amount that increases with temperature. As the base temperature approaches the temperature at which adiabatic escape becomes possible, the escape flux can become arbitrarily large, limited only by the density at the base.

The physical content of the constraint is simple: the escaping atmosphere carries kinetic and potential energy with it, and this outward energy flux must be matched by the supply of radiant energy absorbed within the escaping atmosphere. The amount of energy flux escaping due to mass flow is negligible from a standpoint of planetary energy balance; that energy loss is still by far dominated by infrared emission. However, from the standpoint of the energy budget of the outer atmosphere *alone*, the energy loss due to mass outflow can be the dominant term – at least for gases like H_2 which are poor infrared emitters. Infrared emission, to the extent that it occurs at all, can be thought of as stealing energy from the supply of EUV heating available to sustain escape.

In order to complete the solution, we need to know how the potential temperature varies with radius. Once the potential temperature is known, we have enough thermodynamic information to compute the profiles of pressure and density as well. The radial variation of the potential temperature is obtained from the entropy equation:

$$\frac{d}{dr}c_p \ln \theta = \frac{1}{T}\frac{\delta q}{dr} = \frac{1}{T}\frac{\rho \delta q/dt}{\rho w} = \frac{1}{T}\frac{(r/r_s)^2 \dot{Q}}{\Phi}. \tag{8.61}$$

The heating term $(r/r_s)^2 \dot{Q}$ can be written as the radial gradient of a flux as before, but because of the factor $1/T$ appearing in the entropy equation, this equation cannot be integrated to yield a pointwise relation between entropy and flux the way we did for energy. The entropy change between points r_A and r_B depends on the shape of the heating curve between those points, and not just the amount of heat added; heat added at low temperature has a greater effect on entropy than heat added at high temperature.

We will now exhibit some numerical solutions for the case in which the heating $\dot{Q}$ is a known function of r. Radiative heating depends on the density and temperature so strictly speaking it must be solved for together with the atmospheric structure; the extension to this case is straightforward once one understands how to solve the simpler problem. Given the heating, the numerical solution proceeds as follows. One starts by choosing the critical point position r_c/r_s, from which one can compute the sound speed and temperature at the critical point. Then, one integrates the differential equation (8.61) in a direction toward the planet. Since the escape flux Φ appears as a parameter in this equation, one must guess a value of Φ to carry out the integration. The chosen value of Φ also fixes the density at the critical point since the velocity at the critical point is the local speed of sound. At each step of the integration of Eq. (8.61), one obtains value of $\ln\theta$, but to proceed one also needs to update the value of T. This is done by solving the energy equation, Eq. (8.59) (re-written in terms of Mach number using Eq. (8.56)), for the Mach number, following the subsonic branch. The Mach number, in turn, determines the new temperature and allows the integration to proceed further. As the integration proceeds, a point may be encountered in which solutions to the energy equation no longer exist, in which case the chosen value of Φ is not realizable. If this situation does not arise, the integration is continued until the base r_b is reached. One now knows the value of θ there, which for positive heating will be less (usually *much* less) than the value at the critical point. The integration has already completely determined the temperature structure of the atmosphere, and the ratio of potential temperature at r_b to the value at r_c determines the proportionality constant between the density at r_b and the (known) density at r_c. This procedure yields a family of solutions, with r_c and Φ as parameters. When Φ is large, Eq. (8.61) says that the potential temperature becomes constant, in which case we recover the adiabatic escape solutions which typically have very high temperatures at the base. As Φ is made smaller, heating causes the potential temperature at the base to be much smaller than the potential temperature at the critical point, which results in cooler temperatures at the base. When Φ is made too small, however, the temperatures are driven to excessively low values and one encounters at some point a supersonic transition at a radius where the transonic rule is not satisfied, whereupon the solution ceases to exist. Carrying out this procedure requires only the integration of a first order differential equation and the solution of the energy equation at each step using Newton's method. It is quite straightforward to implement.

Before showing a specific family of solutions obtained using the above procedure, it is useful to identify a few relevant non-dimensional parameters. The energy-limited escape flux identified earlier provides a convenient scale for non-dimensionalization of Φ. Let's call the limiting flux Φ_*. It can be written $\frac{1}{2}F_\oplus/w_{esc}^2$, where w_{esc} is the escape velocity from the surface, namely $\sqrt{2g_s r_s}$. We can define a characteristic temperature T_* such that $c_p T_* = w_{esc}^2$. The heating can be written $\dot{Q} = (d/dr)(F \cdot r^2/r_s^2)$, so it can be non-dimensionalized by multiplying both sides of the entropy equation by r_s (amounting to taking the planetary radius as the unit of length) and then writing $F = F_\oplus \cdot (F/F_\oplus)$. If the temperature is non-dimensionalized against T_*, then the entropy equation depends on the EUV flux, the escape velocity, and the escape flux only through the dimensionless combination Φ/Φ_*. Specifically, the non-dimensional entropy equation becomes

$$\frac{d}{d\tilde{r}}\ln\theta = \frac{1}{\tilde{T}}\frac{\tilde{r}^2\tilde{Q}}{\tilde{\Phi}} \tag{8.62}$$

where $\tilde{r} \equiv r/r_s$, $\tilde{\Phi} \equiv \Phi/\Phi_*$, $\tilde{T} \equiv T/T_*$, and $\tilde{Q}$ is the non-dimensional heating profile. Similarly, if one non-dimensionalizes the energy equation by dividing by w_{esc}^2, one finds that the same three quantities appear in the energy equation only in the form Φ/Φ_*. Thus, for any given *shape* of the EUV heating profile, Φ/Φ_* determines the basic behavior of the atmospheric structure; changing the values of any of the terms in this non-dimensional parameter just uniformly rescales the temperature and density profiles.

In Fig. 8.8 we show a family of solutions for a hydrodynamically escaping H_2 atmosphere on Earth. In this calculation, r_c/r_s is held fixed at 30 while the non-dimensional escape flux is varied. The calculations were carried out with $r_b/r_s = 1.1$. The EUV heating profile was chosen such that

$$\frac{r^2}{r_s^2}\dot{Q} = \frac{d}{dr}\frac{r^2}{r_s^2}F = Q_0\exp-\left(\frac{r-r_b}{r_{EUV}}\right). \tag{8.63}$$

The heating profile determines the flux $(r/r_s)^2F(r)$, which is needed for use in the energy equation. The flux integration shows that when the heating is shallow, $Q_0 \approx \frac{1}{4}F_\odot/r_{EUV}$. The calculations shown in the figure were done with $r_{EUV}/r_s = 1$, yielding a shallow, low-level heating.

Looking at the results, we see that when Φ/Φ_* is made large, the temperature is monotonically decreasing, and at very large escape fluxes the curve looks like the adiabatic escape solution as expected. As $\Phi_*/\Phi \to 1$, however, the temperature at the base falls toward zero, and the temperature rises from cool values to a maximum before decaying approximately adiabatically toward the critical point. At lower values of Φ/Φ_*, solutions fail to exist in the purely radiatively heated case. Although the energy constraint expressed in Eq. (8.60) permits lower escape fluxes when the velocity at infinity is non-zero, the resulting flows cannot be made to satisfy the transonic rule in the configuration under investigation.

The cool-base solutions constitute the desired regime for hydrodynamic escape of hydrogen from Earth or Venus. In these solutions, EUV heating takes the hydrogen entering at the cool temperatures of the lower atmosphere, and heats it as it flows outward, to the point where it is hot enough and far enough out to escape. The EUV heating region is like the

Figure 8.8 Temperature structure of an escaping H_2 Earth atmosphere subject to radiative heating at low levels. The critical point is held fixed at $r/r_s = 30$. The numbers on curves give the value of the non-dimensional escape flux, Φ/Φ_*.

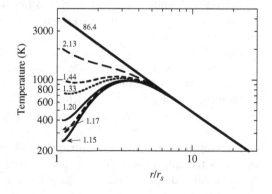

boost phase of a rocket, where burning of fuel launches the rocket and accelerates it to escape velocity.

What we have achieved by introducing radiative heating is the ability to sustain hydrodynamic escape with realistic low-level temperatures. For the cool-base solutions, the escape flux is on the order of Φ_*, so we can use this quantity to estimate the significance of the escape flux. For Earth conditions, where $F_\oplus \approx 0.004\,\mathrm{W/m^2}$, $\Phi_* = 1.6 \cdot 10^{-11}\,\mathrm{kg/m^2 s}$, corresponding to a flux of hydrogen atoms of $10^{16}/\mathrm{m^2 s}$. This greatly exceeds the Jeans escape rate we computed earlier, and is more than sufficient to keep up with the volcanic outgassing. For the Venus case, where $F_\oplus \approx 0.008$ owing to the closer orbit, we find a mass flux of $3.72 \cdot 10^{-11}\,\mathrm{kg/m^2 s}$, or an atomic hydrogen escape flux of $2.24 \cdot 10^{16}/\mathrm{m^2 s}$. In a billion years, this could get rid of the hydrogen from a 10 km deep ocean (of course, one still needs to find something to do with the resulting oxygen). The reader should be cautioned that these numbers should not be taken as definitive; a more sophisticated calculation taking into account heat diffusion and a more realistic model of deposition of radiation could reduce the estimated fluxes. Still we have shown that hydrodynamic escape easily has the potential to get rid of the required amount of hydrogen on Earth or Venus.

In the presence of radiative heating, we can match the desired temperature boundary condition at the base of the escaping atmosphere by varying the escape flux with fixed r_c, but the problem now is that we cannot independently specify the density there. For any given base temperature, this is fixed by the other parameters in the problem. Examination of the density solutions (not shown) reveals that in the family of curves given in Fig. 8.8, the density decreases rapidly as the base temperature increases. In dimensional terms, at 250 K the density is $2.5 \cdot 10^{21}/\mathrm{m^3}$, falling to $3.6 \cdot 10^{20}/\mathrm{m^3}$ at 300 K, and to $2.3 \cdot 10^{19}/\mathrm{m^3}$ at 400 K. The strong temperature dependence arises mainly because the density scale height is smaller for low temperatures, so that the atmosphere has to start with higher density in order to match with the density needed at the critical point. The additional degree of freedom needed to match the density condition at the base while keeping the temperature fixed at some desired value is provided by varying the critical point position r_c. Some numerical experimentation, backed up by a bit of tedious algebra applied to the energy and entropy equations, reveals the following behavior for the density at the base:

- Except for the case of an ideal monatomic gas with $\gamma = \frac{5}{3}$, the density at the base can be made arbitrarily large for fixed base temperature by moving r_c outward; in fact, for large r_c/r_s the base density increases linearly with r_c, with a slope that increases as the base temperature increases.
- For any gas, the base density can be made small by moving r_c toward the surface, though for ideal monatomic gases the decrease in density only becomes effective as r_c moves into the heating region. However, if the base density one is trying to match becomes too small, then the required r_c approaches the surface, meaning that the atmosphere escapes only because it is supersonic already at low levels – i.e. that it escapes by virtue of having been given high kinetic energy at low levels. These solutions are of little physical interest, as they do not patch to an atmosphere in hydrostatic balance at low altitudes.
- For ideal monatomic gases, solutions satisfying the transonic rule cease to exist once r_c is moved moderately outside the heating region. This feature is connected with the insensitivity of T_b to the critical point position, expressed in Eq. (8.53) (applied to the base of the *adiabatic* region rather than all the way to the base of the heating region). As a result, monatomic gases appear to be subject to a *maximum* allowable base density. The implications of this peculiarity for the important case of monatomic hydrogen escape

seem not to have been remarked on or explored in the literature. We speculate that when the base density is "too large," the flow tries to become subsonic out to infinity, whereafter upstream influence in some way acts to alter conditions at the base. Because of the way the EUV heating affects the transonic condition, the behavior of density in the monatomic case is likely to be highly sensitive to the radial distribution of the heating, and in particular to the extent to which some heating persists out to great distances from the planet's surface.

- For low base temperatures, the escape flux remains near the limiting flux Φ_*, and does not increase significantly as the base density increases. Increasing the base density in this regime moves the critical point outward, but does not increase the flux. This behavior contrasts with both the high-temperature regime and with Jeans escape, and comes about because the escape flux is constrained by the energy supplied by EUV heating.

Our understanding of hydrodynamic escape so far can be summarized in the following recipe for estimating the escape flux due to this mechanism. First compute the characteristic temperature T_*, defined by $c_p T_* = 2g_s r_s$. Is the temperature at the base comparable to or greater than this value? If so, you can use the adiabatic escape formulation as an estimate, which yields the proportionality constant between the escape flux and the atmospheric density at the base, which can take on whatever value is dictated by the lower boundary condition. The proportionality constant was computed in Exercise 8.9. On the other hand, if the temperature at the base is much less than T_* your atmosphere is in the low-temperature limit, and EUV absorption plays a critical role in sustaining escape. In this case there is a threshold base density which must be exceeded in order to permit hydrodynamic escape from a level in the atmosphere which is nearly in hydrostatic balance. The threshold density must be determined by integrating the entropy equation inward from some chosen critical point, and moving the critical point inward until the velocity at the base becomes unacceptably large. This is the only hard part of the calculation. However, once you know that the base density exceeds the threshold, the escape flux in the low-temperature regime can be estimated by simply evaluating the limiting flux, $\Phi_* \approx \frac{1}{4} F_\odot / g_s r_s$, where $F_\odot$ is the EUV flux impinging on the planet.

A rather startling aspect of the radiatively heated escape calculation is that the limiting escape flux Φ_* is independent of the molecular weight of the substance making up the escaping atmosphere. From this, it would appear that a given amount of absorbed EUV could make a kilogram of oxygen escape just as easily as a kilogram of hydrogen. Continuum hydrodynamics appears to know nothing about the fact that the fluid is made up of discrete particles, so it is not immediately obvious how the molecular weight should enter the problem. In fact, the microscopic nature of the fluid enters through c_p, which, through equipartition of energy, *does* know about the existence of particles. A kilogram of H_2 has more degrees of freedom than a kilogram of O_2, and this is reflected in a higher value of c_p for H_2. The specific heat enters into the determination of the density at the base, and it is this density boundary condition which tells us that it is essentially impossible for heavy species to escape hydrodynamically, unless the temperature becomes very large. Using dimensional analysis, we can use our H_2 escape calculations to make some inferences about what happens if we try to make a heavier species escape. The value of c_p for atomic oxygen is nearly 10 times smaller than the value for H_2, so that the characteristic temperature T_* for oxygen is more than 10 times that of H_2. That means that to achieve the same particle density for oxygen as was achieved at 250 K in the H_2 case, the base temperature has to be raised to over 2500 K. An escape calculation with atomic oxygen shows that one can indeed achieve a temperature of 250 K at the base by making Φ/Φ_* sufficiently close to the limiting value, but when one does so the base particle density is $10^{32}/m^3$ and the pressure there is over *three million bars*! It is not likely that any planet can meet such a high density threshold.

Effects of heat diffusion

EUV heating is not the only diabatic effect of importance in the outer atmosphere. The other main heating and cooling effect that needs to be taken into account is the redistribution of heat by diffusion. Heat diffusion is much like diffusion of mass, and becomes strong when the mean free path is large. When the temperature gradient is dT/dr, the flux of heat is $\kappa dT/dr$, where κ is the same kind of thermal conductivity we encountered in earlier chapters, only this time applied to a gas rather than a solid. One of the many important effects of heat diffusion is that it allows heat from deposition of EUV at high levels to be diffused down to the lower atmosphere, which is often necessary to sustain hydrodynamic escape. Indeed, the radiative escape calculations we presented above in some sense implicitly assume some diffusion, since it is unlikely that enough EUV would penetrate to r_b to allow the fairly strong heating we assumed there in our calculation.

In contrast with radiative heating, heat diffusion fundamentally changes the mathematical character of the problem. With radiative heating only, the energy equation (8.59) is an algebraic equation that is solved for the Mach number. It does not involve any derivatives. However, with heat diffusion, the heat flux F appearing in the equation is no longer just a function of r, but instead involves a term $\kappa dT/dr$. Thus, diffusion *changes the energy equation from an algebraic equation for the Mach number to a differential equation for temperature*. This is true no matter how small the thermal conductivity, though the equation will become very ill posed when κ is small. One can also see the singular effect of diffusion by examining the entropy equation. With only radiative heating, the highest order derivative is on the left hand side, and constitutes a first order differential equation for θ. With diffusion, the heating term on the right hand side has a contribution from diffusion, which introduces the gradient of $\kappa dT/dr$. The entropy equation thus becomes instead a second order differential equation for T. The extra derivative can provide some additional scope for meeting both a temperature and density boundary condition at the base of the atmosphere, though this advantage is somewhat constrained by the fact that the heat diffusion is weak near the base, where the density is relatively large.

The second order diffusion equation based on the entropy equation can in principle be used as the basis of a solution method in the diffusive case. A more common approach to incorporating heat diffusion proceeds from a variation on the reasoning we used in deriving the transonic rule. Instead of relating the pressure gradient to the potential temperature, as we did in our derivation of (8.49), we can instead write $p = \rho RT$ and use the same manipulations as previously to eliminate $d\rho/dr$ using conservation of mass. In that case, we obtain the following alternate form of Eq. (8.49)

$$\left(1 - \frac{c_T^2}{w^2}\right) w \frac{dw}{dr} = -R\frac{dT}{dr} + c_T^2 \frac{d\ln(A)}{dr} - g_s\frac{r_s^2}{r^2} \tag{8.64}$$

where $c_T^2 \equiv RT$. The left hand side looks similar to the previous form, except that the *isothermal* sound speed c_T (Problem 8.24) appears in place of the adiabatic sound speed. From this equation one would be tempted to draw the paradoxical conclusion that the critical point should be defined with regard to the isothermal sound speed rather than the adiabatic sound speed. This would be a fallacy. It is true that the right hand side must vanish where w equals the isothermal sound speed, but this does not give us a useful transonic rule since we do not *a priori* know the value of dT/dr. This contrasts with the original form, Eq. (8.49), where we know $d\theta/dr$ from the entropy equation, and in particular that it vanishes if there is no heating near the critical point. But that's not the end of the story. If the diffusion is non-zero everywhere, then the energy equation, Eq. (8.59) can be solved for dT/dr in terms of w, T, r, and the radiative flux; the resulting expression has no derivatives, and so when substituted into Eq. (8.64) it does not change the location of the singular point of the equation where the coefficient of the derivative vanishes. It would appear that the introduction of even infinitesimal heat diffusion discontinuously changes the mathematical character of the problem, so that we are left with a transonic rule that is applied at the critical point defined by isothermal rather than adiabatic sound speed. In the diffusive case, the most common approach to obtaining a steady solution is to simultaneously integrate Eq. (8.64) and Eq. (8.59), which jointly define a coupled set of differential equations for the pair (w, T). The integration is supplemented by a transonic condition applied at the point defined by the isothermal sound speed. This is a technically correct approach to the problem, but as a numerical scheme it is sure to become badly behaved in one way or another when the heat diffusivity becomes small. This may account for the difficulty some researchers have reported in finding consistent solutions to the escape equations in the presence of both heating and diffusion.

There is one special case, however, where one can take a shortcut to avoid the complexities and the problematic features of the integration sketched out above. When the heat diffusion is so strong that it keeps the outer atmosphere isothermal, then dT/dr vanishes, and Eq. (8.64) provides a usable constraint on the conditions at the critical point defined with reference to the isothermal sound speed. The isothermal case is worth pursuing, as it provides the opposite extreme to the no-diffusion case we considered previously, and thus serves to highlight the effects of diffusion. However, numerical solutions to the full problem for Earth and Venus have temperature variations that somewhat resemble the cool-base non-diffusive cases shown in Fig. 8.8, so the isothermal limit cannot be relied on for an accurate estimate of the actual escape flux.

As for the adiabatic case, the isothermal case admits a very simple explicit solution for the escape flux. When the atmosphere is isothermal, we do not need to invoke the First Law of Thermodynamics to derive a conservation law from the momentum equation. Since T is constant, substituting $p = \rho RT$ immediately yields the result

$$\frac{1}{2}w^2 + c_T^2 \ln \rho - \frac{1}{2} 2g_s r_s \frac{r_s}{r} = \text{const.} \tag{8.65}$$

Note that this energy expression is a constant even though no explicit heating term appears in the equation. Moreover, c_T is a constant as well, because the atmosphere is isothermal. By equating the values of this expression at the base and at the critical point, we find

$$\ln \frac{\rho(r_c)}{\rho(r_b)} = -\frac{1}{2} \frac{2g_s r_s}{c_T^2} \left(\frac{r_s}{r_b} - \frac{r_s}{r_c} \right) - \frac{1}{2} = -2 \left(\frac{r_c}{r_b} - 1 \right) - \frac{1}{2}. \tag{8.66}$$

As usual, we have assumed that the kinetic energy is small at the base. The second equality arises from application of the isothermal form of the transonic rule. Equation (8.66) links the density at the base to the density at the critical point once the critical point position is specified. The result implies that $\rho(r_c)/\rho(r_b)$ gets exponentially small as the critical point is removed to infinity. Further, the escape flux is $\Phi = \rho(r_c)c_T(r_c)(r_c/r_s)^2$, so Φ is known once the position of the critical point and the density there are known. So far, the nature of the isothermal solution is rather similar to the adiabatic case, in the sense that the base temperature and base density can be fixed by adjusting r_c and $\rho(r_c)$, whereafter the escape flux is also determined. An important difference with the adiabatic case is that the base temperature can be made quite cool, given that the temperature is uniform and its value is set by the (low) speed of sound at the distant critical point. The quantitative behavior of the escape flux in a few illustrative cases is explored in Problem 8.27.

Because diffusion only redistributes heat and is not in itself a source of energy for escape, it would be bizarre if introduction of strong diffusion were to allow an atmosphere to escape at low temperature without any further conditions being met. Indeed, we are not quite done with this problem. The solution must be completed by making use of the energy constraint linking escape flux to EUV heating. For the isothermal case, this comes in via the entropy equation. For an isothermal medium $\ln \theta = -(R/c_p)\ln \rho + $ const. Further, since the temperature is constant, when Eq. (8.61) is integrated over an interval of r the heating term integrates out to a difference in the fluxes at the endpoints. Putting the two results together, integrating the entropy equation from the base to the critical point, and dividing through by R yields the expression

$$-\Phi \cdot \ln \frac{\rho(r_c)}{\rho(r_b)} = \frac{1}{4} \frac{r_{abs}^2}{r_s^2} \frac{F_\odot}{c_T^2} - \frac{\kappa}{c_T^2} \frac{r_c^2}{r_s^2} \frac{dT}{dr}\bigg|_{r_c} + \frac{\kappa}{c_T^2} \frac{r_b^2}{r_s^2} \frac{dT}{dr}\bigg|_{r_b} \qquad (8.67)$$

where we have assumed that the net radiative flux vanishes at the base and becomes constant above r_{abs}. We also assume r_{abs} to be below r_c. Although the system is nearly isothermal, the diffusive heat fluxes out of the boundaries are not necessarily small, since the small (but non-zero) gradients are multiplied by a large diffusivity. If, however, we assume that the diffusion only redistributes heat provided by radiation and does not import heat from the lower atmosphere or export it through the critical point, then the last two terms on the right hand side of Eq. (8.67) can be dropped. Further, the isothermal transonic rule says that $2c_T^2 = g_s r_s^2/r_c$, while Eq. (8.66) allows us to re-write the log of the density ratio in terms of r_c as well. With these substitutions, the flux becomes simply $\Phi = \frac{1}{4}F_\odot/g_s r_s$, assuming $r_b/r_s \approx r_{abs}/r_s \approx 1$ and $r_c/r_b \gg 1$. The latter assumption is in fact necessary to assure that the kinetic energy at the base is indeed negligible, as was assumed in the derivation of Eq. (8.66). This flux is precisely the same as the low-temperature limiting flux derived earlier for the non-diffusive case. This satisfactory result provides a consistency check on our reasoning; that the two results should be the same was a foregone conclusion, given the requirements of energy conservation and the assumptions made regarding the energy available to drive the escape.

As in the low-temperature non-diffusive case, then, Φ is fixed by the gravity and EUV heating, whence evaluating Φ at the critical point implies that $\rho(r_c)$ decreases algebraically (specifically, like $r_c^{-3/2}$) as the critical point is moved outwards. Equation (8.66) then implies that the density at the base can be made arbitrarily large by moving the critical point outward, since the exponential growth of $\rho(r_b)$ trumps the algebraic decay of $\rho(r_c)$. Conversely, there are limits to how small the base density can be made by moving the critical point inward, since one ultimately violates the condition that the flow be nearly at rest near the base. The qualitative behavior is the same as for the non-diffusive case, in that the EUV flux fixes the escape rate, but that one must have enough density at the base if the escaping solution is to exist at all. Unlike the non-diffusive case, though, nothing special happens for monatomic gases.

Wrap-up and summary of hydrodynamic escape

Except for very low mass or very hot bodies, hydrodynamic escape is fueled by energy input due to EUV heating in some elevated layer of the atmosphere. EUV is absorbed very high in the atmosphere, and provides the energy needed for escape, while the much greater far-UV flux penetrates to deeper layers where it provides the energy needed to dissociate heavier species such as water into lighter constituents which may then be transported upward and escape. The atmosphere starts cold below the EUV heating layer and very subsonic. As it enters the heating layer it is boosted to escape velocity, and then launched into space. Some of the required heating is not provided by *in situ* EUV heating, but is rather communicated downward by diffusion of EUV heating from aloft. This increases the thickness of the heating layer. For a cold base, the escape flux is given by the EUV-limited flux Φ_*, but there is a critical density at the base of the escaping flow, below which hydrodynamic escape cannot be sustained. This critical density, all other things being equal, is an exponentially increasing function of the molecular weight of the escaping gas; it also depends on the thickness of the heating layer, and therefore on diffusive heat transport. In the circumstances typically of most interest, such as water loss from a runaway greenhouse on a fairly massive planet, meeting the critical density condition is the main impediment to hydrodynamic escape.

An important feature of the blowoff state is that the escaping flow of a light gas like hydrogen can carry heavier minor constituents along with it, if the outward velocity of the hydrogen is greater than the characteristic fall speed of the heavy constituent. This works only if the concentration of heavy constituents is small enough that it does not increase the mean molecular weight of the gas sufficiently to choke off escape. To determine which species can escape, we note that the fall speed (relative to the background current) of a species with molecular weight M is $w_f = (Mg/R^*T)(b/n)$, where b is the binary diffusion parameter for the heavy species diffusing through hydrogen and n is the number density of the hydrogen. For a heavy gas diffusing through a much lighter gas, the binary parameter is nearly independent of M, so that the fall speed is proportional to M. Species heavier than a threshold value do not escape at all, whereas lighter ones escape at rates which depend linearly on M – in contrast to Jeans escape, which depends exponentially on molecular weight. This differential escape implies a characteristic pattern of enrichment, most readily detected in noble gases like xenon, which are not complicated by chemical reactions. The effect may not be very important for climate evolution, but it provides the main means of determining whether an atmosphere ever experienced a blowoff state in its past.

The one-dimensional treatment of hydrodynamic escape we have given may be elegant, but it suffers from a glaring physical inadequacy: The escape is energized by EUV absorption from the planet's star, which illuminates only the dayside. Yet, despite the fact that the thin outer atmosphere has little thermal inertia to even out the dayside/nightside contrast, the escaping flow has been modeled as spherically symmetric. In reality, the escape is likely to take the form of a complex three-dimensional flow, with an outward directed jet centered on the subsolar point. Some of the outward directed mass flux leaving the dayside will fall back on the cold nightside exosphere instead of escaping to space. While more comprehensive treatments of hydrodynamic escape have added much sophistication in terms of atmospheric chemistry and radiative transfer, none at the time of writing have taken on the grand challenge of modeling the three-dimensional structure of an escaping atmosphere. Another challenge is that the traditional hydrodynamic escape formulation assumes local thermodynamic equilibrium and treats the fluid as a continuum, whereas the actual dynamics become nearly collisionless when one goes sufficiently far out in the atmosphere. If the transition to nearly collisionless dynamics occurs past the transonic point, this may matter little, given that information cannot propagate upstream in supersonic flow. However, if the transition occurs below the transonic point, the very notion of "speed of sound" breaks down and the controlling role of the transonic rule is likely to change substantially. Dealing with the transition from continuum to collisionless flow in this case is a considerable challenge. Given the nonlinearities and thresholds in the escape problem, it is likely that major revisions in our conception of escape rate are in store once somebody rises to these challenges.

8.7.5 Erosion by solar wind

Solar wind erosion is a form of non-thermal escape energized by solar wind particles instead of EUV photons. The corona is basically the exosphere of the Sun, and the solar wind is nothing more nor less than hydrodynamic escape of the solar atmosphere, which is primarily hydrogen ionized to protons. The mechanism is general and applies to virtually all stars, though we will not attempt to discuss here how the stellar wind characteristics vary from star to star, nor the way the stellar wind changes as a star proceeds through its lifecycle.

The solar wind consists almost entirely of protons (95% for the Sun) and this is likely to be the case for all Main Sequence stars. Solar wind particles are extremely energetic; they fly out with speeds of 300 to 600 km/s or even more, having energy between $8 \cdot 10^{-17}$ J and $3.2 \cdot 10^{-16}$ J. The solar wind is very tenuous; at the Earth's orbit, it has a density ranging up to 10^7 particles per cubic meter, leading to a particle flux of $4 \cdot 10^{12}/m^2s$, though it also fluctuates down to values as low as a tenth as much. The flux at other orbits can be estimated using the inverse-square law. The energy flux in the solar wind is on the order of 10^{-3} W/m^2 at Earth's orbit, which is comparable to the total EUV energy flux. Stellar winds are stronger when a star is young, and decrease over time, with the most pronounced decay occurring in the first billion years of the star's life for G-class stars like the Sun. The precise nature of the time evolution is not definitively settled, and ties in with many of the same issues that determine the long-term evolution of EUV output.

Based strictly on energetic considerations, it would appear that the solar wind has the potential to cause a great deal of atmospheric erosion even for planets as massive as Earth or Venus, and still more so for Mars. The escape energy for an oxygen atom from Earth or Venus is about $1.7 \cdot 10^{-18}$ J, so a single solar wind proton has enough energy to knock loose about 120 oxygen atoms if it could be optimally deployed. Taking into account the flux of solar wind protons (reduced by a factor of 4 to allow for averaging over the surface of the planet), this would lead to the loss of 20 bars of oxygen on Venus in the course of a billion years, or about half that much from Earth where the solar wind flux is weaker. For Venus, even this upper bound is only sufficient to remove the oxygen in a 200 meter deep ocean. The oxygen loss estimate for Earth is only relevant to the time after the atmosphere was appreciably oxygenated, but it does indicate that solar wind erosion cannot be a priori ruled out as a factor in evolution of the atmosphere. On Mars, we are principally interested in the process that could lead to the loss of a dense primordial CO_2 atmosphere; the solar wind flux is lower at the orbit of Mars, but the escape energy is less. A purely energetic estimate suggests that solar wind erosion could potentially lead to a loss of 8 bars of CO_2 in the course of a billion years, which would be more than sufficient to leave Mars in its present state.

However, there are a number of factors that greatly limit the ability of collision with solar wind protons to directly erode the heavier species of an atmosphere. The first is that, in a collision between a proton and a heavier particle, only a small part of the proton's energy is transferred to the heavy particle. For in-line collisions, conservation of energy and momentum during a collision between a light particle of mass m_1 and a stationary heavy particle of mass m_2 implies that the fraction of incident energy transferred to the heavy species is only $4(m_1/m_2)$. For protons colliding with CO_2, this amounts to about 10%. Moreover, the light proton reverses in direction in a collision, and therefore tends to escape the planet's gravity well before it can lead to additional escape. Since the typical solar wind proton has enough energy to cause 217 CO_2 molecules to escape Mars, even the limited energy deposited could cause nearly 22 molecules to escape, but only if the subsequent collisions of CO_2 or its components with the rest of the atmosphere caused 100% efficient escape. This is an unrealistic upper bound, since much of the energy of subsequent collisions will be lost to multiple collisions with the rest of the atmosphere, and will simply heat the atmosphere a bit. Thus, if solar wind protons could collide with the Martian atmosphere, the actual loss of CO_2 over a billion years would be somewhere between 0.8 bars and 0.08 bars. The former, based on the upper bound, would still be significant, but the latter would relegate solar wind erosion to the role of a minor player. Similar considerations reduce the estimates of oxygen loss from Earth or Venus.

In reality, the effects of planetary and solar electromagnetic fields render the problem far more complicated. For planets with a strong planetary magnetic field, such as Earth, the planetary field very effectively shields the atmosphere from impacts with solar wind protons, and limits erosion to small values. Titan has no magnetic field of its own, but it benefits from shielding by Saturn's magnetic field during the part of its orbit when it is effectively in the solar wind shadow caused by Saturn's field. The same phenomenon is likely to apply generically to moons orbiting gas giants. Venus has essentially no planetary magnetic field, and Mars has only a very weak one arising from remnant magnetization of the crust. However, buildup of ionized species in the outer atmosphere still leads to electromagnetic fields that are sufficient to deflect most incident protons and keep them from interacting significantly with the atmosphere. For this reason, direct collision with protons is not generally considered to be a significant source of erosion of heavy species even for non-magnetized planets like Mars or Venus.

Heavier ions are less deflected by magnetic fields, and so can penetrate more deeply into the atmosphere. Collisions with heavy ions are also more effective at transferring energy to heavy species. For this reason, it is not the solar wind itself, but secondary acceleration of heavy ions by the solar wind that play the greatest role in solar wind erosion. This is where things get complicated. First of all, you need a supply of heavy ionized species; these are created by ionization due to the EUV flux. On Mars and Early Venus, the heavy ions of principal interest are ionized oxygen atoms, in the Martian case arising from decomposition of CO_2 and in the Early Venus case arising from decomposition of H_2O if the planet is in a runaway state. The second step is accelerating the oxygen ions to an energy where they can cause escape. This step is not done by collisions with solar wind protons, but rather by the forces exerted by the electromagnetic field carried by the solar wind. The energy still ultimately comes from the energy of the solar wind, but it is transferred by the intermediary of large-scale electromagnetic interactions. The calculations necessary to determine the flux and energy of heavy ions are very complex, since they require a model of the electromagnetic field of the solar wind as well as the tracking of a large shower of ions injected into the field. The final stage of the problem is figuring out what happens when an accelerated oxygen ion collides with the atmosphere. When the shower of ions hits the atmosphere, a certain fraction of the target will be splashed out backward with sufficient energy to escape the gravitational well. This process is known as *sputtering*. Sputtering is not nearly 100%

efficient at converting incident energy to escaping particles, and the computation of sputtering efficiency is difficult, depending, among other things, on the degree to which the collisions cause the target molecules to dissociate.

Because of the complexities and considerable uncertainties of such calculations, we are in the regrettable position of not being able to guide the reader through any simple robust estimates of the actual likely mass loss due to solar wind erosion; it is a subject for experts and cognoscenti, but a very important one. The articles cited in the Further Reading section for this chapter will provide some introduction to the subject, and the range of numbers that have emerged to date. Most estimates of loss of CO_2 from Mars due to solar wind sputtering suggest that 0.1 to 0.2 bars could be lost in this way, though the somewhat controversial calculations of Kass and Yung put the number as high as 1 bar (see the discussion in Luhmann *et al.*, listed in the Further Readings). It thus appears that solar wind erosion is a potentially significant factor for loss of a primordial dense Martian atmosphere, but the general sentiment is that most of the mass loss would have to occur by other means (probably impact erosion, to be discussed shortly). Solar wind erosion does not seem to provide a viable mass loss mechanism for either the hydrogen or oxygen in a runaway Venus state, despite the planet's proximity to the Sun and even assuming that the Early Venus, like that of today, had no planetary magnetic field. The only situation in which solar wind erosion could have contributed to loss of a Venus ocean is if the loss occurred in the first hundred million years of the life of the Solar System, during which time it has been speculated that the solar wind may have been over a thousand times as intense as it is today. Even then, the loss occurs not by direct erosion or sputtering, but rather by providing enough additional heating to cause hydrodynamic escape of hydrogen with sufficient velocity to drag oxygen along with it. This mechanism bears more resemblance to hydrodynamic escape, with solar wind substituted for EUV as an energy source, than it does to the sputtering mechanisms that have been discussed in connection with Mars. The shielding effect of the Earth's magnetic field limits solar wind erosion to very small values, so that it is not a major factor in atmospheric evolution for Earth.

As in the case of escape driven by other processes, the trickle of escape of heavy species driven by solar wind interactions is pertinent to a rich variety of interesting questions bearing on the observed structures of outer planetary atmospheres today. Many of these questions are important in interpreting the isotopic composition of atmospheres in terms of past atmospheric evolution. However, for the purposes of this book we are principally interested in those mechanisms that cause enough escape to substantially affect the evolution of a planet's climate or habitability. The best guideline we can provide at this point is that, for bodies orbiting G-class stars like the Sun, solar wind erosion should be kept in mind as a significant factor in climate evolution for bodies the size of Mars or smaller, which are unshielded by a planetary magnetic field. M-dwarfs, despite being cool, have much higher stellar wind fluxes than the Sun, and could therefore sustain significant atmospheric escape from somewhat larger bodies, especially if they are in close orbits.

8.7.6 Impact erosion

The two main cases in the Solar System where theories of atmospheric evolution call for massive atmospheric loss are the problem of water loss on Venus and the loss of a hypothetical dense CO_2 atmosphere on Mars. For Titan the question is the converse – accounting for the lack of atmospheric N_2 loss, and that is plausibly accounted for by the solar wind shielding provided by Saturn's magnetic field and the low EUV flux at such large distances from the Sun. There may have been blowoff early in Titan's history, but on an icy body there are plenty of volatile reservoirs available to restock an atmosphere. Hydrodynamic escape of hydrogen from photodissociated water provides a plausible mechanism for the Venus case, but we still do not have a good way to get rid of a 2 bar Early Mars atmosphere, except for the remote possibility that solar wind erosion could do the trick. If there is no way to lose such a dense greenhouse atmosphere from Early Mars, then explanations for the apparently warm and wet early climate of that planet must be sought elsewhere. It is likely, however, that impact erosion could provide much of the needed loss mechanism. Impact erosion could be similarly important for Mars-sized bodies elsewhere in the Universe. One should be cautious about concluding that the way things happened in the Solar System is the way things must happen elsewhere, but still the fact that Earth and Venus retain thick atmospheres while Mars does not suggests that Mars-sized bodies may be susceptible to loss of atmosphere. One would like to know what the possible mechanisms are, and whether there are circumstances in which a Mars-sized body could retain an atmosphere at temperatures warm enough to be habitable and without the shielding effect of a nearby giant planet.

The impact history in the inner portion of a planetary system, where rocky planets form, can be divided up into five broad stages:

- The *Early Accretion* stage, in which small planetesimals are colliding to form larger objects, which in turn aggregate into a broad spectrum of still larger objects;
- The *Late Accretion* stage, in which most of the aggregation is complete, but there are still a number of planet-sized bodies in nearby orbits. According to simulations, lunar to Mars-sized bodies are common at this stage, and one or more giant impacts are likely. In our own Solar System, there is evidence that Earth, Venus, and Mars all experienced a giant impact at some point;
- The *Sweep* stage, in which each planet has attained nearly its ultimate size, and is sweeping up much smaller debris in the vicinity of its orbit. The impacts in this stage are still frequent, but involve collisions with bodies much smaller than the planet. There are no giant impactors left;
- The *Late Heavy Bombardment*, in which a second swarm of impactors originating from perturbations of mass stored in more distant orbits encounters the inner system. In our Solar System the Late Heavy Bombardment occurred around 3.8 billion years ago. It is not known what caused this bombardment, where the mass came from, or whether such bombardments are a generic feature of the late stages of planetary system formation. Indeed, it is not even completely clear whether the Late Heavy Bombardment is really distinct from the Sweep stage. In any event, the period of impacts that could substantially erode atmospheres came to a close with the Late Heavy Bombardment, about one billion years after the beginning of the formation of our Solar System. This number can be taken as a rough guide to the time scale of impact erosion in other planetary systems, though in systems without a Late Heavy Bombardment the period when inner planets are subject to impact erosion probably would end up to 300 million years earlier;
- The *Steady State Bombardment* stage, in which nearby sources of impactor mass have been used up. During this stage there are infrequent collisions by objects drawn from the pool of cometary and asteroidal material. Large objects of this class are only occasionally flung into planet-crossing orbits by long-term chaotic gravitational interactions, though inconsequential encounters with very small objects (as in meteor showers) occur on a nearly continuous basis.

Impact erosion is a crucial factor in the process by which a planet forms and retains an atmosphere as it accretes by collision of planetesimals, but in the following we will be principally concerned with impacts that occur in the later stages of the process – the Sweep stage and the Late Heavy Bombardment (if it is present). In this stage the planet has attained close to its ultimate mass and most of the impactors are much less massive than the target planet, but there is still enough impactor mass available to cause significant atmospheric erosion. We will also offer a few remarks on the consequences of giant impacts which precede the Sweep stage. The long period of steady state impacts which follow the Sweep stage and continue to this day can have cataclysmic consequences such as the mass extinction triggered by the Cretaceous–Tertiary impact, but these impacts are too small and too infrequent to cause much atmospheric erosion.

In impact erosion, the source of energy available to accelerate parts of the the atmosphere to escape velocity is the kinetic energy of the impactor. This energy depends on the mass of the impactor and the velocity with which it encounters the planet. When an impactor is dropped onto a much more massive planet from a great distance, it will strike the planet with a velocity equal to the planet's escape velocity, provided the impactor is initially at rest relative to the planet. This is a simple consequence of conservation of kinetic plus potential energy. It can be shown that because objects in neighboring orbits travel at different speeds, gravitational interactions cause the typical impact speed to be somewhat greater than the escape velocity. However, the enhancement is seldom as much as 20%, so throughout the following we will simply take the escape velocity as the characteristic impact speed. The scaling of impact velocity with escape velocity eliminates much of the intuitive effect of planetary size on impact erosion: though the atmosphere of a small planet is less gravitationally bound than that of a large planet, the typical impactor energy is also less for a small planet.

The case of satellites is quite different, since there we need to take into account the gravity of the parent body as well as that of the satellite. Consider the case of Titan, which orbits Saturn at a distance of $1.22 \cdot 10^9$ m. At this distance, the gravitational acceleration of Saturn is 0.025 m/s^2. The corresponding escape velocity from Saturn for an object starting at this orbit is 7.9 km/s, which is also the speed an object dropped from infinity reaches under the gravitational influence of Saturn, upon reaching Titan's orbit. In comparison, the escape velocity from Titan is only 2.6 km/s. Under the joint influence of both gravitational fields, the impactor speed would be approximately $\sqrt{2.6^2 + 7.9^2}$, or 8.3 km/s. In reality, the encounter speed could be somewhat greater, depending on the geometry of the impact, since one should also take into account the orbital speed of Titan, which is 5.5 km/s. The enhancement of impactor velocity for satellites allows the atmosphere to be eroded by a much smaller total mass of impactors than would be required if the body were a planet in an orbit of its own. This effect is overwhelmed, however, by the fact that a satellite must compete for impacts with the much larger planet it is orbiting. Impacts are received in proportion to the cross-section areas of the bodies, so the satellite receives many fewer impacts than it would if it were a planet in its own orbit, all other things being equal. For Titan, the ratio of impacts is 0.002, so this effect ups the required mass of available impactors in the orbit by a factor of 500. In the estimates to follow, we will use the term "Titan-P" to refer to a Titanlike body in a planetary orbit of its own, reserving the name "Titan" for the actual satellite subject to the effects of Saturn.

In a similar vein, the classic theory of impact erosion from planets assumes impactors in fairly circular orbits originating not far from the planet being impacted, with the planet itself assumed to be in a fairly circular orbit. Recent simulations of planet formation indicate that there can be stages in which planetesimals have highly eccentric orbits. Indeed the catalog of extrasolar planets is rife with systems having very eccentric orbits. When the impactor or planet has an eccentric orbit, the impact velocity can considerably exceed the planet's escape velocity, much as was the case for impacts on satellites. In this case, the planet's parent star plays much the same role as a satellite's parent planet. If eccentric impacts are common, they tend to tilt the balance in favor of greater atmospheric erosion from small planets, since the eccentric impacts decouple impact energy from the strength of a planet's gravity. The understanding of this effect is rapidly evolving, and so we will merely flag it here as a subject worthy of the reader's attention. Our discussion in the following will be mostly based on the classic scaling of impact energy.

The typical impact speed is much greater than the typical speed of sound. For example, an Earthlike body would have an impact speed of over $11\,000$ m/s, whereas the speed of sound in air at 300 K is only 347 m/s. Since pressure and density modifications can travel no faster than the speed of sound, the impactor carves a cylinder into the atmosphere leaving a near-vacuum in its wake. The walls of the cylinder have little time to close in, and the flow just ahead of the impactor cannot know about the presence of the planet's surface far below. The situation is not like a piston slowly compressing a column of air. Instead, the drag force on the impactor is solely dependent on the instantaneous density and temperature just ahead of its position at any time, and on the velocity and cross-section area of the impactor.

Any amount of energy delivered to the planet could, in principle, cause some amount of atmosphere to be accelerated to escape velocity. The key question in determining how much atmosphere *actually* escapes is the volume of atmosphere over which the energy is diluted. Striking a match releases about 1000 J of energy, which is enough to cause 16 milligrams of air to escape from Earth. It doesn't happen, though, because the energy released is too diluted to cause any gas to escape. The fluid dynamics governing partitioning of energy upon impact is complicated and difficult to simulate accurately. Nonetheless, simulations and theory point to a few simple principles upon which estimates of impact erosion can be based.

We can distinguish three classes of impacts:

- Small impacts, which dissipate their energy in the atmosphere. Simulations indicate that these do not cause much atmospheric escape;
- Intermediate-sized impacts which dissipate most of their energy when striking the ground or ocean, but which are still small compared with the target planet and therefore do not significantly accelerate the planet as a whole. In this case, the impactor still has a great deal of energy left upon striking the solid or liquid surface of a planet. The energy is released in a concentrated burst which can lead to considerable erosion of the portion of the atmosphere in the vicinity of the impact;
- Giant impacts, in which the impactor has mass comparable to the target planet. These can cause total blowoff of an atmosphere since the shocked planet acts like a piston which can accelerate nearly the entire mass of the atmosphere to escape velocity.

The energy required for an impactor to reach the surface with most of its energy left depends on how massive the atmosphere is. Any impactor at all will reach the surface of the Moon, but hardly anything gets through to the surface of Venus – which is why the surface of Venus is characterized by just a few but very large impact craters. The drag force exerted on a high speed sphere of radius r is $C_D \rho_a \pi r^2 U^2$, where U is the speed of the body, ρ_a is the atmospheric density, and C_D is a dimensionless number which asymptotes to values near unity for very supersonic flow. When passing through an atmosphere of thickness H

(estimated as the density scale height), the work done against the object is $C_D \rho_a \pi r^2 U^2 H$. We will assume that the initial velocity is high enough that the additional work done by the force of gravity after the projectile encounters the atmosphere can be neglected. Then, equating the work done by drag to the kinetic energy $\frac{1}{2} M U^2$ for a body of mass M, we find that the condition for the body to reach the surface with some of its initial energy left is $M > 2 C_D \rho_a \pi r^2 H$. Since $C_D \approx 1$, this condition says that the mass of the body must exceed twice the mass of the cylinder of atmosphere having the same cross-section area as the body. Writing $M = \frac{4}{3} \pi r^3 \rho_i$, where ρ_i is the impactor density, we find $r > \frac{3}{2} (\rho_a / \rho_i) H$ for $C_D = 1$. For Earth's present atmosphere, this yields a critical radius of a mere 3.5 m, assuming a silicate impactor having a density of $3000 \, kg/m^3$. For the atmosphere of Venus, the critical radius is a more impressive 340 m.

Exercise 8.12 Estimate the critical radius for a silicate impactor to penetrate the 7 mb atmosphere of present Mars.

Let's focus next on the effect of intermediate impacts. When the impactor reaches the surface, most of its energy is turned into heat, which creates a shock wave which travels outwards from the point of impact, accelerating the atmosphere. Some of the energy also goes into vaporizing the solid surface, which adds mass to the gas that must be ejected, and steals energy that could otherwise be used to feed escape; this is a refinement we shall not pursue quantitatively. Fluid mechanical simulations and analytic shock wave solutions indicate that once a critical energy is reached, essentially all the atmosphere in a narrow cone above the impact is ejected, as shown in the left panel of Fig. 8.9. As the mass of the impactor is increased, the angle of the cone widens, until it reaches the point where all the atmosphere above a plane tangent to the sphere at the point of impact is blown off. Further increases in the mass of the impactor cause little or no additional escape, until the size class of giant impacts is approached. There is a narrow range of impactor masses between the mass where a narrow cone is first blown off and the mass at which the whole tangent slice is blown off. Therefore, one can get a reasonable estimate of impact erosion by non-giant impacts by simply assuming that all impactors with a mass below that required to blow off

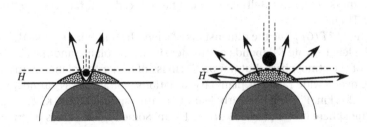

Figure 8.9 Portion of atmosphere subject to impact erosion by non-giant impacts. The left panel shows the portion potentially eroded by small impacts; very small impactors dissipate their energy before reaching the ground and cause little or no erosion. The right panel shows the limiting erosion by a larger impactor, which can erode all the atmosphere above the tangent plane to the planet's surface at the point of impact. Increasing the mass of the impactor does not yield much further erosion, until the giant-impact class, able to significantly accelerate the entire target planet, is reached.

a tangent slice cause no loss, whereas all impactors with mass above this critical mass cause loss of one tangent slice of the atmosphere.

Now we are prepared to answer the key question of how the susceptibility to impact erosion scales with the size of the planet and the thickness of its atmosphere. There are two ingredients to this estimate: the critical impactor mass needed to blow off a tangent slice of the atmosphere, and the fraction of atmospheric mass eroded by each such impact. The critical mass is important because it says how big an impactor has to be to cause significant erosion; this affects the erosion rate because there are many more small impactors than there are big impactors. To estimate the mass of atmosphere above a tangent plane, we represent the atmosphere as a uniform density layer with density ρ_a and depth H equal to the scale height at a mean atmospheric temperature T. With this approximation, the mass above a tangent plane is simply $\rho_a H^2 a$, where a is the radius of the planet. If v_e is the escape velocity, the energy needed to cause this amount of mass to escape to space is $\frac{1}{2} v_e^2 \rho_a H^2 a$. The energy of an impactor of mass m is $\frac{1}{2} m v_i^2$ where v_i is the speed of the impactor. However, since $v_i \approx v_e$, the velocity terms cancel and we find that the critical impactor mass is $m_c \approx \rho_a H^2 a$, i.e. the mass of atmosphere above the tangent plane. Note that $\rho_a H$ is the mass of atmosphere per unit area of the planet's surface. We will find it convenient to use this quantity as our basic measure of the amount of atmosphere remaining on the planet. In terms of the surface pressure $\rho_a H \approx p_s/g$, according to the hydrostatic relation.

For fixed $\rho_a H$, the critical mass scales with $Ha = RTa/g$, but since $g = \frac{4}{3} G \rho_p a$ where ρ_p is the mean density of the planet, we find that the critical mass is

$$m_c \approx \frac{3}{4} \frac{RT}{G \rho_p} (\rho_a H). \tag{8.68}$$

Thus, *for fixed atmospheric mass per unit area, the critical impactor mass is independent of the size of the planet.* While the critical mass formula does not discriminate by size of planet, it does say that massive atmospheres (in the sense of large $\rho_a H$) are more difficult to erode than tenuous atmospheres, because larger impactors are needed to trigger erosion in the more massive case, but there are fewer large impactors than small impactors. For a given mass spectrum of impactors, erosion of a massive atmosphere is initially slow and intermittent, accelerating and becoming more steady as erosion proceeds and the atmosphere becomes less massive. For a satellite, m_c must be reduced by a factor of $(v_e/v_i)^2$, which is about 0.1 for Titan.

The quantity $\frac{3}{4} RT/G\rho_p$ has the dimensions of a length squared; we'll use the symbol ℓ^2 to refer to it. The length scale ℓ depends on the density of the planet and the composition and temperature of the atmosphere, but not on the mass of either the planet or the atmosphere. It varies little over a wide range of planetary situations. For a 1 bar N_2 atmosphere at 280 K on Earth, $\ell = 233$ km. For a 2 bar Early Mars CO_2 atmosphere at 280 K, $\ell = 220$ km. For a 1.5 bar N_2 atmosphere on Titan at 80 K, $\ell = 213$ km. Some typical values of critical impactor mass and size are given in Table 8.7. Note that the critical mass is mostly controlled by the mass path $\rho_a H$, rather than the size of the planet. The Early Mars case with a 2 bar atmosphere and the Titan-P case actually require larger impacts to erode than the present Earth case. The massive atmosphere of Venus is hard to erode, but if Venus ever went through a period when it had a 1 bar atmosphere, it would be essentially as easy to erode as Earth. Generally speaking, erosion of Earthlike atmospheres is sustained by impactors with radii of a few kilometers or more, and somewhat larger impactors are required for the Early Mars case. As the Early Mars atmosphere erodes down to surface pressures of 100 mb, one can

	m_c (kg)	r_c, silicate (km)	r_c, ice (km)	N_e	m_{tot}/m_{Earth}
Earth, 1 bar N_2, 280 K	$5.5 \cdot 10^{14}$	3.5	5.2	3003	$1.0 \cdot 10^{-3}$
Mars, 2 bar CO_2, 280 K	$2.6 \cdot 10^{15}$	5.9	8.7	951	$0.7 \cdot 10^{-3}$
Mars, 100 mbar CO_2, 220 K	$1.0 \cdot 10^{14}$	2.0	2.9	1210	$0.18 \cdot 10^{-3}$
Venus, 90 bar CO_2, 700 K	$9.2 \cdot 10^{16}$	19.4	28.3	1623	$7.1 \cdot 10^{-3}$
Venus, 1 bar N_2, 280 K	$6.4 \cdot 10^{14}$	3.7	5.4	2582	$0.94 \cdot 10^{-3}$
Titan-P, 1.5 bar N_2, 80 K	$5.0 \cdot 10^{15}$	7.4	10.8	585	$0.6 \cdot 10^{-3}$
Titan, 1.5 bar N_2, 80 K	$5.0 \cdot 10^{14}$	3.4	5.0	585	$94. \cdot 10^{-3}$
Super-Earth, 1 bar N_2, 280 K	$3.2 \cdot 10^{14}$	3.0	4.3	8778	$2.3 \cdot 10^{-3}$

In the table, m_c is the critical mass required to blow off a tangent slice, and the r_c columns give the corresponding impactor radii (in km) for silicate or icy impactors. N_e is the number of impacts with $m > m_c$ which are required to deplete most of the atmosphere, and m_{tot} is the total available impactor mass in orbit required to yield this number of impacts. The calculations of m_{tot} depend on the mass distribution; results in this table assume a power law distribution with $q = 1.5$ and the maximum impactor mass m_+ equal to one-tenth the mass of Earth's Moon. The "Titan-P" case gives results for a hypothetical Titanlike body in an orbit of its own, while the "Titan" case includes the effects of Saturn, taking into account the competition with Saturn for available impactor mass. The "Super-Earth" case is for a planet with the same density as Earth and with a mass of 5 Earth masses.

Table 8.7 Table of impact erosion parameters for various bodies.

make do with impactors of somewhat over a third the size, or one-twentieth of the mass. The table also includes a Super-Earth case based on the extrasolar planet Gliese 581c, which has a mass about five times that of Earth. The critical impactor size is slightly lower than for Earth, because the surface gravity is higher and hence a 1 bar atmosphere on the Super-Earth has less mass per unit area than a 1 bar atmosphere on Earth.

For each impactor exceeding the critical mass, the fraction of atmosphere eroded is

$$\pi a \rho_a H^2 / (4\pi \rho_a H a^2) \tag{8.69}$$

which is $\frac{1}{4}H/a$ or equivalently $\frac{1}{4}\ell^2/a^2$. Thus, *the fraction of atmosphere eroded per super-critical impact decreases quadratically with the radius of the planet, all other things being equal.* This is the main reason that small planets are more susceptible to impact erosion than large planets. The characteristic number, N_e, of supercritical impacts needed to erode the atmosphere substantially is $4a^2/\ell^2$. If the impactors arrive in sequence and the atmosphere has a chance to adjust back to uniform coverage of the planet between impacts, then this number of impactors would erode the atmosphere down to a mass of $1/e$ of its initial mass, given that after each impact there is less mass left to erode and each supercritical impact just takes away a fixed fraction of what is there. The values of N_e for some planets of interest are given in Table 8.7. This number is primarily controlled by the size of the planet, ranging from about 600 for Titan to about 3000 for Earth and about 9000 for a Super-Earth. The colder Mars case with a thin atmosphere requires more impactors than the hot Mars case with a 2 bar atmosphere because the scale height is smaller in the former case.

To complete the story, we must estimate the total mass of impactors that must hit the planet in order to get the number of supercritical impacts (N_e) required for substantial erosion. We will carry out this estimate for a late stage in planetary formation, when the planet in question is by far the largest thing near its orbit and the remaining debris near the orbit is all small compared with the planet; our planet is at this stage the big kid on the block, subject to small to intermediate impacts as it sweeps up a late veneer of the remaining debris. We can define a catchment basin for the planet, consisting of the range of orbits for which the debris is more likely to impact the planet under investigation rather than some other planet. As time goes on, essentially all of the mass in this catchment basin will eventually impact the planet. (This part of the story will be slightly modified for satellites.) It is this total mass we shall estimate. For given total mass in the catchment basin, a small planet like Mars will take longer than a larger planet such as Earth to sweep up the debris, in proportion to the relative cross-section areas. Thus, for a small planet, erosion by intermediate impacts will carry on for a longer time than for a large planet. For Mars, the late stage of the erosion process will last 3.5 times longer than it would for Earth, all other things being equal. This time scale comparison may be a significant factor in accounting for the present tenuous atmosphere of Mars, since Mars can regenerate an atmosphere if it loses it early on when it is still tectonically active, but not if the loss occurs later, when the planet's interior has frozen out and has ceased outgassing volatiles.

To proceed further we must make some assumption about the mass distribution of impactors, because of the role of the critical mass m_c in determining how much atmosphere gets blown off by an individual impact. The optimal distribution for erosion would have all the impactors of equal size m_c. Making impactors smaller reduces the erosion because the impactor is unable to blow off a tangent slice, and making the impactors larger wastes impactor mass because a large (but not giant) impact cannot blow off more than a tangent mass. Information about the mass distribution and total available mass of impactors comes to us mainly from the cratering record of rocky planets, and among those primarily from the Moon and Mars (which have well-preserved surfaces not much subject to erosion). The estimates are highly uncertain, and uncertainties in the impactor distribution almost certainly overwhelm uncertainties in the detailed fluid dynamics governing how much mass is blown off by an individual impact. The mass spectrum of impactors can also be estimated from the mass distribution in today's asteroid belt. These estimates are generally compatible with estimates derived from the cratering record. Information about the *timing* of the impacts, and the rate of decay in the inner Solar System, comes from looking at the cratering record in younger resurfaced terrain, and in the lunar case, from direct radiometric dating of crater samples returned to Earth for analysis.

Let $N(m)$ be the number of impactors with mass greater than m. This is the function we need to know. The mass spectrum $n(m)$ is given by $dN/dm = -n(m)$. Equivalently, $n(m)dm$ is the number of impactors in the mass range between $m - dm/2$ and $m + dm/2$. The distribution of crater radii on any individual body has been found to approximately obey an r^{-3} power law, where r is the crater radius. This power law captures the basic crater distribution for moons of Jupiter and Saturn, as well as for the inner planets, though the details of the deviations from the ideal power law are different between the outer Solar System and the inner Solar System. The corrections to the r^{-3} law are strikingly similar between Mercury, the Moon, and Mars. This provides strong evidence that the entire inner Solar System was subject to the same population of impactors. Because crater radius scales with a power of impact energy, the crater distribution implies a power law distribution of impactor energy. Since impact velocity is approximately constant for any given body, the impact energy is proportional to the impact mass for a given body, implying a power law distribution for impactor mass. Specifically, numerical simulations and study of thermonuclear bomb craters imply that crater radius scales approximately with $E^{1/3}$ (i.e. $m^{1/3}$) for impactors. If the crater diameter distribution is $n(r)$, we get the corresponding mass distribution by writing

$$n(r)dr = n(m^{1/3})d(m^{1/3}) = \frac{1}{3}n(m^{1/3})m^{-2/3}dm \tag{8.70}$$

from which we identify $\frac{1}{3}n(m^{1/3})m^{-2/3}$ as the mass spectrum. The r^{-3} crater radius power law thus implies an $m^{-5/3}$ power law for impactor mass. Use of more detailed fits to the crater data along with alternate crater-size models, as well as direct fits to asteroidal mass distribution, yields exponents between 1.5 and 1.8.

Suppose now that $n(m) \propto m^{-q}$ for some exponent q. The total mass of impactors is $\int m \cdot n(m)dm$, and the blowup of $n(m)$ at small m does not cause the total mass to diverge as long as $q < 2$. On the other hand, the *number* of small impactors is infinite for $q > 1$, as is the case for the observed distribution. However, for $q > 1$ the total mass of *large* impactors diverges if the power law continues out to infinite mass. Thus, to make physical sense, the power law must be truncated at some mass m_+, which represents the largest mass impactor in the population. With this assumption, the total mass in the distribution is finite, and we can write a normalized distribution as

$$n(m) = \frac{2-q}{m_+}\frac{m_{tot}}{m_+}\left(\frac{m}{m_+}\right)^{-q} \tag{8.71}$$

where m_{tot} is the total mass of impactors. It is presumed that $n = 0$ for $m > m_+$.

Exercise 8.13 Verify that m_{tot} is the total mass of impactors implied by the distribution in Eq. (8.71). Find the cumulative distribution $N(m)$ and discuss how this behaves for $m_+ \to \infty$ with m_{tot} fixed.

The total number of impactors with mass greater than m_c is

$$N(m_c) = \int_{m_c}^{m_+} n(m)dm = \frac{m_{tot}}{m_+}\frac{2-q}{q-1}\left(\left(\frac{m_c}{m_+}\right)^{(1-q)} - 1\right). \tag{8.72}$$

From this, we set $N(m_c)$ equal to N_e, which tells us the required total mass m_{tot}, given q and m_+. The special case of satellites is treated by reducing m_c according to the estimate of impactor velocity enhancement, and multiplying the value of m_{tot} by the ratio of area of the primary to area of the satellite, so as to take into account the proportion of total available impactors that hit the satellite rather than the primary.

The results of this calculation of m_{tot} are given in the final column of Table 8.7. These calculations were carried out with $q = 1.5$ and $m_+ = 7.35 \cdot 10^{21}$ kg, which is one-tenth the mass of the Earth's Moon. It takes rather little impactor mass in the late stage veneer to deplete the atmosphere of an Earthlike planet – only a tenth of a percent of Earth's mass, which is not an unreasonable amount to be left over after the assembly of an Earth-sized planet. An important result is that if Mars were to start out with a 2 bar CO_2 atmosphere (as suggested by some climate calculations based on evidence for warm, wet early conditions), its atmosphere would not be much more subject to erosion than Earth's. The mass of available impactors required to erode such a Martian atmosphere would be fully 70% of the corresponding mass for Earth. The main reason the estimates are so similar is that a 2 bar atmosphere on Mars has much more mass per unit area than Earth's atmosphere, requiring a higher critical mass of impactor as compared with Earth. A more tenuous Martian atmosphere is much more erodable than Earth's, as illustrated by the 100 mb Mars case in the table. Similarly, if Venus had an Earthlike atmosphere, its atmosphere would be essentially as erodable as Earth's, whereas the actual dense Venus atmosphere requires about seven times as much available impactor mass to erode. The hypothetical Super-Earth case is only a bit less subject to erosion than Earth, in this case because a 1 bar atmosphere on a large planet has less mass per unit area than Earth's atmosphere. The importance of the atmospheric mass effect shows also in the hypothetical planetary Titan case, which, owing to its very massive atmosphere, requires nearly as much available impactor mass to erode as does the 2 bar Early Mars case. The real Titan, in contrast, is very difficult to erode, requiring an available impactor mass of nearly a tenth of Earth's mass, owing to the competition with Saturn for impacts.

The essential puzzle posed by the results of Table 8.7 is that it looks quite plausible that Earth's atmosphere would be subject to loss by impact erosion in the Sweep stage, and that a dense Early Mars atmosphere would not be appreciably less erodable than Earth. How, then, to account for the present tenuous Martian atmosphere, while Earth has a substantial atmosphere remaining? One potential scenario is that Earth's atmosphere was indeed lost by impact erosion, but was regenerated by outgassing from the interior. Consistent with this picture, we note that while Mars requires nearly as much available impactor mass as Earth, this impactor mass is delivered over a much longer time, owing to the smaller cross-section of Mars. Combined with the relatively early shutdown of tectonic activity and hence outgassing on Mars (owing to its small size) it could be that the essential difference between the planets resides not so much in ability to *hold* an atmosphere as in ability to *regenerate* an atmosphere. A severe difficulty with this picture, however, is the abundance of N_2 in Earth's atmosphere. A CO_2 or water vapor atmosphere could be easily regenerated, but it is not easy to hide enough N_2 in the mantle to allow this component to be regenerated. And recall that Venus has even more N_2 in its atmosphere than Earth, suggesting that even if Venus went through an early stage with far less CO_2 in its atmosphere, it did not suffer total atmosphere loss by impacts during that stage. Could it be that there is an ability to sequester a bar or two of N_2 in a planet's mantle? Could it be that Earth started out with much more N_2 in its atmosphere and that what we have today is the small bit left over after substantial impact erosion? Or could it be that the mass of impactors was not in fact sufficient to deplete Earth's atmosphere and that the tenuous Martian atmosphere has some other explanation? Perhaps it never generated a dense atmosphere, because it never received enough oxygen-bearing material to turn carbon into carbonate and CO_2. Perhaps Mars lost its atmosphere in a chance giant impact which got rid of Martian N_2, whereas Earth's Moon-forming impact was not big enough to get rid of all the N_2. If a giant impact removed most of the primordial N_2 on Mars, then perhaps the rest could have been lost by non-thermal escape and solar wind erosion. But if Mars lost its atmosphere too early then it becomes hard to account for the large, extensive water-carved channels on Mars, some of which suggest persistence of active surface hydrology up to 3.5 billion years ago, with episodic recurrence of less extensive river networks extending billions of years later. More precise dating of these hydrological features, which will come ultimately with sample return missions from Mars, will go far to help resolve these puzzles. Still, the Mystery of the Missing Martian Atmosphere is likely to remain one of the Big Questions for a long while to come.

How do giant impacts fit into the picture? Giant impacts do not come in a continuous stream, but lunar to Mars-sized bodies are common enough in the late stages of planetary formation that it is likely that one or more giant impact occurs before the planet attains its final size. The very existence of the Moon provides evidence that Earth experienced a giant impact, while the anomalous retrograde rotation of Venus has been taken as evidence that a giant impact occurred there as well. The Martian crust exhibits a striking dichotomy between rugged thick-crusted and heavily cratered southern hemisphere highlands and smoother, thinner northern hemisphere lowlands; this has sometimes been taken as having resulted from a giant impact, though one smaller in relative scale than Earth's Moon-forming impact. A single giant impact can blow off an entire atmosphere, but this is not inevitable; depending on the energy of the impactor, there can be a substantial proportion of the original atmosphere left. The issues in reconciling the histories of Earth and Mars are essentially the same as for impact erosion at the Sweep stage: how do we account for the story of N_2 on Earth (or Venus, for that matter)? And how are we to account for the hydrology of Early Mars if a giant impact blew off the primordial Martian atmosphere but the planet was unable to regenerate a new CO_2 atmosphere by outgassing?

It should be kept in mind that impacts can also be an important *source* of volatiles. Comets can directly bring in volatiles such at CO_2, water, and methane. Moreover, the high pressure and temperature during the impact shock can cook water vapor out of hydrated minerals such as serpentine. Similarly, impacts can release CO_2 from carbonates in the crust. Carbonates are not primary planet-forming substances, but could be formed in the process of accretion through reaction between primary forms of carbon, oxygen, water, and silicates. If this happens, CO_2 can be retained within carbonate even when some atmospheric loss event has blown away an earlier atmosphere, and this CO_2 can be released as a result of later impacts at a rate that can be much faster than the release associated with volcanic activity.

8.8 WORKBOOK

8.8.1 Silicate weathering

Problem 8.1 Based on the information shown in Fig. 8.2, how high does the temperature have to be in order to inhibit the formation of iron carbonate, if the surface silicate rocks are in contact with an atmosphere having a CO_2 partial pressure of 100 mb? *Python tips:* The Chapter Script EbelmenUreyEq.py carries out the calculations needed for Fig. 8.2.

Problem 8.2 It has been estimated that the current CO_2 outgassing rate on Earth is about $8 \cdot 10^9$ Mole/yr. Assuming that this is entirely balanced by reaction with $CaSiO_3$ to form carbonate, and that suitably weatherable rocks have a surface area amounting to $\frac{1}{6}$ of the Earth's surface, how thick is the layer of rock that needs to be weathered away over the course of a thousand years?

The density of $CaSiO_3$ is about 3000 kg/m^3. In reality, it is a fairly uncommon mineral in the Earth's crust, and weathering reactions involving the feldspar family of silicates are far more important. One member of this family is *anorthite* ($CaAl_2Si_2O_8$), with density of 2700 kg/m^3. How would your result change if this were the mineral weathering? (As a crude approximation to the chemistry, assume that each C from a CO_2 molecule substitutes for one Si in the weathering products. The actual weathering products include a bewildering array of silicate-bearing clay minerals in addition to bicarbonate ion, which reduces the efficiency of the weathering reaction.)

Problem 8.3 *Artificial* CO_2 *outgassing*
Our own planet at present is plagued with too much CO_2 entering the atmosphere, owing to fossil fuel combustion. However, a planet which had cooled off to the point that tectonics had stopped and outgassing of CO_2 had ceased would have the opposite problem, in that CO_2 would be taken out of the atmosphere by silicate weathering without being replaced by a source. This would lead to global cooling, and ultimately a Snowball state on a planet with an ocean. How hard would it be to sustain a technological replacement for outgassing?

The present weathering rate on Earth consumes about $5 \cdot 10^{-5}$ Mole/m^2 yr of carbon over continental crust. The weathering rate should be similar for any vegetated planet with an Earthlike climate and surface mineralogy. Suppose there is a fossil fuel reservoir containing 4000 Gt of carbon. How long could the silicate weathering be offset by judicious burning of this fuel? How many kilowatt-hours of electricity would be produced by doing so? Compare that with current world electricity production.

But what can be done after the fossil fuels are gone? All is not lost. Use the equilibrium properties of the Ebelmen–Urey reactions to develop a strategy for cooking CO_2 out of various carbonate rocks by using solar energy to raise the rocks to some high temperature, and then letting them cool down. This could be done using mirrors to focus solar energy. You may assume that at sufficiently high temperatures the reaction proceeds so quickly that the time to reach equilibrium is not a significant limitation. Determine how many kg of $CaCO_3$ rock would have to be processed over each square meter of land each year. Estimate the energy required to do the processing. Compare this with a rough estimate of the solar energy falling on the area in the course of the year. What proportion of the solar energy needs to be tapped in order to save this planet from freezing? How might your answer differ if the rocks being processed were iron or magnesium carbonates instead?

The specific heat of calcium or magnesium carbonate is about 1000 J/kgK, and that of iron carbonate (siderite) is only slightly less at 700 J/kgK.

Problem 8.4 Re-do the calculation of Fig. 8.3 for a range of different outgassing rates relative to the present value W_0. How much would the outgassing rate have to be increased in order to make the Early Earth temperature similar to the present value?

Python tips: This and subsequent silicate weathering thermostat problems can be done by modifying the Chapter Script WHAK.py.

Problem 8.5 *Precipitation limitation and silicate weathering*
In Section 6.8 we showed that at sufficiently high temperatures the precipitation rate depends explicitly on the surface absorbed solar radiation, and tends to level off at a limiting value. Idealize this precipitation–temperature relation by assuming that the precipitation increases linearly with temperature at the rate assumed in the text until the implied surface latent heat flux equals the surface absorbed solar radiation, and thereafter remains constant.

For simplicity, we will assume that the atmosphere does not absorb any shortwave radiation, though it may reflect some back to space before it can reach the ground, and we will also assume that the albedo of the surface is zero, so that it absorbs all the shortwave radiation incident upon it. In this case, the atmospheric albedo is also the planetary (i.e. top of atmosphere) albedo, so the absorbed solar radiation at the surface or top-of-atmosphere is simply $\frac{1}{4}(1 - \alpha)L_\odot$. Compute the clear-sky *OLR* using the polynomial *OLR* fit for an Earthlike atmosphere.

First, assuming that the atmosphere does not absorb or reflect any incoming shortwave radiation, plot the temperature as a function of $L_\odot$ for a range of different outgassing rates relative to Earth's present value. Discuss how these results differ from the case in which precipitation can increase indefinitely, paying particular attention to the implications for planets in orbits with low values of $L_\odot$.

Now suppose that the atmosphere has clouds in it. If the clouds are low, they just change the albedo, and the effects are equivalent to just changing the value of $L_\odot$. As an opposite extreme, suppose that the clouds have a height such that their albedo effect is exactly canceled by the reduction in *OLR* caused by the clouds, much as happens for deep clouds in the Earth's tropics today. The clouds will then leave the temperature unchanged for any given CO_2 concentration, but they will alter the precipitation. What does this effect do to climate once silicate weathering comes into equilibrium?

Carry out these calculations both for the biotic-land case in which the weathering rate is not directly dependent on pCO_2, and the abiotic case in which the weathering has a power law dependence on pCO_2.

Problem 8.6 *Effect of methane on silicate weathering*
The effect of CH_4 in an atmosphere can be crudely incorporated into the silicate weathering thermostat by changing the *OLR* relation to

$$OLR(T, pCO_2, pCH_4) = OLR(T, pCO_2) - 2.1 \ln pCH_4/10 - 5.5 \qquad (8.73)$$

where the first term on the right hand side is the *OLR* formerly used for the no-methane case and both pCH_4 and pCO_2 are measured in ppmv. Discuss how the addition of methane to an atmosphere affects the temperature and pCO_2 of a planet, assuming the silicate weathering thermostat to be in operation. Discuss the differences between the abiotic case in which weathering depends directly on pCO_2, and the land-biotic case in which weathering rate depends on pCO_2 only indirectly through temperature and precipitation. You may assume the precipitation to increase linearly with temperature, and ignore the limitations on precipitation that may come into play for very warm climates.

What happens to temperature if you bring the CH_4 to very high values and then suddenly oxidize it to CO_2 in the atmosphere, as might happen if the source turns off abruptly or the atmosphere gets suddenly oxygenated? What happens if instead 80% of the extra CO_2 added by methane oxidation is taken up by the ocean? Scenarios of this sort have been proposed as means of triggering a Snowball glaciation.

Problem 8.7 *Silicate weathering on Super- and Mini-Earths*
The size of a rocky planet affects the silicate weathering thermostat in several ways. It affects the CO_2 outgassing rate per unit area, possibly affects the atmospheric inventory of background gases such as N_2, and (through gravity) affects the relation between the partial pressure of CO_2 and the greenhouse effect of the atmosphere. This problem explores just the last of these effects. If we adopt a linearized form of $OLR(T, pCO_2)$, $a + b \cdot (T - 220)$, then holding the total surface pressure fixed at 1 bar the coefficients are $a = 111.75 - 2.24 \ln(pCO_2/300)$ and $b = 1.87 - 0.045 \ln(pCO_2/300)$ for Mars gravity, and $a = 117.4 - 2.17 \ln(pCO_2/300)$ and $b = 2.52 - 0.052 \ln(pCO_2/300)$ for a Super-Earth with surface gravity $20\,m/s^2$. In these formulae, pCO_2 is measured in ppmv. The water vapor feedback has been included in the calculation of the coefficients.

Using these *OLR* fits together with the weathering-rate formulae given in the text, discuss how the curves of steady state temperature and pCO_2 vs. $L_\odot$ differ between the two planets. Show results for outgassing rates per unit surface area equal to Earth's, and both half and twice that value. Discuss the differences between the abiotic case in which weathering depends directly on pCO_2, and the land-biotic case in which weathering rate depends on pCO_2 only indirectly through temperature and precipitation. You may assume the precipitation to increase linearly with temperature, and ignore the limitations on precipitation that may come into play for very warm climates.

8.8.2 Ocean–atmosphere partitioning

Python tips: The carbonate/bicarbonate equilibrium problems in the following group can be done with the assistance of the Chapter Script `CarbonateEq.py`.

Problem 8.8 Suppose space aliens suddenly and instantaneously steal all the O_2 from our atmosphere. Photosynthesis would eventually regenerate the oxygen, but how much O_2 would be put back into the atmosphere by outgassing from the ocean inventory alone? Give your answer in the form of the partial pressure of O_2. You may assume that O_2 is non-reactive in the ocean, so that the ocean storage is governed entirely by Henry's law.

Problem 8.9 Suppose a mutant nitrogen-fixing bacterium takes over the Earth and converts all the N_2 in the Earth's atmosphere into NH_3. Using Henry's law, determine the new composition of the atmosphere once it comes into equilibrium with a 50 m oceanic mixed layer. What is the composition if the atmosphere comes into equilibrium with the top kilometer of ocean instead?

Problem 8.10 It has been proposed that Early Mars could have been warmed by a fairly high abundance of SO_2 in the atmosphere, arising from volcanic outgassing in an anoxic atmosphere. Suppose that Mars at that time had a 500 m deep water ocean covering 20% of the surface area of the planet, with a mean temperature of 280 K. Use Henry's law to determine the partitioning of SO_2 between atmosphere and dissolved gas in the ocean. Ignoring any other oceanic SO_2 reservoir or sinks of SO_2, how many Moles of SO_2 need to be outgassed in order to achieve a concentration of 100 ppmv in a 1 bar CO_2 atmosphere? Compare the estimated outgassing rate per unit surface area to the outgassing rate of CO_2 on Earth.

If you are ambitious, you may want to think about the circumstances under which sulfite and bisulfite can make up a substantial part of the oceanic sulfur pool.

Problem 8.11 *Linearized carbonate equilibrium*
Linearize the carbonate/bicarbonate equilibrium equations about an initial equilibrium state with atmospheric partial pressure $(pCO_2)_0$, bicarbonate concentration $[HCO_3^-]_0$, carbonate concentration $[CO_3^{--}]_0$ and a pH value $(pH)_0$. Specifically, write $pCO_2 = (pCO_2)_0 + \Delta_0$, $[HCO_3^-] = [HCO_3^-]_0 + \Delta_b$ and so forth, and solve for each perturbation Δ in terms of Δ_0 assuming the perturbations to be small compared with the unperturbed values. Find an explicit expression for the change in pH and net atmosphere–ocean carbon storage as a function of the change in pCO_2. Consider both the case in which $[Ca^{++}]$ remains fixed at its unperturbed value and the case in which it is kept in equilibrium with a reservoir of solid $CaCO_3$.

Problem 8.12 Using Newton's method, write a routine to determine the pH which satisfies the charge balance relation in Eq. (8.19), given pCO_2, $[Ca^{++}]$, and the equilibrium coefficients K_H, K_1, and K_2. Assuming $[Ca^{++}] = 0$, find the pH of sea water at 298 K in equilibrium with atmospheric CO_2 concentrations of 280 ppmv and 380 ppmv. Do the same for fresh water, for which $K_1 = 7.65 \cdot 10^{-9}$ and $K_2 = 1.01 \cdot 10^{-12}$ for concentrations measured as mole fraction. Find the pH of carbonated water in a sealed bottle, in equilibrium with 2 bar of pure CO_2 gas.

Problem 8.13 In this problem you will compute the equilibrium carbon storage in the ocean mixed layer as a function of the atmospheric pCO_2, assuming that there is no time for additional $CaCO_3$ to dissolve beyond what is already in the mixed layer at some initial time. You may assume the mixed layer to have a uniform temperature of $20\,°C$.

First estimate the number of Moles of H_2O in the mixed layer, assuming a $50\,m$ mixed layer depth covering $\frac{2}{3}$ of the Earth's surface. Next solve the carbonate/bicarbonate equilibrium equations and use charge balance to compute the concentration $[Ca^{++}]$ assuming that the mixed layer is in equilibrium with an atmosphere having CO_2 concentration 280 ppmv (28 Pa partial pressure) and its pH is 8.15.

Next, compute the total atmosphere/mixed-layer carbon storage and the ocean pH as a function of pCO_2, holding $[Ca^{++}]$ fixed at the value computed in the preceding part. Use your results to discuss how much CO_2 from fossil fuel burning would remain in the atmosphere over times scales for which the mixed layer does not significantly exchange water with the deeper ocean.

Problem 8.14 *Where did my trillion tonnes go?*
In the text we computed where the first trillion tonnes of fossil fuel carbon would go, over time scales long enough for surface waters to mix with the deep ocean, but too short for significant additional input of dissolved $CaCO_3$. What happens when more carbon is added? How much does the reduction in oceanic pH reduce the ocean's ability to take up additional carbon?

To address this problem, compute the total atmosphere–ocean carbon content as a function of the atmospheric pCO_2, assuming $[Ca^{++}]$ to remain fixed at a pre-industrial value corresponding to equilibrium with an atmosphere having a CO_2 concentration of 280 ppmv. Present your results in the form of a graph showing the atmospheric CO_2 concentration as a function of the amount of fossil fuel carbon released in Gt, showing results out to 5000 Gt (the estimated amount of carbon existing in worldwide coal deposits).

The outline of the calculation for this problem is essentially the same as for Problem 8.13, except that this time one equilibrates with the entire volume of the ocean rather than just the mixed layer.

Problem 8.15 Suppose that during a Snowball Earth episode, the atmospheric CO_2 concentration builds up to 10%. The ocean is in equilibrium with the atmosphere, and the dissolved calcium carbonate has been maintained at saturation. Since there is little input of carbonate from land weathering during a global glaciation, all of the calcium carbonate needed to maintain saturation needs to have come from sea-floor dissolution. Estimate the number of Moles of $CaCO_3$ per square meter of ocean floor that need to have dissolved in order to maintain saturation. How thick a layer of $CaCO_3$ does this correspond to? (The density of $CaCO_3$ is $2700\,kg/m^3$.)

Problem 8.16 *Ecopoeisis on Mars*
Plot spoiler warning: Read the story *Ecopoeisis* by Geoffrey Landis before reading the rest of this problem.

In Landis' story, an ecosystem has been introduced on Mars which has increased the annual mean surface pressure of a pure CO_2 atmosphere to 100 mb. This has warmed the planet sufficiently to melt the glaciers and create a water ocean at one of the poles. Taking the temperature range indicated in the story as a given (i.e. leaving aside the question of whether 100 mb of CO_2 would warm Mars that much) consider the effects of the polar ocean on the seasonal cycle of CO_2, including the effects of carbonate/bicarbonate equilibrium. At what parts of the seasonal cycle is the ocean taking up CO_2? At what point is it outgassing?

What is the total mass of CO_2 outgassed each year? What is the effect on the surface pressure if the CO_2 remains in a column over the ocean? If the CO_2 is uniformly distributed over the planet?

If the ocean becomes supersaturated during the outgassing season and then releases all of the CO_2 in a 10 day long burst at the end of the warm season, estimate the wind speed if the gas comes out in a radial current 3 km deep having uniform speed. Does the premise of the story work?

Hints: You will need to make an assumption about the mean pH, or equivalently the anion concentration $[Ca^{++}]$. The latter can be assumed constant over the course of the seasonal cycle (why?). Over much of the Martian surface the soils are acidic and carbonate-poor, so the mean pH could be quite low. On the other hand, some carbonate-rich soils have been discovered in polar regions, and so the ocean might be alkaline, with a pH similar to Earth's oceans. Discuss both of these limiting cases.

Assume that the ocean has a mass of 10^{19} kg, spread uniformly over a layer centered on one pole, having area equal to 20% of the Mars surface. Consider the mixed-layer depth to extend through the full depth of the ocean. Model the seasonal ocean temperature cycle at the pole as a sinusoid with specified mean and amplitude. You are not expected to compute the seasonal cycle from a simulation, though in fact you are equipped to do so using the material covered in Chapter 7. Try some different assumptions about the mean temperature and seasonal variation to see how this affects the basic premise of the story.

8.8.3 Atmospheric chemistry

Problem 8.17 *Follow the photons I*
Consider a planet in an orbit about an M-dwarf star with photospheric temperature 3500 K, which has $L_\odot = 2600$ W/m^2 (about the same as Venus). Suppose that the planet is undergoing a runaway greenhouse and has developed a massive pure water vapor atmosphere. Assuming that water vapor is dissociated by far-UV photons with wavelengths shorter than 0.21 μm, estimate the photon flux in this band by assuming that it is half the blackbody value (as for Earth), and determine how long it would take to dissociate all the water originally contained in a liquid water ocean of average depth 2 km. Does the answer depend on the radius and surface gravity of the planet?

Assuming that the absorption cross-section per molecule is 10^{-25} m^2, estimate the thickness of the layer (expressed in Pa) over which the dissociation occurs, given a surface gravity of 20 m/s^2.

Problem 8.18 *Follow the photons II*
Consider a planet whose upper atmosphere consists of a mixture of gas A having initial molar concentration η_A and far-UV cross-section χ_A, with a gas B having initial molar concentration η_B and a smaller far-UV cross-section χ_B. The cross-sections are independent of wavelength. The atmosphere is exposed to far-UV with a given photon flux. The photolysis products do not react to re-form the photolyzed gases, and also do not absorb any far-UV photons. Compute the time evolution of the concentrations of each gas under the following two limiting assumptions about mixing: (a) The top 10 mb of the atmosphere is a mixed layer within which all molar concentrations are kept uniform in height as they change; (b) There is no vertical mixing, so that a gas lost to photolysis is never replaced by mixing in from a layer where the gas is more abundant. Plot the time evolution of the profile of photon flux as well.

Put in some numbers corresponding roughly to an NH_3/CO_2 atmosphere on Early Earth, and discuss the extent to which a CO_2-dominated atmosphere could shield NH_3 from photolysis. The weakest points in your estimate will be the assumption that the cross-sections are wavelength-independent and the assumption that the photolysis products neither react nor absorb photons.

Problem 8.19 *CO_2 photolysis and re-formation*
Re-do the CO_2 photolysis and re-formation problem discussed in connection with Eq. (8.23), but this time for a thick CO_2-dominated atmosphere in which mixing is slow enough that essentially none of the photolysis products reach the ground before reacting. Model the photolysis assuming a wavelength-independent cross-section of 10^{-25} m^2 for wavelengths shorter than $0.17\,\mu m$, and ignore absorption by the photolysis products. Model the mixing by assuming that a layer extending from the top of the atmosphere down to pressure p_{mix} is instantaneously mixed to uniform composition. Compute the molar concentrations of CO_2, CO, and O once the system has reached equilibrium, as a function of p_{mix} and the photon flux. Restrict your attention to cases in which essentially none of the energetic photon flux exits the bottom of the mixed layer. Put in some numbers for the photon fluxes for Earth, Mars, and Venus, and for a range of mixed layer thickness. For computing the rate constants, you may assume the temperature to be $200\,K$, and use the rate constant given in the text.

If you are ambitious, you can replace the mixed-layer assumption with a solution of steady state diffusion equations for each of the chemical constituents. The methods used are precisely the same as those discussed in Chapter 6, except for the need to introduce chemical sources and sinks. You can solve this either as a time-stepping partial differential equation problem or a steady-state coupled second order ordinary differential equation problem. In either case, remember to take into account the different diffusion constants for the various species. With the diffusion formulation, you can also extend the problem to the case in which CO_2 is a minor constituent in an N_2 background gas (assumed transparent to FUV). In that case you would need to diffuse CO_2 as well as the photolysis products.

Note that even in the absence of catalysis by the OH radical, the re-formation process would in reality be significantly affected by the recombination of 2O into O_2, and the photolysis of O_2.

8.8.4 Atmospheric escape

Problem 8.20 Consider an atmosphere which is pure CO_2 below the homopause, but which photodissociates into CO and O at some pressure level p_1 which lies above the homopause. Note that the term "homopause" here refers to the level above which the mixing is dominated by diffusion rather than by turbulent motions, even though a single-component atmosphere will by definition have uniform composition everywhere; this is why some prefer the term "turbopause." For the purposes of this problem, you may ignore the recombination of the products back into CO_2. You may assume the atmosphere below the exobase to be isothermal at a temperature of $250\,K$. You may assume that conditions are such that the exobase lies at a low enough altitude that the gravity is essentially the same as the surface gravity, but you should state how the rest of your results depend on the planet's gravity.

First compute what the exobase altitude and density would be if the CO_2 did not dissociate and the exobase were pure CO_2. In this case one expects a cool exobase temperature, which you can take to be $300\,K$. Then, keeping the exobase temperature the same, determine the exobase height, density, and composition as a function of the dissociation pressure level p_1. When the exosphere becomes dominated by O, the temperature is expected to rise. Say how your results change if you assume the exobase temperature to be $1000\,K$ instead.

Using the profiles of CO and O you computed above, discuss where you think the recombination into CO_2 is most likely to occur.

Problem 8.21 *Jeans escape from M-dwarf planets*

It has been suggested that for planets in close orbits about M-dwarfs, the high EUV output early in the life of the star could cause massive early atmospheric loss, which in the case of N_2 could be irreversible. One way this could happen would be for the EUV flux to be so high that the outer atmospheric temperature becomes very high, enhancing Jeans escape directly and through elevating the exobase to altitudes where the gravity is weaker. In this situation, it is only the outer EUV-heated part of the atmosphere that would become high. Deeper parts of the atmosphere respond mainly to the net shortwave energy flux from the star, and would remain quite cool.

Consider a Super-Earth with radius and surface gravity twice that of Earth. The deep atmosphere, consisting of almost all of the mass of the atmosphere, is represented by an isothermal layer with a relatively cool temperature T_0; this layer is shallow enough that gravity is essentially constant within it. Above this lies an EUV-heated layer which begins at some specified pressure p_1 and has a high temperature T_{ex}. Supposing the atmosphere to be pure N_2 between p_0 and the ground, dissociating to N at higher altitudes, compute the exobase density and altitude as a function of T_{ex}. How large does this temperature need to be in order for the reduction of gravity to be significant? Compute the Jeans escape flux, and determine how large T_{ex} needs to be in order to lose one bar of N_2 in a billion years. How does your answer depend on p_0?

Note that T_{ex} can substantially exceed the photospheric temperature of the star, since EUV radiation is non-thermal and comes in with an energy flux much greater than what would be expected from the photospheric blackbody temperature. Make an estimate of p_0 in terms of the EUV photon flux using the number and absorption cross-section of molecules at altitudes above p_0, by estimating the number of molecules needed to use up the available photon supply. (Determining the temperature is much more complicated, since one needs to know something of how the N atoms in a nearly collisionless gas lose energy by radiation.)

Problem 8.22 *Diffusion-limited escape on Titan*

Estimate the diffusion-limited escape rate of hydrogen from Titan, assuming that hydrogen escapes by diffusion of H_2 through N_2 and then dissociates to H just before it escapes to space at the top of the atmosphere. Estimate how long it would take to lose one hydrogen atom from each CH_4 molecule in Titan's atmospheric CH_4 inventory, assuming that the photochemical reactions that split off the H are not rate-limiting.

Estimate the Jeans escape rate of H and compare it with the diffusion-limited rate, to see if the Jeans escape rate is high enough to make the escape diffusion-limited. Note that there are escape mechanisms that can be more effective than Jeans escape. To compute the Jeans escape rate you will need an estimate of the H mixing ratio at the exobase. This is difficult to do precisely, so a crude estimate of the likely range will suffice. You may assume that H is a minor constituent at the exobase.

Problem 8.23
In the text the diffusion-limited escape problem was solved assuming that the removal from the exobase was so rapid that the concentration of the escaping constituent

could be considered to be zero at the exobase. Re-do the problem using the actual flux boundary condition at the exobase, namely that $-D d n_A/dz = w_* n_A$ at the exobase, where n_A is the number density of the escaping constituent, D is the diffusivity of this constituent at the exobase, and w_* is the escape flux coefficient, which has dimensions of a velocity. Show that it reduces to the result in the text in the limit of large w_*, and that the escape flux is limited by w_* rather than diffusion when w_* becomes small. Define the notions of "large" and "small" precisely, through formulation of a suitable non-dimensional parameter.

In this problem, you may assume that the diffusive layer is thin compared with the radius of the planet, so that the acceleration of gravity can be considered constant.

Problem 8.24 *Speed of sound*

First show that the wave equation $\partial_{tt}\phi - c^2\partial_{xx}\phi$ is satisfied by any function $\phi(x \pm ct)$, representing a disturbance propagating with speed c. Next, derive the wave equation for sound waves by showing that the momentum equation is $\partial_t u' = -\rho_0^{-1}\partial_x p'$ and the mass conservation equation is $\partial_t \rho' + \rho_0 \partial_x u' = 0$ for small disturbances of velocity, density, and pressure (u', ρ', p') of a state of rest with uniform density and pressure. Assume that pressure can be written as a function of density $p(\rho)$, and use the relation $p' = \rho' dp/d\rho$ to combine the two equations into a wave equation, from which you will be able to conclude that $c^2 = dp/d\rho$. Then, use the ideal gas equation of state and the definition of potential temperature to show that: (a) $c^2 = RT_0$ if the gas is constrained to remain isothermal, as in the case of very strong heat diffusion, and (b) $c^2 = \gamma R T_0$ if the flow remains adiabatic.

Problem 8.25 *The de Laval nozzle*

Consider compressible flow in a tube with circular cross-section, whose radius varies with distance r along the axis of the tube. The velocity in the axial direction is uniform at each r, with value $u(r)$, so that the mass flux is $\rho(r)u(r)A(r)$. The fluid is not subject to any external heating, and the effect of gravity is negligible, as is the diffusion of heat. The gas satisfies the ideal gas equation of state. The equations describing this flow are identical to those used for spherically symmetric adiabatic hydrodynamic escape, except that $A(r)$ replaces $4\pi r^2$ and the gravitational potential has been dropped.

Solve the energy and entropy conservation equations to determine the characteristics of flow in this tube, given the conditions on u, ρ, and T at the inlet. Show that if the inlet flow is subsonic, then the only way to cause the outlet flow to be supersonic is to arrange things such that there is a constriction of the tube such that the transonic point occurs at the position of the minimum of area. What do you think happens if an engineer tries to accelerate the flow to supersonic speeds using a conical nozzle with area decreasing monotonically toward the outlet?

Problem 8.26 *Adiabatic hydrodynamic escape from small bodies*

Consider a spherical body with a single-component atmosphere which is transparent to all incoming shortwave radiation as well as to outgoing infrared. It is a rocky body with a density of $3000\,kg/m^3$. The surface is isothermal and has a specified temperature; once the temperature is given, it does not matter for the purpose of this problem whether the surface temperature is maintained by shortwave absorption, internal heat, or some combination of the two.

For what size of body can you sustain adiabatic hydrodynamic escape of H_2 if the surface temperature is 300 K? What answer do you get if the atmosphere is N_2 instead? Estimate the lifetime of the atmosphere as a function of the size of the body in each of these cases.

For a G star with luminosity equal to the Sun, find the maximum size of body for which adiabatic hydrodynamic escape of an N_2 atmosphere can be sustained as a function of the orbital radius, assuming zero albedo of the body and that the surface is heated by shortwave absorption.

Problem 8.27 *Isothermal hydrodynamic escape flux*
Compute the isothermal hydrodynamic escape flux as a function of the density at the base of the escaping atmosphere and the critical (i.e. transonic) point position. State the corresponding temperature of the layer, which is the same as the critical point temperature since the flow is isothermal. The "base" of the escaping flow can be taken to be at any convenient point of reference where the kinetic energy of the radial velocity is negligible. Put in numbers corresponding to the escape of atomic hydrogen on Earth and on Venus, and say how the results depend on the assumed critical point position. What happens if you try to make a heavier species escape? Consider the cases of atomic oxygen on Venus and atomic carbon on Mars as an example of this. Think hard about the question of whether you can get around the requirement of very high base densities by moving the altitude you choose for the base outward. *Hint:* The answer is that you can't.

Use the entropy equation to derive an expression estimating how much radiative flux is needed to sustain the escape. How does the molecular weight of the escaping species affect the result? *Hint:* Recall that the specific heat is largely determined by molecular weight.

To complete the calculation of escape flux, you need to put in a number for the base density. One gets the same answer regardless of where one takes the base to be so long as the kinetic energy of the radial velocity is negligible there, since one is in the end solving for the structure using hydrostatic balance, whether that is done directly or via the radial momentum equation. Let's restrict attention to the H escape case. A precise calculation requires solving the full photochemistry problem simultaneously with a representation of vertical transport, but we'll take some shortcuts. To get the base density, assume that the homopause (i.e. turbopause) is pure H_2O, but that the water vapor decomposes into H_2 and O_2 just above the homopause. Then, assuming the temperature of the layer above the homopause to be the temperature you computed on the basis of the critical point position above, compute the scale height of each species, and find the altitude at which H_2 is 90% dominant, by Mole fraction. Assume that just above this point, the H_2 dissociates into H, and take the result as your base density of H.

Using this estimate, compute the escape flux for Venus, Earth, and Mars, and determine how long it would take to lose the H in a 2 km deep ocean for each planet. Estimate the EUV flux needed to sustain the escape, and compare this to the typical EUV fluxes encountered in the Solar System. Recall that the young Sun could have put out an EUV flux a factor of ten or more greater than is observed at present.

Problem 8.28 *Impact erosion basics*
First, reproduce the Mars and Earth lines in Table 8.7 to make sure you understand how the calculation of impact erosion is done. Then compute the analogous results for a 1 bar N_2 atmosphere on the Moon assuming a surface temperature of 260 K. Do the calculation first for the case where the Moon orbits the Earth in its present orbit, and then for the case in which the Moon orbits the Sun directly without sharing the orbit with Earth.

Then, re-do the estimates of m_{tot} for Earth and Mars assuming that the mass spectrum of impactors has the exponential form $n(m) \sim \exp(-m/m_1)$ instead of a power law. In this case, do you need to truncate the distribution at a maximum mass m_+?

Problem 8.29 *Alternate assumptions about impactor energy*
In the traditional view of impact erosion, the velocity of impactors scales with the escape velocity of the target planet. More recent models, particularly of the Late Heavy Bombardment, suggest that the population of impactors may instead be drawn from a pool of planetesimals in orbit beyond Jupiter, whose orbits are perturbed and become eccentric enough to cross the inner Solar system. This is known as the "Nice model" (after the town in Southern France, though the model is indeed a nice one as well).

As a simple exploration of this situation, consider a target planet with mass M_p and radius a in a circular orbit at distance r_p from its star. The star has mass $M_\odot$. The population of impactors are in eccentric orbits which take them from a maximum distance r_+ from the star to a minimum distance $r_- < r_p$, which can collide with the target planet when they cross its orbit. The impactors and the target orbit the star in the same direction, so collisions will be rear-end rather than head-on. Use conservation of energy and angular momentum to find an expression for the impactor speed at the point where it crosses the target orbit at distance r_p from the star. Use the expression for the ellipse to find the angle at which the impactor's orbit crosses the planet's orbit, and then use this angle and the planet's orbital speed to determine the velocity of the impactor *relative* to the planet. Under what circumstances is this relative velocity large in comparison with the escape velocity of the planet (which gives the impact speed of an impactor initially at rest relative to the planet, dropped from a great height)? Put in some numbers corresponding to Earth and Mars.

Now re-do the derivation of the formula relating mass of atmosphere lost to net impactor mass, this time assuming the typical velocity of the impactor to have a fixed value v_i independent of the size of the target planet. This assumption is valid when the relative approach speed calculated in the preceding paragraph is large compared with the escape velocity of the planet. Assume the mass distribution of impactors to follow the same power law assumed in the text. Discuss how the tendency of a planet to lose its atmosphere scales with the size of the planet in this case. Note that if two planets are competing for the same population of impactors, then each will capture impactor mass in proportion to its cross-sectional area, so that the smaller planet will get less total mass of impactors than the larger planet.

Apply these results to the relative erodability of atmospheres on Early Earth and Mars, assuming the impactor population to have eccentric orbits extending from that of Jupiter to that of Venus.

8.9 FOR FURTHER READING

The general idea of silicate weathering as a control on Earth's climate has a long history in Earth science, dating back as far as Ebelmen's original but long-neglected work. The first quantitative treatment of the silicate weathering thermostat, particularly in the context of the Faint Young Sun problem, can be found in

- Walker JCG, Hayes PB and Kasting JF 1981: A negative feedback mechanism for the long-term stabilization of Earth's surface temperature, *J. Geophys. Res.* **86**, 9776–9782.

The model introduced there is generally referred to as the *WHAK* model, after the initials of the authors (plus a conjunction to make the acronym easier to pronounce). The model discussed in the text is a variant on the WHAK model.

The following papers contain information on the UV (especially FUV and EUV) output of the Sun and other stars, and how this varies over time:

- Lean J 1990: A comparison of models of the Sun's extreme ultraviolet irradiance variations, *J. Geophys. Res.* **95**, 11933–11944.
- An implementation of Tobiska's *EUV*91 model cited in the preceding reference (SERF2.f) can be retrieved from `http://ccmc.gsfc.nasa.gov/modelweb/sun/serf.html`
- Ribas I, Guinan EF, Gudel M and Audard M 2005: Evolution of the Solar activity over time and effects on planetary atmospheres I, *Astrophys. J.* **622**, 680–694.
- Selsis F, Kasting JF, Levrard B *et al.* 2007: Habitable planets around the star Gl581? *Astron. Astrophys.* **476**, 1373–1387.

The paper by Lean describes instrumental observations of the present Sun. Ribas *et al.* describes results from the *Sun in Time* project, which makes use of observations of other stars to infer how the output of the Sun and other G-class stars vary over their lifetime. This paper is one representative of a considerable series of results coming out of the *Sun in Time* project. Selsis *et al.* includes a discussion of the situation for other spectral classes of stars, notably M-dwarfs.

Comprehensive data on reaction rates and photolysis rates relevant to chemistry of planetary atmospheres can be found in

- Yung Y and DeMore WB 1999: *Photochemistry of Planetary Atmospheres*, Oxford University Press.

This reference is the source for the quantitative data used in our abbreviated discussion of atmospheric chemistry.

The following references provide additional information about atmospheric escape:

- Watson AJ, Donahue TM and Walker JCG 1981: The dynamics of a rapidly escaping atmosphere: applications to the evolution of Earth and Venus, *Icarus* **48**, 150–166.
- Kasting JF and Pollack JB 1983: Loss of water from Venus I. Hydrodynamic escape of hydrogen, *Icarus* **53**, 479–508.
- Luhmann JG, Johnson RE and Zhang MHG 1992: Evolutionary impact of sputtering of the Martian atmosphere by O^+ pickup ions, *Geophys. Res. Lett.* **19**, 2151–2154.

Watson *et al.* provides an exceptionally clear introduction to the basic physics of hydrodynamic escape; Kasting and Pollack couple this physics into a comprehensive photochemical model of the Early Venusian atmosphere. The paper by Luhmann *et al.* gives an example of the kind of calculation needed to quantify solar wind erosion.

A peek at dynamics

9.1 OVERVIEW

So far, we have studiously avoided discussing the circulations of atmospheres or oceans, or indeed fluid mechanics of any type, with the exception of a brief foray into compressible one-dimensional hydrodynamics in Section 8.7.4. This is not because the subject is unimportant, but rather because the subject is *too important* to be relegated to the kind of superficial discussion we could accord it while doing justice to the rest of the physics governing the fluid envelopes of planets. This chapter provides a glimpse at what the reader has been missing. It highlights what the reader needs to keep in mind when learning atmosphere/ocean fluid dynamics, and, for the student who has already acquired some familiarity with that subject, connects fluid mechanical effects with the key planetary climate phenomena that have been the subject of this book. It is, in essence, a sampler of some of the many ways that large-scale fluid dynamics affects planetary climate.

This being the final chapter (for now) of a long journey, we will also take stock of how well we have done at coming to an understanding of the Big Questions introduced in Chapter 1. We wrap up with a reminder of the great breadth of largely unexplored problems the reader is already equipped to take on. The universe of problems becomes all the richer once planetary fluid dynamics is brought into the picture.

9.2 HORIZONTAL HEAT TRANSPORT

From the standpoint of planetary climate, the most important effect of large-scale fluid flow in a planet's fluid envelope is that it carries heat from one place to another, helping to even out the large geographical temperature variations that would otherwise be caused by variations in insolation between poles and tropics, or dayside and nightside.

There are two limits in which one does not need a detailed model of the horizontal heat transport. If the heat transport is very effective, then the temperature can be considered independent of geographical position and so the temperature profile can be determined using a single-column radiative-convective model driven by global mean top-of-atmosphere insolation. Venus is in this category. At the opposite extreme, the atmosphere may be so ineffective at transporting heat that each geographical position can be considered to be in local radiative-convective equilibrium. Again, the climate can be determined using one-dimensional radiative-convective calculations, but this time one must run one such calculation for each geographical position and time period for which the temperature is desired. The weak-transport limit most commonly occurs for thin atmospheres, which simply do not have enough mass to transport heat effectively. Mars is an example of such a planet. Many planets of interest lie between the extremes, however. Earth is one such. For the intermediate cases, it is necessary to model the atmospheric and oceanic heat transport. Note that even in the weak-transport limit, the atmospheric heat transport will significantly affect the *atmospheric* temperature structure, even if its influence on surface temperature is weak. In the long term, this can have an indirect effect on surface temperature, for example through influencing the slow redistribution of ice and other volatiles on the surface.

The discussion in this section will emphasize atmospheric transports. Oceans also transport energy, and oceanic transports are often represented in simplified models in ways quite similar to the treatment of atmospheric transports to be given below. However, the nature of energy-transporting circulations in Earthlike oceans is different in key respects from the atmospheric case. The features that give oceanic circulations a distinct character include the near-incompressibility, the contribution of both salinity and temperature to buoyancy, primary driving by wind-stress and thermal exchange at the top boundary, and the presence of continental barriers to flow. One should not be too hasty to generalize from the nature of Earth's oceans, though. On a waterworld, there would be no continental barriers, and it is not inevitable that salinity should have an important effect on density. Moreover, deep atmospheres which are optically thick both in the infrared and solar spectrum may share some features of the top-driven circulation of the ocean, at least so far as thermal exchanges go. To top it off, the very distinction between liquid ocean and gaseous atmosphere disappears when the pressure exceeds the critical point discussed in Chapter 2. The special features peculiar to oceanic circulations will be left for the reader to pursue elsewhere. The heat-transporting circulations to be discussed below are abstracted from those occurring in the Earth's atmosphere. They typify the major heat-transporting mechanisms found in the atmospheres of all the Solar System bodies with substantial atmospheres, including Jupiter, Saturn, Venus, and Titan; even Mars, with its thin atmosphere, exhibits basically the same features in one form or another. Similar features would apply to a great variety of hypothetical exoplanet atmospheres.

For atmospheres containing condensable substances, as is the case for water vapor on Earth, precipitation is an inevitable byproduct of horizontal heat transport, which occurs in part through transport of latent heat between the portion of the planet where the substance is picked up by evaporation and the portion where it is deposited through condensation. This behavior is a horizontal version of the precipitation that occurs in connection with vertical transport of latent heat in a convecting column. Thus, the problem of determining precipitation distribution is intimately connected with – though not identical to – the problem of horizontal energy transport. The challenge there is to figure out what portion of the heat transport is due to sensible heat transport, and what part is due to latent heat.

The first thing to note about atmospheric energy transport is that the energy content of most known or contemplated atmospheres is dominated by the heat content, with transports of energy stored in the form of kinetic energy playing at best a minor role. For example, the energy per unit mass involved in changing the temperature of a dry Earth air parcel by ΔT is $c_p \Delta T$, while the entire kinetic energy content is $\frac{1}{2} U^2$ where U is the root-mean-square wind speed in the air parcel. The ratio of kinetic to heat energy is $\frac{1}{2} U^2 / c_p \Delta T$, which works out to 0.006 if $U = 20 \, \text{m/s}$ and $\Delta T = 30 \, \text{K}$. Another way of putting it is that, if all the energy in a 20 m/s jet were dissipated and converted to heat, it would only raise the temperature of the air parcel by 0.2 K. Even this is an overestimate of the significance of heating by kinetic energy dissipation, since friction is weak throughout most of the atmosphere, so that it would take quite a long time to frictionally dissipate the kinetic energy. It is possible to envision circumstances where substantial parts of the energy transport were in the form of kinetic energy, but the following discussion will adopt the conventional line of thinking, which neglects kinetic energy terms in the energy budget. *This does not mean that atmospheric and oceanic currents are themselves unimportant.* The kinetic energy stored in the form of organized winds may be negligible, but it is important that some of the planet's energy is in this form, since it is the winds that carry heat from one place to another.

9.2.1 A little fluid mechanics

In order to discuss the horizontal redistribution of heat by fluid motions, it is necessary to introduce a few basic fluid mechanical concepts. The development of this material requires more extensive use of partial differential equations and multivariable vector calculus than we have employed previously. It is useful background, but the reader who feels unequipped for this foray into the world of fluid mechanics can skip ahead to Eq. (9.11) and pick up from there.

We will adopt longitude λ and latitude ϕ as horizontal coordinates, and pressure p as the vertical coordinate. The latitudes and longitudes appearing in the fluid equations below are measured in radians. Since we have adopted pressure as our vertical coordinate, all partial derivatives with regard to latitude, longitude, and time are to be understood as occurring at constant pressure. At each point on the sphere, we decompose the horizontal velocity into a component u which points along the local latitude circle (eastward positive), and v which points along the local longitude circle (northward positive); u is called the *zonal wind*, and v the *meridional wind*. Note that u and v are conventional wind speeds measured in distance moved in the respective directions per unit time; they are *not* angular velocities such as latitude or longitude change per unit time, though they can easily be related to angular velocities. Since pressure is the vertical coordinate, the vertical motion is characterized by the *pressure velocity* ω, which gives the rate of change of pressure with regard to time, as seen by an observer moving along with a fluid parcel. Negative values of ω denote upward motion. Note that even for a flat surface, ω is not generally zero at the ground, since surface pressure generally is time-dependent.

The first expression we need is for the *material derivative*. Let $B(\lambda, \phi, p, t)$ be any quantity that depends on space and time; it could, for example, be an energy density, or the concentration of an atmospheric constituent. Then, its material derivative dB/dt is defined as the rate of change of B as seen by an observer who is blown along by the three-dimensional time-dependent wind field, meaning that the observer's velocity at any given point in space and time is equal to the wind velocity at the observer's current location. A straightforward application of the chain rule for differentiation shows that

$$\frac{d}{dt} \equiv \partial_t + \frac{1}{a\cos\phi}u\partial_\lambda + \frac{1}{a}v\partial_\phi + \omega\partial_p. \tag{9.1}$$

We will also need an equation which represents the conservation of mass – the *mass continuity equation*. For hydrostatic flow, the time derivative drops out of this equation when it is written in pressure coordinates, leading to a simple and convenient form. When there is no mass loss due to condensation, the mass continuity relation reads

$$\frac{1}{a\cos\phi}\partial_\lambda u + \frac{1}{a\cos\phi}\partial_\phi v \cos\phi + \partial_p \omega = 0. \tag{9.2}$$

This expression says that the flow is non-divergent when written in pressure coordinates. When the relative mass loss due to condensation is small, as is the case for the present Earth, this formula remains approximately valid. When there is more substantial condensation, as for CO_2 on Early Mars and CH_4 on Titan today, then a source and sink term due to formation of precipitation and evaporation/sublimation of falling rain or snow needs to be put on the right hand side, making the flow divergent. The divergence introduced by strong condensation can introduce novel fluid mechanical effects which are essentially unexplored, but in the following we will stick to the simpler case in which condensation takes away minimal atmospheric mass.

Using Eq. (9.2), the expression for dB/dt can be recast in a flux form as follows:

$$\begin{aligned}
\frac{dB}{dt} &= \partial_t B + \frac{1}{a\cos\phi}u\partial_\lambda B + \frac{1}{a\cos\phi}v\cos\phi\partial_\phi B + \omega\partial_p B \\
&= \partial_t B + \frac{1}{a\cos\phi}\partial_\lambda uB + \frac{1}{a\cos\phi}\partial_\phi vB\cos\phi + \partial_p\omega B.
\end{aligned} \tag{9.3}$$

Suppose now that B has a source $\mathbb{S}_B$, so that $dB/dt = \mathbb{S}_B$. Then, the vertically integrated conservation equation reads

$$\begin{aligned}
\int_0^{p_s} \partial_t B \frac{dp}{g} + \frac{1}{g}&\left[\omega(p_s)B - \left(\frac{1}{a\cos\phi}u\partial_\lambda p_s + \frac{1}{a\cos\phi}v\cos\phi\partial_\phi p_s\right)B\right] \\
&+ \frac{1}{a\cos\phi}\partial_\lambda\int_0^{p_s}uB\frac{dp}{g} + \frac{1}{a\cos\phi}\partial_\phi\int_0^{p_s}vB\frac{dp}{g}\cos\phi \\
&= \int_0^{p_s}\mathbb{S}_B\frac{dp}{g}
\end{aligned} \tag{9.4}$$

in which we have divided both sides by g to emphasize that the vertical integral is in fact a mass-weighted integral, given the hydrostatic assumption. Now we make use of the fact that, by definition, $\omega(p_s) = dp_s/dt$, which makes the term in brackets reduce to $B\partial_t p_s$. This term is just what is needed to cancel the term from the time dependence of the limit of integration p_s, which appears when one brings ∂_t outside the integral. The conservation equation can then be re-written as

$$\begin{aligned}
\partial_t\int_0^{p_s}B\frac{dp}{g} + \frac{1}{a\cos\phi}\partial_\lambda\int_0^{p_s}uB\frac{dp}{g} + \frac{1}{a\cos\phi}\partial_\phi\int_0^{p_s}vB\frac{dp}{g}\cos\phi \\
= \int_0^{p_s}\mathbb{S}_B\frac{dp}{g}.
\end{aligned} \tag{9.5}$$

Finally, we take the zonal mean of the equation – i.e. the average along latitude circles – denoting zonal mean quantities with angle brackets. The resulting zonal mean conservation equation is

$$\partial_t \left\langle \int_0^{p_s} B \frac{dp}{g} \right\rangle + \frac{1}{a \cos \phi} \partial_\phi \left\langle \int_0^{p_s} v B \frac{dp}{g} \right\rangle \cos \phi = \left\langle \int_0^{p_s} \mathcal{S}_B \frac{dp}{g} \right\rangle. \tag{9.6}$$

To obtain the zonal mean energy equation, we simply apply Eq. (9.6) to the moist static energy per unit mass, $\mathfrak{M}$, defined in Section 2.7.3. We will restrict attention to the case in which the concentration of the condensable component is small, so that the simplified form of the moist static energy applies and we need not worry about the loss of moist static energy through precipitation. The source term of the column integral of $\mathfrak{M}$ takes on a simple form, because $\mathfrak{M}$ is an exact differential. Specifically, suppose that the net vertical flux of energy by all means is $F(p)$ at pressure level p. Then, the energy source per unit mass is $\mathcal{S}_{\mathfrak{M}} \equiv -g dF/dp$, with the convention that downward fluxes are positive. The vertically integrated source appearing in the transport equation is then simply the difference between F at the bottom and top of the atmosphere. Another way of putting it is that, as discussed in Section 2.7.3, the change in the moist static energy content of a column is given by the net energy put into the column through the upper and lower boundaries. Recall, however, that if we weren't working in the dilute limit, it would also be necessary to allow for the loss of energy through precipitation.

At the top of the atmosphere, let $F_{\odot,top}(\phi, t)$ be the net zonal mean solar flux into the atmosphere (downward positive), and $OLR(\phi, t)$ be the zonal mean outgoing thermal infrared (upward positive, as usual). At the surface, let $F_{\odot,s}(\phi, t)$ be the zonal mean solar flux exiting the bottom of the atmosphere; the difference with $F_{\odot,top}(\phi, t)$ gives the atmospheric solar absorption. The heat flux out of the bottom of the atmosphere consists of turbulent as well as infrared radiative terms; call this net flux simply F_s, with the convention that a positive value indicates a transfer of heat from the atmosphere to the underlying surface. With these definitions the vertically integrated source term, which gives the rate of change of moist static energy in a column, becomes

$$\left\langle \int_0^{p_s} \mathcal{S}_B \frac{dp}{g} \right\rangle = F_{\odot,top} - OLR - F_{\odot,s} - F_s. \tag{9.7}$$

Substituting this into Eq. (9.6), and using the moist static energy in place of B, we obtain the energy balance equation

$$\partial_t E_{atm} + \frac{1}{\cos \phi} \partial_\phi \Phi_{atm} \cos \phi = F_{\odot,top} - OLR - F_{\odot,s} - F_s \tag{9.8}$$

where we have defined the atmospheric energy flux as

$$\Phi_{atm} \equiv \frac{1}{a} \left\langle \int_0^{p_s} v \mathfrak{M} \frac{dp}{g} \right\rangle \tag{9.9}$$

and the mean heat storage in the atmospheric column as

$$E_{atm} \equiv \left\langle \int_0^{p_s} \mathfrak{M} \frac{dp}{g} \right\rangle. \tag{9.10}$$

As defined above, Φ_{atm} has the units of an energy flux (W/m^2), because the geometric factor $1/a$ has been absorbed into the definition of the flux. The rate of meridional energy transport per unit length along a latitude circle is $a \cdot \Phi_{atm}$, whence the net rate at which energy is transported across the latitude circle is obtained by multiplying this quantity by the length of the latitude circle, $2\pi a \cos \phi$. Thus, the net rate (in W) at which energy is transported across a given latitude circle is $2\pi a^2 \Phi_{atm} \cos \phi$.

It is convenient to introduce the coordinate $y = \sin \phi$, since then the element of area is simply $2\pi a^2 \, dy$. The atmospheric energy budget equation then becomes

$$\partial_t E_{atm} + \partial_y \Phi_{atm} \cos \phi = F_{\odot,top} - OLR - F_{\odot,s} - F_s. \tag{9.11}$$

Note that $\cos \phi$ can be expressed as $\sqrt{1 - y^2}$. The derivation of this equation in terms of the moist static energy budget has been somewhat long and tedious, but in the end, the only new term we have added to the familiar column budget is $\partial_y \Phi_{atm} \cos \phi$. If we are content to deal with the meridional heat flux in the abstract without relating it to wind and moist static energy density, the same term can be derived immediately from the simple atmospheric column budget shown in Fig. 9.1. The change in energy in a column is obtained by summing up the fluxes entering or leaving each side of the box. Solar and infrared radiative fluxes enter and leave the top of the column, while radiative and turbulent fluxes enter or leave the bottom of the column. A certain amount of dynamically carried flux enters the right hand boundary of the column while a different amount exits from the left. The surface area of the planet is proportional to Δy, so dividing the summed-up fluxes by this quantity yields the budget per unit surface area, resulting in Eq. (9.11) in the limit of small Δy.

To complete the analysis, the atmospheric budget must be coupled into the surface budget. The simplest case is the one in which the surface has negligible thermal inertia, as in the case of a solid. Typically, in this case the surface budget can be considered to be in equilibrium, in which case $F_{\odot,s} + F_s = 0$. When this assumption is valid, Eq. (9.11) involves only the top-of-atmosphere radiation budget and the atmospheric heat storage.

For a planet with an ocean, we need to allow for the thermal inertia of the surface and oceanic heat transports, both of which can throw the surface budget out of local equilibrium. As a simple example of this situation, let's suppose that the oceanic thermal inertia can be represented by a mixed layer of uniform depth H. Suppose further that ocean currents transport heat meridionally at a rate Φ_{ocean}, defined analogously to Φ_{atm}. Most of this heat transport actually occurs below the mixed layer, and is communicated from the deep ocean to the mixed layer by vertical exchange of water. We will not go into the oceanic heat transport processes in any detail, and will be content with merely summarizing the effects on the heat transport, which will be sufficient for the present purposes. Given that the ocean is heated and cooled only at the surface, the energy balance equation for the ocean becomes

$$\partial_t c_{p,w} H T_{mix} + \partial_y \Phi_{ocean} \cos \phi = F_{\odot,s} + F_s \tag{9.12}$$

Figure 9.1 The energy budget of a column of atmosphere having meridional extent Δy. The column shown is a cross-section of a ring of fluid extending along an entire latitude circle.

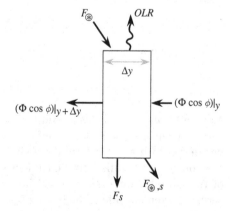

where $c_{p,w}$ is the specific heat of the liquid making up the ocean and T_{mix} is the mixed layer temperature.

Adding the ocean and atmosphere budgets, the surface exchange terms drop out, and we obtain the net column energy budget

$$\partial_t(E_{atm} + c_{p,w}HT_{mix}) + \partial_y\Phi\cos\phi = F_{\odot,\,top} - OLR \qquad (9.13)$$

where the net horizontal heat flux is $\Phi \equiv \Phi_{atm} + \Phi_{ocean}$. Note that the energy input and loss is now written entirely in terms of the top-of-atmosphere radiative exchange. This equation says that any top-of-atmosphere imbalance must be compensated by some combination of horizontal heat flux and change in the atmosphere/ocean energy storage. When the ocean thermal inertia dominates over the atmosphere, then any imbalance that is not accommodated by horizontal heat transport goes into changing the mixed layer temperature. When the ocean thermal inertia is small compared with that of the atmosphere, we are back in the same regime as the solid-surface case, in which the residual imbalance instead goes into changing the atmospheric heat storage.

9.2.2 Some observations

Since the top-of-atmosphere energy budget is purely radiative, the net top-of-atmosphere imbalance in the vicinity of each latitude and longitude point of a planet can be observed by satellite-borne instruments. If such data are averaged over a sufficiently long time period that the energy storage in each ocean–atmosphere column can be considered stationary, then the imbalance provides a direct measure of how much heat must be imported into the column or exported from the column in order to balance the budget. While it gives the net ocean–atmosphere dynamic heat flux, it does not say how the required fluxes are partitioned between the two fluids. Since the seasonal cycle repeats almost exactly from one year to the next, an annual average usually suffices to determine the annual mean dynamic heat flux, even on a planet with substantial thermal inertia in the atmosphere or ocean. The rapid increase in atmospheric CO_2 on Earth due to industrial CO_2 emissions has thrown off the Earth's radiation budget by a few W/m^2, but this is not a large enough effect to seriously compromise estimates of dynamical heat fluxes based on annual mean observations. Top-of-atmosphere data is not in itself sufficient to determine the seasonal cycle of heat transport, however. For example, the January mean energy imbalance for Earth would show a large net input of energy into the southern hemisphere, and a large net deficit in the northern hemisphere. This imbalance is not made up by a massive energy transport between the hemispheres; instead, it mostly reflects transient uptake of heat by the ocean in the summer hemisphere, and transient heat release from the ocean in the winter hemisphere.

So far, geographically resolved top-of-atmosphere radiation budget measurements are only available for the Earth. Zonal mean data from the ERBE satellite observations are shown in Fig. 9.2. The data are plotted as a function of the coordinate $y = \sin\phi$ because $dy = \cos\phi d\phi$ is proportional to the element of area on the sphere, so that the area under the energy imbalance curve between two values of y is the net energy imbalance integrated over that region. The net imbalance is very nearly symmetric about the equator. At the equator, the atmosphere receives about $60\,W/m^2$ more solar energy than it emits locally as infrared. At each pole, the situation is reversed, with the polar regions emitting about $100\,W/m^2$ more than they receive as sunlight. There is a net atmosphere ocean transport from the regions with $|y| < 0.6$ to the higher latitude regions. In terms of latitude, the

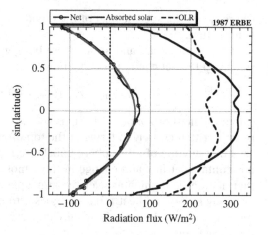

Figure 9.2 Top of atmosphere energy balance from the ERBE satellite dataset, annually averaged for 1987. The gray solid line is a quadratic fit to the net top-of-atmosphere imbalance.

division between the two regions occurs at $\pm 37°$. The net top-of atmosphere imbalance, F_{toa}, can be quite well fit by the parabola $F_{toa} = 62.216 - 7.112y - 169.96y^2$. The global mean of this quantity is about $5\,\text{W/m}^2$ out of balance, which is comparable to the measurement error of the ERBE instruments. Note that the interhemispheric asymmetry is very weak. This is a quite mysterious result, and surprising in view of the considerable asymmetry in the continental distribution between the hemispheres. This behavior is as-yet unexplained, and the range of circumstances under which one should expect symmetric heat transport are not known.

From Eq. (9.13) the meridional flux $\Phi\cos\phi$ can be obtained by integrating the polynomial expression for F_{toa} with respect to y, assuming the annual mean system to be in a steady state. The constant of integration is determined by requiring that $\Phi\cos\phi$ vanish at one of the poles. (This is not a trivial requirement despite the vanishing of $\cos\phi$ at the poles, since Φ itself could become infinite at the pole.) If the system is truly in a steady state, then the integral of F_{toa} with respect to y must vanish, guaranteeing that the meridional flux will vanish at the opposite pole as well; observational errors or deviations from annual mean equilibrium would prevent the latter condition from being met exactly. Instead of integrating the flux from pole to pole, here we will take the simpler approach of neglecting the interhemispheric asymmetry, zeroing out the small term in the polynomial fit which is linear in y. In this case, the boundary condition on $\Phi\cos\phi$ is that it vanish at the equator, and the integration can proceed from there. Carrying out this procedure, we find that the maximum value of $\Phi\cos\phi$ is approximately $25\,\text{W/m}^2$, corresponding to a poleward heat transport of $6.4 \cdot 10^{15}\,\text{W}$ across the $37°$ north or south latitude circle.

Detailed analysis of surface fluxes indicates that over most of the planet, the heat transport implied by Fig. 9.2 is dominated by the atmosphere. Moreover, modeling studies indicate that if the oceanic heat transport is suppressed, the atmosphere takes up much of the slack. Hence, we are somewhat justified in focusing attention on the atmospheric transport mechanisms, but the reader should be aware that the oceanic heat transport can also have a significant effect on the climate, though we shall not discuss it here.

There are two main styles of atmospheric heat-transporting circulations, both of which occur on Earth, and both of which to varying degrees account for meridional heat transport on other planets as well. In the low latitudes, the heat transport is dominated by a zonally symmetric overturning *Hadley circulation*, which varies gradually over the course of

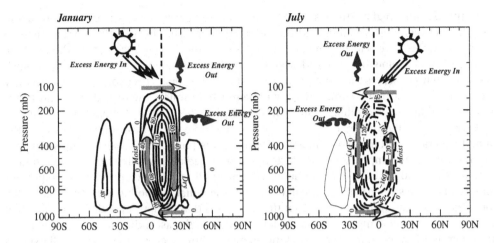

Figure 9.3 The January and July Hadley cell structure, based on 1960 to 1970 data for the Earth. The contours give the mass streamfunction in units of 10^9 kg/s. Dashed contours are negative, and the arrows indicate the sense of the circulation.

the seasonal cycle. At higher latitudes, the heat transport is dominated by wavelike mobile *synoptic eddies* which have spatial scales on the order of a few thousand kilometers and propagate from west to east at a rate of 10-20 m/s. The synoptic eddies are evident on conventional weather maps as mobile high/low pressure systems, with their familiar systems of warm and cold fronts and associated weather.

Figure 9.3 shows the climatological mean January and July Hadley circulation for the years 1960 to 1970, diagnosed from zonally averaged horizontal wind observations. The circulation is characterized by a mass streamfunction in the latitude–height plane. The winds blow along the streamlines, and the mass flux between any two contours (in units of mass per unit time) is given by the difference between the streamfunction values on the two contours. Thus, the atmospheric currents transport mass more rapidly where the contours are more closely packed.

On the Earth, the Hadley cell is confined to the latitude range from 30N to 30S, and effectively defines the tropics. The circulation reverses between the solstices. In each solstice, the summer hemisphere portion of the tropics receives more solar radiation than it radiates away locally, while the winter hemisphere portion of the tropics suffers a net loss of energy, both through an excess of infrared emission relative to local solar absorption, and because synoptic eddies transport heat out of the winter tropical regions. Heat transports by the Hadley cell make up the difference, carrying energy from the summer to the winter hemisphere across the equator. For climates like that of the present Earth, moisture is crucial to the way the Hadley cell transports heat. Observations show that above the boundary layer the temperature profile is very nearly identical between the ascending and descending branch of the Hadley circulation; the profile in both regions remains very close to the moist adiabat. The horizontal homogeneity of temperature in the tropics is rooted in dynamical affects associated with the relatively weak effect of the Earth's rotation when one is sufficiently near the equator. It is a generic feature of planetary circulations, though the distance from the equator over which the temperature is uniform will depend on the rate of the planet's rotation, among other things.

In the Earthlike regime, the transport of moist static energy comes from the contrast in humidity between the ascending and the descending branch. The ascending branch is near saturation, while there is little or no deep convection in the descending branch, whence there is little supply of moisture to the descending air. In consequence, the descending branch approximately conserves the moisture mixing ratio of the upper troposphere, and is very dry when it nears the ground. Ordinarily, dry descent would follow the dry adiabat, and hence be warmer than the ascending branch upon reaching the ground. Instead, the dynamical heat transport in the Hadley cell is efficient enough to keep the descending branch on the same temperature profile as the moist adiabat. This is possible because the subsiding air, which warms as it compresses, loses heat by infrared radiative cooling at the rate required to balance the budget. Indeed, the radiative cooling on the dry branch provides an important constraint on the strength of the circulation, since the radiative cooling must be balanced by compressional heating. The net result of this process is that the low level air in the subsiding branch has the same temperature as the low level air in the ascending branch, but is much drier; hence it has much lower moist static energy. By this process, the Hadley cell exports air with high moist static energy in the upper-level outflow, while it imports low level air with lower moist static energy. The Hadley cell would behave similarly for any planet that had a significant condensable component in its atmosphere, notably for the case of methane on Titan.

On a dry planet – for example Snowball Earth, which is dry because it is cold – the energy transport in the Hadley circulation must work differently. It must work by transporting dry static energy, and that requires that there be a difference in dry static energy between the upper level outflow and the lower level inflow. This requires a horizontal temperature gradient, at least at the outflow and inflow levels. On a dry planet, the circulation itself looks much the same as that shown in Fig. 9.3, but the thermal structure of the tropics is rather different from Earth's present tropics.

For a watery planet like Earth, the thermal inertia of the ocean also affects the amount of heat that must be transported by the Hadley circulation, and hence its strength. That is because the ocean is out of equilibrium in the course of the seasonal cycle. In the winter hemisphere, it is able to release stored energy to help make up the energy deficit of the atmospheric column, and this reduces the amount of dynamical heat transport required to balance the budget.

Figure 9.4 shows an example of the meridional transport of heat by synoptic eddies. The figure shows a typical three-day sequence of the evolution of the lower-tropospheric temperature pattern on Earth. The reader's attention is directed to the southern hemisphere, where the influence of the synoptic eddies is most clearly revealed; similar phenomena occur in the northern hemisphere, but are less immediately evident to the eye because they are distorted by the planetary-scale undulations forced by the extensive northern hemisphere topography. In the vicinity of latitude $45°$ S, one sees that the midlatitude temperature gradient is not zonal, but rather is disturbed by a series of undulations having a horizontal scale of a few thousand kilometers. A detailed examination of the wind fields shows that the cold tongues are being blown equatorward, while warm tropical air is being transported poleward. One such system has been highlighted in the figure, and it can be seen that the system propagates towards the east – in the direction of the prevailing upper air jet – as it evolves. Such systems ultimately draw their energy from the potential energy associated with the meridional temperature gradient, since energy can be released by bringing warm, light tropical air poleward and upward while moving cold, dense polar air downward and equatorward to replace it. Much is understood about the intricate fluid dynamics governing such motions,

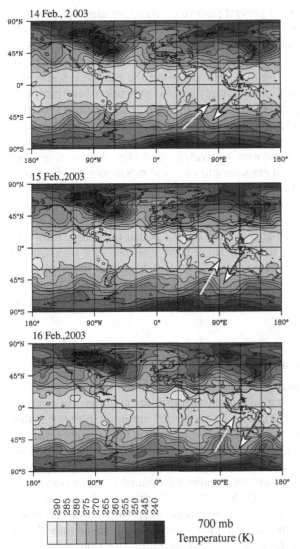

Figure 9.4 Three-day sequence of the temperature pattern on the 700 mb pressure surface. The large arrows indicate the general direction of the wind in the vicinity of one particular southern hemisphere synoptic system. Note the eastward propagation of the system.

but a complete understanding of the nonlinear fluid system, such as would be required to determine the meridional heat transport without carrying out a full numerical simulation, has proved elusive.

For an atmosphere such as Earth's, with an appreciable concentration of a condensable substance, synoptic eddies transport energy via latent heat as well as sensible heat. In accordance with Clausius–Clapeyron, the warm equatorward-moving tongues of air also tend to be moist, whereas the returning poleward cold currents tend to be dry. One cannot in general, however, infer that a moist atmosphere will tend to transport heat more rapidly than a dry atmosphere, because the resolution of that issue requires one to know how the presence of moisture affects the typical wind speed and spatial scale of the eddies. In fact, the partitioning of heat transport between latent and sensible forms is currently a subject of very active research, and is likely to remain so for a long time.

Heat-transporting synoptic eddies are a general feature of planetary atmospheres, and occur wherever the atmosphere supports a sufficiently large meridional temperature gradient. In places where the effects of a planet's rotation are weak, then temperature gradients tend to be weak since the pressure gradients arising from temperature variations cause strong winds which redistribute mass efficiently so as to wipe out the gradients. Something like this occurs near the equator on any planet, implying that synoptic eddy transports tend to become more dominant as one moves away from the equator. On a slowly rotating planet, the weak-gradient regime can expand to fill essentially the entire planet, in which case synoptic eddy heat transport becomes insignificant. This is the case on Venus and Titan. Synoptic eddies are prominent in the winter hemisphere of Mars, however, and are ubiquitous on Saturn and Jupiter. Detailed consideration of fluid mechanics is required to determine precisely what is meant by "fast" vs. "slow" rotation and "far from" vs. "close to" the equator, and to say how such notions depend on the size of the planet and the vertical structure of the atmosphere.

An interesting feature of the heat imbalance data in Fig. 9.2 is that the energy transport is seamless at the edge of the Hadley circulation. There is no evidence of a break in the transport mechanism, and the synoptic eddy transport picks up smoothly at the edge of the tropics as the Hadley transport dies out. This is probably because synoptic eddies penetrate significantly into the subsiding branch of the Hadley circulation, and indeed are an integral part of the heat and moisture budget there.

9.2.3 Scale analysis of atmospheric heat transport

There is no general theory of atmospheric heat flux, but a little scale analysis applied to Eq. (9.9) sheds some light on the factors that influence this important quantity. The analysis is similar to the analysis of turbulent surface layer fluxes carried out in Chapter 6, but this time the fluid motions involved are large-scale, including things such as Hadley cells and synoptic eddies. The great degree of organization of these motions and the diversity of energy sources they can tap is in large measure the source of the complexity of the subject.

First note that $\Phi/L_\odot$ is a dimensionless quantity because Φ is defined in such a way that it has units of W/m^2. An examination of Eq. (9.11) shows that $\Phi/L_\odot$ provides the key measure of how effective the meridional transport is at wiping out meridional temperature variations due to the variation in insolation between pole and equator, since all the shortwave fluxes in the equation scale with $L_\odot$. When $\Phi/L_\odot$ is small, the contribution of transport to the local energy budget is small, and local equilibrium applies. When $\Phi/L_\odot$ is order unity or larger, then the transport term is comparable to the variation in insolation, and we begin to approach the well-mixed regime. Earth is in the order-unity range.

Let V be the typical magnitude of meridional velocity and $\Delta\mathfrak{M}$ be the typical magnitude of fluctuation of the moist static energy per unit mass. Then the atmospheric heat flux given in Eq. (9.9) has the scaling

$$\Phi_{atm} \approx \frac{1}{a} \frac{p_s}{g} V \Delta\mathfrak{M} \tag{9.14}$$

where p_s is the surface pressure and g is the acceleration of gravity. From the second factor in the scaling, we infer that, all other things being equal, a low-mass atmosphere will transport heat less rapidly than a high-mass atmosphere. Thus, the thin atmosphere of Mars would need to have a much more vigorous circulation in order to transport heat at the same rate as Earth's more dense atmosphere, and a very sluggish circulation on Venus can still transport vast amounts of heat. For related reasons, the Earth's ocean transports heat at a rate only somewhat less than the atmosphere, despite the fact that ocean currents are two orders of magnitude weaker than atmospheric winds; the relatively large heat transport is possible because the oceans have well over a hundred times the mass of the atmosphere.

From the first factor in Eq. (9.14) we infer that, all other things being equal, the typical magnitude of Φ will be lower for a large planet than for a small planet, in inverse proportion to planetary radius. The actual rate of energy transport across a given latitude circle goes up as a planet is made larger, since that is given by $2\pi a^2 \Phi \cos \phi$. However, when considering the effect of meridional heat transport on the temperature variation, the increase in heat transport is more than offset by the increase in the area of the planet, which means that a given difference between high-latitude and low-latitude insolation accumulates over a greater area. Thus, the scaling of Φ with a means that all other things (including $L_\odot$) being equal, a larger planet will tend to have a greater temperature variation between pole and equator than a smaller planet.

The estimate of $\Delta \mathfrak{M}$ depends on the nature of the circulation and the relative importance of condensable substances. Hadley cells work by exporting air with high moist static energy from the tropics at high altitudes, and importing air with lower moist static energy at low altitudes. In this case, it is the vertical contrast of $\mathfrak{M}$ which enters the scaling. Hadley cells can transport heat even if there is relatively little horizontal contrast in temperature. Insofar as the vertical variations of $\mathfrak{M}$ tend to be large, Hadley cells can transport heat efficiently even if V is quite small. In the synoptic eddy regime, the meridional contrast of $\mathfrak{M}$ is more relevant; for cold, dry cases, this contrast is dominated by $c_p \Delta T$, whereas for hot, moist cases it is dominated by the latent heat term $L \Delta q$. In accordance with Clausius–Clapeyron, latent heat should become increasingly dominant as temperature increases, but it is not easy to determine how the net heat flux should behave, since V might systematically change as latent heat comes to assume a more dominant role.

The estimate of $\Delta \mathfrak{M}$ in the synoptic eddy regime depends further on the characteristic horizontal length scale of the synoptic eddies. Taking the dry case as an example, if the eddies are of planetary scale, then ΔT would be on the order of the entire pole to equator temperature gradient. However, if the eddies are small compared with the planet, they tap into only a smaller portion of the total temperature gradient to create fluctuations, and ΔT would be correspondingly smaller. In the moist case, there is the additional complication that synoptic eddies tap into vertical as well as horizontal contrasts, and q has very strong vertical variations; thus, the appropriate value of Δq depends also on the typical vertical slope of the air parcel trajectories driven by the synoptic eddies. Generally speaking, Δq will always be greater than the value expected on the basis of the horizontal gradients of q.

The determination of the velocity scale V is the hardest part of a generally hard problem. How would V change if the Earth had 10 times its present mass of N_2 in its atmosphere? How would V change if it were 20 K warmer and had a correspondingly higher water vapor content? What would V be for a planet which was in an Earthlike orbit about a G star, but which had twice Earth's rotation rate, or twice its radius? Theory provides no easy answers to these questions, though considerable progress on the general phenomenology has been made from detailed numerical simulations. The essential difficulty is that, as we saw at the outset of this chapter, the kinetic energy is only a tiny fraction of the total heat energy available to drive the circulation on a given planet. Thus, energetics provides no significant intrinsic constraint on the circulation of a planet. The whole name of the game is in the fluid dynamics which determines the extent to which the circulation can configure itself so as to tap into some of the heat energy and convert it to kinetic energy. On top of the issue of extraction of kinetic energy, there is the issue of how it is dissipated, and there are many ways of doing that. Since dissipation is weak, a small source of kinetic energy can build up over a very long time yielding strong winds. Uranus provides an illustration of this: it has the strongest large-scale winds in the Solar System, despite the fact that it experiences very weak thermal driving. Closer to home, the balance that determines V in the Hadley circulation is very different from that which determines V in the synoptic eddy regime, and the difference between the nature of the balance in the two cases determines the rate at which V increases with increasing horizontal temperature gradient or moist static energy gradient. Even to discuss such things sensibly requires far more fluid dynamics than we wish to bring into the picture at this point.

The influence of planetary rotation on heat transport provides *prima facie* evidence of the importance of fluid dynamical details, since the rotation enters the energetics only indirectly via fluid dynamics. The rotation rate of a planet is one of the most important factors in governing the nature of the heat-transporting circulations. It enters into both V and $\Delta\mathfrak{M}$. For a very slowly rotating planet, any appreciable horizontal temperature gradient leads to horizontal pressure gradients by the hydrostatic relation. This would drive a flow which redistributes mass so as to rapidly eliminate the temperature gradient. For a rapidly rotating planet, meridional currents are inhibited by the Coriolis force, and instead pressure gradients can be balanced by the Coriolis force due to zonal winds. In that case, it is the instability of the resulting jets (tapping into the potential energy of the meridional gradient) which creates synoptic eddies and transports heat. But in order to distinguish between "rapid" and "slow" rotation, the angular velocity of rotation Ω must be compared to some other intrinsic time scale, so as to yield a non-dimensional number. One can obtain such a time scale from the velocity $c_0 \equiv \sqrt{gH}$ where H is the scale height of the atmosphere and g is the acceleration of gravity. The quantity c_0 is the speed of the pressure waves which would equalize temperature on a non-rotating planet – the analogy of the speed of propagation of long waves on the surface of the ocean or of water sloshing in a bathtub. If a is the radius of the planet, then one can form the time scale a/c_0 and hence the non-dimensional rotation rate $\Omega a/c_0$. When this is large, the planet is in a dominantly synoptic eddy regime like Jupiter. When this is small, the planet is in a Hadley regime with weak temperature gradients, like Titan or Venus. When it is order unity, the planet exhibits some features of both cases, as in Earth or Mars. The waves which equalize temperature often are internal waves living on the atmospheric internal stratification rather than waves displacing the entire mass of the atmosphere, so one can get a better classification of the circulation regime by using a modified form of c_0 based on the potential temperature stratification. Nonetheless, $\Omega a/c_0$ provides a useful first look at the circulation regime. It yields the correct insight that for fixed rotation rate, smaller planets, planets with high gravity, or planets with deep atmospheres are less dominated by rotation effects.

As an example of the subtlety and perversity of planetary fluid dynamics, though, one should keep the atmosphere of Venus in mind. Although the solid planet has such slow rotation as to be dynamically insignificant, the upper bar or so of the atmosphere develops a rotation of its own which gives the upper atmospheric circulation a character that has more affinities with the significantly rotating regime than one would expect on the basis of the slow rotation rate of the planetary surface. The mechanism and generality of this phenomenon (which also occurs in the upper atmosphere of Titan) is currently a matter of intense inquiry.

9.2.4 Formulation of energy balance models

In some sense, every climate model is an energy balance model. Full general circulation models could be described as energy balance models in which the heat transports are computed by solving the three-dimensional equations governing fluid motion. What is generally meant by the term "energy balance model," however, is a model in which some assumption is invoked to allow the heat transport to be computed rapidly without actually solving for the underlying flow. As part of this simplification, it is usually assumed that the thermal state of the atmosphere and ocean at each latitude can be represented by a single temperature T, most commonly taken to be the surface temperature. Various more elaborated forms of energy balance models are also in use. For example, instead of using a single temperature to characterize both the atmosphere and the surface, a separate temperature can be retained for each, with their corresponding energy budgets. For that matter, two-dimensional energy balance models which retain the variation of temperature in longitude have also been written, though in that case it becomes still harder to justify simple representations of the energy transport. We will not pursue either of these elaborations here; in skillful hands, and used judiciously and with due humility with regard to their limitations, such models can be illuminating, but once the models become so complicated that they are no longer easy to understand, they lose much of their utility.

The most common form of energy balance model represents the atmosphere–ocean heat transport as a thermal diffusion, namely

$$\Phi \cos \phi = D \cos \phi \partial_\phi T = (1 - y^2) D \partial_y T. \tag{9.15}$$

The diffusivity D may depend on y, either directly or through the intermediary of T and its gradient. The assumption that the moist static energy transport can be written in terms of the temperature gradient is dubious, but this prescription at least has the reasonable behavior that it relaxes the system towards a state of uniform surface temperature.

To close the problem, it is necessary to write the OLR as a function of T, which requires making an assumption about the vertical distribution of temperature and humidity. For the atmosphere, this is typically done by assuming that the atmosphere is on the moist adiabat and that the relative humidity has some fixed value, as we did in computing the $OLR(T)$ fits in Table 4.2. This also permits the atmospheric part of the vertical heat storage to be written as a function of T; the mixed layer part of the oceanic heat storage is by its nature proportional to T, if the mixed layer temperature doesn't deviate too much from the overlying air temperature. More problematically, one needs to make some assumption about the effects of clouds on albedo and OLR. There is really no satisfactory way to do this within an energy balance model, and so we will not discuss cloud effects in any great detail. Clouds are often taken into account through an adjustment of the albedo, but this can lead to erroneous conclusions if one tunes the adjustment to one climate and then applies it to an altered climate.

With the above simplifications, the energy balance model now reads

$$\partial_t E(T) = \partial_y[(1 - y^2) D \partial_y T] + (1 - \alpha) L_\odot f(y, t) - OLR(T) \tag{9.16}$$

where α is the planetary albedo, and f is the flux distribution factor discussed in Chapter 7. For a rapidly rotating planet, the diurnally averaged form of f would typically be used, but for a planet whose day is comparable to its year the zonal mean of the instantaneous value would generally be more appropriate. The albedo in general will depend on y, either directly, or indirectly through the effect of temperature on ice cover. If we write $\mu(T) \equiv dE/dT$, then the left hand side of Eq. (9.16) becomes $\mu \partial_t T$. The equation thus takes on the form of a one-dimensional time-dependent diffusion equation with a source term, and can be solved numerically using the methods discussed in Chapter 7. In the following, we will confine attention to steady solutions.

An appropriate value for diffusivity in the present Earth's climate can be estimated from the ERBE data. The maximum observed value of $\Phi \cos \phi$ is about $25 \, \mathrm{W/m^2}$, so we may estimate D using $25 \, \mathrm{W/m^2} \approx D \Delta T / \Delta y$. The annual mean pole to equator surface temperature difference for the northern hemisphere is $43 \, \mathrm{K}$, and we argue that it is better to use this for ΔT instead of the larger southern hemisphere value, which is strongly affected by the altitude of Antarctica; focusing on the Arctic polar temperature allows us to skirt the difficult issue of how to allow for topographic effects in diffusive energy balance models. Then, since $\Delta y = 1$ between the pole and equator, we have $D \approx 0.6 \, \mathrm{W/m^2 K}$. It would be a mistake, however, to apply this value to climates which differ significantly from that of the present Earth. We have no direct proxy for the meridional heat transport in Earth's past climates, but full fluid dynamical simulations suggest that the effective diffusivity is higher in warm ice-free climates like the Eocene and lower in cold icy climates like the Last Glacial Maximum or Snowball Earth. This is in part due to the changing role of water vapor in these climates, but there are many other dynamical influences in play as well.

9.2.5 Equilibrium solution of diffusive energy balance models

In a steady state, the zonal-mean diffusive energy balance model becomes

$$\frac{d}{dy}(1 - y^2)D\frac{dT}{dy} = OLR(T) - (1 - \alpha)L_\odot f(y). \tag{9.17}$$

We have replaced the partial derivatives with ordinary derivatives, since this is now an ordinary differential equation (ODE). There are two distinct cases in which the steady state form of the equation is appropriate. First, steady state conditions apply approximately if the seasonal temperature fluctuations are weak, as would be the case for a planet with a deep mixed layer ocean or one with low obliquity in a nearly circular orbit. In that case, the flux factor $f(y)$ would be taken to be the annual mean flux factor such as displayed in Fig. 7.6. Even if the seasonal fluctuation are not small, the steady state equation will still govern the annual mean temperature if $OLR(T)$ and α are approximately linear in T. Alternately, steady state conditions apply instantaneously at each point of the seasonal cycle if the response time of the system is short compared with the planet's year, as would typically be the case for a planet with an all-solid surface having an atmosphere which is not too thick. In that case, $f(y)$ would be the diurnally averaged flux factor at any given point in the seasonal cycle, if the planet is rapidly rotating. (If the response time is very short, or the planet's day is very long, the use of a zonally averaged model becomes questionable.) The flux factor varies with time in the course of the seasonal cycle, but because the response time of the system is short the temperature has time to come into equilibrium with the solar flux received at each individual point of the seasonal cycle.

Equation (9.17) is a second order equation and therefore requires two boundary conditions. Given that the physical problem is on a sphere, which has no physical boundaries, the notion of boundary condition may seem somewhat peculiar. However, although the sphere itself has no boundaries, our *coordinate system* for the sphere has boundaries at the poles, $y = \pm 1$. As boundary conditions, we require that no energy leak out of the system through the poles, namely that $(1 - y^2)D dT/dy$ vanish at $y = \pm 1$. It might seem that this condition is tautologically satisfied in view of the factor $(1 - y^2)$, but the condition can be violated if dT/dy blows up at one or more pole. The no-flux condition is thus equivalent to the requirement that dT/dy be well-behaved at the poles. In the special case where the forcing is symmetric about the equator, one of the polar conditions can be replaced by the requirement that $dT/dy = 0$ at $y = 0$.

With the specification of the boundary conditions, Eq. (9.17) becomes a nonlinear two-point ODE boundary value problem. The nonlinearity creeps in through $OLR(T)$, and perhaps also through the temperature dependence of the albedo. One needs to find a solution that simultaneously satisfies both boundary conditions, but numerical integration starting at one of the boundaries requires the specification of *two* initial conditions. In this situation, iterative methods can be used to find a solution. One first defines the auxiliary variable $Y \equiv (1-y^2)D\frac{dT}{dy}$, and writes the problem as a pair of first-order equations in the two variables Y and T. The equation for T is $dT/dy = D^{-1}Y/(1 - y^2)$, which becomes singular at $y = \pm 1$. The simplest way to deal with this problem is to start the numerical integration at a value y_0 which is very close to, but not at, the pole; one must use a correspondingly small numerical integration step in order to resolve the near-singularity. With that trick in mind, one carries out the integration starting at y_0, using $Y = 0$ and a *guess* T_p of the value of temperature at the pole. Upon integrating to the other pole, it will generally be found that Y is not zero there. One completes the solution by iterating on T_p using Newton's method until the value of Y at the opposite pole is acceptably small. In the case of a symmetric climate, the procedure is the same except that one integrates only to the equator and imposes the condition $Y = 0$ there instead. In this case, one can alternatively start at the equator and integrate toward the pole.

If the OLR is written in the linearized form $OLR(T) = a_0 + a_1 T$ and the albedo is independent of T, then Eq. (9.17) becomes a linear system. Note that it remains linear even if the albedo depends on y. Ice-albedo feedback would make the problem nonlinear, but we will see shortly how one can deal with that through iterating on the ice margin. In the linear case, the problem can be solved by superposing two suitably chosen solutions, without the need for an iteration. The solution method is a standard one for linear second order boundary value problems, and the only numerical tool required is an integration routine for systems of ordinary differential equations given initial conditions at a specified point. With Y defined as above, the first order system with the assumed linear form of OLR can be written

$$\frac{dY}{dy} - a_1 T = a_0 - (1 - \alpha)L_\odot f(y), \quad \frac{dT}{dy} - \frac{1}{(1 - y^2)D}Y = 0. \qquad (9.18)$$

We will illustrate the rest of the solution method for the symmetric case, where we can impose the boundary condition $Y = 0$ at $y = 0$. First we numerically find a particular solution (Y_p, T_p) satisfying Eq. (9.18) subject to $Y_p = 0$ and $T_p = 0$ at $y = 0$. Then we numerically find a homogeneous solution (Y_h, T_h) satisfying Eq. (9.18) with the right hand side of the first equation zeroed out, and subject to the conditions $Y_h = 0$ and $T_h = 1$ at $y = 0$. One then completes the solution by adding to the particular solution the correct multiple of the homogeneous solution needed to make the superposition satisfy the boundary condition at the pole. This procedure breaks down for very low diffusivity, owing to exponential blowup of the solutions, but one rarely needs to deal with diffusivity low enough that this is a problem.

Some numerical solutions of the linearized energy balance model are shown in Fig. 9.5. The OLR coefficients are chosen to correspond to a moist Earthlike atmosphere with 50% relative humidity and a CO_2 concentration of 300 ppmv, and a diurnally/annually averaged

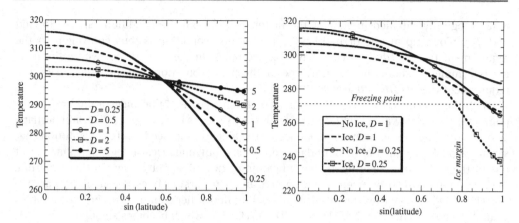

Figure 9.5 Solution to the equilibrium energy balance model for a fixed albedo of 0.15 (left panel) and for a case with ice where $|y| > 0.8$ (right panel), where $y = \sin \phi$. The ice albedo is 0.6. Results are shown for several different values of the diffusivity coefficient D (in $W/m^2 K$), indicated on the curves. The meridional profile of insolation was computed for Earth's present obliquity.

form of f corresponding to the Earth's present obliquity was assumed. The left panel shows results for a waterworld with uniform albedo. With $D = 0.25\,W/m^2\,K$, the tropical temperature is much warmer than observed on Earth, and the gradient in temperature between pole and equator is large. As D is increased, the temperature becomes more uniform. For $D = 0.5\,W/m^2\,K$, the temperature gradient is fairly realistic, but but the tropics is still too warm, and the pole somewhat less so. For $D = 1\,W/m^2\,K$, the tropics is somewhat cooler, but the poles are now too warm and the temperature gradient too weak, as compared with observations. When polar ice is introduced in the right panel of Fig. 9.5, the polar temperature drops, but because of heat diffusion the tropical temperature is dragged down as well. With polar ice, the case $D = 1\,W/m^2\,K$ has a realistic tropical temperature, though the polar temperature is somewhat too warm and hence the pole to equator temperature difference is somewhat too small. Still, it is encouraging that a fairly realistic profile can be achieved with a diffusivity that is in the general vicinity of the value $D \approx 0.6\,W/m^2\,K$ deduced from ERBE data. A more exact fit to observations can be achieved by adjusting the assumed ice albedo and the implicit cloud effects, but we shall not indulge in any such fine-tuning exercise here.

Note that the heat diffusion smooths out the discontinuity of temperature one would otherwise get at the ice margin. When diffusivity is weak, one sees a stronger temperature contrast across the ice margin. For weak diffusivity, the polar ice also has less effect on tropical temperatures, because the two regions are more thermally decoupled.

The results for the polar ice case shown in Fig. 9.5 are inconsistent, because the ice margin temperature is not at the freezing point. For $D = 0.25\,W/m^2\,K$ the ice margin temperature is below freezing, and the ice margin will tend to advance and cover more of the globe. For $D = 1\,W/m^2\,K$ the ice margin temperature is above freezing, so the ice margin will tend to retreat toward the poles. In order to find the equilibrium ice margin, one must perform a series of calculations, and determine the ice margin temperature as a function of the ice margin position; the equilibria are determined by the intersection of the curve with the freezing point temperature. There are also two limiting cases where the ice margin temperature can consistently be different from the freezing point. If the ice "margin" is at

$y = 0$, then the planet is in a Snowball state, and a state with the temperature at $y = 0$ below freezing is consistent since the ice has no more room to advance. At the opposite extreme, if the ice "margin" is at $y = 1$ the temperature is free to be above freezing, since that corresponds to an ice-free world with no more room for the ice to retreat.

A calculation of the ice margin temperature curve is shown in Fig. 9.6. For the highest diffusivity case, $D = 2 \, \text{W/m}^2 \, \text{K}$, the curve is monotonically increasing in y_{ice}. There is a stable Snowball state for $y_{ice} = 0$ and a stable ice-free state for $y_{ice} = 1$. There is also an additional partially ice-covered equilibrium state, but it can readily be seen that this is an unstable equilibrium. A displacement of the ice margin poleward causes the ice margin temperature to be above freezing, leading to further retreat until the planet falls into the ice-free state. Similarly, a displacement toward the equator causes the planet to fall into the Snowball state. As the diffusivity is reduced toward $0.5 \, \text{W/m}^2$, the curve develops a local maximum, but the dip does not cause further intersections with the freezing line until D is decreased further. For the low diffusivity case $D = 0.25 \, \text{W/m}^2$, the ice-free state no longer exists, but there is now a stable state with a small polar ice cap.

The general behavior is qualitatively like that we saw in the ice-albedo bifurcation diagram based on the zero-dimensional energy balance model in Chapter 3. Now, however, we have the advantage that the behavior is more closely tied to the actual distribution of solar radiation over the planet's surface, and the magnitude of the heat transport. Note that the case $D = 0.5 \, \text{W/m}^2$, which is in some sense the most like the real Earth in terms of its temperature structure, just misses having a stable small polar ice cap. This suggests that our

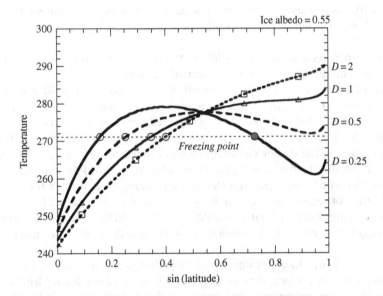

Figure 9.6 The temperature at the ice margin as a function of the ice margin position, computed for four different values of the diffusivity (values indicated on curves, in $\text{W/m}^2\text{K}$). The ice albedo was taken to be 0.55 and the oceanic albedo was fixed at 0.15. Unstable equilibrium points are indicated with open circles, while the stable equilibrium point is indicated with a shaded circle. In all cases the Snowball state ($y_{ice} = 0$) is also a stable equilibrium. Except for the lowest diffusivity case, the ice-free state is likewise a stable equilibrium.

present climate state may be rather fragile, with modest effects from clouds or ocean heat fluxes pushing the bifurcation behavior one way or another. Seasonal effects, which we have also left out of the picture, may also play a decisive role.

Based on general energy balance reasoning, we can anticipate that high obliquity states will be less favorable to polar ice, because the poles receive more solar energy in those cases; conversely, low obliquity states will be more favorable to polar ice. For any given obliquity, low diffusivity will favor polar ice if the planet is fairly warm, since less tropical heat invades the polar region. On the other hand, if the planet is quite cool, the situation is reversed, and high diffusivity can cause the planet to fall into a Snowball state, since some of the heat from the tropics can be diffusively bled off into the bitterly cold polar regions, causing the tropics to freeze over. A more complete exploration of the ice-albedo bifurcation diagram associated with the diffusive energy balance model is carried out in Problem 9.12.

As a slight refinement of the preceding approach, one sometimes requires that the ice margin temperature be a few degrees below the freezing point of sea water, rather than at freezing. This is meant to allow for the effects of the seasonal cycle, which can allow sea ice to melt back severely in the summer even if the annual mean temperature is below freezing.

9.2.6 Limitations of diffusive energy balance models

Diffusive energy balance models provide a powerful tool for exploratory work in planetary climate. They allow one to examine the qualitative effect of local energy imbalances within a framework that is at least energetically consistent. However, in using such models it is important to keep their limitations in mind. These include the following:

- It is difficult to determine how the diffusivity should vary with temperature gradient, planetary rotation rate, and vertical structure of the atmosphere.
- It is difficult to determine the effect of latent heat transport on the diffusivity. It is likewise difficult to treat the hydrological cycle (notably precipitation) within energy balance models. The hydrological cycle is sometimes treated through diffusion of the water vapor mixing ratio, but there are reasons to believe this to be an unreliable formulation.
- The basic physics of the Hadley circulation are not well represented by thermal diffusion.
- Water vapor radiative feedback and analogous radiative feedback from other condensable substances can only be incorporated through an assumption of fixed relative humidity, as in radiative-convective models. There is no good way to represent the influence of atmospheric motions on the relative humidity (see Section 9.3). Similarly, because clouds also are largely controlled by dynamics, there is no reliable way to incorporate cloud feedbacks.
- There is no reliable way to determine the surface wind field, which is needed to determine evaporation and sensible heat flux, as well as wind stresses needed to drive ocean circulations. Similarly, the atmospheric heat transport due to large-scale wind fields is difficult to assess.

Energy balance models are most useful when used in conjunction with a general circulation model, which can help constrain parameters such as relative humidity or diffusivity which are hard to set a priori. They can also be used in a diagnostic fashion to build an understanding of a given observed climate state, setting parameters according to observations rather than according to general circulation model behavior. Diffusive energy balance models have little predictive value on their own, however.

The quest for better models of dynamical heat transport continues. As already noted, in the well-mixed limit, one can avoid the problem of heat flux representation completely, and simply determine it diagnostically by letting it take on whatever value is needed to keep the temperature horizontally uniform. This approach has been productively used in the Earth's tropics, and it should prove fruitful for the tropical regions of many other planets as well. For sufficiently slowly rotating planets, including most tide-locked cases, the uniform-temperature region can expand to encompass the entire planet, making the well-stirred approximation globally applicable. Aside from this limit, there are interesting proposals to formulate models based on diffusion of moist static energy rather than temperature, as well as models based on mixing of various fluid mechanical quantities, especially potential vorticity (about which the reader will hear much, upon further pursuit of the fluid mechanical topics not covered in this book).

Until fairly recently, computational efficiency was one of the major motivations for pursuing simplified energy balance models, but as computers become faster many of the tasks traditionally explored within such models can be easily taken on with moderate-resolution general circulation models which explicitly resolve the fluid dynamics. This does not eliminate the need for simplified models, though, since these will always be needed if one is to have a chance to understand what the complex general circulation models are doing. As the emphasis shifts away from computational efficiency, however, one must keep in mind that the chief goal in formulating simplified energy balance models is that they be understandable and subject to analysis in terms of basic physical principles. Models with many *ad hoc* empirical parameters no longer have any legitimate role in the subject of planetary climate, if indeed they ever did.

9.3 DYNAMICS OF RELATIVE HUMIDITY

Throughout this book, we have seen that water vapor feedback, and analogous feedback due to other radiatively active condensable substances, has an important bearing on planetary climate. In the extreme, this includes the runaway greenhouse phenomenon. In the models of water vapor feedback we have discussed, it has been assumed that the relative humidity is a known constant, since that is essentially the best one can do without bringing in dynamics. In reality, relative humidity is a dynamically controlled quantity of considerable subtlety.

A snapshot of mid-tropospheric relative humidity is shown in Fig. 9.7. The first thing one notes is that the atmosphere is not saturated, and that there is considerable variability in the relative humidity. The variability is not random, but shows a pattern that can be linked to the large-scale dynamics. In the tropics, one sees relatively saturated air in the convection zones corresponding to the ascent in the Hadley cell and similar large-scale overturning tropical circulations; conversely, one sees very dry air in large-scale subsiding regions, with vast regions of the tropics having relative humidities of 20% or less. In the midlatitudes one sees a more small-scale pattern, corresponding to the mobile upward and downward air parcel trajectories driven by synoptic eddies. The moist regions are concentrated in rather thin filaments, and in a movie one would see that these are not geographically

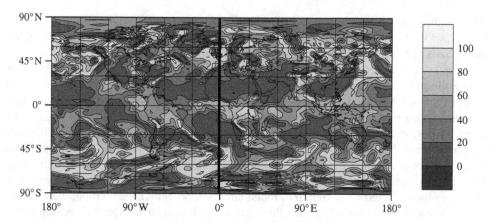

Figure 9.7 A snapshot of relative humidity on the 600 mb surface, for 1 January, 2001. The data is taken from the NCEP reanalysis, which is a blend of observations with model results, but the general patterns are the same as those seen in pure satellite data.

fixed (except in a statistical sense) but rather propagate along with the synoptic eddies.

The main thing to understand is the degree and pattern of subsaturation in the atmosphere. Why is there so much subsaturated air, and should we expect this on other planets as well? Subsaturation is caused by mixing through an environment providing temperature inhomogeneities. When a saturated parcel of air is brought to a cold place, it rains out its moisture so as to prevent supersaturation. When the parcel is then warmed either by adiabatic compression or radiative or turbulent heating, it will then be subsaturated. The distribution of subsaturation depends on the competition between the rate at which this process produces subsaturation and the rate at which the moisture supply resaturates air parcels. The cold air regions that dehydrate an atmosphere can be the cold regions aloft, or the cold regions at the poles, or some combination of the two. In the absence of moisture resupply, which comes from convection and from synoptic eddies bringing air parcels near the subtropical ocean surface, the entire atmosphere would relax toward the saturation mixing ratio of the coldest point in the entire troposphere, since all air parcels will eventually experience that point given enough time. The moisture source and sink processes are quite well represented in general circulation models, and some progress has been made in the direction of distilling these processes into simple stochastic representations which can be used in idealized climate models.

Because clouds form in air that becomes saturated, the cloud formation process is subject to all the same dynamical influences as water vapor, and is impossible to treat without bringing in dynamics. The cloud problem is more challenging than water vapor in general, because it is controlled by the extreme tail of the humidity distribution that brings it near saturation, and developing theories for extreme and sparse events is intrinsically difficult. On top of that, the processes that dissipate clouds, as well as those that govern their longwave and shortwave radiative effects, are affected by microphysical processes happening on time scales of minutes and spatial scales down to a few microns. It is no wonder that clouds have proved the chief impediment to quantitative calculations of planetary climate.

9.4 DYNAMICS OF STATIC STABILITY

Throughout this book we have seen the critical role played by the vertical temperature structure of the atmosphere in determining planetary climate. It is impossible to discuss the radiation budget of a planet except in the context of the vertical structure. The troposphere is the portion of an atmosphere (usually in the deeper regions) within which mixing by fluid motions strongly affects the vertical structure. In this book we have treated this mixing by assuming that the end state of the mixing is an adiabat, because that is the best that can be done without bringing in detailed fluid dynamics. Adjustment to an adiabat is a reasonable approximation when mixing is dominated by vigorous convection, and indeed observations have shown that in the Earth's tropics it provides a very accurate description of vertical structure even where convection is not active. It is not clear that Hadley cells would be able to maintain this state in all planetary circumstances, particularly for non-condensing atmospheres where one loses the essential asymmetry between upward and downward motions. An even more serious complication is that mixing in the midlatitudes of planets is dominated by large-scale synoptic eddies rather than convection, provided the planet is rotating rapidly enough. Dynamical effects on the vertical temperature structure of the atmosphere, and particularly the degree to which the troposphere maintains a statically stable rather than moist-adiabatic profile, could potentially have a profound impact on climate.

To illustrate the extent of deviation from adiabatic control of vertical structure, we show a few vertical profiles from Earth's atmosphere in Fig. 9.8. It can be seen that the sounding for the summer continental interior conforms quite closely to a moist adiabat, indicating that convection is in control. Even there, however, there are notable differences which demand an explanation, in that the observed summer sounding is slightly more unstable than the moist adiabat (though it is still stable compared with the dry adiabat). This

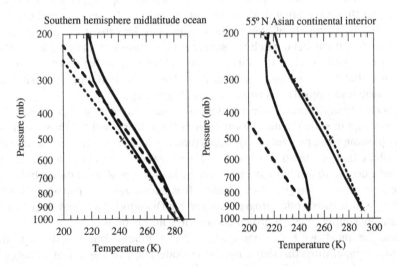

Figure 9.8 Monthly mean temperature soundings for January and July at selected individual extratropical points on the Earth. Solid lines give the observed temperature profiles, and the summer results in each case correspond to the warmer sounding. Dashed lines give the corresponding moist adiabats starting from the observed low level air temperature. Data is taken from the NCEP dataset for the year 2003.

deviation probably has something to do with the intermittency of condensation, combined with the fact that the midlatitudes can support greater horizontal variability in temperature than the tropics. The deviations from the moist adiabat are dramatic in winter continental interiors, with the observed atmosphere being vastly more stable than the moist adiabat, correspondingly reducing the magnitude of the greenhouse effect there. The southern hemisphere is mostly ocean, and gives a good example of what is expected on a waterworld, or over marine points in general where the seasonal cycle is not as extreme. The two soundings shown are in the midlatitude Southern Ocean. In both the winter and summer profiles, the atmosphere is significantly more stable than the moist adiabat.

In midlatitudes, synoptic eddies maintain a statically stable state by transporting cold air in the polar regions equatorward and downward, sliding it underneath warm tropical air which is transported poleward and upward. The process forms "bottom air" at the poles, much as the oceans form cold, dense "bottom water" in the North Atlantic and around Antarctica. This effect is particularly pronounced in the wintertime over continents, where the horizontal temperature contrasts are most extreme. The deep-winter continental interior profiles are also strongly influenced by local radiative relaxation to a stable profile; in some sense, there is no troposphere in these cases, and the stratosphere can almost be considered to come right down to the ground. There has been some progress in developing simple theoretical models of these extratropical processes, at least for non-condensing atmospheres, but a completely satisfactory closure of the problem has not yet been achieved outside comprehensive general circulation models.

9.5 AFTERWORD: ENDINGS AND BEGINNINGS

The Lapland journey of Carl Linnaeus, who put in a brief appearance at the outset of this book, took him through a great deal of unfamiliar territory. It was long and often arduous, but punctuated by moments of sheer delight – as when he caught his first glimpse of the flora of Taradalen. It has been much the same for the author of this book, and perhaps also for the reader who has made it this far. Linnaeus returned from his journey with a treasure trove of ideas that were to change the course of European thinking about natural history. When he returned to Uppsala, his work was just beginning.

There is clearly much more to learn about how planetary climate operates. A lifetime is not nearly enough to stale the infinite varieties of the subject, and for those working in this field, every passing week presents some opportunity to learn some new bit of physics, chemistry, or biology that opens up new avenues of exploration. It would be a mistake, however, to think that one had to attain to an encyclopedic knowledge of such things before embarking on creative research on planetary climate. It is encouraging to pause and contemplate the extent to which substantial progress on the Big Questions has been made on the basis of the relatively simple physical principles set forth in this book.

First and foremost, we have done quite well at fulfilling Fourier's vision of determining planetary temperature through a precise consideration of the planet's energy balance. Indeed, that has been most of what this book has been about. For cases where the infrared absorption by the atmosphere is dominated by atmospheric gases, the understanding is essentially complete, though a few details pertaining to lapse rate and to undersaturation of condensable gases remain to be settled. The basic greenhouse theory provides a satisfactory account of the differences between the climates of the Solar System planets, and explains a great deal about Earth's current and past climate variations. The greenhouse theory provides the essential tool used to think about the possible range of climates of extrasolar planets. The development and refinement of this theory surely constitutes one of the great triumphs of the past two centuries of science.

We have also done well at explaining the seasonal cycle on Earth and other planets, in terms of the interplay of the distribution of insolation over the surface of the planet and the thermal inertia of the planet's atmosphere and surface. The variations in the orbital parameters that govern the insolation ("Milankovic cycles") go a long way towards explaining the Pleistocene glacial–interglacial cycles on Earth, and provide some tantalizing hints as to what might have been going on in the past climate of Mars. For Earth, significant gaps remain in the understanding of response of both sea ice and land glaciers. The lack of a viable theory of the glacial–interglacial CO_2 variations is an even more gaping hole, though it is virtually certain that the answer to this problem lies somewhere in the ocean's uptake and release of carbon.

With regard to the Big Question of anthropogenic global warming, we have managed to understand the geochemical factors limiting ocean carbon uptake well enough to know that burning known or likely fossil fuel reserves would bring atmospheric CO_2 concentrations to four times their pre-industrial values, if not more. Moreover, the projected rate of emissions is great enough to cause a doubling in well under a century. Based on real gas radiative transfer and the thermodynamics of water vapor, we also know that the radiative effect of doubling CO_2, amplified by water vapor feedback, is sufficient to cause a global mean warming of about 2 K. We also know how to add in the radiative forcing due to other greenhouse gases (notably methane) whose concentrations are affected by human activities. A precise determination of the effect of cloud feedbacks has proved elusive, but we have understood enough about the effect of Earth's clouds to know that changes in cloud properties may have the potential to moderate the warming, but they have an even greater potential to amplify it. The behavior of paleoclimate – in particular the degree of cooling in the southern hemisphere during the Last Glacial Maximum and the degree of warming during the PETM – if anything tend to rule out a low climate sensitivity.

Stimulated by the problem of understanding the climate and atmosphere of Venus, we have also come to a pretty good understanding of the factors governing the inner edge of the habitable zone for a water-rich planet. The central phenomenon is the runaway greenhouse, which provides a plausible scenario for creating a water-vapor-dominated atmosphere. In the "dry runaway" variant, all of the ocean goes into the atmosphere, whereas in the "wet runaway" a hot liquid ocean remains, but enough has still gone into the atmosphere to achieve high water vapor concentrations aloft. In either scenario, water vapor photolyzes; our understanding of hydrodynamic escape provides a plausible mechanism for losing the resulting hydrogen. Getting rid of the oxygen is still somewhat problematic, and is tied up with unresolved issues concerning crustal recycling and crustal geochemistry on Venus and similar planets. Once liquid water is lost, our understanding of silicate weathering provides a clear means by which outgassing of CO_2 would lead to the thick CO_2 atmosphere we see on Venus. It is far from certain, however, that Venus really had a wet start; it could have been water-impoverished from the very beginning.

The problem of determining the outer edge of the habitable zone, and in particular of explaining the dichotomy between the apparently warm, wet climate of Early Mars and its present Moonlike nearly airless climate, has proved more recalcitrant. Given that the example of Venus shows that an almost unlimited supply of CO_2 can outgas and accumulate in the atmosphere of a rocky planet, one would think that such a planet could be made habitable even if it is quite far from its star, by virtue of simply letting the greenhouse effect from accumulated CO_2 grow until it compensated for the faint illumination. This pathway to habitability is defeated by the onset of CO_2 condensation, which can halt the accumulation of CO_2 before its gaseous greenhouse effect is sufficient to bring the surface temperature above freezing. Using a generalization of the runaway greenhouse calculation to CO_2, we have been able to compute the threshold absorbed stellar radiation below which this becomes a problem. The calculation also shows that the threshold has a rather weak dependence on the planet's surface gravity. Mars itself is quite close to the limit, so a little nudge from some additional warming effect can push the planet over the threshold of habitability. We have shown that the scattering greenhouse effect of CO_2 ice clouds is one means by which this nudge could be realized, but there could also be a role for other, less condensable greenhouse gases such as CH_4 or SO_2. For the latter two gases to play a significant role, however, one would have to come up with a source that is vigorous enough to offset the sink due to atmospheric photochemistry. An additional issue with CH_4 as a means of making faintly illuminated planets habitable is its quite strong absorption of incoming shortwave radiation from the star; this can give rise to an anti-greenhouse effect which limits warming. Besides that, a very CH_4-rich atmosphere tends to form organic haze clouds such as those of Titan, and these, too, limit warming. Even for gases like CO_2 which are less effective shortwave absorbers, the Rayleigh scattering of a thick atmosphere can seriously interfere with greenhouse warming.

Even if we can account for a warm, wet Early Mars through a 2 bar CO_2 atmosphere nudged warmer with some additional exotic effect, there remains the question of how to account for the loss of atmosphere. The fact that Mars has little N_2 left, in contrast to Earth and Venus, provides an important clue. Impact erosion may be able to do the trick, but it is not completely clear that Mars is much more subject to this kind of loss than Earth or Venus. Would Mars have remained habitable if it were as massive as Earth or more massive? These questions remain to be answered. The answer has a great bearing on the outer limit of habitability in extrasolar planetary systems, and in particular on the habitability of M-dwarf Super-Earths such as Gliese 581d. For M-dwarfs, which typically have very active phases, one has the additional question of whether atmospheres in the otherwise habitable zone tend to get blown away by periods of extreme EUV flux. Depending on the timing and duration of the blowoff, a planet may be able to regenerate a CO_2 atmosphere or water ocean by outgassing, but regeneration of atmospheric N_2 (necessary for life as we know it) is more problematic.

We have caught a glimpse of the means by which the Earth maintained its generally equable climate over billions of years, despite evolution of the Sun, likely variations in outgassing, and biological innovations that radically altered the composition of the atmosphere. The answer comes down to a feedback involving silicate weathering, and seems likely to apply equally well to any planet which has a silicate-rich continental crust, a supply of CO_2 due to outgassing, and adequate liquid water rainfall. It is only a glimpse of the solution, however, because a great deal about the quantification of silicate weathering rates remains to be understood. For Earth, chief among these is the effect of vegetation and microbial activity on continental weathering. More broadly, there is the question of how the differing mineralogy of planets elsewhere in the Universe might affect the operation of the silicate weathering thermostat, and what happens on a planet which has such deep oceans that silicate weathering is dominated by sea-floor weathering. And what happens on a silicate-poor waterworld? Or on a Super-Titan with an icy crust and interior? The study of alternate geochemical climate regulatory mechanisms, such as those involving SO_2, is in its infancy. Climate regulatory feedbacks are critical to planetary habitability in space and time. What are the broadest conditions under which they can operate? What does it take for them to break down? When the do break down, is habitability gone for good, or does a planet have a certain capacity to recover?

We have come to a quite good understanding of one of these habitability crises: the globally glaciated Snowball Earth. The ice-albedo bifurcation and hysteresis provides a generic pathway for getting into a Snowball, and highlights the difficulty of getting out. We have understood this as a general habitability crisis for watery worlds, regardless of whether or not our own planet ever succumbed to a Snowball. Through a study of the radiative physics in a Snowball state, we have understood why it is hard to get out by a gaseous greenhouse effect of accumulating CO_2 alone, though our study of silicate weathering and ocean carbon uptake provides crucial background for assessing the possible buildup during the Snowball, and fate of CO_2 in the aftermath of deglaciation. One possible exit strategy involves high clouds, and we have at least understood enough about cloud radiative effects to see why this is a possibility. Another exit strategy involves accumulation of dust on the ice surface, and while we have not gone much into the transport of dust to and by ice, we at least know how much albedo change could trigger a deglaciation. Our study of the surface energy budget also has told us much about the effect of solar penetration and sublimation on ice thickness, and about the typical snowfall rate to be expected during a Snowball. One of the outstanding Big Questions, though, is why the Proterozoic consisted of a billion years of ice-free conditions bookended at the beginning and end by massive glaciations that could well have been true Snowball states. And could a Snowball state happen again, or are we done with all that?

The opposite extreme from a Snowball Earth (short of the runaway greenhouse) is the ice-free hothouse climate such as the Earth undoubtedly experienced during the Cretaceous, Eocene, Paleocene and indeed through much of its history. Here, we have come to a good understanding of how high the concentrations of CO_2 or some other greenhouse gas would have to rise in order to keep the polar regions ice-free. We even have some idea of the kinds of fluctuations in the long-term inorganic carbon cycle that could give rise to the required variations in CO_2 over geological time scales, though the story here is far from complete. The general link between CO_2 decline and the long slow cooling that led from the hothouse climate of the Eocene to the onset of Antarctic glaciation in the Oligocene, and later the Pleistocene Arctic glaciations, seems secure. Changes in silicate weathering are almost certain to be behind the gradual CO_2 decline, though the precise mechanism for the decline has not yet been identified.

The outstanding problem of hothouse climates, however, is to account for the apparently low pole to equator temperature gradient characteristic of hothouse climates. A related and even more perplexing problem is how to account for the absence of cold continental interior winters. Both issues are unresolved, and both engage fluid dynamical phenomena which are outside the scope of this book. It is also quite possible, maybe even likely, that clouds might play a role in smoothing out the geographical variations in radiative forcing that would normally lead to cold winters and cold poles or hot tropics. Whatever the clouds are doing, they may also play some role in resolving the difficulty of reconciling the large warming caused by the PETM CO_2 release with the high CO_2 thought to be necessary to explain the already-warm Paleocene climate.

Clouds are the great unfinished business of climate physics. They introduce uncertainty into virtually every important planetary climate problem one wishes to pose, at least for planets with a significant atmosphere. Clouds are the main source of uncertainty in the sensitivity of climate to anthropogenic increases of CO_2. They enter into a possible resolution of the nature of Early Mars climate, though in a way that at present leaves it hard to come to a definitive resolution. They enter into the conditions needed to enter or exit a Snowball state. They introduce uncertainty into the water vapor runaway greenhouse, and can make the difference as to whether Venus experienced a wet or dry runaway. They almost certainly have something to do with enigmatic low-gradient hothouse climates such as those of the Cretaceous and Eocene. Through application of basic physical principles we have come to a good understanding of the factors governing the albedo and greenhouse effect of a cloud in terms of its specified properties. We at least know what factors are important. What has remained elusive is the understanding of what determines cloud fraction, condensate content, particle size, and the vertical distribution of these quantities. These quantities can be modeled, with varying degrees of confidence, in comprehensive general circulation models, but one keenly feels the lack of idealized but useful theoretical constructs that can convey real understanding of the effect of clouds on climate.

The student who has mastered the building blocks of planetary climate as presented in this book is already equipped to tackle a rich variety of original research topics. To mention a few:

- What happens to the climate of Venus if you change the properties of its sulfuric acid clouds, or take them away altogether? If you couple this to volcanic outgassing of SO_2 and reactions at the surface of Venus, what kind of climate cycles result?
- Does every planet that undergoes a runaway greenhouse turn into something like Venus? Does such a planet inevitably lose its water? What if the planet were more massive, as is the case for a Super-Earth? What happens if the planet does not irreversibly lose its water? Even if it does lose water, does it inevitably develop thick high-altitude sulfuric acid clouds like those of Venus? For that matter, if one sees a high albedo object in an extrasolar system, how does one know whether one is looking at a Venus or at a Snowball?
- Carbon dioxide ice clouds may have helped to keep Early Mars warm, and may play a similar role at the outer edge of the habitable zone for extrasolar systems as well. The formation of such clouds is in some sense self-limiting, because the latent heat released by CO_2 condensation must be balanced by infrared cooling to space, yet the clouds themselves can reduce the cooling rate. Since the cloud condensate path results from a balance between loss by precipitation and gain by condensation, the coupled system fixes the condensate path. What is the behavior of the coupled system? Does it reach a stable equilibrium, does it oscillate, or does it do something more complex? How does all that interact with the seasonal cycle, and how does it depend on latitude? What is the effect of the stellar spectrum on the capacity for CO_2 clouds to warm the planet? Could CO_2 clouds push a planet like Gliese 581d into the habitable zone?
- Present Mars does not form permanent CO_2 glaciers, but under conditions with a denser CO_2 atmosphere it is possible that such glaciers might form. In that case, there is a novel form of ice feedback, in that the formation of the glaciers has an albedo feedback, but also has a greenhouse feedback, as the loss of atmospheric CO_2 to the glaciers has an additional cooling effect. What is the behavior of this system? Does it exhibit hysteresis? How would the response be affected by Marslike extreme obliquity cycles? Note that to do this properly, you would need to learn a bit about glacier dynamics, since the flow of glaciers from cold regions into warm regions where they melt or sublimate is an important part of the story.
- The effect of CO_2 condensation on deglaciation of a Snowball is a variation on the preceding theme, and involves CO_2 glaciers on top of water glaciers. The difference here is that the feedback is almost entirely due to the greenhouse feedback of the glacier cycle, rather than the albedo feedback.

- During high-obliquity stages of the Martian obliquity cycle, polar conditions could go well above freezing, especially at a time when there is a dense CO_2 atmosphere. How deep is the layer of water permafrost that could be melted in these conditions? Could it account for some of the fluvial features of Early Mars? How does this seasonal melt affect silicate weathering on Mars? Could this process account for the carbonates in polar Martian soils found by the Phoenix lander?

- It has been speculated that life could emerge in the cold hydrocarbon lakes of Titan. The existence of such lakes (which could be whole oceans on a planet with a greater methane inventory) requires sufficiently low temperatures, and can be limited by a form of methane runaway greenhouse. One can define a hydrocarbon lake "habitable zone" as the range of orbits for which Titanlike methane/ethane lakes can persist. What is the inner and outer edge of this zone, and how does it depend on the surface gravity of the body and the stellar spectrum? To what extent do organic haze clouds limit the methane runaway?

- How cold would Titan get if you took away the methane? How much of the N_2 in the atmosphere would condense out? For what range of orbits around a star can a planet with an N_2/CH_4 inventory maintain liquid hydrocarbon lakes such as Titan's? How is this affected by the surface gravity of the planet, the planet's size (which affects ability of hydrogen to escape), and the spectrum of the star? This question is of considerable interest, since it has been proposed that biochemistry can occur in Titanlike hydrocarbon lakes. The main stumbling block for life in such conditions seems to be finding a cell membrane that does not dissolve in liquid hydrocarbons.

- At pressures of a few bars, at an altitude where the gravitation acceleration is a few Earth gravities, Jupiter maintains Earthlike temperatures even though it absorbs only $9\,W/m^2$ of solar radiation, averaged over its surface. The problem with Jupiter as an environment for life is that it has no surface at this level which could provide a stable environment for life to emerge. Could a rocky Super-Earth in an orbit with insolation like Jupiter's maintain habitable conditions with a Jovian-type H_2-dominated atmosphere with a surface pressure of a few bars? How long could such a planet hold on to the H_2 in its atmosphere?

- What is the effect of seasonality on the Snowball bifurcation diagram? What happens if the planet is tide-locked? What if it is in a highly eccentric orbit and is in a quasi-tide-locked state? What if it is an a 3:2 spin state like Mercury? Under what circumstances can a mostly glaciated planet in such an orbit maintain perpetual open water conditions in some region?

- Does the reddish spectrum of an M-dwarf significantly inhibit a planet's ability to recover from a Snowball state by accumulating CO_2 in the atmosphere? How is this affected by surface gravity and planet size?

- How would weathering cycles based on the sulfur cycle and the greenhouse effect of SO_2 affect the habitable zone in time and space? How does the competition between the greenhouse effect of SO_2 and the cooling effect of sulfate aerosols play out? How does the FUV and EUV environment about an M-dwarf alter the photochemistry of SO_2?

- For planets orbiting M-dwarfs, what is the effect of the reddish stellar spectrum on the runaway greenhouse? To what extent is the runaway greenhouse inhibited? Does the steam atmosphere form a troposphere, or is there a deep isothermal layer?

- What is the climate evolution path of a silicate-poor waterworld with a very deep ocean, for which there is little CO_2 outgassing and little reaction with the rocky crust? If such a planet undergoes a runaway greenhouse, does it lose hydrogen by hydrodynamic escape, or does the accumulation of oxygen choke off the escape? If hydrogen escape occurred in an extrasolar system, would the escaping hydrogen flux be detectable from Earth?

- How would the silicate weathering thermostat work on a waterworld without any continental crust, in which case the main sink of CO_2 is from sea-floor weathering? How does the answer depend on the nature of sea-floor crustal recycling?

Work on some of these topics has already appeared in the literature, and more will appear over time. But the possibilities of these topics have been no means exhausted, and of course there are a whole lot more questions in the well from which these were drawn.

Linnaeus himself was quite deeply interested in applying his botanical knowledge in the kitchen, and became an accomplished cook. No doubt his success there came from a deep understanding of his ingredients. Indeed, the best cooking often happens when you know your ingredients so well that you don't need to adhere to a recipe. I hope this book has helped the reader to understand the basic ingredients of planetary climate. The possibilities of these ingredients are limitless, and for the student who goes on to learn a bit of fluid dynamics to complement the physics that has been presented here the possibilities are beyond limitless. Wherever this may lead the reader, remember: profound ideas grow out of simple models. And always remember to think deeply of simple things.

9.6 WORKBOOK

9.6.1 Basic concepts

Problem 9.1 *Mechanical energy dissipation at surface*
On a planet with a solid or liquid surface, most of the dissipation of kinetic energy usually occurs at the ground, where the atmosphere rubs up against the stationary surface. Dissipation occurs even if the near-surface wind is steady, since the surface exerts a drag force on the wind, but the wind is moving while the force is being exerted; the work done against the wind is the force applied times the distance the air moves in a given time, and therefore the energy dissipation rate (in W/m^2) is the drag force per unit area times the wind speed. The drag force per unit area is given by the bulk exchange formula, Eq. (6.10), applied to momentum. The result is $\rho_s C_D U^2$, where U is the wind speed at the upper edge of the surface layer. The drag coefficient C_D has essentially the same value for momentum as it does for sensible heat exchange, and can be determined from the data and formulae given in Chapter 6. The energy dissipation rate is $\rho_s C_D U^3$. For this problem, you may use neutral surface layer theory to compute C_D.

Assuming a 5 m/s wind at 2 m above the ground, estimate the surface mechanical energy dissipation for Earth, present Mars, Venus, and Titan, and compare the results to the absorbed solar energy flux, so as to determine how significant this term may be in the atmospheric energy budget.

Note that this is a *diagnostic* result, in that it describes what happens if you are given the surface wind from observations. In reality, the surface drag will affect the surface wind speed, and higher drag would typically lead to lower near-surface wind speeds. The amount of the decrease depends on the nature of the forces maintaining the surface wind against dissipation. This pursuit of that matter quickly leads one into some very intricate areas of fluid mechanics.

Problem 9.2 *Mechanical energy dissipation by falling rain*
Raindrops (of any condensed substance) do not accelerate as they fall, but instead fall at a size-dependent *terminal velocity* because they then experience a drag force which counteracts the force of gravity. This means that the falling rain is doing mechanical work on the

environment, and dissipating mechanical energy. Find an expression for the corresponding energy flux (in W/m^2) in terms of the rainfall rate $\dot{P}$ in kg/m^2s, the terminal velocity w_f, the planet's gravitational acceleration g, the density ρ of the raindrop, and its radius r. Assume the raindrop to be spherical. Put in some numbers corresponding to rain falling with terminal velocity $1\,m/s$ at a rate of $100\,kg/m^2$ per day on Earth. Assume spherical raindrops with a radius of $1\,mm$. Compare the result to the typical absorbed solar radiation. How do you think the answer would differ for methane rain on Titan?

Problem 9.3 *An entropy balance model*
Consider an atmosphere without any condensable component, and for which the only heating of the atmosphere is provided by absorbed solar radiation within the atmosphere and convective heat transport which carries absorbed solar radiation from the solid surface to the atmosphere. The solid surface can be assumed to have zero thermal inertia, and hand off all solar energy immediately to the atmosphere via convection. Derive an equation for the mass-weighted vertically integrated zonal mean dry entropy $c_p \ln \theta$, where θ is the dry potential temperature. This equation is analogous to the equation for mean moist static energy derived in the text. Recall that the mass-weighted vertical integral is proportional to the integral with respect to pressure.

Show that in this case, the source term for the mean entropy requires knowledge of the vertical distribution of temperature and heating, and no longer reduces to a top-of-atmosphere flux as was the case for moist static energy

Problem 9.4 Suppose that the annual mean observed top-of-atmosphere imbalance between absorbed incoming shortwave and outgoing longwave flux for a planet with radius a is found to have the form $b_0 - b_1 y^2$ where $y = \sin \phi$. Find an expression for the meridional moist static energy flux Φ defined in the text.

Problem 9.5 The pure CO_2 atmosphere of Mars currently has a surface pressure on the order of $10\,mb$. Recall that $\Phi/L_\circledast$ is the key non-dimensional quantity determining the effect of meridional heat flux on the planetary temperature gradient. Making use of the scaling of Φ with planetary radius and atmospheric mass path p_s/g, how strong would typical Martian winds need to be in order for the meridional heat transport to have an effect on Martian temperature variations of similar magnitude to that in Earth's atmosphere? To carry out this estimate you need an estimate of ΔT. For that, you may assume that under the conditions where Mars is given an Earthlike temperature gradient, ΔT has a value similar to that on the present Earth. Don't forget to take into account the specific heat of CO_2 vs. that of Earth air.

You may do this estimate ignoring the effects of moisture in the Earth case. In reality, about half of the heat transport in the present Earth's atmosphere is due to moisture. What does this fact do to your estimate of the winds required on Mars?

Problem 9.6 *Does planetary size matter?*
All other things being equal, should a giant planet be more in the uniform-temperature regime than a smaller planet, or should it be more in the regime where temperature is determined by a local balance at each individual latitude? To explore this, consider a planet with radius a, an active heat transporting layer of thickness Δp in pressure, a gravitational acceleration g within that layer, and a specified typical velocity scale V within that layer. Discuss how the non-dimensional transport $\Phi/L_\circledast$ scales with the size of the planet, and use this to determine whether a giant planet with a given V tends to be more or less isothermal

than a smaller planet. You may assume that the planet is rapidly rotating, so that the scaling of Φ is determined by the synoptic eddy regime rather than the Hadley cell regime. Note that if the eddies are small in scale compared with the planet, then the ΔT entering into the scaling of Φ will be proportionately smaller than the full pole-to-equator temperature gradient. You should take this possibility into account in your discussion. Note also that for fixed composition a larger planet will tend to have higher g in the heat-transporting layer. What effect does this have on your result? Put in some numbers corresponding to Jupiter.

9.6.2 Diffusive energy balance models

General hints: For many of the problems in the following group you will need the annual-mean flux factor as a function of latitude for a circular orbit. You can compute this by numerically evaluating an integral using the formulae given in Chapter 7 (or using the ready-made Python routines provided as part of the courseware). Alternately, to keep things simpler, you can make use of the fact that the flux factor is well fit by the polynomial $f(y) = a_0 + a_2 y^2 + a_4 y^4$ where $y = \sin \phi$. The coefficients (a_0, a_2, a_4) have the values $(0.311, -0.0873, -0.162)$ for $10°$ obliquity, $(0.308, -0.134, -0.0682)$ for $20°$ obliquity, $(0.300, -0.155, 0.0116)$ for $30°$ obliquity, and $(0.284, -0.137, 0.0563)$ for $40°$ obliquity.

To keep the energy balance model linear, you will also need a linearized form of the function $OLR(T)$. For an Earthlike atmosphere at 50% relative humidity with a CO_2 concentration of 300 ppmv, a reasonable fit to the actual OLR curve in the range of 220 K to 310 K is $OLR(T) = a + b \cdot (T - 220)$ where $a = 113 \, W/m^2$ and $b = 2.177 \, W/m^2 K$. For other values of CO_2 concentration in the range between 50 and 5000 ppmv, a fairly good approximation to reality can be obtained by changing the value of a to $a' = a - 2.2 \ln(pCO_2/300)$ and b to $b' = b - 0.05 \ln(pCO_2/300)$, where pCO_2 is measured in ppmv. These coefficients assume a 1 bar total surface pressure and Earth gravity. Because of the effects of water vapor on radiation and lapse rate, it is difficult to give a simple relation allowing one to scale these results to other surface pressures and gravities. As a general basis for planetary explorations we provide the following examples of computations for other conditions. For Mars with a 1 bar atmosphere dominated by N_2 with CO_2 as a minor constituent and water vapor at 50% relative humidity, the coefficients are $a' = 111.75 - 2.24 \ln(pCO_2/300)$ and $b' = 1.87 - 0.045 \ln(pCO_2/300)$. For a Super-Earth having the same kind of atmosphere but a surface gravity of $20 \, m/s^2$ the coefficients are $a' = 117.4 - 2.17 \ln(pCO_2/300)$ and $b' = 2.52 - 0.052 \ln(pCO_2/300)$. Note that the smaller slope in the Mars case arises from the enhancement of water vapor feedback owing to the increased mass path of water in a low-gravity atmosphere.

Python tips: These coefficients can be computed for any desired pressure, gravity, and temperature range using the function `OLRcoeffs` in the module `ccmradFunctions.py` in the Chapter Scripts for Chapter 4.

Python tips: The basic steady-state linear diffusive energy balance model is implemented in the Chapter Script `SteadyLinearDiffusiveEBM.py`. Most of the numerical problems in the following group can be done by modifying this script.

Problem 9.7 For the general steady diffusive EBM described in Eq. (9.17), show that in the limit of very small D, the temperature is determined by local energy balance. Show that in the limit of very large D the temperature is uniform. Compute this temperature, by considering the integral of Eq. (9.17) over the entire planet. Show that when $OLR(T)$ is linear in T this procedure gives the exact correct result for the global mean T regardless of the magnitude of the diffusivity.

In non-dimensional terms, what is meant by "small" diffusivity? *Hint:* Build a number with the same dimensions as diffusivity, based on the typical pole to equator variation in solar absorption.

Problem 9.8 Following on from the results of Problem 9.7, find an approximate expression for the variation of T over the globe when D is large but not infinite. Specifically, write $T(y)$ as $T_0 + D^{-1}T_1(y)$ where T_0 is constant, plug into Eq. (9.17), collect terms of order D^{-1}, and analytically find an expression that solves the resulting equation. You should be able to do this problem without assuming that $OLR(T)$ is linear.

Problem 9.9 Discuss the behavior of a diffusive energy balance model in the limit of small but non-zero D. In particular, consider a situation with a discontinuity of albedo at y_0, and discuss the characteristic distance over which the temperature response makes the transition from the low-albedo to the high-albedo value.

Hint: You should be able to do this problem analytically, without the use of numerical simulations. The trick is to make use of the fact that for small D the basic solutions take the form of exponentials that decay or grow very rapidly as a function of y.

Problem 9.10 *Numerical solution for uniform albedo*
Numerically solve the steady linear diffusive energy balance model equations for a planet having uniform albedo of 0.2, forced by annual-mean insolation for a circular orbit. The planet has an Earthlike atmosphere and gravity, so a linear $OLR(T)$ fit based on Earth conditions can be used. Explore the behavior as a function of obliquity, $L_\odot$, and the atmospheric CO_2 concentration. Determine the conditions under which the planet can be expected to be ice-free, assuming it to be a waterworld. You make keep the large-scale diffusion parameter D fixed at $1\,W/m^2\,K$. Leaving aside the effects of gravity, how would the size of the planet affect the results (if at all)?

Hint: If you re-write the model in terms of the shifted temperature variable $T' \equiv (T - 220 + a'/b)$ then $L_\odot$ can be scaled out of the problem by using $L_\odot/b$ as a temperature scale. Thus, results for arbitrary $L_\odot$ can be obtained from a single numerical solution.

Problem 9.11 *Numerical solution for solstice conditions*
Consider a planet with a uniform albedo solid surface within which there is no horizontal heat transport. The thermal inertia of the surface is negligible on seasonal time scales. This situation applies equally well to a rocky planet without significant snow cover, or to a Snowball. The planet is rotating rapidly enough that the diurnal-mean insolation can be used in the calculation. Use the linearized expressions for OLR given above.

Numerically compute the response of a diffusive energy balance model to the instantaneous solstice insolation. Discuss how the results depend on obliquity, $L_\odot$, and the large-scale diffusivity D. Use the results to discuss the solstice temperature pattern on a Snowball Earth. Use the results to discuss the chances for getting above-freezing summer temperatures during a high-obliquity state of Mars, assuming a moderate greenhouse effect comparable to that for an Earthlike atmosphere.

Problem 9.12 *Ice-albedo bifurcation revisited*
Using a steady linear diffusive energy balance model with the ice latitude determined consistently with the freeze temperature as described in the text, recompute the ice-albedo

bifurcation diagram using $L_\odot$ as the control parameter, which we earlier computed using a zero-dimensional globally averaged model (Fig. 3.10). For this problem, you may assume that the atmosphere is transparent to infrared radiation, but is nonetheless massive enough to transport significant amounts of heat. Explore the behavior as a function of the diffusivity used to represent large-scale heat transport, and as a function of the obliquity of the planet, assuming the orbit to be circular and that the length of the day is short compared with the length of the year, as on Earth. Display your results in two different forms: first using the global mean temperature as the response variable (as in Fig. 3.10) and then using the ice margin y_{ice} as the response variable.

You may assume that the freeze temperature of sea water is 271 K and that the ocean albedo is 0.1. Show results for ice albedos of 0.5 and 0.7, corresponding to different degrees of snow cover on the ice. For *OLR*, use a linearization of σT^4 about the freezing temperature.

To implement the calculation, you first write a routine that computes the ice-margin temperature as a function of the ice margin position, making use of the differential equation integration described in the text. Then, you hand this function to a Newton's method solver or some other root-finder, which will find the positions of the equilibria; you need to re-run the root-finder many times with different initial conditions in order to find all the solutions, since there may be more than one. One way to make sure you find them all is to generate guesses by computing a rather coarse array of $T_{ice} - T_{freeze}$ and look for the values of y where there is a zero-crossing.

Problem 9.13 *Ice-albedo bifurcation with* CO_2

Re-do Problem 9.12, this time including the effects of CO_2. First re-do the bifurcation diagram using $L_\odot$ as the control parameter as before, but this time showing results for a range of CO_2 values with fixed D. Then, fix $L_\odot$ and D and compute a bifurcation diagram using the CO_2 concentration as the control variable. Fix D at some suitable value, and show results for a few different values of $L_\odot$. For some values of $L_\odot$, you will run into the problem that the CO_2 concentration needed to deglaciate is so high that it is beyond the valid range of the polynomial fit for the coefficients $a(CO_2)$ and $b(CO_2)$. Handle this by just showing the portion of the bifurcation diagram for which you can make a reasonable estimate of the *OLR* coefficients.

Problem 9.14 *Ice-albedo bifurcation for a tide-locked waterworld*

Numerically determine the ice-albedo bifurcation diagram for a tide-locked waterworld in a circular orbit. In this case, if you orient the axis of the spherical coordinate system so that it goes through the subsolar point, then the insolation pattern is simply $L_\odot \sin \phi$ on the dayside and zero on the nightside. Assume that there is just one iceline separating the glaciated and unglaciated portions of the planet.

Problem 9.15 *A weak temperature gradient model*

On a slowly rotating planet, the atmospheric heat transport is so efficient that the atmospheric temperature can be considered horizontally uniform. However, because the surface energy balance allows the surface temperature to deviate from the low-level air temperature, the surface temperature variations can still be considerable.

Carry out a calculation for the annual mean surface temperature in this case, assuming that the surface albedo is uniform and that the surface material does not transport any heat horizontally, so that the surface budget is locally in equilibrium. In this case, the atmospheric temperature is determined by imposing the requirement of global mean top-of-atmosphere energy balance. For simplicity, you may assume that the atmosphere is optically thick enough that the surface temperature does not affect the *OLR*. Once you have the atmospheric temperature, you can determine the surface temperature using the surface flux formulae discussed in Chapter 6.

First consider a planet with sufficient thermal inertia that the diurnally averaged insolation formula can be used despite the slow rotation. This is the regime Titan is in. Discuss how the temperature varies with latitude assuming Earthlike surface flux parameters. If the surface is water, under what circumstances does ice form at the poles? In the tropics? Then put in insolation and surface flux parameters corresponding to Titan, including the effects of methane evaporation. How much does the surface temperature vary if the surface is uniformly damp with methane? If the tropics are dry but the polar regions are damp? For the Titan case, compute the *OLR* by linearizing σT^4 about 95 K.

Finally, do the calculation for a tide-locked waterworld with an Earthlike atmosphere. Under what circumstances is the nightside ice-free? Under what circumstances is a globally glaciated state possible?

Problem 9.16 *An EBM based on moist static energy diffusion*
Derive a steady energy balance model based on the assumption that the moist static energy flux is proportional to the gradient of mean moist static energy E_{atm} defined in Eq. (9.10), instead of being proportional to the gradient of the zonal mean surface temperature T. In other words, derive an alternate energy balance model based on diffusion of moist static energy instead of temperature. In order to close the problem it is necessary to be able to write E_{atm} as a function of T, since the *OLR* (and perhaps also the albedo) is a function of T. Show that this can be done provided that the following two conditions are satisfied: (1) The atmosphere is saturated everywhere; and (2) The atmosphere is on the moist adiabat, so that the moist static energy is uniform in each column. State the function $E_{atm}(T)$ in terms of the saturation vapor pressure function.

Using the chain rule on the y derivative of $E_{atm}(T)$ appearing in the diffusion equation, show that the equation can be re-cast as a temperature diffusion energy balance model with a temperature-dependent diffusivity. Discuss the behavior of this diffusivity as a function of temperature, paying particular attention to the cold, dry limit and the very hot moisture-dominated limit.

What kind of additional equation do you need to close the problem if the vertical temperature profile is described by the saturated moist adiabat, but there are nonetheless unsaturated regions? This situation is an accurate description of the Earth's tropics, and a somewhat worse approximation for the extratropics.

Throughout this problem, you may assume moisture is dilute, so that it is valid to use the approximate dilute form of moist static energy.

9.7 FOR FURTHER READING

The formulation of the atmosphere–ocean energy budget and diagnosis of heat transport from data is discussed in

• Trenberth KE and Caron JM 2001: Estimates of meridional atmosphere and ocean heat transports, *J. Clim.* **14**, 3433–3443.

There are many excellent books on atmospheric and oceanic fluid dynamics. A good, current introduction to the subject can be found in

- Vallis GK 2006: *Atmospheric and Oceanic Fluid Dynamics: Fundamentals and Large-scale Circulation*. Cambridge University Press.

Among other things, this book has a good discussion of the fluid dynamics of Hadley circulations.

The dynamics of synoptic eddies and the instabilities that give rise to them are discussed in

- Pierrehumbert RT and Swanson KL 1995: Baroclinic instability, *Ann. Rev. Fluid Mech.* **27**, 419–467.

The general processes governing relative humidity are discussed in

- Pierrehumbert RT, Brogniez H and Roca R 2007: On the relative humidity of the atmosphere. In *The Global Circulation of the Atmosphere*, T Schneider and A Sobel, eds. Princeton University Press.

The principles introduced in this work apply equally well to any condensable substance in a planetary atmosphere.

The diaries of Carl Linnaeus' journey to Lapland have been most recently reprinted in

- Linnaeus, C 2003: *Iter Lapponicum 1732*, Kungliga. Skytteanska Samfundets Handlingar, Stockholm.

The one English translation of this magnificent work has been out of print for decades and is exceedingly difficult to come by. Even Swedish editions are rather scarce at present.

And because (when it comes to cookery) one does not live by planetary climate alone, the following book is heartily recommended

- David, Elizabeth 1977: *English Bread and Yeast Cookery*. Penguin Books.

Index